2nd International Conference on Numerical Modelling in Engineering (NME 2019)

IOP Conference Series: Materials Science and Engineering
Volume 657

Beijing, China
19 – 22 August 2019

ISBN: 978-1-5108-9850-9
ISSN: 1757-8981

Printed from e-media with permission by:

Curran Associates, Inc.
57 Morehouse Lane
Red Hook, NY 12571

Some format issues inherent in the e-media version may also appear in this print version.

Institute of Physics
Dirac House, Temple Back
Bristol BS1 6BE UK

Phone: 44 1 17 929 7481
Fax: 44 1 17 920 0979

techtracking@iop.org

Additional copies of this publication are available from:

Curran Associates, Inc.
57 Morehouse Lane
Red Hook, NY 12571 USA
Phone: 845-758-0400
Fax: 845-758-2633
Email: curran@proceedings.com
Web: www.proceedings.com

TABLE OF CONTENTS

IOP Conf. Series: Materials Science and Engineering **657** (2019) 011001 doi:10.1088/1757-899X/657/1/011001

Proceedings of 2nd International Conference on Numerical Modelling in Engineering (NME), 19-22 August 2019, Beijing, China

Chairman

Prof. Magd Abdel Wahab
Ghent University, Belgium

International Scientific Committee

Prof. D. Ribeiro, School of Engineering, Polytechnic of Porto (ISEP-IPP), Portugal
Prof. J. Santos, University of Madeira, Portugal
Prof. J. Toribio, University of Salamanca, Spain
Prof. B.B Zhang, Glasgow Caledonian University, UK
Prof. V. Silberschmidt, Loughborough University, UK
Prof. T. Rabczuk, Bauhaus University Weimar, Germany
Prof. L. Vanegas Useche, Universidad Tecnológica de Pereira, Colombia
Prof. S. Bordas, University of Luxembourg, Luxembourg
Prof. N. S. Mahjoub, Institut Préparatoire aux Etudes d'Ingénieurs de Monastir, Tunisia
Prof. A. Cheknane, Amar Telidji University of Laghouat, Algeria
Prof. E. N. Farsangi, Kerman Graduate University of Advanced Technology (KGUT), Iran
Prof. N. A. Noda, Kyushu Institute of Technology, Japan
Prof. K. Oda, Oita University, Japan
Prof. S. Abdullah, Universiti Kebangsaan Malaysia, Malaysia
Prof. C. Zhou, Nanjing University of Aeronautics and Astronautics, China
Prof. B. Bhusan Das, National Institute of Technology Karnataka, India
Prof. R. V. Prakash, Indian Institute of Technology, India
Prof. H. N. Xuan, Hutech University, Vietnam
Prof. Giuseppe Carbone， University of Calabria, Italy
Dr. H. T. Ali, University of Bristol, UK
Dr. A. San-Blas, Miguel Hernández University of Elche, Spain
Dr. G. Minafo, University of Palermo, Italy
Dr. A. Caggiano, Technische Universität Darmstadt, Germany
Dr. S .Khatir, Ghent University, Belgium
Dr. T. Yue, Ghent University, Belgium
Dr. P. Phung Van, Ghent University, Belgium
Dr. A. Rudawska, Lublin University of Technology, Poland
Dr. L. V. Tran, Sejong University, South Korea
Dr. Y. L. Zhou, National University of Singapore, Singapore
Dr. X. Zhuang, Leibniz Unversität Hannover, Germany
Dr. I. Hilmy, International Islamic University Malaysia, Malaysia
Dr. C. Wang, Liaocheng University, China
Dr. M. Mirrashid, Semnan University, Iran
Prof. A. G. Correia, University of Minho, Portugal
Dr. M. Wang, Los Alamos National Laboratory, USA
Dr. Filippo Genco, Adolfo Ibáez University, USA

Editorial

This volume contains the proceedings of the 2nd International Conference on Numerical Modelling in Engineering (NME), 19-22 August 2019, Beijing, China. Numerical Modelling in Engineering 2019 is the 2nd NME conference, whereas the 1st NME conference was celebrated in Ghent (Belgium, 2018).

The overall objective of the conference is to bring together international scientists and engineers in academia and industry in fields related to advanced numerical techniques, such as FEM, BEM, IGA, etc., and their applications to a wide range of engineering disciplines. The conference covers industrial engineering applications of numerical simulations to Civil Engineering, Aerospace Engineering, Materials Engineering, Mechanical Engineering and Biomedical Engineering, etc. The presentations of NME 2019 are divided into two main sessions, namely 1) Mechanical and Materials Engineering and 2) Civil Engineering. The papers that are presented in NME 2019 include stress analysis, thermal analysis, solid structures, fluid mechanics, contact mechanics, biomechanics, Acoustic analysis, finite element analysis and boundary element analysis. The conference gathers international scientists from more than 20 different countries.

The organising committee is grateful to keynote speakers; Prof. Guirong Liu, University of Cincinnati, USA, for his keynote lecture entitled 'Numerical Methods: Indispensable Tools for Engineering Applications' and Professor Hussam Mahmoud, George T. Abell Professor in Infrastructure, the Department of Civil Engineering, Colorado State University (CSU), USA, for his keynote lecture entitled 'Advances in Computational Methods for the Assessment of Structures under Fires and Fire Following Earthquakes'.

Special thanks go to members of the Scientific Committee of NME 2019 for reviewing the articles published in this volume and for judging their scientific merits. Based on the comments of reviewers and the scientific merits of the submitted manuscripts, the articles were accepted for publication in the conference proceedings and for presentation at the conference venue. The accepted papers are of a very high scientific quality and contribute to advancement of knowledge in all research topics relevant to NME conference.

Finally, the organising committee would like to thank all authors, who have contributed to this volume and presented their research work at the confrence venue in Beijing, China.

Professor Magd Abdel Wahab

Chairman of NME 2019

Peer review statement

All papers published in this volume of *IOP Conference Series: Materials Science and Engineering* have been peer reviewed through processes administered by the proceedings Editors. Reviews were conducted by expert referees to the professional and scientific standards expected of a proceedings journal published by IOP Publishing.

IOP Conf. Series: Materials Science and Engineering **657** (2019) 012001 doi:10.1088/1757-899X/657/1/012001

The true working performance inversion simulation analysis of Jinping level-I arch dam

G X Zhang[1], Y Liu[1], L Zhang[1,2], Y Z Liu[1] and S H Li[1]

[1]State Key Laboratory of Simulation and Regulation of Water Cycle in River Basin, China Institute of Water Resources and Hydropower Research, Beijing 100038, China

E-mail:3040252@qq.com

Abstract. Jinping Level-I Super-high Arch Dam, which is up to 305 m high, ranks the highest among all the arch dams that are built up and under construction in the world. But various technical problems have appeared in the design and construction of the super-high arch dam. Therefore, it is necessary to precisely evaluate and predict the temperature, stress and deformation and address the working performance well during the construction, which is critical for the dam construction quality and safe operation of the dam. By performing a whole-dam whole-process simulation analysis, the paper discusses the true load simulation, the simulation of the transverse joint, adiabatic temperature rising model and efficient technologies to resolve difficulties for the super-high arch dam. The whole-dam whole-process simulation analysis platform – SapTis is used to conduct simulation computation in order to track the true working performance of Jinping Level-I super-high arch dam during its construction. Through analysis of the monitoring data and inversion simulation of the computation parameters, the paper simulates the current working performance of the dam, predicts late-stage response, researches the various problems occurring in the construction to come up with effective solutions. The use of the inversion simulation analysis helps have a whole picture of the true working conditions and performance of the dam during its construction, water impounding and operation and guarantee the dam construction quality and operation safety, while creating a new technology solution for the control of high arc dam construction quality.

1. Introduction

The maximum dam height of a Double Curvature Arch Dam of Jinping Level-I hydropower station is 305 m, which ranks the highest arch dam under construction in the world. With the completion of Jinping Level-I Extra-high Arch Dam, the construction of high arch dam in our country is being realized the leap from the following to the leading. There is a significant performance difference between the Level-300 m extra-high arch dam and the Level-200 m arch dam [1], first, the stress level of Level-300 m extra-high arch dam is high, its main tensile stress control is different from that of ordinary arch dam, both tensile and compressive stress of extra-high arch dam reach the critical value of standard allowable stress, the overall safety margin may be less than ordinary arch dam; second, working performance and stress in construction period will have a great influence on the operation later state, so that the true working performance of the dam is different from the initial design state greatly, third, because extra-high arch dam bottom width is bigger, later concrete temperature rise greatly, difficulty of temperature control and crack prevention increase [2], its requirement on spatial gradient when water cooling is high, result in the increase of cantilever height, it may lead to the deterioration of working condition of arch dam during construction period. In some cases, working performance may occur safety control condition during the construction period.

The article takes Jinping Level-I Extra-high Arch Dam as the study object, considering the complex topographic and geological conditions in dam area and various boundary conditions of arch dam construction period, from the beginning of dam pouring and carry out simulation analysis and back analysis and according to actual construction progress, environmental boundary, monitoring data, construction quality and other factors of the arch dam to veritably, reliably understand the working performance of arch dam and further guide the construction and ensure the safety operation of the dam.

2. Whole dam whole process simulation analysis method and software

The whole dam whole process simulation analysis method is an important method to study the true working performance of Jinping Level-I Extra-high Arch Dam [3], and there are three basic characteristics of this method, first is the whole dam, carry out simulation for the whole dam including all transverse joint, fillet, orifice, pier, complex geological conditions; second is the whole process, need to simulate the working performance of the dam construction period, the initial filling and the operation period from the pouring of first hopper of concrete of the dam, third is simulation, get close to the true state as far as possible from the aspect of model, boundary condition, construction process, calculation parameters and others. For Jinping Level-I Arch Dam, mainly from the following aspects to ensure the implementation of working performance inversion simulation during construction period.

2.1. Fine finite element grids

Carry out finite element modeling according to actual excavation on site, foundation treatment, structure design and the latest geological data, the model reflects the accurate figure, detail structure, and various joints of pedestal and dam, to fit the temperature field analysis requirements during pouring, when dividing grids, pay attention to fine degree of grids and gradient change grids in the position of upstream and downstream surface of dam, foundation cushion, baseboard and others contacting with water and air; the basic model reflects the complex geological structure, Including partitions of various rocks (Type II, Type III1, Type III2, Type IV1, Type IV2 and Type V1), complex fault (f2, f18, f5, f8, f42-9, f13, f14 and X), A variety of treatment measures for the ground (anti - shear hole, grouting, backfill, etc.), various treatment measures for foundation (anti-shear tunnel, grouting, backfill, etc.). The models are shown in the figure 1.

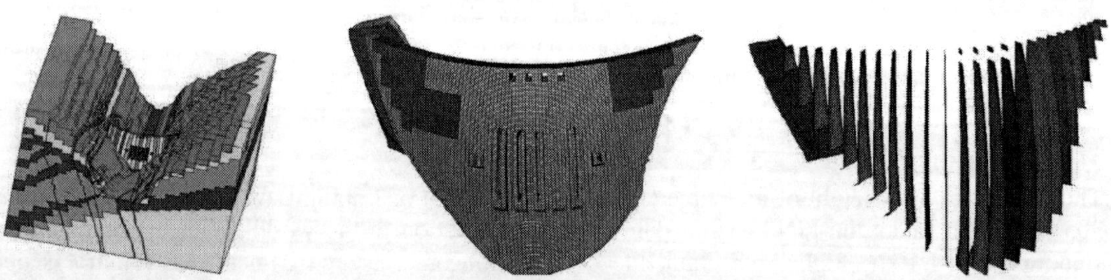

(a) (b) (c)

Figure 1. Finite element computational grid model of Jinping level-I arch dam. (a) Foundation, (b) Dam body and (c) Various joints.

2.2. True load simulation

At present, the loads playing the important role in dam deformation including self-weight, water pressure and temperature. The load simulation adopts whole process simulation analysis method, simulate the construction process of the dam, Stage by stage simulate the function of self-weight and water pressure, simulation of temperature load starts from the day of concrete pouring, accurately considering the various environmental temperature and temperature change of concrete itself, considering the whole process of temperature load, at the same time, according to the monitoring data during the construction process to carry out feedback on internal temperature process of dam, to make the load simulation get closer to the actual situation.

2.3. Simulation model of transverse joints true performance

Transverse joints performance has important influence on arch dam performance. In the early stage of pouring, due to temperature rising transverse joints in compacting state, transverse joints can pass compacting and shearing action, is of less tensile strength; during the cooling process, the temperature drops and it shrinks, the transverse joints opens, and the transverse joints have a certain degree of opening; after closure grouting of dam, the transverse joints interim is filled, transverse joints can pass compacting and shearing action, with a certain tensile strength. The Simulation of true performance of transverse joints, innovatively through the study of the contact surface opening and closing iteration algorithm, to establish concrete contact surface model which can simulate the actual opening and closing process of transverse joints. When disposing the opening and closing iteration of contract surface elements, to prevent the above problems caused by the contact surface embedding, the state variable of the contact surface model is introduced, record the opening and closing state of contact surface before and after each iterative step, when the unit state is considered no longer changed the iteration converge; at the same time, in each iteration step adopt the method of pre-judging the status of the contact surface, promptly increasing decreasing stiffness springs, addition and subtraction force, greatly speed up the convergence rate, under the premise of ensuring the reasonable calculation results, reduce the computational cost, which is suitable for large scale numerical calculation.

Additionally, through the derivation of stress-strain relationship of contact surface element with zero thickness key bed, establish the contact surface model which can simulate the trapezoid, triangle and spherical key bed. When establishing finite element model, the key bed in vertical and horizontal joints shall be simulated carefully, the workload is huge, and it is difficult to operate at the same time, Therefore, it is very effective to simulate mechanical behavior of key bed with contact surface equivalent key bed model. As for the spherical key bed shown in figure 2, the contact stiffness can be described by the following formula:

$$[D]' = \begin{bmatrix} \alpha_h^x K_s & 0 & 0 \\ 0 & \alpha_h^y K_s & 0 \\ 0 & 0 & \alpha_n K_n \end{bmatrix} \tag{1}$$

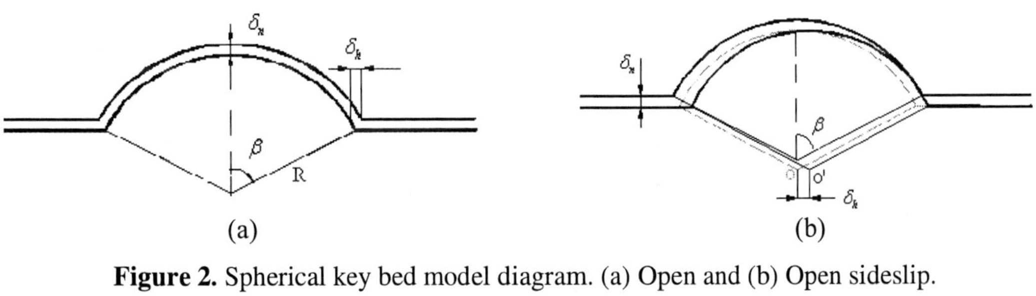

Figure 2. Spherical key bed model diagram. (a) Open and (b) Open sideslip.

In formula (1), Kn, Ks is normal, tangential contact stiffness, α_h,α_n are related to the following three states:

- $\delta_n \le 0$, close, $\alpha_h = 1$, $\alpha_n = 1$
- $\delta_n > 0$, $|\delta_h| \ge R\sin\beta - \sqrt{R^2 \sin^2\beta - 2R\delta_n \cos\beta - \delta_n^2}$, open: $\alpha_h = 0$, $\alpha_n = 0$
- $\delta_n > 0$, $|\delta_h| \ge R\sin\beta - \sqrt{R^2 \sin^2\beta - 2R\delta_n \cos\beta - \delta_n^2}$, open the sideslip, shear of lateral

contact: $\alpha_n = \xi \cos\beta \ \alpha_h = 2.5\xi \sin\beta$

2.4. Post heating adiabatic temperature rising model

In recent years, high arch dam after being water cooling and temperature decrease, there is a big rebound in temperature; the temperature of some is still rising even in the age of six months to a year. We propose a combined function model on the basis of conventional formula [4] for it, as shown in the formula (2), which can more flexibly accurately simulate the actual hydration heat process of concrete, so as to calculate the temperature field and the temperature stress of the whole process.

$$Q(\tau) = Q_1\left(1 - e^{-\alpha_1\tau^{\beta_1}}\right) + Q_2\left(1 - e^{-\alpha_2\tau^{\beta_2}}\right)$$
$$Q(\tau) = Q_1\tau^{\beta_1} / \left(\alpha_1 + \tau^{\beta_1}\right) + Q_2\tau^{\beta_2} / \left(\alpha_2 + \tau^{\beta_2}\right) \quad (2)$$
$$Q(\tau) = Q_1\left(1 - e^{-\alpha_1\tau^{\beta_1}}\right) + Q_2\tau^{\beta_2} / \left(\alpha_2 + \tau^{\beta_2}\right)$$

2.5. Efficient parallel algorithm of equation set solution

Considering the huge simulation calculation of whole dam whole process, an efficient parallel algorithm is developed; considering the flexibility and extensibility of the software, equation paralleling solvers are developed respectively based on OMP and MPI platform. Adopt parallel algorithm for sparse direct solution based on incomplete LU factorization and efficient Krylov subspace iteration method combined with pre condition technique and other methods to realize efficient solution of linear algebraic equations of super-large scale massive computation.

2.6. Whole dam whole process simulation analysis platform - SAPTIS

SapTis [5,6] is a large-scale structure multi-field simulation and nonlinear analysis software with more than 20 years of development and application history, has been equipped with a variety of functions such as pouring of concrete dam, excavation of cavern, temperature, seepage, deformation, stress and other multi-field simulation, structural nonlinear analysis and others, which can be used in simulation analysis of temperature stress, seepage, stress, deformation and safety degree and others of simulating the concrete dam from foundation excavation, pouring, filling, until the whole process of long run. The software entire process consider the construction process, temperature boundary conditions, concrete hydration heat, water cooling process, elastic modulus variation process, creeping degree and others in complex concrete structure. By real-time simulation, grasp the spatial and temporal distribution of temperature, deformation and stress of concrete structure and true working performance of engineering structure. So far, it has been applied in Three Gorges, Ertan, Longtan, Xiaowan, Xiluodu, Jinping, Danjiangkou, and other dozens of engineering, and achieved good results. SapTis software platform is the core computing platform of true working performance inversion simulation analysis of Jinping Level-I Arch Dam.

3. Inversion of foundation and dam deformation modulus

The thermal and mechanical parameters of materials are the basis of the simulation analysis; it is short of monitoring data in early construction phase, generally according to the design parameters and the test parameters of site and test room, to reflect the dam and basic real thermal and mechanical properties as far as possible. With the accumulation of monitoring data, shall according to the site actual temperature, deformation and other monitoring data, through the whole process simulation calculation, carry out inversion and feedback analysis of physical and mechanical parameters of dam and foundation, calibration analysis model input parameter, to ensure that temperature field and stress field calculation result can truly reflect the actual situation of the engineering.

For Jinping Level-I Arch Dam, first, the feedback of the temperature field is carried out, make the calculated temperature process coincide with actual measured temperature process by adjusting the parameters and conditions of temperature calculation, then take advantage of temperature field obtained by feedback, and based on the dam deformation monitoring data, to carry out further feedback analysis for material parameters of dam and bedrock. In the feedback analysis, considering the little change of density, poisson ratio and linear expansion coefficient and other parameters and other thermal parameters, therefore, its value is selected according to the design value or the test value, and assumed that the value is fixed, only change variable modulus values of dam body and foundation model. The basic idea of feedback analysis [7,8] is to make the calculated dam displacement of finite element model get close to actual measured dam displacement as far as possible by continuously adjusting variable modulus values of each partition of dam concrete and different types of rock of foundation in finite element model. When the calculated results and the actual measured deformation mean square error take the minimum values, the taken model modulus value is considered to be more realistic elastic and deformation modulus value of dam concrete and bedrock.

The geological condition of Jinping Level-I Arch Dam is complex, the foundation surface layer is mainly Type Ⅲ2 and Type Ⅳ1 rock mass, the interior is mainly Type Ⅲ1 and Type Ⅱ rock mass, the deep part is mainly Type Ⅲ1 rock mass. According to the vertical monitoring results and gallery settlement monitoring results, from the geological survey data, the actual measured data of surface rock mass is more adequate, due to the technical condition of deep rock mass, its geological prospecting data is not enough, therefore, when foundation deformation modulus inversion, adopt actual test values for surface layer with Type Ⅲ2 and Type Ⅳ1 rock mass, carry out inversion for variable modulus of interior Type Ⅲ1 and Type Ⅱ rock mass, for the deep rock mass (below 1400 m), its rock mass deformation modulus is determined by the empirical relationship between ground stress and deformation modulus, no more inversion here. In addition, according to the law of the dam gallery settlement observation value in the monitoring data, subdivide local area of dam foundation, carry out variable modulus inversion. Carry out inversion for the elastic modulus of dam body and the pedestal on the basis of the designing elastic modulus. Due to the large number of parameters of the dam and the foundation material, it is necessary to carry out repeated trial calculation to approach the correction target of the dam and the foundation material, therefore, use orthogonal test method to determine inversion sample, then combine the monitoring data with effects of dam and foundation variable modulus, to finally determine each inversion calculation conditions. Inversion ideas based on vertical displacement and horizontal displacement are as follows:

- Vertical displacement

The relationship between the vertical displacement and the dam body, the foundation variable modulus parameter is close, vertical displacement of measuring point in different position can respectively reflect deformation characteristics of dam body and foundation. Because the low-elevation gallery is close to the dam foundation, its vertical deformation data mainly reflect the influence of bedrock deformation, take the deformation of this section as feedback target, can carry out vertical variable feedback correction for the Grade Ⅲ2 rock mass, Type Ⅱ rock, weak structural plane and other materials in foundation bottom; galleries vertical displacement difference of other galleries of upper elevation and other adjacent elevation is less affected by the foundation variable modulus, it can basically reflect the influence of the dam concrete modulus of the elevation section, take each gallery displacement difference as feedback target, preliminary estimate dam body deformation modulus value; also according to the deformation difference of multi-point displacement meter, can estimate the foundation deformation modulus value; then combine vertical data, gallery level and other monitoring data, to carry out feedback correction for dam concrete and foundation modulus.

- Radial displacement

The radial displacement of dam is also affected by dam body, foundation variable modulus parameters. Take radial displacement of dam as feedback target; on the basis of roughly estimate dam body variable modulus according to adjacent elevation gallery deformation difference can carry out feedback correction for dam concrete variable modulus.

Monitoring data reflect that, with the increase of dam body pouring elevation, under the overhang effect, the dam is gradually shifted to the upstream; with filling to 1710 m elevation, 1760 m elevation, 1800 m elevation and 1840 m elevation, the dam is gradually deformed toward downstream, middle and low elevation displacement is larger than middle and high elevation, at present, the maximum radial displacement is about 27 mm; gallery level monitoring data show that the whole dam is in a state of subsidence, the subsidence of middle section is larger, the subsidence of both sides is less, of basic symmetric distribution, local section on both sides is of slightly elevated vertical displacement change trend. Figures 3 and 4 are the radial displacement of the dam and the vertical displacement monitoring curve graph at the initial filling stage.

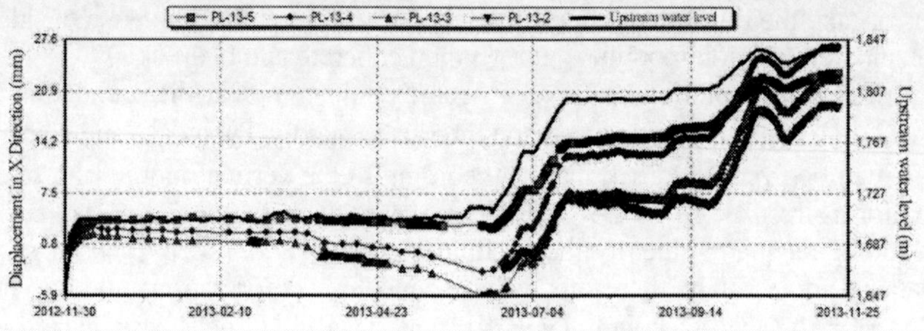

Figure 3. 13# dam section vertical radial displacement actual measured process line.

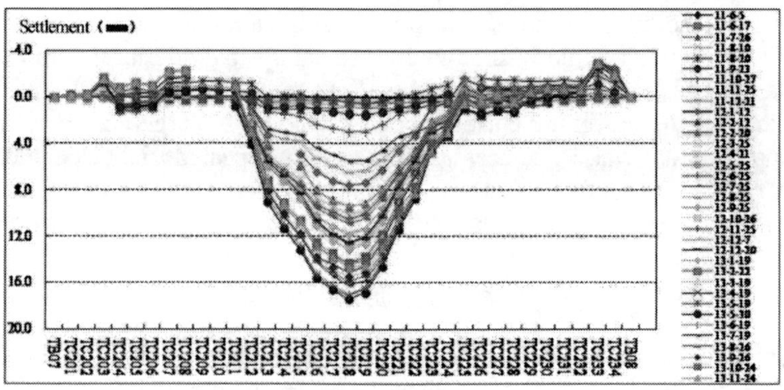

Figure 4. Vertical displacement of dam 1664 m elevation distribution map.

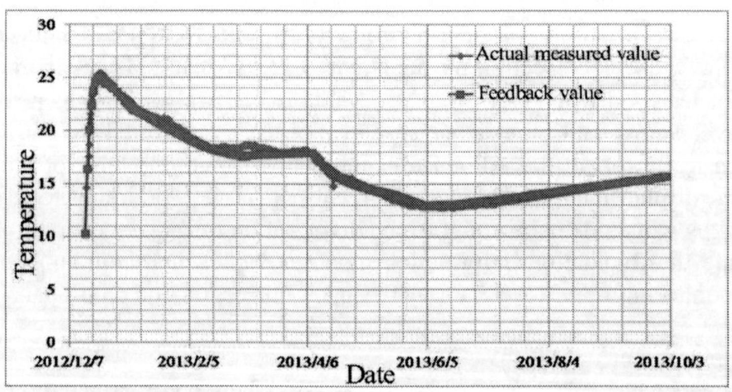

Figure 5. Typical warehouse temperature process feedback curve.

Figure 5 is the contrast curve of feedback temperature and actual measured temperature of typical warehouse, make the calculated temperature process coincide with actual measured temperature process by adjusting the parameters and conditions of temperature calculation, based on this, according to the dam deformation performance reflected by monitoring data, carry out inversion for dam and foundation elastic modulus by the method of whole process simulation analysis. Figure 6 is settlement diachronic curve comparison of 1778 m elevation gallery in some period of 17# dam section and 1601 m elevation gallery of 13# dam section, figure 7 is vertical radial displacement and calculated value contrast curve of typical parts of typical dam section, it can be seen that the calculated and measured processes are consistent, and the deformation modulus obtained by inversion can reflect the true performance of the dam.

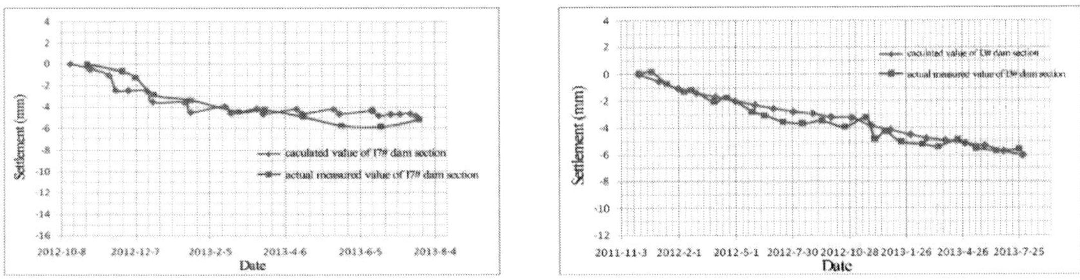

(a) (b)

Figure 6. Relative settlement of gallery level of different dam sections diachronic comtrast grap. (a) 1778 m gallery of 17# dam section (b) 1601 m gallery of 13# dam section.

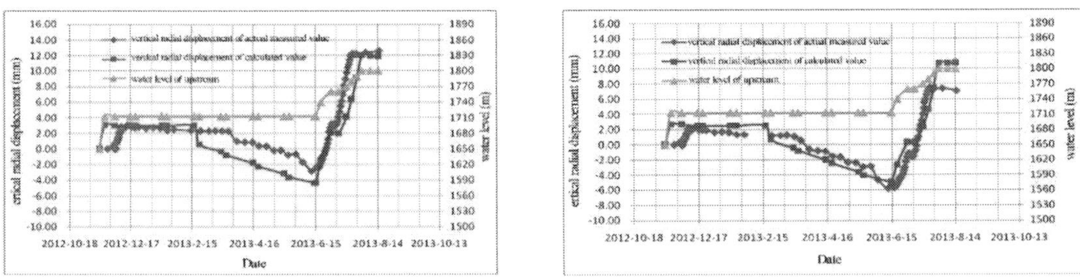

(a) (b)

Figure 7. Working performance analysis of transverse joints during construction. (a) 1730 m elevation and (b) 1601 m gallery.

4. Working Performance Analysis of Transverse Joints during Construction

4.1. Monitoring data analysis of transverse joint opening degree during construction

The monitoring data of Jinping Level-I Arch Dam during construction period shows that Most of the opening process of transverse joints is very well correlated with the temperature, early concrete temperature rise, transverse joints in compacting state, with water cooling the temperature of internal concrete of dam body drop, the transverse joints opening degree gradually increase and reach the maximum transverse joints opening degree before groution, the maximum value of the transverse joint opening degree is generally less than 3 mm, which conform to the general rule, after grouting, the opening degree of the transverse joints basically remain unchanged. But the transverse joints opening degree of some dam sections is larger, and there is obvious abrupt change, mutation time is close to the grouting time, the maximum transverse joints opening degree of some parts is more than 8 mm, which does not conform to the general rule of the transverse joints opening degree. Figure 8 is the abnormal joints opening degree monitoring data diagram.

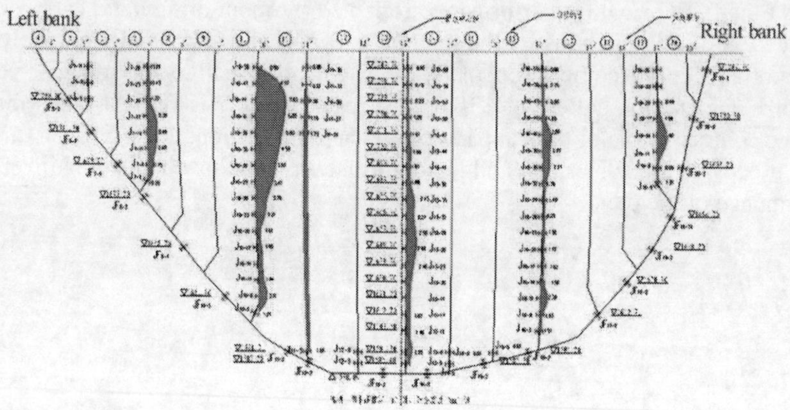

(a)

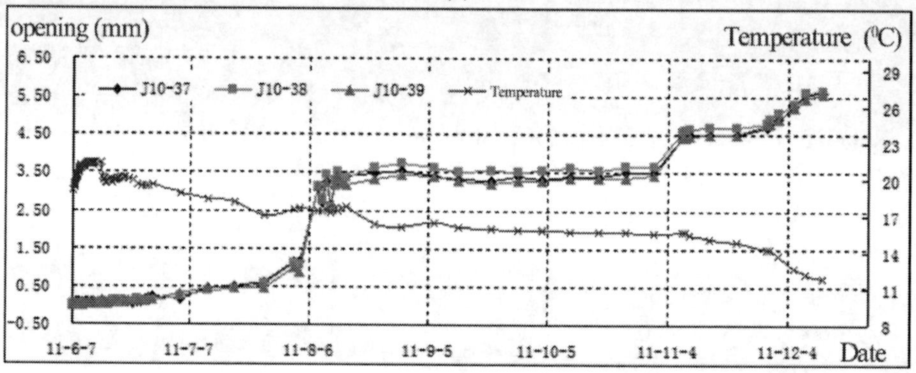

(b)

Figure 8. Actual measured transverse joints opening degree. (a) Transverse joints opening degree diagram and (b) typical transverse joints opening degree process line.

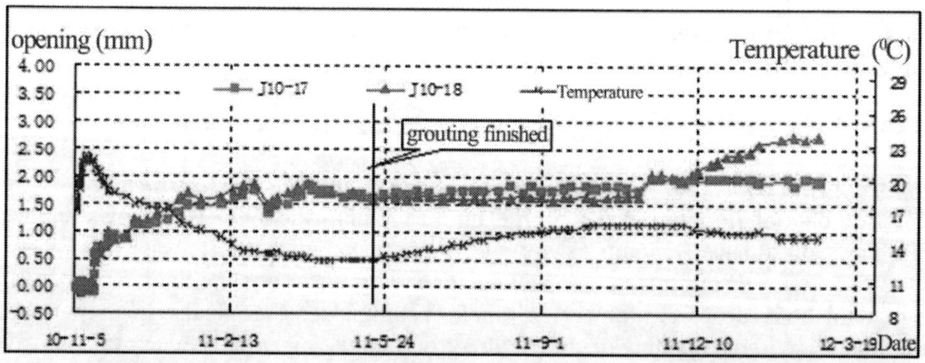

Figure 9. Transverse joint opening degree process line after grouting.

In addition, the joint meter shows that there are different degrees of opening in the transverse joints in already grouting area of the lower part of the arch dam during construction period, As shown in the figure 9. after arch dam closure till June 20, 2012, the opening degree of 10# transverse joint in EL1620 m-EL1680 m is larger, the opening degree reached 0.4 mm-0.9 mm; opening degrees in the rest elevation are smaller, the opening degree is about 0.1 mm. 13# transverse joints opening degrees above EL1620 m are less than 0.2 mm; its opening degrees below EL1620 m are larger, opening degree in the upstream of EL1585.7 m reaches 2.42 mm, and there is abrupt change of opening degree.

Most of the elevation opening degrees of 16# transverse joints are less than 0.15 mm, individual larger elevation opening degree reaches 0.31 mm. Therefore, analyze the causes of abnormal phenomenon of transverse joint opening degree, and to analyze its harmfulness, and further take the corresponding control measures.

4.2. Problem analysis on oversize partial transverse joint opening degree in construction period

For the problem of oversize partial transverse joint opening degree in the construction period, through the method of whole process inversion simulation analysis, on the basis of the inversion of actual temperature field and main mechanical parameters, carry out simulation in accordance with the actual construction process, temperature control process and closure grouting process of dam, the calculation parameters consider the concrete elastic modulus change with time, concrete creep, concrete autogenous volume deformation and other factors, respectively analyze the influence on transverse joint opening degree under the synthetic effect of self weight, temperature, autogenous volume deformation and other factors and various factors. In addition, through the analysis of monitoring data, it is found that the time of occurrence of the opening degree is consistent with the transverse joint grouting time, and the transverse joint positions with abrupt change are all above grouting area, the higher the elevation, the greater the influence, it is speculated that the transverse joint opening mutation abrupt change is related to transverse joints grouting according to time anastomosis; therefore, carry out corresponding simulation for grouting pressure degree, grouting method, analyze the influence of grouting factor on transverse joint opening degree.

The simulation results show that [9], the effect of the self weight on the average joint opening degree is about 0.2 mm; under the effect of autogenous volume deformation, under the condition of design parameters, the transverse joint opening degree is 0.18-0.53 mm, the average value is 0.39 mm; under the condition of temperature and autogenous volume deformation, the average value of transverse joint opening degree is 1.93 mm, the maximum is 2.9 mm. Under the synthetic effect of self weight, temperature, concrete autogenous volume deformation and other factors, the change law of calculated transverse joint opening degree is the same as monitoring value, early concrete temperature rise and expansion, transverse joint is in compacting state, with the decrease of the temperature of the dam, the transverse joints gradually open, and the maximum value is reached when the temperature reaches the minimum. In the abrupt change position of transverse joint opening degree, the difference between the calculated value and the monitoring value is large; the calculated value is close to the monitoring value after the removal of abrupt change. Therefore, in order to analyze the influence of grouting, carry out sensitivity analysis respectively for whether through horizontal pressure or not, grouting pressure degree, pressure impose height and others, simulation calculation results shows that: The grouting pressure has little effect on the transverse joint opening degree when through horizontal pressure, and the influence scope is limited to the grouting elevation nearby, and the effect of the joint opening degree in grouting area is less than 0.3 mm; when do not through horizontal pressure, the grouting pressure affects the transverse joint opening degree of upper part, and it is of increasing trend with the increase of height, the upper part reach 1.39 mm in 9 m under the pressure of 0.5 MPa, the greater the pressure area, the greater the transverse joint opening degree, the maximum value is 5.98 mm in 18 m grouting area under the pressure of 0.5 MPa. The results are shown in the figure 10 and figure 11. Combine monitoring data and simulation results, it can be seen that grouting method is the main reason affecting transverse joint opening degree, shall strictly in accordance with the relevant requirements of the grouting operation in the actual construction process.

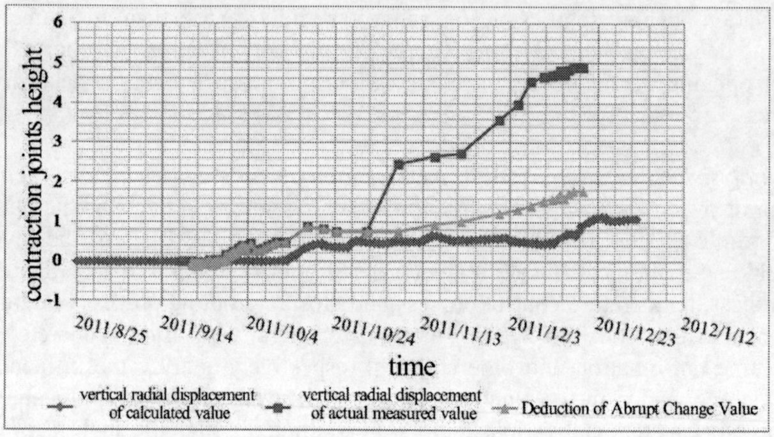

Figure 10. Contrast of process line of simulation calculation and actual measurement and joint opening degree after deduction of abrupt change value.

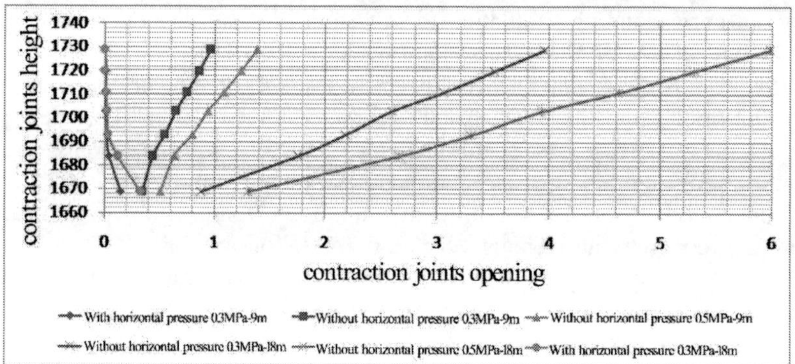

Figure 11. Influence of grouting on joint opening degree curve.

4.3. Partial opening problem of transverse joint after grouting during construction period

For opening problem of the transverse joint after grouting, also through the method of whole process inversion simulation analysis, on the basis of the analysis of the monitoring data and on the basis of the inversion of the actual temperature field and the main mechanical parameters, carry out simulation analysis according to actual pouring process and the closure process of the dam, focusing on the cause of transverse joint opening after closure, carry out quantitative analysis from self weight, temperature and other aspect affecting factors, the analysis results show that [10]:

- Take10#, 13# and 16# transverse joints as the research objects, considering the combined effects of weight and temperature, by June 20, 2012, the average value of calculated the transverse joint added opening degree after closure was 0.26 mm, the monitoring value was 0.19mm, the calculated values and monitoring values were slightly different, but the law of the two is basically the same, the 13# transverse joints added opening degree in dam section of river bed upstream is that the upstream is bigger than that of downstream, 10# , 16# transverse joints added opening degree in dam section of slope is that the downstream is bigger than upstream. Considering the effect of self weight alone, the average value of transverse joints added degree is 0.2 mm, considering temperature effect alone, he average value of transverse joints added degree is 0.06 mm, self-weight plays a major role in transverse joint opening after closure. The results are shown in the figure 12.
- For the position of transverse joint added opening degree with abrupt change after grouting,

through the analysis of the monitoring data, there is pressurized water operation before joint grouting at the moment of abrupt change in upper grouting area, affected by it; there are abrupt changes in the transverse joint open degree of already grouting lower part.

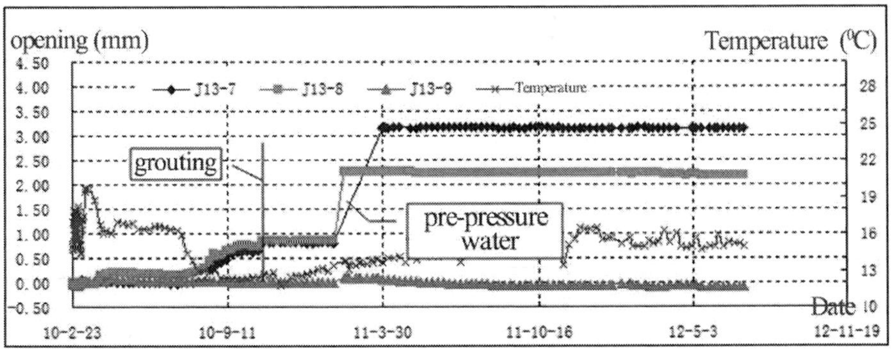

Figure 12. 13# dam section transverse joint opening degree process line.

4.4. Impact analysis on oversize transverse joint opening degree and added opening problem

The oversize transverse joint opening degree during construction process, and there is the trend of development after grouting, therefore, it is necessary to study the influence of oversize transverse joint opening degree on the dam's stress.

Before closure grouting, the parts above closure elevation are separate bearing load of each dam section, when designing in accordance with the consideration of the working condition, and calculate the height of the cantilever, the oversize transverse joint opening degree will not affect the stress at this moment. While closure grouting, big transverse joint opening degree is beneficial for the entry of grouting slurry. As long as the design is elaborate, the construction is meticulous, ensure grouting quality, guarantee the grouting compaction, after the hardening of grouting slurry, it is basically equivalent to concrete body elastic modulus. After grouting, the dam will become a complete as a whole, which does not affect the dam stress. After closure grouting, the already grouting transverse joint opening degree continue to develop under the effect of upper self weight, and grouting and various factors, if cause opening of the already grouting transverse joint, it will have a negative impact. The article through the method of whole process simulation analysis to simulate the actual the actual opening degree of the transverse joint and the open state after grouting, from the construction period simulation calculation to the dam filling, calculation results show that: after filling, under the effect of water pressure, transverse joint need to be close first so that it can arch bear hydro-thrust, displacement along river and tensile stress of dam heel will increase. According to the transverse joint opening situation reflected by monitoring data, after the filling to 1880 m, displacement along river growth rate is 3.07%; the dam body stress distribution law remain unchanged, the extreme value of tensile stress of dam heel will increased by about 5%, tensile stress range will increased by 0.2 m.

Therefore in the process of construction, shall strictly in accordance with the technical requirements of grouting to carry out operation, especially to achieve "when carry out grouting for transverse joint in an elevation, shall through horizontal pressure for no grouting adjacent transverse joint of the same elevation [11]". At the same time in order to reduce the influence of self weight on the transverse joint opening, shall take timely measures such as upstream water filling and others.

5. Construction period overhang and control of whole arch dam

Overhang is one of the main features of hyperbolic double curvature arch dam, in the process of construction, overhang has some effects on the stress and deformation of dam, poor control will bring the stress state contrary to the normal load of the design, even produce adverse effects on the dam structure. Jinping Level-I Arch Dam is higher than 300 m, the problem of overhang is more prominent during construction, especially the overhang of upstream surface, if the control is poor, it is easy to occur larger tensile stress on downstream surface, or even occur cracks; in addition, upstream deformation brought by overhang effect may bring about the opening of transverse joint in local area, so it is necessary to study overhang problem of Jinping Level-I Arch Dam during construction, and put forward corresponding measures.

In order to study the working performance of dam in overhang state, according to existing specification requirements, adopt whole process simulation analysis method, establish the model of whole dam, considering the influence of self weight, water pressure and temperature load, carry out calculation for various overhang working conditions which may occur during construction period by using finite element equivalent stress method, carry out evaluation in accordance with the corresponding stress control standards. The stress state at different stages of the dam under the condition of empty reservoir and various water filling conditions is mainly studied.

Simulation calculation results show that: under the effect of self-weight in empty reservoir the Jinping Level-I Arch Dam will produce a certain range of tensile stress near the foundation surface of downstream surface, because the dam overhang has the trend of tilting to riverbed and the trend of tilting to upstream, the first principal stress of downstream is greater than that of the upstream, As shown in the figure 13, the direction of principal stress makes an acute angle with foundation surface, which explains that the first principal stress is produced by the combined effects of high shear stress of foundation surface along the steep slope and downstream tensile stress caused by overall overhang. The large tensile stress in the position may lead to the foundation surface disengaging, or occurring cracks in the nearby area, which needed to be concerned about. Take the image empty reservoir state in May 2012 for example, after making equivalent upstream surface tensile stress is 1.1 MPa, which do not exceed the standard, but the downstream surface tensile stress reaches 2.1 MPa, the area exceeding the standard area is about 20 m^2, the depth is about 12 m. Considering the upstream water filling scheme, taking the upstream filling to the 1660 m elevation as example, the maximum downstream surface tensile stress is reduced to 1.8 MPa, the results are shown in the figure 14. In addition, the effect of the upstream water filling impoundment seam is very small, which can be ignored. Therefore, proper filling is favorable to reducing tensile stress near the downstream surface of foundation surface.

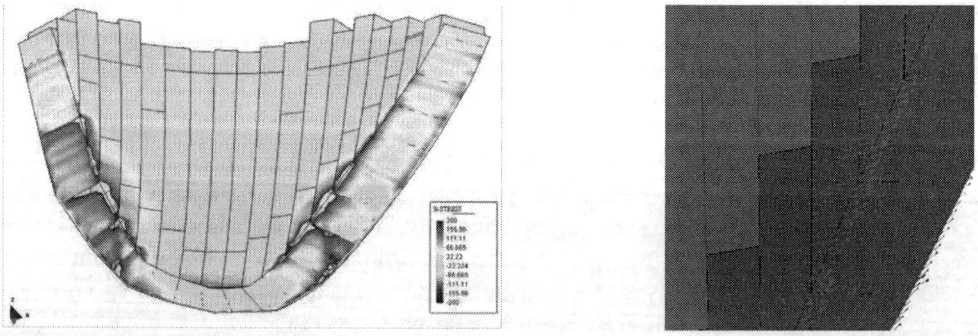

Figure 13. Cloud image and its vector diagram of downstream surface tensile stress under effect of overhang.

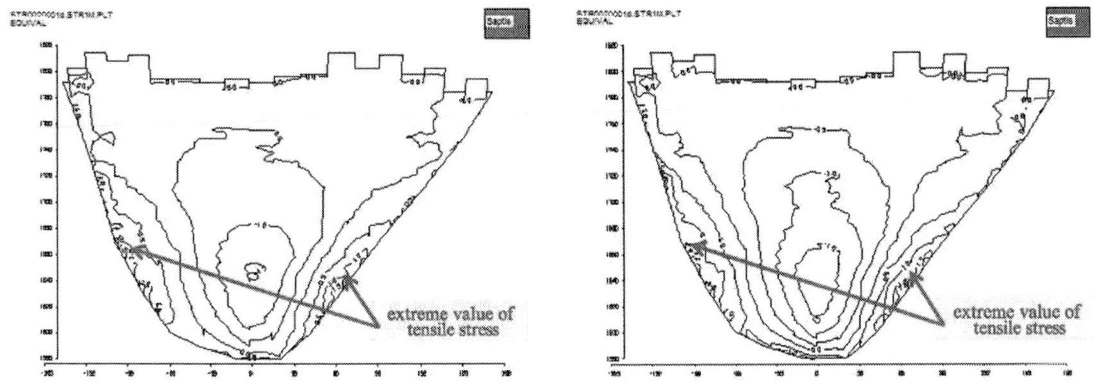

(a) (b)

Figure 14. Equivalent stress of downstream surface. (a) Empty reservoir state and (b) upstream filling to 1660 m elevation.

Therefore, in order to reduce the adverse effects of the dam overall overhang, high dam empty reservoir state shall be avoided. It can be adopted that select appropriate time to do upstream water filling, considering the river water temperature is lower than the temperature, water filling time shall be avoided in the low temperature season. During the construction of Jinping Level-I Arch Dam, Summer upstream appropriate water filling scheme is adopted, which guarantee the construction quality of dam and the safety during construction period.

6. Study on working performance of dam in initial filling stage

Jinping Level-I hydropower station started filling from guiding tunnel gate in November 2012, it is planned to filled to the normal water level of 1880 m by September 2014, according to the construction schedule of engineering, the dam pouring was still continuing in the initial filling, joint grouting was not completed, especially in October 2013 after guiding bottom gate, when the reservoir filling from 1800 m water level to 1840 m water level, the joint grouting elevation of the dam was about 1859 m, there was still 26 m of dam without joint grouting, in order to make the arch ring grouting of upper dam arch avoid the adverse effects of high water level pressure impoundment seam on the transverse joint grouting as far as possible, need to study the best opportunity of transverse joints closure grouting according to the reservoir water level and the project schedule.

In the initial filling stage dam is under construction, various design parameters may vary with the loading history and loading path, the compatibility between the strength of the dam itself and the actual boundary around the dam is not exactly consistent with the design expectation; and according to the dam accident statistics, it showed that, almost 60% of the dam accidents occurred in the initial filling stage or initial operation stage. Therefore, according to the monitoring data of each stage during the initial filling, carry out tracking and inversion simulation analysis for working performance of the dam, promptly grasp the current operation safety state of the dam, and make predictions for next safety status, to provide the decision basis for making safe operation monitoring standard and reservoir scheduling operation mode.

6.1. Analysis on impoundment seam problem of the initial filling

In the initial filling stage, the upper elevation closure grouting is not yet completed; the water pressure will produce impoundment seam effect on the no grouting transverse joint of upper different elevations. Simulate filling process, refer to the actual monitoring value of transverse joint, predict the effect of filling and arch closure scheme on the grouting of dam.

The simulation calculation results show that [12], in the same position, the overall transverse joint opening degree is in gradually compacted state with the rise of the water level, the higher the water level, the more severe the impoundment seam. The open degree of the downstream side of the arch crown cantilever shows an open trend under the condition of lower water level. The change law of transverse joint opening degree near the riverbed is: severe impoundment seam is most severe on the upstream side at the same elevation, impoundment seam value decreased gradually from upstream side to the downstream side. The change law of transverse joint opening degree near the two sides is: severe impoundment seam is most severe on the downstream side at the same elevation, impoundment seam value increased gradually from upstream side to the downstream side.

According to the existing construction progress and filling progress, when the water level is below the 1800 m elevation, the total impoundment seam value is not large, and the average impoundment seam value is less than 1 mm; when the water level is elevated from 1800 m to 1840 m elevation, the average impoundment seam value is up to 1.35 mm, calculate according to current average joint opening degree of 1.83 mm based on monitoring data statistics, the remaining average joint opening degree is 0.48 mm. When the water level reaches 1840 m elevation, impoundment seam is larger is small scale of transverse joint, which is more than 2 mm, it is mainly concentrated in the parts above 1859 m elevation, so when the initial filling reaches to 1840 m elevation, it may have an effect on the closure grouting of these parts.

If considering reducing the water level to 1800 m by the end of February 2012, and then carrying out closure grouting in March-May, the total impoundment seam value would have a significant improvement compared with filling to 1840 m elevation, it is close to impoundment seam situation of first filling to 1800 m, the maximum impoundment seam value is generally less than 1 mm. So it can be considered that after the water level is reduced, select machine to closure grouting according to the actual monitoring joint opening degree situation. The current engineering is also drawing lessons from on the results of this research program.

6.2. Analysis on current working performance of dam

The finite element calculation results show that, in the process of filling, with the water level increased, the dam displacement along the river show an increasing trend toward downstream; at the same time, with the increase of delivery lift, the displacement increment becomes larger, increment of low elevation of dam is smaller, the higher the elevation is the upper point of the dam, with the increase of the water level, the more obvious the increment; left and right bank deformation is basic symmetrical , because considering the step by step pouring process of dam, at the same time, apply water pressure load according to normal filling time in the pouring process, so dam deformation contour is of stepped distribution. On tangential deformation, the left and the right bank of the dam are respectively pointing to the corresponding shore direction, and the left and right bank deformation is basic symmetrical. The whole deformation law of other points in time is the same as that of the water level, but with fluctuation of time and water level, the deformation value is different.

The vertical stress is basically compressive stress, and the maximum compressive stress value appeared in the bottom section of the dam, compressive stress gradually decreased from bottom to top, with the increase of the water level, the compressive stress of upstream surface is in decreasing trend, the compressive stress of downstream surface is in increasing trend. Except foundation surface local stress concentration point the upstream surface's arch stress is all compressive stress, the stress is gradually increased from the periphery to the middle part, the maximum value of compressive stress occurs in the middle elevation of the arch crown cantilever, the maximum compressive stress is increased with the increase of the water level; when the water levels are1810 m, 1820 m, 1830 m and 1840 m, the maximum stresses in dam heel (beam) are 8.5 MPa, 8.2 MP, 7.9 MPa and 7.2 MPa, the maximum compressive stress of the dam heel pressure actual measured is 6.66 MPa, appear in the 12# dam section, as shown in the figure 14. The simulation results are slightly different from the actual measurement [13], but the overall basically reflect the actual stress of dam heel.

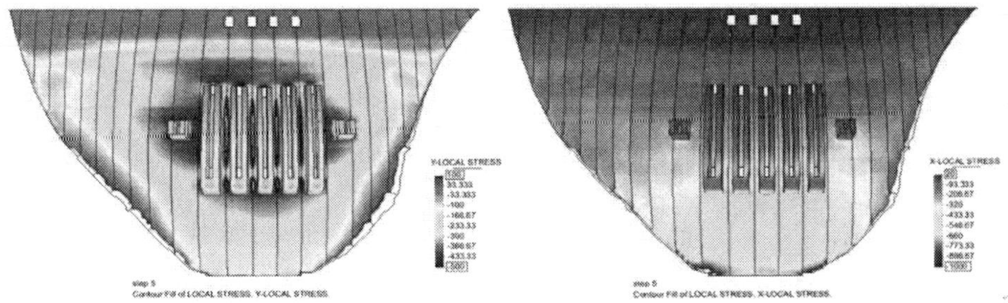

(a) Arch (b) Beam

Figure 15. Cloud image of current stress calculation of dam (1839 m water level).

Figure 16. Diagram of vertical stress distribution of dam heel (November 20, 2013).

The general deformation and stress of the dam are shown in table 1 and figure 15, the overall displacement is consistent with the monitoring value, the dam body stress is in accordance with the general law, except the individual stress concentration point significantly exceeds the tensile stress, the tensile and compressive stresses were less than concrete tensile and compressive strength. In summary, the results of inversion tracking calculation analysis show that the current working performance of arch dam is generally normal.

Table 1. Comparison between calculated deformation values and monitoring values.

Elevation (m)	Monitoring Instrument	Water Level 1800 m-1839 m Radial Deformation Increment (mm)		Current Radial Deformation (mm) By November 20, 2012	
		Calculated value	Monitoring value	Calculated value	Monitoring value
1778	PL9-3	12.03	9.40	19.97	17.25
1730	PL9-4	7.23	6.38	15.76	15.96
1664	PL9-5	1.41	2.34	5.31	7.99
1778	PL11-3	15.99	13.94	27.42	25.16
1730	PL11-4	11.88	11.43	25.30	28.47
1664	PL11-5	4.84	6.13	14.50	19.06
1778	PL13-3	16.65	14.08	29.24	22.43
1730	PL13-4	13.73	12.16	28.63	26.53
1664	PL13-5	7.64	7.42	19.18	23.22
1778	PL16-3	11.35	9.95	22.09	19.68
1730	PL16-4	9.72	9.05	22.95	24.01
1664	PL16-5	5.23	6.13	16.20	22.01
1778	PL19-3	4.36	5.39	9.69	10.62
1730	PL19-4	2.77	4.50	9.25	12.75
1664	PL19-5	0.93	2.42	5.89	8.93

7. Conclusion

In the article, the key technologies of the whole dam whole process simulation analysis method are discussed, the whole dam whole process simulation analysis platform - SAPTIS is introduced from the true load simulation, the true performance simulation of the transverse joint, the post heating adiabatic temperature rising model, the efficient parallel algorithm of the equation set solution and other aspects. Use Saptis as the basic tool, and adopt the simulation analysis method to carry out inversion simulation analysis for the true working performance of Jinping Level-I Extra-high Arch Dam. (1) carry out inversion analysis for the foundation and mechanical parameters of dam concrete according to actual measured vertical and radial displacement, then carry out whole process simulation for the overall dam; (2) carry out simulation analysis for oversize partial transverse joint opening degree problem, the abrupt change problem of the transverse joint opening degree, the partial opening problem of the transverse joint after grouting during the construction period, study the causes, effects and improvement measures of these problems; (3) carry out detailed analysis and research for excessive tensile stress and possible crack problem caused by the overall overhang of arch dam during construction period, put forward suggested measures of advance filling to reduce overhang; (4) study the impoundment seam problem of the initial filling, put forward water level scheduling scheme which avoid affecting the quality of the upper closure grouting; (5) carry out whole process tracking inversion simulation analysis for the true working performance during filling process of dam, guide the simulation according to monitoring data, make accurate prediction for future performance, put forward proposed scheme which ensure the safety of dam.

The whole process tracking inversion and simulation analysis research and its application results show that, the simulation inversion method can grasp true working performance of the whole process of dam construction and filling, ensure the safety of the dam during construction and initial filling, the proposed technical measures and control measures have been adopted in the construction process, and achieved good results, guarantee the construction quality and construction, operation safety of the dam, At the same time, it created a new model of technical support for the high quality construction of the Level-300 ms extra-high arch dam.

There are many contents in the working performance of the dam construction period and the initial filling period, limited to the length of the article, only describes the above aspects, a preliminary study has been done on temperature rise, the dam heel stress, tilting deformation, aging deformation, Initial seepage field effect and others, which will be introduced in the following articles.

Acknowledgments

The research was supported by the National Key Research and Development Project of China (Grant NO. 2016YFB0201000, 2018YFC0406700, 2016YFC0401608), The research was supported by the National Natural Science Foundation of China (Grant NO. 51579252, 51779277, 51439005, 51578544), the Special Scientific Research Project of the China Institute of Water Resources and Hydropower Research (Grant NO. SS0145B392016, SS0145B612017), the Special Scientific Research Project of the State Key Laboratory of Simulation and Regulation of Water Cycle in River Basin and the Special Scientific Research Project of the China Institute of Water Resources and Hydropower Research (Grant NO. 2015TS05, 2016ZY10, Huaneng Group Science and Technology Project (HNKJ16-H16).

Conflict of Interests

The authors do not have any conflict of interest with the content of the manuscript.

References

[1] Zhang G X, Zhu B F and Jin F 2013 Summary report of study on the real working form of high arch dam and engineering application results (Beijing: Tsinghua University, China Institute of Water Resources and Hydropower Research)

[2] Zhang G X, Liu Y Z, Liu Y *et al* 2010 Discussion of the problem of extra-hig arch dam

temperature control and crack prevention *J. Hydroelectr. Eng.* **29** 125-31
[3] Zhang G X, Liu Y, Zhu B F and Wang R K 2008 Theory and method for the simulation of real working performance of high arch dam *Research Progress of Dam Technology and Long Term Performance* **2008** 24-9
[4] Zhu B F 2009 *New Development of Theory and Technology of Concrete Dam* (Beijing: China Water Power Press)
[5] Zhang G X 2013 SapTis: Development and application of multi-field simulation and nonlinear analysis software (one) *Water Resour. Hydr. Eng.* **44** 31-5
[6] China Institute of Water Resources and Hydropower Research 2013 Stress deformation of dam structure and safety evaluation during the construction period and initial filling of Jinping Level-I hydropower station (phase report Ⅱ)
[7] China Institute of Water Resources and Hydropower Research 2013 Stress deformation of dam structure and safety evaluation during the construction period and initial filling of Jinping Level-I hydropower station (phase report Ⅲ)
[8] China Institute of Water Resources and Hydropower Research 2012 Special research report of transverse joint opening degree analysis of 10# dam of Yalong Jinping Level-I hydropower station
[9] China Institute of Water Resources and Hydropower Research 2012 Analysis research report of causes and effects of opening of transverse joints after grouting in Yalong Jinping Level-I hydropower station
[10] Chengdu Survey Design and Research Institute of China Hydropower Consulting Group 2009 The joint grouting construction technology requirements of a Jinping Level-I hydropower station in Yalong River Sichuan Province
[11] China Institute of Water Resources and Hydropower Research 2013 Research of dam filling process control and selection of closure opportunity of Jinping Level-I hydropower station
[12] Zhang G X and Zhou Q J 2013 Causal analysis of the difference of the high arch dam heel stress actual measurement and elastic calculation results *J. Hydraul. Eng.* **44** 640-7

Structural design and strength analysis of multi-ring packed flywheel with heavy metal alloy segments

L Jiang[1,2], W Zhang[2], C W Wu[2,3], L P Zhang[1], Y X Zhang[1] and Z Y Liu[1]

[1]National Key Laboratory of Science and Technology on Reactor System Design Technology, Nuclear Power Institute of China, Chengdu 610041, China
[2]State Key Laboratory of Structural Analysis for Industrial Equipment, Dalian University of Technology, Dalian 116024, China

E-mail: cwwu@dlut.edu.cn

Abstract. Flywheel, as a kinetic energy storage and conversion device, has promising applications due to its high efficiency, no pollution and long lifetime. In particular, with a three-layer structure packed with heavy metal alloy middle layer, a flywheel can achieve both a large rotational inertia and a high energy density. To guarantee structural strength safety of this type of multi-ring packed flywheel operating at high temperature, structural and thermal stress analysis for a flywheel with a tungsten alloy layer sandwiched between two stainless steel layers is carried out using finite element method (FEM). Based on plane stress assumption, flywheel stress characteristic along radius under assembly force, centrifugal force and thermal loading is demonstrated. The results show that, the rise of temperature causes inconsistent thermal expansion between neighbouring layers and results in a large thermal stress. Particularly, a high circumferential tensile stress occurs in the middle layer. It is also found that, the segments of the middle tungsten alloy layer can efficiently relieve the inconsistent thermal expansion and lower the thermal stress. Moreover, the number of segments significantly influences the efficiency of thermal stress releasing and the reserved gap between adjacent segments significantly improves the contact pressure between inner hub and tungsten alloy layer. This work may provide reference for structural design of the multi-ring packed flywheels under thermal loading.

Nomenclature

r_1: Inner radius of inner hub [mm]
r_2: Outer radius of inner hub [mm]
r_3: Outer radius of tungsten alloy layer [mm]
r_4: Outer radius of outer retainer [mm]
E: Elastic modulus of material [GPa]
υ: Poisson's ratio
ρ: Mass density [kg/m^3]
σ_y: Yield strength [MPa]
α: Thermal expansion coefficient [/K]
σ_r: Radial stress [MPa]
σ_θ: Circumferential stress [MPa]
ω: Angular velocity [rad/s^2]
δ: Interference fit magnitude [mm]
T: Temperature [K]
p: Contact pressure [MPa]

N: segment number
δ_{gap}: gap dimension [mm]

1. Introduction

With the advantages of high efficiency, no pollution and long lifetime, flywheel has broad applications in power system, vehicles and aerospace [1-3]. The flywheel stores enormous kinetic energy at rotating condition and releases this energy when needed, e.g., it can be used to punch the workpiece in punching machine, stabilize the output power in power station, and provide the delayed flow in reactor coolant pump (RCP) system [4-6]. Theoretically, the amount of energy stored is linearly proportional to the square of rotation speed and the rotational inertia of rotor about the rotation axis [7,8]. However, when the rotation speed is fixed by machine system, a rotor with large enough rotational inertia is desired to ensure the ability of energy storage. To enhance rotational inertia of a rotor, one may design a larger geometry or select a higher density material [9]. Nevertheless, for traditional single disk rotor, the large geometry encounters high circumferential stress in inner radius and high density material often results in an overweight exceeding load-carrying ability of supporting bearings [10-13]. On the other hand, in terms of materials utilization, the single disk rotor is not efficient as the mass element adjacent to rotation axis has much less contribution to rotational inertia than the mass near outer rim. To improve such an old design, three-layer structure flywheel packed with a high density layer (middle layer) was recently proposed [14-16]. Owing to its large rotating radius and high density middle layer, this multi-ring packed flywheel can achieve not only a larger rotational inertia but also a higher energy density, which provides novel design guidance for large rotational inertia flywheels.

To date, little is known about the stress characteristics of this flywheel under thermal loading. However, in reality, during early design phase, thermal stress evaluation is very essential to guarantee safety operation under temperature effect [17-19]. With respect to this multi-ring packed flywheel, the inner layer is responsible for connecting the shaft with flywheel, and the middle layer supplies the major rotational inertia of flywheel. As for the outer layer, it plays the role of packing the inner and middle layers. The outer layer is assembled outside the middle layer using an interference fit to provide prestress to keep inner layers closed. When this flywheel operates under high temperature condition, apart from assembly prestress and rotating stress, thermal stress occurs due to strain induced by thermal expansion [20,21]. As pointed out by Lippmann [22] and Kovacs [23], temperature rising reduces the assembly prestress in the interface of interference fit and increases the risk of assembly failure. Antoni [24] considered a rotating two-cylinder assembly under thermal loading and found that the reasonable magnitude for interference fit is necessary to prevent the occurrence of separation due to temperature effect. On the other hand, the multi-ring packed flywheel confronts inconsistent thermal expansion due to mismatch in expansion coefficients between neighboring layers. You [25] showed that the mismatch in thermal deformation often causes the fiber breaks, debonds and cracks at the interfaces of fiber-reinforced composite cylinder. Clearly, effect of thermal loading needs to be studied for the multi-ring packed flywheel operating at high temperature.

In this communication, structural and thermal stress analysis for the multi-ring packed flywheel in RCP is carried out using finite element method (FEM). Two-dimensional finite element analysis (FEA) model is established through introducing plane stress assumption and the stresses along radius under assembly force, centrifugal force and thermal loading are discussed. To understand the coupling effect of these loads, load condition in the FEA process is divided into three steps: assembly, rotating and thermal loading conditions. Under assembly condition, the interference fit is studied to determine assembly prestress, while under rotating condition both the interference fit and rotational speed are considered to obtain rotating stress. Under thermal condition, the interference fit, rotating speed and thermal loading are all taken into account. For a deep understanding of the multi-ring packed structure, structural and thermal stress characteristics of the flywheel packed with a number of tungsten alloy segments middle layer is investigated. Analysis processes are accomplished in Abaqus which is specialized in simulating coupled fields and contact problems [26-29].

2. Numerical analysis

2.1. Model construction

As shown in figure 1(a), three layers are described as inner hub, heavy metal alloy layer, and outer retainer. They play the roles of connecting rotor with shaft, providing a large rotational inertia, and packing the inner and middle layers, respectively. The shaft and inner hub are made of stainless steel, and heavy metal layer is a type of tungsten alloy with twice the density of stainless steel. The outer retainer is made of high strength maraging steel so as to withstand the assembly prestress and rotating stress. Flywheel is assumed to rotate in horizontal plane, i.e., the rotating shaft is in vertical direction. Such arrangement is commonly seen in RCP system [30] and wind power generation system [31].

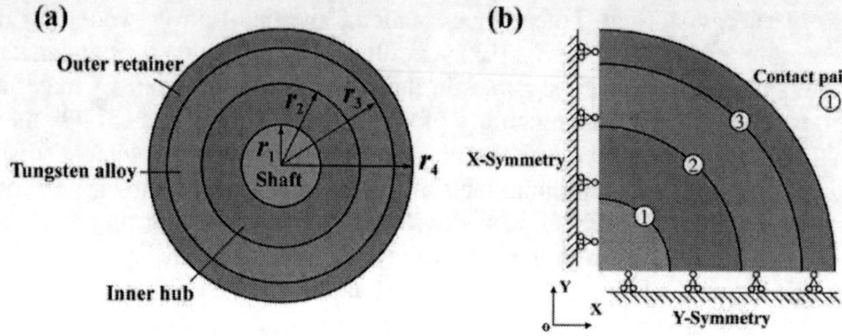

Figure 1. Multi-ring packed flywheel with tungsten alloy layer: (a) structure diagram, (b) FEA model.

The load consists of three parts: assembly force, centrifugal force, and the stress induced by thermal expansion. The gravity force is negligible as it has much less effect on stress relative to these forces. Considering the smaller axial thickness relative to the flywheel radial radius, we give concentration on the stress distribution along radial direction using plane stress model [32]. To further reduce calculation cost, 1/4 structure is taken to construct the FEA model due to its symmetry characteristic as used in literature [15]. As shown in figure 1(b), symmetry boundary conditions are defined, respectively. Contact relationships between the shaft, inner hub, tungsten alloy layer, and outer retainer are considered. Surface to surface contact is used, and friction coefficients are all set as 0.15 in Abaqus. Final model established is displayed in figure 1(b). Tables 1 and 2 give respectively the geometry parameters and mechanical properties for flywheel.

Table 1. Geometry parameters of the multi-ring packed flywheel.

Description	Value
Inner radius of inner hub, r_1	150 mm
Outer radius of inner hub, r_2	300 mm
Outer radius of tungsten alloy layer, r_3	430 mm
Outer radius of outer retainer, r_4	500 mm

Table 2. Mechanical properties of the flywheel [33].

Components	E (GPa)	ρ (kg/m^3)	σ_y (MPa)	α (/K)
Shaft, inner hub	198	7900	275	1.12×10^{-5}
Tungsten alloy	325	17500	520	4.5×10^{-6}
Outer retainer	198	7900	1500~1700	1.12×10^{-5}

2.2. Rotating stress

Different initial stress fields may express distinct responses to thermal loading. Here, rotating stress of this multi-ring packed flywheel is studied in advance before thermal stress analysis. To look into its stress characteristic, the load condition is divided into two steps: assembly condition and rotating condition. Under assembly condition, the interference fit between tungsten alloy layer and outer retainer is studied to determine assembly prestress, while under rotating condition both the interference fit and rotational speed are considered to obtain rotating stress. In this paper, interference fit magnitude of δ=1.0 mm and rotational speed of n=1800 rpm is studied as example. Centrifugal force is applied by defining a rotational body force with the given angular velocity in Abaqus (angular velocity, ω=188 rad/s^2). Since the stress keeps constant around the circumference, the radial stress, σ_r, and circumferential stress, σ_θ, along the radius are evaluated.

In figure 2(a) we can see that, the radial stresses of flywheel along radius show characteristic of continuous distribution. Clearly, owing to the assembly prestress, effective contacts between layers are established. The contact pressures in three contact interfaces are defined respectively as p_1, p_2 and p_3 (see figure 3) and they show the relationship of $0<p_1<p_2<p_3$. The centrifugal force makes all layers expand outward releasing part of the contact pressures. Specifically, this rotating condition significantly lowers p_1 but has less impact on p_3 (see red line with circle symbols in figure 2(a)). With respect to circumferential stresses (see figure 2(b)), the inner hub and tungsten alloy layer both show low compressive stress levels rather than high tensile stress in inner radius for traditional single disk flywheel [34]. Apparently, this outer retainer plays a critical role in packing the flywheel and making inner layers stay at low stress levels.

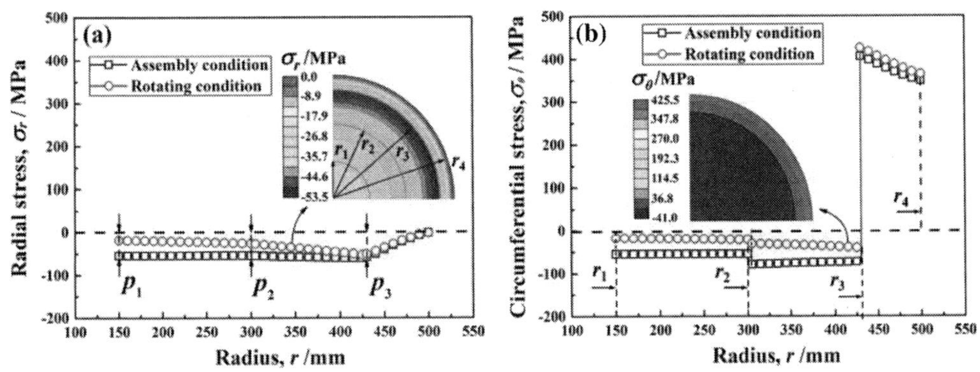

Figure 2. Stress distribution along radius: (a) radial stress, (b) circumferential stress under assembly and rotating conditions.

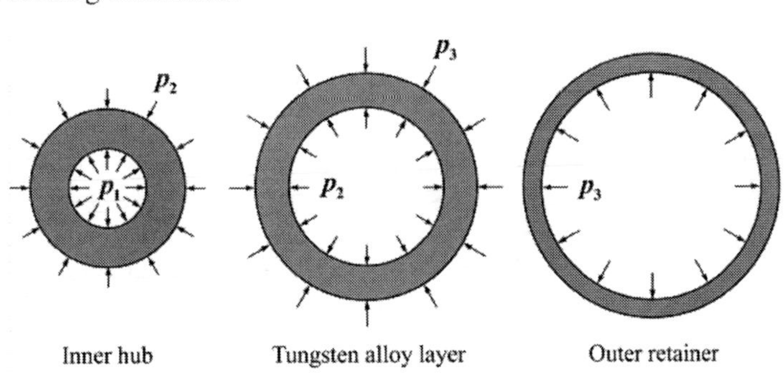

Figure 3. Schematic of the contact pressures between layers.

2.3. *Thermal effect*

As known, thermal stress takes place due to either non-uniform temperature fields or mismatch in thermal expansion coefficients of components. Since the thermal expansion coefficient of tungsten alloy layer is about one half of that of inner layer and outer retainer, thermal stress occurs due to inconsistent thermal expansion between layers. To investigate its effect, stress analysis under interference fit magnitude of δ=1.0 mm, rotational speed of n=1800 rpm and constant temperature of T=561 K (steady temperature caused by coolant flow through the reactor core under normal working condition [35]) for RCP flywheel is carried out. Room temperature of T=294 K is taken as initial temperature field.

Figures 4(a) and 4(b) give the radial and circumferential stress distributions along the flywheel radius. It is seen that temperature effect leads to remarkable rising of stresses in the inner hub and tungsten alloy layer (see red lines with circle symbols). Particularly, considerable high stress turns up in the inner surface of tungsten alloy layer. In fact, considering the thermal expansion, the tungsten alloy layer has a smaller radial deformation than inner layers (shaft and inner hub) due to its smaller thermal expansion coefficient. Consequently, this inconsistent radial deformation builds an interference fit relationship between tungsten alloy layer and inner hub. As presented in figures 4(a) and 4(b), this additional interference fit results in much larger contact pressures between layers and the considerable high stress in inner surface of tungsten alloy layer. With the same mechanism, the outer retainer has a larger radial deformation than tungsten alloy layer which leads to the decrease of circumferential stress in outer retainer.

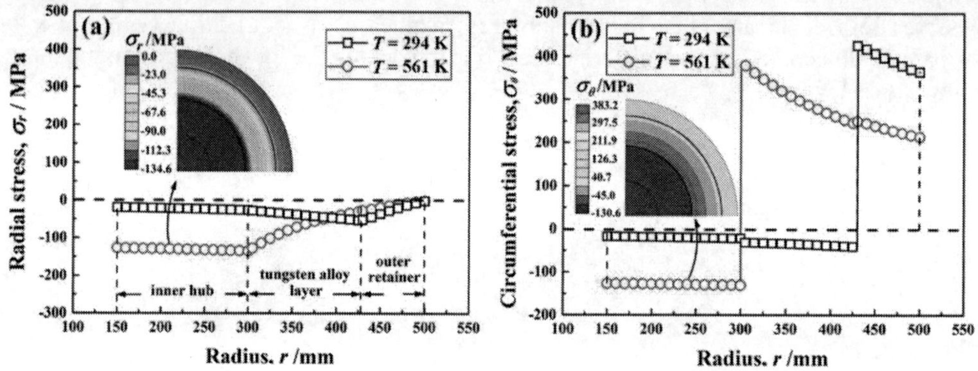

Figure 4. Temperature effect on: (a) radial stress, (b) circumferential stress distributions along radius.

3. Segment design for the middle layer

3.1. Thermal stress characteristic

To deal with the thermal stress problem pointed out, an improved design for multi-ring packed flywheel is proposed. The integrated tungsten alloy layer is replaced by 12 tungsten alloy segments evenly around the circumference (see figure 5(a)). Flywheel becomes a typical cyclic symmetric structure about the rotation axis and contact relationships between tungsten alloy segments need to be considered. Based on this symmetry characteristic, 1/12 flywheel structure is taken to construct the FEA model. Four contact pairs are established between the shaft, inner hub, tungsten alloy segments and outer retainer. To make a comparison with the integrated tungsten alloy layer, we also carry out stress analysis process under assembly and rotating conditions with δ=1.0 mm and n=1800 rpm for this type of design. Considering that stress changes around the circumference, stress along the radial path crossing a single segment center (see $\vec{r}$ in figure 5(c)) is observed. Comparing results in figures 5(c) and 5(d) with those in figures 2(a) and 2(b), we can find that two types of flywheel designs present almost the same stress distributions.

Further, to learn the thermal stress characteristics, radial and circumferential stress along the radius under temperature of T=561 K is given in figures 6(a) and 6(b). It can be found that, stresses in inner hub and tungsten alloy layer are both under compression and show low stress levels. Though at high temperature, they both maintain a low stress level rather than the significant rising presented in figure 4 To understand this phenomenon, firstly, we recognize that the segment design of tungsten alloy layer substantially breaks the state of high tensile stress around the circumference and thus releases the

IOP Conf. Series: Materials Science and Engineering **657** (2019) 012002 doi:10.1088/1757-899X/657/1/012002

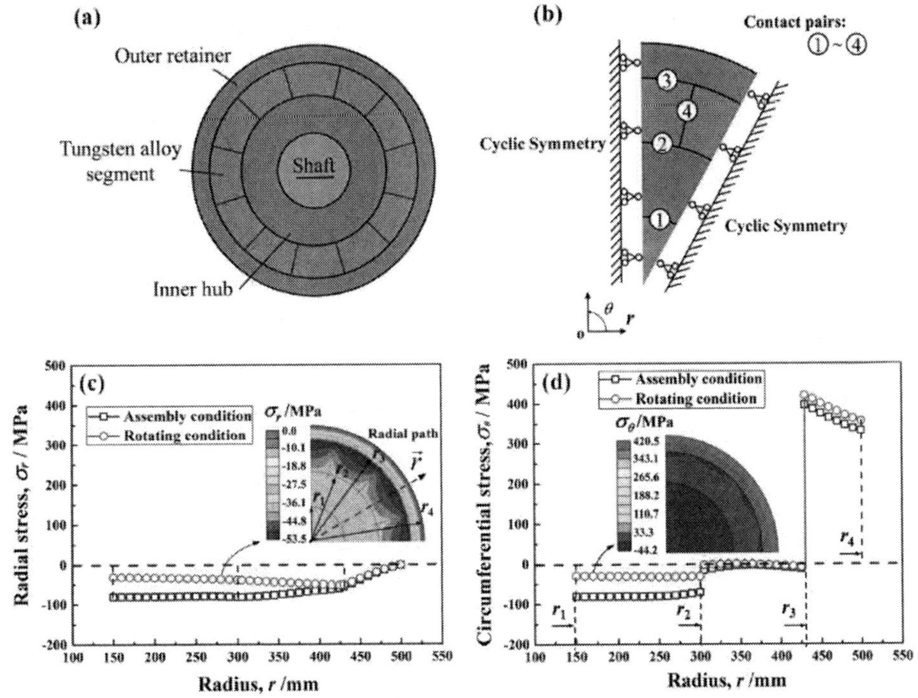

Figure 5. Multi-ring packed flywheel with tungsten alloy segments: (a) structure diagram, (b) FEA model, (c) radial stress, and (d) circumferential stress distributions along specified radial path.

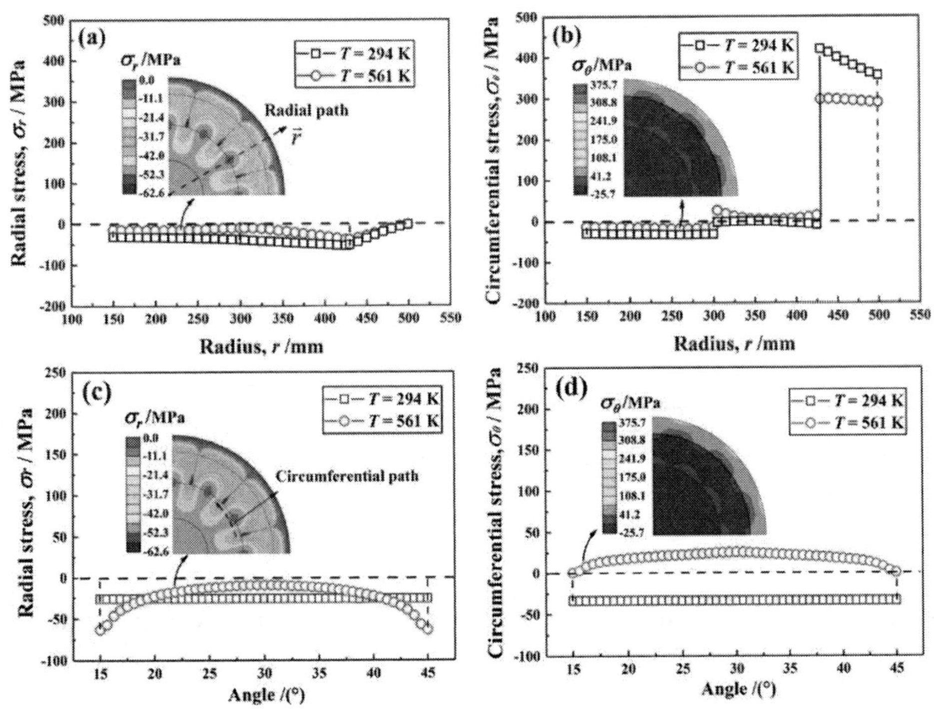

Figure 6. Temperature effect on multi-ring packed flywheel with tungsten alloy segments: (a)-(b) radial and circumferential stresses along specified radial path, (c)-(d) radial and circumferential stresses along specified circumferential path.

circumferential stress. Secondly, instead of an integrated tungsten alloy layer, those tungsten alloy segments enhance the deformation ability of tungsten alloy layer around the circumference. Thus, previously stated inconsistent thermal expansion between layers will be greatly weakened and thus thermal stress will be highly relieved. Consequently, these two factors contribute to these low stresses in inner hub and tungsten alloy layer. This improved flywheel design with tungsten alloy segments middle layer can successfully solve the thermal stress problem and maintain the low stress environment for packed inner components. However, what we need to acknowledge is that segment design break the uniform contacts between the inner hub, middle layer, and outer retainer. Figures 6(c) and 6(d) show the radial and circumferential stresses along a specified circumferential path in the inner surface of a single segment (see figure 6(c)). It is seen that, within the single segment, pressure in middle contact area is higher than that in both sides. In terms of this non-uniform contact, effect of segment number will be studied in the following subsection.

3.2. Effect of the segment number for middle layer

Segment design of middle layer releases the high tensile stress around circumference and weakens the interference fit relationship built by inconsistent thermal expansion. To understand the segment number design for tungsten alloy layer, effect of segment number on relieving the circumferential stress is explored. Flywheel stresses with different segment numbers are studied. Similarly, making full use of the structural cyclic symmetry characteristic, stress analysis processes are carried out using FEM. In figure 7, the stress contour with N=1 represents integrated tungsten alloy layer of which high circumferential stress is displayed. When integrated tungsten alloy layer is divided into a number of segments, the high circumferential tensile stress is highly released. Specifically, it presents a fast declining with increasing segment numbers. But this decrease slows down quickly with large segment numbers. This performance reveals that different segment numbers contribute to different degrees of stress relieving in middle layer, i.e., too small a number cannot guarantee a low stress. In other words, increasing the segment number can help improve the efficiency of relieving the circumferential stress and weakening the inconsistent thermal expansion. However, we should put in mind that too large a number actually will bring difficulties during practical manufacturing and assembly process. From the curve in figure 7, it can be concluded that segment number of around 10-14 can not only contribute to a high degree relieving of thermal stress, but also maintain simplicity for manufacturing and assembly process.

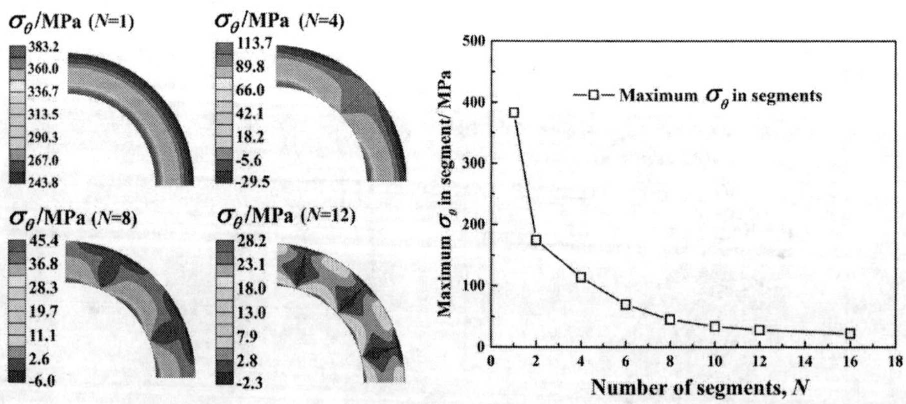

Figure 7. Effect of different segment numbers (N=2, 4, 6, 8, 10, 12, 16) on relieving the circumferential tensile stress, σ_θ, in middle layer (N=1 represents the integrated tungsten alloy layer).

In figure 7, we also noticed that segments break the uniform contacts between inner hub, middle layer and outer retainer. Stress concentration appears in the inner surface of outer retainer and the

IOP Conf. Series: Materials Science and Engineering **657** (2019) 012002 doi:10.1088/1757-899X/657/1/012002

maximum peak stress occurs in the contact area between two adjacent segments. Due to mismatch in interface deformation, interface friction force occurs. This friction force gives shear effect on the inner surface of outer retainer and hence results in non-uniform contact. Stress contour graphs with different segment numbers are given in figure 8. Based on cyclic symmetry characteristic, we look into the stress around circumference in the inner surface of outer retainer within a periodic structure. It is seen that, they all present an inverted funnel shape distribution and the maximum peak stress presents a declining trend with increasing segment numbers.

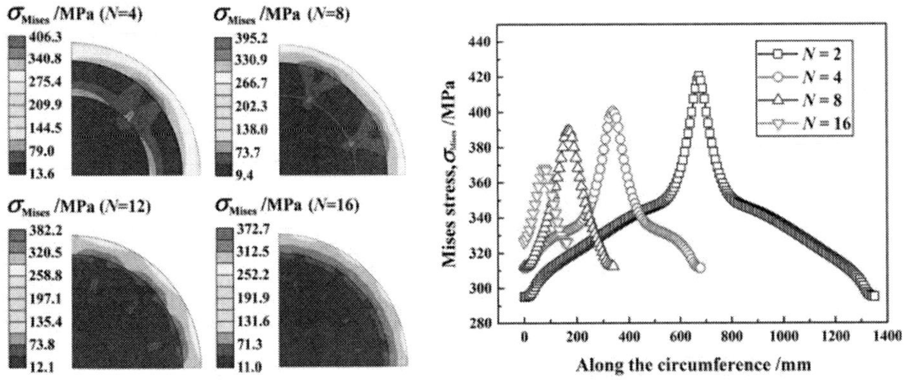

Figure 8. Effect of segment number on stress distribution around circumference in the inner surface of outer retainer.

3.3. Effect of gap dimension on thermal stress

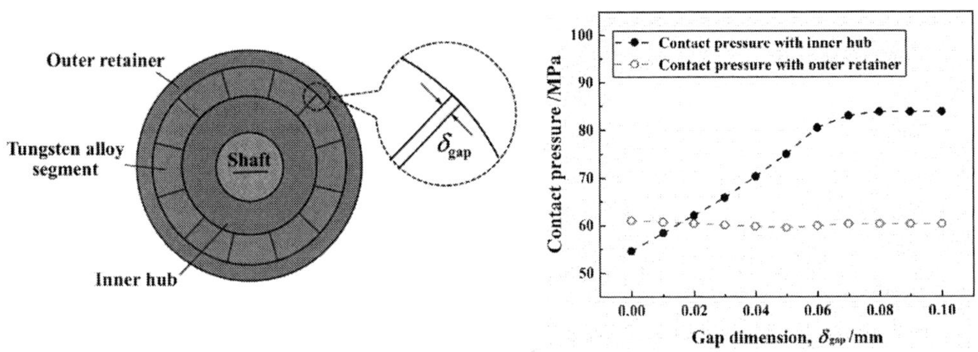

Figure 9. Effect of gap dimension on contact pressures.

Figure 9 shows the gap structure between adjacent segments. This gap actually provides space for the circumferential deformations of segments which will accommodate the thermal expansion and affect the compatible deformations between layers. To explore the effect of gap dimension (δ_{gap}) on thermal stress, contact pressures between neighboring layer (see figure 9) and between adjacent segments (see figure 10) are studied. In figure 9, the maximum contact pressure between tungsten alloy layer and inner hub shows a fast growth with the increase of gap dimension (0~0.08 mm). Apparently, the gap enhances the contact between segments and inner hub. Compared with no gap (δ_{gap}=0 mm), the maximum pressure improvement rate reaches 53.5% when δ_{gap}=0.08 mm. In figure 10, the contact pressure between adjacent segments is 0 MPa when δ_{gap}=0.08 mm. It indicates that adjacent segments are in open state and this gap dimension completely accommodates the circumferential deformation. Thus, the contact pressure becomes stable when the gap dimension exceeds about 0.08 mm. With respect the contact pressure between tungsten alloy layer and outer retainer, it almost keeps a constant versus the gap dimension (see figure 10) as this contact pressure is mainly dominated by the assembly prestress between tungsten alloy layer and outer retainer.

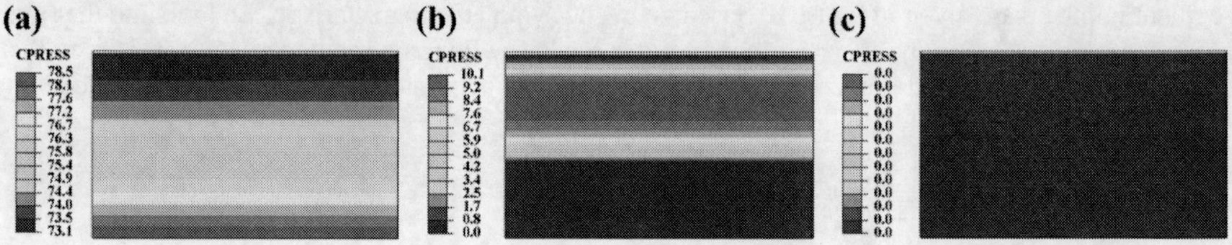

Figure 10. Contact pressures between adjacent tungsten alloy segments under different gap dimensions: (a)δ_{gap}=0 mm, (b)δ_{gap}=0.06 mm, (c)δ_{gap}=0.08 mm.

4. Conclusions

In this paper, structural and thermal stress analysis for a multi-ring packed flywheel composed of inner hub, tungsten alloy layer and outer retainer is carried out. Results show that, for this type of three-layer structure, temperature effect gives rise to considerable thermal stress because of different thermal expansion coefficients of neighbouring layers. Specifically, inconsistent thermal expansion causes high circumferential stress in the middle layer. Fortunately, adopting segment design for middle layer can effectively accommodate and weaken the inconsistent thermal expansion and thus relieve the high circumferential stress. Moreover, a proper segment number contributes to a high stress relieving and also a significant improvement of stress concentration induced by non-uniform contact. Reserving gaps between adjacent segments helps adapt to circumferential deformation and enhances the contact pressure, which is beneficial to keep inner layers closed. This work may provide reference for structural design of the multi-ring packed flywheels under thermal loading.

Acknowledgments

This work was supported by National Basic Research Program of China (Grant No. 2015CB057306), National Natural Science Foundation of China (Grant No. 11572080, 11772086), and Fundamental Research Funds for the Central Universities of China (Grant No. DUT17ZD229).

Conflict of Interest

The authors declare that they have no conflict of interest.

References

[1] Bitterly J G 1998 Flywheel technology: Past, present, and 21st century projections *IEEE Aero. El. Sys. Mag.* **13** 13-6

[2] Arani A A K, Karami H, Gharehpetian G B *et al* 2017 Review of flywheel energy storage systems structures and applications in power systems and microgrids *Renewable Sustainable Energy Rev.* **69** 9-18

[3] Christopher D A and Beach R 1998 Flywheel technology development program for aerospace applications *IEEE Aero. El. Sys. Mag.* **13** 9-14

[4] Tso P L and Li C H 2008 Study of servo press with a flywheel *J. Adv. Mech. Des. Sys.* **2** 1-11

[5] Gayathri N S, Senroy N and Kar I N 2016 Smoothing of wind power using flywheel energy storage system *IET Renew. Power Gen.* **11** 289-98

[6] Gao H, Gao F, Zhao X *et al* 2011 Transient flow analysis in reactor coolant pump systems during flow coastdown period *Nucl. Eng. Des.* **241** 509-14

[7] Jiang L and Wu C W 2017 Topology optimization of energy storage flywheel *Struct. Multidiscip. O.* **55** 1917-25

[8] Jiang L, Zhang W, Ma G J *et al* 2017 Shape optimization of energy storage flywheel rotor *Struct. Multidiscip. O.* **55** 739-50

[9] Dasseh M, Samer Y and Webb J F 2015 A study on maximizing the energy density of a system by choosing a suitable flywheel *Global J. Tech. Optim.* **6** 1-6
[10] Belingardi G, Genta G and Gola M 1979 A study of the stress distribution in rotating orthotropic discs *Compos.* **10** 77-80
[11] Genta G and Gola M 1981 The stress distribution in orthotropic rotating disks *J. Appl. Mech.* **48** 559-62
[12] Genta G 1989 Some considerations on the constant stress disc profile *Meccanica* **24** 235-48
[13] You L H and Zhang J J 1999 Elastic-plastic stresses in a rotating solid disk *Int. J. Mech. Sci.* **41** 269-82
[14] Finegan J R. Kreke F J and Casamassa J J 2013 Reactor coolant pump flywheel *U.S. Patent* 8,590,419
[15] Wang Y, Wang D Z, Yin J *et al* 2014 The use of experimental design for the shrink-fit assembly of multi-ring flywheel *2014 22nd Int. Conf. Nucl. Eng., Prague* V001T03A012 1-6
[16] Wang Y J, Wang D Z, Guo W Z *et al* 2017 The study on the shrink-fit multi-ring flywheel with fall-off function *Nucl. Sci. Tech.* **28** 28-42
[17] Liew K M, Kitipornchai S, Zhang X Z *et al* 2003 Analysis of the thermal stress behavior of functionally graded hollow circular cylinders *Int. J. Solids Struct.* **40** 2355-80
[18] Zhou S S, Gao X L and Griffith G W 2012 Stress analysis and structural optimization of a three-layer composite cladding tube under thermo-mechanical loads *J. Eng. Mater. Technol.* **134** 031001
[19] Vedeld K and Sollund H A 2014 Stresses in heated pressurized multi-layer cylinders in generalized plane strain conditions *Int. J. Pressure Vessels Piping* **120** 27-35
[20] Çallioğlu H 2007 Thermal stress analysis of curvilinearly orthotropic rotating discs *J. Thermoplast. Compos. Mater.* **20** 357-69
[21] Pei Y C, Tan Q C, Zhuang Q W *et al* 2013 Elastic–plastic stresses in rotating connection disk under temperature rise and transmitted torque *Int. J. Mech. Sci.* **69** 141-9
[22] Lippmann H 1992 The effect of a temperature cycle on the stress distribution in a shrink fit *Int. J. Plast.* **8** 567-82
[23] Kovacs A 1994 Thermal stresses in a shrink fit due to an inhomogeneous temperature distribution *Acta Mech.* **105** 173-87
[24] Antoni N 2013 Contact separation and failure analysis of a rotating thermo-elastoplastic shrink-fit assembly *Appl. Math. Model.* **37** 2352-63
[25] You L H 2003 Thermal stress distributions in axisymmetrically orthotropic fiber-reinforced composites with multilayered interfaces *J. Mater. Sci.* **38** 2963-70
[26] Ozturk F 2011 Finite-element modelling of two-disc shrink fit assembly and an evaluation of material pairs of discs *Proc. Inst. Mech. Eng., Part C* **225** 263-73
[27] Qayyum F, Guk S, Kawalla R *et al* 2018 Experimental investigations and multiscale modeling to study the effect of sulfur content on formability of 16MnCr5 alloy steel *Steel Res. Int.* 1800369 1-9
[28] Ullah M, Wu C S and Qayyum F 2018 Prediction of crack tip plasticity induced due to variation in solidification rate of weld pool and its effect on fatigue crack propagation rate (FCPR) *J. Mech. Sci. Tech.* **32** 3625-35
[29] Qayyum F, Kamran A, Ali A *et al* 2017 3D numerical simulation of thermal fatigue damage in wedge specimen of AISI H13 tool steel *Eng. Fract. Mech.* **180** 240-53
[30] Park J S, Song H C, Yoon K S *et al* 2004 A structural integrity evaluation for the interference-fit flywheel of reactor coolant pump *ASME/JSME 2004 Pressure Vessels Piping Conf.* (USA: San Diego) 73-80
[31] Daoud M I, Abdel-Khalik A S, Massoud A *et al* 2012 On the development of flywheel storage systems for power system applications: A survey *2012 XXth Int. Conf. on Electrical Machines, Marseille* 2119-25
[32] Alexandrova N, Alexandrov S and Real P M M V 2004 Displacement field and strain distribution

in a rotating annular disk *Mech. Based Des. Struc.* **32** 441-57
[33] Curtiss-Wright Electro-Mechanical Corporation 2006 Structural analysis summary for the AP1000 reactor coolant pump high inertia flywheel (Pittsburgh)
[34] Pedersen N L 2016 On optimization of interference fit assembly *Struct. Multidiscip. O.* **54** 349-59
[35] Wang Y J, Wang D Z, Guo W Z *et al* 2016 The effect of smaller turbulent motions on heat transfer in the annular gap flow of flywheel *Ann. Nucl. Energy* **91** 1-7

IOP Conf. Series: Materials Science and Engineering **657** (2019) 012003 doi:10.1088/1757-899X/657/1/012003

Effect of clearance jet on aerodynamic performance of centrifugal fan

G J Zhu[1], X Y Bo[1], J J Feng[1,2] and G K Wu[1]

[1]Institute of Water Resources and Hydro-electric Engineering, Xi'an University of Technology, Xi'an 710048, China

E-mail: jianjunfeng@xaut.edu.cn

Abstract. The change of the depth of the inlet collector inserted into the impeller will change the gap shape between the collector and the impeller, and then affect the performance of the ventilator. Taking the small centrifugal ventilator as the object, the influence of the depth of the impeller inserted into the collector on the aerodynamic performance of the ventilator is studied in this paper. Firstly, the depth of the collector inserted into the front disk of the impeller is adjusted to 2 mm, 3 mm and 4 mm respectively to change the geometry of the clearance. Then, based on the computational fluid dynamics analysis method, the influence of the insertion depth of the above three different concentrators on the aerodynamic performance of the ventilator is analyzed. The analysis results show that the gap leakage is positively correlated with the depth of the impeller inserted into the collector, and the aerodynamic performance of the ventilator can be significantly improved by reducing the insertion depth of the collector under the conditions of design flow rate and large flow rate, but under the condition of small flow, the decrease of insertion depth weakens the effect of gap jet on the flow field in impeller, resulting in the decrease of fan efficiency and static pressure.

1. Introduction

Centrifugal fan is widely used in the field of ventilation and cooling in industrial production. The external rotor centrifugal fan usually includes three parts: the inlet collector, the impeller and the outer rotor motor. There is a gap between the inlet collector and the impeller to separate the dynamic and static parts. The volume loss caused by the gap is the main reason for the decrease of the efficiency of the ventilator. Therefore, it is of great significance to understand the clearance leakage mechanism of centrifugal fan and the influence of leakage on the aerodynamic performance of centrifugal fan to improve the aerodynamic performance of centrifugal fan.

The influence of the inlet clearance between the collector and the impeller on the aerodynamic performance of the centrifugal fan cannot be ignored. The internal flow of centrifugal fan is very complicated, and it is difficult to reflect the real characteristics of the internal flow of the centrifugal fan only for the numerical calculation of one blade passage or section of the centrifugal fan [1]. The air flow parameters near the front plate of the fan impeller vary greatly, and are affected by the vortex area on the back of the collector. It is easy to form a large vortex area, which varies with the relative axial position between the collector and the impeller and the flow rate of the fan, which constitutes a main source of fan loss and noise. Kind and Tobin [2] tested the flow field of the squirrel cage fan with a five-hole probe, the results show that the impeller flow state under low flow conditions is different from that of the impeller under high efficiency conditions. Fukutomi *et al* [3] and Wang *et al* [4] designed the inlet nozzle of a fan, and improved the aerodynamic performance of fan by affecting the effective flow area of impeller and weakening the flow vortex in blade channel at a specific position.

In centrifugal turbomachines, the leakage of impeller inlet clearance is one of the main reasons for its efficiency reduction. Dong *et al* [5] investigated the effect of gap leakage jet on the flow field structure of centrifugal ventilator under different relative length of inlet clearance. Bunker [6] has found that the leakage flow of the top clearance of the turbine blade is one of the important factors leading to the increase of the flow loss. In addition, other scholars have studied the effect of ring clearance on leakage loss and mechanical efficiency of centrifugal pump [7-9]. CFD has become a main method to calculate the internal flow field of turbomachine and analyze its internal flow mechanism [10].

The pressure difference flow caused by the pressure difference at both ends of the gap is the main cause of the formation of the gap jet. Rain [11] uses the leakage model to analyze the pressure difference flow in the compressor, and accurately forecasts the gap loss of the compressor. Baskharone *et al* [12] used finite element analysis method to simulate the leakage jet flow in the internal clearance of multistage pump, and compared it with the existing impeller leakage analysis model. The fan clearance has an influence on the adjacent flow channel. When the radial clearance is large enough at any flow rate, the influence can be extended to the whole flow channel area [13]. Therefore, it is very important to find a suitable inlet clearance to control the flow field at the inlet of the ventilator.

The depth change of the current collector into the front disc of the impeller will affect the geometry and size of the gap between the current collector and the impeller, thus affecting the performance of the ventilator. In order to explore the influence of the depth change of the collector inserted into the impeller front disk on the internal flow characteristics and energy external characteristics of the centrifugal fan. In this paper, different schemes by adjusting the depth of the collector inserted into the impeller front disk have been obtained, and then the aerodynamic external characteristics and internal flow field characteristics of the fan have been compared and analyzed under different schemes based on the computational fluid dynamics method. Finally, according to the analysis results, the variable influence of the insertion depth of the collector on the fan performance is revealed, which provides a reference for the relative position design between the collector and the centrifugal fan impeller in the future.

2. Calculation method

2.1. Geomery and grid generation

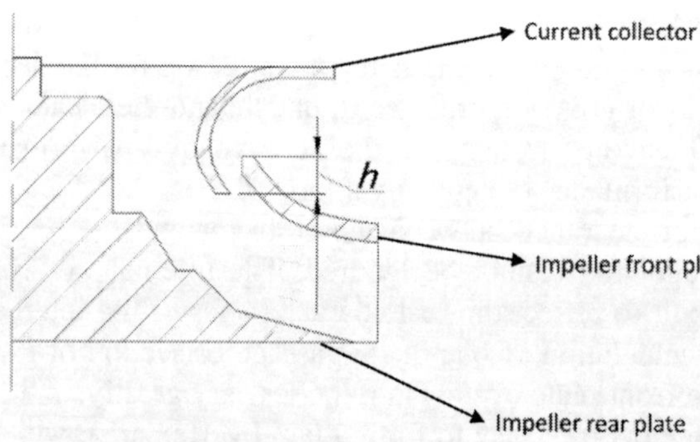

Figure 1. The depth of the collector inserted into the front disk of the impeller is *h*.

Taking a small centrifugal fan as the research object, the main parameters of the fan are as follows: the design speed is n= 3300 rpm, the design flow rate is Q_v= 364 m^3/s, the inner diameter of the impeller is D_1= 134.6 mm, and the outer diameter of the impeller is D_2= 196 mm, blade number z = 6. The 3D modeling software centrifugal fan is used to model the fan. Figure 1 shows the profile of fan impeller, and the insertion depth *h* is 2 mm, 3 mm and 4 mm, respectively. The model is divided into three parts: the inlet section, the impeller section and the exit section.

The whole fluid region is divided into a rotating impeller area and two static regions, and the fan impeller is inside the wind tunnel, as shown in figure 2. In order to fully simulate the flow field in the impeller, the inlet region of the model is hemispherical, and the outlet section is the modeling of the wind tunnel.

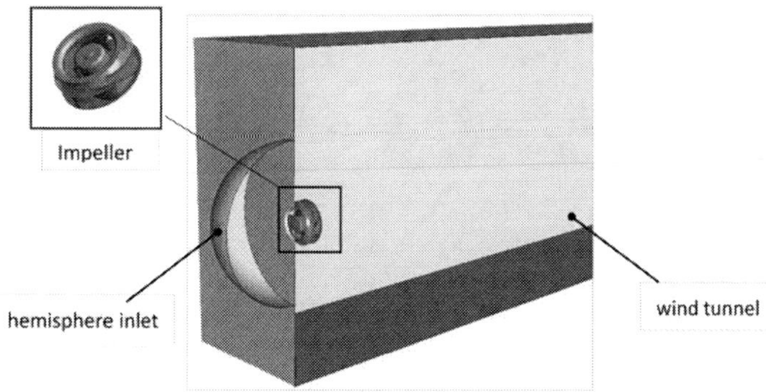

Figure 2. Three-dimensional modeling of the ventilator.

The full flow channel of the ventilator is divided by hexahedral structured grid, as shown in figure 3, the total grid number is verified by the static pressure and efficiency of the ventilator. When the total number of fan model grid is 5.29 million, the number of grid has no effect on the calculation results.

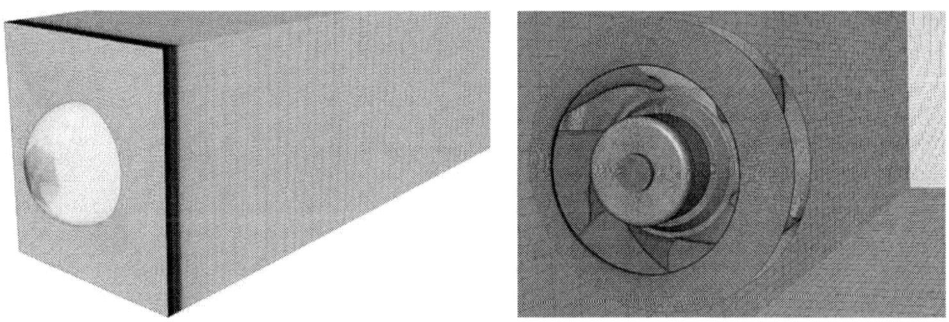

(a) (b)

Figure 3. Grid division of ventilator. (a) Grid view and (b) impeller grid.

The aerodynamic performance of the prototype fan was tested by the large wind tunnel rear side method. The aerodynamic performance was tested on the test platform with reference to the international standard AMCA-210-2007 [14]. Figure 4 shows the sketches and photos of the experimental equipment.

(a) (b)

Figure 4. Aerodynamic performance experimental device. (a) Fan experimental wind tunnel and (b) Inner impeller diagram of wind tunnel.

The static pressure P_S and static pressure efficiency η at the outlet of the test fan can be calculated according to the test data according to the following formula:

$$P_s = P_t - \frac{1}{2}\rho(\frac{Q_v}{A})^2 \tag{1}$$

$$\eta = \frac{Q_v.P_s}{W} \times 100\% \tag{2}$$

Where Q_v represents the volume flow rate of the ventilator, A indicates the area of the outlet of the wind tunnel, and ρ is the density of the air, P_t, W is the total pressure and total power consumed by the test fan.

2.2. Boundary condition settings

In the numerical calculation, the inlet boundary condition of the calculation domain is set to the static pressure boundary condition and total pressure is 0, the outlet boundary is given the mass flow rate, and the solid wall is set to the non-slip wall. In the steady calculation, the interface between the rotating part and the static part is connected by the frozen rotor method. In the process of solving the N-S equation, the convective term in the governing equation is discretized by the second order difference scheme, and the SST k-ω turbulence model is used for numerical simulation. In order to verify the reliability of the numerical calculation, five working conditions are selected for steady calculation in and around the rated working condition of the unit, and the working conditions are shown in table 1.

Table 1. Working conditions and parameters of centrifugal ventilator.

Working condition number	Airflow rate /(m^3*h^{-1})
Condition 1	150
Condition 2	313
Condition 3	364
Condition 4	500
Condition 5	626

The experimental results of static pressure efficiency for the clearance depth of the prototype fan of 3 mm are compared with the results of numerical simulation, as shown in figure 5 below. It can be seen that the results of numerical simulation are the same as the experimental results, and the agreement between them is very good, which indicates that the results of numerical simulation are reliable. The efficiency obtained by numerical simulation is slightly larger than the experimental value, which is due to the fact that the surface roughness of fan impeller is not taken into account in the process of numerical simulation, so the numerical efficiency is slightly higher than the actual efficiency.

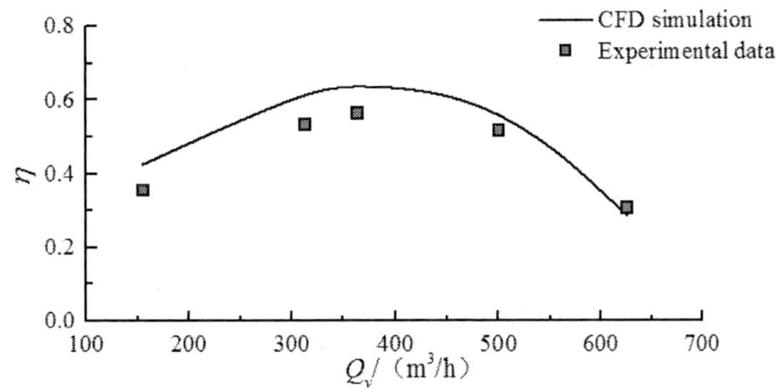

Figure 5. Comparison between numerical simulation results and experimental values.

3. Results and discussion

According to the different insertion depth of the collector, five calculation points are calculated with the help of Ansys software CFX. Through the treatment and analysis of the numerical results, the external characteristic curve and the internal flow field structure of the centrifugal fan are obtained.

3.1. Analysis of external characteristic curves

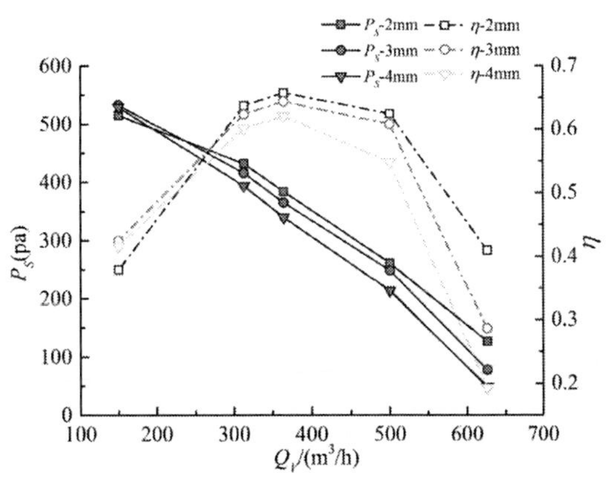

Figure 6. Static pressure and efficiency curve of centrifugal fan at different insertion depth.

The static pressure and efficiency of centrifugal fan are selected as the external characteristic parameters of centrifugal fan. Figure 6 shows the static pressure and efficiency curves of centrifugal fan at different depth of collector insertion. It can be seen that with the increase of the depth of the front disk of the impeller inserted into the collector, the static pressure shows a decreasing trend. In the small flow area, with the increase of the gap depth, the static pressure changes little, but the static pressure decreases obviously in the large flow area. As far as the efficiency curve is concerned, the efficiency change of different inlet clearance fans is basically the same, showing the trend of increasing at first and then decreasing. When the clearance depth increases from 2 mm to 3 mm, the fan efficiency decreases by 2% under rated working conditions, and the fan efficiency decreases more obviously in large flow area. To sum up, when the gap insertion depth is 2 mm, the fan has obvious advantages near rated working conditions and in large flow area.

Figure 7 shows the variation of gap leakage in different sizes. With the increase of the depth of the front disk of the impeller inserted into the collector, the leakage of the gap increases, which is caused by the increase of the cross section area of the gap. With the increase of inlet flow rate, the total pressure obtained by the gas decreases gradually, and the pressure difference between the two sides of the clearance becomes smaller with the decrease of the total pressure of the fan, so the leakage loss will also decrease, and the trend is more obvious with the increase of the insertion depth. This shows that when the insertion depth of the front disk of the impeller is 2 mm, the leakage of the fan is reduced, the internal flow field of the fan is improved, and the efficiency of the fan is also improved.

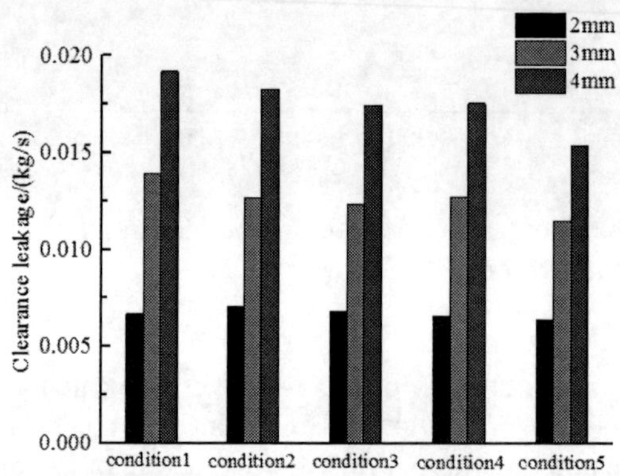

Figure 7. Leakage quantity of gap of different dimensions.

3.2. Internal flow analysis

The axial cross-section of the flow field is selected, as shown in figure 8, there is a significant swirl flow in the space area between the air inlet, the casing and the front disc of the impeller, and the flow from the impeller outlet to the gap between the impeller and the air inlet experiences a process of generation, development and dissipation. The change of the insertion depth of the front disc of the impeller is bound to influence the flow field dynamics in the fan, which in turn leads to the change of the air performance of the fan.

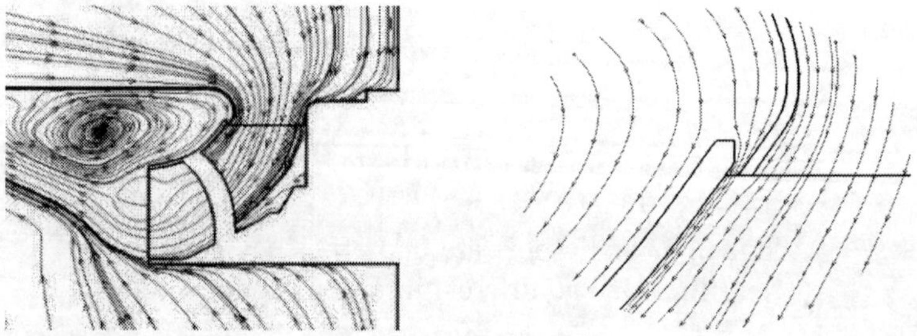

(a) (b)

Figure 8. Axial profile of fan flow field. (a) Fan impeller internal flow field and (b) Internal reflux in fan clearance.

The flow rates of 150 m^3/h, 313 m^3/h and 626 m^3/h are selected as the small flow, design flow and large flow of centrifugal ventilator for detailed analysis. Figure 9 shows the three-dimensional flow distribution diagram at 3 different gaps under 3 working conditions. Under the condition of small flow rate, the internal flow pattern of the blade flow channel is disordered, and the effect of the gap jet on the internal flow pattern is not obvious. With the increase of the inlet flow rate of the fan, the fluid distribution in the single flow channel is more uniform, but the flow pattern in the fan impeller with different clearance is not much different.

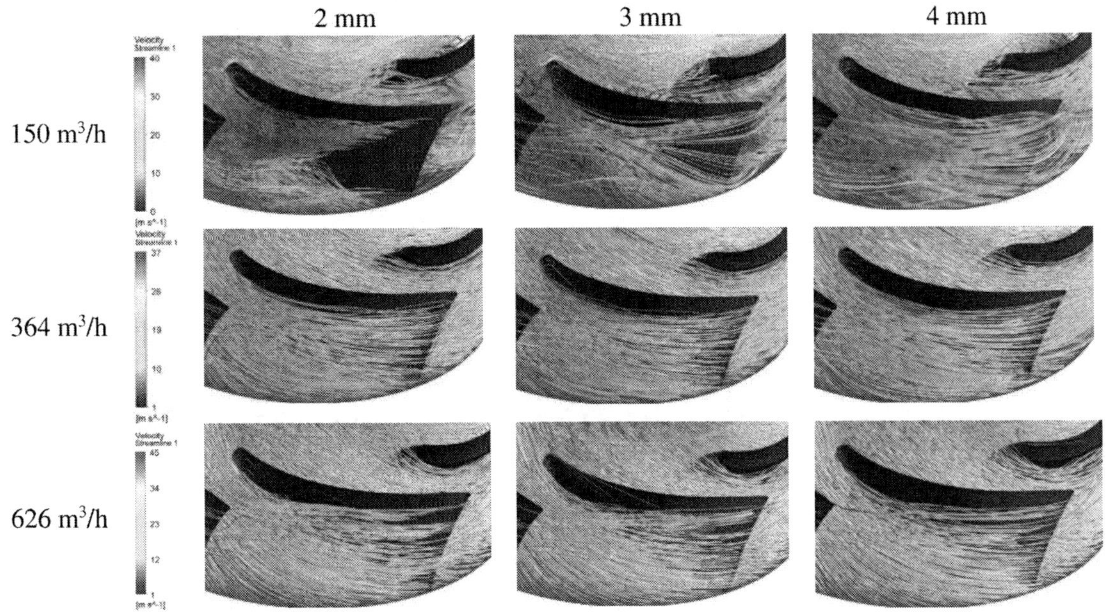

Figure 9. Three-dimensional streamline distribution map at different clearance of fan.

Figure 10 shows the pressure distribution of blade airfoil under different working conditions. In order to investigate the effect of gap jet on the pressure distribution of the blade, the blade position near the gap, that is, the blade airfoil surface at 0.98 times the blade height, is selected for analysis. Where the transverse coordinate z/c denotes the relative chord length of the airfoil, z/c is 0, 1 represents the head of the airfoil, the tail position, and the longitudinal coordinate is the static pressure coefficient Cp, which is defined as the static pressure coefficient.

$$\mathbf{C_p} = \frac{\boldsymbol{P_s}}{\frac{\mathbf{1}}{\mathbf{2}}\boldsymbol{\rho v^2}} \tag{3}$$

Where Ps is static pressure, ρ is density, v is the circumferential velocity of the outer edge of impeller.

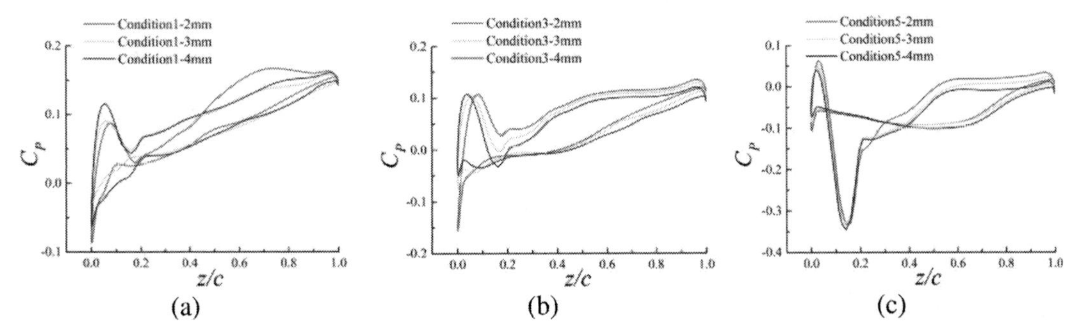

Figure 10. Pressure distribution of airfoil on the front disk surface of fan impeller. (a) Q_v=150 m³/h, (b) Q_v =362 m³/h and (c) Q_v=626 m³/h.

It can be seen from the figure that with the change of the size of the gap, the pressure change trends of the two ends of the wind wing type are basically the same. The pressure difference between the blades near the hub is the largest, along the chord length direction of the airfoil, the pressure difference between the two ends of the blade is first reduced and then increased. The pressure difference of the two ends of the airfoil at the strength of 0.2 times is the smallest, and the pressure distribution of the airfoil surface of the blade is found to have no significant difference from other cross-sectional analysis. The inlet flow rate of the fan has a great influence on the pressure distribution on the airfoil surface, especially under the condition of large flow rate, the pressure surface pressure of the blade near 0.2 times chord length is much smaller than that near the suction surface, which leads to the uneven force on the blade and the serious decrease of the fan efficiency. Therefore, the fan should avoid running under the condition of large flow rate.

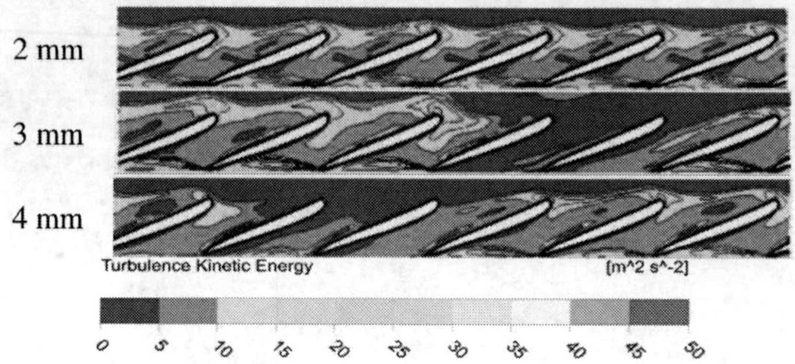

Figure 11. Distribution of kinetic energy.

Figure 11 shows the kinetic energy distribution of three clearance impellers at 73% of height of the fan under the condition of small flow. It can be seen from the diagram that when the clearance depth of the impeller is 2 mm, the internal flow is unstable, the turbulent kinetic energy at the inlet of the impeller and the outlet of the impeller is slightly larger. With the increase of the insertion depth of the collector, the influence of the gap jet on the inner flow channel of the impeller is enhanced, which makes the flow in part of the impeller smooth and the turbulent kinetic energy in the impeller decreases.

4. Conclusions

- In the small flow area, with the increase of the gap, the static pressure changes slightly. However under the design flow and large flow conditions, the pneumatic performance of the ventilator can be significantly improved by the decrease of the insertion depth of the collector, and the operation advantage of the fan with the clearance depth of 2 mm is obvious.
- With the increase of gap depth, the amount of gap leakage increases. Under the condition of small flow rate, the clearance jet can improve the flow field in the impeller obviously. When the clearance depth is 3 mm, the turbulent kinetic energy in the impeller decreases obviously, and the static pressure and efficiency of the fan increase.
- The inlet flow rate of the fan has a great influence on the surface pressure distribution of the airfoil, and the pressure difference between the two sides of the airfoil is the smallest at 20% chord length. Under the condition of large flow rate, the pressure surface pressure of the blade near 20% chord length is much smaller than that of the suction surface, which leads to the uneven force of the blade and the serious decrease of fan efficiency.

Acknowledgments
This work was partially supported by Natural Science Basic Research Plan in Shaanxi Province of China (Program Nos. 2019JQ-044 and 2018JM5102) and Scientific Research Plan Projects of Shaanxi Education Department (Program No. 19JK0570), and also be funded by the National Natural Science

Foundation of China (Grant No. 51679195).

References

[1] Li X H, He H W *et al* 2002 Numerical analysis of whole steady flow field for a centrifugal fan *J. Eng. Therm.* **23** 453-6

[2] Kind R J and Tobin M G 1990 Flow in a centrifugal fan of the squirrel-cage type *J. Turbomach.* **112** 84-90

[3] Fukutomi J, Kuwauchi T, Itabashi A *et al* 1999 A study ofsirocco fan with eccentric inlet nozzle *Trans. JSME: Ser. B* **65** 2023-9

[4] Wang J B, Ou Y D and Wu K Q 2005 Analysis on the vortices flow in the blade passages of the fan for air-conditioner and the influence of the eccentric fan inlet *J. Eng. Thermophys* **26** 951-3

[5] Dong Y S, Yang A L, Chen E Y *et al* 2016 Influence of clearance leakage jet flow in wear-rings on flow characteristics of a centrifugal fan *Chin. J. Power Eng.* **36** 286-93

[6] Bunker R S 2006 Axial turbine blade tips: Function, design, and durability *J. Propul. Power.* **22** 271-85

[7] Tam L T, Przekwas A J, Muszynska A *et al* 1987 Numerical and analytical study of fluid dynamic forces in seals and bearings *J. Vib. Acoust.* **110** 112-9

[8] Childs D W 1991 Centrifugal-acceleration modes for incompressible fluid in the leakage annulus between a shrouded pump impeller and its housing *J. Vib. Acoust* **113** 209-18

[9] Wang Y and Zhang X 2008 Effects of clearance of impeller wear rings on efficiency of low specific speed centrifugal pump *J. Drain. Irrig. Mach. Eng.* **26** 27,30

[10] Coutierdelghosa O, Régiane F P, Reboud J L *et al* 2003 Experimental and numerical studies of a centrifugal pump in cavitating conditions *J. Fluid Eng.* **125** 492-7

[11] Rains D A 1954 Tip clearance flows in axial compressors and pumps (Pasadena: California Institute of Technology)

[12] Baskharones E A and Wyman 1999 Primary/leakage flow interaction in a pump stage *J. Fluid Eng* **121** 133-8

[13] Xi G and Wang S J 1997 Numerical analysis of the three-dimensional turbulence flow in an unshroud impeller of centrifugal compressors *J. Xi'an Jiaotong Univ.* **1997** 93,98

[14] AMCA/ASHRAE 2007 ANSI/AMCA standard 210-07/ANSI/ASHRAE standard 51-07, laboratory methods of testing fans for certified aerodynamic performance rating (Arlington Heights, IL: Air Movement and Control Association and Atlanta: American Society of Heating, Refrigerating and Air-Conditioning Engineers, Inc.)

IOP Conf. Series: Materials Science and Engineering **657** (2019) 012004 doi:10.1088/1757-899X/657/1/012004

Application of the zero thickness interface element approach to the upper bound limit analysis of shallow tunnel excavation

Y L Lee[1,2]

[1]Department of Civil Engineering, Chung Hua University, 707, Sec. 2, WuFu Rd., Hsinchu, 30012, Taiwan

E-mail: rosalee@chu.edu.tw

Abstract. The aim of this study is to apply the zero thickness element approach to investigate the upper bound value of limit analysis with the different failure mechanism of a shallow tunnel excavation in soft clay. This interface is modelled as a contact element with zero thickness and with the Coulomb's friction behaviour. The procedure of numerical analysis adopted the zero thickness element particularly developed and installed into the finite element program in the laboratory. A series of finite element analysis is used to investigate the solutions of stability and simulate the soil progressive failure mechanism due to a shallow tunnel excavation. Four geometry of failure mechanism are taken into account the effects of overburden depth and critical collapse ratio. According to the results obtained, the upper bound values between the limit equilibrium analysis and the numerical calculation are theoretically coincidence, and also the progressive failure mechanism of cohesive soil can be practically predicted. The numerical analyses show that the predicted stability ratio is in reasonable agreement with the analytical solutions in the limit analysis.

1. Introduction

A basic engineering decision to be made in designing a tunnel in soft clay is whether or not the tunnel can be excavated without internal support. The safety of constructing a shallow heading in soft clay can be assessed in terms of the fluid support pressure which may be required to maintain stability. This can be estimated by means of the lower and upper bound theorems of plasticity. In this paper, only upper bound stability solutions are derived for collapse under undrained conditions. The results obtained of limit analysis are compared by finite element analysis that using the zero thickness interface elements is particularly investigated. Engineering literature contains a variety of interface element formulations; however, they can generally be classified as either a stiffness approach (e.g. directionally stiff elements) or a constraint approach (e.g. Lagrange multipliers). A hybrid of these two approaches has also been developed. Several types of finite elements have been proposed for the modeling of joints and interfaces. They can be described and classified in the following categories.

A common interface element is proposed by Goodman *et al* [1] with relative displacements being the nodal unknowns. The thickness is assumed to be zero and to avoid overlapping and penetrating a very high normal stiffness of the interface. An element considering relative motions within the interface element as independent degree of freedom is developed [2]. A similar element, based on an isoparametric formulation has been proposed for two- or three-dimensional applications [3,4]. The linkage elements in which only the connections between opposite nodes which are connected by discrete springs is proposed by Frank [5]. Although the desired effect may be obtained, data preparation, mesh generation and interpretation of results turn out to be very involved. Desai *et al* [6] proposed a thin-layer element based on an isoparametric formulation. The main features are the introduction of an independent shear modulus in the elastic range and the possibility of taking into account stick, slip, debonding and rebonding modes. Although some applications have been reported, the performance of this element is not fully investigated. The application of zero thickness elements to the analyses of a retaining wall and the numerical instability caused by ill-conditioning of the stiffness matrix and high stress gradients was also investigated by Day and Potts [7]. Recently, the developments on zero-thickness interface elements for coupled hydro-mechanics problems are described [8].

The approach used in this paper used the idea of the zero thickness elements described by Carol and Alonso [3] and Beer [4] for the analysis of rock joints, the main advantage being an easier implementation into existing finite element codes. The influence of interface stiffness, stability ratio, the failure mechanism and the deformed soil surface profile is studied. The numerically calculated stability ratio is compared with approximate analytically derived values used in limit analysis. Solutions to the stability problem of a shallow tunnel in cohesive material applying the proposed elements are briefly presented.

2. Interface element formulation

2.1. Geometry, stiffness and constitutive law

For the definition of the interface elements, an isoparametric formulation is chosen that it can be easily used in conjunction with the continuum elements. The interface element with four or six nodes is fully compatible with four and eight-node isoparametric 2D soil element. The use of an element based on an isoparametric formulation has the advantage of being easily implemented into existing finite element codes. Furthermore, the data preparation and the mesh generation can be done in a routine manner.

To obtain the basic element stiffness matrices, the interface stress is characterized by the normal and shear stresses is assumed. The normal stress σ and the shear stress τ are related by the constitutive law to the normal and tangential interface element strain, ε and γ as shown the following:

$$\begin{Bmatrix} \sigma \\ \tau \end{Bmatrix} = [D] \begin{Bmatrix} \varepsilon \\ \gamma \end{Bmatrix} \tag{1}$$

where [D] is the material stiffness matrix, and represented by

$$[D] = \begin{bmatrix} K_n & 0 \\ 0 & K_s \end{bmatrix} \tag{2}$$

where K_s and K_n are the shear and normal stiffness of elasticity, respectively (in units of F/L3).

The linear elastic-perfectly-plastic model using a Mohr-Coulomb failure criterion as the yield surface is investigated in the analyses. The formulation of the constitutive behavior was based on plasticity theory. The Mohr-Coulomb failure criterion is defined by the yield function f, and represented as

$$f(\sigma,\tau) = \tau - \sigma \tan(\phi) - c = 0 \tag{3}$$

where ϕ is the internal friction angle, c is the cohesion.

2.2. Numerical stability

In the sense of simple formulation of the failure criterion, when the shear stress reaches and satisfies the failure criterion, the value of shear stiffness is set to zero, and the normal stiffness remains unchanged. The tensile stress in the interface element is generally not permitted. If tensile stress occurs, both the normal and shear stiffness are set to zero, and the tensile stress is redistributed by the solution algorithm. If the interface element moves such that the maximum normal tensile strength is exceed $c/\tan\phi$, the interface is allowed to subsequently open and close, and the residual tensile stress is redistributed. When the interface is open the normal stress remains equal to $c/\tan\phi$ and shear stress remains equal to zero. This is essentially the same as setting both normal and shear stiffness to zero. The amount of opening of the interface is recorded. When the interface recluses and reforms contact, the constitutive model again defines the interface behaviors.

Although the zero thickness interface elements have been successfully used by many others without reporting particular problems, but it also necessary to be taken into account that the Ill-conditioning can be reduced by careful selection of the size of the 2D elements adjacent to the interface, and the problem of steep stress gradients is entirely one of inadequate mesh design. For the convergence of solutions, an accelerated modified Newton-Raphson scheme with a sub-stepping stress point algorithm was employed to solve the non-linear finite element equations in this study.

3. Limit analysis – upper bound theorem

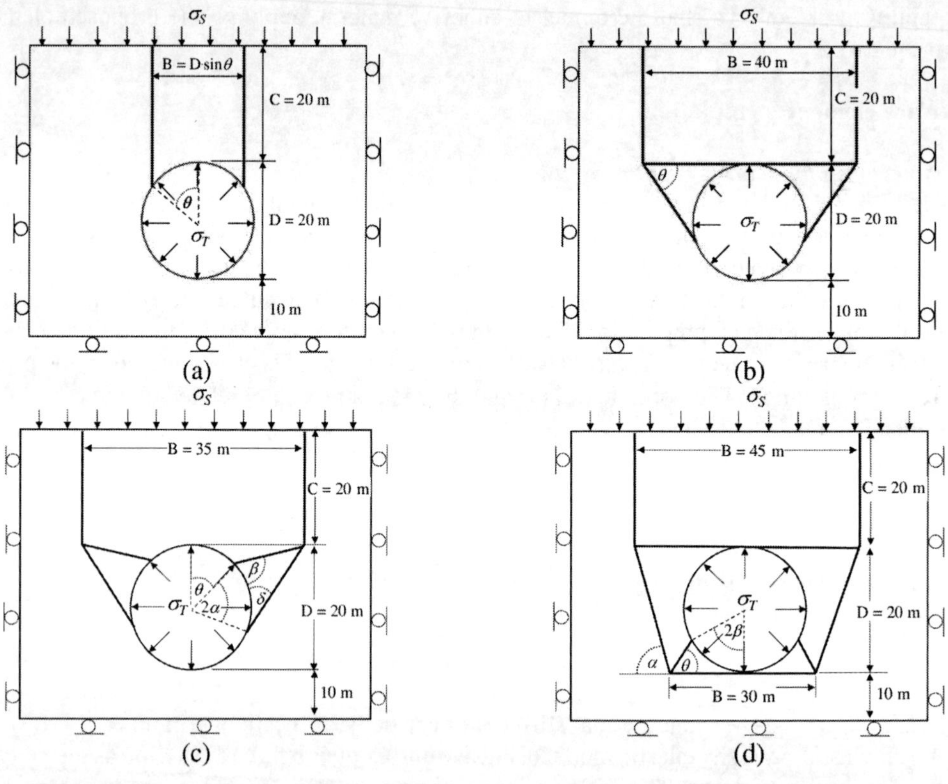

Figure 1. Schematic illustration of four upper bound mechanisms A, B, C and D.

As the working face of tunnel is advanced, a means of supporting the ground close the face may be needed. In soft ground with immediate support provided by such mean as shield, the use of compress air or clay slurry under pressure may be required; otherwise collapse may occur as a result of gross plastic deformation of the soil, possibly accompanied by flooding. Under these circumstances, it is possible to idealize the process of tunneling as shown in figure 1. in which a circular un-lining tunnel of diameter *D*, which there acts a uniform fluid pressure σ_T, is shown being constructed with a depth of cover *C*. The ground has a unit weight γ and a uniform pressure σ_S acts on the soil surface. The tunnel pressure is necessary to maintain the stability of the heading for different values of the parameters that have been defined (*C*, *D*, σ_S) and the strength of ground is investigated. The collapse of the tunnel heading will usually be a sudden event and hence it is appropriate to characterize the strength of the ground by its undrained shear strength c_u. In following analysis, it is assumed that c_u is constant with

IOP Conf. Series: Materials Science and Engineering **657** (2019) 012004 doi:10.1088/1757-899X/657/1/012004

depth.

In the point of view of stability of un-lining tunnel, Broms and Bennermark [9] defined a stability ratio N, equal to the difference between the total overburden stress in the ground at the axis of the tunnel and the tunnel pressure divided by the undrained shear strength c_u.

$$N = \left[\sigma_s - \sigma_T + \gamma(C + D/2)\right]/c_u \tag{4}$$

They concluded that if N is less than 6 then the opening will be stable. The upper bound theorem states that if a work calculation is performed for a kinematically admissible collapse mechanism then the loads thus deduced will be higher than (or equal to) those for collapse.

For a complete solution the requirements of equilibrium, compatibility, material behavior and the boundary conditions, both load and displacement, must all be satisfied. All four of the above requirements are satisfied only for a limit analysis solution in which both upper and lower bound calculations lead to the same results. Stability solutions will be obtained using the limit theorems of plasticity.

4. Applications – The plane strain circular tunnel

An upper bound to the true collapse load can be found by selecting any kinematically admissible collapse mechanism and performing an appropriate work rate calculation. It may be noted that the solution will lead to an unsafe or low value for the tunnel pressure. The accuracy of an upper bound calculation will depend, in part, on the closeness of the assumed mechanism to the real one, and it is in the selection of a collapse mechanism that the models tests have proved to be of value. In practice, it is important to establish the minimum tunnel pressure necessary for stability.

Four upper bound mechanisms are shown in figure 1. The mechanism A and B are simple 'roof' and 'roof and sides' mechanisms each containing one variable dimension (or angle θ). The mechanism C has four variable angles (θ, 2α, β, δ) in its specification and includes mechanism A and B as the special case. The mechanism D is a 'roof, sides and bottom' mechanism with three variable angles (θ, α, 2β). The work described in this paper has been limited to plane strain circular tunnel. The soil is idealized as a linear elastic material. The interface is assumed as an elastic-perfectly-plastic material with a cohesion c_u. In addition, the input data of calculation is represented in table 1. The boundary conditions are constrained by the rollers in both sides and bottom of meshes proposed. The procedure for determining the critical collapse load is to derive an expression for (σ_S - σ_T) (involving the variable dimension or angle) and then to minimize the value of (σ_S - σ_T) with respect to the variable dimension or angle.

Table 1. Input data of calculation.

Cohesive soil and interface		Failure mechanism	
Parameter	Value	Parameter	Value
Elastic modulus, E (MPa)	300.0	Mechanism A (degree)	$\theta = 30 \sim 60$
Poisson's ratio, ν	0.25	Dimension (B, C, D) (m)	20sinθ, 20, 20
Assumed unit weight, γ (MPa/m)	0.0	Mechanism B (degree)	$\theta = 30 \sim 60$
Undrained cohesion, c_u (MPa)	0.1	Dimension (B, C, D) (m)	40, 20, 20
Friction angle, ϕ (°)	0.0	Mechanism C (degree)	$2\alpha = \beta = \delta = \theta = 30 \sim 60$
Interface elements		Dimension (B, C, D) (m)	35, 20, 20
Elastic modulus, E (MPa)	300.0	Mechanism D (degree)	$\alpha = 2\beta = \theta = 30 \sim 60$
Poisson's ratio, ν	0.0	Dimension (B, C, D) (m)	45, 20, 20

After finite element numerical modeling, the deformed meshes of four upper bound mechanisms are shown in figure 2. The reason that σ_S and σ_T appear only in the form (σ_S - σ_T) in the upper bound calculations is that since a kinematically permissible mechanism for cohesive material involves no volume change then the decrease in area of the tunnel must equal the area of ground loss at the surface. Hence the work done by the pressures in the work calculation will be (σ_S - σ_T) multiplied by that area.

The sliding block mechanisms also suggest that a movement of soil inwards near the tunnel is accompanied by an (approximately) equal settlement of the ground surface.

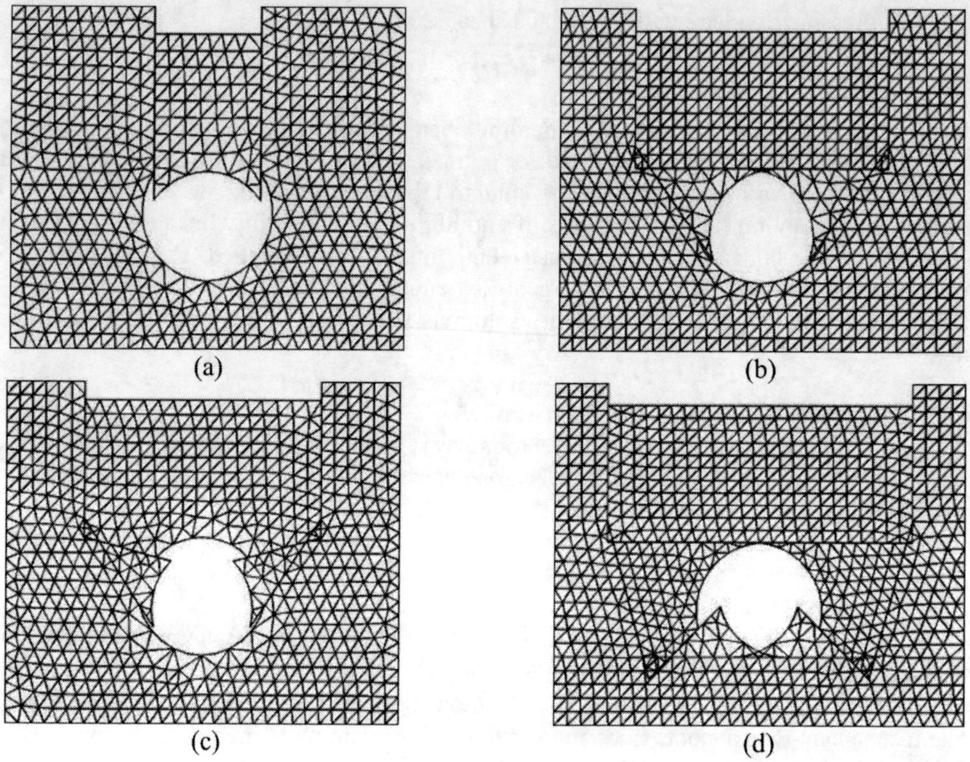

Figure 2. Deformed meshes of four upper bound mechanisms by the finite element analysis.

According to the results obtained by the graphic expressions, figure 3 shows that the failure mechanism is indicated by the vectors of incremental displacement in the last increment of the analysis. The failure mechanism for mechanism A to D in which the interface has been fully mobilized is also observed. All soil movement occurs within a different shape failure.

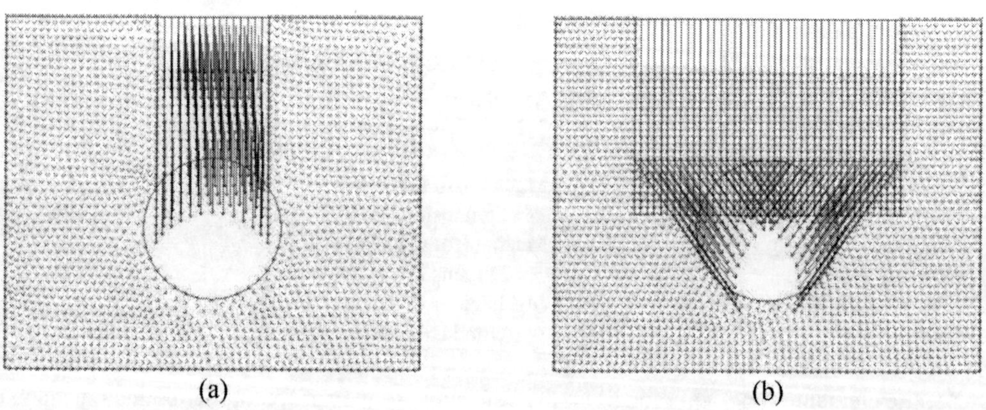

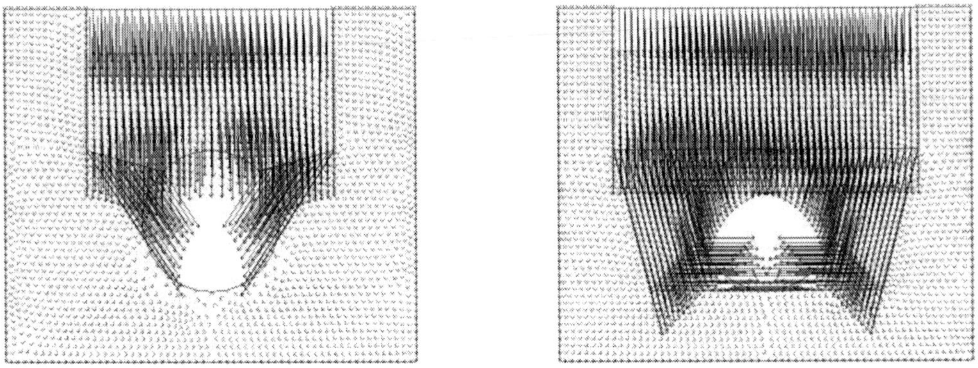

Figure 3. Displacement fields of four upper bound mechanisms by the finite element analysis.

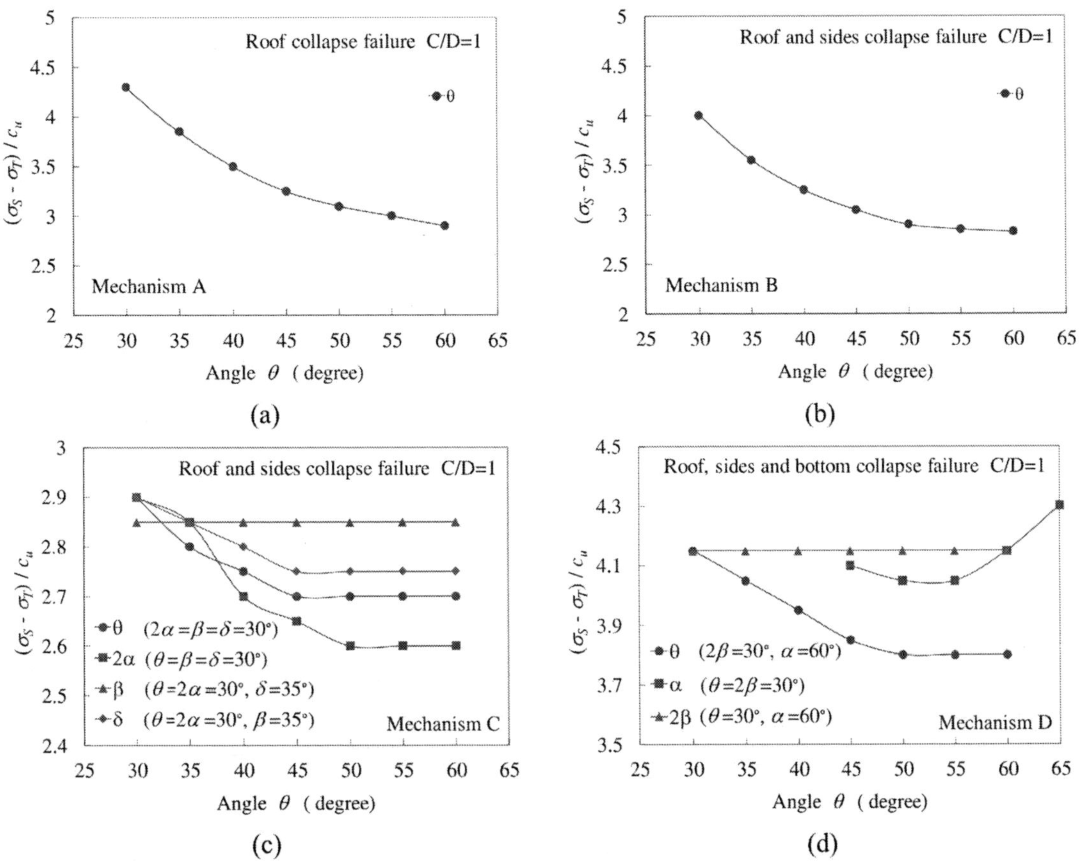

Figure 4. Parametric studies of four upper bound mechanisms A, B, C and D.

The parametric studies of variable angles of each mechanism are particularly examined. Figure 4 show that in the neighborhood of the optimum upper bound, changes in the variable angles lead to small changes in the collapse load. For obtaining such optimum upper bound, the variable angle θ is approximately equal to 60 degrees for the mechanism A and B. On the other hand, for the mechanism C and D, the optimum value of θ is about 50 degrees. The results of a numerical optimization to discover the critical mechanisms are only for $\gamma D/c_u$=0 as a criterion for deciding the tunnel pressure should be a safe procedure. This is because it always corresponds to a lower value of N, or a higher value of σ_T.

Examining the results obtained, the comparison between analytical solution and numerical calculation is investigated. Figure 5 clearly indicate the mechanism C is more critical for low values of *C/D* and is superseded by mechanism D for high values of *C/D*. It also shows that the ability of the simple upper bound solutions to predict the stability of shallow tunnel in soft ground with satisfying accuracy. The numerical results presented in this paper can be used for the calculation of the undrained stability of tunnels when *C/D*<3.

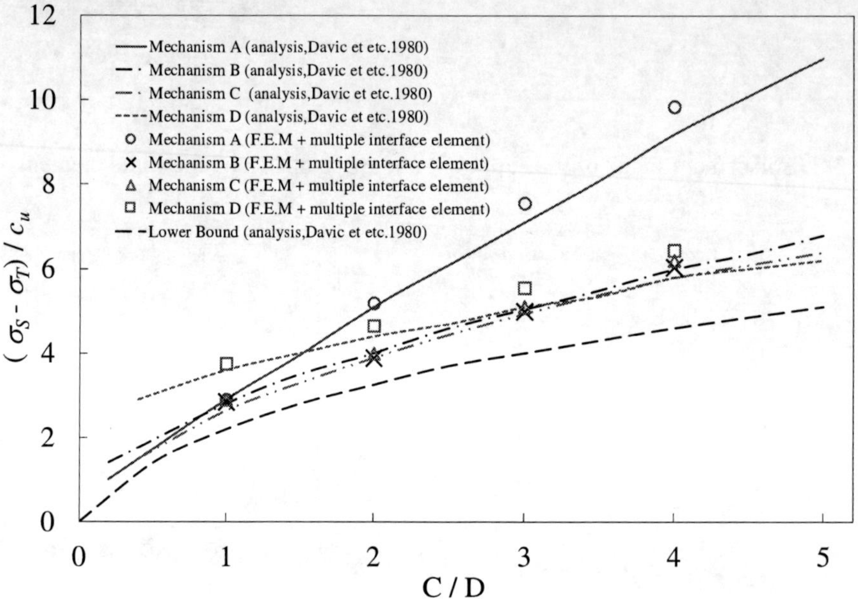

Figure 5. Comparison of results obtained between analytical solution and numerical calculation for a circular tunnel ($\gamma D / c_u = 0$).

There is no much difference between mechanisms B, C and D for practical purposes (i.e. in estimated collapse load) although the mechanisms of deformation very different. In general, it seems that a mechanism with one variable will yield an adequate upper bound (i.e. close to a good lower bound) providing an appropriate pattern of collapse is chosen. The stability solutions of analytical limit analysis methods are compared with numerical results which indicate the zero thickness interface elements provided a useful means to model cohesive material in finite element analysis of a shallow tunnel. The simple elastoplastic Mohr-Coulomb constitutive model appears to be satisfactory for the modeling of interface behavior in this type of analysis.

5. Conclusion

The numerical application of zero thickness interface elements considered in the finite element program with a simple elastoplastic Mohr-Coulomb failure criterion has predicted the upper bound of stability ratio for various types of a shallow tunnel failure mechanism in reasonable agreement with analytical solutions. These analyses have provided numerical confirmation of the analytical solutions such as Davis *et al* [10] that used the limit theorems of plasticity to estimate the tunnel pressures necessary for the support of shallow tunnels and underground openings in soft ground.

The results obtained in the analysis of the failure modes of a shallow tunnel modeled with zero thickness interface elements assuming associated plastic flow are presented. In addition, the upper bound of stability ratio is independent of the elastic stiffness.

IOP Conf. Series: Materials Science and Engineering **657** (2019) 012004 doi:10.1088/1757-899X/657/1/012004

References

[1] Goodman R E, Taylor R L and Brekke T L 1968 A model for mechanics of jointed rock *J. Soil Mech. Foundation Division* **94** 637-59
[2] Ghaboussi J, Wilson E L and Isenberg J 1973 Finite element for rock joints and inter faces *J. Soil Mech. Foundation Division* **99** 833-48
[3] Carol I and Alonso E E 1983 A new joint for the analysis of fractures rock *5th International Congress on Rock Mechanics* (Melbourne) pp 147-51
[4] Beer G 1985 An isoparametric joint/interface element for finite element analysis *Int. J. Numer. Methods Eng.* **21** 585-600
[5] Frank R, Guenot A and Humbert P 1982 Numerical analysis of contacts in geomechanics *Proc. 4th Int. Conf. on Numerical Methods in Geomechanics* (Rotterdam) pp 37-42
[6] Desai C S, Zaman M M, Lightner J G and Sirwardane H J 1984 Thin-layer elcment for interface and joint *Int. J. Numer. Analyt. Methods Geomech.* **8** 19-43
[7] Day R A and Potts D M 1994 Zero thickness interface elements – numerical stability and application *Int. J. Numer. Analyt. Methods Geomech.* **18** 689-708
[8] Garolera D, Aliguer I, Segura N M, Carol I, Lakshmikantha M R and Alvarellos J E 2013 Zero-thickness interface elements with hydro-mechanics coupling, formulation and applications in geomechanics *XII International Conference on Computational Plasticity. Fundamentals and Applications, COMPLAS XII* E Oñate, D R J Owen, D Peric and B Suárez edis pp 1-12
[9] Broms B B and Bennermark H 1967 Stability of clay in vertical openings *J. Soil Mech. Foundation Division* **93** 71-94
[10] Davis E H, Gunn M J, Mair R J and Seneviratne H N 1980 The stability of shallow tunnels and underground openings in cohesive material *Géotechnique* **30** 397-416

IOP Conf. Series: Materials Science and Engineering **657** (2019) 012005 doi:10.1088/1757-899X/657/1/012005

Research on vibration characteristic simulation of the 220 kV gas insulated metal enclosed switchgear equipment

X P Wang[1,4], X P Jiang[2], Q Wang[2], Y K Long[2], J W Tan[3] and J Hao[1]

[1]State Key Laboratory of Power Equipment & System Security and New Technology, Chongqing University, Chongqing, China
[2]State Grid Chongqing Electric Power Company, Chongqing Electric Power Research Institute, Chongqing, China
[3]Chongqing Pulse Science Technology Co., Ltd., Chongqing, China

E-mail: xpwang@cqu.edu.cn

Abstract. GIS equipment (gas insulated switchgear) is more and more popular in power system, and people often pay more attention to its discharge fault than other mechanical faults. But in the operation of GIS, besides discharge faults, mechanical faults are also a major cause of accidents. Therefore, we should give more research to abnormal mechanical vibration. At the same time, compared with discharge signal, mechanical vibration fault is more difficult to study. In this article, based on the size of real GIS equipment, we first built a three-dimensional model through COMSOL software to study the propagation characteristics of vibration signals. Simply considered, the bus air chamber is two-dimensional axisymmetric. Firstly, the natural vibration frequency of the equipment is calculated by finite element method, and the vibration propagation characteristics of GIS shield cover loose of the GIS equipment are calculated and analyzed. The simulation calculation shows that the longer the shell is, the lower the natural vibration frequency is. Therefore, we should pay more attention to the natural vibration frequency far away from 100 Hz, when designing and manufacturing GIS equipment, so as to avoid serious consequences caused by resonance. Because the basic vibration frequency caused by the loosening of shield cover is twice as high as the power frequency of 50 Hz. Then the 100 Hz simulation source is added to COMSOL model as the vibration source of loosening fault, and the propagation rule and characteristics of vibration fault about GIS are received. After the vibration fault occurs, the loosening vibration signal at the connection between the GIS shell and the basin insulator is the strongest. These conclusions by simulation also conform to objective rules and facts. Meanwhile it can guide us to arrange sensors reasonably and improve the signal-to-noise ratio of vibration detection signals.

1. Introduction

GIS equipment has many advantages, such as compact structure, reliable operation, less impact on the environment and less maintenance workload. It is an important manifestation of technological progress in power equipment manufacturing. In recent years, with the development of high voltage power transmission and the need of urbanization construction, GIS has been more and more widely used [1-3].

Although GIS has high reliability and low failure rate, with the extensive application of GIS, the failure of GIS has become a problem that can not be ignored. Existing research methods at home and abroad are mainly aimed at GIS discharge fault, but in fact, the phenomenon of mechanical fault is

IOP Conf. Series: Materials Science and Engineering **657** (2019) 012005 doi:10.1088/1757-899X/657/1/012005

increasing. Therefore, it is necessary to systematically study the mechanical vibration fault of GIS [4,5]. At present, in most literatures, the research on vibration of GIS equipment is mainly carried out by experiments. Undoubtedly, the test data are quite real, but it is time-consuming and laborious to carry out the test, and it also needs a more expensive test platform. Therefore, more and more researchers begin to carry out the corresponding simulation research. Through the simulation research, we can quickly understand the generation and transmission characteristics of GIS vibration signal, and then guide the development of the experiment. Through the combination of simulation and experiment, the abnormal sound fault of GIS equipment can be studied efficiently.

In this paper, the finite element simulation software COMSOL is used to model and simulate 220 kV GIS real equipment. Firstly, the natural vibration frequency of the equipment is obtained, and then the vibration propagation characteristics are obtained by adding the excitation source to simulate the vibration. The simulation results are used to guide the follow-up experimental research, which is of great significance to the safe operation of GIS equipment.

2. Calculation and simulation of natural frequency of GIS equipment

The study of the vibration frequency of GIS is generally divided into the natural vibration of GIS conductor and shell. The main vibration frequency of this type of GIS buses is close to the first-order natural frequency f of a single span. If the span length of the bus is changed, the natural frequency f will also change.

The theoretical formula for natural frequencies of bus is as follows [6]:

$$f_0 = \frac{\pi}{2L_{ks}^2}\sqrt{\frac{EJ}{\rho s}}\ (Hz) \tag{1}$$

Among them, E is the elastic modulus of conductor material; J is the moment of inertia of a conductor; s is the cross-sectional area of the conductor; ρ is the mass density of conductor materials; L_{ks} is the pot insulator spacing (span).

The ratio of expected thickness radius of GIS shell is very small, generally less than 2.5%. For the structure of GIS, flanges are often clamped on both sides of the pot insulator and the shell is reinforced. A real conductor - shell system can be regarded as a beam to approximate the natural frequency of the shell.

The natural frequencies of the shell can be calculated according to the following formula [6]:

$$f_{ks} = \frac{k_s}{2\pi}\sqrt{\frac{2EC_2}{(1-\mu)^2\times(0.5D_2)^2}}\ (Hz) \tag{2}$$

Among them, k_s is related to C_2, D_2 and L_{ks}; C_2 is the thickness of the shell; D_2 is the diameter of the shell; μ is the Poisson's ratio of the shell material; E is the elastic modulus of the shell material.

However, real GIS equipment is often not such a simple geometry, so it is not simple and accurate to calculate the natural frequency in engineering. In this paper, a simple GIS model is built by COMSOL, and the natural frequency simulation is carried out [1].

As shown in figure 1, the GIS bus model in COMSOL consists of a 2.5-meter-long GIS shell, two basin-shaped insulators and a central conductor. The outer diameter and wall thickness of GIS shell are 620 mm and 20 mm respectively. The two ends are sealed with a 20 mm thick cover plate, and the radius of the central conductor of GIS is 50 mm. By simplifying the processing, a three-dimensional model of GIS structure is established.

In this model, GIS shell is made of steel. Copper is used as the central conductor. The material of the basin insulator is polypropylene composite material. The parameters of SF_6 are set at 0.4 MPa pressure. The shield is made of aluminium. The material properties are shown in table 1.

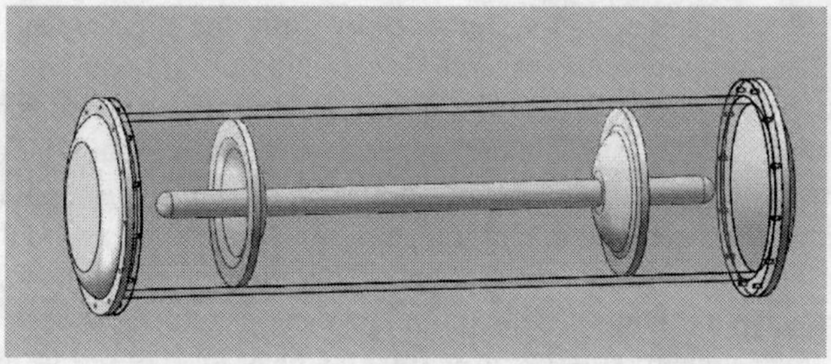

Figure 1. The three-dimensional model.

Table 1. Material properties.

Material	Density (kg/m^3)	Young's modulus (Pa)	Poisson's ratio
4340 standard steel	7850	205E9	0.28
Copper	8960	110 E9	0.35
Polypropylene composite material	2203	73.1 E9	0.17
SF6 (0.4MPa)	25.25	/	/
Aluminum	2700	70E9	0.33

In order to get the characteristics and changing rules of natural vibration frequency, we calculated the following data by changing the length of bus air chamber in GIS. It is worth mentioning that there are many natural frequencies, here only five lower natural frequencies are listed, as shown in table 2.

Table 2. Five natural frequencies of the low frequency section.

Length	Natural vibration frequency (Hz)				
2 m	143.33	190.89	210.43	238.24	270.63
2.5 m	135.70	136.37	143.34	190.68	238.21
3 m	91.76	95.56	143.32	181.02	190.86

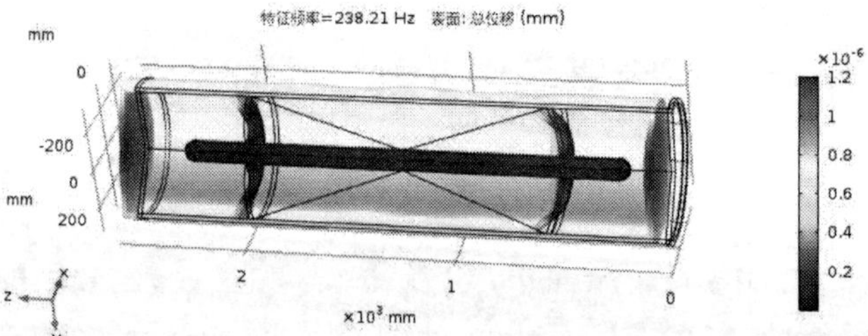

Figure 2. Vibration displacement distribution for 238.21 Hz.

From the above data, we can find that the natural vibration frequency of GIS bus air chamber is closely related to its length. The simulation results show that in a certain range, the longer the bus air chamber is, the lower its natural vibration frequency is, the shorter the bus air chamber is, and the higher its natural vibration frequency is. This is in line with our sensory perception. At the same time, we know that the common mechanical fault vibration of GIS equipment is caused by electric force, and the frequency of electric force caused by poor contact is twice the working frequency, that is 100 Hz. Therefore, we should pay attention to the natural vibration frequency of GIS equipment and keep it away from the common fault frequency of 100 Hz, so as to avoid causing more serious problems of resonance. Figure 2 show the vibration displacement distribution for 238.21 Hz, which is a typical example through simulation.

3. Simulation of vibration propagation characteristics

Based on the above model, in order to simplify the simulation and calculation, a 100 Hz vibration source is added to the simulation model to simulate the fault signal. The propagation characteristics of the vibration signal in the gas chamber of GIS bus are obtained by simulation[7-10]. Taking 2.5 m bus air chamber length as an example, vibration sources are added to the central conductor, shield cover and basin insulator to simulate the mechanical fault. Through simulation calculation, the surface pressure distribution of GIS equipment shell can be obtained, as shown in figures 3 to 5.

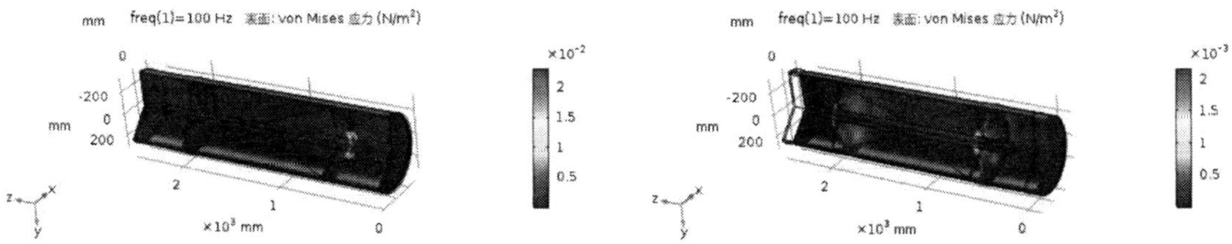

(a) (b)

Figure 3. Vibration excitation is applied to the connection between the basin insulator and the central conductor. (a) R directional vibration excitation is applied to the connection between the basin insulator and the central conductor and (b) Z directional vibration excitation is applied to the connection between the basin insulator and the central conductor.

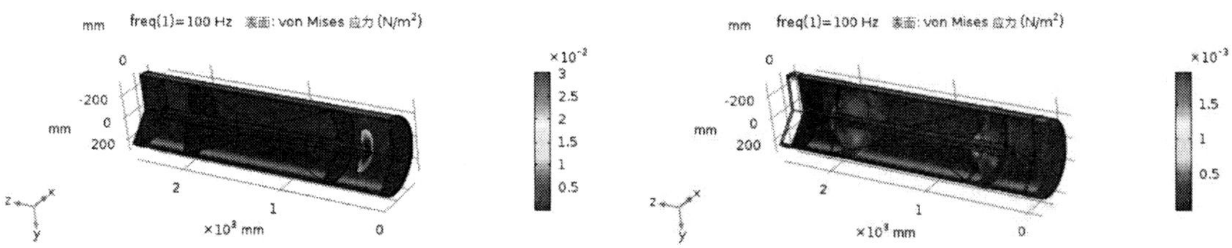

(a) (b)

Figure 4. Vibration excitation is applied to the shield. (a) R directional vibration excitation is applied to the shield and (b) Z directional vibration excitation is applied to the shield.

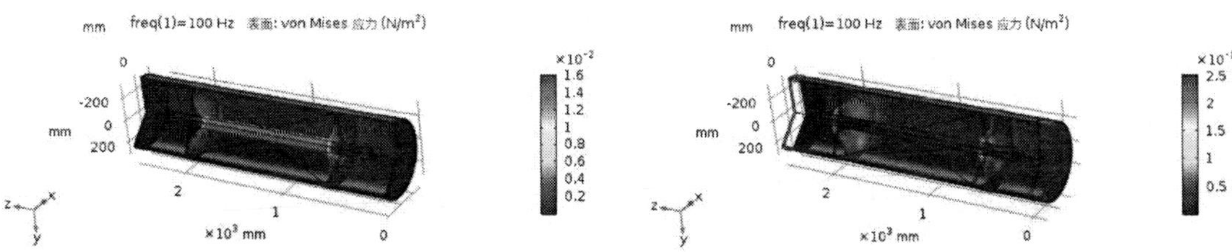

(a) (b)

Figure 5. Vibration excitation is applied to the central conductor. (a) R directional vibration excitation is applied to the central conductor and (b) Z directional vibration excitation is applied to the central conductor.

As shown in the six figures, it could be clearly seen that the vibration signal is strongest at the connection between the GIS shell and the basin insulator when the fault occurs inside the GIS equipment and produces the vibration signal. Because the central conductor is directly connected to the insulator, the vibration signal propagates along the solid to the surface of the GIS shell. The vibration of the central conductor causes pressure changes in the gas chamber, and then transmits to the GIS shell through gas pressure or acoustic vibration. Obviously, the resistance of this propagation path is huge, and the signal will be attenuated to a large extent. Therefore, when measuring the mechanical fault signal, the vibration sensor should be as close as possible to the basin insulator to get a higher signal-to-noise ratio signal, improve the effectiveness of the detection signal and test sensitivity.

In addition, the analysis of the above six graphs shows that the vibration in R direction is more intense than that in Z direction. This is because the gas chamber of GIS bus is similar to a cylinder, with smaller diameter in the R direction and longer size in the Z direction. Therefore, for the same degree of vibration, the R direction will be more sensitive and the vibration will be more intense.

Finally, we can also find that loosening of the connection between the basin insulator and the central conductor can lead to more serious vibration. This is also evident, because the basin insulator is directly installed at the joints of the GIS shell, the loose signal of the basin insulator itself is transmitted to the shell and captured by the sensor almost without attenuation.

4. Conclusion

The natural vibration frequency of GIS bus air chamber is negatively correlated with its length. Therefore, when designing the size of GIS equipment, the common frequency of mechanical fault should be avoided at 100 Hz to avoid causing more serious consequences of resonance.

Mechanical fault signal is the largest in the area where the basin insulator is connected with the GIS shell. Vibration sensors should be arranged as close as possible to the insulator to improve the signal-to-noise ratio and obtain more effective data. In addition, special attention should be paid to the loosening fault along the radius direction and the loosening signal of the basin insulator itself.

Acknowledgment

All the authors thank the State Grid Chongqing Electric Power Company Chongqing Electric Power Research Institute (Programm 2018 Yudian Keji 5#) for the financial support.

References

[1] Wang Q, Jiang X P, Tan J W, Hao J, Zheng L and Wu G L 2018 Simulation study on the vibration characteristics and vibration fault propagation of insulated metal-enclosed switchgear (GIS) busbar *IEEE International Conference on High Voltage Engineering and Application (ICHVE)* (Athens, Greece)

[2] Li K, Xu H H, Chen B B, Cui Y L and Ma H Z 2017 Research on vibration mechanism and natural frequency in GIS *Electrical Measurement & Instrumentation* **54**(3) 14-8

[3] Lv F C and Zhang B 2014 Recognition of GIS discharge types based on the S_kohonen network *Electrical Measurement & Instrumentation* **51**(20) 21-4

[4] Huang Q, Wei X, Xu J G and Zhou Z C 2017 Research of GIS bus loose contact fault diagnosis techniques based on mechanical vibration *High Voltage Apparatus* **53**(11) 0099-104

[5] Chang G, Zhang Z Q and Wang Y 2011 Review on mechanical fault diagnosis of high-voltage circuit breakers based on vibration diagnosis *High Voltage Apparatus* **47**(8) 85-90

[6] Xing J 2016 Research on mechanical fault vibration detecting of GIS (Shengyang: Shenyang Institute of Engineering)

[7] Li C, Sánchez R, Zurita G *et al* 2016 Fault diagnosis for rotating machinery using vibration measurement deep statistical feature learning *Sensors* **16** 895

[8] Kim J K, Hahn S C, Park K Y *et al* 2005 Temperature rise prediction of EHV GIS bus bar by coupled magnetothermal finite element method *IEEE Trans. Magn.* **41** 1636-9

[9] Qixiao M, Dengwei D, Zhihang X *et al* 2016 Abnormal vibration analysis of GIS based on time frequency detection *International Conference on Condition Monitoring & Diagnosis* (Xi'an, China)

[10] Jian M J, Zhang G C and Zhang D Q 2014 Finite element analysis of the shell of GIS busbar tube *Adv. Mater. Res* **889-890** 1406-9

IOP Conf. Series: Materials Science and Engineering **657** (2019) 012006 doi:10.1088/1757-899X/657/1/012006

Numerical modelling for the simulation of nonlinear ultrasound in liquids with gas bubbles

C Vanhille[1,2] and M T Tejedor Sastre[1]

[1]NANLA, Universidad Rey Juan Carlos, Tulipán s/n, 28933 Móstoles, Madrid, Spain
[2]Author to whom any correspondence should be addressed

E-mail: christian.vanhille@urjc.es

Abstract. Several numerical models have been developed in different configurations to simulate the behaviour of finite-amplitude ultrasound when interacting with tiny gas bubbles in a liquid. Since this interaction is highly nonlinear, specific models must be developed to understand the propagation of the waves in this kind of dispersive media for which their nonlinear and attenuation coefficients, as well as the sound speed, are extremely dependent on the ratio of the driven frequency to the bubble resonance. The bubble volume variation is mathematically modelled in the time domain through a Rayleigh-Plesset equation with terms up to the second order, whereas the time-dependent acoustic field relies on the wave equation in one or several dimensions. Both differential equations are coupled and auxiliary conditions are imposed. The differential systems are solved by the developed numerical models. In this paper we study in a three-dimensional resonator with axial symmetry how new harmonics obtained by nonlinear distortion can be enhanced by taking the nonlinear resonance effect into account, and we show that the generation of new frequency components by nonlinear frequency mixing exists. We also analyse the stable cavitation phenomenon in a three-dimensional focused field with axial symmetry by considering a nonlinear dependence of bubble generation in the liquid and the existence of primary Bjerknes forces.

1. Introduction

Bubbly liquids are biphasic media for which a very small amount of gas bubbles can produce very high nonlinear effects on ultrasound without changing the liquid density and with relatively small pressure amplitudes [1-4]. Bubbles also produce dispersive effects and attenuation [1-4]. These media find huge interest in many industrial and medical applications [5-9]. However, the experimental study of such media is delicate since a stable population of homogeneous gas bubbles is not easy to obtain. The development of numerical tools able to simulate the behavior of finite-amplitude waves in bubbly liquids is thus necessary [1-4,6,10-12]. Moreover, many high-power ultrasound processes are based on acoustic cavitation, which includes two different regimes, stable and inertial [2,4,13,14]. Very active theoretical and experimental works have been performed [2-4,6], but a higher understanding of all the phenomenon involved in this framework is still needed. The complexity of the physical problem is such that the mathematical models must include restrictions to idealize the solution obtain after the solution procedure.

Linear responses of the system are obtained when infinitesimal acoustic pressure amplitudes are considered in a liquid, but nonlinear distortion affects the wave when finite amplitudes are assumed, i.e., harmonic components of the fundamental frequency are created and nonlinear combinations of signals of different frequencies generate other components [1-4,10-14]. Moreover, if those amplitudes are above the cavitation threshold, the generation of stable cavitation bubbles from the nuclei present in the liquid

IOP Conf. Series: Materials Science and Engineering **657** (2019) 012006 doi:10.1088/1757-899X/657/1/012006

is triggered, and these bubbles oscillate nonlinearly [4,13,14]. For even higher amplitudes, inertial cavitation takes place, and the bubbles grow and collapse [4,13]. Primary Bjerknes forces are net radiation forces due to an acoustic wave on oscillating bubbles [3]. These phenomena have been studied in the literature [2-4,6,12,13,15].

In this paper we chose to study two aspects of ultrasound propagation in bubbly liquids, in three-dimensional volumes with axial symmetry. On the one hand, we propose to analyze the possibilities of a resonator to enhance the power of harmonics by using the nonlinear resonance effect [3]. We also aim to show that a difference component is generated from a dual-frequency source by nonlinear frequency mixing [16,17]. On the other hand, we perform a study of the stable cavitation process in a focused ultrasonic field, by considering a threshold-based nonlinear dependence of void fraction in the liquid on rarefaction acoustic pressure [4,13,18], and how primary Bjerknes forces act on these bubbles to form agglomerates [1,2,4,13,19]. These analysis are carried out by numerical simulations after adapting our ad-hoc numerical models [19-21]. To this end we consider a nonlinear coupled differential system formed by the wave equation and a Rayleigh-Plesset equation for which the bubble volume variation and the acoustic pressure are nonlinear dependent variables [2,3,11,21].

This mathematical model solved here considers the nonlinear mutual interaction of pressure and bubble vibration variables. In this sense, the nonlinearity of the bubbly medium is due to the presence of bubbles [3,17]. Also, these bubbles imply dispersion and attenuation of acoustic waves. Thus, the nonlinearity and sound speed in the bubbly liquid depends on the ratio of the driving frequency to the bubble resonance. This feature is extremely important for nonlinear waves with multi-frequency components [2,3,21].

Section 2 describes the physical problem we consider and the mathematical model we solve. It also defines how we simulate the stable cavitation process, including the nonlinear dependence law, and the equations we solve to evaluate the bubble motion. Section 3 presents the numerical simulations and discusses our results.

2. Models

2.1. Physical problem

A three-dimensional cylindrical volume of liquid of radius R and length Z in which a population of gas bubbles is evenly distributed is considered. The study is performed from time $t = 0\,s$ to T. Cylindrical coordinates with axial symmetry around $\overrightarrow{e_z}\,(0 \leq \theta \leq 2\pi)$ are used, so that the coordinates (z, r, t) are employed. The bubbly liquid volume is excited by an ultrasonic source at $z = 0\,m$. The problem we solve is the nonlinear interaction of oscillating bubbles and ultrasound of finite amplitudes. The void fraction of gas in the liquid leads to high nonlinearity of the medium and to dispersion of features such like acoustic attenuation and nonlinearity.

2.2. Differential system

The system we describe here models the nonlinear interaction of ultrasound and gas bubble oscillations in the most general configuration: three-dimensional space domain and variable density of bubbles in the liquid. Both fields are defined by the bubble-volume variation $v(z, r, t)$, i.e., its perturbation compared to its initial volume v_0 (corresponding to the initial radius R_0), and the acoustic pressure $p(z, r, t)$. The wave equation and a Rayleigh-Plesset equation are coupled to simulate the nonlinear mutual influence of bubbles and ultrasound, accounting for nonlinearity, dispersion, and attenuation due to the bubbles in the liquid [3,11,21]. In addition to the coupled differential system, auxiliary conditions are added to model the absence of acoustic and bubble perturbations at the outset, the axial symmetry, and the boundary conditions. The differential problem is then the following, in which partial derivatives are indicated by subscript on z, r and t:

$$
\begin{aligned}
& v_{tt} + \delta\omega_0 v_t + \omega_0^2 v = av^2 + b\left(2vv_{tt} + v_t^2\right) - \eta p, \ (z,r,t) \in [0,Z]\times[0,R]\times(0,T), \\
& p_{rr} + p_r/r + p_{zz} - p_{tt}/c_l^2 = -\rho_l N(z,r) v_{tt}, \ (z,r,t) \in (0,Z)\times(0,R)\times(0,T), \\
& p(z \neq 0) = v = p_t(z \neq 0) = v_t = 0, \ t = 0, (z,r) \in [0,Z]\times[0,R], \\
& p_r = 0, \ r = 0, (z,t) \in [0,Z]\times(0,T], \\
& p_z = -p_t/c_l \text{ or } p = 0, \ z = Z, (r,t) \in [0,R]\times(0,T], \\
& p_r = -p_t/c_l \text{ or } p = 0, \ r = R, (z,t) \in [0,Z]\times(0,T], \\
& p = s, \ z = 0, (r,t) \in [0,R]\times(0,T].
\end{aligned} \tag{1}
$$

ω_0 is the first bubble resonance frequency, $\delta = 4\mu_k/\omega_0 R_0^2$ is the viscous damping coefficient of the bubble-liquid mixture, where μ_k is the kinematic viscosity of the liquid, ρ_l is the equilibrium density of the liquid, c_l is the small-amplitude sound speed of the liquid, $N(z,r)$ is the bubble density at the point (z,r), $s(r,t)$ is the function defining the excitation at the source, $a = (\gamma+1)\omega_0^2/2v_0$ is the nonlinear coefficient related to the adiabatic gas law, where γ is the specific heat ratio of the gas, $b = 1/6v_0$ is the nonlinear coefficient accounting for the dynamic response of bubbles, and $\eta = 4\pi R_0/\rho_l$ is a constant.

For standing waves in cavities a bubbly liquid is considered and free-wall conditions are assumed. With a single-frequency source the following function is used: $s(t) = p_0 \sin(\omega t)$, where p_0 is the amplitude and $\omega = 2\pi f$ is the frequency, whereas when a dual-frequency source is considered the following function is used: $s(t) = p_0 \sin(\omega_1 t) + p_0 \sin(\omega_2 t)$, where $\omega_1 = 2\pi f_1$ and $\omega_2 = 2\pi f_2$ are the primary frequencies.

When the stable cavitation process is assumed, a population with homogeneous density N_- of very tiny spherical gas bubbles of radius R_{0-} is present in the liquid at the outset, simulating the gas nuclei from which the cavitation process can take place. A phased array composed by S spherical harmonic pressure sources, $p_s(r,t) = p_0 g_s(r) \sin(\omega t + \varphi_s)$, where g_s is the spatial distribution and φ_s is the phase, excited with phase delays is considered to create a focused pressure field. The whole contribution of the source is denoted by $s(r,t)$.The finite-difference numerical model developed in [21] is adapted to account for bubble generation in such a way that the acoustic pressure field and the bubble-volume variation field are calculated at each space point on the fast time scale, i.e., on the interval $[0,T]$, whereas the bubble density in the liquid is evaluated on the slow time scale from the minimum rarefaction pressure $p_-(z,r)$, in absolute value, reached at each space point during Θ (several acoustic cycles), following the nonlinear law [18,22]:

$$
N(z,r) = \begin{cases} N_+ \tanh\left(\dfrac{p_-(z,r) - \Phi_c}{p_-^+ - \Phi_c}\pi\right) \text{ if } p_-(z,r) \geq \Phi_c, \\ N_- \text{ if } p_-(z,r) < \Phi_c, \end{cases} \quad (z,r) \in (0,Z)\times(0,R). \tag{2}
$$

No new bubbles are created for p_- lower than Φ_c, the cavitation threshold of the medium. However, once Φ_c is reached, new bigger gas bubbles are generated in the liquid with a density that follows the above nonlinear law, where N_+ is the maximum possible density of new bubble and p_-^+ is the maximum p_- able to change the bubble density. This process simulates the generation of bubbles, which modify ultrasound, and the bubble creation is then modified, and so on [18,22]. We also consider the primary Bjerknes forces B_1 acting on the stable cavitation bubble newly created in the three-dimensional space. B_1 are evaluated on the slow time scale by averaging the radiation force on

$\Theta = [t_1, t_2]$. The bubble speed is then calculated. The motion of the nth bubble due to B_l is tracked using the time evolution of its position vector in the (z,r) plane at time t, $\overrightarrow{u_n}(t) = (\overline{z_n}(t), \overline{r_n}(t))$, i.e., the solution of the system formed by the dynamic equation, the initial position, $\overrightarrow{u_{n0}} = \overrightarrow{u_n}(t = t_1)$, and the null initial speed,

$$\begin{aligned} & M\overrightarrow{u_{n\,tt}} = \langle \overrightarrow{F_{ln}} \rangle,\ t \in (t_1, t_2), \\ & \overrightarrow{u_n} = \overrightarrow{u_{n0}},\ t = t_1, \\ & \overrightarrow{u_{n\,t}} = 0,\ t = t_1. \end{aligned} \tag{3}$$

$\langle \overrightarrow{F_{ln}} \rangle$ is the average on Θ of $\overrightarrow{F_{ln}}(t) = -V_n(t)\vec{\nabla} p_n(t)$, i.e., B_l applied to the nth bubble of mass $M = \langle V_n \rangle (\rho_g + \rho_l/2)$, and for which $V_n(t)$ is its volume, $\overrightarrow{u_{n\,t}}$ is its speed, $\vec{\nabla}$ is the gradient operator, $p_n(t)$ is the acoustic pressure at $\overrightarrow{u_n}(t)$, and ρ_g is the equilibrium density of the gas. The solution of this differential system is obtained by applying a finite-difference scheme [19].

The main physical restrictions of the mathematical model used are described in [1-3,11].

2.3. *Numerical models*

Two different numerical models have been developed to solve the coupled set of differential equations shown in Section 2.2. For standing waves, a finite-volume based numerical model was developed in [20] and is used in Section 3.1. For stable cavitation bubbles, a finite-difference based numerical model was developed in [21] and is used in Section 3.2. In these references the discretized equations are described in details.

3. Results and discussion

We consider that the liquid is water, $c_l = 1500\,\mathrm{m\,s^{-1}}$, $\rho_l = 1000\,\mathrm{kg\,m^{-3}}$, $\mu_d = 1.43 \times 10^{-3}\,\mathrm{kg\,s^{-1}m^{-1}}$, $\mu_k = 1.43 \times 10^{-6}\,\mathrm{m^2\,s^{-1}}$, and the bubble gas is air, $\gamma = 1.4$, $\rho_g = 1.29\,\mathrm{kg\,m^{-3}}$. Two types of numerical simulations are performed to illustrate the use of the models developed in Section 2. On one hand, we study the nonlinear resonance effect in a three-dimensional cavity with axial symmetry to enhance the nonlinear harmonics of distorted ultrasound and the generation of new frequency components by nonlinear frequency mixing. On the other hand, we analyse the stable cavitation process in a three-dimensional focused field with axial symmetry and the effect of primary Bjerknes forces.

3.1. *Nonlinear resonance and nonlinear frequency mixing in a standing-wave field*

We consider a cylindrical cavity of dimensions $0.00175 \times 0.0035\,\mathrm{m^2}$ in the (z,r) plane (figure 1) excited at its resonance frequency $f = 258\,\mathrm{kHz}$ in the bubbly liquid defined by $R_0 = 2.5\,\mu\mathrm{m}$ and $N = 5 \times 10^{11}$ bubbles $\mathrm{m^{-3}}$, and for which the sound speed at f is $c = 987.63\ \mathrm{ms^{-1}}$ [3]. A frequency sweep is applied around the resonance at two amplitudes at the source, $p_0 = 1\,\mathrm{Pa}$ and $p_0 = 15\,\mathrm{kPa}$. The maximum pressure amplitude reached in the cavity is presented in figure 2. For low amplitudes (linear regime), the highest value is obtained at the source frequency $f_L = 258\,\mathrm{kHz}$, which matches the resonance. For high amplitudes (nonlinear regime), the highest value is obtained at the source frequency $f_{NL} = 255.01\,\mathrm{kHz}$, which is the nonlinear resonance of the cavity. This frequency does not match the linear resonance, a shift of the resonance of the cavity is produced when the amplitude is raised: $\Delta f = 2.99\,\mathrm{kHz}$. This means that a softening is undergone by the bubbly liquid when amplitudes are increased, i.e., the sound speed is lowered [23]. Figure 3 represents the difference of amplitude of the fundamental, second and third harmonics obtained in the resonator when the excitation is carried out at

$f_{NL} = 255.01\,\text{kHz}$ and at $f_L = 258\,\text{kHz}$. The enhancement of the response in the cavity when the nonlinear resonance phenomenon is taken into account can be seen.

Figure 4 shows the amplitude of the difference frequency component obtained in the same cavity with the same bubbly medium by nonlinear mixing of two signals from the source, for which the primary frequencies are $f_1 = 6\times10^5$ Hz and $f_2 = 9.1758\times10^5$ Hz , and the amplitude at the source $p_0 = 12\,\text{kPa}$. The difference frequency is $f_d = f_2 - f_1 = 3.1758\times10^5$ Hz , for which the sound speed in the bubbly liquid is $c = 1215.7\,\text{ms}^{-1}$.

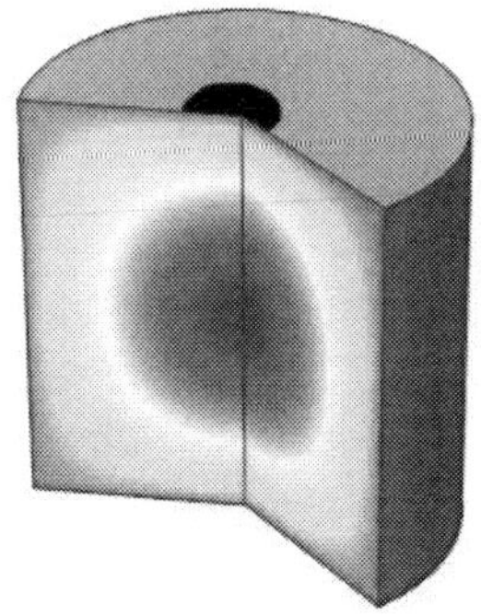

Figure 1. Cylindrical cavity excited at its resonance frequency.

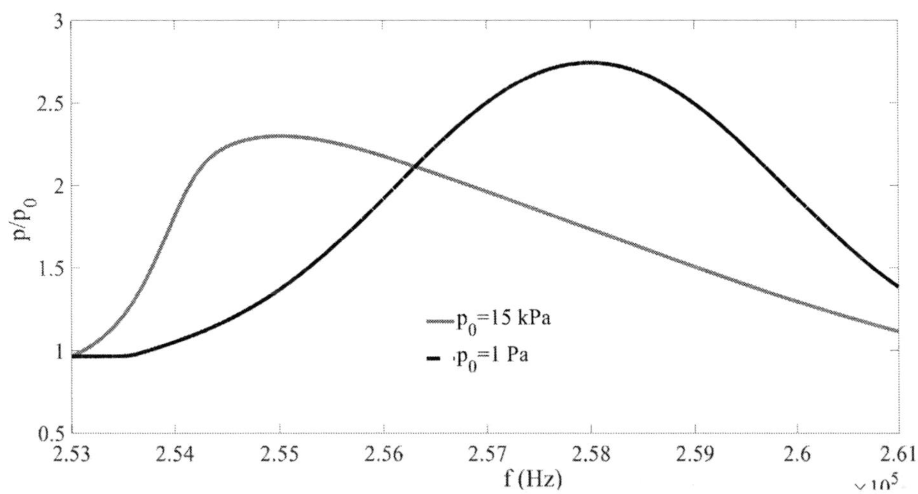

Figure 2. Maximum pressure amplitude reached in the cavity with two amplitudes at the source.

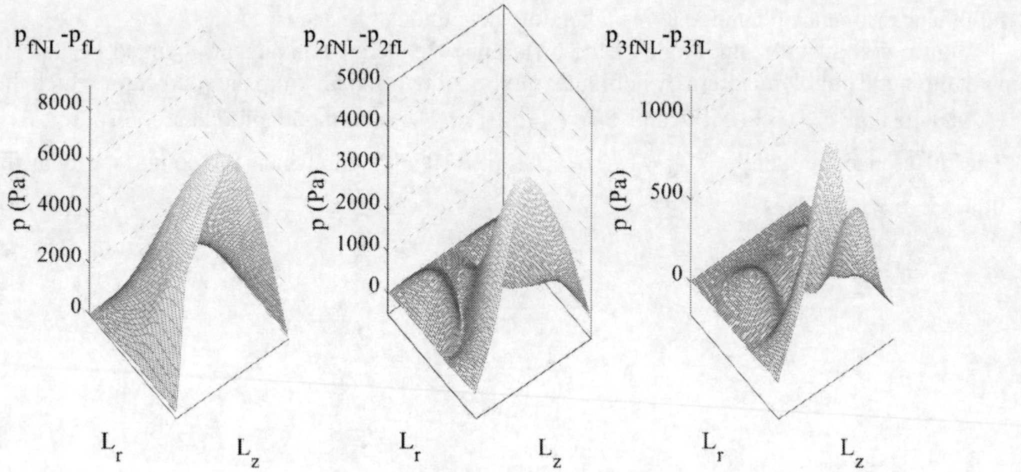

Figure 3. Difference of amplitude of fundamental, second and third harmonics obtained with f_{NL} and f_L.

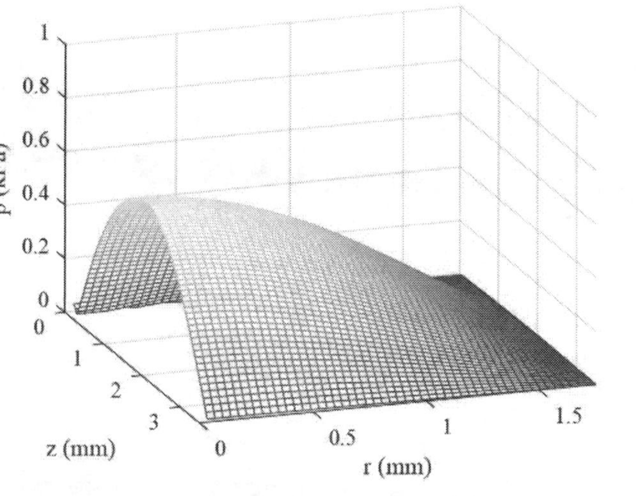

Figure 4. Amplitude of the difference frequency component by nonlinear mixing.

3.2. *Motion of stable cavitation bubbles in a focused field*

We consider the source at $f = 1\text{MHz}$ emitting up to $T = 8\times10^{-6}\,\text{s}$ in the liquid-nuclei mixture for which $R_{0-} = 0.01\mu\text{m}$, $N_- = 1\times10^{10}$ bubbles m^{-3}, $p_-^+ = 3.5\text{MPa}$ and $\Phi_c = 1.5\text{MPa}$, confined in the cylinder of dimensions $Z = 0.0075\text{m}$ and $R = 0.0045\text{m}$. We assume that $\Theta = 2/f$. For the medium with new stable cavitation bubbles we assume $N_+ = 2.5\times10^{10}$ bubbles m^{-3}, $R_0 = 35\mu\text{m}$, and the maximal void fraction is $V_f = 0.449\%$. Figure 5a shows p_- in the (z,r) plane for $\Theta\#1-\Theta\#4$. Since $p_- > \Phi_c$ in some area from $\Theta\#2$, new big stable cavitation bubbles are created. Figure 5b represents N in the (z,r) plane at the end of $\Theta\#1-\Theta\#4$, which is used for the evaluation of acoustic pressure at $\Theta\#2-\Theta\#5$. These cavitation bubbles form a cloud around the focus that changes the propagation of ultrasound. The nonlinearity of the bubbly liquid raises in the cloud and the pressure signal is distorted

IOP Conf. Series: Materials Science and Engineering **657** (2019) 012006 doi:10.1088/1757-899X/657/1/012006

nonlinearly. Figure 6 displays the frequency decomposition of p on the symmetry axis ($\Theta\#5$). Higher harmonics appear in the spectrum: 2 MHz, 3 MHz, 4 MHz, which means that the effect of the nonlinearly oscillating big gas bubbles on ultrasound is the transfer of acoustic energy from the fundamental to higher harmonics. Note that the corresponding linear diagram at low amplitude would show one single frequency (the source frequency), since no new cavitation bubbles would be created.

Figure 7 represents the distribution in the (z,r) plane at $\Theta\#5$ of $<F_{1z}>$ (a) and $<F_{1r}>$ (b) that affect the new cavitation bubbles, and their speed field (c). Around $z=0.002\,\mathrm{m}$ the bubbles are pushed toward the source and some are pushed away from the axis. Around $z\approx 0.0045\,\mathrm{m}$ the bubbles agglomerate. This result, due to the combined effect of stable cavitation process and primary Bjerknes forces, is quite coherent with experimental data [24,25].

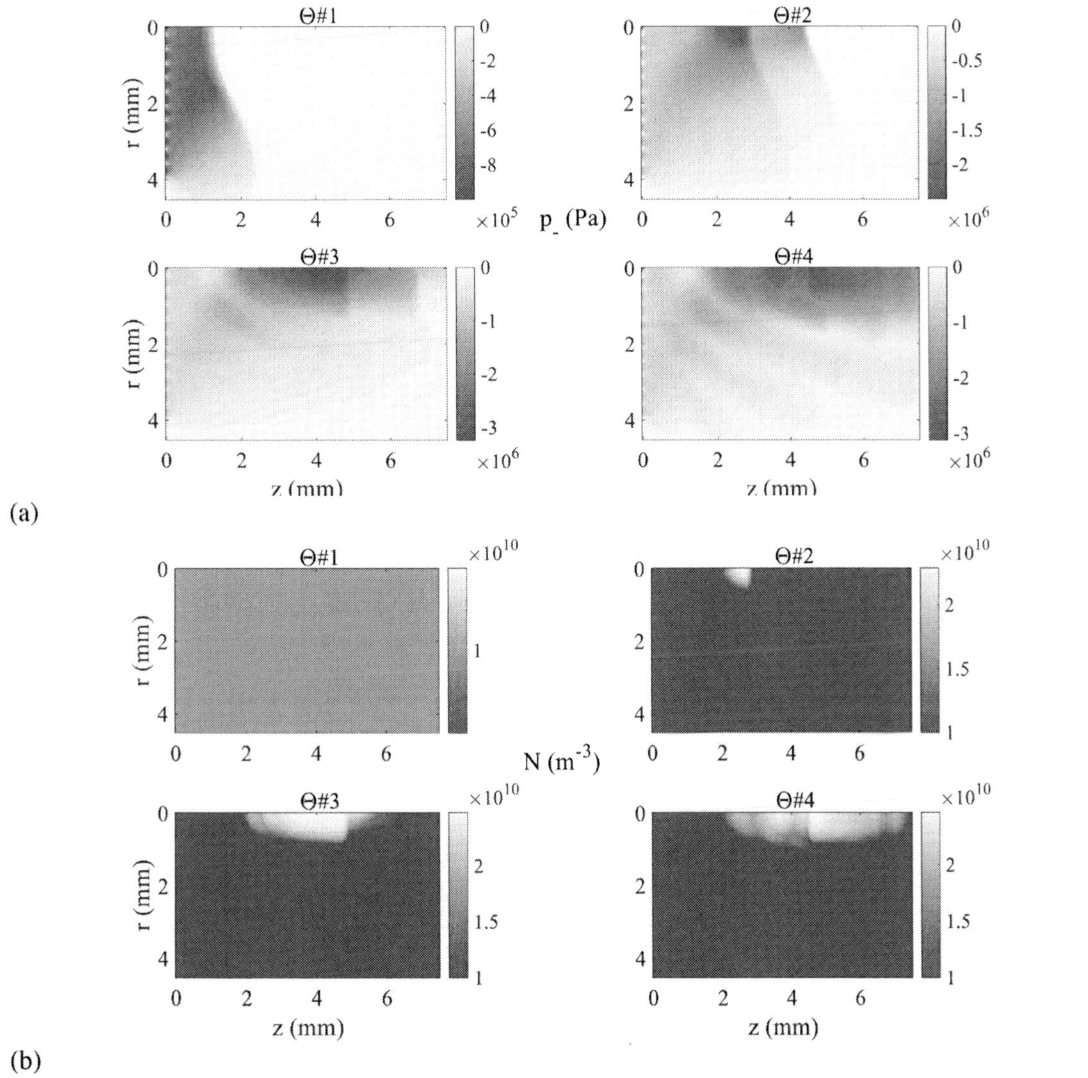

Figure 5. (a) p_- in the (z,r) plane for $\Theta\#1-\Theta\#4$. (b) N in the (z,r) plane at the end of $\Theta\#1-\Theta\#4$.

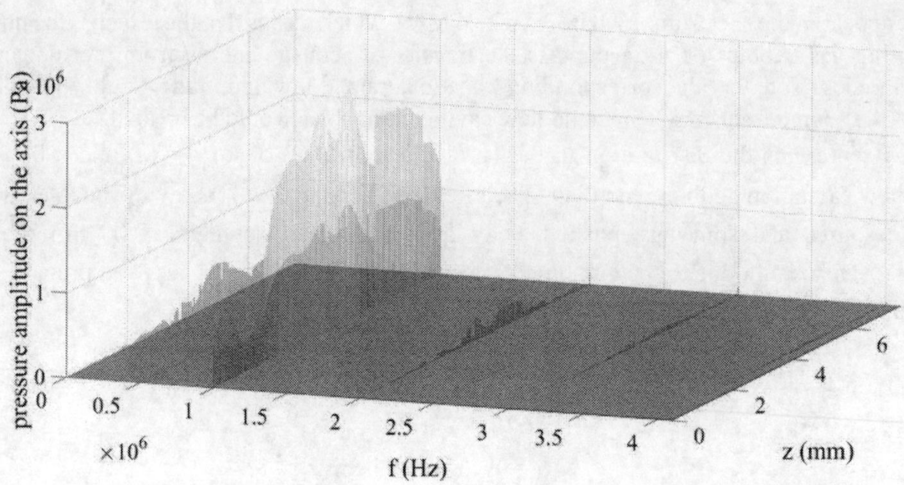

Figure 6. Frequency decomposition of p on the symmetry axis (Θ#5).

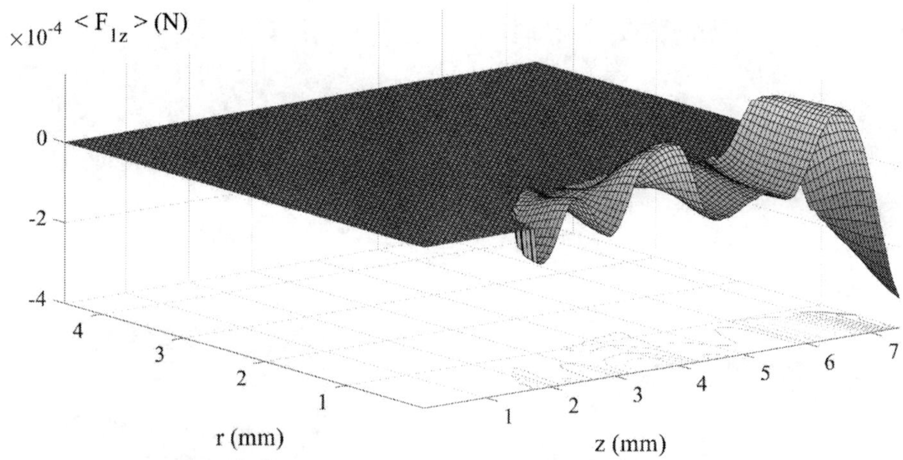

(a)

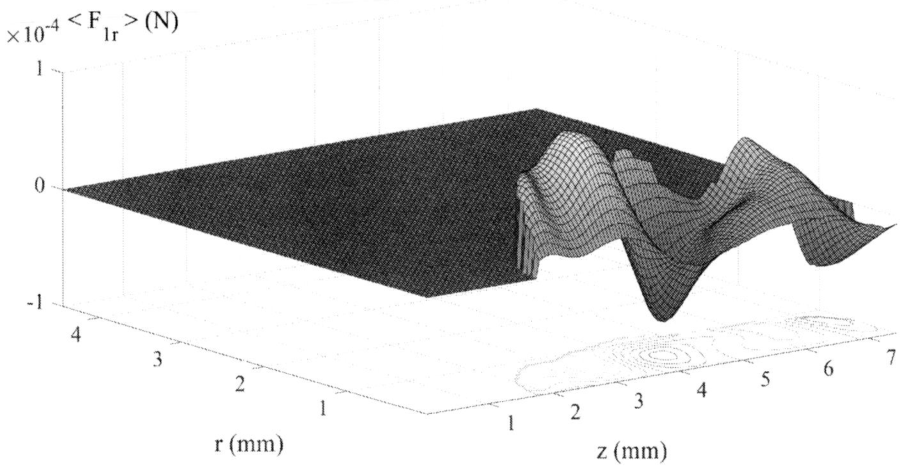

(b)

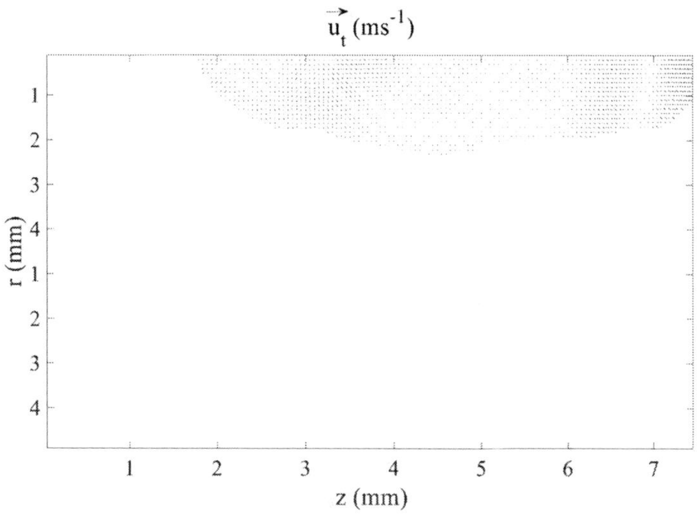

(c)

Figure 7. (a) Distribution of $< F_{1z} >$ in the (z,r) plane (Θ#5). (b) Distribution of $< F_{1r} >$ in the (z,r) plane (Θ#5). (c) Distribution of stable cavitation bubble speed in the (z,r) plane (Θ#5).

4. Acknowledgments

This work is supported by the National Agency for Research of Spain and the European Regional Development Fund through the grant DPI2017-84758-P.
Christian Vanhille dedicates this work to Dr. Cleofé Campos-Pozuelo.

References

[1] Naugolnykh K and Ostrovsky L 1998 *Nonlinear Wave Processes in Acoustics* (Cambridge University Press, New York)
[2] Grieser F, Choi PK, Enomoto N, Harada H, Okitsu K and Yasui K (Eds.) 2015 *Sonochemistry and the Acoustic Bubble* (Elsevier, Amsterdam)
[3] Hamilton MF and Blackstock DT (Eds.) 1998 *Nonlinear Acoustics* (Academic Press, San Diego)
[4] Brennen CE 1995 *Cavitation and Bubble Dynamics* (Oxford University Press, New York)

[5] Mason TJ and Lorimer JP 2002 *Applied Sonochemistry: the Uses of Power Ultrasound in Chemistry and Processing* (Wiley-VCH, Weinheim)
[6] Gallego-Juárez JA and Graff KF (Eds.) 2015 *Power ultrasonics: applications of high-intensity ultrasound* (Woodhead Publishing Series in Electronic and Optical Materials, vol. 66, Elsevier, Amsterdam)
[7] Matsumoto Y, Allen JS, Yoshizawa S, Ikeda T and Kaneko Y 2005 *Exp. Therm. Fluid Sci.* **29** 255-65
[8] Abramov OV 1998 *High-Intensity Ultrasound: Theory and Industrial Applications* (Gordon and Breach, London)
[9] Wu J and Nyborg WL (Eds.) 2006 *Emerging therapeutic ultrasound* (World Scientific, New Jersey)
[10] Yasui K 2011 Computational methods in sonochemistry *Computational Methods in Nonlinear Acoustics: Current Trends* ed C Vanhille and C Campos-Pozuelo (Kerala: Research Signpost) pp 137-150
[11] Zabolotskaya EA and Soluyan SI 1973 *Sov. Phys.-Acoust.* **18** 396-8
[12] Lauterborn W 2002 Nonlinear acoustics and acoustic chaos *Sound-flow interactions (Lecture Notes in Physics 586)* ed Y Aurégan, A Maurel, V Pagneux and JF Pinton (Berlin: Springer-Verlag) pp 265-284.
[13] Young FR 1989 *Cavitation* (McGraw-Hill, London)
[14] Vaughan PW 1968 *J. Sound Vib.* **7** 236-46
[15] Tudela I, Sáez V, Esclapez MD, Díez-García MI, Bonete P and González-García J 2014 *Ultrason. Sonochem.* **21** 909-19
[16] Tejedor Sastre MT and Vanhille C 2017 *Ultrason. Sonochem.* **34** 881-8
[17] Vanhille C, Pantea C and Sinha DN 2017 *Shock Vib.* **2017** 2518168 1-7
[18] Vanhille C 2016 *Ultrason. Sonochem.* **31** 631-6
[19] Vanhille C 2016 *Wave Motion* **63** 23-31
[20] Tejedor Sastre MT and Vanhille C 2017 *Ultrason. Sonochem.* **39** 597-610
[21] Vanhille C and Campos-Pozuelo C 2013 *Ultrason. Sonochem.* **20** 963-9
[22] Vanhille C and Campos-Pozuelo C 2012 *Ultrason. Sonochem.* **19** 217-20
[23] Tejedor Sastre MT and Vanhille C 2018 *Shock Vib.* **2018** 1570508 1-9
[24] Dubus B, Vanhille C, Campos-Pozuelo C and Granger C 2010 *Ultrason. Sonochem.* **17** 810-8
[25] Vanhille C, Campos-Pozuelo C, Granger C and Dubus B 2015 *Phys. Proc.* **70** 1070-3

IOP Conf. Series: Materials Science and Engineering **657** (2019) 012007 doi:10.1088/1757-899X/657/1/012007

Acoustic response of thin-walled, orthogonally stiffened cylinders

D Zhao[1,2], N S Ferguson[1] and G Squicciarini[1]

[1]Institute of Sound and Vibration, University of Southampton, UK

E-mail: Dong.Zhao@soton.ac.uk

Abstract. A semi-analytical model is developed to predict the acoustic response of thin-walled, orthogonally stiffened cylinders. The free vibration model is solved by the approximate assumed-modes method. The excitation of the normal modes is considered in terms of the Joint Acceptance Function (JAF) for incident acoustic harmonic plane waves. The surface pressure distribution is obtained by the shallow shell approximation, where the blocked excitation pressure is double the incident pressure, and by the 2-D rigid cylinder approximation considering the sound scattering. The acoustic response is then compared with that of a fully coupled FE-BE model. An efficient estimation of the response can be obtained by the two models over a certain range of accuracy. Numerical results for different structural configurations are then given. The effect of the stiffeners on the modal characteristics is discussed. It also illustrates the modal contribution to the total displacement response and addresses both circumferential and axial orthogonally stiffened cylinder arrangements.

1. Introduction

Thin-walled, stiffened cylinders are representative of widely used aerospace structures. When subjected to high-level acoustic loading, typically from rocket or jet engines, such structures are likely to experience considerable vibration and stress reversals, which may lead to fatigue failure. An efficient and accurate prediction of acoustically induced vibration (AIV) is thus required.

Much progress has been made in the past to predict the dynamic response of stiffened structures subjected to sound pressure. Often, flat structures are considered, and this significantly simplifies the structural modelling and acoustic response prediction [1-3]. The acoustically excited modes can be characterized by a structure-acoustic coupling ratio, the Joint Acceptance Function (JAF), which effectively relates the modal forces to the spatial matching between the acoustic field and the structural mode [4]. The acoustic response of stiffened plates has been analysed in a recent study by the authors [5]. However, these assumptions are not necessary applicable for curved structures. In comparison to stiffened flat plates, the stiffeners produce different effects on the modal characteristics of the cylinders due to the coupling of extension/bending motion around the circumference [6,7]. In addition, scattering effects must be considered for cylinders; for flat structures modelling of the excitation assumes that the blocked pressure is the twice that of the incident pressure and it is even appropriate for panels with small curvatures [8], but this assumption is invalid for a structure with significant curvature, whose dimension is comparable to the acoustic wavelength [9]. The JAFs of diffuse field has been investigated for cylindrical geometries based on a suitable spatial correlation function [10]. To the best of authors' knowledge, the acoustic scattering has not been explicitly addressed in the AIV within the aerospace domain. An exact solution for this problem requires a fully coupled vibroacoustic analysis, usually provided using a model that combines the structural model (FEM) and the acoustic domain (BEM). However, this approach requires high computational resources and it is not always practical for preliminary investigation and parametric study. Alternatively, the vibroacoustic response can be predicted by developing pre-processing tools compatible with commercial FE codes to reduce the computational cost. In this sense, Seon and Roy [11] developed a pre-processing tool to perform vibroacoustic analysis in Nastran for the Thermal Protection Panels (TPS) in a launch vehicle. For more generalized structures, like stiffened cylinders, an efficient model incorporating features like stiffener configurations and scattering is necessary for the preliminary studies.

This paper presents a semi-analytical approach to predict the linear response of stiffened cylinders to harmonic plane wave excitation. The free vibration problem is formulated by the assumed-modes method. The subsequent modal forces are derived based on two approximations for the pressure distribution: (i) blocked pressure around the cylinder and (ii) 2-D scattered pressure. These approaches are then compared with the results of a fully coupled vibroacoustic analysis from a FE/BE model.

2. The free vibration

The formulation adopted in this work to express the energy of the vibrating stiffened cylinder follows the procedures in reference [6]. Figure 1 shows the geometry of a thin-walled cylinder which is internally stiffened by a set of rings (around the circumference) and by a set of stringers (along the length). The flexure and extension of the cylindrical shell are considered. As a result of these deformations, the stiffeners are subjected to flexure (about two perpendicular axes), extension and torsion. From the linear strain-displacement relation [12] and the equation of state for the plane stress, the strain energy for the cylinder and the stiffeners can be formulated as a function of the displacement. The displacement of the stiffeners is uncoupled using the elastic axis and referenced to that of the line of attachment. Then, the compatibility conditions between the cylinder and the stiffeners are adopted to express the total energy in terms of the displacement of the cylinder middle surface. The detailed derivation can be found in reference [6]. The end boundary conditions of the cylinder are assumed to be shear diaphragms (S-D) [13]. The displacement of the middle surface of the cylinder can be expressed as,

$$
\begin{aligned}
u &= \sqrt{2}\sum_{m}\sum_{n}(\bar{u}_{mn}\cos n\theta + \bar{u}'_{mn}\sin n\theta)\cos m\pi x/a \\
v &= \sqrt{2}\sum_{m}\sum_{n}(\bar{v}_{mn}\sin n\theta - \bar{v}'_{mn}\cos n\theta)\sin m\pi x/a \qquad (1) \\
w &= \sqrt{2}\sum_{m}\sum_{n}(\bar{w}_{mn}\cos n\theta + \bar{w}'_{mn}\sin n\theta)\sin m\pi x/a
\end{aligned}
$$

where $\bar{u}_{mn}$, $\bar{v}_{mn}$ and $\bar{w}_{mn}$ are the generalized coordinates of the symmetric circumferential modes and $\bar{u}'_{mn}$, $\bar{v}'_{mn}$ and $\bar{w}'_{mn}$ are those of the antisymmetric modes.

The homogeneous equations of motion have been obtained by substituting the energy expression and equation (1) into Lagrange's equation [14]. The resulting frequency equation have been numerically solved in MATLAB to obtain the natural frequencies and mode shapes of the stiffened cylinder.

3. The acoustic response

Figure 2 shows that an acoustic harmonic plane wave impinging on the outer surface of a cylinder with an incident angle β to the x axis. Without losing any generality, it is convenient to assume that the incident acoustic wave component in the $y-z$ plane only travels along the y axis. In this paper it is assumed that the two closed ends of the cylinder have no influence to the acoustic field, i.e., the diffraction and scattering due to the finite length of cylinder are neglected. To simplify the analysis, the 3-D incident plane wave is written in cylindrical coordinates as,

$$p_i(\theta, s, t) = p_0 \sum_{q=-\infty}^{\infty} (-i)^q J_q(kR\cos(\beta)) e^{iq\theta} e^{ikRs\sin\beta} e^{i\omega t}, \quad (2)$$

where J_q is the Bessel function of first kind of order q and ϵ_q is the Neumann factor. p_0 is the amplitude of the free field incident pressure and $s = x/R$ is a normalized coordinate.

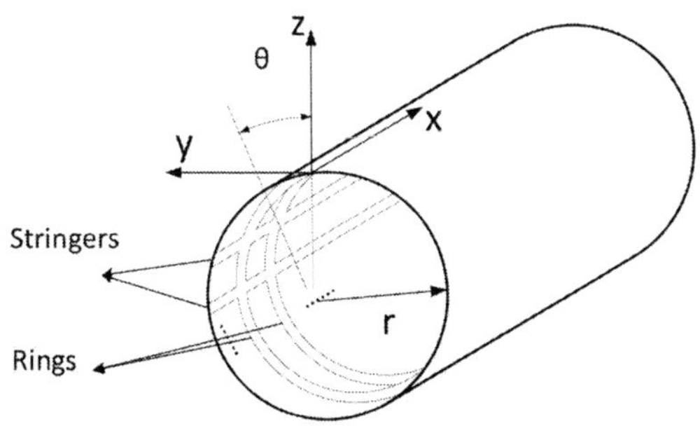

Figure 1. Geometry of a stiffened cylinder with the coordinate system.

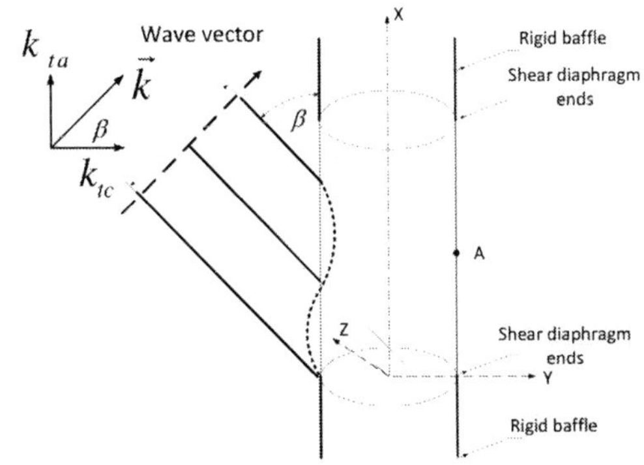

Figure 2. Diagram of an incident plane wave. k is the acoustic wavenumber and k_{ta}, k_{tc} are respectively the acoustic trace wavenumber (projection) along and normal to the shell axis.

3.1. *Joint acceptance function (JAF) for the shallow shell assumption*

In the blocked pressure assumption, the total pressure distribution can be represented as

$$p(\theta, s, t) = 2p_i(\theta, s, t). \quad (3)$$

Note that this is only a rough approximation that maybe valid only for some points of the cylinder and in a frequency range where the acoustic wavelength is small compared to the radius of curvature of the cylinder. From the modal analysis [14], the modal forces are,

$$L_{mn}(t) = \int_0^{2\pi} \int_0^{l/R} p(\theta, s, t)\phi_{mn}(\theta, s) \mathrm{d}sR\mathrm{d}\theta = F_0 j_{mn}(\omega) e^{i\omega t} \quad (4)$$

where $F_0 = p_0 S$ is defined as an equivalent static force acting on the cylinder surface with a normalized area $S = 2R\sqrt{\pi R l}$. The area has been normalized to simplify the expression for the JAF. $j_{mn}(\omega) = j_{Rn}(k_{tr}R) j_{Am}(k_{ta}l)$ can be regarded as the JAF for the $(m,n)^{th}$ mode of the cylinder for this type of incident pressure field. j_{Rn} and j_{Am} are therefore its circumferential and axial components respectively. They are given by the expressions

$$j_{Cn}(k_{tc}R) = (-i)^n J_n(k_{tc}R) \quad (5)$$

$$j_{Am}(k_{ta}l) = \frac{(-1)^m e^{-ik_{ta}l} - 1}{(k_{ta}l)^2 - (m\pi)^2} m\pi, \quad (6)$$

where k_{tc} and k_{ta} are the trace wavenumbers in the circumferential and axial directions, namely

$$k_{tc} = k\cos\beta, \; k_{ta} = k\sin\beta.$$

3.2. JAF using a 2-D rigid cylinder assumption for the scattered pressure

When the radius of the cylinder is comparable to the acoustic wavelength, the sound scattering due to the curvature will be significant. Assuming a rigid cylinder, the total pressure is in this case [15]:

$$p_t(\theta, s, t) = p_0 \sum_{q=-\infty}^{\infty} (-i)^q \big(J_q(k_{tc}R) - R_q(k_{tc}R)\big) e^{iq\theta} e^{ik_{ta}Rs} e^{i\omega t}. \tag{7}$$

$$where\ R_q(k_{tc}R) = \left(\frac{J_{q-1}(k_{tc}R) - J_{q+1}(k_{tc}R)}{H_{q-1}^{(2)}(k_{tc}R) - H_{q+1}^{(2)}(k_{tc}R)}\right) H_q^{(2)}(k_{tc}R). \tag{8}$$

$H_n^{(2)}(kR)$ are Hankel functions of the second kind.

It can be seen that only the pressure distribution in the circumferential direction is influenced by the scattering effect. The axial JAF remains the same to equation (6). The circumferential JAF becomes

$$j_{Cn}^{t}(k_{tc}R) = (-1)^n [J_n(k_{tc}R) - R_n(k_{tc}R)]. \tag{9}$$

4. Numerical results and discussion

Four types of structural configurations are considered for the aluminium cylinders. They are a bare cylinder, a cylinder stiffened by 9 equally spaced rings, a cylinder stiffened by 8 equally spaced stringers and a cylinder stiffened by both rings and stringers of the same number and positions. The dimensions and the material properties are given in table 1. Stringers and rings have rectangular cross section. In the numerical calculation, hysteric damping is assumed by using a damping loss factor of 1%. The numerical simulation is performed in the MATLAB.

Table 1. Parameters for numerical simulation of an aluminium cylinder.

Parameters	Symbol	Value	Parameters	Symbol	Value
Cylinder radius (m)	r	0.9	ring width (m)	r_w	0.002
Cylinder length (m)	l	0.3	ring depth (m)	r_d	0.004
Cylinder thickness (m)	t	0.002	Stringer width (m)	s_w	0.002
Density (kg/m^3)	ρ	2.7×10^3	Stringer depth (m)	s_d	0.004
Poisson's ratio	ν	0.3	Elastic modulus (N/m^2)	E	71×10^9
Incident angle (rad)	β	$\pi/4$	Pressure amplitude (Pa)	p_0	1

Figure 3 gives the natural frequencies of the lowest 20 modes for each structural configurations denoted by different types of symbols. The natural frequencies of adjacent modes appear in pairs for each structural configuration. This is because all the cylinders are symmetric. Each pair of frequencies corresponds to the symmetric/anti-symmetric modes. They are equal for the bare cylinder and slightly different for the stiffened ones. Compared to the bare cylinder, the rings increase the natural frequencies in each mode while the stringers decrease them slightly. This is because these modes are dominated by the circumferential deformation. The rings constrain the extensional or the flexural motions for these modes, which effectively stiffens the cylinder. The stringers put no constraint but apply mass to these modes. The natural frequencies of ring-stiffened cylinders also tend to occur in clusters over a small frequency range. This is because the stiffness effect of the rings makes the cylinder more like a finite periodic structure characterized by the 'frequency passband'.

Figure 4 plots the absolute values of the circumferential and the axial JAFs against the non-dimensional parameters, $k_{tc}R$ and $k_{ta}l$. These are for the bare cylinder and are based on the blocked pressure approximation. The results show a series of maxima for each mode. The maximum value occurs in the mode with the lowest order, corresponding to the 'breathing' mode for the circumferential direction and to the 'beam' bending mode in the axial direction. However, these modes occur at relatively high frequency for this geometry, being at 907 Hz (breathing mode) and 897 Hz (beam bending mode) and corresponding to mode number 135 and 131. For the other modes, the maxima in the axial JAF occur when the acoustic trace wavenumber matches the structural bending wavenumber, which is known as coincidence phenomenon. The maxima of the circumferential JAF results from the combination of pressure distribution around the circumference and the mode shapes. The pressure distribution around the circumference due to the incident plane waves does not result in a constant acoustic wavelength. It is therefore not straightforward to associate maxima in the circumferential JAF to the same coincidence phenomena found in the axial direction (or in flat plates). The dips in the JAFs correspond to the zero modal force, which occur when the integrations of the production of pressure and mode shape on the cylinder surface equal to zero in equation (4).

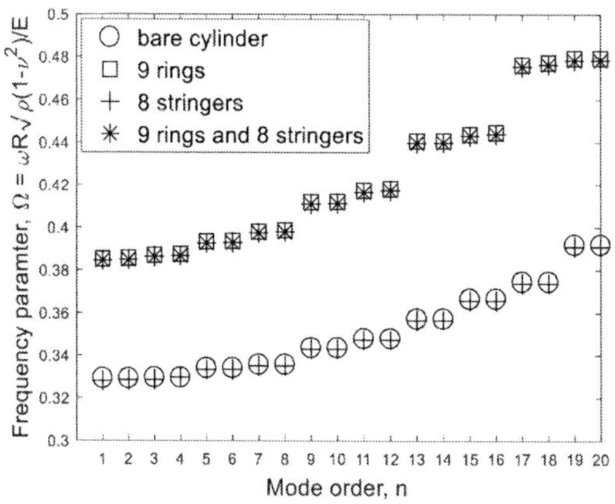

Figure 3. The natural frequencies of the lowest 20 modes for each case, plotted versus non-dimensional frequency $\Omega = \omega/\omega_R$ where $\omega_R = \sqrt{\rho(1-\nu^2)/E}/R$ is the ring frequency for a cylindrical bare shell.

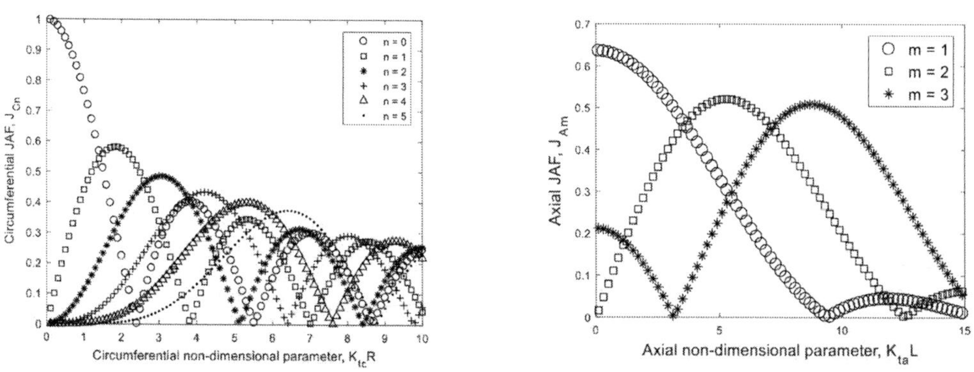

(a) (b)

Figure 4. The circumferential (a) JAFs for lowest order axial modes and (b) JAFs for breathing modes ($n = 0$) against the non-dimensional parameters equal to the Helmholtz number (kR or kL) and k_t is the trace wavenumber in the normal to the shell axis (k_{tc}) or axial (k_{ta}) directions.

Figures 5(a) to 5(d) present the spectra for the displacements predicted at a reference point for each cylinder. The assumptions of blocked pressure is compared with 2-D scattering. Point A depicted in figure 2 is chosen as an example point. Scattering is significant at this position and the point is always located away from stringers and rings. The frequency range is from 200 Hz to 1000 Hz and all the

IOP Conf. Series: Materials Science and Engineering **657** (2019) 012007 doi:10.1088/1757-899X/657/1/012007

modes up to 1200 Hz have been included in the calculations. For all the different structures, figure 5(a) to 5(d), three distinct frequency ranges can be observed. Below 300 Hz the response is mostly constant and this is due to the absence of vibration modes. Between 300 and 500 Hz the structures start to show response with peaks corresponding to the first few vibration modes. Above 500 Hz the response increases even further showing a broad peak superimposed by various other peaks.

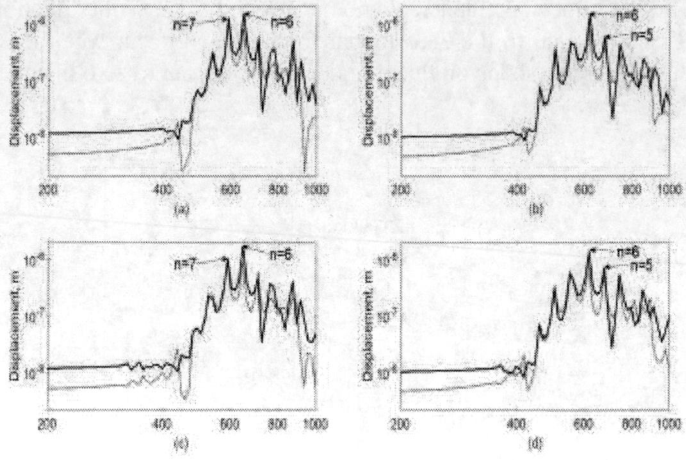

Figure 5. The radial displacement FRF (per unit pressure harmonic excitation) solid lines for blocked pressure results and dotted lines for scattering pressure results in upper figures) at a reference point on the furthermost side of the bare cylinder (a), the stiffened cylinder with 9 rings (b), with 8 stringers (c) and with both rings and stringers (d).

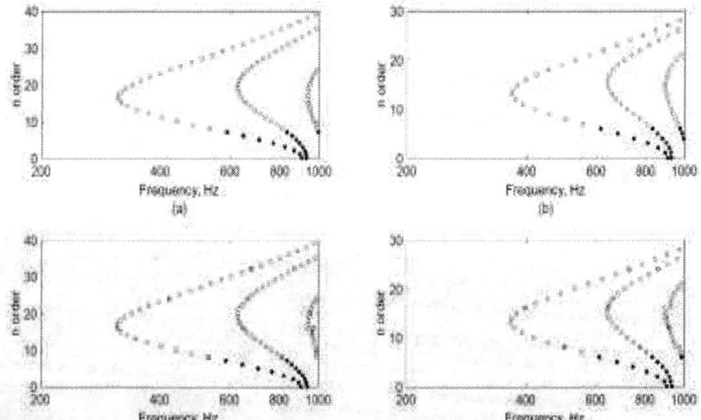

Figure 6. The acoustically excited modes (the predominant modes which corresponds to the peaks in figure 5 are marked by solid circles) at a reference point on the furthermost side for the bare cylinder (a), the stiffened cylinder with 9 rings (b), with 8 stringers (c) and with both rings and stringers (d).

Figures 6(a) to 6(d) show the acoustically excited modes and their frequency response dependency corresponding to the displacement responses in figures 5(a) to 5(d) for each cylinder. The marked solid circles which are the pronounced dominant modes corresponding to the resonant peaks in figure

IOP Conf. Series: Materials Science and Engineering **657** (2019) 012007 doi:10.1088/1757-899X/657/1/012007

5. These figures confirm the absence of modes below 300 Hz and show that below 500 Hz the order around the circumference (n) is relatively high. From equation (4) it was found that high order n results in smaller JAF (see figure 4). The response of the structure becomes larger for $n < 8$, which occurs at frequencies above 500 Hz. In this instance and for all the four structures considered, the highest peak in the response correspond to mode $m = 1, n = 6$. Additionally, all the figures show that blocked pressure gives a more conservative prediction of the response compared to the scattering pressure. The difference between the two decreases at the higher frequencies, when the acoustic wavelength is much smaller than the circumferential dimension of the cylinder. For this chosen response point, and for these geometries, the effect of stiffeners is limited.

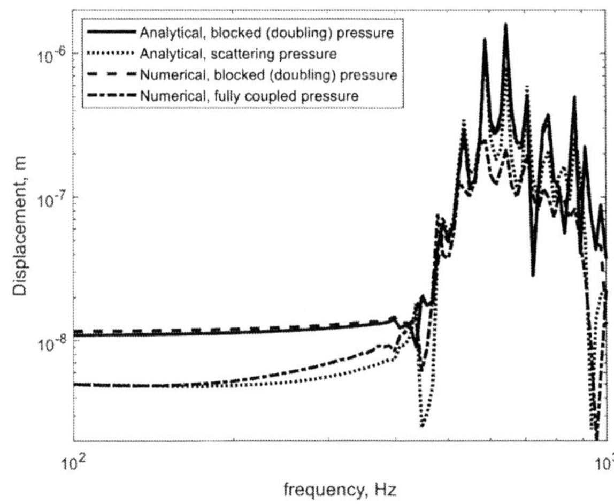

Figure 7. Displacement of the reference point for the bare cylinder: blocked pressure (solid line), 2-D scattered pressure (dotted line), FE blocked pressure (dashed line) and the FE-BE fully coupled pressure (dash - dot line). This shows the effect of the assumed acoustic pressure excitation on the bare cylinder response.

Figure 7 compares analytical predictions with those from a numerical model. This is calculated with both a blocked pressure assumption and with a full FE-BE coupling. Both have been developed in COMSOL and only the case of a bare cylinder has been analysed. The size of structural and acoustic mesh guarantees at least six elements per wavelength in each domain. The prediction based on blocked pressure overlaps completely with that in the FE model and for this reason it is not visible in the figure. In the FE-BE model, two finite cylinders longer than the longest acoustic wavelength in the simulation approximate the semi-infinite rigid cylinders at the ends of the structures. The result shows that in the frequency range away from the pronounced resonances ($\leq$ 400 Hz), the result of the 2-D scattering pressure approximation agrees well with that of the fully coupled analysis, while the prediction of the response based on the blocked pressure approximation is higher. This is because at low frequency the structural vibration is so small that it has negligible re-radiation effect, but the interaction between the sound and the structure geometry is significant in this case. At high frequencies, where resonances occur, the fully coupled model response is lower than the two approximate models. This is likely to be due to radiation damping, the effect of which appears above 400 Hz. The resonant frequencies in the fully coupled response are slightly lower than those in the two approximate models, which is likely to be due to the mass effect of the fluid loading from the surrounded air. The run time for the FE-BE model of the bare cylinder is about 40 hours on a standard desktop computer, while that of the two approximate models is about 5 minutes. For the other stiffened cylinders, the fully coupled FE-BE simulation can be more demanding and has not yet been addressed by the authors.

5. Conclusions

This paper presented an analytical model to predict the response of cylinders with and without stiffeners due to acoustic plane wave excitation. Results are produced to show the effect of stiffeners on the vibration modes, the modal contribution in the JAFs and acoustic response based on two different incident pressure approximations. The natural frequency grouped behaviour is introduced by the stiffeners, making the cylinder behave like a finite periodic structure. The modes with lower circumference mode order give a larger contribution to the total response. Both of the two approximate pressure assumptions lead to an overestimate of the response. They can significantly reduce the computational time at the cost of sacrificing some accuracy but are still conservative predictions. The analytical model could in the future be improved by including the effect of radiation damping.

Acknowledgments

The authors would like to acknowledge the BISE-FEE Joint Laboratory, who funded the project of acoustic fatigue prediction for advanced aerospace structures.

References

[1] Wijker J 2018 Miles' Equation in Random vibrations: Theory and applications in spacecraft structures design *Solid Mechanics and Its Applications no. 248* (Switzerland: Springer)

[2] Joshi P, Mulani S B and Kapania R K 2015 Multi-objective vibro-acoustic optimization of stiffened panels *AIAA/ASME Structural Dynamics and Material Conference* **2011-1750**

[3] Davis R B and Fulcher C W 2011 Framework for optimizing the vibroacoustic response of component loaded panels *AIAAJ* **53** 265-72

[4] Division E P 1996 Response prediction by deterministic methods *Structural Acoustics Design Manual* (The Netherlands: ESA Publications Division)

[5] Zhao D, Squicciarini G, and Ferguson N S 2019 The acoustic response of stiffened plates *J Physics: Conf Series* **1264** 012041

[6] Egle D M and Sewall J L 1968 An analysis of free vibration of orthogonally stiffened cylindrical shells with stiffeners treated as discrete elements *AIAAJ* **6** 518-26

[7] Kataoka S and Hida T 2016 Evaluation of stress in vibrating cylindrical shells due to acoustic loading based on theory of shells *ASME 2016 Pressure Vessels and Piping Conference* (Canada) p V003T03A019

[8] Cunningham P R, Langley R S and White R G 2003 Dynamic response of doubly curved honeycomb sandwich panels to random acoustic excitation Part 2: Theoretical study *J. Sound Vib.* **264** 605-37

[9] Fahy F J 1994 Acoustically induced vibration of structures *Sound and Structural Vibration: Radiation, Transmission and Response* (London; San Diego; New York: Academic Press) p 221

[10] Coyette J P, Lielens G, Robbe M, and Neple P 2005 An efficient method for evaluating diffuse field joint acceptance functions for cylindrical and truncated conical geometries *J. Acous. Soc. Am.* **117** 1009-19

[11] Seon G and Roy N 2011 Vibro-acoustic response to diffuse acoustic field using a modal approach and application to the IXV re-entry vehicle design justification *EUCASS 2011*

[12] Flügge W and Persen L N 1961 Stresses in shells *J Appl Mech* **28** 477-8

[13] Leissa A W 1973 Thin circular cylindrical shells *Vibration of shells United States: Scientific and Technical Information Office* (National Aeronautics and Space Administration) p 49

[14] Meirovitch L 1967 *Analytical Methods in Vibrations* (New York: Macmillan)

[15] Skudrzyk E 1971 The wave equation in cylindrical coordinates and its applications *The Foundations of Acoustics: Basic Mathematics and Basic Acoustics* (Wien: Springer Vienna) p 446

Refinement construction simulation research of BIM technology on pumping concrete

Y Li[1], C Y Song[1,2], Z G Wang[1] and Z G Li[1]

[1]College of Architecture and Civil Engineering, Beijing University of Technology, Chaoyang, Beijing, China

E-mail: ysong19@126.com

Abstract. With the gradual increase of pumping construction of medium and high-rise buildings in China, the construction requirements for concrete pumping process are gradually improved. Based on a project in Wuhan, this paper applied BIM technology on the pumping concrete construction process. Through the use of several softwares, the optimization of pumping pipeline layout, simulation of construction schedule, and priority grouping of marked bends was achieved. The details of the concrete construction of concrete pumping using BIM technology were shown, providing a reference for the application of BIM technology in the field of pumping concrete.

1. Introduction

The BIM is abbreviation of Building Information Modeling, which was first proposed by Chuck Eastman in the last 20th century. It set the foundation of the current "Building Information Modeling" for the concept of Chuck Eastman's "Building Description System" [1]. At the same time, other scholars Jerry Laiserin [2] and McGraw-Hill Building Information Company[3] also defined the BIM concept. BIM was a Shared knowledge resource which contained digital representations of all physical and functional characteristics throughout the entire life cycle of a construction project. Furthermore, the BIM was usually used to provide a reliable basis for decision-making through the lifecycle from concept to dismantling. At the same time, during different stages and periods of the project, various stakeholders involved in the project could store, extract, transfer and modify information through BIM at any stages of the project, so as to support and map the collaborative work and the interaction between different participants and responsibilities more quickly and conveniently.

From the top on, we could see that BIM covered the integration, sharing and transmission of the whole life cycle information and achieved the purpose of fine management and macro management through the dynamic monitoring platform.

The development of BIM technology is nearly 40 years. In early 2007, the American Academy of Building Sciences introduced the relatively BIM completed standard NBIMS (National Building Information Modeling Standard), which clearly stated the connotation of BIM [4,5]. In the United Kingdom, the application of BIM technology began in 2010. The standard NBS (National Building Specification), as a Government Construction Strategy document, proposed that the collaborative 3D-BIM should be finished fully by 2016, and the whole of information documents should be managed and the goal of different stage needed to complete was formulated [6]. Subsequently, a lot of BIM technology research in different fields have been conducted by a large number of domestic and foreign scholars. Li [7] studied the way to create building model components using BIM technology in traditional residential design methods. Yan [8] analyzed the advantages of introducing BIM technology from the perspective of life cycle, combining with the application of BIM technology in Shanghai's actual fabricated buildings. Kang *et al* [9] studied the network-based BIM-4D construction visualization technology and found that the asynchronous communication method through the network can greatly improve the decision-making effectiveness and optimize data management among project participants.

In recent years, BIM technology has been gradually promoted and applied in China's civil engineering design and construction fields. For example, the application of BIM technology in projects such as China-Zun, the Metro in Guangzhou and the East Station in Hangzhou, have greatly reflected the BIM technology needs in the domestic market. However, there were few cases that the BIM technology was taken advantage to solve the construction problems of concrete in the pumping process with the purpose of refining construction,

In response to the above problems, relied on a building in Wuhan, China, this paper conducted the research as follows:

- Three-dimensional visual modeling of a floor and pumping pipes components in Wuhan was completed utilizing BIM technology.
- Adopt Project software to realize construction organization planning; import the formed documents and the 3D model in the first step into Navisworks [9,10] software, finish simulating pumping construction progress, optimize pump tube pipeline simulation and group-marking bending tube.

A scheme of applying BIM technology in this paper is shown in figure 1.

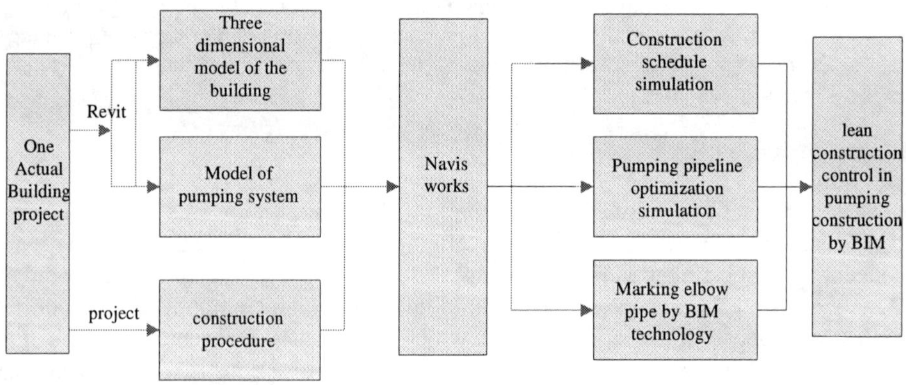

Figure 1. A scheme of BIM application.

2. Project overview

The location of a certain building is in Wuchang zone, Wuhan, China. Because the underground soil layer is soft soil, the pile foundation structure was adopted, and the upper building style of structure adopted the reinforced concrete frame structure. The construction method of the floor concrete utilized the transport-form of pumping concrete by high-pressure pumps. The pump tube was iron steel pipe with inner diameter of 150 mm, the horizontal part and the vertical part of the pump tubes were both 2 m, and the curvature at the elbow was 1/30 of the diameter of the pump tubes. The fixed base was a support form fixed by wrapping the pump tubes. The number of layers to floor body was 11, the pumping concrete was made of C60 type concrete, the column size was 600×600 (mm^2), the beam size was 400×800 (mm^2), and the connection diagram of pumping truck and pumping tube is shown in figure 2 below.

Figure 2. The connection layout size of pump and pipe.

3. Establishment of 3D model and optimization of pump tube layout

This paper chose Revit included in series of BIM softwares to carry out three-dimensional visual modeling of the main structure and pumping pipeline system of some specific project for a building in Wuhan. The 3D visualization model established by Revit software can visually and clearly display the dimensional relationship, positional relationship, apparent material of every components, and virtually and realistically express the specific objects in the real scene. For example, in this paper, the connection joints and support component sets of pumping pipes are shown below (figures 2 and 3), created by the way of self-creation "family" library and created according to the size of the real connecting and supporting components where truly reflecting the specific position of joint and support of the pumping pipes, providing the basis for the optimization of the pumping piping arrangement.

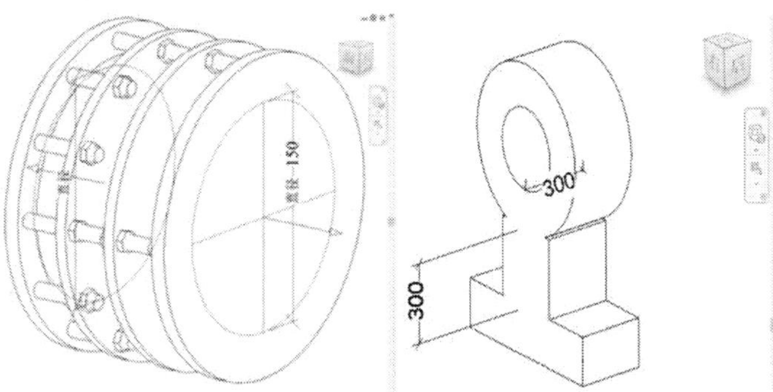

Figure 3. Pumping connection joints and support sets.

According to the actual project, the spatial position of structural sets created in Revit family via the self-built family and the system family function in 3D space was usually modified by software instructions such as move, offset, alignment, mirror, copy, rotate and construct under modify- button in the elevation and the grid system. Firstly, the three-dimensional modeling of the pump piping system was completed, and different construction material stacking positions and driving routes were divided in the site model. The on-site pumping layout model was rationally designed and optimized, thereby reducing the numbers of bending of the pumping pipelines and the total length of the pumping pipes with achieving the purpose of using BIM technology to reduce pumping pressure loss, civilized operation and fine construction.

The data before and after the optimization of the number and length of the horizontal pump tubes on the ground are listed in table 1.

Table 1. The data of pumping tubes in horizontal layout.

Category	Numbers of straight pipes	Numbers of bend pipes	Numbers of joint pipes	straight pipes length /m	Total length of pump lines /m
Unoptimized	35	6	41	67	74.85
BIM-layout	31	4	35	59	65.28

The article completed the creation of basic structures' component such as walls, beams, slabs, columns, piles and pile caps, furthermore completed the construction of this building in Wuhan, including the doors, windows, roofs, slabs, stairs, railings, curtain walls and so on. The finally Revit model of 3D visualization is shown in figure 4. In the figure, the different enclosed areas indicate different stacking positions of types of construction materials, the dark circles on the outer ring indicate the driving route, and the light bars in the inner circle indicate the routes in the construction. The establishment of 3D visualization model provided the necessary foundation for realizing construction progress simulation of the next step work.

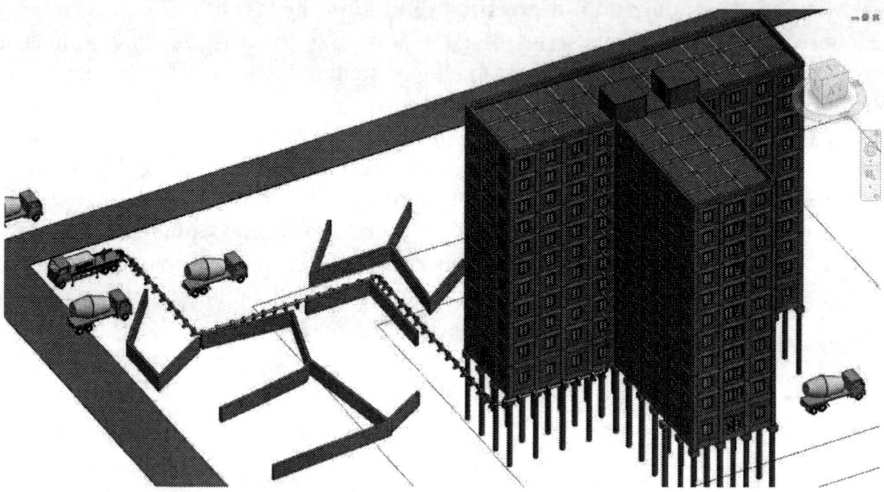

Figure 4. Revit model of body and pump tubes system.

4. 3D visualization of construction progress simulation and priority marking

The 3D visualization construction progress simulation is also called BIM-4D modeling information technology in the BIM field. It is a dynamic correlation between the 3D visualization construction model and the schedule of project constructions using a series of BIM softwares, which provides help for the pre-project and construction of the project. And also it provides supporting data timely and accurately about the ordering, distribution and utilization of materials in the process with improving engineering efficiency.

This paper adopts Microsoproject2010 as a construction process programming software in BIM softwares to compile and optimize the construction schedule. This software is developed by Microsoft to assist project managers in developing plans, scheduling and task allocation. It can efficiently track project progress, manage engineering budget and calculate the size of the engineering quantity.Using the Project2010 software, the different components in the Revit 3D visualization model were hierarchically organized into names and serial numbers in order of categories and time completion. Then the theoretical and actual start time and completion time of each component of the created model in the software were input, according to the rules and orders of preparation, which was determined in advance and the specific arrangement process of the engineering project time. Through the divided construction task also named Work Breakdown Structure (WBS), the paper created a different level of construction schedule, as shown in figure 5 below. The mpp formal document created by Microsoproject provided the basis for the completion of the next 3D visualization construction progress simulation.

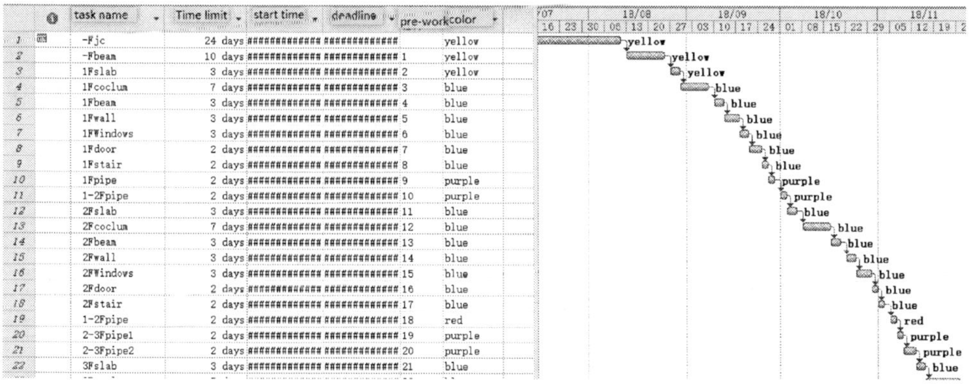

	task name	Time limit	start time	deadline	pre-work	color
1	-Fjc	24 days	#############	#############		yellow
2	-Fbeam	10 days	#############	#############	1	yellow
3	1Fslab	3 days	#############	#############	2	yellow
4	1Fcoclum	7 days	#############	#############	3	blue
5	1Fbeam	3 days	#############	#############	4	blue
6	1Fwall	3 days	#############	#############	5	blue
7	1FWindows	3 days	#############	#############	6	blue
8	1Fdoor	2 days	#############	#############	7	blue
9	1Fstair	2 days	#############	#############	8	blue
10	1Fpipe	2 days	#############	#############	9	purple
11	1-2Fpipe	2 days	#############	#############	10	purple
12	2Fslab	3 days	#############	#############	11	blue
13	2Fcoclum	7 days	#############	#############	12	blue
14	2Fbeam	3 days	#############	#############	13	blue
15	2Fwall	3 days	#############	#############	14	blue
16	2FWindows	3 days	#############	#############	15	blue
17	2Fdoor	2 days	#############	#############	16	blue
18	2Fstair	2 days	#############	#############	17	blue
19	1-2Fpipe	2 days	#############	#############	18	red
20	2-3Fpipe1	2 days	#############	#############	19	purple
21	2-3Fpipe2	2 days	#############	#############	20	purple
22	3Fslab	3 days	#############	#############	21	blue

Figure 5. Construction schedule (partial).

As a simulation software in the BIM softwares, AutodeskNavisworks is a building engineering management software developed by Autodesk, with which engineers usually optimize structural and architectural design and control the quality of construction projects. This article imported the two things into Autodesk Navisworks, one was the rvt formal model created by Revit and the other was the mpp formal document created by Microsoproject. Through the "select tree" and viewpoint function of Navisworks software, it was conveniently to save the 3D model created by Revit according to certain rules. Then, by using different colors for different floors and functional components, the simulation growth problems in the model can be clearly observed so that the designer can timely modify and optimize the structural design to avoid unnecessary waste. Figure 6 below is a simulation diagram of column hierarchical displayed by nw software. The column of each layer of color represents the columns of different layers. It was more convenient to find out whether the numbers and position of components in the simulation construction process were correct through the simulation of different components, thereby achieving a refined simulation construction process.

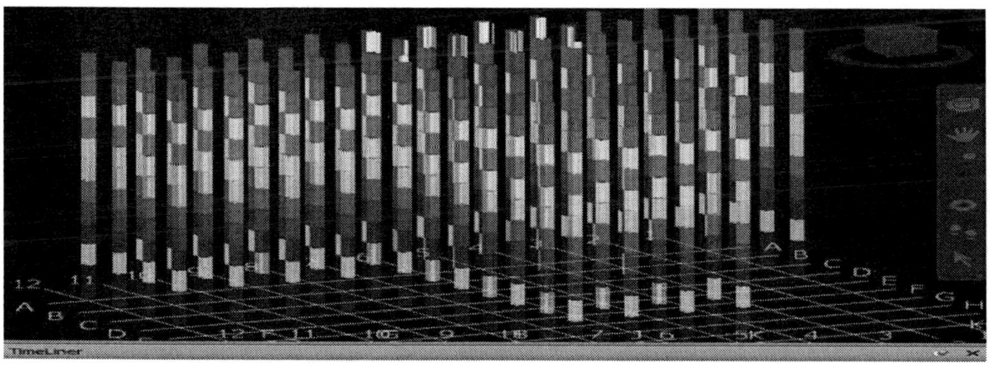

Figure 6. Different floor column simulation diagram.

As to the concrete transportation and pouring operation, it was carried out continuously in the pumping process. With the three-dimensional visual construction progress simulation implemented, the three things could be found in advance, that was problems in the construction process, the construction machinery arrival sequence and the construction progress. Thus, it could meet the needs of the project party to accurately grasp the construction situation during the construction process at any time. In this paper, the Timeliner function in Navisworks software was used to simulate the 3D visualization construction progress. The resulting effect is shown in figure 7. Using NW software, not only realized the simulation of construction progress, but also reproduced the progress of the project according to the requirements of the construction process at any time. This three-dimensional visual construction simulation was corresponded to the actual construction progress node, which greatly facilitated the engineer to compare the construction progress of the site with the progress of Navisworks 4D model so as to find the deviation of the construction process, arrange the construction personnel and determine the order of the construction machinery.

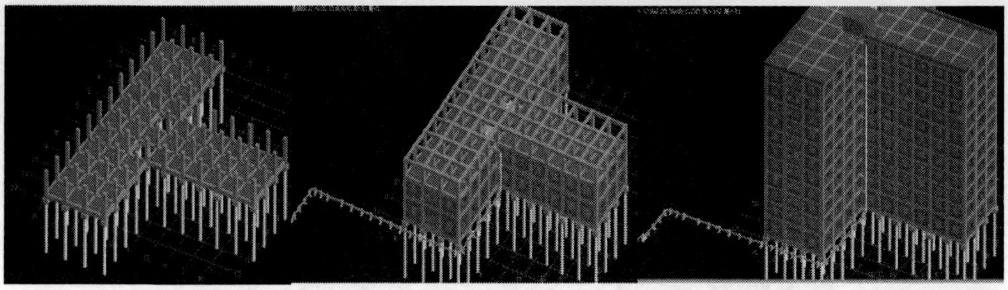

Figure 7. 3D visualization construction progress simulation.

At the same time, for the pumping elbows in the pumping system, priority grouping were performed in a manner similar to that of figure 6, depending on the phenomenon of trapping of the elbow during the pumping process. Level1: Horizontal and vertical conversion elbows > Level2: Top elbows > Level3: Horizontal elbows. With the help of the grouping function of the software, when pumping problem occurs, the construction workers could use the priority order of the marking to conduct the inspection, which makes the construction process simpler and clearer, with achieving the purpose of using the BIM technology to realize the fine construction of the pumping process and improving the construction efficiency in pumping construction.

5. Conclusion

- In this paper, we optimized the arrangement of the pumping pipeline by revit three-dimensional modeling, reducing the length of the overall pumping pipeline, thus reducing the pressure loss during pumping.
- Through detailed demonstration of the refinement simulation construction process using revit, project and nw software, the problems of construction works' arrangement, mechanical approach sequence and so on were solved with improving the efficiency of pumping construction by BIM technology.
- The BIM technology was used to model, prioritize and mark the pumping elbows, which provided a reference for field personnel to check the pumping and blocking problems in order with the efficiency of the pumping process improved.

Acknowledgments
The authors would like to acknowledge the financial support provided by National Key R & D Program of China-Key materials and preparation technology of high crack resistant ready-mixed concrete (2017YFB0310100), National Natural Science Foundation of China (51678011); Beijing Natural Science Foundation (8162005).

References

[1] Jerry and Lether W X 2011 *Arch. Creation* **2011** 146-50
[2] Guo H and Li H S M 2010 *J. Eng. Manage.* **26** 41-7
[3] McGraw-Hill 2009 *J. Infor. China. Cons.* **2009** 10-1

[4] He Q H and Qian L L 2012 *J. Eng.Manage.* **26** 12-6
[5] USN 2007 *National Institute of Buiding Sciences* Version1-P1
[6] Liu H G 2015 *J. High Rail. Tech.* **2015** 59-66
[7] Liu Y J 2008 *J. Chongqing Univ.* p 14
[8] Min L and Liu L 2014 *J. Hous. Sci.* **34** 53-6
[9] Kang J H, Anderson S D and Clayton M J 2007 *J. Cons. Eng. & Manage.* **133** 447-61
[10] Jiang A M and Huang S 2014 *J. Cons. Tech.* **43** 86-9

IOP Conf. Series: Materials Science and Engineering **657** (2019) 012009 doi:10.1088/1757-899X/657/1/012009

Particle swarm optimization particle filter denoising algorithm with mutation operator

H X Chen[1,2], L Fang[1], D L Fan[1], W J Huang[1], G Y Zhang[1], C H Cao[1], J M Huang[1] and L Yang[1]

[1]School of Mechanical and Electrical Engineering, Wuhan Institute of Technology, Wuhan 430205, China

E-mail: 449392297@qq.com

Abstract. In this paper, a novel particle filter (NPSO-PF) algorithm is proposed, which is called particle cluster optimization particle filter algorithm with mutation operator, which is used for real-time filtering and noise reduction of nonlinear vibration signals. Because of its introduction of mutation operator, this algorithm has no problem that particle swarm optimization (PSO) algorithm is easy to fall into local optimal value and the calculation accuracy is not high. At the same time, through the mutation operation, the distribution and diversity of particles in the sampling process are improved, and the particle filter (PF) algorithm is solved in which the particles are poor and the utilization rate is not high. The mutation control function makes the particle set optimization process in the early and late stages, and improves the convergence speed of the particle set, which greatly reduces the running time of the whole algorithm. Simulation experiments show that compared with PF algorithm and PSO-PF algorithm, the proposed new particle swarm optimization particle filter (NPSO-PF) algorithm has lower root mean square error, shorter running time, higher signal to noise ratio and more stable filtering performance. It is proved that the algorithm is suitable for real-time filtering and noise reduction processing of nonlinear signals.

1. Introduction

Denoising is an important part of signal processing. The effect of denoising directly affects the accuracy of subsequent analysis results. The denoising method for studying nonlinear signals is the focus of attention in recent years. Ananth *et al* [1] used a PSO based blur filter to remove high density image impulse noise from the signal. These algorithms can achieve the denoising of nonlinear signals. However, using the PSO algorithm structure, it is easy to fall into the local optimal solution. Processing of signals with multiple local extrema, the convergence speed and accuracy are not high, the convergence speed and accuracy are not high. In order to improve the defects of the PSO algorithm, Dai J et al. introduced the rough set theory into the PSO algorithm in [2], and constructed a discrete PSO method to enhance the search ability. Kar *et al* [3] proposed a particle swarm optimization algorithm based on gravitational search algorithm (GSA), which is used for the optimization design of two common analog circuits and achieved good results.

The unique advantage of PF algorithm in dealing with parameter estimation and state filtering of nonlinear non-Gaussian time-varying systems , which makes the PF method attract attention at home and abroad in recent years .However, the conventional PF algorithm adopts the suboptimal importance function in the resampling process, which makes the conventional particle filtering method have the problems of low particle and low computational efficiency. The intelligent optimization algorithm is integrated into particle filter, which becomes the hotspot of particle filter at present. The particle filter algorithm based on particle swarm optimization algorithm is one of the representatives of intelligent particle filter algorithm. In [4], Ramazan *et al* proposed a dual estimation method for joint parameters and state estimation based on edge particle filter and particle swarm optimization, which improves the performance of PF algorithm. However, the algorithm tends to be locally optimal in the middle and late stages, and the operation results are unstable and the denoising effect is poor. In order to improve the performance of the PSO-PF algorithm, Tao *et al* [5] introduced chaotic sequences into the PSO-PF algorithm, and the literature [6] and [7] combined the genetic resampling method with the PSO-PF algorithm. These measures all improve the filtering performance of the PSO-PF algorithm. However, as the structure of the algorithm becomes more complicated, the amount of calculation will increase, which is not suitable for real-time denoising processing.

Inspired by the above, this paper proposes a NPSO-PF algorithm for filtering noise reduction. The algorithm passes the genetic variation control function after sampling, so that the particle optimization process is divided into the front and the end, which accelerates the convergence of the particle set and improves the operation speed. At the same time, through the mutation exercise, the particle diversity is controlled, and the problem of low particle consumption and low utilization rate is avoided. The algorithm is used for noise reduction of nonlinear signals with the advantages of stability, speed and effectiveness. This paper first introduces the principles of PF, PSO and PSO-PF algorithms and analyzes their defects, and then constructs the NPSO-PF algorithm. In order to verify the filtering performance of the NPSO-PF algorithm, the state estimation experiment was carried out. Finally, the denoising simulation experiment was carried out to verify the filtering noise reduction of the NPSO-PF algorithm.

2. Conventional particle filter algorithm

The essence of the PF algorithm is to use the Monte Carlo method to realize the integral operation of the optimal Bayesian estimation. The sample particles are used to simulate the current state of a given system and determine the particle samples corresponding to the system in different states at different times. Thereby a posterior probability distribution approximated to the state of the system is obtained. The key is that the weight of the core particles can be constantly corrected and adjusted, and finally an accurate estimate is obtained [8]. The state equation and observation equation of nonlinear dynamic system are selected as follows:

$$\begin{cases} x_k = f(x_{k-1}) + v_{k-1} \\ y_k = h(x_k) + n_k \end{cases} \tag{1}$$

Where x_k is the system state vector at k moment, y_k is the observation output. And the v_{k-1} is the system noise. The n_k is the observation noise. The $f(\cdot)$ and $h(\cdot)$ are non-linear functions respectively.

The implementation steps of particle filter are as follows:

Step 1: Initialization. At $k=0$ moment, $\{x_0^i, i=1,2,..., N\}$ is obtained by sampling according to $P(x_0)$ distribution.

Step 2: Sampling

- For $i=1,2,\ldots,N$. Sample x_k^i from the proposal distribution $q\left(x_k \middle| x_{k-1}^i, z_k\right)$.
- Evaluate the importance weights

$$\omega_k^i = \omega_{k-1}^i \frac{P(z_k|x_k^i)P(x_k^i|x_{k-1}^i)}{q(x_k^i|x_{k-1}^i, z_k)}, i=1,2,\ldots,N \tag{2}$$

- Normalize the weights

$$\tilde{w}_k^i = w_k^i * \left(\sum_{i=1}^{N} w_k^i \right)^{-1} ,i=1,2,3....N \tag{3}$$

Step 3: Resampling. The set $\{x_0^i ,i=1,2,..., N\}$ of new particles at k time is obtained by re-sampling from set $\{x_0^i ,i=1,2,..., N\}$ according to the importance weight $\tilde{\omega}_k^i$, and the weight of particles is redistributed to $\omega_k^i = \tilde{\omega}_k^i = 1/N$.

Step 4: Output. State estimation

$$\tilde{x}_k = \sum_{i=1}^{N} \omega_k^i \tilde{x}_k^i \tag{4}$$

3. PSO-PF algorithm

3.1. The principle of PSO algorithm

The PSO algorithm is a kand of optimization algorithm for simulating swarm intelligence behavior proposed by Kennedy and Eberhart in 1995. It is an effective global optimization algorithm by optimizing the search through the group intelligence guidance generated by the cooperation and competition of particles in the group [9]. Specifically, it can be expressed as: randomly initialize a particle group (the quantity is m) in an N-dimensional search space. The particl i position is $X_i = (x_{i1}, x_{i2}, ..., x_{in})$, its velocity is $V_i = (v_{i1}, v_{i2}, ..., v_{in})$, its individual extreme value $P_i = (p_{i1}, p_{i2}, ..., p_{in})$, and the population's global extreme value is $G_i = (g_{i1}, g_{i2}, ..., g_{in})$. After determining the above two extremes, you can update the speed and position of the particle according to the following equation:

$$\begin{cases} V_i^{k+1} = \omega * V_i^k + c_1 * rand() * (P_i^k - X_i^k) + c_2 * rand() * (G_i^k - X_i^k) \\ X_i^{k+1} = X_i^k + V_i^{k+1} \end{cases} \tag{5}$$

Where: $rand()$ is a random number between interval $(0,1)$, and c_1, c_2 are collectively referred to as a learning factor or an acceleration factor, generally $c_1 = c_2 = 1.4692$. ω is the inertia coefficient.

3.2. PSO-PF algorithm

The PSO algorithm is integrated into the PF algorithm by first introducing the latest observations into the sampling process of the particle samples and defining the fitness function in the PSO algorithm as:

$$fit = \exp\left\{ -\frac{1}{2R_t} (Y_{new} - Y_{pred}) \right\} \tag{6}$$

Where R_t is the observed noise variance, Y_{new} is the latest observation, an Y_{pred} dis the predicted observation.

During the sampling process, if the particle sets are distributed near the real state, the fitness of the particles will be high. Particles that are not in the real state will have a low fitness value. At this point, the PSO algorithm optimization algorithm is used to continuously update the velocity and position of each particle according to the optimal value and equation (5), so that the particles are constantly approaching the real state.Thereby, the omission of important particles is avoided, which not only ensures the diversity of the particles, but also improves the utilization rate of the effective particles.

4. PSO-PF algorithm with variation operator (NPSO-PF)

4.1. PSO algorithm with mutation operator

The conventional PSO algorithm is prone to fall into the local optimal value in the early iteration. In order to improve this defect, in view of the variation idea of the genetic mutation algorithm, the genetic algorithm operation operator [10] is introduced into the PSO algorithm to keep the diversity within the population. Furthermore, the algorithm avoids premature convergence to the local optimal value. The specific method is to introduce a mutation operator control function to control the mutation rate of the mutation operator.The mutation control function is shown in equation (7):

$$y(epoch)=\left(1-(epoch/epoch_{max})^{\alpha}\right)^{\beta} \tag{7}$$

Where $epoch$ represents the current number of iterations, $epoch_{max}$ represents the maximum number of iterations, α、β represents the control coefficient.

According to equation (7), the control rate of the mutation operator is:

$$u=m\cdot y(epoch) \tag{8}$$

Where u is the mutation rate of the mutation operator, and m is the preset mutation rate, which is unchanged after setting.

According to the mutation rate, the number of particles undergoing the mutation operation is:

$$M=[N*u] \tag{9}$$

M particles are randomly selected for mutation operation, and real numbers are used. Considering that particle k is selected for mutation operation, and element j of the $X_k=(x_{k1},x_{k2},\cdots,x_{kD})$ is mutated, the operation strategy is:

$$x_{kj}=x_{kj}+rand*y(epoch) \tag{10}$$

Where $rand\in(-a,a)$ is a random number.

In this paper, the dynamic decreasing inertia coefficient is used, and the dynamic inertia coefficient is calculated as:

$$w=w_{min}+(w_{max}-w_{min})\cdot y(epoch) \tag{11}$$

Where w_{max} represents the inertia coefficient and w_{min} represents the minimum inertia coefficient.

The mutation operator controls the operation of the PSO algorithm through the particle's mutation rate, dynamic inertia coefficient and mutation operation. The mutation rate is determined by equation (7), the dynamic inertia coefficient is determined by equation (11), and the mutation operation is performed by equations (8)-(10), which are all related to the number of iterations of the algorithm. In the early stages of algorithm iteration, the control variation rate and the dynamic inertia coefficient take a large value, and the variation operation of the particles is large, which maintains the diversity of the internal particles of the population, and makes the PSO algorithm search in a large range of the solution space. In the later stage of the algorithm, the control variation rate and the dynamic inertia coefficient take a small value, and the variation operation of the particles is very small. At this time, the particles inside the population can quickly become uniform, making the PSO algorithm within a small range of the solution space. Search. Thereby, the convergence speed and convergence precision of the algorithm are improved.The specific implementation steps of the NPSO algorithm are as follows:

Step 1: Initialize group size N, learning factors C_1 and C_2, inertia coefficient w, position X_i and velocity V_i of each particle, maximum iteration number $epoch_{max}$, preset mutation rate m.

Step 2: The control function $y(epoch)$is calculated according to equation (7), and the inertia coefficient w is calculated according to equation (11).

Step 3: Calculate the mutation rate $y(epoch)$ according to the equation (8), and randomly select the M particles to perform the mutation operation according to the equation (10).

Step 4: For each particle, the fitness solution $f_{it}[i]$ is expressed according to equation (6) and compared with its individual optimal value P_i, if $f_{it}[i] > P_i$ then $P_i = f_{it}[i]$.

Step 5: Find the global optimal value G_i, execution procedures such as if $P_i > G_i$ then $P_i = G_i$.

Step 6: Update the velocity V_i and the position X_i of the particle according to equation (5).

Step 7: When the number of iterations or the optimal value of the particle group meets a certain threshold ε, the result is output, otherwise it returns to the second step.

4.2. NPSO-PF algorithm

The principle of the NPSO-PF algorithm is to optimize the sampling process of particle filter by using the fast and accurate convergence performance of the PSO algorithm with mutation operator. The PSO algorithm with mutation operator is used to calculate the fitness to move all particles to the optimal particle. Thereby improving the distribution of the particle set, so that the particle set is in the vicinity of the real state, thereby improving the precision and efficiency of the particle filtering, the specific steps are:

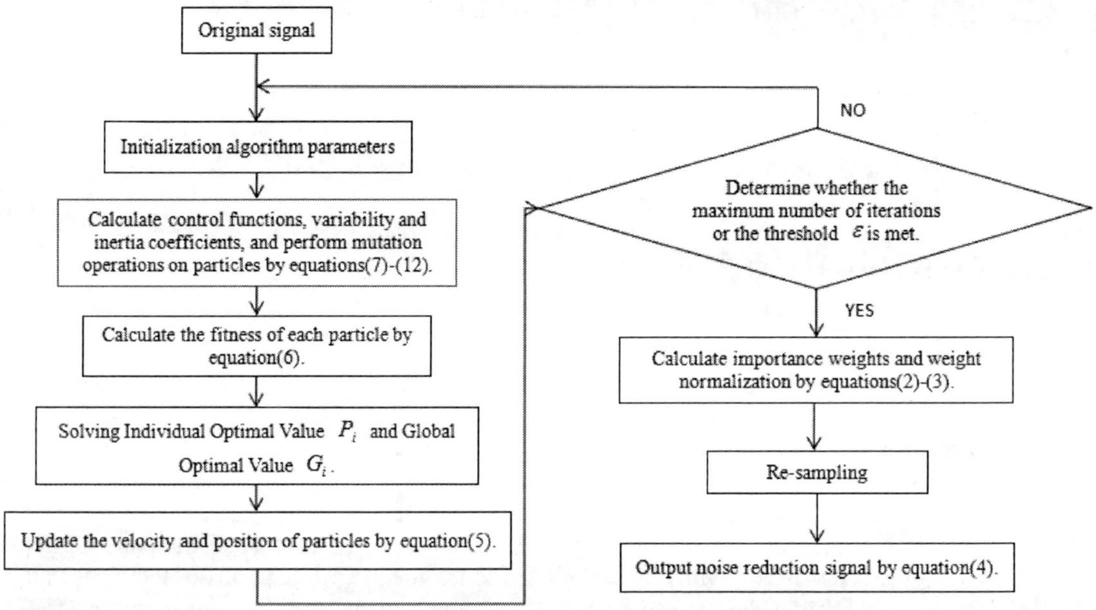

Figure 1. The flowchart of NPSO-PF algorithm.

Step 1: initialization. At time $k = 0$, $\{x_0^i, i = 1, 2, ..., N\}$ is obtained from the $P(x_0)$ distribution sample.

Step 2: NPSO optimizes the particle set distribution. Combine the NPSO implementation steps described in the Section 4.1 to optimize the particle swarm. When the number of iterations is reached or the optimal value of the particle swarm meets a certain threshold ε, the particle is output to obtain a new particle sample $\hat{x}_k^i$.

Step 3: Importance weight calculation. Sampling $\{\hat{x}_k^i, i = 1, 2, ..., N\} \sim q(x_k | x_{k-1}^i, y_k)$, calculating a new importance weight according to equation (2), and performing weight normalization according to equation (3).

Step 4: Resampling. Resampling from set $\{\hat{x}_k^i, i = 1, 2, ..., N\}$ based on new importance weights, the process is consistent with traditional particle filtering algorithms.

Step 5: Output. The updated state estimate is obtained according to equation (4).

The general steps of filtering and noise reduction are shown in figure 1.

5. Test simulation and result analysis

5.1. Particle filter simulation test and comparison

The state equation of the dynamic non-linear system selected by the test is:

$$x_k = 0.5 x_{k-1} + \frac{25 x_{k-1}}{1 + x_{k-1}^2} + 8\cos(1.2(k-1)) + v_{k-1} \tag{12}$$

The observation equation is:

$$y_k = \frac{x_k^2}{20} + n_k \tag{13}$$

Where v_{k-1} and n_k are zero mean Gaussian noise. The PF algorithm, PSO-PF algorithm and NPSO-PF algorithm are used to estimate and track the state of the nonlinear system respectively. To compare the performance of the three algorithms, N=50, N=500 and N=1000 respectively. Three different kinds of particle numbers are selected, taking the initial state probability density function as $N(0,2)$, measuring noise variance and process noise variance are $R=1\times10^{-3}$ and $Q=1\times10^{-2}$, respectively, and the parameters of the PSO algorithm are the same as above. The Root of Mean Square Error (RMSE) is calculated as follows:

$$RMSE = \left[\frac{1}{T} \sum_{k=1}^{T} (x_k - \hat{x}_k)^2 \right]^{1/2} \tag{14}$$

Where T is the number of samples, x_k is the true value of the sample, and $\hat{x}_k$ is the average value of the sample.

In order to compare the overall performance of the three algorithms,the simulation experiment was carried out 30 times, and the average of 30 root mean square errors and program running time of the three algorithms under three different particle numbers were measured, as shown in tables 1 and 2:

Table 1. RMSE of three algorithms.

Algorithm	Mean of Root Mean Square Errors		
	N=100	N=500	N=1000
PF	0.2627	0.1101	0.0974
PSO-PF	0.1547	0.0966	0.0800
NPSO-PF	0.0797	0.0572	0.0561

Table 2. Run time of three different algorithms.

Algorithm	Program Running Time (S)		
	N=100	N=500	N=1000
PF	0.8608	0.8930	1.0748
PSO-PF	0.7597	1.0519	1.1111
NPSO-PF	0.7744	0.7904	0.9247

From the results of tables 1 and 2, the average estimation error and running time of the NSPO-PF algorithm are lower than the other two. With the increase of the number of particles, the effect is more obvious. The PSO-PF algorithm is only slightly better than the PF algorithm, and the performance is basically the same as the number of particles increases.In summary, the NPSO-PF algorithm has better and more stable estimation performance than the PSO-PF and PF algorithms.

5.2. Noise reduction experiment simulation

Using the above three methods to perform noise reduction simulation experiments on the following signals, select an original signal equation as follows:

$$Y(t)=0.01\cos(800\pi t-5)e^{-0.5\left(\frac{t-0.01}{0.004}\right)^2}+0.003\cos(1000\pi t-5)e^{-0.5\left(\frac{t-0.03}{0.005}\right)^2}+0.01\cos(1200\pi t-5)e^{-0.5\left(\frac{t-0.05}{0.006}\right)^2}$$
$$+0.005\cos(1400\pi t-5)e^{-0.5\left(\frac{t-0.07}{0.005}\right)^2}+0.002\cos(1600\pi t-5)e^{-0.5\left(\frac{t-0.09}{0.004}\right)^2} \quad (15)$$

Add a random noise signal of the same size as the original signal:

$$Z(t)=Y(t)+0.005R(t) \quad (16)$$

In the above equation, the original signal equation (15) and the noisy signal equation (16) are used to replace the state equation (12) and the observation equation (13) in the upper section, respectively, to form a spatial state model among the three algorithms. The initial probability density function is $N(0,2)$, the initial state value is 1, the measurement noise variance $R=1\times10^{-4}$, the process noise variance $Q=1\times10^{-5}$, and the PSO algorithm related parameters are the same as above, and the three kinds of particles in the upper section are experimentally simulated respectively.

The signal-to-noise ratio is the ratio of the original signal to the pure noise signal. The larger the signal-to-noise ratio is, the smaller the noise is. Therefore, the signal-to-noise ratio is an important indicator to measure the noise reduction effect. In order to better compare the noise reduction performance of the three algorithms, the signal to noise ratio (SNR) analysis is performed on the above experimental results. Taking the 50 noise reduction experiments, the three algorithms analyze the signal-to-noise ratio mean values of the three particle numbers. The results are shown in table 3. The SNR calculation formula is:

$$SNR=10\lg\left(\sum x^2/\sum(z-x)^2\right) \quad (17)$$

Table 3. Comparison of simulated noise reduction results of three algorithms.

Particle Number	SNR before filtering	SNR after filtering			Improved SNR Value		
		PF	PSO-PF	NPSO-PF	PF	PSO-PF	NPSO-PF
N=100	-0.3200	1.0314	1.8626	3.0332	1.3514	2.1826	3.3532
N=500	-0.2256	4.2526	4.3995	6.0405	4.4782	4.6251	6.2661
N=1000	-0.2427	4.3535	4.4469	6.7677	4.5962	4.6896	7.0104

It can be seen from table 3 that the signal-to-noise ratio filtered by the NPSO-PF algorithm is higher than the other two algorithms under the test of the three particle numbers, and the signal-to-noise ratio after filtering by PSO-PF and PF algorithm is basically the same. It shows that the noise reduction performance of NPSO-PF algorithm is better than the other two algorithms, and the noise reduction performance of PSO-PF and PF algorithm is basically the same. The improved value of signal-to-noise ratio (SNR) of NPSO-PF algorithm is generally higher than that of the other two algorithms, and it continues to rise with the increase of the number of particles, while the PSO-PF and PSO algorithms are basically saturated. In summary, the NPSO-PF algorithm has better and more stable filtering noise reduction performance.

6. Conclusions

In this paper, the mutation operator is introduced into the PSO algorithm, which improves the structure of the PSO algorithm, optimizes the problem that the traditional PSO algorithm is easy to fall into the local optimal solution, and improves the performance of the PSO algorithm. Combining it with particle filter algorithm constitutes NPSO-PF algorithm, which not only has better estimation performance, but also has better effect in filtering noise reduction. The simulation results show that the NPSO-PF algorithm is suitable for real-time filtering and noise reduction of nonlinear signals.

Acknowledgment

The experimental data is provided by the Reliability Research Lab in the Department of Mechanical Engineering at the University of Alberta in Canada. This work was supported by the Special Major Project of the Ministry of Science and Technology of Hubei Province of China (Grant No. 2016AAA056), Major project of Hubei Provincial Department of Education (Z20101501), Research foundation for the returned Overseas Chinese Scholars of State Education Ministry (20091001) and the National Natural Science Foundation of China (Grant 51775390).

References

[1] Ananth C, Vivek T, Selvakumar S *et al* 2017 *Impulse Noise Removal Using Improved Particle Swarm Optimization* (Social Science Electronic Publishing Sarasota Florida USA)
[2] Dai J, Han H, Hu Q *et al* 2016 Discrete particle swarm optimization approach for cost sensitive attribute reduction *knowl. based syst.* **102** C 116-26
[3] Mallick S, Kar R, Mandal D *et al* 2015 Optimal sizing of CMOS analog circuits using gravitational search algorithm with particle swarm optimization *Int. J. Machine Learning and Cybernetics* **8** 1
[4] Havangi and Ramazan 2018 Joint parameter and state estimation based on marginal particle filter and particle swarm optimization *Circ. Syst. Signal Pr.* **38**(8) 3558-75
[5] Ershen W, Tao P, Pingping Q U *et al* 2016 Improved particle filter algorithm based on chaos particle swarm optimization *J. Beijing University of Aeronautics & Astronautics* **42**(5) 885-90
[6] Xue H, Qifeng C, Tingting Z *et al* 2016 Improved particle swarm optimization based on re-sampling of particle filter and mutation *J. Comput. Applications* **36**(4)1008-14
[7] Haibo L, Jingjing K E, Yi Z and Rao-Blackwellized 2016 Particle filter algorithm combined particle swarm optimization and genetic re-sampling *Comput. Eng.***42**(11) 295-9
[8] Fang Z, Geng G F and Xu X H 2007 Particle swarm optimization particle filter *J. Contr. Decision* **22**(3) 273-7
[9] Li Y, Bai B D and Zhang Y K 2010 Improved particle swarm optimization algorithm for Fuzzy mufti-class SVM *J. Systems Eng. and Electronics* **21** 509-13
[10] Yu R B, Zhao X P and Meng F L 2016 A PSO algorithm with mutation operator *Ship Electronics Eng.* **36** 26-9

IOP Conf. Series: Materials Science and Engineering **657** (2019) 012010 doi:10.1088/1757-899X/657/1/012010

Investigating mechanical properties of purplish-red siltstones and mudstones

T Wen[1,2], L Huang[3,4,5]

[1]The Key Laboratory of Well Stability and Fluid & Rock Mechanics in Oil and Gas Reservoir of Shaanxi Province, Xi'an Shiyou University, Xi'an, Shaanxi 710065, China
[2]School of Geosciences, Yangtze University, Wuhan, Hubei 430100, China
[3]Three Gorges Research Center for Geohazards, Ministry of Education, China University of Geosciences, Wuhan, Hubei 430074, China
[4]Department of Civil Engineering, Clemson University, Clemson, South Carolina 29634, United States

E-mail: huanglei@cug.edu.cn

Abstract. We investigate mechanical properties and energy conversion patterns of special rock types (purplish-red siltstones and mudstones) broadly distributed in the Three Gorges Reservoir (TGR) area, China. The mechanical properties deteriorate under unloading conditions. For loading conditions, the cumulative dissipation energy is slightly lower than elastic strain energy before crack propagation stage, while after crack propagation stage the elastic strain energy converts into cumulative dissipation energy. Similar energy conversion patterns are found for unloading conditions. This work offers a new perspective into the failure mechanism of the two special rock types.

1. Introduction

Rock mass associated with underground engineering is conditioned by triaxial ground stress before excavation [1]. Certainly, the rock mass excavation is usually an unloading process, which significantly alters the original stress state [2]. As the excavation continues, the rock mass mechanical properties are weakened, thereby destabilizing the rock mass itself [3,4]. The rock mass under unloading conditions differ considerably from that under loading conditions in behaviors, such as the constitutive relationships, mechanical evolution patterns and deformation and failure patterns [5].

Laboratory triaxial loading tests and triaxial unloading tests allow to reveal the deformation properties and the energy conversion patterns of rocks. In the past decade, progresses have been made in evaluating the properties of rock mass under loading conditions or under unloading conditions [6-8] and the loading theory is therefore formed for estimating the mechanical parameters [9,10]. This widely used loading theory, however, is not well suited to the unloading conditions [11,12]. Therefore, it is becoming very necessary to understand their mechanical properties under loading and especially unloading conditions.

Recent perspective states that the failure phenomena of rocks are caused by energy-driven instability, which is an irreversible process of energy dissipation experiencing local dissipation, followed by local failure and finally overall destruction (e.g., [13]). Previous researches on rocks under different stress paths from the perspective of energy concentrate on the variation in physical properties [14,15], yet their differences are rarely revealed from this perspective. The energy dissipation tends to lead to the rock degradation and strength reduction, while sudden overall destruction is partly due to the energy release [1,16]. At present, the researches about influences of the confining pressure on energy input/output remain lacking, which is of great importance for determining strength yielding criteria during a complete deformation and failure process. During the process, the energy conversion is dynamic. There is a mutual balance and conversion between various energies [17]. The effects of loading path on the energy conversion of rocks would be different from that of unloading path. Therefore, understanding the energy conversion patterns during a complete deformation and failure process are of significance [13].

In this study, we conduct triaxial loading tests and unloading tests of purplish-red siltstones and purplish-red mudstones. Mechanical properties are analyzed and the energy conversion patterns are investigated.

2. Preparation of siltstone and mudstone specimens and test/analysis methodologies

2.1. Specimens preparation

The rock samples were collected from a slope located in Dongrangkou Town, Badong County, in the TGR area. The first type of lithology is purplish-red siltstones from Badong Formation Section Two (b2), Middle Triassic. The second type of lithology is purplish-red mudstones from Badong Formation Section Four (b4), Middle Triassic. Triaxial loading tests and triaxial unloading tests of purplish-red siltstones and purplish-red mudstones were conducted in Chinese Academy of Sciences. The relative error in the axial strain measurement was no more than 0.6%, and the relative error in lateral strain measurement was no more than 1.3%.

2.2. Test method

In order to reveal the mechanical properties and the energy conversion patterns of purplish-red siltstones and mudstones under loading and unloading conditions, the initial σ_3 is set to be four levels, that is 2, 4, 8 and 16 MPa.

In the loading tests, σ_3 is unchanged but the axial stress is increased. Moreover, the axial stress-controlled method is employed to guarantee unchanged loading rate. Afterwards, keeping σ_3 stable, the axial stress is loaded until the rock sample is broken.

In the unloading tests, the axial stress is unchanged but σ_3 is decreased. Initially, the axial stress-controlled method is also employed to guarantee unchanged loading rate before the axial stress reaches 80% of peak strength (PS). Then, the control method is transformed into the strain control. Specifically, the axial stress and σ_3 is loaded. Afterwards, keeping the axial stress stable when it reaches 80% of the PS, σ_3 is unloaded until rock sample is destroyed.

2.3. Energy parameters determination method

The formation and coalescence of microcracks that cause rock deformation and failure are energy consuming, the dissipation energy (DE) represents the reduction degree of the initial rock strength. The energy distribution patterns based on the stress-strain relationships are obtained, which is referred in reference [18]. The total absorption energy is simplified as TAE. The elastic strain energy is simplified as ESE. Moreover, the cumulative dissipation energy is simplified as CDE. According to the energy conservation theorem and the first law of thermodynamics, the equations of TAE, ESE and CDE can be obtained by [18].

3. Test/analysis results

3.1. Mechanical properties

3.1.1. Loading conditions. Figure 1 presents the measured curves under loading conditions. With the increase of σ_3, the ductility of b2 rocks are more obvious, while the ductility of b4 rocks are not distinct. Moreover, when σ_3=16 MPa, the brittle characteristic of b4 rocks is more obvious than that under low σ_3 (2, 4 and 8 MPa). For b2 rocks, the PS in cases of σ_3=16 MPa is two times larger than that in cases of σ_3=8 MPa. During the period from 80% of the PS to the PS, the damage of the rock is aggravated and it will soon reach the critical point of the destruction of the rock. After the PS, the stress of b2 rocks decreases rapidly with a multilevel failure characteristic, which embodies in three strength reductions in the σ_3 of 2 MPa, and two strength reductions in the σ_3 of 4 MPa and 8 MPa, with the exception of the σ_3 of 16 MPa.

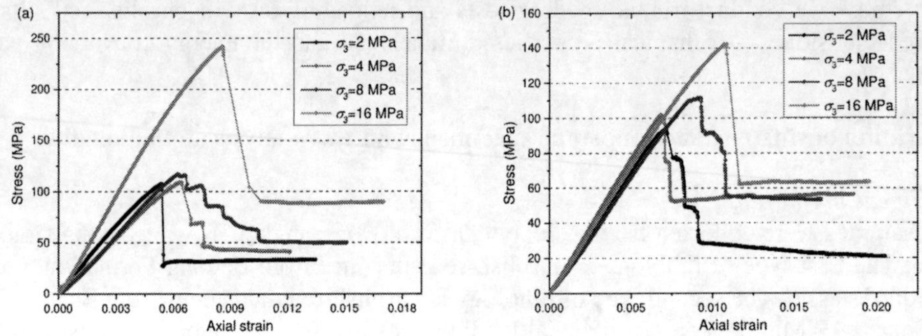

Figure 1. The curves of red bed rocks for triaxail loading tests. (a) b2 rock. (b) b4 rocks.

Under loading conditions, the mechanical parameters of red bed rocks are shown in table 1. As σ_3 increases, the PS and the RS gradually increase. The ratios of the RS to the PS are approximately 0.5, except for b4-2-1 rock. The range of the elastic modulus of b4 rocks is 12-14 GPa, while the Poisson's ratio is approximately 0.3 under different σ_3, which are less than those of b2 rocks. The results also illustrate that c of b4 rocks at the PS or at the RS are higher than that of b2 rocks, while φ at the PS or at the RS are less than that of b2 rocks, indicating that the mechanical parameters of rocks are deteriorated. Furthermore, for b2 rocks, the angle of rupture β decreases when σ_3 increases from 2 MPa to 8 MPa, and then increases with increasing of σ_3.

Table 1. The mechanical parameters of red bed rocks under loading conditions.

Rock type	PS parameters		RS parameters	
	c (MPa)	φ (°)	c (MPa)	φ (°)
b2	10.0	57	6.0	37
b4	25.6	32	10.9	22

3.1.2. Unloading conditions. The measured curves under unloading conditions is shown in figure 2. With σ_3 increased, the ductility of b2 rocks are more obvious. b2 rock failure is rapid and abrupt. However, for b4 rocks, the axial strain after the PS is also greater than the axial strain for loading conditions, while the axial strain at the PS is lower than the axial strain for loading conditions. With the increase of σ_3, the plastic characteristic is slightly less visibility.

It can be seen from figure 2, as σ_3 increases, the PS and the RS gradually increase. The PS and the RS are lower than those under loading conditions, indicating that the destruction of the rocks is more serious and the bearing capacity is lower than the loading condition. For b2 rocks, σ_3 after the failure of the rock is very small or even close to 0 at initial σ_3 of 2 MPa, 4 MPa and 8 MPa, which hints that the final σ_3 is affected by the unloading rate. Nevertheless, σ_3 after rock destroys is approximately 5 MPa at initial σ_3 of 16 MPa. For b4 rocks, the range of the ratio of the RS to the PS is approximately 0.1-0.2, which is obviously lower than the loading condition. Remarkably, σ_3 is relatively small when the rock sample is destroyed.

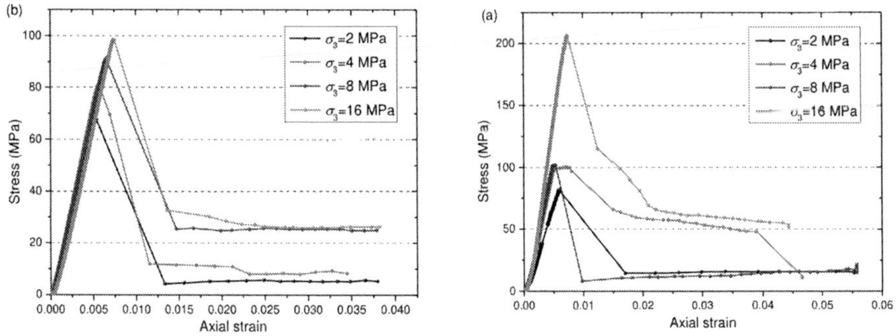

Figure 2. The curves of red bed rocks for triaxail unloading tests. (a) b2 rock. (b) b4 rock.

3.2. Energy evolution

3.2.1. Loading conditions. To investigate the energy evolution at different σ_3, some related curves of red bed rocks are plotted in figure 3. Due to the similar trend of the energy conversion at the same stress path, the b2-2-1 rock and the b4-2-1 rock are selected as examples (figure 3). The ESE of b2 rock accounts for a large proportion of the TAE before crack propagation stage while after crack propagation stage, the increasing rate of the ESE reduces. Nevertheless, the CDE is lower than the ESE before crack propagation stage while the CDE increases abruptly after crack propagation stage. The variation pattern of the CDE during the post-peak failure stage is consistent with that of the TAE. Moreover, the ESE-strain curves of b2-2-2 and b2-2-3 rocks are characterized by multistep descent, which is almost identical with the stress-strain curves.

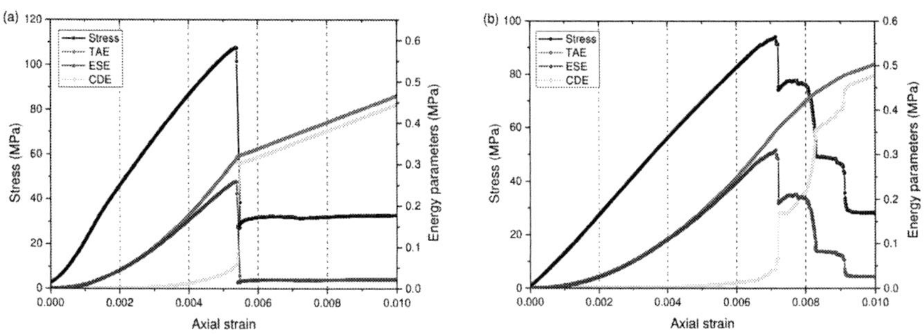

Figure 3. The energy evolution patterns of red bed rocks under loading conditions. (a) b2 rock. (b) b4 rock.

For b4 rocks, the increasing trend of the ESE before crack propagation stage is basically consistent with that of the TAE, while the difference between the ESE and the TAE is getting bigger and bigger after crack propagation stage. The change is due to the generation the microcracks and the production of plastic deformation. Furthermore, after the PS, the release of most of the ESE instantaneously is used to accelerate the propagation of the microcracks by transforming into the CDE. It is very obvious that the ESE-strain curves have multistep descent characteristic, which is almost identical with the stress-strain curves, except for the b4-2-4 rock. In contrary, the CDE before crack propagation stage is approximately 0, then the CDE gradually increases as the axial strain climbs. After the PS, the increasing trend of the TAE-strain curves and the CDE-strain curves have similar trends with characterizing of multistep rise. Under high σ_3, much energy is required to destroy the rock, which is owing to the presence of σ_3 that restrains the formation of microcracks. Therefore, more external work should be done to destroy the rock sample under high σ_3.

3.2.2. Unloading conditions. Figure 4 plots the related curves of red bed rocks under unloading conditions. The energy parameters curves of b2 and b4 rocks are similar under different σ_3, hence, the b2-3-1 rock and the b4-3-1 rock are just selected to take as examples in figure 4. Before the stress of b2 rocks reaches the PS, the ESE approximately accounts for 90% of the TAE while the increase rate of the ESE reduces. The ESE is up to the maximum at the PS. Afterwards, a large amount of the ESE begins to release and transform into the CDE, which is used to accelerate the development of microcracks for the rock. The evolution patterns of the energy parameters of b2 rocks are basically identical with loading conditions, except for the small CDE before the PS. After the PS, CDE/TAE is larger than the CDE/TAE under loading conditions, and the CDE raises abruptly.

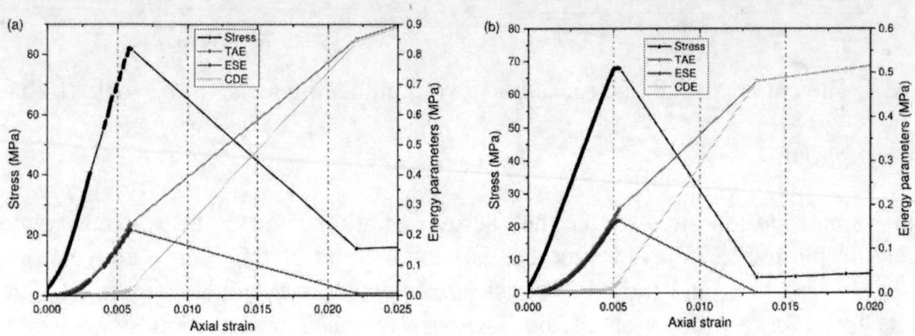

Figure 4. The energy evolution patterns of red bed rocks under unloading conditions. (a) b2 rock. (b) b4 rock.

For b4 rocks, the growth rate of the ESE gradually decreases before the PS. The value of the ESE still increases. At the PS, the ESE that approximately accounts for 90% of the TAE is up to the maximum. Afterwards, a large number of the ESE is released to aggravate the development of the microcracks. Finally, the ESE is very small at the RS, especially in σ_3 of 2 MPa and 4 MPa. However, the evolution patterns of the CDE are opposite to that of the ESE. At the PS, the CDE increases sharply. The increase rate of the CDE during the post-peak failure stage is basically stable until the stress reaches the RS. Compared with the results under loading conditions, the evolution patterns of the energy parameters under unloading conditions are almost identical.

3.3. Deterioration of deformation parameters

According to the analysis mentioned above, under unloading conditions, the mechanical properties of b2 and b4 rocks are obviously different with loading conditions. The deformation parameters deteriorate chiefly for the expansion, sliding and instability of cracks inside rock during the reduction of σ_3. Finally, the quality of rock mass will deteriorate. The deterioration of deformation parameters of these rocks is studied in this study by taking the unloading quality that represents the degree of unloading as a parametric variable, then the relationship between the deformation parameters and the unloading quality, defined as follows:

$$m = \frac{\sigma_{30} - \sigma_3}{\sigma_{30}} \times 100\% \tag{1}$$

where m is the unloading quality, σ_{30} is the initial confining pressure.

Deformation parameters (E and μ) are calculated using the methods proposed in [19]. The Poisson's ratio in post-peak failure stage is greater than 0.8, which has no original meaning. This is because the surrounding rebound unloading leads to the formation of the microcracks. In addition, the direction of the propagation of the microcracks is perpendicular to the direction of unloading, which results in the transverse expansion intensely. Therefore, only the region from the starting point of unloading to the PS is considered.

The relationships between the deformation parameters and the unloading quality are presented in figure 5, taking b2 rocks as an example. E and μ of b2 and b4 rocks deteriorate during the unloading,

IOP Conf. Series: Materials Science and Engineering **657** (2019) 012010 doi:10.1088/1757-899X/657/1/012010

and the degree of deterioration of the deformation parameters for b2 rocks is larger than that for b4 rocks. Generally, the deterioration effect of the Poisson's ratio shows a non-linear characteristic with the increase of the unloading quality, while the deformation modulus tends to linearly change. When the unloading quality is small, the deterioration effect of the deformation parameters is not obvious. When the unloading quality reaches 80%, E and μ change sharply, in which E decreases while μ increases. This will lead to the decrease of the rock integrity and the increase of the transverse deformation, which is prone to tensile failure. In addition, under higher σ_3 (i.e. 8 MPa and 16 MPa), the variation of the deformation parameters at the unloading yield stage is larger than that under lower σ_3. Therefore, the unloading quality can be used to characterize the deterioration of the deformation parameters under unloading conditions.

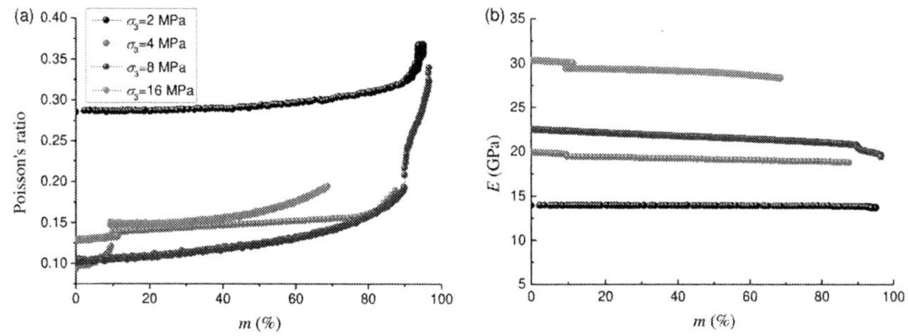

Figure 5. The relationships between the deformation parameters and the unloading quality for b2 rocks. (a) Poisson's ratio. (b) Deformation modulus.

Due to the non-linear characteristic of the deterioration of μ, the relationship between μ and the unloading quality can be fitted by polynomial formula $\mu=g(m)=B_0+B_1*m+B_2*m^2$. In this formula, B_0, B_1 and B_2 are fitting coefficients, respectively, and the fitting results are great, as shown in table 2. However, the deterioration of E presents linear characteristic, so the linear formula $E=f(m)=k*m+b$ can be used to fit their relationship, in which k and b are fitting coefficients. The fitting coefficient k reflects the deterioration degree of E with the change of the unloading quality, while the fitting coefficient b represents the deformation modulus before unloading. The fitting coefficients k and b are shown in table 3.

Table 2. The fitted parameters between Poisson's ratio and unloading quantity.

Coefficient	b2				b4			
	σ_3=2	σ_3=4	σ_3=8	σ_3=16	σ_3=2	σ_3=4	σ_3=8	σ_3=16
B_0	0.30	0.10	0.13	0.14	0.24	0.26	0.195	0.198
B_1	-0.001	0.002	-0.002	2.18E-4	2.97E-6	2.04E-4	-0.001	-1.37E-4
B_2	19.5E-5	-10.07E-5	30.07E-5	7.07E-6	5.44E-6	2.94E-6	2.24E-5	4.76E-6
R^2	0.89	0.87	0.82	0.88	0.99	0.99	0.85	0.99

Table 3. The fitted parameters between deformation modulus and unloading quantity.

Coefficient	b2				b4			
	σ_3=2	σ_3=4	σ_3=8	σ_3=16	σ_3=2	σ_3=4	σ_3=8	σ_3=16
k	-0.003	-0.013	-0.022	-0.021	-0.005	-0.006	-0.008	-0.014
b	14.02	19.79	22.60	29.92	13.33	15.12	15.32	14.78
R^2	0.83	0.92	0.90	0.86	0.98	0.99	0.88	1.00

From the above analysis, it can be seen that the deformation parameters deteriorate during unloading process, characterizing by approximately non-linear behavior for the deformation modulus and linear behavior for μ with the change of unloading quality. The unloading quality is a quantity to indicate the unloading degree, and its expression includes the confining pressures before and after unloading. However, volume expansion, brittle failure and parameter degradation are sensitive to the initial confining pressure and the unloading degree. Therefore, it is of practical and theoretical significance to utilize the unloading quality to express the degradation of the deformation parameters during unloading of rocks.

4. Discussion

As mentioned earlier, the energy conversion and mechanical properties of b2 and b4 rocks under unloading condition are essentially different from loading condition. The above results make it possible to estimate the mechanical properties and the energy conversion patterns. For example, the mechanical properties of b2 and b4 rocks in TGR area may be determined by dividing the complete stress-strain curves into several different deformation stages, if the test conditions are similar to those in the paper. The energy conversion of b2 and b4 rocks in TGR area may offer the possibility to expose the failure patterns of reservoir slopes from the energy perspective, and determine whether the deformation in reservoir slopes is still in progress or is failure.

The specific application of the results mentioned above is as follows. The Section 2 and Section 4 of Badong formation in the TGR area are typical slippery strata [20]. The instability of reservoir slopes in the urban area of Badong County is closely related to stratum lithology and slope structure. This study about the energy conversion and mechanical properties of b2 and b4 rocks in the TGR area can be conductive to the deformation and failure patterns of reservoir slopes. Due to the difference of the quantity of energy dissipation and the quantity of deformation of b2 and b4 rocks, it is easy to slide along the contact interface between hard rocks and soft rocks, which results in the development of bedding slip zone within the rock mass in Badong urban area. Before the penetration of the sliding surface, the rocks maintain good structures without obvious collapse and fracture occurred [21,22], and the energy dissipation of the rocks that leads to rock damage is relatively small. The results mentioned above show that a large amount of ESE is stored in the rocks before sliding. However, the ESE are suddenly released when forming a through sliding surface near the contact interface, which result in severe sliding. Therefore, this study is of significance in the deformation and failure patterns and the stability assessment of reservoir slopes.

5. Conclusions

Major conclusions regarding the mechanical properties and energy conversion patterns of purplish-red siltstones and mudstones are as follows:

- For loading tests, b2 and b4 rocks under different σ_3 destroy and the axial stress decreases abruptly, characterizing multilevel descent after the PS. The occurrence of ESE of b2 rock taking up a large proportion of the TAE before crack propagation stage and the CDE increasing sharply can be a criterion of the rock failure.
- For unloading tests, b2 rock failure is rapid and abrupt. Relative to the loading conditions, the evolution patterns of several parameters about energy are mainly identical. However, the mechanical parameters of b2 and b4 rocks are deteriorated. Thus, the failure of the rock under loading conditions is mainly caused by the loading at the axial direction, while the failure of the rock under unloading conditions is due primarily to the expansion at the circumferential direction.

Acknowledgments

This work was supported by the Key Laboratory of Well Stability and Fluid & Rock Mechanics in Oil and Gas Reservoir of Shaanxi Province, Xi'an Shiyou University (Grant number WSFRM20190101001), the China Postdoctoral Science Foundation (Grant number 2018M642799) and the National Key Research and Development Program of China (Grant number 2017YFC1501305).

References

[1] Li Y R, Huang D and Li X A 2014 Strain rate dependency of coarse crystal marble under uniaxial compression: strength, deformation and strain energy *Rock Mech. Rock Eng.* **47** 1153-64
[2] Chandler N A 2013 Quantifying long-term strength and rock damage properties from plots of shear strain versus volume strain *Int. J. Rock Mech. Min. Sci.* **59** 105-10
[3] Ribacchi R 2000 Mechanical tests on pervasively jointed rock material: Insight into rock mass behaviour *Rock Mech. Rock Eng.g* **33** 243-66
[4] Yuan S and Harrison J P 2004 An empirical dilatancy index for the dilatant deformation of rock *Int. J. Rock Mech. Min. Sci.* **41** 679-86
[5] Torres-Suarez M C, Alarcon-Guzman A and Berdugo-De Moya R 2014 Effects of loading-unloading and wetting-drying cycles on geomechanical behaviors of mudrocks in the colombian andes *J. Rock Mech. Geotech. Eng.* **6** 257-68
[6] Guo Y T, Yang C H and Mao H J 2012 Mechanical properties of Jintan mine rock salt under complex stress paths *Int. J. Rock Mech. Min. Sci* **56** 54-61
[7] Sirdesai N N, Singh T N and Pathegama Gamage R 2017 Thermal alterations in the poro-mechanical characteristic of an Indian sandstone-a comparative study *Eng. Geol.* **226** 208-20
[8] Singh A, Kumar C, Kannan L G *et al* 2018 Estimation of creep parameters of rock salt from uniaxial compression tests *Int. J. Rock Mech. Min. Sci.* **107** 243-8
[9] Lau J S O and Chandler N A 2004 Innovative laboratory testing *Int. J. Rock Mech. Min. Sci.* **41** 1427-45
[10] Trippetta F, Collettini C, Meredith P G *et al* 2013 Evolution of the elastic moduli of seismogenic triassic evaporites subjected to cyclic stressing *Tectonophysics* **592** 67-79
[11] Riazi N, Clarkson C R, Ghanizadeh A *et al* 2017 Determination of elastic properties of tight rocks from ultrasonic measurements: examples from the montney formation (alberta, canada) *Fuel* **196** 442-57
[12] Yan C B, Xu X and Huang L 2018 Identifying the impact factors of the dynamic strength of mudded intercalations during cyclic loading *Adv. Civ. Eng.* **2018** 1-9
[13] Munoz H, Taheri A and Chanda E K 2016 Rock drilling performance evaluation by an energy dissipation based rock brittleness index *Rock Mech. Rock Eng.* **49** 3343-55
[14] Nara Y and Kaneko K 2006 Sub-critical crack growth in anisotropic rock *Int. J. Rock Mech. Min. Sci* **43** 437-53
[15] Bagde M N and Petroš V 2009 Fatigue and dynamic energy behaviour of rock subjected to cyclical loading *Int. J. Rock Mech. Min. Sci* **46** 200-9
[16] Akdag S, Karakus M, Taheri A *et al* 2018 Effects of thermal damage on strain burst mechanism for brittle rocks under true-triaxial loading conditions *Rock Mech.Rock Eng.* **51** 1657-82
[17] Meng Q B, Zhang M W, Han L J *et al* 2016 Effects of acoustic emission and energy evolution of rock specimens under the uniaxial cyclic loading and unloading compression *Rock Mech. Rock Eng.* **49** 3873-86
[18] Wen T, Tang H, Wang Y *et al* 2019 Mechanical characteristics and energy evolution laws for red bed rock of badong formation under different stress paths *Adv. Civ. Eng.* **2019** 1-16
[19] Wen T, Tang H M, Ma J W *et al* 2019 Energy analysis of the deformation and failure process of sandstone and damage constitutive model *KSCE J. Civ. Eng.* **23** 1-12
[20] Wen T, Tang H M, Ma J W *et al* 2018 Evaluation of methods for determining crack initiation stress under compression *Eng. Geol.* **235** 81-97.
[21] Tang H, Huang L, Juang C H *et al* 2017 Optimizing the Terzaghi estimator of the 3D distribution of rock fracture orientations *Rock Mech. Rock Eng.* **50** 2085-99
[22] Huang L, Tang H, Wang L *et al* 2019 Minimum scanline-to-fracture angle and sample size required to produce a highly accurate estimate of the 3-D fracture orientation distribution *Rock Mech. Rock Eng.* **52** 803-25

IOP Conf. Series: Materials Science and Engineering **657** (2019) 012011 doi:10.1088/1757-899X/657/1/012011

Finite element numerical simulation analysis based on ultrasonic phased array

H X Chen[1,2], D L Fan[1], L Fang[1], G Y Zhang[1], C H Cao[1], W J Huang[1], J M Huang[1] and L Yang[1]

[1]School of Mechanical and Electrical Engineering, Wuhan Institute of Technology, Wuhan 430205, China

E-mail: pg01074075@163.com

Abstract. In this paper, the ultrasonic phased array technology is used to detect the defects in the welded seam test specimens. Using pulse echo method to detect a specimen with defects and establishing the corresponding numerical model. At the same time, the finite element is used to simulate the propagation of ultrasonic waves in the defective test specimens, and figures of the sound pressure amplitude are obtained. Compared with the theoretical values, the experimental values and the simulated values show good agreement, and the errors are only -1.37% and 4.11%, respectively. The results show that the ultrasonic phased array with finite element simulation is effective.

1. Introduction

Ultrasonic non-destructive testing (NDT) techniques have been remarkably used in test and evaluation of critical components of defence, space and in industrial applications [1]. NDT is a method of detecting and testing the size, quantity, position, type and other characteristics of defects on the surface or inside the material without affecting the performance of the object to be inspected and the internal organization of the material [2]. Ultrasonic testing is one of the most commonly nondestructive testing methods, its main principle is based on the characteristic of ultrasonic wave propagation in the specimen. Using the ultrasonic propagation characteristics in order to evaluating the specimen itself or its internal defects and the features of defect [3].

In the past 30 years, acoustic emission has been widely used for the non-destructive evaluation of damage and the study of fracture behavior of materials [4]. Ultrasonic phased array detection is widely used as an important acoustic emission detection technology. The transducer array of this technology is designed based on the Huygens principle. The transducer element has a small area and can be regarded as an ultrasonic emission source to emit an ultrasonic beam to the inside of the detected work-piece. When the sound beam is emitted, the excitation delay sequence of each array of wafers in the transducer is controlled by computer software, so that the sound beams emitted by the array elements generate phase superposition, constructive interference and form a focused sound field at a set depth position, thereby controlling the shape and direction of the transmitted ultrasonic beam, and control the angle, focus position and focus size of the generated beam to achieve focusing and deflection of the ultrasonic beam. Subsequently, the ultrasonic transducer echo signal containing the internal defect or tissue structure information of the detected work-piece is received by the array transducer, and then the signal processing is performed to evaluate and image the internal defect or structure of the detected work-piece. It has the advantages of high detection efficiency, high signal-to-noise ratio, convenience and flexibility, and easy detection of complex shaped work-pieces [5].

With the rapid development of electronic technology, the ultrasonic phased array has also developed rapidly. In 2000, A Erhard [6] of BAM College in Germany used a phased array system to detect austenite welds with coarse grain structure and severe ultrasonic attenuation, and enhanced the intensity of the sound beam in the detection area by dynamic focusing. Further, the IZFP non-destructive testing institute in Germany [7] successfully realized the miniaturization of the piezoelectric ultrasonic phased array, which could complete the axial and radial high-quality imaging of the pipeline. There are a lot of experimental studies on the sound field numerical simulation technology of ultrasonic phased array. Zhao *et al* [8] proposed an approximate multi-Gaussian beam model, and used this to show the distribution of the sound field of the phased array when focusing on the axis of the array element. Satyanarayan [9] used 2-D time domain finite difference method to model and simulate the pipeline structure with artificial defects. The simulation error was large and the results were not intuitive enough. Zhu et al [10] proposed a finite element reconstruction algorithm based on the frequency domain Helmholtz equation for ultrasonic tomography, which was used to monitor grouting defects in reinforced concrete structures. The numerical simulation results were compared with the experimental results to verify the reliable and effective phased array ultrasonic testing. The finite element method has rich material elements and flexible meshing, and it is close to the exact solution by approximation calculation, which has unique advantages for the calculation of complex structures.

2. Experimental research

2.1. The experimental system composition and model establishment

The equipment used in this experiment includes a computer, a Muti 12000 portable ultrasonic phased array device, a 64-wafer phased array linear array transducer with center frequency of 5 MHz(array element spacing of 0. 6 mm), a phased array control system, and a 54.7° wedge, an artificial defect test specimens. As shown in figure 1.

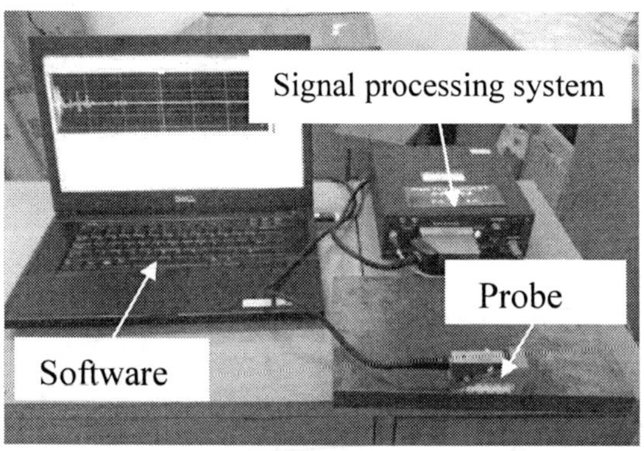

Figure 1. Ultrasonic phased array inspection system.

Figure 2. Weld seam test block with defects.

Modeling plays an important role in the understanding of any physical process. This is particularly true of non-destructive testing and evaluation where a variety of energy forms are used to inspect engineering components. In the process of establishing a two-dimensional model of a defect test block according to the purpose of the experiment, for the defect test piece of the crack hole defect type, we directly perform such a hole defect by performing a Boolean operation on the defect test block by ANSYS software. For irregular weld defects such as unfused at the sidewall and root, model the Auto-CAD software and import it into ANSYS software.

This experiment uses a natural weld defect with a length, width and height of 300 mm*300 mm*16 mm made of Q235. As shown in figure 2, the phased array probe is placed above the work-piece under test during the experiment, and then the wave-forms are transmitted by the computer to observe the A-scan and S-scan. The A-scan is a graph of the amplitude and propagation time of an ultrasonic signal during propagation. The abscissa represents the propagation time of the wave, and the ordinate represents the signal amplitude. S-scan, also known as fan-shaped scan, is two-dimensional images of the probe's delay and refraction angles that have been corrected and superimposed on all A-scans in a particular channel. The S-scan reflects the actual coordinate position of the reflector and is very helpful for intuitive display of the test results and figures comparison. As shown in figure 2.

2.2. Experimental methods and results

The phased array probe is placed on the surface of the work-piece to be tested, and the excitation sequence of the array element is adjusted by the computer to control the shape and direction of the emitted ultrasonic beam, so that the ultrasonic beam is focused and deflected to detect the defect without moving the probe.

When the work-piece is tested by adjusting the excitation rules, when there is no defect, there will be no defect echo as shown in figure 3(a), and no defect display on the right S-scan. When a defect is detected, the situation shown in figure 3(b) will appear. Both the A-scan and the S-scan have defects. By pulling the coordinate pointer, you can clearly see the amplitude and geometric position of the coordinate position.

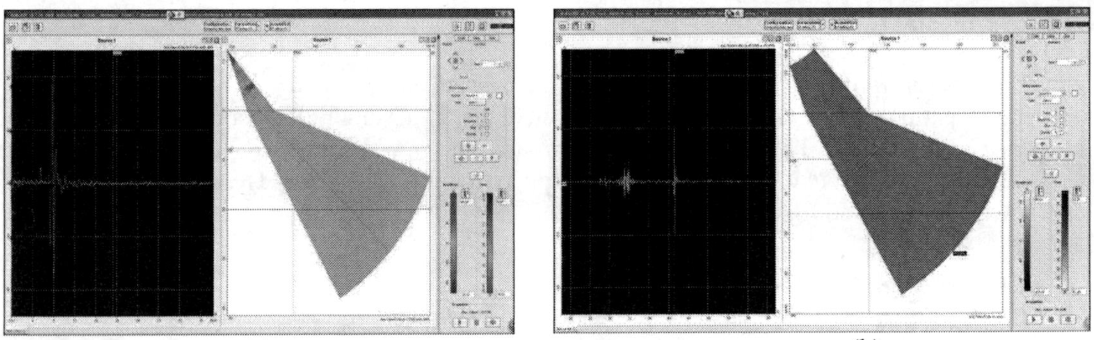

(a) (b)

Figure 3. (a) Defect-free situation and (b) defective situation.

3. Simulation section

3.1. Ultrasonic detection model establishment

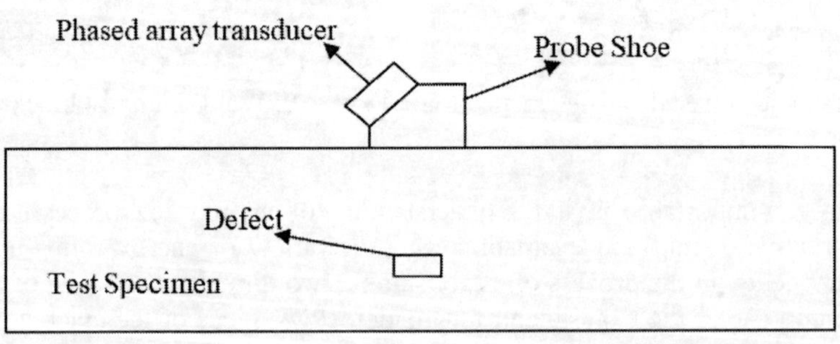

Figure 4. Conventional ultrasonic detection by phased array technique.

The ultrasonic testing model generally includes the following three parts about the phased array transducers, probe shoes and a defective test specimen as shown in figure 4. The role of the phased array transducer is to transmit and receive ultrasound signals. The probe shoe is used that the ultrasonic wave is obliquely incident on the detecting surface. The function of the probe shoe is to control the beam angle to form a strict angle between the wafer of the probe and the surface of the work-piecc. The defective test specimen is detected by the object.

The role and effects of the coupling layer and the wedge are ignored during the simulation. The processing technique for this point is that the simulation does not change. In the experiment, it is discriminated that the ultrasonic wave is injected into the work-piece, and then the ultrasonic wave is framed by the gate and finally collected. The simulation process is equivalent to simulating different pulse signals emitted by the ultrasonic probe according to the focus rules.

3.2. Ultrasonic detection model establishment

In this paper, the ultrasonic finite element analysis model by ANSYS software is established to investigate the ultrasonic propagation process mechanism and the effects of the weld slag on the ultrasonic diffraction. In the experimental part, the selected transducer probe contains 64 crystals. A part of the crystal can be selected to apply a time pulse containing a corresponding delay to control the emission direction of the ultrasonic wave. In the ultrasonic simulation of the model, the loading of the ultrasonic signal is mainly to introduce a waveform function similar to the experimental ultrasonic function, and then apply the waveform function as a load to the nodes on the surface. Therefore, in the simulation, the ultrasonic wave source can be regarded as excited directly at the node of the model surface. However, in the experimental process, the ultrasonic wave is excited at the probe, passes through the coupling layer, wedge, and finally enters the work-piece. According to the formula of the delay rules, the mathematical relationship between the angle (∂) of the sound beam, the velocity (V), the time delay (Δt) of the adjacent wafer, and the distance (d) between adjacent wafers are:

$$\sin\partial = \frac{V\Delta t}{d} \tag{1}$$

In the ultrasonic inspection model, the simulation of the ultrasonic wave source is the most important. The ultrasonic waves from the phased array transducer are modeled by applying appropriate transient excitation pulses to different nodes on the surface of the specimen model. The transient excitation pulse function equation commonly used for ultrasonic simulation is shown in the equation (2):

$$Y_{(t)} = \begin{cases} \cos(2\pi ft)[1-\cos(\frac{2\pi f}{N}t)], 0 \le t \le \frac{N}{f} \\ 0 \end{cases} \tag{2}$$

Which f represents the pulse excitation frequency and N represents the number of waves in the excitation pulse waveform. The probe frequency is defined by the pulse width and the number of the cycles of the pulse. The frequency in the model is 5 MHz and $N = 3$.

Figure 5 is a comparison of the experimental ultrasonic source and the simulated wave source in the simulation. Part (a) is the load waveform of equation (2), taken in ANSYS software. Part (b) is the waveform of the wave source in the experiment, which is extracted from the configuration section of the experimental software. Comparing the wave source diagrams of the experimental and simulation models, it can be seen that the phase of the simulated wave source is exactly opposite to the experiment, and the rest show a good consistency.

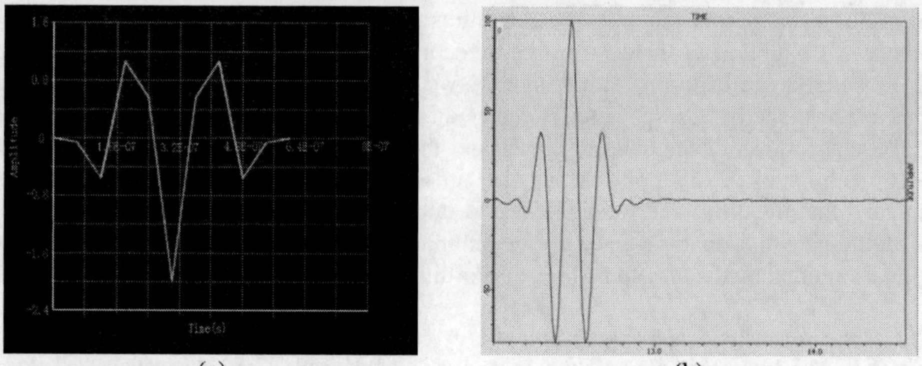

(a) (b)

Figure 5. Comparison of simulated wave sources in ultrasonic wave source and simulation: (a) Load waveform diagram; (b) Experimental waveform.

3.3. Finite element simulation of ultrasonic propagation in defective specimens

As described in Section 2.2, a specimen with groove is used for the model of the conventional ultrasonic detection for the defect by phased array technique. The geometric size and material parameters in simulation models are the same as the experiments. A two-dimensional flat specimen modeling is developed to simulate the ultrasonic wave propagation. The longitudinal wave is used for the detection of the defect because the wave propagation propagates in the vertical direction, which coincides with the position of the defect.

According to the various parts of the ultrasonic detection in equation (2), a 2-D numerical simulation model is established to simulate the ultrasonic B-scan of the specimen and the A-scan is obtained by the displacement of the special nodes. The four ultrasonic pulse snapshots shown in figure 6 are all generated according to the specific parameters. These figures are obtained at different time points when the ultrasonic pulse passes through the defective test specimen, and the degree of the color indicates the value of the displacement. The letters L, S, H and R represent the longitudinal, transverse, head and Rayleigh waves, respectively. L-L represents a longitudinal wave generated by waveform conversion of a longitudinal wave, and S-L represents a transverse wave generated by waveform conversion of a longitudinal wave. These color images clearly show that the longitudinal, shear, head and Rayleigh waves are generated from the nodes excited by the source and propagate through the model. They can be easily distinguished based on the difference in wave velocities.

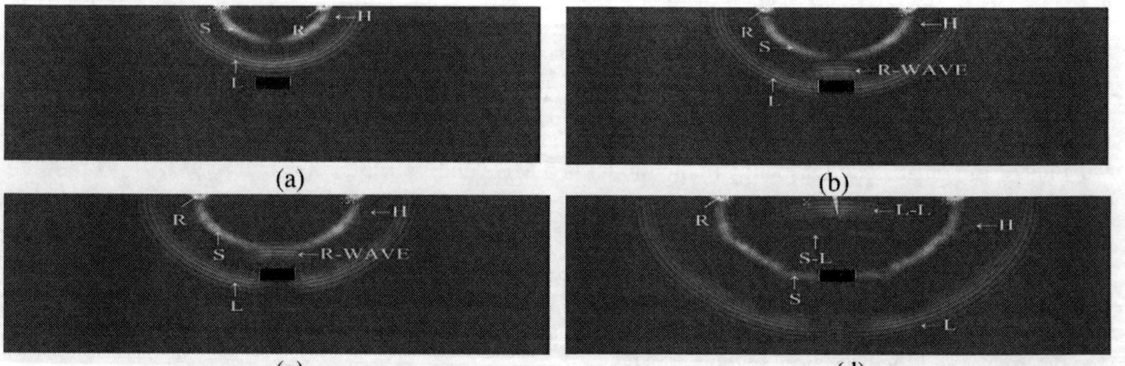

(a) (b) (c) (d)

Figure 6. Snapshot results of four different moments in the defect test simulation: 2.1 us; (b) 2.4 us; (c) 2.8 us; (d) 3.8 us.

Figure 6 shows the snapshot diagrams extracted at the four different times of 2.1 us, 2.4 us, 3.1 us, and 3.8 us in the numerical simulation results of the defect test. The figure 6(a) below shows the simulation results, and the blue cuboid part shows the test block. The central black cuboid represents the stomata defect. The grayish white ripple indicates the propagation of ultrasound in the test block. In figure 6(b), the longitudinal wave propagates to the defect at 2.4 us, which will reflect in the opposite direction and produce a reflected wave (R-Wave). In figure 6(c), at 2.8 us, the defect reflection wave propagates in the opposite direction of the incident direction. Since the waveform conversion occurs, the entire reflected wave contains the reflected longitudinal wave and the reflected transverse wave. Since the longitudinal wave velocity is twice the transverse wave, it is also It is also possible to distinguish them by the wave speed (or the time of wave propagation). In figure 6(d), it is observed that the longitudinal wave in the reflected wave is about to reach the incident point at 3.8 us, and the transverse wave propagates along the longitudinal wave but has a certain distance from the incident point.

4. Comparative analysis and results

The node to which the load is applied in the specimen model is taken as a special node, and the time displacement data of the node is derived as a simulated A-scan view. Extract the A-scan data of the test results in the experiment, and compare the simulation results with the experimental results. The amplitude map corresponding to the two methods is shown in figure 7. The abscissa is the time of ultrasonic propagation, and the ordinate is the defect echo percentage of amplitude.

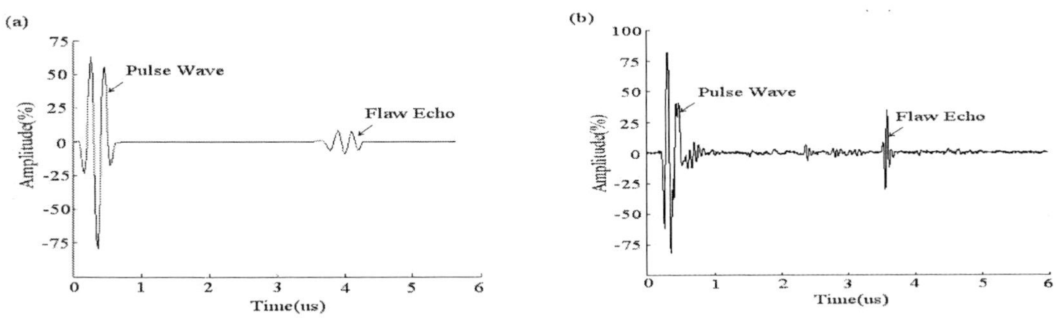

Figure 7. A-scan simulation and experimental results for defect detection: (a) Simulated signals; (b) Experimental signals.

Figure 7 compares the finite element analog signal with the A-scan experimental signal for a clear comparison of the defect echo amplitude and time. Figure 7(a) represents the simulation signal and figure 7(b) represents the experiment signal. Figure 7(a) shows that the pulse node receives the reflect wave when the time is 3.9 us and the amplitude of the flaw echo signal is observed clearly. Figure 7(b) shows the probe received the flaw echo when the time 3.7 us and the amplitude of flaw echo is large enough to be observed. The echo time by the defect in the simulation wave is the same as that in the experimental signal. With the theoretical value as the reference standard, it can be found that in the column of the defect in the echo time, the theoretically calculated receiving probe should receive the defect echo at 3.65 us, and the experimental value and the simulated value both show a relatively good consistency.

Table 1 compares the defect echo time and the echo amplitude in the A-scan data of the defect test piece simulation and experimental results. The errors are only -1.37% and 4.11%, respectively. In the case of echo amplitude, the experimental value is 50.1%, and the simulation value is only 15.6%, which is far lower than the experimental value. The distortion of the experimental result is more serious. The reasons for this phenomenon are as follows: (1) Error exists. As an approximate calculation method, the finite element method itself has a certain error range. (2) Calculation accuracy. Although the finite element method has constant error, it can greatly improve the accuracy by increasing the calculation conditions (such as reducing the mesh size and shortening the time step). In theory, as the calculation conditions improve, the simulation results will become closer to the actual situation.

Table 1. Comparison between the two methods and theoretical calculations.

	Echo Time (us)	Error(%)	Echo Amplitude(%)
Theory value	3.65	0	\
Experimental value	3.6	-1.37	50.1
Simulation value	3.8	4.11	15.6

5. Conclusion

The finite element numerical simulation of ultrasonic phased array is the main part of this paper. It is very effective and novel to simulate the propagation of ultrasonic waves through load loading, and it has certain accuracy. Compared with the theoretical values, the experimental values and the simulated values show good agreement, and the errors are only -1.37% and 4.11%, respectively. The results show that the ultrasonic phased array with finite element simulation is effective.

Acknowledgments

The experimental data is provided by the Reliability Research Lab in the Department of Mechanical Engineering at the University of Alberta in Canada. This work was supported by the Special Major Project of the Ministry of Science and Technology of Hubei Province of China (Grant No. 2016AAA056), Major project of Hubei Provincial Department of Education (Z20101501) , Research foundation for the returned Overseas Chinese Scholars of State Education Ministry (20091001) and the National Natural Science Foundation of China (Grant 51775390).

References

[1] Upadhyay A, Sherpa B B, Kumar S, Kumar P D, Srivastava N and Agarwal A 2019 Phased array-based ultrasonic testing of explosively welded aluminium and stainless steel plates *Trans. Indian Inst. Met.* **72** 1521-5

[2] Li X M 2007 *Nondestructive Testing* (Beijing: Mechanical Industry Press)

[3] Zhong W B, Xiao Y N and Chen S D 2006 Phased array ultrasonic testing system and its application in industrial nondestructive testing *Weld Technol.* **35** 41-3

[4] Ben B S, Yang S H, Ratnam C and Ben B A 2013 Ultrasonic based structural damage detection using combined finite element and model lamb wave propagation parameters in composite materials *Int. J. Adv. Manuf. Technol.* **67** 1847-56

[5] Wang Y M 2014 *Ultrasonic Phased Array Detection Technology and Application* (Beijing: National Defence Industry Press)

[6] Pohl R, Erhard A, Montag H J, Erhard A and Thomas H M 2004 NDT techniques for railroad wheel and gauge corner inspection *NDT & E International* **37** 89-94

[7] Rieder H, Spies M, Licht R and Kreier P 2003 Development and optimization of a rotating phased array inspection system *AIP Conf. Proceedings.* **657** 785-92

[8] Zhao X Y, Gang T and Zhang B X 2008 Sound field calculation of phased array transducers with non-paraxial approximation multi-Gaussian beam model *Acta Acousatactics Sinica* **5** 475-80

[9] Satyanarayan L, Kumar A, Jayakumar T, Krishnamurthy C V, Balasubramaniam K and Raj B 2009 Sizing cracks in power plant components using array based ultrasonic techniques *J. Nondestructive Eval.* **28** 111-24

[10] Zhu Z Q, Yu B, Li Y N, Xiao J Y and Zhou Y 2015 Simulation study on ultrasonic tomography for grouted reinforced concrete by finite element *J. Cent. South Univ.* **22** 2791-9

IOP Conf. Series: Materials Science and Engineering **657** (2019) 012012 doi:10.1088/1757-899X/657/1/012012

Detection and analysis of coherent structures in the near-field of a turbulent annular swirling jet: an unsteady Reynolds-averaged Navier–Stokes (URANS) simulation

Y Zhang[1] and M Vanierschot[1,2]

[1]KU Leuven, Mechanical Engineering Technology Cluster TC, Group T Leuven campus, Leuven, Belgium

[2]Author to whom any correspondence should be addressed

E-mail: yang.zhang2@student.kuleuven.be, maarten.vanierschot@kuleuven.be

Abstract. This paper analyses the capabilities of unsteady Reynolds-averaged Navier–Stokes (URANS) simulations to predict the coherent structures found in a swirling jet undergoing vortex breakdown. Recently, tomographic particle image velocimetry experiments of an annular swirling jet at a Reynolds number of 8500 at moderate swirl showed the presence of a double helical structure in the flow field (Vanierschot et al., Physical Review Fluids, 2018). This structure corresponds to the double helix vortex breakdown mode and is rarely observed in turbulent flows. In this study, the same flow topology is simulated using Computational Fluid Dynamics (CFD). Turbulence is modeled using the unsteady RANS methodology with a RNG k–ε turbulence model. The coherent structures in the flow field were analysed using the spectral proper orthogonal decomposition technique. Despite the fact that it is known that steady-state RANS simulations using two-equation isotropic turbulence models have problems in accurately describing swirling flows which are highly anisotropic, this study shows that the unsteady variant was able to predict the large scale flow structures and their associated dynamics reasonable well. In particular, similar to the experiments, a double helical structure was found in the flow field. The structure has windings in the counter-swirl direction and it is wrapped around the central breakdown bubble. To the authors knowledge, this study is the first one to show the ability of unsteady RANS to predict not only the presence of the double helix vortex breakdown in the flow field, but also the spatial and temporal structure of it.

1. Introduction

Turbulent swirling flows are ubiquitous in nature, such as in tornadoes and typhoons and in engineering technical applications, for instance, the vortices shedding from the leading edge of aircraft and the rotational motion in combustion chambers. In combustion systems, swirl is added to the flow to enhance the fuel-air mixing rate and to reduce the emission of nitrogen oxides through the recirculation zone, which is caused by the centrifugal effects or adverse pressure gradients above a certain level of swirl intensity [1]. Other prevalent features of turbulent swirling jets are the so-called 'precessing vortex core' (PVC) and 'vortex breakdown' (VB), which is referred to as the abrupt change in the structure of a vortex core. Both coherent structures have been intensively studied during the past several decades and different theories

IOP Conf. Series: Materials Science and Engineering **657** (2019) 012012 doi:10.1088/1757-899X/657/1/012012

have been introduced to explain their occurrence. However, there is currently not a unifying theory which is accepted by the community [2].

Although Reynolds-averaged Navier-Stokes (RANS) modeling is not as accurate as large eddy simulation or direct numerical simulation, it is widely adopted to model turbulent flows as it has low computational cost and can be used for highly complex flow geometries. Investigations have been conducted to assess the capability of different RANS turbulent models to simulate turbulent swirling flows, both in steady and unsteady cases [3–10]. Their results show that the k–ε family and Reynolds stress model are capable of capturing the highly unsteady flow dynamics such as the PVC and its precessing frequency.

Recently, Vanierschot et al. [11] conducted experimental studies on a turbulent annular swirling jet by tomographic particle image velocimetry, and not only single helix but also double helix vortex breakdown was identified. In this work, an unsteady RANS simulation is used to model the turbulent annular swirling jet flow firstly. Then, we apply a modal decomposition to identify the coherent structures and their dynamics and compare the results with those from literature.

2. Computational setup

2.1. Governing equations and turbulence model

The governing equations are the three-dimensional incompressible Reynolds-averaged Navier–Stokes equations for averaged velocity $\bar{u}_i$ and averaged pressure $\bar{p}$:

$$\frac{\partial \bar{u}_i}{\partial x_i} = 0, \tag{1}$$

$$\frac{\partial \bar{u}_i}{\partial t} + \bar{u}_j \frac{\partial \bar{u}_i}{\partial x_j} = -\frac{1}{\rho}\frac{\partial \bar{p}}{\partial x_i} + \nu \frac{\partial^2 \bar{u}_i}{\partial x_j \partial x_j} - \frac{\partial \overline{u'_i u'_j}}{\partial x_j}, \tag{2}$$

where x_i, t, ρ and ν denote distance, time, density and molecular viscosity, respectively. The Reynolds stress tensor $\overline{u'_i u'_j}$ is modeled through the turbulent-viscosity hypothesis introduced by Boussinesq:

$$\overline{u'_i u'_j} = \frac{2}{3} k \delta_{ij} - \nu_t \left(\frac{\partial \bar{u}_i}{\partial x_j} + \frac{\partial \bar{u}_j}{\partial x_i}\right), \tag{3}$$

which assumes that the deviatoric Reynolds stress is proportional to the mean rate of strain. Here, ν_t is the turbulent or eddy viscosity. In the present work, the RNG k–ε model is chosen to simulate the flow. Compared to the standard k–ε model, the effect of swirl on turbulence is included in the RNG k–ε model, which improves accuracy for swirling flows. The eddy viscosity coefficient is computed from turbulent kinetic energy and dissipation rate $\nu_t = C_\mu k^2/\varepsilon$, where $C_\mu = 0.09$ is a model constant.

2.2. Computational domain, boundary conditions and control parameters

The cross section of the computational domain is shown in figure 1. The working fluid flows through an annular tube firstly and then mixes with the ambient medium in a large cylindrical domain whose length and diameter are $15D_o$ and $16D_o$, respectively. Here, $D_o = 27mm$ is the outer diameter of the annular tube, whereas the length L and inner diameter D_i are $54mm$ and $18mm$, respectively.

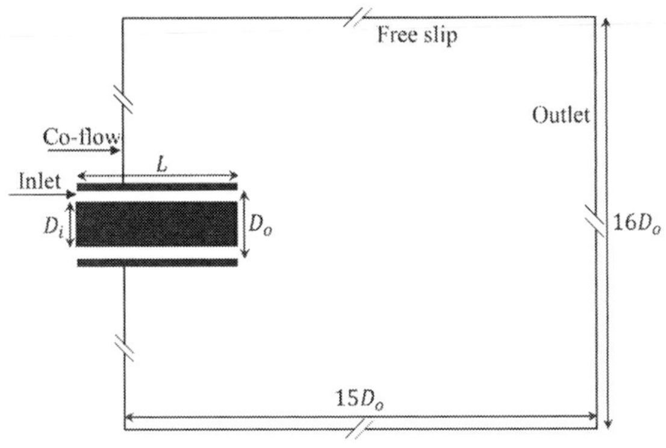

Figure 1. The sketch of the computational flow domain.

The swirling flow is generated by setting an axial-tangential fluid entry, and the constant inlet axial and azimuthal velocities are given by U_0 and U_θ, respectively. The no-slip velocity boundary condition is applied to the inner and outer wall of the annular pipe, and the end wall of the central bluff-body. The entrained fluid, a co-flow, is fed in by a mild velocity $U_c = 4\%U_0$. As the boundaries of the jet are sufficiently far from the nozzle outlet, symmetry boundary conditions are employed at the lateral wall of the domain and zero gradient conditions for fluid leaving the computational domain [5, 12]. For the pressure boundary conditions, ambient pressure is imposed at the far field boundary, whereas zero gradient boundary conditions are applied to all other boundaries. The turbulent kinetic energy and dissipation rate at the inlet are calculated from the turbulence intensity (15%) and integral length scale (half the hydraulic diameter, $(D_o - D_i)/2$) and zero gradient conditions across the faces of the computational domain are applied. The behavior close to solid walls is modeled with scalable wall functions.

The flow is characterized by two control parameters. The first one being the Reynolds number $Re = \frac{U_0 D_h}{\nu}$ based on the axial inlet velocity $U_0 = 0.94m/s$, and the hydraulic diameter $D_h = 9mm$ of the annular tube. The working fluid is water with kinematic viscosity $\nu = 1.0048 \times 10^{-6} m^2/s$. The other parameter is the swirl number $S = \frac{M_\theta}{M_x \frac{D_o}{2}}$, which is the ratio between axial flux of tangential momentum $M_\theta = \int_{\frac{D_i}{2}}^{\frac{D_o}{2}} 2\pi\rho \overline{U_x}\,\overline{U_\theta} r^2 dr$ and axial flux of axial momentum $M_x = \int_{\frac{D_i}{2}}^{\frac{D_o}{2}} 2\pi\rho(\overline{U_x})^2 r dr$, where $\overline{U_x}$ and $\overline{U_\theta}$ are the mean axial and tangential velocity measured at the end of the annular pipe, respectively. The computed Reynolds number Re is 8500, and the swirl number S is calculated as 0.46.

2.3. Grids and discretization methods

The numerical grid is constructed using the software Pointwise V18.0. Figures 2(a) and 2(b) display the computational grid close to the nozzle exit. All grid cells are hexahedral and the grid is finer in the core region. The SIMPLEC (Semi-Implicit Method for Pressure Linked Equations-Consistent) algorithm is employed to solve the governing equations. The transient term is discretised using second order implicit time differencing. The least square method is used for the gradient terms and the QUICK (Quadratic Upstream Interpolation for Convective Kinematics) scheme is employed for spatial discretization. We use scalable wall functions to model the flow field close to solid walls. The computations are done by the commercial software package ANSYS Fluent 18.2.

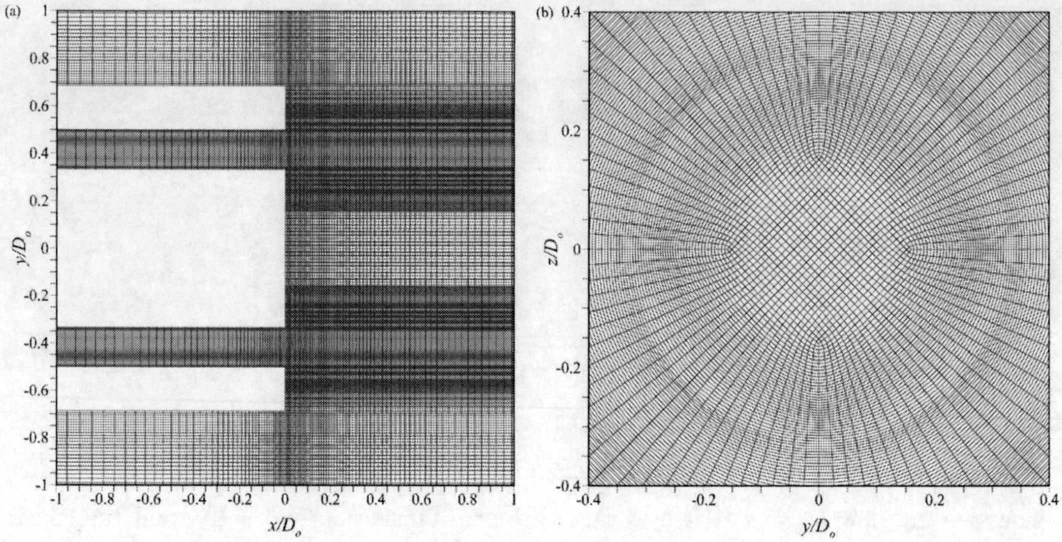

Figure 2. Computational grid close to the nozzle exit: (a) cross section in the $x - y$ plane, (b) symmetry plane perpendicular to the axis of the jet.

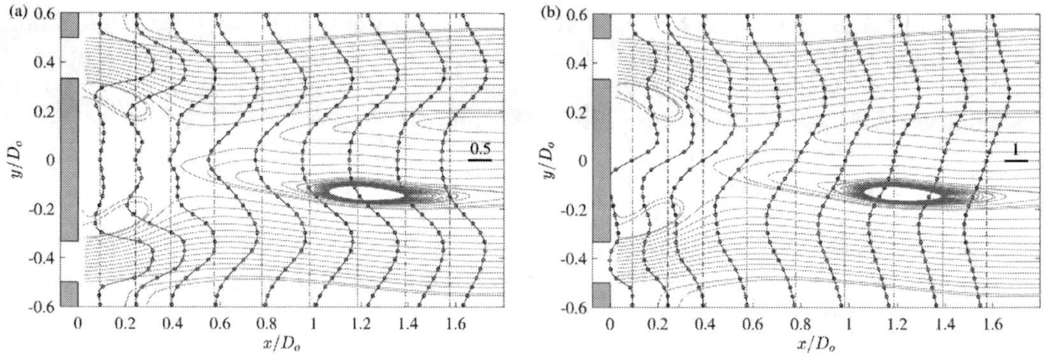

Figure 3. Time-averaged velocity profiles at different axial locations normalized by U_0: (a) axial velocity and (b) tangential velocity. The streamlines are depicted by the grey lines in the background.

3. Results and discussion

3.1. Mean flow fields

Figures 3(a) and 3(b) provide the mean axial and azimuthal velocity profiles at different axial locations. It is seen that the axial velocity profiles are wake-like in the near-field of the annular jet, which is similar to the observation in experimental measurements [13]. The first reason is due to flow separation behind the central bluff-body, which causes a central toroidal recirculation zone. The other one comes from the bubble caused by vortex breakdown further downstream at $x/D_o > 0.4$.

3.2. Spectral proper orthogonal decomposition analysis
One of the key issues in this research field is to identify the coherent structures from flow fields, which involves the construction of empirical modes that dominate the flow dynamics. The most popular methods, such as energy-ranked proper orthogonal decomposition (POD) and frequency-ranked fast Fourier decomposition or dynamic mode decomposition (DMD), are not applicable when coherent structures occur with low energies or fluctuating frequencies. However, the recently introduced spectral proper orthogonal decomposition (SPOD) approach allows to detect the coherent structures both in spatial and in frequency domains especially when the flow dynamics is rather complicated [14].

Figure 4 displays the extracted SPOD mode pairs and their energy content. The first four ranked pairs with the highest harmonic correlation are labeled with numbers. Each dot corresponds to one mode pair and the size of it represents the harmonic correlation between the modes. Here, the Strouhal number St is defined as $St = fD_h/U_0$, where f is the numerically obtained precessing frequency. The SPOD results show that the first two mode pairs account for 23.6% and 6.7% of the total turbulent kinetic energy, respectively. Hence, they represent the dominant coherent structures in the flow field. The evolution of their temporal coefficients is given in figures 5(a) and 5(b), which have the same content except for a phase difference of $\pi/2$. The power spectral density (PSD) of the harmonic mode coefficient shown in figures 5(c) and 5(d) indicates that their spectra have a peak at Strouhal numbers 0.23 and 0.53, corresponding to frequencies of 24.4Hz and 54.8Hz, respectively. Both POD analysis of the pressure fields [15] and phase-averaged velocity fields [11] in turbulent annular swirling jet have shown similar results.

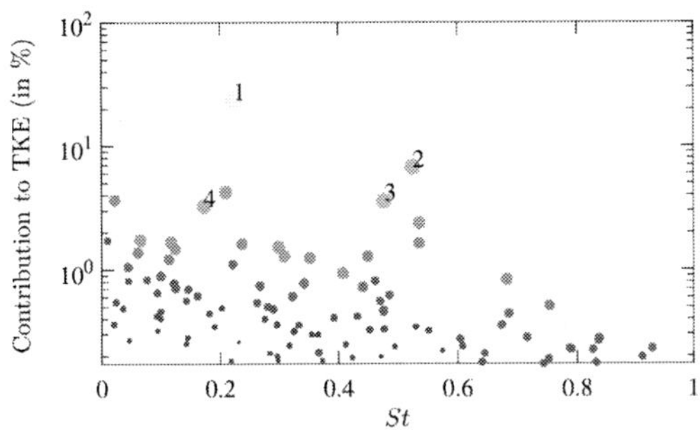

Figure 4. The contribution to the total turbulent kinetic energy (TKE) by the identified SPOD mode pairs.

3.3. Single and double helical structures
The reconstructed velocity fields are presented in figures 6(a) and 6(b). Here, the Q criterion [16] is applied to identify vortical structures in the flow field. The results show that not only a single helical but also a double helical vortical structure is found, and they both correspond to the spiral type of vortex breakdown. These spiral-shaped structures have high vorticity wrapped around the vortex breakdown bubble, and they are not always present in the flow field according to the mode coefficients shown in figures 5(a) 5(b). It should be noted that the double helix structure is rarely observed in swirling jets, and it is in line with the one detected in a similar turbulent annular swirling jet by phase averaging PIV measurements [11].

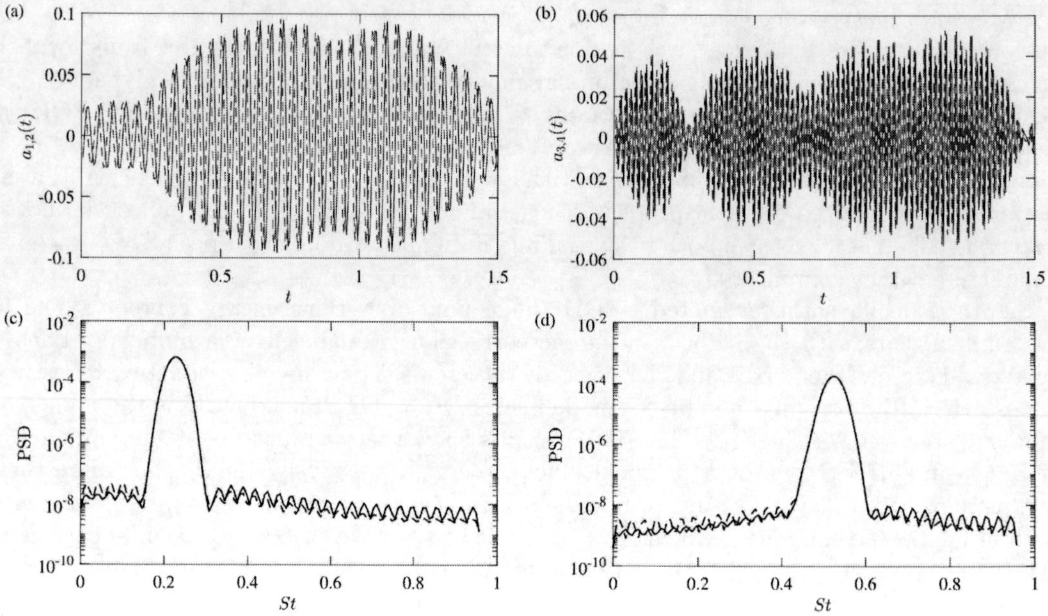

Figure 5. The time evolution (a-b) and power spectral density (c-d) of the first two ranked SPOD mode pairs. The time t is given in seconds.

Figure 6. Reconstructed velocity field with the fist (a) and second (b) ranked mode pair: slices with counters of axial velocity in $x - y$ plane, and Q criterion vortices visualization with isosurfaces of positive $Q = 4 \times 10^4$ $1/s^2$.

4. Summary and conclusions

In this paper, a turbulent annular swirling jet at Reynolds number $Re = 8500$ and swirl number $S = 0.46$ is simulated using the unsteady Reynolds-averaged Navier-Stokes approach, where the RNG k–ε turbulence model is used. The spectral proper orthogonal decomposition methodology is applied to investigate the coherent structures in the flow field. It is found that, apart from the single helical vortical structure usually reported in turbulent swirling flows, a double helical vortical structure is also identified. These single and double spiral type coherent structures with excessive vorticity are wrapped around the vortex breakdown bubble in the counter-swirl direction and rotate in the swirl direction with Strouhal numbers equal to 0.23 and 0.53, respectively. This result is in agreement with the recent experimental observations by Vanierschot et al. [11]. Although it is widely known that the k–ε model has disadvantages such as over-prediction of eddy viscosity and is not capable to capture the anisotropic turbulence of swirling flows, the findings in this work indicate that the unsteady RNG k–ε two-equation model can predict the periodic large scale coherent structures in turbulent annular swirling jet flow both temporally and spatially with good accuracy.

Acknowledgements

This work was supported by the China Scholarship Council (CSC) under Grant No. 201706890016.

References

[1] Gupta A K, Lilley D G and Syred N 1984 *Tunbridge Wells (Kent, English: Abacus Press)*
[2] Lucca-Negro O and O'doherty T 2001 *Prog. Energ. Combust.* **27** 431–481
[3] Wegner B, Maltsev A, Schneider C, Sadiki A, Dreizler A and Janicka J 2004 *Int. J. Heat Fluid Fl.* **25** 528–536
[4] Jochmann P, Sinigersky A, Hehle M, Schäfer O, Koch R and Bauer H J 2006 *Int. J. Heat Fluid Fl.* **27** 192–203
[5] Vanierschot M and Van den Bulck E 2009 *Int. J. Numer. Meth. Fl.* **59** 577–592
[6] Dunham D, Spencer A, McGuirk J J and Dianat M 2009 *J. Eng. Gas Turb. Power* **131** 011502
[7] Ramirez J A and Cortes C 2010 *Numer. Heat Tr. B-Fund.* **58** 98–120
[8] Javadi A and Nilsson H 2015 *Flow Turbul. Combust.* **95** 755–774
[9] Miltner M, Jordan C and Harasek M 2015 *Appl. Therm. Eng.* **89** 1117–1126
[10] Rukes L, Paschereit C O and Oberleithner K 2016 *Eur. J. Mech. B-Fluid* **59** 205–218
[11] Vanierschot M, Percin M and van Oudheusden B 2018 *Phys. Rev. Fluids* **3** 034703
[12] Vanierschot M and Van den Bulck E 2007 *Int. J. Numer. Meth. Fl.* **54** 313–324
[13] Vanierschot M and Van den Bulck E 2008 *Phys. Fluids* **20** 105104
[14] Sieber M, Paschereit C O and Oberleithner K 2016 *J. Fluid Mech.* **792** 798–828
[15] Percin M, Vanierschot M and van Oudheusden B 2017 *Exp. Fluids* **58** 166
[16] Jeong J and Hussain F 1995 *J. Fluid Mech.* **285** 69–94

IOP Conf. Series: Materials Science and Engineering **657** (2019) 012013 doi:10.1088/1757-899X/657/1/012013

Impact characteristics of gustiness debris flow on check dam

X G Jiang[1,2,3], F W Liu[1], K H Hu[2], T Liu[1], H J Wang[1], L Li[1], S Y Zheng[1], Y Chen[1], S J Song[1] and Y D Wang[1]

[1]College of Civil Engineering, Sichuan Agricultural University, Dujiangyan 611830, China

[2]Key Laboratory of Mountain Hazards and Surface Process, Institute of Mountain Hazards and Environment, Chinese Academy of Sciences, Chengdu 610041, China

E-mail: jxgjim@163.com

Abstract. The gustiness is a common motion form of debris flow, especially viscous debris flow. However, few researches have considered the gustiness into the impact characteristics of debris flow, which needs further research. In this study, the check dam was used as the load receptor to study the impact characteristics of gustiness debris flow. The results show that, under the same general flow depth, the dam stress considering the impact of gustiness debris flow is far less than that of disposable continuous debris flow. The impact force increases with the increase of the times of gustiness debris flow, and the impact torque increases linearly. Under the impact of gustiness debris flow, the displacement and stress of the dam vary in the same rule. The safety factors of anti-sliding and anti-overturning of the dam under the effect of gustiness debris flow are larger than those of disposable continuous debris flow, and the decrease rate of the safety factors increases with the increase of the times of gustiness debris flow. And under the condition of the same total flow depth, the increase of the debris flow depth can significantly decrease the stability of dam. The conclusions can provide reference for the design of blocking structures of gustiness and frequent debris flow.

1. Introduction

Debris flow is a common natural disaster all over the world, which occurs suddenly, fiercely and rapidly, and usually causes collapse, landslide and flood damage. Therefore, the damage degree of debris flow is more serious than a single collapse, landslide or flood. For example, in 1999, a massive mudslide in Venezuela killed more than 30,000 people and left 140,000 people homeless [1]. In 2010, debris flow occurred in the Zhouqu County, resulting in more than 1,600 deaths and about 300 homeless [2].

According to previous research data, debris flows, especially viscous debris flows, mostly moved in the form of gustiness flow [3-7], such as the gustiness debris flow in Jiangjia gully as shown in figure 1. According to years of observation and research about debris flow in Jiangjia Gully, it is known that the movement of viscous debris flow has thixotropy [8], which can affect the gustiness and bed-making process of debris flow [9]. Pierson [10] observed that debris flows moved in waves in the trench of mount Thomas, generating many standing waves in the process. Li *et al* [11] found that the movement of debris flows usually exists in the form of surge waves and consists of multiple waves that are temporally and spatially separated and have different patterns. Arai [12] found that surge intermittent is a common feature of viscous debris flow and muddy debris flow. According to on-site detection, Kean *et al* [13] found that debris flow has obvious surge characteristics.

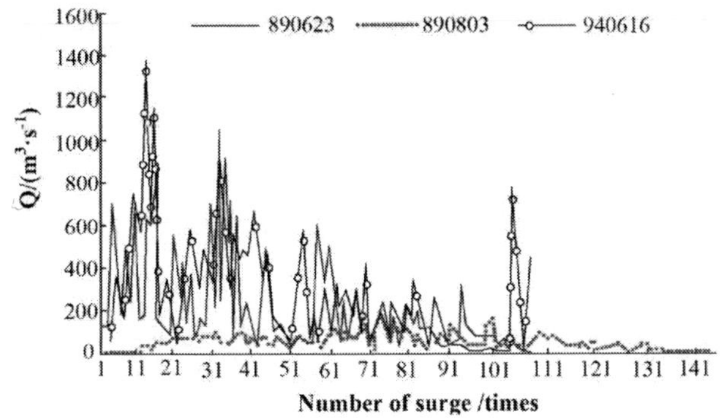

Figure 1. The discharge of three gustiness debris flows[7]. The data is recorded in Jiangjia gully by Li[7]. The number 890623, 890803, and 940616 are the issue numbers of three gustiness debris flow.

In order to prevent and control debris flow, researchers have conducted a series of studies and designed kinds of blocking structures. Check dams are mostly used to prevent debris flow. When designing such structures, most of the designers considered the impact of debris flow slurry and coarse particles on such devices [14]. And the impact force addition of slurry and coarse particles are supposed as the design value of the debris flow impact force. Other researchers made coupling analysis between debris flow and its interception project to calculate the stress and deformation of retaining structure under impact of debris flow [15,16]. However, most of the above studies simplified the effect of debris flow into a single impact process, ignoring the impact of gustiness on the blocking device, which was often inconsistent with the actual situation. Therefore, considering the gustiness of debris flow is helpful to design more reasonable debris flow prevention engineering.

This study firstly obtained the real dynamic parameters of gustiness debris flow through field observation. Then, combined with the dynamic parameters of debris flow, the interaction between debris flow and interception structures under gustiness conditions was analyzed. Finally, the influence of the impact force of the debris flow on the different parameters of the check dam was analyzed. The conclusion of this study can provide some guidance and suggestions for structural design and theoretical research of prevention and control engineering of debris flow.

2. Methodology

In order to discuss the impact of gust impact characteristics in detail, the check dam was chosen as the load receptor. According to reference in "Handbook of debris-flow prevention" [17], the dam is set to a length of 10 m, a width of 7 m and a height of 10 m. The width of dam crest was 2 m, the bottom width of the dam was 7 m, the dam material was concrete with a per unit weight of 24 kN/m^3. Midas/ Gts Nx was used to simulate the impact influence of debris flow on check dam. Midas/Gts Nx is widely used in Gts Nx engineering field which could analyse dynamics process efficiently and accurately.

In the finite element model, the dimension was set the same as that of the check dam, and adopting tetrahedral meshes, there were 3,201,281 dividing units (figure 2). The bottom of the check dam was embedded in a trench bed, and both sides of the sediment storage dam were embedded in mountains, the bottom and both sides of the model were set as fixed boundary, as shown in figure 2. To reflect the gustiness of debris flow, it was assumed that the dam experienced five times of debris flow, each flow depth was 2 m (figure 3). The simulation experiments were deigned in table 1. Each time the mudslide acted on the dam, and then the static earth pressure was generated, which combined with the next flow to act uniformly on the back of the dam.

Debris flow moves surging, the impact force of the mudslide on the check dam varies with different flow depths. Even for mudslide with the same depth, the impact force on the blocking structure is different. Therefore, this study selected the maximum impact forces at different flow depths from the actual measured debris flow impact data [18], as the impact force of the corresponding depth in numerical analysis. The details are as follows.

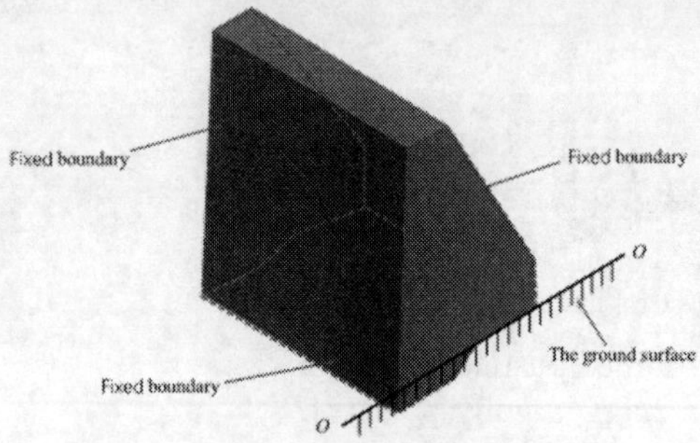

Figure 2. Check dam simulated model: the set of simulated model.

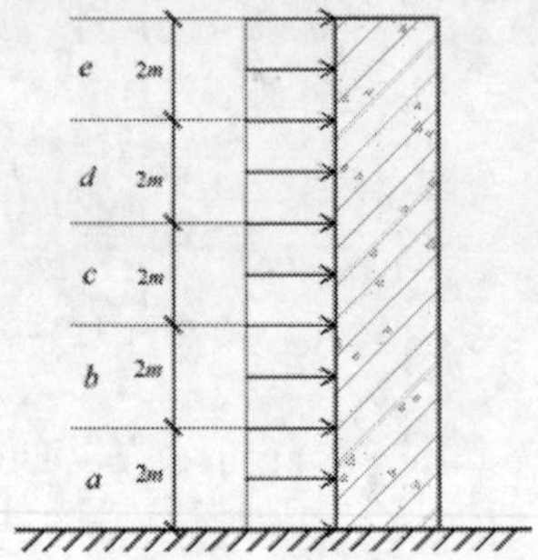

Figure 3. The set of debris flow surges.

The impact force of the actual measured debris flow is divided into slurry and rock impact force, both of them are highly correlated with space. Therefore, the functions of the slurry impact force and rock impact force with position height are established respectively. Based on the above two functions, the impact load of slurry and rock can be simulated when the debris flow hits the sediment storage dam. At the same time, another function between static earth pressure and position height is built. When the second dynamic load is applied, the third function (function between static earth pressure and position height) will be added as the static soil pressure in the first dynamic load application region; when the third dynamic load is applied, the first and second dynamic load application region will add the third function as static earth pressure, and so on.

Table 1. Design parameters.

Surge Name	Flow Depth /m	Distance of between surge debris flow surface and ground
Flow a	2	2
Flow b	2	4
Flow c	2	6
Flow d	2	8
Flow e	2	10
Disposable continuous debris flow	10	10

3. Results

3.1. Characteristics of stress in the check dam affected by gustiness debris flow

Figure 4 shows the stress contour of the dam under the action of five gustiness debris flows and disposable (flow depth of 10 m) continuous debris flow. It can be seen from the figure that the stress distribution of the dam caused by five gustiness debris flows was similar, that was, the compressive stress was near the impact load action area, the tensile stress was in the other region, and the stress in the middle part of the impact zone was larger, while the stress on both sides gradually decreased. With the increase of the times of gustiness debris flow, the area of the compressive stress region increased gradually, the area of the tensile stress region decreased gradually, and the maximum stress value also increased. Compared with the disposable impact of debris flow depth of 10 m on the dam, it can be seen that the stress generated by the disposable continuous impact was larger than the maximum impact stress of gustiness debris flow. For example, the maximum compressive stress generated by the disposable continuous impact was about 1.5 times of the other.

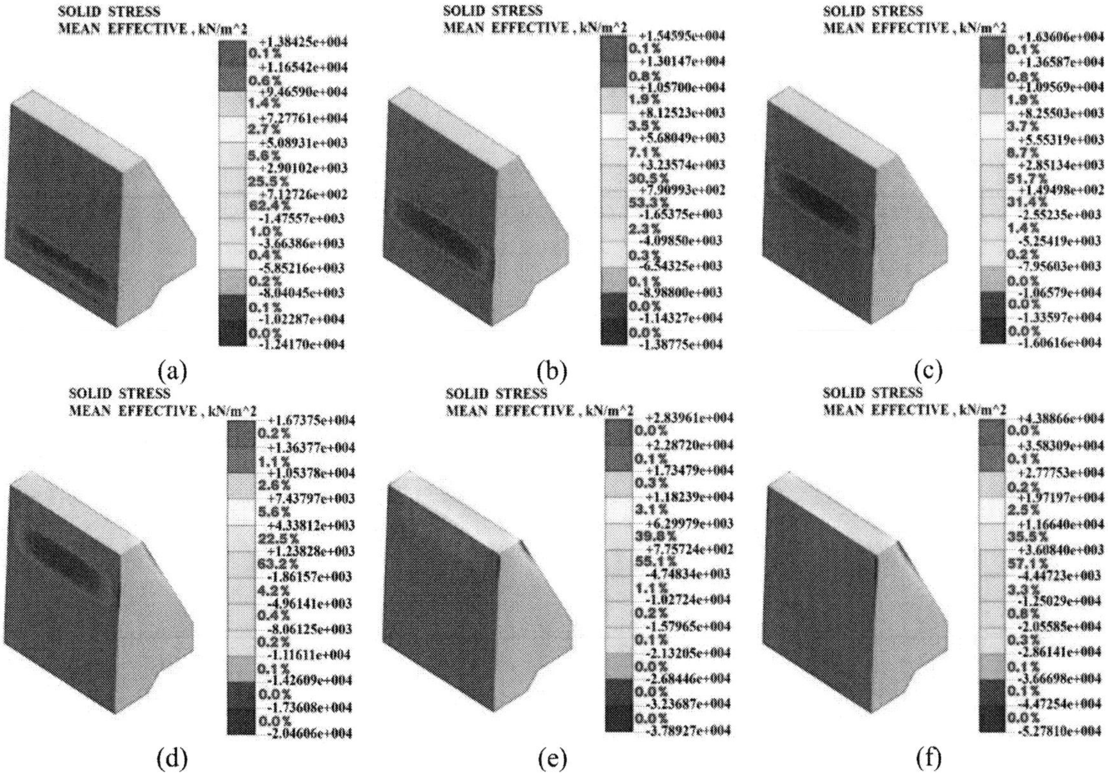

Figure 4. Stress contour of the dam: (a)-(e) effect of gustiness debris flow; (f) effect of disposable continuous debris flow.

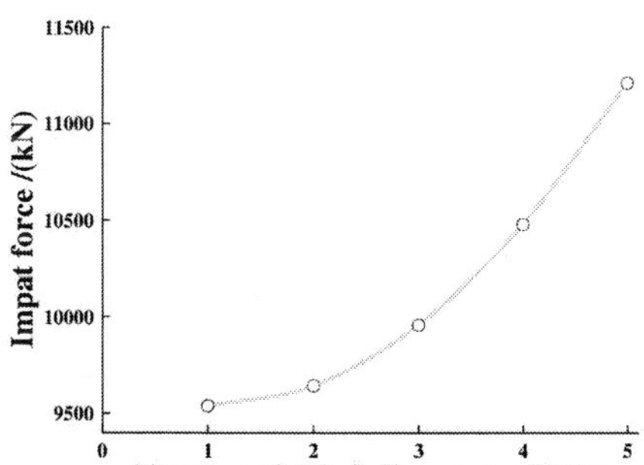

Figure 5. The relationship between the impact force and the gustiness debris flow sequences of surges. The impact force in the figure is the resultant force along the debris flow direction, including the dynamic load and soil pressure generated by the gustiness debris flow.

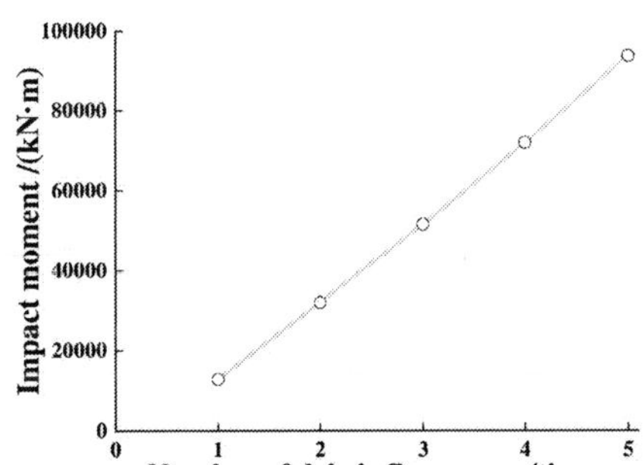

Figure 6. The relationship between the impact moment of debris flow and the numbers of debris flow surges.

Figure 5 shows the change characteristics of impact force generated by five times of gustiness debris flow. We can see from the figure that the horizontal impact force increased in a nonlinear form with the increase of times of gustiness debris flow. Moreover, with an increase in times of gustiness debris flow, the difference of impact forces between two adjacent flows increased gradually. Figure 6 shows the relationship between the moment of the horizontal impact force to the ground and the number of flows. We can get the conclusion that the impact moment increased linearly with the number of flows. The difference of moments between two adjacent flows increased with the increase of the times of flow, but the difference was not big. According to calculation, the horizontal force and moment generated by the disposable continuous impact of debris flow at a depth of 10 m were about 1.1×10^5 kN and 5.4×10^5 kN·m, which were respectively 10 times and 6 times of the fifth (the last) flow impact.

3.2. Characteristics of displacement in the check dams affected by gustiness debris flow

Figure 7 shows the general displacement cloud diagram of the dam under the action of five gustiness debris flows and disposable continuous debris flow. We can see from the figure that the total displacement was larger in the impact load area of debris flow, and the further away from the impact load area of debris flow, the smaller the displacement was. Furthermore, the maximum total displacement occurred on the upper surface of the impact load (distributed load).

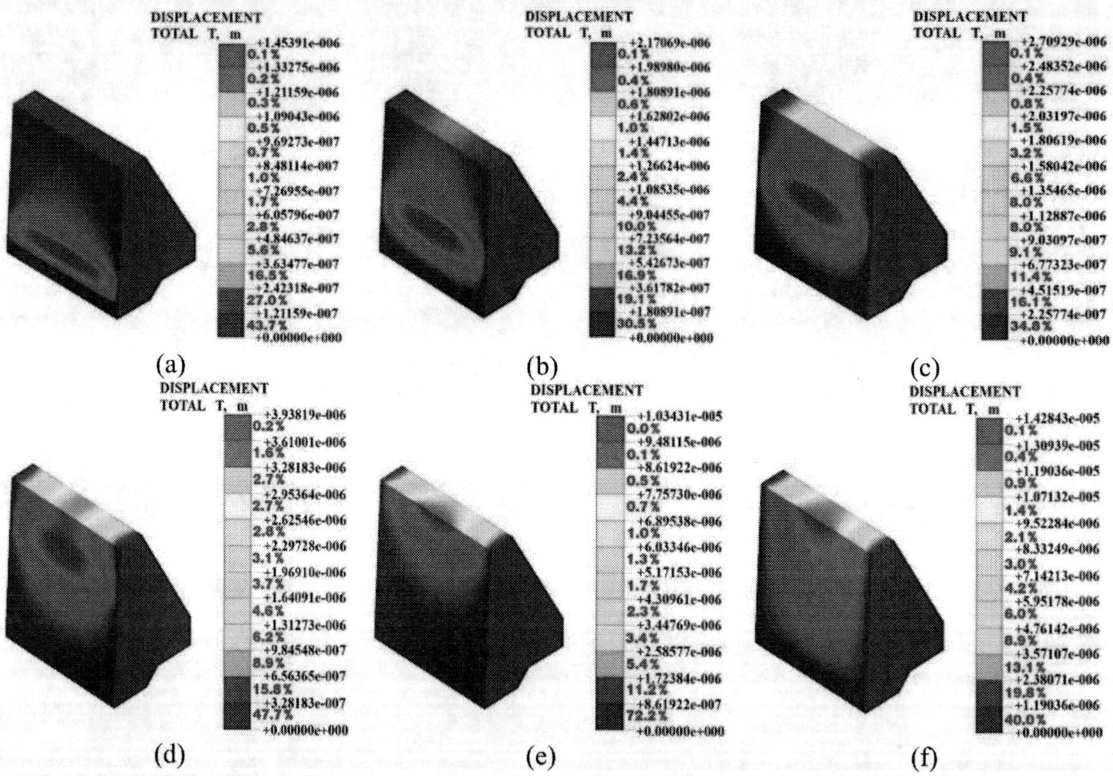

Figure 7. Displacement nephogram of the check dam: (a)-(e) effect of gustiness debris flow; (f) effect of disposable continuous debris flow.

The relationship between the maximum displacement of the check dam and the times of debris flow is shown in figure 8. It can be seen from the figure that the maximum displacement increased with the increase of the number of debris flows, that is, it increased with an increase in the height of debris flow. Furthermore, the difference of maximum displacement between two adjacent flows increased with the increase of the number of debris flows. For example, the maximum displacement difference generated by the first and second debris flows was 7.1397×10^{-3} mm, and the maximum displacement difference generated by the fourth and fifth flows was 5.87401×10^{-3} mm. In addition, from the comparison of the effect of gustiness debris flow and disposable continuous debris flow, it can be seen that the displacement generated by disposable continuous impact was larger than that generated by gustiness impact under the same total flow depth condition.

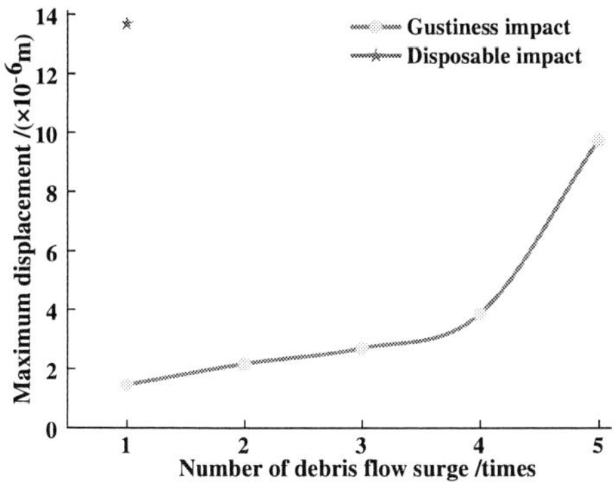

Figure 8. Relationship between the maximum displacement and the gustiness debris flow surges.

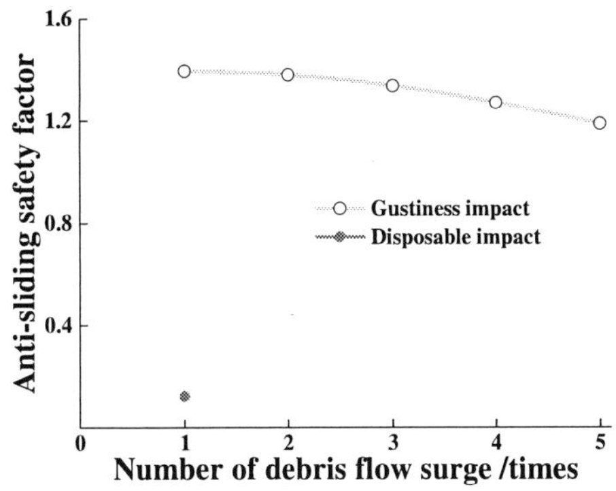

Figure 9. The relationship between the anti-sliding safety factor and the gustiness debris flow surges.

3.3. The safety of check dam impacted by gustiness debris flow

The whole stability of the dam must be ensured during the normal operation. Under the action of design load or combination of dangerous load, the anti-sliding coefficient is a parameter to measure the anti-sliding stability of the dam. The anti-sliding coefficient of the dam is expressed in equation (1) [20].

$$k_\sigma = \frac{f\sum W}{\sum Q} \tag{1}$$

Where k_σ is the anti-sliding safety factor of the check dam, f is the friction coefficient between masonry and foundation of the dam, $\sum W$ is the sum of the vertical forces acting on the unit length of the dam, $\sum Q$ is the sum of the horizontal forces acting on the unit length of the dam.

Considering the impact of debris flow, earth pressure, deadweight and friction of the dam, the anti-sliding coefficient of the dam is shown in figure 9. It can be seen from the graph that the anti-sliding safety factor decreased gradually with the increase of the number of debris flows, and the minimum value was about 1.2, larger than 1. While the safety factor of disposable continuous debris flow was about 0.1, which was much smaller. Therefore, when the total flow depth of debris flow was the same, the impact load of gustiness debris flow was much less dangerous than that of disposable continuous debris flow.

The anti-overturning coefficient is an index to measure the anti-overturning stability of the dam, which can be calculated in equation (2) [20].

$$k_y = \frac{\sum M_y}{\sum M_o} \tag{2}$$

Where k_y is the anti-overturning safety factor of the dam, $\sum M_y$ is the anti-overturning moment of the dam, $\sum M_O$ is the overturning moment of the dam.

The toe of the downstream slope of the dam was chosen as the rotation shaft, and the relationship between the calculated anti-overturning safety factor and the times of gustiness debris flow is shown in figure 10. It shows that the anti-overturning safety factor decreased with the increase number of debris flows, especially when debris flow accumulated to 2-4 meters, the anti-overturning safety factor decreased sharply, it might be the anti-overturning ability of the dam reached to a critical point due to the accumulation of a certain height of debris flow. The anti-overturning safety factor of the dam under the impact of gustiness debris flow was larger than that of disposable continuous debris flow, indicating that under the same total flow depth condition, the impact of gustiness debris flow was less dangerous than that of disposable continuous flow.

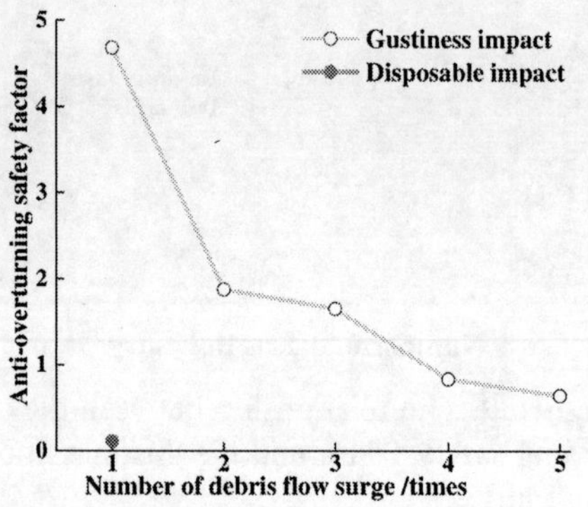

Figure 10. The relationship between the anti-overturning safety factor and gustiness debris flow surges.

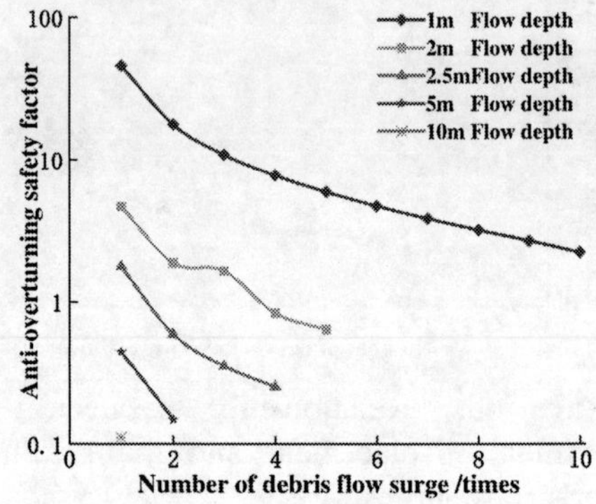

Figure 11. Influence of gustiness debris flow depth on: anti-overturning safety factor.

3.4. The influence of gustiness debris flow depth to check dam

Under the condition of the same general flow depth, shallower depth of a single gustiness debris flow means more times of the gustiness debris flows. The above analysis all depended on the situation of single flow depth of 2 m, in order to analyze the influence of flow depth on impact characteristics, five different flow depths (1 m, 2 m, 2.5 m, 5 m, 10 m) were selected in this study, it is worth mentioning that the flow depth of 10 m meant the gustiness of debris flow was not considered. The safety factors of anti-overturning under different flow depths were calculated as shown in figure 11.

As can be seen from figure 12, when the total flow depth of debris flow remained unchanged, shallower depth of single gustiness debris flow resulted in greater anti-sliding and anti-overturning safety factors. The results show that under the impact of a single flow, the low-depth debris flow had little influence on the safety of the dam, especially after considering the gustiness of debris flow, the anti-sliding and anti-overturning safety factors were obviously larger than those of disposable continuous debris flow.

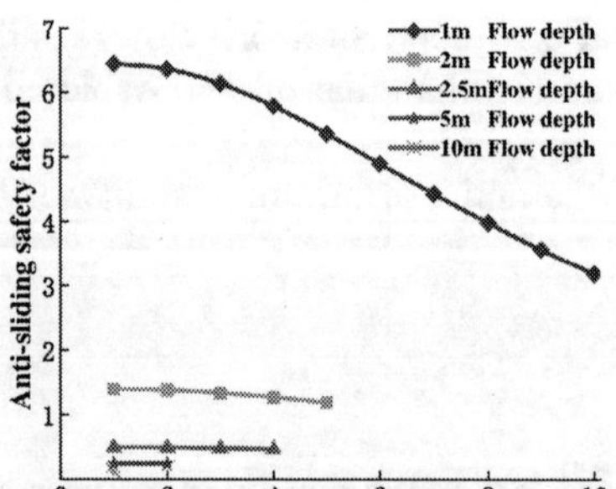

Figure 12. Influence of gustiness debris flow depth on: anti-sliding safety factor.

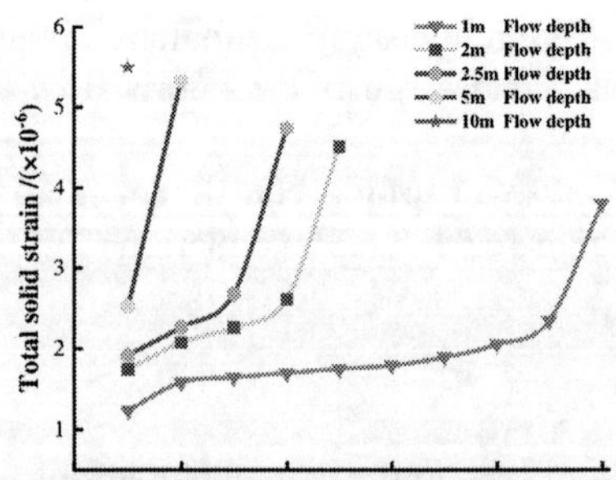

Figure 13. Influence of gustiness debris flow depth on strain of the check dam.

The strain of the check dam under the effect of load is an intuitive parameter to evaluate the anti-safety performance of the check dam. Figure 13 shows the strain of gravity check dam when it was first impacted by debris flows with flow depth of 1 m, 2 m, 2.5 m, 4 m, 5 m, respectively. Conclusions can be drawn from figure 13 that the growth rate of total strain increased with the increase of times of gustiness debris flow, and the greater the depth of a single debris flow, the faster the strain increased. Overall, keep the general depth of debris flow unchanged, the shallower the depth of a single debris flow, the smaller the final strain of the dam body would be.

4. Conclusions

Combined with the debris flow impact measurement data in Jiangjiagou gully, Yunnan and numerical analysis, the following conclusions were drawn.

- Under the condition of the same total flow depth, the impact force and moment of disposable continuous debris flow are far greater than those of gustiness debris flow. The impact force of gustiness debris flow has a nonlinear relationship with the times of gustiness debris flow, and the growth rate of impact force increases with the increase of the times of gustiness debris flow. While the impact moment has a linear relationship with the number of debris flows.
- Under the condition of the same general flow depth, the displacement of the dam under the impact of gustiness debris flow is smaller than that of continuous debris flow. Under the effect of gustiness debris flow, the displacement has a nonlinear relationship with times of gustiness debris flow, and the displacement growth rate increases with the increase of the times of flows.
- Under the condition of the same general flow depth, the anti-sliding safety factor and anti-overturning factor of continuous flow are much smaller than those of gustiness debris flow.
- As the total flow depth keeps the same, the stability of the dam gets worse with the increase of the gustiness debris flow depth.

In actual situation, the front of the dam is rarely filled by debris flow in one time, but by repeated debris flow. Therefore, multiple debris flow events can be simplified into gustiness flows of a long time series. The impact characteristics of gustiness debris flow discussed in this paper can provide a reference for the design of blocking structure in debris flow prone-to-occur area.

Acknowledgments

The authors acknowledge financial support provided by The National Natural Science Foundation of China (No. 41807289); Sichuan province fundamental scientific research (No. 18ZA0374); Open fund of Key Laboratory of mountain hazards and surface processes, Chinese Academy of Sciences (No. KLMHESP-17-05) and Innovative Training Program for College Students in Sichuan Province (No. 201810626152).

References

[1] Francisco L P 2001 Matrix granulometry of catastrophic debris flows (December 1999) in central coastal Venezuela *Catena* **45** 163-83
[2] Hu K H, Ge Y G, Cui P, Guo X J and Yang W 2010 Preliminary analysis of extra-large scale debris flow disaster in Zhouqu County of Gansu Province *J. Mount. Sci.* **28** 628-34
[3] Sharp R P and Nobles L H 1953 Mudflow of 1941 at Wrightwood, southern California *Geol. Soc. Am. Bull* **64** 547-60
[4] Liu J, Li Y, Su P, Cheng Z and Cui P 2009 Temporal variation of intermittent surges of debris flow *J. Hydrol* ***365*** 322-8
[5] Davies T R H 1986 Large debris flows: a macro-viscous phenomenon *Acta. Mech* **63** 161-78
[6] Major J J 1997 Depositional processes in large-scale debris-flow experiments *J. Geol* **105** 345-66
[7] Li Y 2014 Spatial and temporal characteristics of Yangjiagou debris flow shower *Chin. J. Nat* **36** 319-24
[8] Wang Y Y and Fang T N 1997 Discussion on the phenomenological model of debris flow burst flow and thixotropy *J. Nat. Disaster* **6** 38-47
[9] Ng C O and Mei C C 1994 Roll waves on a shallow layer of mud modelled as a power-law fluid

J. Fluid. Mech **263** 151-84

[10] Pierson T C 1980 Erosion and deposition by debris flows at Mt Thomas, north Canterbury, New Zealand *Earth Surf. Proc. and Landf* **5** 227-47

[11] Li Y, Jing J L, Hu K H, Peng C S 2012 Probability distribution of measured debris-flow velocity in Jiangjia Gully, Yunnan Province, China *Nat. Hazards* **60** 689-701

[12] Arai M 2010 Influence of nonuniform solid concentration on surge occurrence of muddy debris flow and viscous debris flow *EGU General Assembly Conference Abstracts* **12** 7740

[13] Kean J W, McCoy S W, Tucker G E, Staley D M and Coe J A 2013 Runoffgenerated debris flows: Observations and modeling of surge initiation, magnitude, and frequency *J. Geophys. Res-Earth* **118** 2190-207

[14] Li J J, Wang X L and Zhu Y P 2014 Experimental study on dynamic responses of a new type of debris flow dam under solid impact load *Applied Mechanics and Materials* (Shanghai, China) pp 299-308

[15] Liang Y X, Wang X L, Wu C and Lv Z G 2014 Study on the impact properties of a new debris flow dam based on steel-concrete composite structure *In Applied Mechanics and Materials* (Shanghai, China) pp 2056-9

[16] Li K, Chen X Q, Chen J G, Wang F and Wang X J 2016 Numerical simulation of the slit-trap dam based on fluid-structure interaction *J. Donghua Univ (Eng. Ed.)* **33** 740-4

[17] Zhou B F, Li D J and Luo D F 1991 *Handbook of Debris-Flow Prevention* 1st ed (China: Beijing Sciences Press) pp 132-41

[18] Chen H K, Tang H M, Xian X F and Zhang Y P 2010 Experimental model of debris flow impact features *J. Chongqing Univ* **33** 114-9

IOP Conf. Series: Materials Science and Engineering **657** (2019) 012014 doi:10.1088/1757-899X/657/1/012014

Numerical study of the reflected elastic waves using Rayleigh diffraction integral

N Maghlaoui[1,4], H Djelouah[2], A Oudina [2], M Ourak[3] and F Benmeddour[3]

[1]Ecole Supérieure en Sciences Appliquées d'Alger, BP 474, Place des Martyrs 16001, Alger, Algérie
[2]Université des Sciences et de la Technologie Houari Boumediene, Faculté de Physique, Laboratoire de Physique des Matériaux, BP 32 El Alia 16111, Algiers, Algeria
[3]Univ. Polytechnique Hauts-de-France, CNRS, Univ. Lille, YNCREA,Centrale Lille, UMR 8520 - IEMN, DOAE, F-59313 Valenciennes, France

E-mail : maghlaoui.n79@gmail.com

Abstract. In this work, the transient ultrasonic waves radiated by a linear phased array transducer in a liquid then reflected at a liquid solid interface is studied. A model based on the Rayleigh integral is used where the reflection at the plane interface is considered by using the reflection coefficients for harmonic plane waves. The transient field is obtained by an inverse Fourier transform of the harmonic field. The obtained results highlight the different components of the ultrasonic field: the direct and edge waves as well as the longitudinal head waves or leaky Rayleigh waves. The temporal representation of these waves has been analyzed and discussed by the rays' model. Instantaneous cartographies allowed a clear description of all the waves which appear at the liquid-solid interface. The results have been compared to those obtained by using a finite element method.

1. Introduction

The propagation study of the ultrasonic waves at the interface liquid solid radiated by a multi-element linear transducer has been the subject of many publications [1-9].
The advantage of multi-element transducers is to allow the focusing and the steering of the beam in both liquid and solid [1- 10].

Generated acoustic waves by elementary sources composing a transducer can be modeling by superposing the produced fields. Each source is modeled by point source superposition applied to the Rayleigh integral, the spatial impulse response model, and other analytically equivalent integral approaches.

A good review of the earlier developments of the acoustic wave modeling generated by multi-element linear transducers in front of a planar liquid solid interface can be found in [8, 11-21].
In this research work, the Rayleigh integral has been used to simulate the propagation of the ultrasonic wave emitted in water by a multi-element linear transducer in presence of a plane water-solid interface. This model starts from the knowledge of the particle velocity at the radiating surface. Then, the harmonic velocity potential is derived through the Rayleigh integral on the surface. The transient incident wave is then obtained by an inverse time Fourier transform of the harmonic results. In the presence of a fluid solid interface the echo-graphic response in the liquid is calculated by taking into

IOP Conf. Series: Materials Science and Engineering **657** (2019) 012014 doi:10.1088/1757-899X/657/1/012014

account the reflection coefficient for harmonic plane waves. The obtained results highlighted the different components of the acoustic field in the liquid: the bulk waves as well as the surface waves radiated in the water. The temporal representation of these waveforms has been studied. The arrival time of each contribution is analytically calculated and the physical origin of each of them is clearly determined (Table 1). Time domain snapshot of different wave front propagation before and after reflection allowed a physical description of all waves. These results are compared with those obtained by a finite element method [22].

2. Theoretical model

2.1. Analytical formulation

2.1.1. Emission in the water

A multi-element linear transducer comprising N small sources is considered. The latter emits acoustic waves in water, the propagation velocity of these waves is c_f. The width of each source is $2a$ and the distance between the center of two successive source is denoted d. Let $\mathcal{R}(\overrightarrow{Ox},\overrightarrow{Oz})$ be the reference frame where $\overrightarrow{Ox}$ is taken as the reference line defining the $z = 0$ plane and the second axis $\overrightarrow{Oz}$ perpendicular to the previous axis (figure 1).

Assuming that the radiating surface is a set of elementary sources and each elementary surface dS vibrates with a uniform normal particle velocity $v_n(t)$.

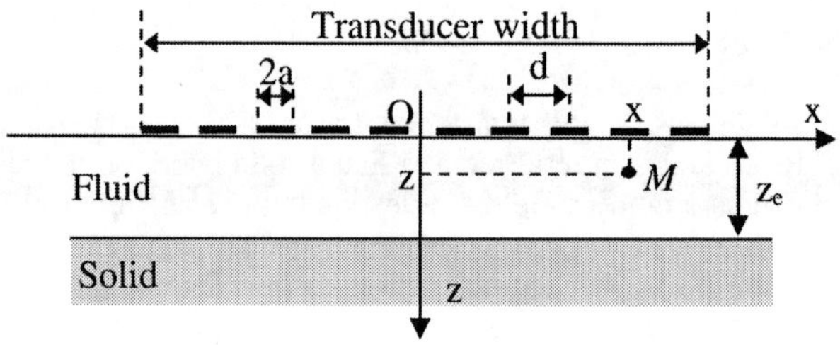

Figure 1. Geometry of the problem

At a point $M(x,z)$ in a water, the pressure $p_j(x,z,t)$ radiated by the jth element of the multi-element source can be expressed is:

$$p_j(x,z,t) = \rho_f \frac{d\varphi_j(x,z,t)}{dt}, \quad (1)$$

$\varphi_j(x,z,t)$ is the transient velocity potential at the field point (x,z).

$$\bar{\varphi}_j^*(k_x,z,\omega) = \int_{-\infty}^{+\infty} \varphi_j^*(x,z,\omega)\, exp(-ik_x x)\, dx. \quad (2)$$

Equation (2) shows that the velocity potential $\varphi_j(x,z,t)$ at any point inside the medium is a harmonic synthesis of monochromatic plane waves $\bar{\varphi}_j^*(k_x,z,\omega)$.

$\bar{\varphi}_j^*(k_x,z,\omega)$ is the spatial Fourier transform of the velocity potential $\varphi_j^*(x,z,\omega)$ calculated in the monochromatic case by the Rayleigh integral:

$$\varphi_j^*(x,z,\omega) = \frac{v_{nj}^*(\omega)}{2\pi} \iint_S \frac{exp(ikr)}{\sqrt{r}}\, dS, \quad (3)$$

where dS is the array of the elementary sources constituting the jth element, r is the distance from the point $M(x,z)$ to the elementary source, $k = \omega/c_f$ is the wave number, ω is the angular frequency. Each transducer element is constituted from elementary sources which vibrate with a velocity $v_{nj}(x,t)$.

$$v_{nj}^*(\omega) = \int_{-\infty}^{+\infty} v_n(t-\tau_j)\, exp(i\omega t)\, dt, \quad (4)$$

where $v_{nj}(x,t)$ and τ_j are respectively the particle velocity and the time delay associated with the jth element of the transducer.

The wave field at the point $M(x,z)$ is the sum of the wave fields emitted by the N elements of the multi-element transducer:

$$p(x,z,t) = \sum_{j=1}^{N} p_j(x,z,t) \tag{5}$$

In order to obtain a focusing beam, the time delay applied to the jth element is:

$$\tau_j = \frac{F - \sqrt{\left(x_f - x_j\right)^2 + z_f^2}}{c_f} \tag{6}$$

where $F\left(x_f, z_f\right)$ is the focal point.

2.1.2. Water and aluminium media separated by a plane interface
In the presence of the water aluminium plane interface at $z = z_e$. The incident wave at the interface defined by $z = z_e$ plane, is given by:

$$p_i(x,z_e,t) = \sum_{j=1}^{N} p_{ij}(x,z_e,t), \tag{7}$$

with

$$p_{ij}(x,z_e,t) = \rho_f \frac{d\varphi_{ij}(x,z_e,t)}{dt}, \tag{8}$$

where φ_{ij} is the incident velocity potential radiated by the jth element of the source.
The space time Fourier transform of the velocity potential radiated by the jth element of the source and reflected at the interface is:

$$\bar{\varphi}^*_{rj}(k_x,z_e,\omega) = \bar{\varphi}^*_{ij}(k_x\,,z_e,\omega)\boldsymbol{R}(k_x\,,\omega), \tag{9}$$

where $\boldsymbol{R}(k_x,\omega)$ is the reflection coefficient at the water-aluminium plane interface.

The reflection coefficient $\boldsymbol{R}(k_x,\omega)$ is calculated by taking into account the specular and non specular reflection. Considering the fact that the reflected pressure must be real valued functions, their Fourier transforms must be Hermitian.

The echo-graphic response of the wave coming back to a z-plane is given by:

$$p_r(x,z,t) = \sum_{j=1}^{N} p_{rj}(x,z,t), \tag{10}$$

where

$$p_{rj}(x,z,t) = \rho_f \frac{d\varphi_{rj}(x,z,t)}{dt}, \tag{11}$$

$p_{rj}(x, z, t)$ is the echo-graphic response of the pressure field emitted by the jth source.

2.2. Numerical implementation
MATLAB programs have been developed to model the acoustic field in presence of a water-aluminium interface using Rayleigh integral method.
The waveform over the active surface of the source is represented by the excitation signal $f(t)$, (figure 2), which is a pulse whose center frequency is f =1 MHz defined by:

$$f(t) = A\cos[2\pi f(t - t_0)]\exp\{-(t - t_0)^2/B\}. \tag{12}$$

The propagation medium is water, characterized by an acoustic velocity c_f = 1485 ms^{-1} and whose density is $\rho_f = 10^3\text{kg/m}^3$. The solid is aluminum for which the velocities of the longitudinal and transverse waves are respectively: c_L = 6420 ms^{-1}, c_T = 3110 ms^{-1} and whose density is $\rho_s = 2.7 \cdot 10^3\text{kg/m}^3$.

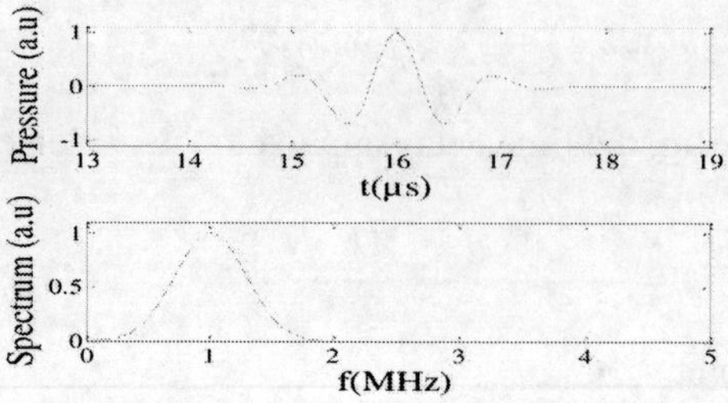

Figure 2. Time waveform $p(0,t)$ and its time Fourier transform. Frequency f = 1 MHz, B = 5 10^{-13} s^2 and $t_0 = 16$ μs.

3. Discussion

3.1. Obtained results

In order to explain the different kind of contribution to the echo-graphic response, the characteristics of the wave emitted by a cylindrical transducer [8] is used. The field emitted forward by the source is the superposition of two contributions: the direct wave emanating from the front face of the source and the edge waves from the edge of the source. When these contributions arrive at the solid surface, they are reflected and transmitted (figure A1 – A2).

Considering that, during the propagating time, the circular edge waves sweep over the solid surface. The reflection phenomenon can be described by simply considering an incident wave emerging from the edge of the multi-element source and having an incidence angle varying with time.

Figure 3 (a) represents the pressure received by the 64th element of multi-element transducer, located at the middle of the phased array transducer. In this case the first impulse corresponds to the superposition between the reflected direct wave and the longitudinal head wave followed by the superposition between the leaky Rayleigh wave (RW) and the reflected edge waves (REW). In this case the reflected direct wave is due to the reflection of the direct wave by the aluminum surface, which arrives to the 64th element of phased array transducer. The leaky Rayleigh wave (RW) is due to the conversion of the direct wave, incident at Rayleigh incidence angle ($\theta = 30.65°$), into a surface wave which propagates at the velocity $c_R = 2917\ \mathrm{ms}^{-1}$ over the solid surface.

Figure 3 (b) represents the reflected wave, received by the elements from 62th to 66th of phased array transducer. This signal is obtained by a simple summation of the received pressure by the elements 62th to 66th. The first wave corresponds to the interference between the reflected direct wave and the longitudinal head wave. The next wave corresponds to the interference between the leaky Rayleigh wave (RW) and the reflected edge waves (REW). It is worth noting that the amplitude of the reflected direct wave (RDW) is significant compared to the other contributions (REW, RW and LW).

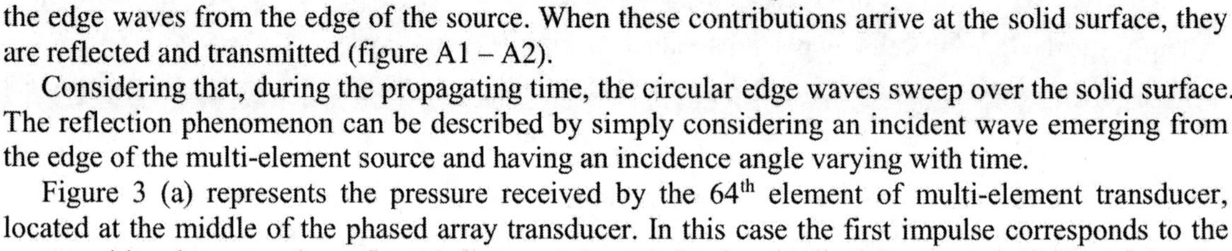

(a) (b)

Figure 3. Representation of the wave (Arbitrary units) received by the transducer. Number of elements N = 128, Width $2\ a$ = 0.27 mm, pitch of the elements d = *0.3 mm*, f = 2.5 MHz, plane interface z_e = 10 mm, Focal point $F(x_f = 0, z_f = 30mm)$. (a) Wave field received by the 64th element of the multi-element source, (b) Wave field received by the elements from 62 to 66 of the multi-element source.

Figure 4 represents the pressure wave fronts at different instant times. At the time t = 2.68 μs (figure 4 (a)) the presence of the longitudinal bulk waves (DW and EW) is observed. The different wave fronts are focused on the field point $F(x = 0\ mm, z = 20\ mm)$ from the multi-element source. After reflection, $t = 12.19$ μs (figure 4 (b)), the presence of the wave front (RDW) and the wave fronts (REW) are noticed. These waves are due to the specular reflection of the direct and edge waves at water solid interface. It is also noted the presence of weak amplitude waves identified by the leaky Rayleigh wave (RW) and the longitudinal head wave (LW) (figure 4 (b)). These waves are characterized by a linear wave front propagating in the direction $\theta = \pm 30.65°$ in the case of the Rayleigh wave (RW) and in the direction $\theta = \pm 13.39°$ in the case of the longitudinal head wave (LW). Inside the radiated zone, the non-specular contributions are due to the non-specular reflection of the direct wave arriving onto the interface with critical incidence angles. Outside the radiated zone, the non-specular contributions are due to the non-specular reflection of the edge wave arriving onto the interface with critical incidence angles. Furthermore, the energy converted to the longitudinal head wave is weaker than that converted to the leaky Rayleigh wave.

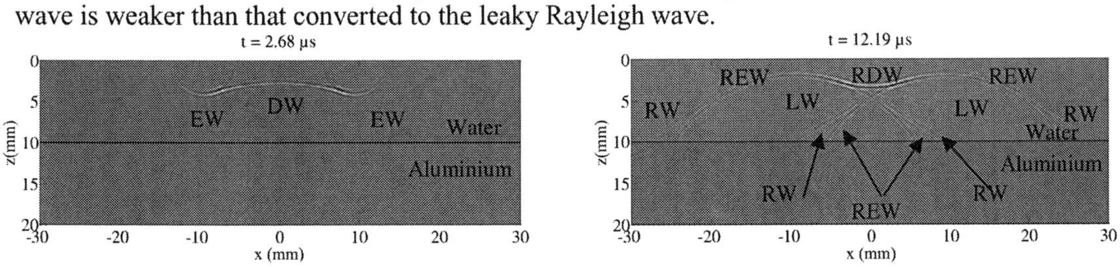

(a) (b)

Figures 4. Representation of the wave field $p(x, z)$ before and after reflection by the water aluminium interface. Transducer parameter (N = 128, $2a$ = 0.13 mm, d = *0.15 mm*, f = 2.5 MHz), z_e = 10 mm, $F(x_f = 0, z_f = 30\ mm)$. (a) Before reflection t = 2.68 μs, (b) After reflection t = 12.19 μs.

3.2. Comparison with finite element method

Finite element method has been used for comparison purposes by using a finite element package [8, 22 - 23].

The choice of the finite element method mesh must be able to reproduce the propagation and the reflection of the elastic wave. The spatial and time steps Δx, Δz and Δt verify the conditions:

$$\max(\Delta x, \Delta z) < \lambda_{min}/10 \qquad \text{and} \qquad \Delta t < 0.7\frac{\min(\Delta x, \Delta z)}{c_L}, \qquad (13)$$

where λ_{min} is the smallest wavelength associated to the maximum frequency spectrum. The used time discretization step is $\Delta t = 0.015$μs.

The incident and reflected pressure fields are computed by the finite element model in the presence of the water aluminum plane interface, the obtained results are shown in figure 5. The results put in evidence the bulk waves (Figure 5 (a)). After reflection, at the instant time $t = 13.60$ μs, the different waves are reflected and generate the reflected bulk wave (RDW and REW) and the surface waves which radiate in water (RW and LW).

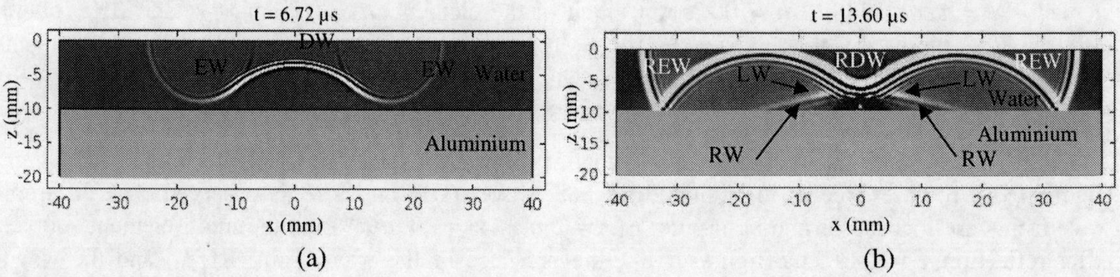

Figure 5. Representation of the wave field *p*(*x*, *z*) before and after reflection by the water aluminium interface. Transducer parameter (N = 128, $2a$ = 0.2 mm, $d = 0.25\ mm$, f = 2.5 MHz), z_e = 10 mm, $F(x_f = 0, z_f = 20mm)$. (a) Before reflection t = 6.72 µs, (b) After reflection t = 13.60 µs.

These contributions are identified by a linear wave front propagating in the direction $\theta = 13.39°$ in the longitudinal head wave (LW) case and the direction $\theta = 30.65°$ in the leaky Rayleigh (RW) case (Figure 5 (b)).

At the frequency $f = 2.5\ MHz$ and for the element number $N = 128$, the computing time is 691.80 s with the finite element method and 126.20 s with the Rayleigh integral method, respectively in a computer with an Intel Pentium Core (TM) i7 2.4 GHz processor.

4. Conclusion

Rayleigh integral method is used to model the ultrasonic wave generated by multi-element linear transducer in front of the water-aluminum interface. This semi analytical model considers the radiation, the specular and non-specular reflection. One major advantage of this method is to use the expression of the reflection coefficient, which reduces the computing time. One other advantage of the Rayleigh integral method is its capability of modeling critical reflection phenomenon.

The computed results show the advantage provided by the linear array transducer in the acoustic microscopy application.

The obtained results agree with those computed with the finite element method.
The method can be extended to the case of more complex transducers or more complex structure.

Appendix

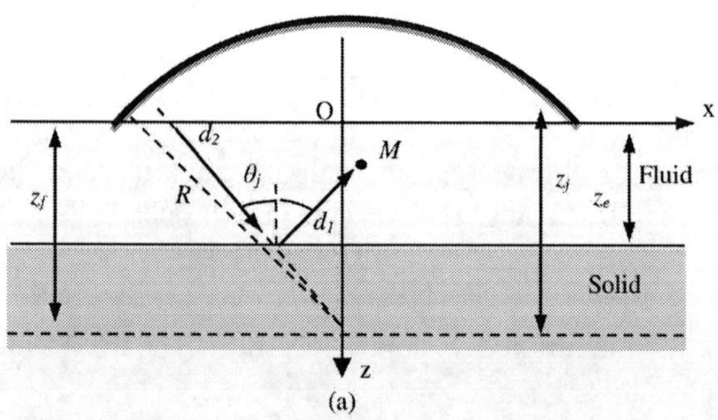

Figure A1. Paths of the reflected Bulk wave contributions.

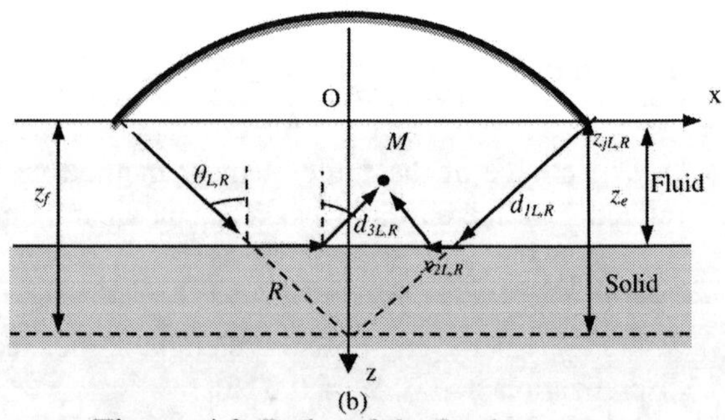

Figure A2. Paths of the Leaky wave contributions.

Table 1. The calculation of the arrival times of the reflected waves are detailed in [8].

Bulk reflected waves (Figure A1)	Leaky surface waves (Figure A2)
$t_{RDW} = \dfrac{d_1 + d_2}{c_f}$	$t_{RW} = \left(\dfrac{d_{1R}}{c_f} + \dfrac{d_{2R}}{c_f} + \dfrac{x_R}{c_R}\right)$
$t_{REW} = \left(\dfrac{\sqrt{(z_e - z)^2 + x_1^2}}{c_f}\right) + \left(\dfrac{\sqrt{{z_e}^2 + x_2^2}}{c_f}\right)$	$t_{LW} = \dfrac{d_{1L}}{c_f} + \dfrac{d_{3L}}{c_f} + \dfrac{x_{2L}}{c_L}$

References

[1] Frénet D, Calmon P and Ouaftouh M 1998 Generation of leaky Rayleigh waves using a conical phased-array transducer : modeling time-domain signals on anisotropic materials *Proc. IEEE Ultrason. Symp Japan* **1** 269-72

[2] Zeng X and McGough R J 2009 Optimal simulations of ultrasonic fields produced by large thermal therapy arrays using the angular spectrum approach *J. Acoust. Soc. Am.* **125**(5) 2967-77

[3] Wu P, Kazys R and Stepinsky T 1995 Calculation of transient fields in immersed solids radiated by linear focusing arrays *Proc. IEEE Ultrason. Symp.* **2** 993-7

[4] Titov S, Maev R and Bogachenkov A 2006 Measurements of velocity and attenuation of leaky waves using an ultrasonic array *Ultrasonics* **44** 182-7

[5] Banerjee S, Kundu T and Alnuaimi N A 2007 DPSM technique for ultrasonic field modelling near fluid–solid interface *Ultrasonics* **46** 235-50

[6] Ahmad R, Kundu T and Placko D 2005 Modeling of phased array transducers *J. Acoust. Soc. Am.* **117**(4) 1762-76

[7] Banerjee S and Kundu T 2007 Ultrasonic field modeling in plates immersed in fluid *Int. J. Sol. Str.* **44** 6013–29

[8] Maghlaoui N, Belgroune D, Ourak M and Djelouah H 2016 Reflection at a liquid–solid interface of a transient ultrasonic field radiated by a linear phased array transducer *Ultrasonics* **71** 98-105

[9] Belgroune D, de Belleval J F and Djelouah H 2008 A theoretical study of ultrasonic wave transmission through a liquid-solid interface *Ultrasonics* **48** 220-30

[10] Song S J and Kim C H 2002 Simulation of 3-D radiation beam patterns propagated through a planar interface from ultrasonic phased array transducers *Ultrasonics* **40** 519-24

[11] Oberhettinger F 1961 On transient solutions of the baffled piston problem *J. Res. Nat. Bur. Stand.* **65B** 1-6

[12] Stepanishen R P 1971 Transient radiation from piston in an infinite planar baffle *J. Acoust. Soc. A.* **49**(5) 1629-38

[13] Harris G R 1981 Review of transient field theory for a baffled planar piston *J. Acoust. Soc. A.* **70** 10-20

[14] Wu P, Kazys R and Stepinski T 1995 Analysis of the numerically implemented angular spectrum approach based on the evaluation of two-dimensional acoustic fields. Part I. Errors due to the discrete Fourier transform and discretization *J. Acoust. Soc. A.* **99** 1139-48

[15] Lerch T P, Schmerr L W, Sedov A 1998 Ultrasonic beam models: an edge element approach *J. Acoust. Soc. A.* **104**(3) 1256-65

[16] Zhang J, Guy P, Baboux J C and Jayet Y 1999 Theoretical and experimental responses for a large-aperture broadband spherical transducer probing a liquid–solid boundary *J. App. Phy.* **86**(5) 2825-35

[17] Fellinger P, Marklein R, Langenberg K J and Klaholz S 1995 Numerical modeling of elastic wave propagation and scattering with EFIT-elastodynamic finite integration technique *Wave Motion* **21** 47-66

[18] Rappel H, Yousefi-Koma A, Jamali J and Bahari A 2014 Numerical Time-Domain Modeling of Lamb Wave Propagation Using Elastodynamic Finite Integration Technique *Shock Vib.* **2014**(6) 1-6

[19] Zhu J, Popovics J S Schubert F 2004 Leaky Rayleigh and Scholte waves at the fluid–solid interface subjected to transient point loading *J. Acoust. Soc. A.* **116** (4) 2101–10

[20] Wu P, Kazys R and Stepinsky T 1995 Calculation of transient fields in immersed solids radiated by linear focusing arrays *Proc. IEEE. Ultrason. Symp.* **2** 993-7

[21] Du Y, Jensen H, Jensen J A 2013 Investigation of an angular spectrum approach for pulsed ultrasound fields *Ultrasonics* **53** 1185–91

[22] http://www.COMSOL.com

[23] Yuan M, Kang T, Zhang J, Song S J and Kim H J 2013 Numerical simulation of ultrasonic minimum reflection for residual stress evaluation in 2D case *J. Mech. Sci. Technol.* **27**(11) 3207-14

IOP Conf. Series: Materials Science and Engineering **657** (2019) 012015 doi:10.1088/1757-899X/657/1/012015

Numerical investigation of plane wave propagation in three soil models and a simple viscoelastic treatment of the boundary considering gravity using MPM

R Y Zhang[1,2] and A Haider[1]

[1]Department of Civil Engineering, Tsinghua University, Beijing, China

E-mail: zry11tsinghua@163.com

Abstract. The importance of plane wave propagation in geotechnical engineering problems considering both small and large deformation is well understood but the theory of wave propagation in the latter is maybe not mature. In this paper, the mechanism of plane wave propagation in constitutive models such as linear elastic model, Mohr Coulomb model (MC) and elastoplastic density-dependent model (DDSM) considering finite deformations is extensively studied using the material point method. The classical laws of wave propagation and reflection in a continuum are found to provide reasonable explanations to the considered problems. The stress propagation mechanism in the three models considered is mainly related to the variation of the material wave impedance. An accurate silent boundary condition is found difficult even impossible to be determined for complicated models, which is verified by a simple deduction. Moreover, a simple viscoelastic silent boundary is proposed to consider the realistic effects of gravity in geotechnical engineering problems and its performance is compared with that of the classical viscous boundary for truncating a 1D linear elastic half-space.

1. Introduction

The study of wave propagation is involved in many practical problems in geotechnics. The theory of elastic wave propagation has already been developed as early as the nineteenth century, but stress wave in many practical problems is plasitc. Though limited analytical solutions have been proposed for plastic wave propagation problems, less work has been made to verify the numerical simulations with these analytical results [1]. Meanwhile, stress wave propagation for finite deformation problems have not gotten enough attention.

When the material domain of the problem is infinite, proper silent boundary conditions are necessary to be imposed on the truncated boundary which account for unobstructed propagation of waves from the near field to the far field, and minimize the reflection of waves on the computational domain boundary. Many formulations for silent boundaries have been proposed in connection with the finite element method. Despite their advantages, silent boundaries have limited applications in the material point method (MPM) [2].

MPM uses dual descriptions of Lagrange and Euler. A continuum is discretized into finite material particles that carry all material information. The motion and deformation states of the material are characterized by these material particles, while the Eulerian background grid is used to solve momentum equations and calculate spatial derivatives. During each time step, the material particles are solidly connected to the background grid. The material information is first mapped from material particles to grid nodes, and the momentum equations are solved on these grid nodes. Then, the information is mapped back to the material particles, and the material information on the grid nodes is set to be zero. By MPM, there are no convection terms, and the grid distortion is also avoided. Hence, it is suitable for dealing with large deformation problems.

In this paper, stress wave propagation mechanism for finite deformation problems in constitutive models such as linear elastic model, Mohr Coulomb model (MC) and Elastoplastic density-dependent model (DDSM) is investigated using numerical simulations. The difficulty in proposing an accurate silent boundary has been highlighted and verified by a simple deduction. As an alternate choice, the applicability of the traditional viscous silent boundary for finite deformation problems is investigated and a simple viscoelastic silent boundary is proposed considering the realistic effects of gravity.

2. Methodology

2.1. Material point method

MPM was proposed by Sulsky *et al* [3] based on researches by Harlow and Evans [4], Dawson [5], Brackbill and Ruppel [6], which combines the advantages of the Lagrangian and Eulerian methods. For the specific control equations, numerical processes and program implementations, readers are referred to the literature cited [7].

For standard MPM method, there is numerical noise caused by material particles crossing the background grid boundary. Zhang *et al* [8] proposed a new method for calculating the shape function gradient, by weighting the shape function gradient based on material particles in the standard material point method and the shape function gradient based on the background grid nodes in the fluid implicit particle method (FLIP) [6], to reduce the stress oscillation caused by material particles crossing the background grid boundary, which is called the dual domain material point method (DDMP).

In DDMP, the modified shape function gradient is as follows:

$$\widehat{N}_{Ip,j} = \alpha(\mathbf{x}_p)N_{Ip,j} + [1-\alpha(\mathbf{x}_p)]\hat{N}_{Ip,j} \tag{1}$$

$$\hat{N}_{Ip,j} = \sum_{J=1}^{n_g} \frac{N_{Jp}}{V_J} \int_V N_J N_{I,j} \mathrm{d}V \tag{2}$$

where the subscript I denotes the background grid node, p denotes the material particle, and j denotes the direction; $N_{Ip,j}$ is the shape function gradient of standard MPM, and $\hat{N}_{Ip,j}$ is the shape function gradient based on the background grid nodes; $\alpha(\mathbf{x}_p)$ is a weight function that is zero on background grid boundaries, to ensure the continuity of the modified gradient [9].

However, there is another problem for DDMP. The background grid node internal force is calculated by the modified shape function gradient, but the background grid node external force is calculated by the original shape function of standard MPM. This inconsistency will cause errors to some extent for the calculation of the deformation and stress near the material boundary. To overcome this small discrepancy, a modified shape function especially used to calculate the background grid node external force is proposed here:

$$\widehat{N}_{Ip} = \alpha(\mathbf{x}_p)N_{Ip} + [1-\alpha(\mathbf{x}_p)]\hat{N}_{Ip} \tag{3}$$

$$\hat{N}_{Ip} = \sum_{J=1}^{n_g} \frac{N_{Jp}}{V_J} \int_V N_J N_I \mathrm{d}V \tag{4}$$

in which N_{Ip} is the shape function gradient of standard MPM, and $\hat{N}_{Ip}$ is the shape function gradient based on the background grid nodes.

For DDMP, the influence domains of a material particle and of a node are enlarged, which will induce the occurrence that a node near the material boundary will get node force from material particles but no mass. To overcome this problem, a simple but effective method is used to distribute the node force of a node with no mass to surrounding nodes with mass [8,10]. If this method is used, when information except node force is mapped from material particles to grid nodes, the original shape function of standard MPM is also effective, which means that only the calculation of grid node external force needs the modified shape function.

2.2. *Silent boundary*

The first implementation of silent boundaries in MPM was carried out by Shen and Chen [11]. They used the standard viscous boundary proposed by Lysmer and Kuhlemeyer [12] to simulate the film delamination. The viscous boundary is not only very easy to implement, but also quite accurate in many applications. Later on, several viscoelastic boundaries have also been proposed, but they are not the focus here.

The viscous damping stress on the boundary in 1-D case is as follows:

$$\sigma_n = -\rho c_p v_n \tag{5}$$

in which ρ is the material density, c_p is the velocity of the stress wave, and v_n is the particle velocity.

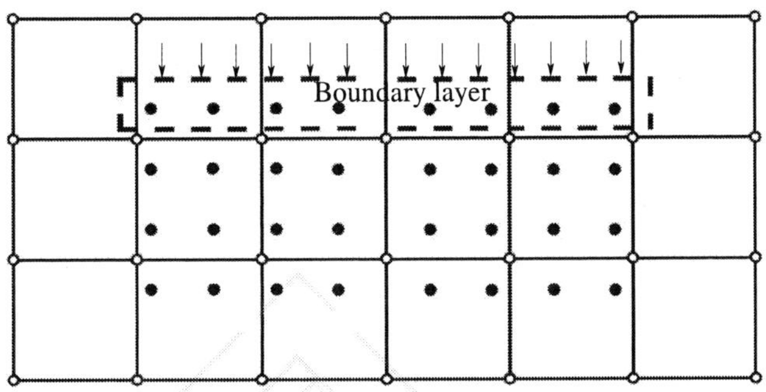

Figure 1. Application of boundary force in MPM.

In MPM, the material particles carry all information such as density, stiffness, velocity, stress, etc., so as shown in figure 1, the boundary surface force is treated as the body force of the boundary layer, seen as the second term on the right side of the following equation:

$$f_{iI}^{\text{ext}} = \sum_{p=1}^{n_p} N_{Ip} m_p b_{ip} + \sum_{p=1}^{n_p} N_{Ip} m_p \bar{T}_{ip}^s h^{-1} \tag{6}$$

in which f_{iI}^{ext} is grid node external force; m_p is the material particle mass; b_{ip} is body force; h is the thickness of the boundary layer; $\bar{T}_{ip}^s$ is the specific boundary surface force (surface force over density); $\bar{T}_{ip}^s h^{-1}$ is the equivalent body force of the boundary surface force.

When applying viscous silent boundary in MPM, the outermost layer of material particles on the boundary is regarded as the boundary layer. The damping force of the silent boundary is approximately determined by the velocity, density and stiffness of the material particles in the boundary layer, then it is converted into the body force of the hypothetical boundary layer. This body force is then mapped to the background grid nodes and processed as node external force.

Equation (5) is accurate for simple linear elastic problems. Now another 1-D problem with two kinds of linear elastic materials is considered. Due to the refection and refraction of the stress wave on the interface, the accurate stress expression here should be as follows:

$$\sigma_n = -\frac{2\rho_1 c_{p1}\rho_2 c_{p2}}{\rho_1 c_{p1} + \rho_2 c_{p2}} v_n \tag{7}$$

where the subscript 1 denotes the near field through which the stress wave passes first; 2 denotes far field through which the stress wave passes last. In this case, if we place a silent boundary just at the interface between these two materials, the accurate viscous boundary conditions should be equation (7) rather than 5.

Equation (7) indicates that an accurate silent boundary should consider the material properties on both sides of the silent boundary. In more realistic geotechnical applications, the nolinearity, plasticity and large deformation all need to be considered, then an accurate silent boundary will be quite complicated, even impossible, due to the reflection of the stress wave caused by variation of material properties. As another choice, the variation of material properties near silent boundary is assumed to be small, and to investigate the applicability of the traditional silent boundary conditions in more realistic constitutive models will be meaningful.

In some dynamic cases, gravity may also have a certain significant impact. Here, a simple viscoelastic silent boundary is proposed considering the realistic effects of gravity:

$$\sigma_n = -\rho c_p v_n - \rho g h \tag{8}$$

In the 1-D problem, in order to avoid tracking the distance from the silent boundary to the surface of soil, the above formula can be converted into:

$$\sigma_n = -\rho c_p v_n - \rho_0 g h_0 \tag{9}$$

in which ρ_0 is the initial material density; h_0 is the initial distance from the silent boundary to the surface of soil.

3. Numerical examples

The program of MPM is based on finite deformation theory, so all examples are actually finite deformation problems in this study. All numerical examples are 1-D here. Stress wave propagations in four material situations are considered. The first is in elastic material (EM). The second is in two different elastic materials, between which there is a interface (EM2M). MC model is used in the third one. An elastoplastic density-dependent model (DDSM) is used in the fourth one. In DDSM, the hardening of soil and the change of elastic modulus with variation of density can be considered, and the specific model is referred to the study of Zhang and Sun *et al* [9].

For convenience, some abbreviations are adopted. ID denotes that infinite material domain is adopted. SB denotes that the silent boundary condition shown as equation (5) is adopted. SB1 denotes that the silent boundary shown as equation (7) is adopted. SB2 denotes that the silent boundary shown as equation (9) is adopted. FB denotes that the fixed boundary is adopted. The depth of the 1-D bar are all 8 m for SB, SB1, SB2, FB, and the data in all figures are from points close to this depth or the boundaries.

3.1. Single elastic material

In order to approximate small deformation condition in MPM, the modulus is large and the load is relatively small. The elastic modulus is 5 Gpa, and the density is 2650 kg/m^3. The input load is 100 kPa.

As shown in figure 2, the effect of SB is very close to that of ID, and the effect of SB is obviously better a lot than that of FB. The result of SB is a little different from that of ID, and this is maybe caused by the adoption of the finite deformation theory, which will be further verified in section 3.2.

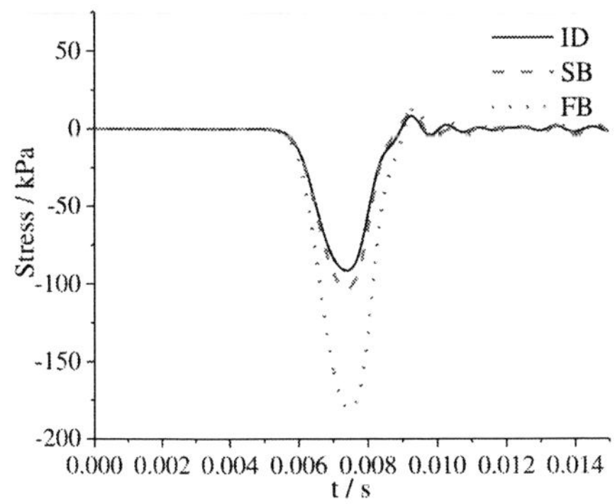

Figure 2. Stress wave propagation and application of silent boundary in EM.

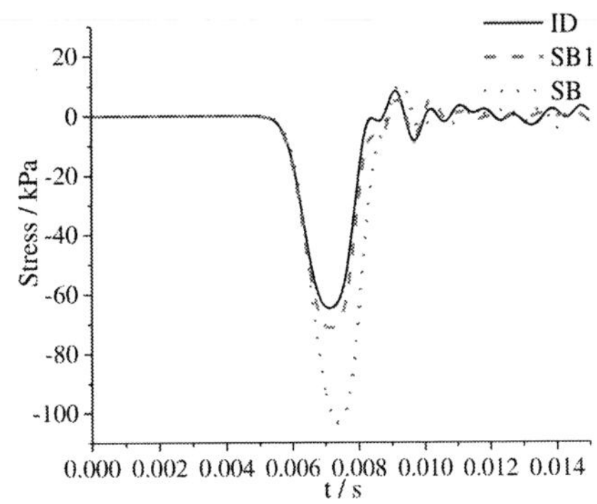

Figure 3. Stress wave progogation and application of silent boundary in EM2M.

3.2. Two different elastic materials

The second material is started from 8 m depth. The parameters for the elastic material through which the stress wave passes first are the same as section 3.1. For the second elastic material, the elastic modulus is 2.5 Gpa, and the density is 1325 kg/m^3. The input load is 100 kPa.

Calculated by equation (7), the stress on the interface should be 67 kPa in small deformation condition. As shown in figure 3, the results of SB1 and ID are close to this value, while that of SB deviate form this value a lot. It means that, if the material properties change along the stress wave propagation path, the accurate silent boundary condition will be very complicated. It is difficult even may be impossible to get a totally accurate silent boundary condition in more realistic soil models, because the material properties on both sides of the silent boundary should be considered.

According to equation (7), if $\rho_2 c_2$ of the second material is smaller than that of the first material, then the total stress on the interface will be smaller than the incident stress because the superposition of the reflection stress to the incident stress. If besides the incident stress the refection stress is absorbed by the silent boundary at the same time, then the total stress close to the silent boundary will increase. ρc is called wave impedance. So the result of SB is larger than that of ID in figure 2, and the result of SB1 is larger than that of ID in figure 3, due to that the finite deformation theory is adopted and the wave impedence in front of the stress wave is smaller than that behind the stress wave.

1-D elastic stress wave theory can be used to discuss the error of the silent boundary due to the absorption of the reflection stress wave. This error is:

$$\Delta\sigma = (k-1)(k+1)^{-1}(\sigma_1 - \sigma_0) = R_\sigma(\sigma_1 - \sigma_0) \tag{10}$$

$$k = \rho_2 c_2 (\rho_1 c_1)^{-1} \tag{11}$$

in which σ_1 is the incident stress; σ_0 is the initial stress of one point. If the variation of the material properties and the difference between the incident stress and the initial stress are all very small, then this error will be very small. This law should be the same for more complicated models. If the energy is absorbed by the plastic deformation, and the silent boundary is placed enough far from the load end, then R_σ and $\sigma_1 - \sigma_0$ will be very small, and the silent boundary will be more accurate.

3.3. MC model

The Young's modulus is 50 MPa; the density is 2650 kg/m^3; the Poisson's ratio is 0.3; the friction angle is 30°; the other parameters are zero. The input load is 100 kPa.

As shown in figure 4, the result of SB is also close to that of ID, and obviously better a lot than that of FB for MC model.

It should be noted that, the input loads for section 3.1, 3.2, 3.3 are all relatively small, so the deformations are close to small deformation, and the variation of the material properties may be not so obvious. Due to space limitations, the influence of large variation of the material properties for single elastic material and MC model is not investigated here and that will be shown directly in simulation of DDSM model.

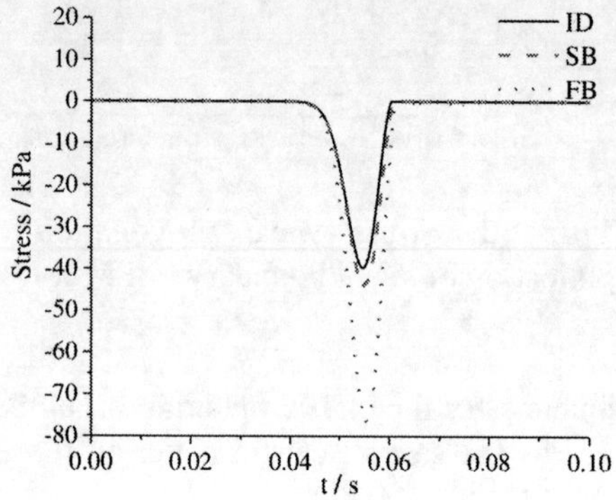

Figure 4. Stress wave propagation and application of silent boundary in MC.

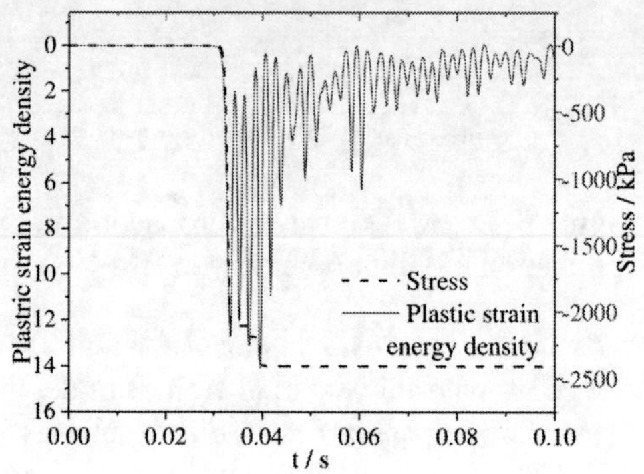

Figure 5. Stress wave propagation and plastic strain energy density in DDSM.

3.4. DDSM model

The specific model is referred to the study of Zhang and Sun *et al* [9]. The parameters are shown in table 1, in which is e_0 the void ratio at pressure 0, h_s is called grain hardness and is a parameter reflecting the grain skeleton characteristics of soil, m is called stiffness pressure correlation index related to the properties of soil, k_{ul} is the ratio of the unloading modulus to the loading modulus, υ is Poisson's ratio, c is cohesion, φ is friction angle, and ψ is dilatancy angle. Because energy is absorbed very quickly in this model, the input load is 40 MPa.

Table 1. Parameters for DDSM model.

Parameters	e_0	h_s / MPa	m	k_{ul}	υ	c / kPa	φ / °	ψ / °
Value	0.5	2600	0.7	8.0	0.3	5	35	0

As shown in figure 5, the stress propagation is quite complicated. But the peak of the stress and the peak of the plastic strain energy occur at the same time, it means that the stress after the peak will not produce a substantial impact, so we can focus on the peak of the stress only.

As shown in figure 6, though the result of SB deviate from that of ID to some extent, but the result of SB still quite better than that of FB.

3.5. Gravity

The single elastic material is adopted and the parameters are the same as that in section 3.1. As shown if figure 7, if gravity is considered, the result of SB will occur rigid motion, while SB2 proposed in this study behave well.

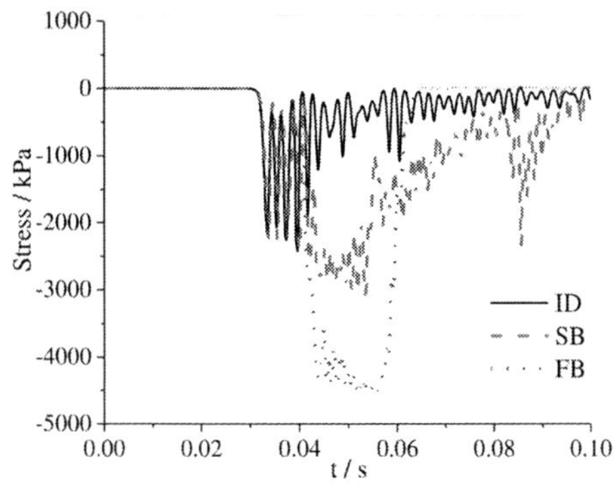

Figure 6. Stress wave propagation and application of silent boundary in DDSM.

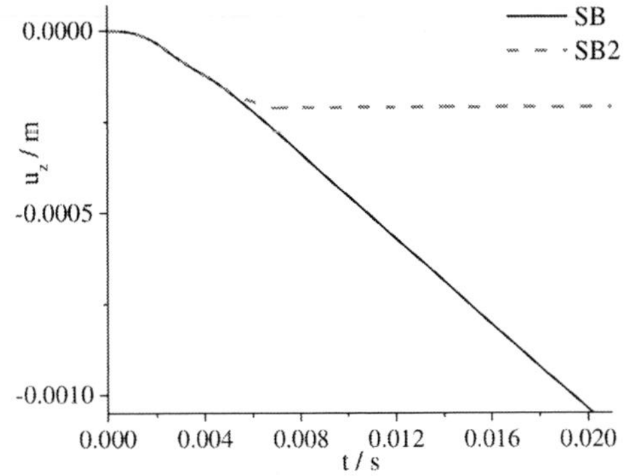

Figure 7. Application of silent boundary considering gravity.

4. Conclusions

In this paper, the mechanism of plane wave propagation in constitutive models such as linear elastic model, MC model and DDSM model considering finite deformations is extensively studied using MPM. The stress propagation mechanism can be regarded as mainly related to the variation of the material wave impedance. A totally accurate silent boundary condition is found difficult even impossible to be determined in complicated models, which has been verified by a simple deduction. The error of the traditional silent boundary in models such as MC and DDSM, due to the absorption of the refection stress caused by variation of the material properties, will be very small if the silent boundary is placed enough far form the load end (of course not too far). A simple viscoelastic silent boundary is proposed considering the realistic effects of gravity in geotechnical engineering problems and it performs well. The results presented in this work are a result of preliminary investigations only and detailed work will follow later.

Acknowledgments

This study was supported by the National Key Fundamental Research Program of China (No. 2014CB047003).

References

[1] Haider A, Song E and Li P 2019 Numerical simulation and absorbing boundary conditions for wave propagation in a semi-infinite media with a linear isotropic hardening plastic model *Soil Dyn. Earthq. Eng.* **125** 105627
[2] Bisht V and Salgado R 2018 Local transmitting boundaries for the generalized interpolation material point method *Int. J. Numer. Methods Eng.* **114** 1228-44
[3] Sulsky D, Chen Z and Schreyer H L 1994 A particle method for history-dependent materials *Comput. Meth. Appl. Mech. Eng.* **118** 179-96
[4] Harlow F H and Evans M 1955 A machine calculation method for hydrodynamic problems *LAMS-1956* (Los Alamos: Los Alamos Scientific Laboratory)
[5] Dawson J M 1983 Particle simulation of plasmas *Rev. Mod. Phys.* **55** 403
[6] Brackbill J U and Ruppel H M 1986 FLIP: A method for adaptively zoned, particle-in-cell calculations of fluid flows in two dimensions *J. Comput. Phys.* **65** 314-43
[7] Sun Y and Song E 2015 Simulation of large-displacement landslide by material point method *Chin. J. Geotech. Eng.* **37** 1218-25
[8] Zhang D Z, Ma X and Giguere P T 2011 Material point method enhanced by modified gradient of shape function *J. Comput. Phys.* **230** 6379-98
[9] Zhang R, Sun Y and Song E 2019 Simulation of dynamic compaction using material point method and analysis of its energy conversion law *Chin. J. Geotech. Eng.* **41** 1208-16
[10] Ma X, Giguere P T, Jayaraman B and Zhang D Z 2010 Distribution coefficient algorithm for

small mass nodes in material point method *J. Comput. Phys.* **229** 7819-33
[11] Shen L and Chen Z 2005 A silent boundary scheme with the material point method for dynamic analyses *CMES-Comp. Model. Eng. Sci.* **7** 305-20
[12] Lysmer J and Kuhlemeyer R L 1969 Finite dynamic model for infinite media *J. Eng. Mech. Div.* **95** 859-78

IOP Conf. Series: Materials Science and Engineering **657** (2019) 012016 doi:10.1088/1757-899X/657/1/012016

Effect of vibration frequency on settlement characteristic of reinforced embankment by geogrid with strengthened nodes using DEM simulation

C C Qiu[1,3], Y L Lin[2] and Y Ma[2], M He[1] and S J hong[1]

[1]School of Civil Engineering, Yancheng Institute of Technology, Yancheng, Jiangsu, China
[2]Department of Civil Engineering, Shanghai University, Shanghai, China

E-mail: craigqiu@163.com

Abstract. Geogrid with strengthened nodes (GSN) is a new reinforcement system that is formed by attaching blocks of a certain thickness to the joints of the longitudinal and transverse ribs of an ordinary biaxial geogrid to form strengthened nodes. A series of discrete element method (DEM) simulations were performed to investigate the dynamic response of GSN-reinforced embankment at various vibration frequencies and compare the reinforcing effect of GSN with that of the ordinary geogrid. It was concluded that the settlement ratio of the loading platen at the top of the unreinforced and reinforced embankment decreases with the increase of the vibration frequency. As the frequency increases, the difference between the settlement ratios decreases, which is especially evident in the reinforced embankment. At the same frequency, GSN performs better in restricting the settlement ratio than the ordinary geogrid, and as the frequency increases, the reinforcing effects of both the ordinary geogrid and GSN decrease. The displacement fields of the simulated embankments were analyzed. It was found that the movement of soil particles below the loading platen is restricted due to the presence of the geogrid, resulting in less settlement of the loading platen and smaller lateral deformation of the embankment slope, compared with the unreinforced embankment. As for the GSN-reinforced embankment, the strengthened nodes provide resistance to soil particles, leading to a more effective reinforcing effect, so that the displacement vectors are even smaller.

1. Introduction

With the development of the earth-reinforcing technique, the reinforcement itself has evolved from metallic materials to polymer materials (i.e. the reinforcement used in the retaining wall develops from steel strips-to-steel mesh-to-geosynthetics [1]). Geosynthetics (such as geotextiles, geogrids, geocells, etc) are widely used in the construction of the roadbed, retaining walls, abutments, and other projects because of their low cost and good reinforcing effect. A large number of laboratory tests and numerical simulations using the finite element method (FEM) have been directed to identify the mechanical behaviors of the geosynthetics-soil system [2-7]. However, the continuum-based FEM simulation does not take into consideration particle rolling and sliding on the geosynthetics-reinforced structure. The discrete element method (DEM) has a distinct advantage in capturing the kinematic behavior of discontinuous media at a microscopic scale [8]. Considerable research on the geosynthetics- reinforced soil behavior has been performed based on DEM simulations [9-13].

In addition to the research on the conventional reinforcing materials, the researchers are actively exploring the reinforcing effect of new materials and innovative reinforcement layout. Zhang et al. [14] proposed a new concept of soil reinforced with three-dimensional (3D) reinforcing elements, from which horizontal-vertical reinforcements and denti-strip reinforcements were derived [9-10,15]. Mosallanezhad *et al* [16,17] introduced a new reinforcement system named "Grid-Anchor", which includes geogrids with anchors attached to them, and found that the anchors can effectively improve the pullout resistance of the geogrid. Horpibulsukt *et al* [18,19] studied the mechanism of a new reinforcement in the retaining wall, which is named "bearing reinforcement", consisting of longitudinal reinforced bars and transverse barrier steel plates. Li *et al* [20] carried out model tests on the reinforced embankment of waste tires and found that the position of the first layer reinforcement had a great influence on the failure mode.

Geogrid with strengthened nodes (GSN) is a new reinforcement system by attaching blocks of a certain thickness to the joints of the longitudinal and transverse ribs of an ordinary biaxial geogrid to form strengthened nodes, which is also derived from the 3D reinforcement proposed by Zhang *et al* [14]. In this paper, a series of DEM simulations were performed to investigate the dynamic response of the GSN-reinforced embankment at various vibration frequencies and the displacement field was analyzed to explain the GSN-soil interaction at a microscopic level.

2. DEM modelling

2.1. Modelling procedure

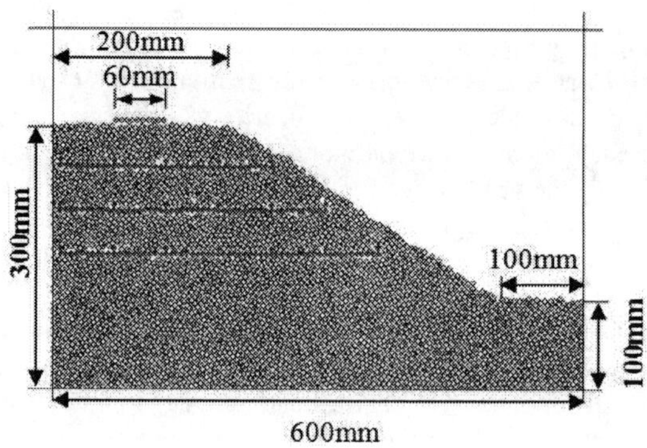

Figure 1. PFC^{2D} model of GSN-reinforced embankment.

The numerical model is established using the DEM software Particle Flow Code (PFC^{2D}), based on a model test. The model test was conducted based on a secondary road by reducing the ratio of the geometric dimensions to 1:20. As the shape of the embankment was symmetric, only half of the embankment is simulated. PFC^{2D} software provides two basic entities: balls and walls. To build the PFC^{2D} model, firstly, four walls were generated to simulate the boundaries of the model box and a large number of ball particles were randomly generated within the boundaries according to the grading curve of soil particles. It is noted that the grading curve of soil particles in the model test has been modified and up-scaled by 3 times with the proved technique of "up-scaling", which can reduce the number of particles to save the computational time [8,10,11]. Then, according to the thickness of the foundation, the angle of the slope and the height of the embankment, the excess particles were deleted to form the sloped embankment model. The calculation cycle stopped when the ratio of maximum unbalanced force to maximum contact force was 1×10^{-6}, indicating the standard of the equilibrium state was reached. After this, the ball particles at the place where the reinforcement should be placed were removed to set aside the position for the formation of the reinforcement. The reinforcement was modeled by some ball particles by assigning the parallel bond between particles to transmit both force and moment. The simulation model also was cycled until the ratio of maximum unbalanced force to maximum contact force reached 1×10^{-6}. Finally, a loading plate was generated at the center of the top of the embankment using the clump function in PFC^{2D}. A clump is a group of slaved particles behaving as a rigid body and will not break apart [21]. The final PFC^{2D} model is shown in figure 1. By controlling the force change on the clumped balls, the loading process can be successfully simulated. In this study, simplifying the traffic load into a half sinusoidal wave, dynamic loading was carried out by applying a half sinusoidal force to each ball in the loading plate at different frequencies (1, 5, 10 Hz). The amplitude value of the dynamic loading was 30 kPa.

2.2. *Reinforcement simulation*

The real reinforcement is a kind of biaxial polyester welded geogrid with an aperture size of 50 mm x 50 mm. Blocks of 6 mm × 6 mm × 6 mm were fixed at the joints of the longitudinal and transverse ribs of the geogrid to form a geogrid with strengthened nodes. As the real biaxial geogrid is a network structure, it should be simplified into a 2D model in PFC^{2D}. Because of the existence of the meshes, the joints of the longitudinal and transverse members of the geogrid are firmly designed. Since the 2D modelling cannot fully reflect the influence of the meshes, the characteristics of the solid joints should be grasped. The geogrid longitudinal member was simulated with one row of bonded ball particles and the joints were simulated by setting some balls into rigid bodies using the clump function. As for GSN, the strengthened nodes were also set into clumped balls according to the thickness of the nodes, as shown in figure 2.

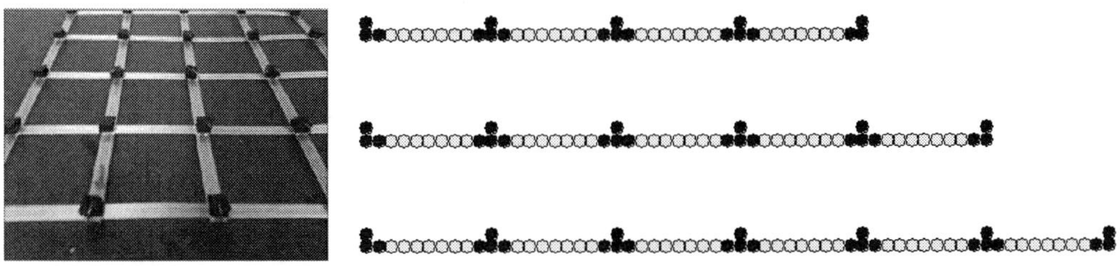

(a) (b)

Figure 2. Geogrid with strengthened nodes: (a) real material, (b) simulation model

2.3. *Input parameters*

The input parameters of the model have a great influence on the accuracy of the simulation results in PFC^{2D}. In this study, the Hertz-Mindlin model provided by PFC^{2D} was used for the contact model of ball particles, as this model can better reflect the non-linear nature of soil under dynamic loading [22]. The parallel-bond model was also assigned to the ball particles of the simulated reinforcement to resist both forces and moments. Through repeated trial calculation and referring to the microscopic parameters of the horizontal-vertical reinforcement [10], the final input parameters of the soil particles, reinforcement and loading plate used are shown in table 1.

Table 1. Input parameters for PFC^{2D} simulation.

Parameters	Soil	Reinforcement	Loading platen
Density ρ(kg/m^3)	1575	1000	2600
Particle radius R (mm)	2~3	3	3
Shear modulus G(Pa)	1×10^{10}	1×10^{10}	1×10^{10}
Poisson's ratio ν	0.3	0.2	—
Friction coefficient f	10	20	25
Parallel-bond radius multiplier *pb_radius*	1	1	—
Parallel-bond shear strength *pb_sstrength* (Pa)	1×10^5	1×10^{10}	—
Parallel-bond normal strength *pb_nstrength* (Pa)	1×10^5	1×10^{10}	—
Parallel-bond shear stiffness *pb_ks* (Pa/m)	5×10^6	1×10^8	—
Parallel-bond normal stiffness *pb_kn* (Pa/m)	5×10^6	1×10^8	—

3. Simulation results

3.1. *Effect of frequencies on the settlement*

Figure 3 shows the settlement ratio *S*/*B* (ratio of settlement *S* to loading platen width *B*) at various frequencies for different cases versus the number of load cycles and table 2 summarizes the cumulative settlement ratio after 1000 cycles for different cases. For both unreinforced and reinforced cases, the settlement ratio developed rapidly in the early stage of vibration at each frequency, but with the increase of cycle number, the settlement ratio developed slowly and tended to be stable. For the same case, the settlement ratio of the small frequency (f = 1 Hz) was more significant, and with the increase of the frequency, the settlement ratio decreased. As the frequency increased (f = 5, 10 Hz), the difference between the settlement ratios became smaller, especially for the reinforced cases, as shown in figure 3(a)-3(c). This may be due to that when the frequency is small, the duration of the vibration period is long, and the soil deformation is sufficient in this period, which leads to a lower modulus of resilience and a larger cumulative deformation under the same cycle number.

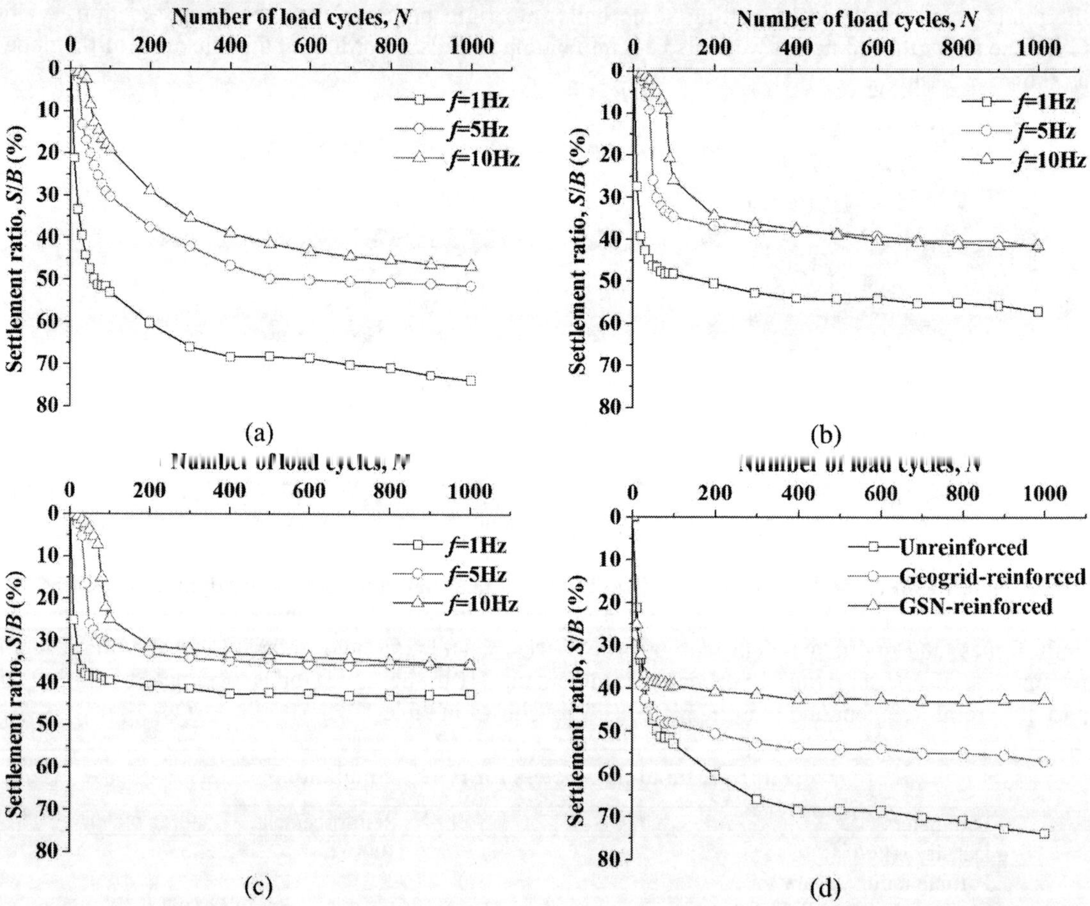

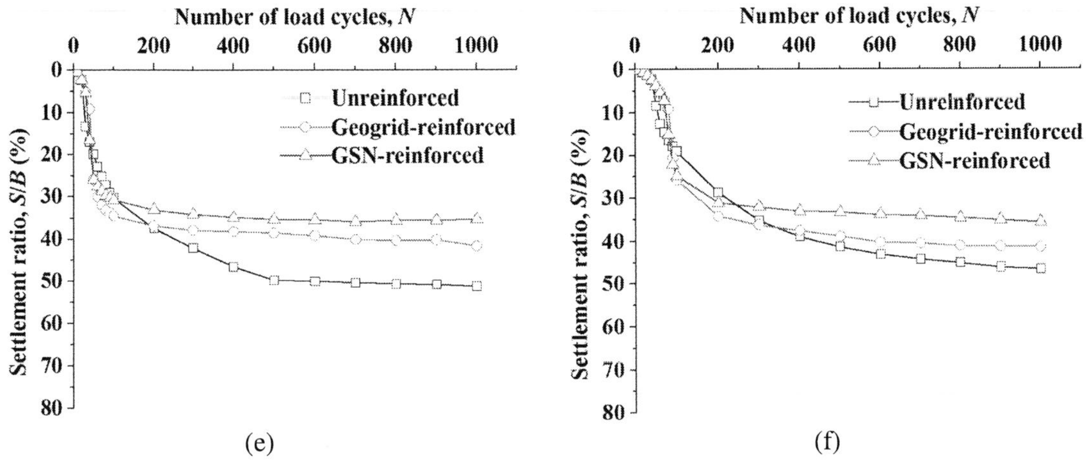

Figure 3. Settlement ratio versus number of load cycles for different cases of (a) unreinforced embankment, embankments reinforced with (b) ordinary geogrid and (c) GSN at various frequencies of (d) 1Hz, (e) 5Hz and (f) 10Hz.

As shown in figure 3(d)-3(f), at the same vibration frequency, the settlement ratio at the top of the embankment was effectively reduced by the geogrid reinforcement, and the effect of GSN was better than that of the ordinary geogrid in restricting settlement. For example, at the vibration frequency of f = 1 Hz, the settlement ratio of the unreinforced embankment was 73.75%, while that of the geogrid-reinforced embankment was 57%, 22.71% less than that of the unreinforced embankment. For GSN-reinforced embankment, the settlement ratio was 42.57%, 42.28% less than that of the unreinforced embankment and 25.32% less than that of the geogrid-reinforced embankment. However, as can be seen from table 2, the reinforcing effects of both geogrid reinforcement and GSN reinforcement decreased with the increase in vibration frequency. Besides, it can be observed that the settlement ratios of the unreinforced, geogrid-, and GSN-reinforced embankment are interlaced from 0 to 200 cycles, at the frequency of 5 and 10 Hz. On the one hand, this may be because when the frequency is large, the vibration time of each cycle is short, and the deformation of the soil is insufficient, which is not enough to mobilize the reinforcing effect of the reinforcement. On the other hand, it may be because the soil in the early stage of vibration is in the stage of vibration compaction, the settlement development is relatively fast, the reinforcement has not yet played a role, and the settlement ratios of the unreinforced and reinforced cases are not much different.

Table 2. Settlement ratios for different cases.

Frequency/Hz	Settlement ratio/%			α^a/%	β^b/%	γ^c/%
	Unreinforced	Geogrid	GSN			
1	73.75	57.00	42.57	22.71	42.28	25.32
5	51.30	42.07	35.37	18.00	31.06	15.93
10	46.62	41.78	35.32	10.37	24.24	15.48

[a]α refers to the reduction rate of the settlement ratio of geogrid-reinforced embankment compared with that of unreinforced embankment.
[b]β refers to the reduction rate of the settlement ratio of GSN-reinforced embankment compared with that of unreinforced embankment.
[c]γ refers to the reduction rate of the settlement ratio of GSN-reinforced embankment compared with that of geogrid-reinforced embankment.

3.2. Displacement field

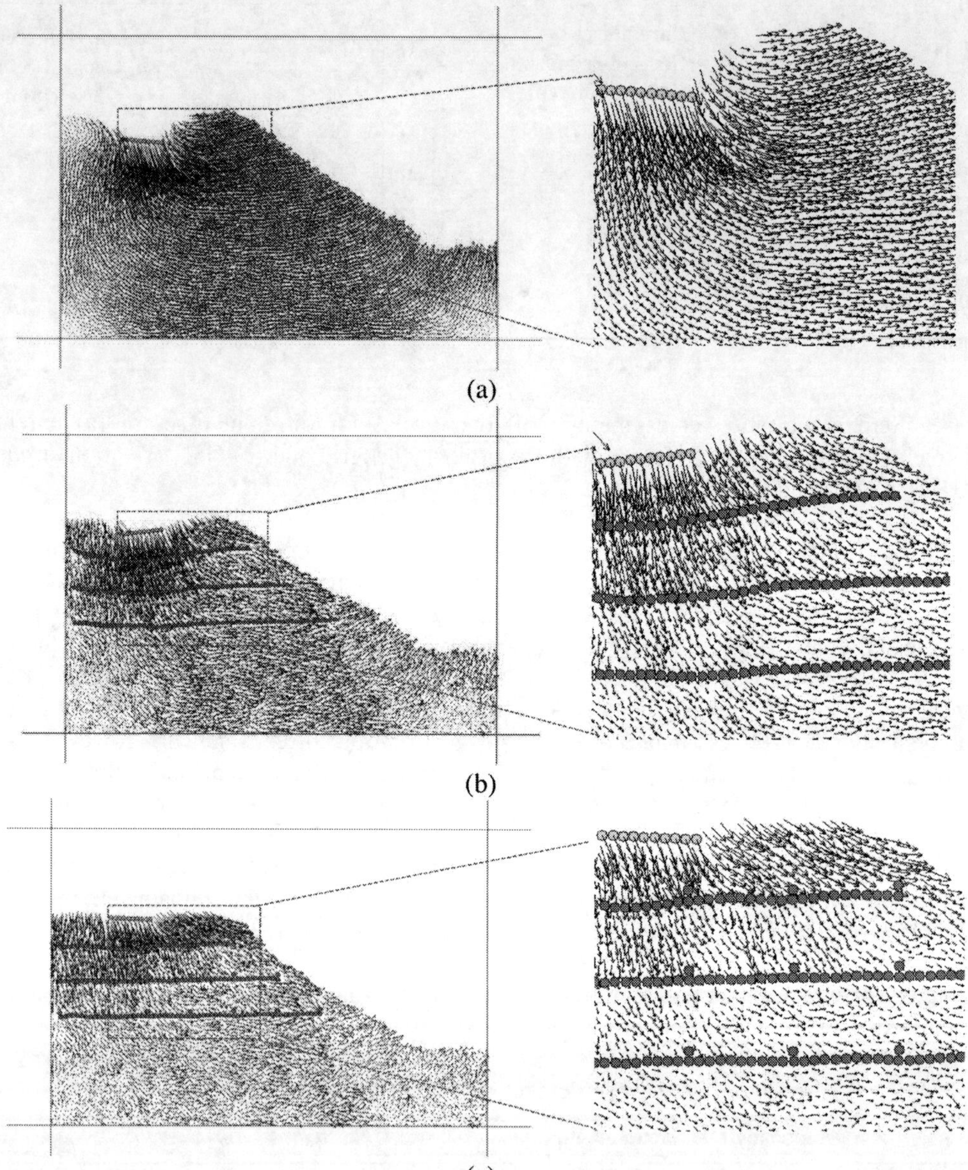

(a)

(b)

(c)

Figure 4. Displacement fields of the embankments after 1000 cycles at the frequency of 1Hz: (a) unreinforced; (b) ordinary geogrid; (c) GSN.

In PFC^{2D}, the displacement field can be represented by the displacement vector of ball particles. The displacement vector is a line from the center of mass to the end of the arrow. The direction of the arrow indicates the direction in which the particle moves at a certain time step. The length of the line is proportional to the displacement magnitude. Figure 4 shows the displacement fields after 1000 cycles for three different cases at the frequency of 1 Hz. It can be seen that the displacement vectors inside the unreinforced embankment were very long, indicating large displacements. The settlement of the loading plate at the top of the unreinforced embankment was large, and the lateral deformation of the slope was obvious, especially the severe uplift in the slope crest. In the geogrid-reinforced embankment, the displacement vectors were obviously smaller than that of the unreinforced embankment. The displacement of the particles right below the loading plate was larger, and the geogrid was bent under the push of the particles. For the presence of the reinforcement, the downward movement of the particles was restricted, resulting in a large reduction in the settlement of the loading platen and a small displacement within the embankment. As for the embankment reinforced with GSN, due to the strengthened nodes, the soil beneath the loading plate was further strengthened, the displacement of particles was reduced, and the degree of bending of the geogrid becomes smaller.

4. Conclusions

The settlement ratio develops rapidly in the initial stage of vibration, either in the unreinforced or reinforced embankment. With the increase of the vibration number, the settlement ratio develops slowly and tends to be stable. At a low frequency, the settlement ratio of the embankment is relatively large, and as the frequency increases, the settlement ratio decreases and the difference between the settlement ratios is getting smaller, which is particularly evident in the reinforced embankments. At the same frequency, the effect of GSN on restricting the settlement ratio is better than that of the ordinary geogrid, but as the frequency increases, the reinforcing effects of both GSN and ordinary geogrid are reduced. Through the displacement field analysis, it is found that the geogrid restricts the movement of the particles under the loading platen, resulting in smaller deformation than that of the unreinforced embankment. For the GSN-reinforced embankment, the strengthened nodes provide resistance to the soil particles, making the reinforcing effect more efficient.

Acknowledgment

The financial assistance from the Natural Science Foundation of Jiangsu Province (Grants No BK20170477 and BK20160437) and the National Natural Science Foundation of China (Grants No 51808481) is herein gratefully acknowledged.

References

[1] Koerner R M and Koerner G R 2018 An extended data base and recommendations regarding 320 failed geosynthetic reinforced mechanically stabilized earth (MSE) walls *Geotext. Geomembr.* **46** 904-12
[2] Rowe R K and Taechakumthorn C 2011 Design of reinforced embankments on soft clay deposits considering the viscosity of both foundation and reinforcement *Geotext. Geomembr.* **29** 448-61
[3] Leshchinsky B and Ling H 2012 Effects of geocell confinement on strength and deformation behavior of gravel *J. Geotech. Geoenviron. Eng.* **139** 340-52
[4] Turker E, Sadoglu E, Cure E and Uzuner B A 2014 Bearing capacity of eccentrically loaded strip footings close to geotextile-reinforced sand slope *Can. Geotech. J.* **51** 884-95
[5] Abdesssemed M, Kenai S and Bali A 2015 Experimental and numerical analysis of the behavior of an airport pavement reinforced by geogrids *Constr. Build. Mater.* **94** 547-54
[6] Mehrjardi G T, Ghanbari A and Mehdizadeh H 2016 Experimental study on the behaviour of geogrid-reinforced slopes with respect to aggregate size *Geotext. Geomembr.* **44** 862-71
[7] Song F, Liu H, Ma L and Hu H 2018 Numerical analysis of geocell-reinforced retaining wall failure modes *Geotext. Geomembr.* **46** 284-96
[8] Wang Z, Jacobs F and Ziegler M 2014 Visualization of load transfer behaviourbetween geogrid and sand using PFC^{2D} *Geotext. Geomembr.* **42** 83-90
[9] Lin Y L, Zhang M X, Javadi A A, Lu Y and Zhang S L 2013 Experimental and dem simulation of sandy soil reinforced with H-V inclusions in plane strain tests *Geosynth. Int.* **20** 162-73
[10] Zhang M X, Qiu C C, Javadi A A, Lu Y and Zhang S L 2013 Discrete-element method simulation of a model test of an embankment reinforced with horizontal–vertical inclusions *Geosynth. Int.* **20** 238-51
[11] Wang Z, Jacobs F and Ziegler M 2016 Experimental and DEM investigation of geogrid-soil

interaction under pullout loads *Geotext. Geomembr.* **44** 230-46

[12] Lai H J, Zheng J J, Zhang J, Zhang R J and Cui L 2014 DEM analysis of "soil"-arching within geogrid-reinforced and unreinforced pile-supported embankments *Comput. Geotech.* **61** 13-23

[13] Miao C X, Zheng J J, Zhang R J and Cui L 2017 DEM modeling of pullout behavior of geogrid reinforced ballast: the effect of particle shape *Comput. Geotech.* **81** 249-61

[14] Zhang M X, Javadi A A and Min X 2006 Triaxial tests of sand reinforced with 3D inclusions *Geotext. Geomembr.* **24** 201-9

[15] Zhang M X, Qiu C C, Javadi A A and Zhang S L 2014 Model tests on reinforced sloped embankment with denti-strip inclusions under monotonic loading *KSCE J. Civ. Eng.* **18** 1342-50

[16] Mosallanezhad M, Hataf N and Ghahramani A 2008 Experimental study of bearing capacity of granular soils, reinforced with innovative grid-anchor system *Geotech. Geol. Eng.* **26** 299-312

[17] Mosallanezhad M, Taghavi S H S, Hataf N and Alfaro M C 2016 Experimental and numerical studies of the performance of the new reinforcement system under pull-out conditions *Geotext. Geomembr.* **44** 70-80

[18] Horpibulsuk S, Suksiripattanapong C and Chinkulkijniwat A 2013 Design Method for Bearing Reinforcement Earth Wall *Geotech. Eng. J. SEAGS AGSSEA* **44** 125-31

[19] Horpibulsuk S, Udomchai A, Joongklang A, Chinkulkijniwat A, Mavong N and Suddeepong A and Arulrajah A 2016 Pullout resistance mechanism of bearing reinforcement embedded in residual clayey soils *Geosynth. Int.* **24** 255-63

[20] Li L H, Xiao H L, Ferreira P and Cui X L 2016 Study of a small scale tyre-reinforced embankment *Geotext. Geomembr.* **44** 201-8

[21] Itasca Consulting Group Inc 2004 PFC^{2D} (Particle Flow Code in 2 Dimensions), Version 3.1, ICG, Minneapolis, MN, USA

[22] Liu H L, Gao D Q, Chen Y M and Li G Y 2009 Simulation of dynamic modulus and damping ratio test on filter material by particle flow code *J. Disaster Prevention and Mitigation Engineering* **29** 361-7 (in Chinese)

Simulation analysis of liquid sloshing under different working conditions of hazardous materials rescue truck

S W Zhou[1] and Z H Liu[1,2]

[1]College of Mechanical Engineering and Automation, Northeastern University, Shenyang Liaoning 110819, China

E-mail: zhenhongneu@163.com

Abstract. The geometric model is built with SOLIDWORKS, and the VOF(Volume Of Fluid) method in the CFD(Computational Fluid Dynamics) software fluent is used to simulate the sloshing characteristics during the brake and steering of the Hazardous Materials Rescue Truck. The braking acceleration, the liquid filling ratio of the vehicle, lateral acceleration, the effect of the baffle during braking and steering are respectively studied. The results show that the steering and braking of Hazardous Materials Rescue Trucks will cause the liquid in the tank to slosh, and the liquid sloshing will cause the wall slosh force of the tank to change. The driving stability of Hazardous Materials Rescue Trucks can be effectively improved by the reduction of braking force, proper liquid-filling ratio, reduction of lateral acceleration, and increase of baffles.

1. Introduction

Hazardous Materials Rescue Truck is a three-in-one functional integrated body that can collect and transport emergency materials in the form of three kinds of substances: toxic, harmful, flammable, explosive and corrosive liquid, and solid liquid mixtures. Due to the sloshing of liquid vehicle go in the non-full load tank, the dynamic load is changed by steering and braking will affect the driving stability of the tanker [1]. At present, methods for studying liquid sloshing in tanks at home and abroad include quasi-static analysis [2], simulation method [3], equivalent mechanical model [4] and test [5]. Literature [6] studied the effects of different baffles and different filling ratios on the driving stability of tankers during braking. In order to improve the driving stability of Hazardous Materials Rescue Trucks, it is important to study the influence of liquid sloshing on stability during steering and braking.Braking force distribution of front and rear axles.

2. Gathering tank parameters and finite element models

2.1. Gathering tank parameters

As shown in figure 1, the two tanks are placed in the longitudinal direction are filter tanks, and one tank is placed in the lateral direction are collection tanks. In this paper, the liquid sloshing in the collection tanks is studied. The collection tank has a nominal diameter of 1800 mm and the total length of the tank is 4000 mm. The two ends are standard elliptical heads, and the area of the baffle is 1.27 m^2.

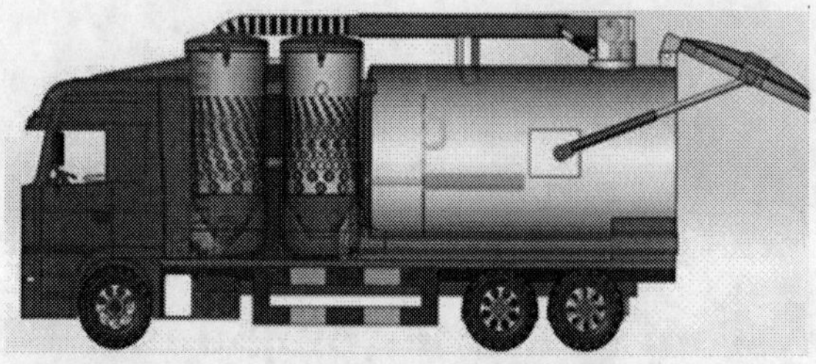

Figure 1. The hazardous materials rescue truck.

2.2. *Finite element models*

In the establishment of the finite element model, in order to simplify the calculation of the model, it is necessary to simplify some of the components of the model, ignoring the role of non-bearing small components. Based on the above principles, a tank model is made up of two ellipse head, a cylinder and baffles. Since the thickness of the baffle is small relative to the length of the entire tank, the effect of thickness is ignored during modeling this paper does not consider the influence of the reinforcement rib and the wall thickness of the tank.

As shown in figure 2, The nominal diameter of the tank is 1800 mm, the length of the cylinder is 4000 mm, and the volume of the tank body is 10.04 m^3. According to the nominal diameter of the tank, the relevant parameters of the selected elliptical head are calculated: the curved surface height is 450 mm, and the straight edge height is 40 mm. The X-axis positive half-axis is set to the vehicle's forward direction, the Z-axis direction is set to the vehicle's lateral acceleration direction, and the Y-axis is set to perpendicular to the ground. The acceleration of the fluid is applied by defining the momentum source term, and the definition of the momentum source term can be setting by UDF.

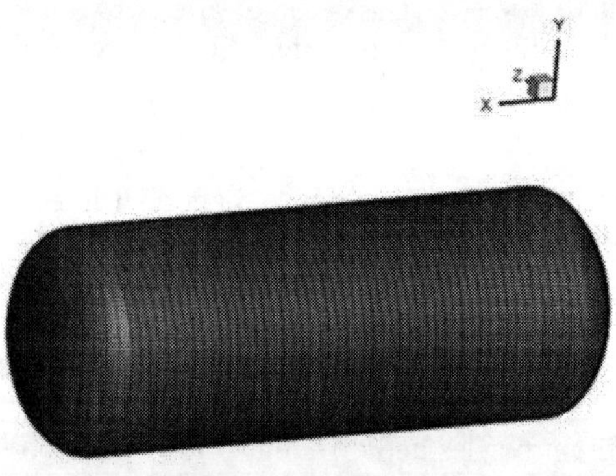

Figure 2. Meshing.

The two phases in the tank are water and air, and the density of water is 998.2 kg/m^3 and dynamic viscosity is 1.003×10^{-3} kg/(m·s). In this paper, the VOF method is used to solve the instantaneous liquid level during liquid sloshing. Changes are accurately captured by VOF method in gas-liquid two-phase interface, the continuity equation and the momentum conservation equation is considered.

Continuity equation:

$$\frac{\partial \rho}{\partial t}+\frac{\partial}{\partial x_i}(\rho u_i)=0 \tag{1}$$

Momentum conservation equation:

$$\frac{\partial \rho_i}{\partial t}+\frac{\partial}{\partial x_j}(\rho u_i u_j)=-\frac{\partial p}{\partial x_i}+\frac{\partial}{\partial x_j}\left[\mu\left(\frac{\partial u_i}{\partial x_j}+\frac{\partial u_j}{\partial x_i}\right)\right] \tag{2}$$

where u and p are velocity and pressure, respectively. ρ and μ are the average density and dynamic viscosity based on the volume friction, respectively.

$$\mu = \mu_w + (1-\alpha_w)\mu_a \tag{3}$$

$$\rho = \rho_w + (1-\alpha_w)\rho_a \tag{4}$$

where α_w is the liquid volume fraction, ρ_w and ρ_a are the liquid and gas densities, respectively, and μ_w and μ_a are the liquid and gas viscosities, respectively.

3. Calculation results and analysis

3.1. Simulation of vehicle steering

In this paper, the liquid-filling ratio is defined as K=h/2R, h is set to the height of liquid level, R is set to the tank radius, assuming a liquid-filling ratio of 0.5 and a lateral acceleration of 0.3 g. At the beginning, there is no sloshing on the liquid and the wall surface, as shown in figure 3, at the initial time the tank is divided into two part of the same area, the upper part of the interface is air, the lower part is water.

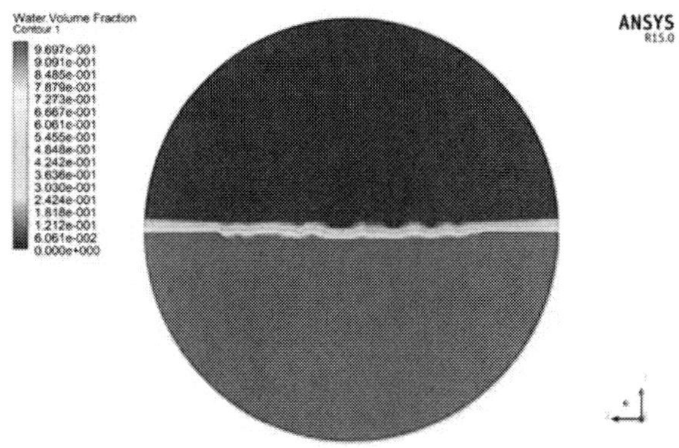

Figure 3. Distributions of gas-liquid phase.

3.1.1. Effect of different liquid filling ratio on the force in direction z. The liquid fill level in the tank truck cannot be too high, so it is necessary to reserve certain space. Especially when the dangerous chemicals storage vehicle is studied in this paper transports dangerous chemicals, the liquid is expanded and the temperature is raised by too much solution sloshing, so the liquid fill level needs to be strictly controlled. The tank truck transport volume is usually above half load, so the fill-to-liquid ratio is set to 0.5, 0.6, 0.7, 0.8, 0.9, and lateral acceleration is set to 0.3 g. In figure 4, the abscissa represents simulation time, the ordinate F_z represents forces in direction z on tank wall.

As shown in figure 4, when the liquid filling ratio is set to 0.5 to 0.6, the crest value of forces in direction z becomes larger as the liquid filling ratio increases, and the magnitude of the wall slosh force is mainly affected by the amplitude of the shaking. As shown in table 1， with the increase of liquid-filling ratio, the inertial force gradually increases, and the liquid in the tank will surge forward at a faster speed, and the arrival time of the crest gradually decreases. When the liquid filling ratio is increased from 0.7 to 0.9, the crest value decreases as the liquid filling ratio increases. Because as the liquid-filling ratio increases, the mass ratio of the liquid to the tank gradually increases, and the inertial force plays a leading role. At this time, the liquid sloshing amplitude also decreases. When the liquid filling ratio reaches 0.9, the solution in the tank is close to full load, and the peak value is much lower than the filling ratio of 0.8.

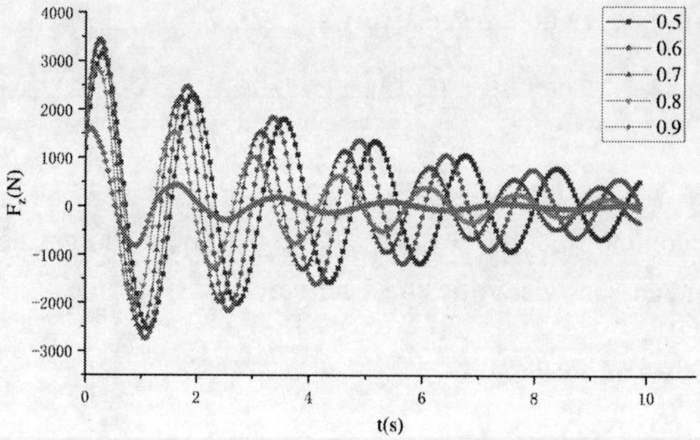

Figure 4. Forces in directions z during steering.

Table 1. First crest value of forces in direction z

Liquid filling ratio	First crest(kN)	First crest time (s)
0.5	3.148	0.42
0.6	3.441	0.38
0.7	3.420	0.36
0.8	2.846	0.32
0.9	1.605	0.22

3.1.2. Influence of lateral acceleration on the force in direction z. The rollover threshold of heavy-duty trucks is 0.4 g~0.6 g [7], and the estimated value is high due to the neglect of the suspension and tire elasticity. Therefore, assuming the upper limit of the study on lateral acceleration is 0.4 g. The lateral accelerations are set to 0.1 g, 0.2 g, 0.3 g and 0.4 g respectively, and the liquid filling ratio was set to 0.5. In figure 5, the abscissa represents simulation time, the ordinate F_z represents forces in direction z on tank wall. The figure 5 show that with the increase of lateral acceleration, the liquid slosh force on the force in direction z become larger and the liquid sloshing is intensified.

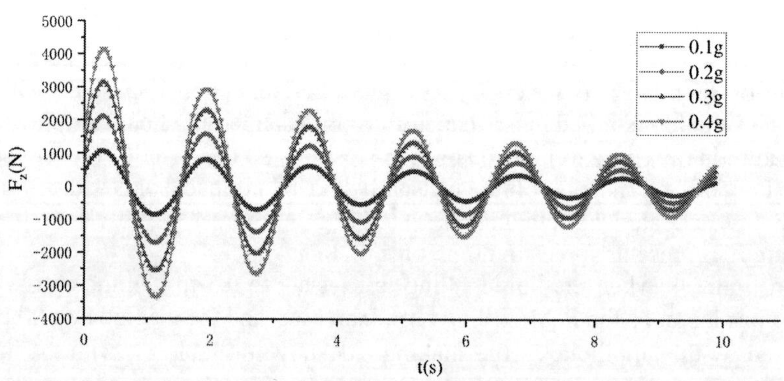

Figure 5. Forces in directions z during steering.

3.2. Simulation of vehicle braking

This paper is assumed that the tank truck is driven in the positive direction of x at an initial speed of 80km/h, and apply braking deceleration for 2 s, acceleration is taken as 6 m/s^2. In figures 6 and 7, the abscissa represents simulation time, the ordinate F_x represents forces in direction x on front head of tank.

After 2 s, the liquid in the tank will oscillate freely under the gravity.

3.2.1. Effect of different braking deceleration on the force in direction x. In this paper, the braking deceleration is set to 5.6 m/s^2 to 7.2 m/s^2, respectively, with an interval of 0.4 m/s^2. To take the braking deceleration 5.6 m/s^2 as an example, as shown in figure 6, F_x increases rapidly with the change of time, reaching the maximum value at about 0.8 s. Because when the vehicle starts braking, due to the sudden effect of braking deceleration, the liquid in the tank rapidly oscillates towards the front wall of the tank body under the action of inertial force and impacts the front wall, so that the tank receives a sharply rising F_x.

When the F_x reaches the maximum value, F_x gradually decreases with the change of time. Because when the all liquid in the tank moves to the front side of the tank under the action of inertial force, due to the reaction force of the front wall of the tank on the liquid, part of the liquid moves backward, causing the F_x to gradually decrease. At 1.28 s, F_x reaches a trough. From 1.28 s to 2 s, F_x gradually increases with the change of time. Because after the liquid moves to the rear side of the tank , the liquid continues to move forward due to the reaction force of the rear wall of the tank body on the liquid. At 2 s, the force on the front head will drop abruptly, because of the sudden disappearance of the brake deceleration, which causes the inertial force of the liquid applied to the tank to suddenly disappear. Comparing with other braking deceleration in figure 6, as the brake deceleration increases, the crest value of the force at the front head is increased. Therefore, it is recommended that when driving the tank truck, the pedal should be slowly pressed to ensure a small brake deceleration when the vehicle brakes.

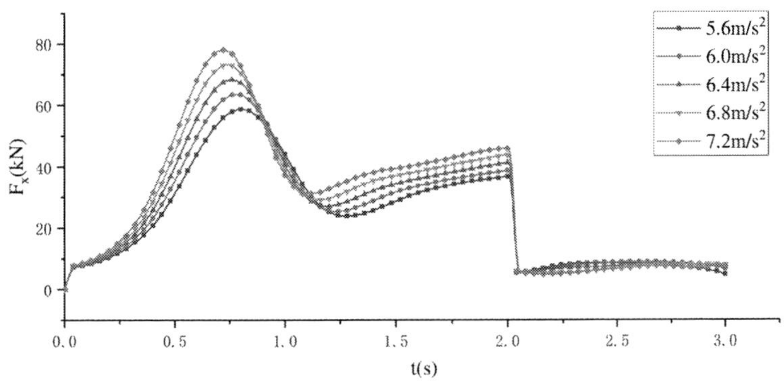

Figure 6. Forces in directions x during braking.

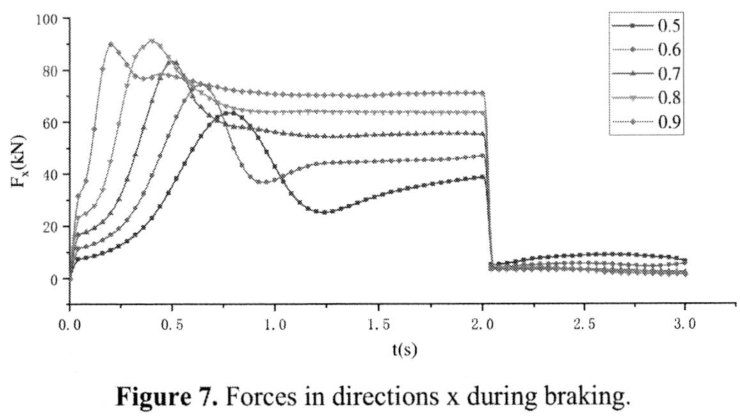

Figure 7. Forces in directions x during braking.

3.2.2. Effect of different liquid filling ratio on the force in direction x. Assuming the vehicle is driven at a constant speed at the initial moment. The liquid ratio is set to 0.5, 0.6, 0.7, 0.8, 0.9 respectively, and the brake deceleration is set to 6 m/s^2. As shown in figure 7, when the liquid filling ratio is set to 0.5 to

0.8, the crest value of the force in direction x increases as the liquid filling ratio increases. When the liquid filling ratio is set to 0.9, the liquid in the tank is close to full load, and the liquid sloshing amplitude is extremely small, so the crest value is lower than the filling ratio of 0.8.

3.3. The effect of the number of baffles on the force in direction x

Since the lateral baffle studied in this paper has little effect on the suppression of the liquid slosh force during steering, only the influence of the change in the number of baffles on the braking is considered. Assuming that the vehicle is driven at the initial speed of 80 km/h, the braking deceleration is set to 6 m/s^2, the number of baffles is 1,3,5,7 and the filling ratio is 0.5 and 0.6 respectively.

As shown in the figure 8, when the number of baffle is one, it is located in the center of the tank. In figure 9, the abscissa represents simulation time, the ordinate Fx represents forces in direction x on front head of tank The results in figure 9 show that after adding the baffles in the tank, the crest value of the force in direction x at the front head is obviously decreased, and the fluctuation range is more gentle than that without the baffles, which indicates that the addition of the baffles to the liquid in the tank sloshing plays a great role in suppressing. The crest value of slosh force at the front head significantly decreases as the number of baffles increases.

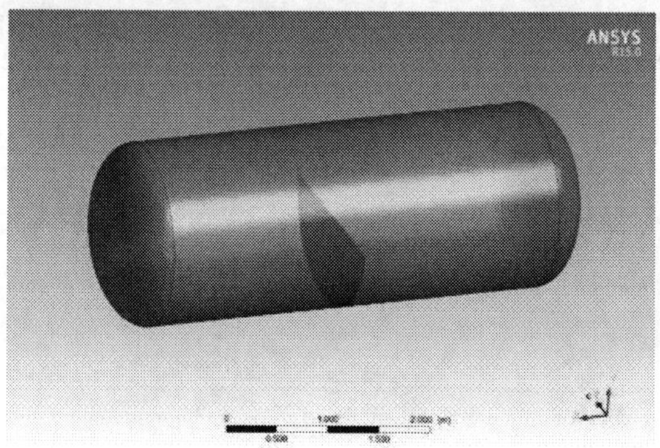

Figure 8. Baffle distribution.

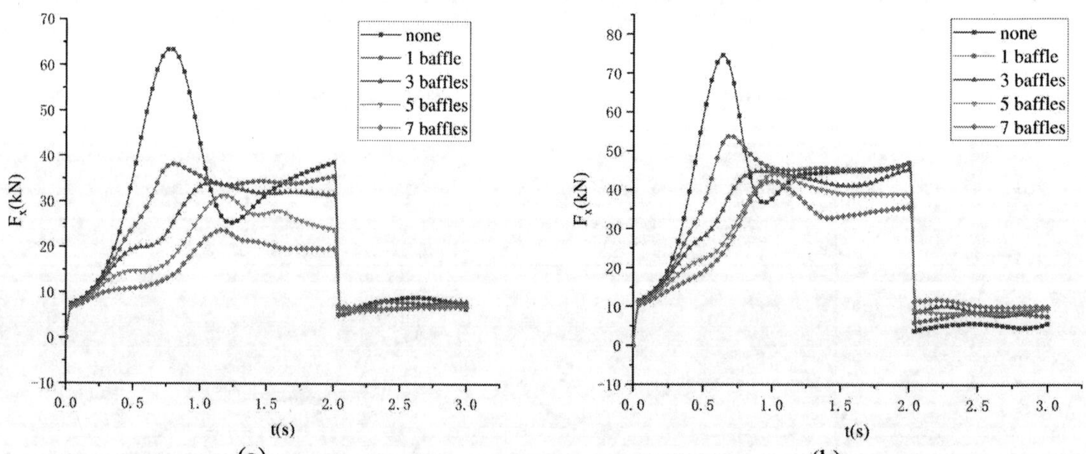

Figure 9. Forces in directions x during braking. (a) K=0.5 and (b) K=0.6.

4. Conclusions

In this paper, the effects of different braking deceleration, liquid-filling ratio, lateral acceleration and the number of baffles on the driving stability of the vehicle during the steering and braking of Hazardous Materials Rescue Trucks are studied. In summary, the following conclusions are offered:

When Hazardous Materials Rescue Truck is turning, maintaining a high liquid filling ratio and a low steering acceleration is advantageous for reducing the liquid slosh force in the tank.

When Hazardous Materials Rescue Truck is braking, as the braking deceleration increases, the crest value of the slosh force of the front head increases. At a fixed brake deceleration, when the liquid filling ratio is 0.5 to 0.8, the change in the force in direction x of the front head is positively correlated with the change in the liquid filling ratio. When the liquid filling ratio continues to increase, the force in direction x of the front head decreasess.

The addition of baffles in the tank has a good effect on suppressing liquid sloshing in the tank. As the number of baffles increases, the driving stability of the vehicle increases significantly.

References

[1] Tian J W, Wang J W and Wang X F 2014 *J. Chongqing Univ. Technol.* **28** 625-30
[2] Zheng X L, Li X S *et al* 2014 *J. Jilin Univ.: Technol. Ed.* **44** 46-51
[3] Wang H P 2014 Study on the influence of liquid sloshing caused by baffle setting in Tank (Jilin, China: Jilin University)
[4] Zheng X L and Li X S 2014 Research on driving stability of tank vehicles based on equivalent mechanical model of liquid sloshing (Jilin, China: Jilin University)
[5] Li S, Gao F Q *et al* 2007 *Nucl. Power Eng.* **28** 54-57
[6] Kang N and Liu K 2010 *J. Zhejiang Univ.* **11** 317-24
[7] Yu Z S 2009 *Automobile Theory* (Beijing: China Machine Press) p195

Simulation research on regenerative braking control strategy of electric vehicle

S W Zhou[1] and Q Y Wang[1,2]

[1]College of Mechanical Engineering and Automation, Northeastern University, Shenyang Liaoning 110819, China

E-mail: qywangedu@163.com

Abstract. In view of the shortage of cruising range of current electric vehicles, the research on the regenerative braking system of electric vehicles has been carried out. Aiming at improving energy utilization rate, combined with the working characteristics of electric motor and battery, a regenerative braking control strategy based on braking intensity is proposed. The control strategy model is built in MATLAB/Simulink firstly, and then the control strategy is verified by the integrated simulation of AVL CRUISE and MATLAB. The simulation results show that the energy utilization rate of the proposed control strategy under custom mild and moderate braking conditions can reach 34.8% and 21.5% respectively.

1. Introduction

Under the current social background of depletion of energy and deteriorating ecological degradation, new energy vehicles represented by electric vehicles have gradually become research hotspots in the automotive field due to their low energy consumption, low ecological damage and high electronic integration [1]. However, due to the technical shortcomings of power batteries, the driving range of electric vehicles is greatly limited, which is also an important reason why today's electric vehicles have not been widely promoted.

Vehicles operating under urban conditions need frequent braking and deceleration, and a lot of energy is consumed in the braking process. For electric vehicles, its unique regenerative braking system can recover part of the energy consumed in the braking process, thus significantly improving the endurance mileage and energy utilization rate of the vehicle. However, the magnitude of the regenerative braking force is limited by factors such as the energy storage device, the motor, and the structure of the brake system [2,3]. This paper focuses on the regenerative braking control strategy of electric vehicles, and a common precursor pure electric vehicle is taken as the research object. Through the control algorithm, the front and rear axle braking force and the relationship between mechanical braking force and regenerative braking force are coordinated in order to achieve stable braking and recover considerable braking energy.

2. Braking force distribution of front and rear axles

In the braking state of the vehicle, the distribution of the braking force of the front and rear axles is related to the steering stability of the vehicle and the degree of utilization of the attachment conditions. In addition, the distribution of braking force and the locking of the front and rear wheels also affect the operation of the regenerative braking system [4].

2.1. *Brake force distribution safety zone*

When the vehicle is in the braking state, if the front and rear axle wheels are locked at the same time, it is beneficial to the utilization of the attachment conditions, and is also beneficial to improving the braking stability of the vehicle. The braking force relationship of the front and rear axles at this time is referred to as ideal braking force distribution curve (I curve).

$$F_f = \frac{1}{2}\left[\sqrt{{L_b}^2 + \frac{4h_g L}{G}F_f} - \left(\frac{GL_b}{h_g} + 2F_f\right)\right] \tag{1}$$

In the expression: F_r is the rear axle braking force; F_f is the front axle braking force; L_b is the distance from the center of mass to the rear axle; h_g is the centroid height; L is the wheelbase; G represents the gravity of the whole vehicle.

Under the road surface with a certain adhesion coefficient, the distribution curve of breaking force between the front and rear axle is called F Line when the front wheels of the car are locked and the rear wheels are not locked.

$$\begin{cases} F_f = \varphi \dfrac{G}{L}\left(L_b + zh_g\right) \\ F_r = Gz - F_f \end{cases} \tag{2}$$

In the expression: φ is the adhesion coefficient of the road surface (0.8 in this paper); z stands for braking strength.

The ECE (Economic Commission of Europe) R13 braking regulation formulated by the United Nations Economic Commission for Europe is formulated from the perspective of meeting the braking efficiency of vehicles. The regulation stipulates the minimum rear axle braking force in the braking process of vehicles. The distribution curve that meets the ECE regulatory line is called the M curve:

$$F_f = \frac{z + 0.07}{0.85}\frac{G}{L}\left(L_b + zh_g\right) \tag{3}$$

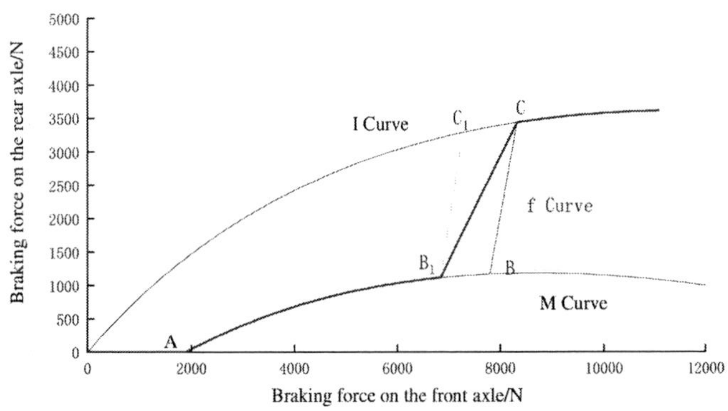

Figure 1. Braking force distribution between front and rear axle.

As shown in figure 1, the range enclosed by the M line f line and the I line and the horizontal axis coordinates is referred to as a safe area for the distribution of braking force between the front and the rear axle during vehicle braking [5].

2.2. *Brake force distribution algorithm*

When braking, a electric vehicle can only recover the energy added to the drive shaft. In order to recover more energy, the braking force distribution curve should be as close as possible to the M curve. On the other hand, in order to pursue better braking performance, it is necessary to increase the braking force of the rear axle so that the distribution curve is close to the I curve [6,7]. Considering the above situation, the following algorithm is obtained:

When the braking strength $z \leq 0.1289$, the braking force of the front and the rear axle are distributed according to the OA line, that is, the braking force of the whole vehicle is provided by the front axle. In the expression: point A is the intersection of the abscissa axis and the M curve. The corresponding braking strength of point A is 0.1289.

When the braking strength $z>0.1289$, The braking force of the front and the rear axle are distributed according to the AB part of the M curve. The point $B_1$is the intersection of the B_1C_1 line and the M curve, the line B_1C_1 is parallel to the BC line, and the braking force of the front axle is 90% of the BC line. Point C is the intersection of the corresponding f curve and I curve under the road surface adhesion coefficient.

When the braking strength $z \geq 0.542$, the braking forces are distributed according to B_1C line.

$$\begin{cases} F_f = 0.3784Gz + 3764.5325 \\ F_r = Gz - F_f \end{cases} \tag{4}$$

When the braking strength is greater than the adhesion coefficient of the road surface, the braking force of the front and the rear axle is distributed according to the I curve.

3. Regenerative brake intervention control strategy

The regenerative braking intervention control strategy is responsible for balancing the regenerative braking force with the friction braking force. When the braking strength $z<0.2$, the braking force of the front axle is preferentially provided by the regenerative braking force; When braking strength $z \geq 0.2$, the share of regenerative braking force in front axle braking force is determined by fuzzy controller.

3.1. Design of fuzzy controller

The control of regenerative braking force has the characteristics of nonlinearity, randomness and time variation, so it is difficult to express the control process with mathematical model. Fuzzy control mimics human reasoning and decision-making process from behaviour : using fuzzy rules to complete fuzzy reasoning. It does not require a mathematical model to achieve precise control. So it is very suitable for controlling the regenerative braking force, and the fuzzy controller shown in figure 2 is proposed.

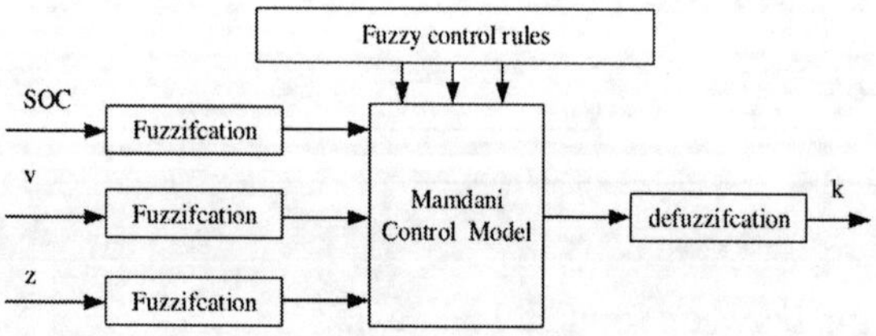

Figure 2. Design of fuzzy controller.

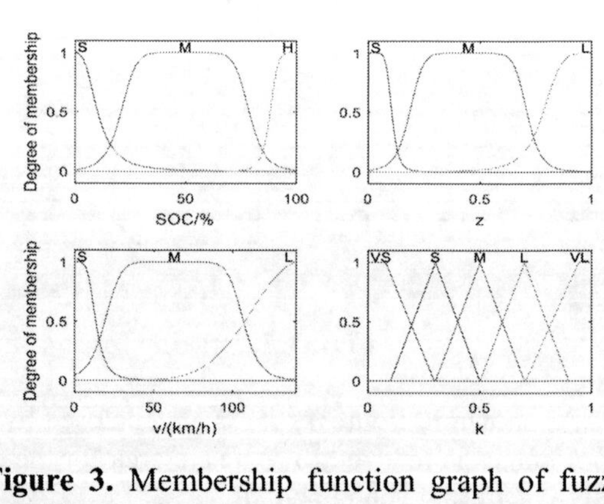

Figure 3. Membership function graph of fuzzy variables.

Firstly, according to the characteristics of the input and output variables, the membership function as shown in the Figure 3. is established, and then combined with practical experience, 27 fuzzy rules shown in table 1 have been developed.

Table 1. Fuzzy rule.

i	1	2	3	4	5	6	7	8	9	10	11	12	13	14
SOC	S	S	S	S	S	S	S	S	S	M	M	M	M	M
z	S	S	S	M	M	M	L	L	L	S	S	S	M	M
v	S	M	L	S	M	L	S	M	L	S	M	L	S	M
k	M	VL	L	M	VL	L	L	VL	L	M	L	M	L	VL
i	15	16	17	18	19	20	21	22	23	24	25	26	27	
SOC	M	M	M	M	L	L	L	L	L	L	L	L	L	
z	M	L	L	L	S	S	S	M	M	L	L	L	L	
v	L	S	M	L	S	M	L	S	M	L	M	M	L	
k	L	L	VL	VL	M	M	S	S	M	S	S	S	S	

3.2. Regeneration braking force limitation and correction

In the above, the regenerative braking coefficient k is obtained by using the fuzzy controller. However, the regenerative braking is restricted by the operating characteristics of the motor and the battery, and it is necessary to limit and correct the k value and the regenerative braking force.

First, it is necessary to specify the conditions for regenerative braking intervention based on the parameters of the battery SOC and the vehicle speed v. When the battery SOC value is high (reaching the threshold value of 95%), in order to prevent the battery from overcharging, the front axle does not use regenerative braking. When the vehicle speed is too low, the recovered regenerative energy is limited too, and the condition for charging the battery is not satisfied, thus the regenerative braking is turned off.

In addition, the maximum braking force of regenerative braking is also limited by the operating characteristics of the motor: when the motor speed is lower than the rated speed, the motor is in the constant torque zone ;when the motor speed is greater than the rated speed, the motor is in the constant power zone as shown in figure 4.

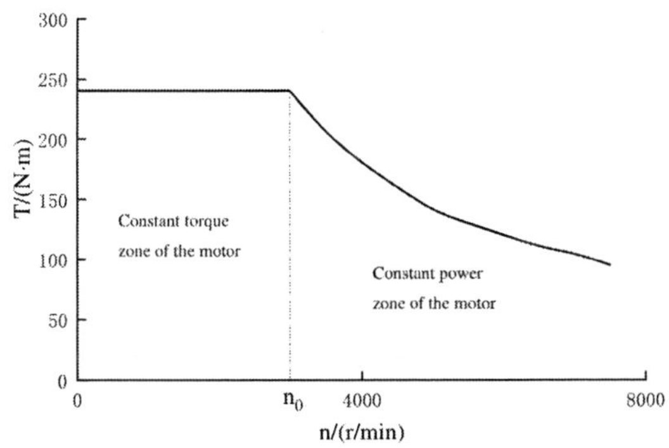

Figure 4. Speed-torque curve of the electric motor.

What' more, when the motor speed is too low, the induced electromotive force generated by the motor is low too, however, the regenerative braking at low speed is not good, so the vehicle should turn off regenerative braking when the motor working at speed less than 500 r/min. To sum up, the model of regenerative braking force limit value shown in formula (5) is obtained.

$$F_{el\max} = \begin{cases} 0 & 0 \le n < n_{\min} \\ T_N i_m / r\eta_T & n_{\min} \le n < n_0 \\ 9550 P_{\max} i_m / nr & n \ge n_0 \end{cases} \quad (5)$$

In the expression: F_{elmax} stands for the largest regenerative braking force; T_N is the maximum braking torque of the motor; P_{max} is the maximum power of the motor; n is the instantaneous speed of the motor; i_m and η_T is the transmission ratio and transmission efficiency of the motor to the front wheel; n_0 is the rated speed of the motor; r represents the radius of the wheel.

4. Simulation calculation

In order to verify the regenerative braking control strategy proposed in this paper, the control strategy developed in this paper is first modeled by MATLAB/Simulink, and the generated SIMULINK control strategy is compiled to generate dynamic link library files (dll file). Then, the generated dll file is imported into the electric vehicle model built by AVL CRUISE software, and the simulation analysis can be performed after the data bus connection is successfully completed. The vehicle model and the simulation data parameters of each component are shown in table 2. The control strategy model built by SIMULINK is shown in figure 5, and the vehicle model built by AVL CRUISE is shown in figure 6.

Table 2. Simulation parameters.

parameter	unit	Parameter value
Vehicle weight	kg	1500
Gravity acceleration	m/s^2	9.8
Centroid height	m	0.56
Wheelbase	m	2.4
Distance from centroid to rear axle	m	1.25
Wheel radius	m	0.279
Maximum torque of the motor	(N·m)	240
Maximum motor power	kW	75
Battery nominal voltage	V	320
battery capacity	(A·h)	110

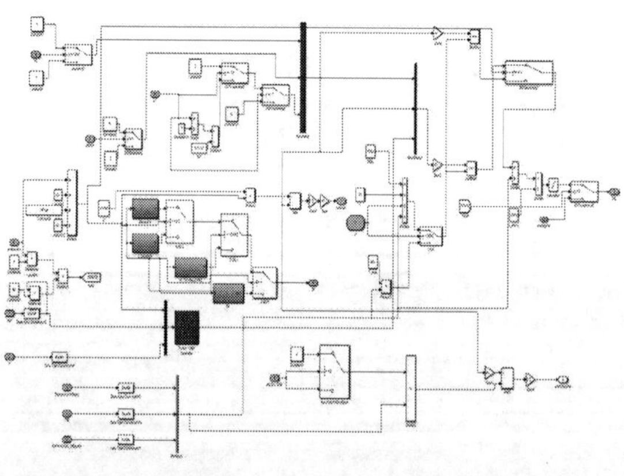

Figure 5. Simulink model of control strategy.

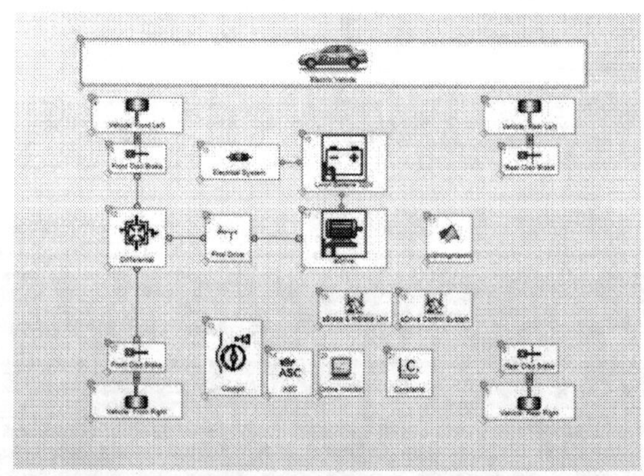

Figure 6. Vehicle model based on AVL CRUISE.

In order to verify the effectiveness of the proposed control strategy, two brake conditions were defined to simulate the driving state of electric vehicle in mild and moderate braking conditions. The energy recovered during the braking process can be obtained by comparing the rate of change of the SOC value in the simulation result with the parameter data of the battery. The energy recovery efficiency during braking can be expressed as expression (6).

$$\eta_{reg} = \frac{\Delta SOC \cdot U_N \cdot Q_N \cdot 3600}{\frac{1}{2} m \Delta v^2} \tag{6}$$

In the expression: η_{reg} represents energy recovery; ΔSOC represents the increment of the SOC; U_N is charging nominal voltage; Q_N is battery capacity; Δv represents the change value of electric vehicle speed during braking.

The simulation results of mild braking are shown in Figure 7. Under mild braking conditions, the electric vehicle started to decelerate from 40 km/h and the braking strength was set to 0.2. The simulation results show that the vehicle speed reduced to 0 after 5.67s braking time, the recovered energy is 35481J, and the energy recovery rate is 34.8%.

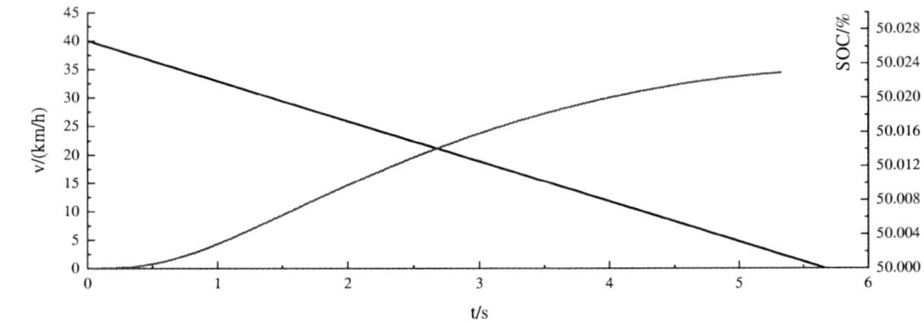

Figure 7. Simulation result of light braking condition.

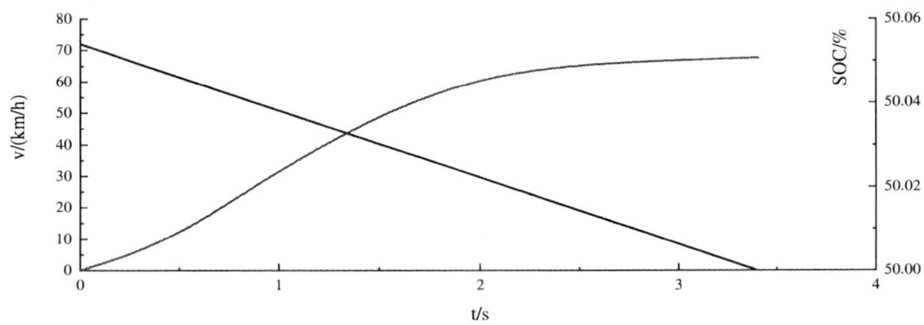

Figure 8. Simulation result of moderate braking condition.

The simulation results of moderate braking are shown in Figure 8. In the case of moderate braking, the vehicle started to brake at 72 km/h, and after 5 seconds, the car stops, recovered 64627J of energy in this process. and the energy recovery rate was 21.5%.

5. Conclusions

In order to improve the cruising range of electric vehicles, a regenerative braking control strategy based on fuzzy control was proposed under the premise of fully considering the working characteristics of the motor and battery. The control strategy and the vehicle model were firstly built using MATLAB/Simulink and AVL CRUISE respectively. After that the simulink model of the control strategy was compiled into a dynamic link library file (dll file) and then embedded in AVL CRUISE. In order to verify the proposed control strategy, simulations were performed in custom moderate and mild braking conditions. The simulation results show that under medium and mild braking conditions, the energy recovery efficiency of the regenerative braking system is 21.5% and 34.8% respectively.

References

[1] Ye M and Guo J W 2013 *Regenerative Braking and Its Control Technology on Electric Vehicle* (Beijing: China Communications Press) pp 52-57
[2] Poursamad A and Montazeri M 2008 *Control Eng. Pract.* **16** 861-73
[3] Chen Z Y, Yang Y *et al* 2016 *J. Northeast Univ.* **37** 1750-63
[4] Yan C Y, Jiang B *et al* 2017 *Electric Vehicle Motor Control and Drive Technology* Zhao H Q ed. (Beijing: Mechanical Industry Press) pp 163-72
[5] Shi Q S 2009 Key technologies research on energy management problems of pure electric vehicles (Shandong, China: Shandong University)
[6] Guo Z J, Yue D D *et al* 2018 *Mach. Des. Manuf.* **1** 173-7
[7] Chen Y, Bei S Y *et al* 2016 *Mod. Manuf. Eng.* **12** 62-7

IOP Conf. Series: Materials Science and Engineering **657** (2019) 012019 doi:10.1088/1757-899X/657/1/012019

Study on dynamic characteristics of horizontal suspension in engineering vehicle seat based on rolling vibration isolation principle

Z H Yan[1], X J Zhang[1], G Q Liu[1], Z Y Wang[1] and Y R Luo[1,2]

[1]School of Mechanical and Aerospace Engineering, Jilin University, Changchun 130022, China

E-mail: luoyr@jlu.edu.cn

Abstract. The low frequency horizontal vibration of engineering vehicles is a part of multi-dimensional vibration, which causes driver's occupational disease. Therefore, the design of low frequency horizontal vibration isolation is of great significance. The principle of rolling vibration isolation is applied to seat suspension for low frequency horizontal vibration isolation in this paper. The natural frequency characteristic model of horizontal suspension and the seat-human system dynamics model with horizontal suspension based on the Lagrangian equation are established. The dynamic characteristics of rolling horizontal suspension are studied on account of the established models. Results show that this horizontal suspension can achieve very low natural frequency so as to isolate the low frequency vibration in horizontal direction by the matching design of the rollers and concave dimensions. The vibration isolation characteristics of the front and back seat–human system are improved with the reduction of the rollers' radius. The vibration isolation characteristic of the front and back seat - human system increases with the increase of the concave surfaces' radius.

1. Introduction

Engineering vehicles work in the mine, site and other harsh environment with complicated working condition. Therefore, there exist severe multi-dimensional, low-frequency and large amplitude vibration, which cause all kinds of occupational diseases (backache, lumbago, stomach problem [1], etc.) and seriously affect the physical and mental health of drivers. The seat system is in direct contact with a large area of the driver's body, and changing its parameters has little effect on the performance of the whole machine. Therefore, it's particularly important to improve the comfort of driving [2] and good vibration isolation system can make it true.

Engineering vehicle seats are mainly composed of suspension, cushion and backrest. All three parts have important influence on the vibration comfort. The suspension is the main part in terms of vibration isolation. The current seat suspension type mainly includes: linear suspension [3], non-linear suspension [4-6], and active and semi-active suspension [7]. The widely used air spring suspension [8] is also a nonlinear suspension due to the nonlinear force-displacement characteristics of the spring.

There has been a lot of research on the engineering vehicle seat, mainly involving the vibration comfort of the seat (seat-man system) [9,10], Static and dynamic characteristics of seat suspension [11] and seat-man system dynamics model [12-14] etc.. The seat suspensions mentioned above mainly plays the role of isolating vertical vibration and most of them are vertical one-dimensional vibration isolation suspension, But the vibration of engineering machinery is multiple dimensions including vertical vibration, horizontal vibration, swing vibration, and so on [15]. The horizontal vibration isolation demand of some types of engineering vehicle seats is higher than that of vertical vibration isolation, such as the seat of excavator).Vibration in multiple directions causes health damage to the driver [16]. In the field of multi-dimensional vibration research, many researchers have long recognized that the difficulty of horizontal isolation design is that there needs to be very low natural frequency of horizontal vibration (0.7 Hz) [17], which is difficult to achieve in structure. In addition, a seat suspension with vertical and horizontal vibration isolation capability is designed [18]. Through a three-layer structure, It realized vibration isolation in three directions (from vertical to left and right). Due to the complexity of the structure, engineering transformation has not been realized in the past years.

Rolling vibration isolation system has a history of more than 100 years, mainly used in buildings, large equipment and other seismic isolation [19,20]. The rolling vibration isolation principle is applied to low frequency vibration isolation of seat horizontal vibration suspension in this paper firstly. It can realize low frequency vibration isolation in horizontal plane to four directions. The working principle is to achieve the isolation of horizontal vibration by rolling the roller in the upper and lower concave surface. The types of rollers are spherical, ellipsoid and cylindrical. The concave surface of a concave part is spherical, conical, inclined plane and variable curvature. The rolling vibration isolation system is divided into single - layer and double - layer systems [21]. The low frequency vibration isolation can be achieved by the reasonable curvature design and roller curvature. This system is applied to seat horizontal vibration suspension in this paper, and the low frequency vibration isolation can be realized in two horizontal directions through a layer structure. The dynamic model of seat - human coupling in front and back direction is established to study the effect of key design parameters of horizontal vibration isolation system on vibration isolation performance of seat - human coupling system in front and back direction.

2. Modeling of natural frequency characteristic

The suspension principle in this paper is shown in figure 1. Four spherical rollers are installed between the top and bottom concave surfaces. The upper and the lower concave structure curvature can be the same or different, and the curvature of the concave from the middle to the edge of the radius of curvature may change. If radius of curvature in the middle of the area is big, the horizontal seat–human system has low natural frequency, and the ability of low-frequency vibration isolation is strong. The rollers can be prevented to roll out the concave structure due to the small curvature radius of edge area, which can limit the rollers positions. The horizontal vibration natural frequency of seat-human system is the key index. When designing horizontal suspension, low-frequency excitation can be isolated, the natural frequency must be lower. The model of seat-human horizontal vibration natural frequency is established in the following discussion. The human body is assumed to be a rigid body, the quality is m.

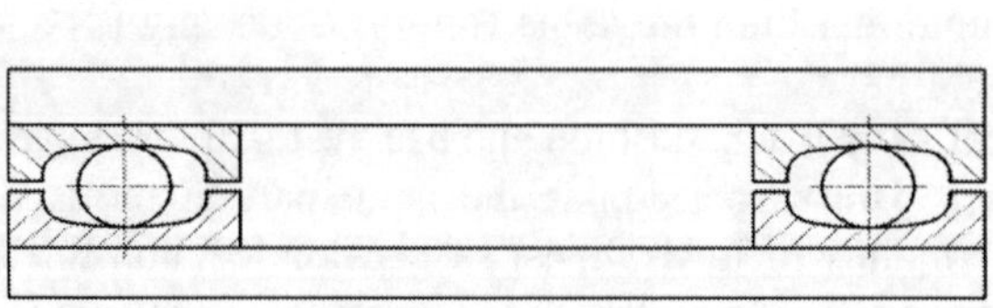

Figure 1. Schematic diagram of horizontal suspension.

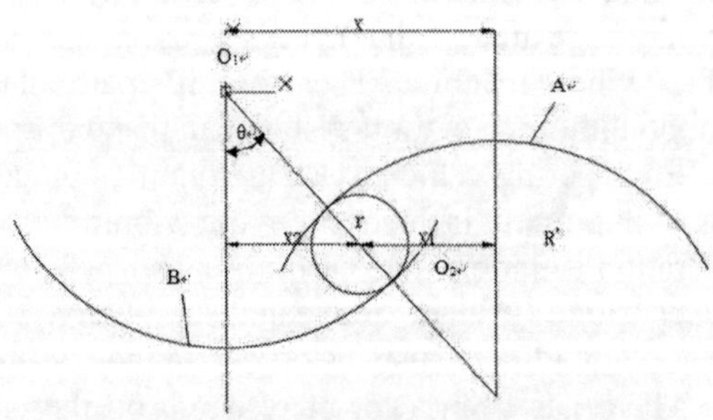

Figure 2. Diagram of force analysis.

Figure 2 shows the static force diagram of the roller at any position. Its gravity is negligible since the roller is small compared with the human body mass. The contact points between the roller center and the upper and lower concave surface form a three-point line. Their extended line c passes through the centers of the concave surface O_1 and O_2.

In the horizontal direction, the equation of free vibration of human body is:

$$m\ddot{x} = ma = mg \sin\theta \tag{1}$$

Where, θ is the Angle between the central connection, the former three-point line, and the vertical direction. $\sin\theta$ can be expressed as a function of the human body's horizontal displacement x, the process is as follows.

A and B are the highest and lowest points of the upper and lower concave surfaces respectively. The horizontal distance between A and B is the horizontal displacement of the seat suspension, which can be obtained by geometric relations:

$$x = x_1 + x_2 \tag{2}$$

Where, x_1 and x_2 are the horizontal displacement of the upper and lower concave surfaces respectively relative to the roller center. And their geometrical relation can be found easily:

$$\frac{x_1}{(R-r)} = \sin\theta \tag{3}$$

$$\frac{x_2}{(R'-r)} = \sin\theta \tag{4}$$

Where, R is the curvature radius of the lower concave surface; R' is the curvature radius of the upper concave surface and r is the radius of the roller.

Substitute equations (3) and (4) into equations (2)

$$x = (R-r)\sin\theta + (R'-r)\sin\theta = (R+R'-2r)\sin\theta \tag{5}$$

Substitute equations (5) into equations (1)

$$m\ddot{x} = \frac{mgx}{(R+R'-2r)} \tag{6}$$

According to equation (6), the natural angular frequency of the horizontal vibration isolation system is

$$\omega_n = \sqrt{\frac{g}{(R+R'-2r)}} \tag{7}$$

Then the natural frequency is

$$f_n = \frac{\omega_n}{2\pi} = \frac{\sqrt{\frac{g}{(R+R'-2r)}}}{2\pi} \tag{8}$$

3. Modeling of seat-human system dynamics

There is multiple natural frequency in human body. The dynamic characteristics of multiple degree of freedom of human body must be considered since human body can not be regarded as rigid during the design of seat suspension. As a result, the front and back seat-human system dynamics model is established in this paper, as shown is figure 3. The dynamic coupling characteristics between horizontal suspension and human body in the front and back directions are also be discussed. The human body model is quoted from literature [22], a 6 degree of freedom model. The parameters related to the human body model and the seat are shown in table 1. M_i is the mass of each mass block in the human body model. Ki is the stiffness of each linear spring; K_{ti} is the stiffness of each torsion spring. C_i is the damping of each linear damper; C_{ti} is the damping of each torsional damper.

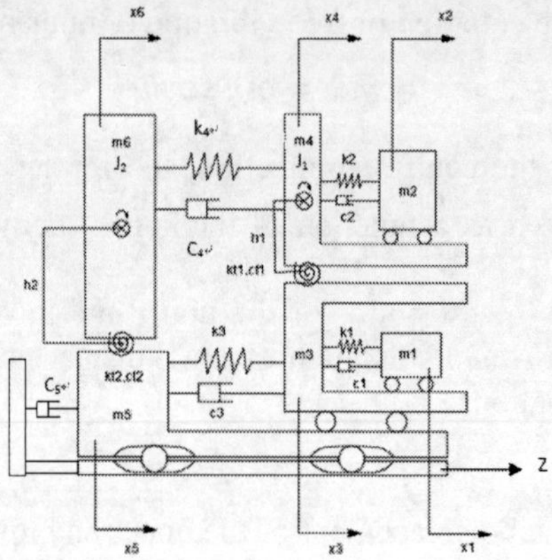

Figure 3. Seat - human system dynamics model in front and back directions.

Table 1. Human body dynamics model and seat parameters in front and back directions.

Related parameters of human body model											
m1 (kg)	m2 (kg)	m3 (kg)	m4 (kg)	J1 (kgm^2)	k1 (N/m)	k2 (N/m)	Kt1 (N/m)	c1 (Ns/m)	c2 (Ns/m)	ct1 (Ns/m)	h1 (m)
20	35	10	8	5	39886	10924	10	359	542	112	0.35
Related parameters of seat model											
m5 (kg)	m6 (kg)	J2 (kgm^2)	k3 (N/m)	k4 (N/m)	Kt2 (N/m)	c3 (Ns/m)	c4 (Ns/m)	c5 (Ns/m)	ct2 (Ns/m)	h2 (m)	
6	2	1.0	26646	24610	10	0.0	0.0	100	1316	0.3	

The dynamic model of the seat-human system in front and back directions is established based on the Lagrange equation.

$$\frac{d}{dt}\left(\frac{\partial T}{\partial x_i'}\right) - \frac{\partial T}{\partial x_i} + \frac{\partial U}{\partial x_i} + \frac{\partial D}{\partial x_i'} = Q_i \tag{9}$$

Where, T is the kinetic energy of the system; U is the system potential energy; D is the dissipated energy of the system; X_i is the generalized degree of freedom of i item; Q_i is the generalized force corresponding to the generalized degree of freedom of i item.

Kinetic energy T of system is

$$T = \sum_{i=1}^{6} \frac{1}{2} m_i x'^2 + \sum_{j=1}^{2} \frac{1}{2} J_j \theta'^2 \tag{10}$$

Where, $\theta_1 = \dfrac{x_3 - x_4}{h_1}$, $\theta_2 = \dfrac{x_6 - x_5}{h_2}$

System potential energy is

$$U = \sum_{i=1}^{4} \frac{1}{2} ki\Delta x_i'^2 + \sum_{j=1}^{2} \frac{1}{2} ktj\theta_j'^2 + Mgy \tag{11}$$

Where, $\Delta x_1 = x_1 - x_3$, $\Delta x_2 = x_2 - x_6$, $\Delta x_3 = x_3 - x_5$, $\Delta x_4 = x_4 - x_6$, $\Delta x_5 = x_5 - z$, $M = \sum_{i=1}^{6} m_i$ (The human body is assumed as a mass block in the calculation since the dynamic characteristics of horizontal direction are considered, and the gravitational potential energy acts in the vertical direction), $y = 2(R-r)\left[1 - \sqrt{1 - \frac{(z-x_5)^2}{4(R-r)^2}}\right]$.

Dissipation energy of the system is

$$D = \sum_{i=1}^{4} \frac{1}{2} c_i \Delta x_i'^2 + \sum_{j=1}^{2} \frac{1}{2} c_{tj} \theta_j'^2 \tag{12}$$

The corresponding generalized forces for each generalized degree of freedom are respectively, $Q_1 = 0$, $Q_2 = 0$, $Q_3 = 0$, $Q_4 = 0$, $Q_5 = -\mu k y$, $Q_6 = 0$.

Substitute equations (10), (11) and (12) into equations (9). The 6 degree of freedom dynamics model of the front and back seat-human system with horizontal suspension can be obtained

$$m_1 x_1'' + k_1(x_1 - x_3) + c_1(x_1' - x_3') = 0 \tag{13}$$

$$m_2 x_2'' + k_2(x_2 - x_4) + c_2(x_2' - x_4') = 0 \tag{14}$$

$$m_3 x_3'' + \frac{J_1}{h_1^2}(x_3'' - x_4'') + k_1(x_3 - x_1) + k_3(x_3 - x_5) + \frac{k_{t1}}{h_1^2}(x_3 - x_4) + c_1(x_3' - x_1') + c_3(x_3' - x_5') + \frac{c_{t1}}{h_1^2}(x_3' - x_4') = 0 \tag{15}$$

$$m_4 x_4'' + \frac{J_1}{h_1^2}(x_4'' - x_3'') + k_2(x_4 - x_2) + k_4(x_4 - x_6) + \frac{k_{t1}}{h_1^2}(x_4 - x_3) + c_2(x_4' - x_2') + c_4(x_4' - x_6') + \frac{c_{t1}}{h_1^2}(x_4' - x_3') = 0 \tag{16}$$

$$m_5 x_5'' + \frac{J_2}{h_2^2}(x_5'' - x_6'') + k_3(x_5 - x_3) + \frac{k_{t2}}{h_2^2}(x_5 - x_6) + \frac{1}{2}\frac{Mg}{R-r}\frac{x_5 - z}{\sqrt{1 - \frac{(x_5 - z)^2}{4(R-r)^2}}} + c_3(x_5' - x_3') + \frac{c_{t2}}{h_2^2}(x_5' - x_6') + c_5(x_5' - z') = -\mu k \,\square\, 2(R-r)\left[1 - \sqrt{1 - \frac{(z - x_5)^2}{4(R-r)^2}}\right] \tag{17}$$

$$m_6 x_6'' + \frac{J_2}{h_2^2}(x_6'' - x_5'') + k_4(x_6 - x_4) + \frac{k_{t2}}{h_2^2}(x_6 - x_5) + c_4(x_6' - x_4') + \frac{c_{t2}}{h_2^2}(x_6' - x_5') = 0 \tag{18}$$

4. Model verification based on Adams

In order to verify this seat - human system dynamics model in this paper, the human body suspension system dynamic model with single degree of freedom is established based on the Adams software. As

IOP Conf. Series: Materials Science and Engineering **657** (2019) 012019 doi:10.1088/1757-899X/657/1/012019

shown in figure 4(a), model parameters are shown in table 2.

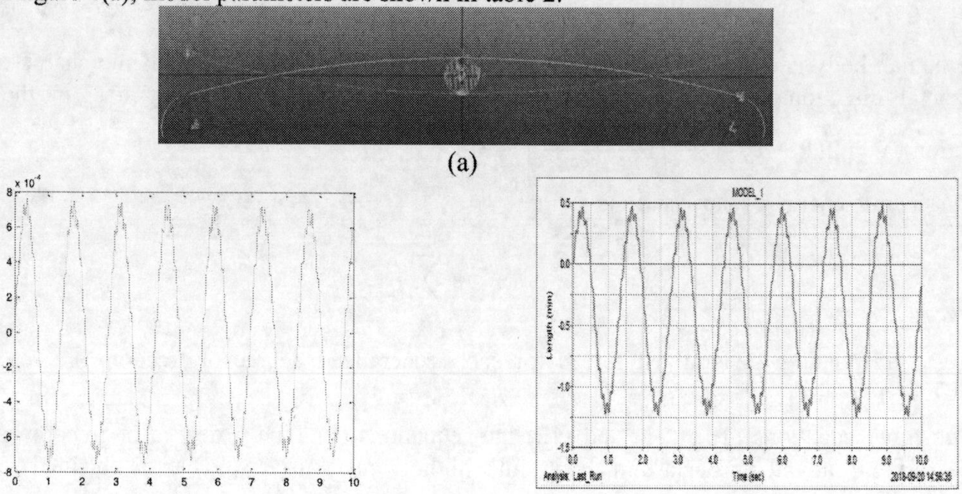

(a)

(b) (c)

Figure 4. Response comparison of human body's forward and backward displacement in time domain: (a) Adams model of single degree of freedom horizontal suspension; (b) The time domain displacement response curve of human body in front and back directions based on the dynamic model of seat - human system established in this paper; (c) The time domain displacement response curve of human body in the front and back directions based on Adams software.

Table 2. Model parameters of Adams.

M (kg)	R (m)	R' (m)	r (m)
m1+m2+m3+m4+m5+ m6	0.25	0.25	0.004

In addition, the spring stiffness parameters between the human body each mass in equations (13)-(18) increase to ten times the original parameters, which holds that the human body is a rigid body, then solve. human body displacement response in time domain can be achieved by two methods respectively, as shown in figures 4(b) and 4(c).

From figure 4, two models for human body response in time domain are consistent, which confirm the correctness of the seat-human system dynamics model. However, human body dynamic model quoted from literature [22] directly has never been verified in this paper. And researching on the dynamic characteristics of seat-human system dynamics model should be followed.

5. Results and discussion

5.1. Natural frequency characteristics

The horizontal natural frequency of the seat-human system is related to R, R' and r, and has nothing to do with the human body quality when human body is regarded as a mass block from expression (8). When R = R', the natural frequency of system horizontal directions (front and back, left and right) are the function of (R - r). Effect of (R - r) on the natural frequency of the system is shown in figure 5. It is similar to a hyperbola function. When it is close to 0, system natural frequency increases rapidly. When it is away from 0, with the increase of (R - r), the natural frequency of system declines slowly. The natural frequency is the key parameter in design of horizontal isolation suspension. To isolate the horizontal low-frequency excitation, the natural frequency of seat-human system in horizontal direction must be lower. Based on the frequency of isolation horizontal excitation, the natural frequency characteristics of seat-human system can be determined and further the following isolation system design parameters.

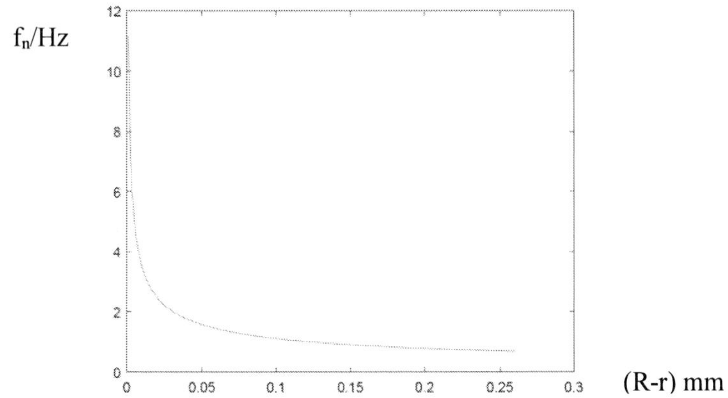

Figure 5. Effect of (R-r) on the natural frequency of the system.

5.2. System dynamic characteristics

Effects of the key design parameters (R, r, c_5) of horizontal vibration isolation suspension on the vibration transfer characteristics is studied on account of seat-human system dynamic model in front and back directions (equations (13)-(18)). Simple harmonic small amplitude sweep excitation is exerted on the floor. In a study of design parameters on the system vibration isolation characteristics, the other parameters are fixed parameter selection of each variable as shown in table 3.

Table 3. Numerical value of R, r and c_5.

	R	r	c_5
1	0.13	0.016, 0.032, 0.064	100
2	0.065, 0.13, 0.26	0.032	100
3	0.13	0.032	10, 100, 1000

- Effect of r on seat-human system vibration isolation characteristics in front and back directions is shown is figure 6(a). with the decrease of r, the natural frequency of system decreases, resonance peak moves to low frequency zone and vibration isolation performance is improved. But limited by structural design, too small value of r is not appropriate. The value of r can be determined by considering the value of R and structural design feasibility.
- Effect of R on seat-human system vibration isolation characteristics in front and back directions is shown is figure 6(b). with the increase of R, the natural frequency of system decreases, resonance peak moves to low frequency zone and vibration isolation performance is improved. Combined with r, the realization of structure(the structural limitation of machining pits on steel plates.) is also considered in design of R. In addition, there must be enough clearance between the upper and the lower concave surfaces.
- Effect of c_5 on seat-human system vibration isolation characteristics in front and back directions is shown is figure 6(c). With the increase of c_5, basically resonance peak frequency remains unchanged, but the peak value decreases. The curve on the right side of the isolation zone increases, which means the decrease of vibration isolation performance. This effect and the influence of damping on the vibration isolation performance of the linear system is consistent. c_5 must be a reasonable value in order to reduce the resonance peak values, even leading to a fall in vibration isolation performance of the isolation zone, since the horizontal excitation frequency may pass through the resonance zone.

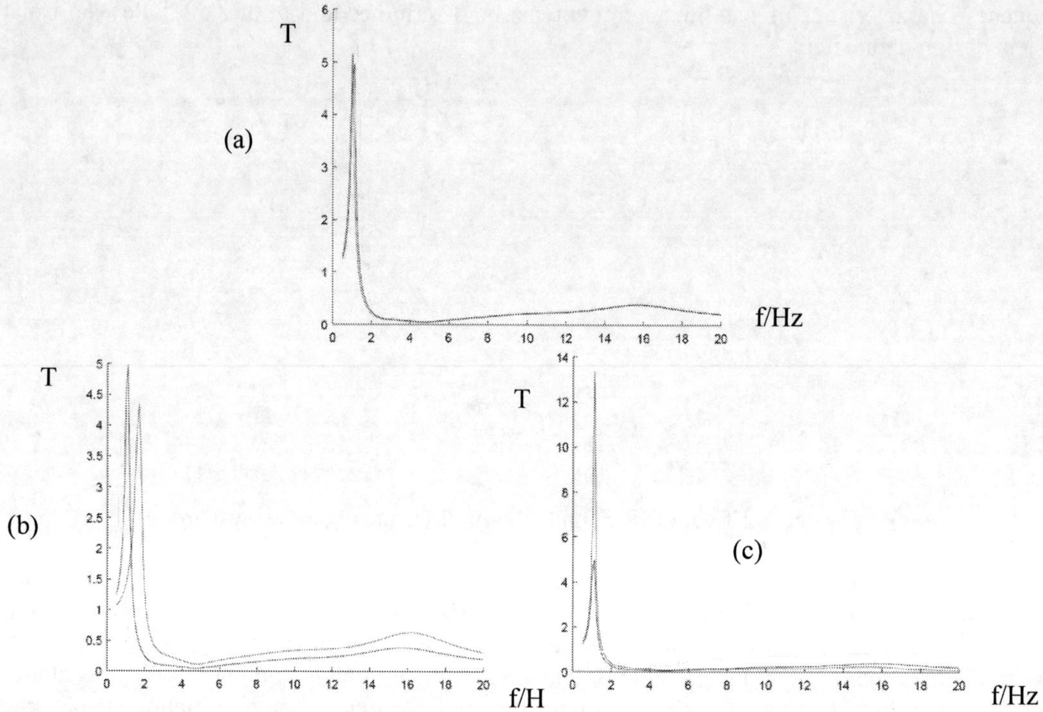

Figure 6. Effect of r, R, c_5 on seat-human system vibration isolation characteristics in front and back directions: (a) Effect of r; (b) Effect of R; (c) Effect of c_5.

6. Conclusion

Aiming at the problem of horizontal low-frequency vibration isolation of engineering vehicles, the design of horizontal suspension based on the principle of rolling vibration isolation is put forward in this paper. This suspension can realize the low frequency vibration in the horizontal plane(front and back, left and right). The natural frequency mathematical model of horizontal vibration isolation is established in the assumption that human body is a mass block. Considering complex dynamic characteristics of the human body, based on the Lagrange equation, the seat -human system dynamics model in front and back directions is deduced. The simulation verification has been done based on Adams software. The horizontal natural frequency of the vibration isolation suspension features and the effect of key design parameters on the vibration isolation characteristics in front and back directions are analyzed. Results show that the horizontal seat suspension based on rolling vibration isolation system can achieve very low natural frequency, and isolate the horizontal low frequency vibration in engineering vehicles. But there are still some problems, such as horizontal vibration limit problem or and the horizontal vibration isolation problems under complex conditions, etc. All the problems will be considered afterwards.

Acknowledgments

This work is partially supported by the National Natural Science Foundation of China (51605184) and the Postdoctoral Science Foundation of China (2018M641777).

References

[1] Kim, Jeong H *et al* 2016 Whole body vibration exposures and health status among professional truck drivers: a cross-sectional analysis *Ann. Occup. Hyg.* **60** 936-48
[2] Stein G J *et al* 2008 Modelling and simulation of locomotive driver's seat vertical suspension vibration isolation system *Int. J. Ind. Ergon.* **38** 384-95

[3] Hill K and Anoop D 2003 Modeling, analysis and optimization of a scissors linkage seat suspension *Eng. Optimiz.* **35** 341-57
[4] Le T D and Kyoung K A 2011 A vibration isolation system in low frequency excitation region using negative stiffness structure for vehicle seat *J. Sound Vibr.* **330** 6311-35
[5] Lee C-M and Goverdovskiy V N 2012 A multi-stage high-speed railroad vibration isolation system with "negative" stiffness *J. Sound Vibr.* **331** 914-21
[6] Loghavi S 2015 Analysis of negative stiffness devices with application to vehicle seat suspensions (The Ohio State University)
[7] Ahn K K 2014 Active pneumatic vibration isolation system using negative stiffness structures for a vehicle seat *J. Sound Vibr.* **333** 1245-68
[8] Holtz Marco W and Johannes L V N 2010 Modelling and design of a novel air-spring for a suspension seat *J. Sound Vibr.* **329** 4354-66
[9] Basri B and Michael J G 2014 The application of SEAT values for predicting how compliant seats with backrests influence vibration discomfort *Appl. Ergon.* **45** 1461-74
[10] Gong W D and Michael J G 2018 Measuring, evaluating and assessing the transmission of vibration through the seats of railway vehicles *Proc. Inst. Mech. Eng. Part F-J. Rail Rapid Transit* **232** 384-95
[11] Yan Z H *et al* 2015 Modeling and analysis of static and dynamic characteristics of nonlinear seat suspension for off-road vehicles *Shock Vib.* **2015** 1-13
[12] Kim S K et al 2003 Simplified models of the vibration of mannequins in car seats *J. Sound Vibr.* **264** 49-90
[13] Tewari V K and Prasad N 1999 Three-DOF modelling of tractor seat-operator system *J. Terramech.* **36** 207-19
[14] Ippili R K *et al* 2008 Nonlinear multi-body dynamic modeling of seat–occupant system with polyurethane seat and H-point prediction *Int. J. Ind. Ergon.* **38** 368-83
[15] Deboli R, Angela C and Christian P 2017 Whole-body vibration: Measurement of horizontal and vertical transmissibility of an agricultural tractor seat *Int. J. Ind. Ergon.* **58** 69-78
[16] ISO 2631-1 1997 *Mechanical Vibration and Shock-Evaluation of Human Exposure to Whole Body Vibration. Part 1: General Requirements*
[17] Rakheja S, Afework Y and Sankar S 1994 An analytical and experimental investigation of the driver-seat-suspension system *Veh. Syst. Dyn.* **23** 501-24
[18] Sankar S and Afonso M 1993 Design and testing of lateral seat suspension for off-road vehicles *J. Terramech.* **30** 371-93
[19] Harvey Jr P S and Karah C K 2016 A review of rolling-type seismic isolation: Historical development and future directions *Eng. Struct.* **125** 521-31
[20] Yang C-Y *et al* 2012 Effectiveness of an eccentric rolling isolation system with friction damping *J. Vib. Control* **18** 2149-63
[21] Harvey Jr P S and Henri P G 2014 Double rolling isolation systems: A mathematical model and experimental validation *Int. J. Non-Linear Mech.* **61** 80-92
[22] Qiu Y and Michael J G 2011 Modelling the fore-and-aft apparent mass of the human body and the transmissibility of seat backrests *Veh. Syst. Dyn.* **49** 703-22

IOP Conf. Series: Materials Science and Engineering **657** (2019) 012020 doi:10.1088/1757-899X/657/1/012020

Composite scattering from complex conductor-medium assembly object above a rough surface

N Li[1], S J Xie[1], G L Tian[1,2] and S J Wei[1]

[1]College of Air and Missile Defence, Air Force Engineering University, Xi'an, 710051, China

E-mail: tianguilong163@163.com

Abstract. To solve the scattering from complex object combining conductor and medium, this paper proposes a hybrid iterated method combining single integral equation (SIE) and Kirchhoff approximation (KA). Firstly, SIE is used to calculate the electromagnetic current in the facet of the conductor-medium assembly object. Then, the electric and magnetic currents on the facet of a rough surface are calculated by KA. The electromagnetic current on the facet of object and a rough surface are updated by mutual iteration. Finally, some examples for composite scattering from object above a rough surface are analysed and discussed in detained.

1. Introduction

Composite scattering from object above a rough surface has been attracted the interest of many researchers in recent years. It has been widely applied in object detection and remote sensing [1,2]. In the study of composite scattering from object above a rough surface, most of the articles are about the perfect electric conductive object above a rough surface or medium object above a rough surface [3,4], and some papers only study the conductor-medium assembly object regardless of the rough surface[5-7]. To solve the scattering from conductor-medium assembly object, the coupled integral equation(CIE) are mostly used with the surface of the conductor analyzed by electric field integral equation(EFIE), magnetic integral field equation(MFIE) or the combined field integral equation(CFIE), while the medium surface analyzed by EFIE[8] or Poggio-Miller-Chang-Harrington-Wu-Tsai(PMCHWT)[9]. However, with the coupled integral equation, the equivalent electric and magnetic currents are needed to be computed, which is very time-consuming. To reduce the computation complexity, the single integral equation(SIE)[10,11] is proposed by substituting the equivalent electric and magnetic currents with the effective current inner the medium, which reduce the unknowns by half. However, there are rare articles about scattering from conductor- medium object above a rough surface, which is widely used in the practical application.

In this paper, a hybrid method combining the single integral equation(SIE) and the Kirchhoff approximation (KA) is proposed for calculating the conductor- medium assembly object above a rough surface. Firstly, the composite model from conductor-medium assembly object above a rough surface is constructed. Next, the electric and magnetic currents on object are calculated by SIE and the electric and magnetic current on the rough surface are calculated by KA. The electric and magnetic currents on the facets of object and a rough surface are updated by mutual iteration. Finally, the total scattering field is obtained.

2. Composite model for conductor-medium assembly object above a rough surface

IOP Conf. Series: Materials Science and Engineering **657** (2019) 012020 doi:10.1088/1757-899X/657/1/012020

The geometry of conductor-medium assembly object above a rough surface is shown in figure 1. The object is the combination of conductor and medium, the surface of the conductor is signed as S_1; the surface of the medium is signed as S_2; the interface of the conducts and the medium is S_3; the surface of the rough surface is S_S. The permittivity of the medium part of the objects ε_r; the magnetic conductivity of the medium part of the object is μ_r; the permittivity of the free space is ε; the magnetic conductivity of the free space is μ. $\boldsymbol{k}_i$ and $\boldsymbol{k}_s$ are the incident electric field direction and scattering electric field direction, respectively.

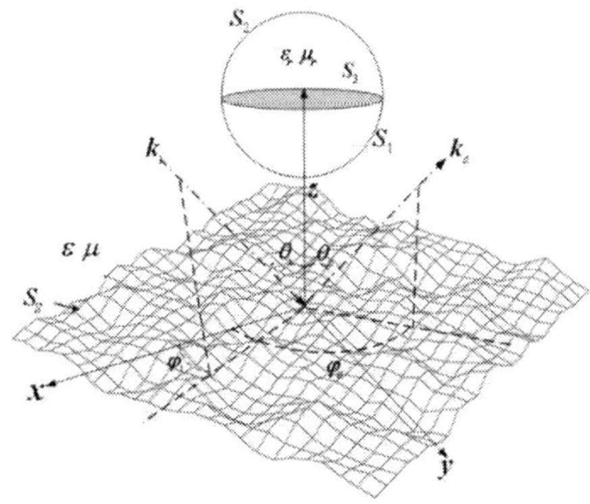

Figure 1. Geometry of composite scattering from conductor-medium assemblyobject above a rough surface.

In composite model, both the object and the rough surface are discrete by RWG function. The surface of the conductive part of the object is divided into N_1 faces. The surface of the medium part of the object is divided into N_2 faces. The interface between the conductive part and the medium part is divided into N_3 faces. The electric current on the facet of conductive part of objects $\eta_0 \boldsymbol{J}_{1e}(\boldsymbol{r})$, the electric and magnetic current on the facet of medium part of object are $\eta_0 \boldsymbol{J}_{2e}(\boldsymbol{r})$ and $\boldsymbol{J}_{2m}(\boldsymbol{r})$, respectively. The electric and magnetic current on the facet of the rough surface are $\eta_0 \boldsymbol{J}_{se}(\boldsymbol{r})$ and $\eta_0 \boldsymbol{J}_{sm}(\boldsymbol{r})$, respectively.

$$\eta_0 \boldsymbol{J}_{1e}(\boldsymbol{r}) = \sum_{j=1}^{N_1} x_{1j} \boldsymbol{f}_{1j}(\boldsymbol{r}), \quad \boldsymbol{r} \in \mathrm{S}_1 \tag{1}$$

$$\eta_0 \boldsymbol{J}_{2e}(\boldsymbol{r}) = \sum_{j=1}^{N_2} c_j \boldsymbol{f}_{2j}(\boldsymbol{r}), \quad \boldsymbol{r} \in \mathrm{S}_2 \tag{2}$$

$$-\eta_0 \boldsymbol{J}_{2m}(\boldsymbol{r}) = \sum_{j=1}^{N_2} d_j \boldsymbol{f}_{2j}(\boldsymbol{r}), \boldsymbol{r} \in \mathrm{S}_2 \tag{3}$$

$$-\eta_1 \boldsymbol{J}_{cd}(\boldsymbol{r}) = \sum_{j=1}^{N_3} x_{3j} \boldsymbol{f}_{3j}(\boldsymbol{r}), \quad \boldsymbol{r} \in \mathrm{S}_3 \tag{4}$$

$$\eta_0 \boldsymbol{J}_{se}(\boldsymbol{r}) = \sum_{j=1}^{N_S} a_j \boldsymbol{f}_{sj}(\boldsymbol{r}), \quad \boldsymbol{r} \in \mathrm{S}_S \tag{5}$$

$$\eta_0 \boldsymbol{J}_{sm}(\boldsymbol{r}) = \sum_{j=1}^{N_S} b_j \boldsymbol{f}_{sj}(\boldsymbol{r}), \quad \boldsymbol{r} \in S_S \tag{6}$$

Where, $\boldsymbol{J}_{1e}$ denotes the equivalent electric current on the facet of the conductive part of the object, $\boldsymbol{J}_{2e}$ denotes the electric current on the facet of the medium part of the object, $\boldsymbol{J}_{2m}$ denotes the magnetic current on the facet of medium part of the object, $\boldsymbol{J}_{cd}$ denotes the electric current on the facet of the interference between the conduct and the medium. $\boldsymbol{J}_{se}$ denotes the equivalent current on the facet of a rough surface S_S, $\boldsymbol{J}_{sm}$ denotes the equivalent magnetic current on the facet of surface S_S, $\boldsymbol{f}_{1j}$, $\boldsymbol{f}_{2j}$, $\boldsymbol{f}_{3j}$, $\boldsymbol{f}_{sj}$ is the RWG function on the facet of $\boldsymbol{S}_1$, $\boldsymbol{S}_2$, $\boldsymbol{S}_3$ and $\boldsymbol{S}_s$, respectively.

The scattering in the free space is generated by the electromagnetic current on the facet of the object,as well as the rough surface:

$$\boldsymbol{E}^s = -\eta_0 \boldsymbol{L}_{01}(\boldsymbol{J}_{1e}) - \eta_0 \boldsymbol{L}_{02}(\boldsymbol{J}_{2e}) + \boldsymbol{K}_{02}(\boldsymbol{J}_{2m}) - \eta_0 \boldsymbol{L}_{0S}(\boldsymbol{J}_{se}) + \boldsymbol{K}_{0S}(\boldsymbol{J}_{sm}) \tag{7}$$

Where, $\boldsymbol{L}$ and $\boldsymbol{K}$ denotes the electric and magnetic field integral operator, respectively. The first subscript denotes that the integration domain in free space or medium area, the second subscript denotes that integration on which faces. The specific definition of $\boldsymbol{L}$ and $\boldsymbol{K}$ are:

$$\begin{aligned} \boldsymbol{L}_{0q}(\boldsymbol{X}) &= jk_0 \int_{S_q} [\boldsymbol{G}_0 \cdot \boldsymbol{X}(\boldsymbol{r}') + \frac{1}{k_0^2} \nabla \boldsymbol{G}_0 \cdot \nabla' \cdot \boldsymbol{X}(\boldsymbol{r}')] dS' \\ \mathrm{K}_{0q}(\boldsymbol{X}) &= \int_{S_q} [\boldsymbol{X}(\boldsymbol{r}') \times \nabla \boldsymbol{G}_0(\boldsymbol{r},\boldsymbol{r}')] dS' \qquad (q=1,2) \end{aligned} \tag{8}$$

Where, $G_0 = e^{-jk_0 R} / 4\pi R$ is the green function in the free space, $\eta_0 = \sqrt{\mu_0 / \varepsilon_0}$ is the wave impedance in free space.

2.1. *Coupled integral equations for scattering from conductor-medium assembly object*

As we see, the geometry of conductor-medium assembly object is shown in figure 2. The medium of object is isotropic media.

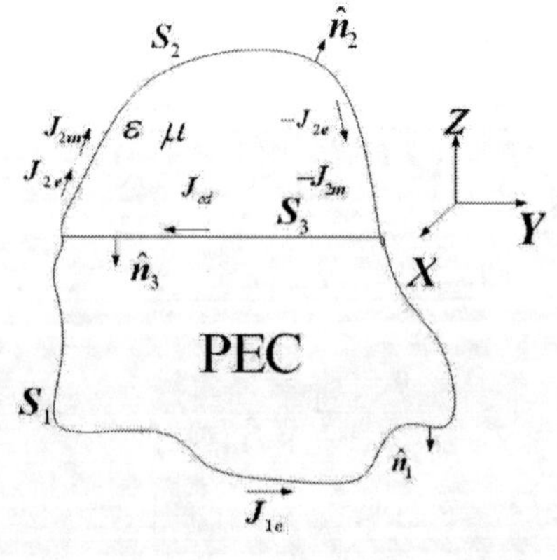

Figure 2. Geometry of 3D arbitrary shape conductor-media assembly object.

The electric field integral equation on the surface $\boldsymbol{S}_1$ and $\boldsymbol{S}_2$ can be expressed as:

$$S_1 : \boldsymbol{n}_1 \times \boldsymbol{E}^i = \boldsymbol{n}_1 \times \boldsymbol{L}_{01}(\eta_0 \boldsymbol{J}_{1e}) + \boldsymbol{n}_2 \times \boldsymbol{L}_{02}(\eta_0 \boldsymbol{J}_{2e}) - \boldsymbol{n}_1 \times \tilde{\boldsymbol{K}}_{02}(\boldsymbol{J}_{2m}) \tag{9}$$

$$S_2 : \boldsymbol{n}_2 \times \boldsymbol{E}^i = \boldsymbol{n}_2 \times \boldsymbol{L}_{01}(\eta_0 \boldsymbol{J}_{1e}) + \boldsymbol{n}_2 \times \boldsymbol{L}_{02}(\eta_0 \boldsymbol{J}_{2e}) - \frac{1}{2}\boldsymbol{J}_{2m} - \boldsymbol{n}_2 \times \tilde{\boldsymbol{K}}_{02}(\boldsymbol{J}_{2m}) \tag{10}$$

Where, $\tilde{\boldsymbol{K}}$ is the principal value except the singular integral at point $\boldsymbol{r}'=\boldsymbol{r}$ .The field in the region of medium is expressed by the electric current $\boldsymbol{J}_{cd}$ on the interface of conductor- medium and the electromagnetic current $-\boldsymbol{J}_{2e}$ and $-\boldsymbol{J}_{2m}$ in the medium. Thus, the inner field in the medium can be expressed as:

$$\boldsymbol{E}^t = -\eta_1 \boldsymbol{L}_{13}(\boldsymbol{J}_{cd}) - \eta_1 \boldsymbol{L}_{12}(-\boldsymbol{J}_{2e}) + \boldsymbol{K}_{12}(-\boldsymbol{J}_{2m}) \tag{11}$$

Where, $\eta_1 = \sqrt{\mu_1/\varepsilon_1}$, $k_1 = \omega\sqrt{\mu_1\varepsilon_1}$; when the observation point is internally close to the object surface, the electric field integral equation on the inner surface of S_2 and S_3 is:

$$S_2 : 0 = \boldsymbol{n}_2 \times \boldsymbol{L}_{13}(-\eta_1 \boldsymbol{J}_{cd}) + \frac{\eta_1}{\eta_0}\boldsymbol{n}_2 \times \boldsymbol{L}_{12}(\eta_0 \boldsymbol{J}_{2e}) + \frac{1}{2}\boldsymbol{J}_{2m} + \boldsymbol{n}_2 \times \tilde{\boldsymbol{K}}_{12}(-\boldsymbol{J}_{2m}) \tag{12}$$

$$S_3 : 0 = \boldsymbol{n}_3 \times \boldsymbol{L}_{13}(-\eta_1 \boldsymbol{J}_{cd}) + \frac{\eta_1}{\eta_0}\boldsymbol{n}_3 \times \boldsymbol{L}_{12}(\eta_0 \boldsymbol{J}_{2e}) + \boldsymbol{n}_3 \times \tilde{\boldsymbol{K}}_{12}(-\boldsymbol{J}_{2m}) \tag{13}$$

By substituting the equation(1)-(4) into equations (9), (10), (12) and (13), and the $\boldsymbol{n}\times\boldsymbol{RWG}$ function $\boldsymbol{g}_{1i} = \boldsymbol{n}_{1i} \times \boldsymbol{f}_{1i}$ and $\boldsymbol{g}_{3i} = \boldsymbol{n}_{3i} \times \boldsymbol{f}_{3i}$ are used to equations (9) and (10), respectively. The function $\boldsymbol{g}_{2i} = \boldsymbol{n}_{2i} \times \boldsymbol{f}_{2i}$ and $\boldsymbol{g}_{3i} = \boldsymbol{n}_{3i} \times \boldsymbol{f}_{3i}$ are used to test equations (12) and (13). Finally, the matrix equation is obtained:

$$\begin{bmatrix} \boldsymbol{Q}_{11} & \boldsymbol{Q}_{12} & \boldsymbol{P}_{12} & 0 \\ \boldsymbol{Q}_{21} & \boldsymbol{Q}_{22} & \boldsymbol{U}_{22}+\boldsymbol{P}_{22} & 0 \\ 0 & \tilde{\boldsymbol{Q}}_{22} & -\boldsymbol{U}_{22}+\tilde{\boldsymbol{P}}_{22} & \tilde{\boldsymbol{Q}}_{23} \\ 0 & \tilde{\boldsymbol{Q}}_{32} & \tilde{\boldsymbol{P}}_{32} & \tilde{\boldsymbol{Q}}_{33} \end{bmatrix} \begin{Bmatrix} \boldsymbol{x}_1 \\ \boldsymbol{c} \\ \boldsymbol{d} \\ \boldsymbol{x}_3 \end{Bmatrix} = \begin{Bmatrix} \boldsymbol{b}_1 \\ \boldsymbol{b}_2 \\ 0 \\ 0 \end{Bmatrix} \tag{14}$$

Where

$$\begin{aligned}
\boldsymbol{Q}_{pq}(i,j) &= \int_{T_{pi}} \boldsymbol{f}_{pi}(\boldsymbol{r}) \cdot \boldsymbol{L}_{0q}(\boldsymbol{f}_{qj}) dS, p=1,2; q=1,2 \\
\boldsymbol{P}_{pq}(i,j) &= \int_{T_{pi}} \boldsymbol{f}_{pi}(\boldsymbol{r}) \cdot \tilde{\boldsymbol{K}}_{0q}(\boldsymbol{f}_{qj}) dS, \quad p=1,2; q=2 \\
\boldsymbol{U}_{pq}(i,j) &= \frac{1}{2}\int_{T_{pi}} \boldsymbol{g}_{pi}(\boldsymbol{r}) \cdot \boldsymbol{f}_{qj}(\boldsymbol{r}) dS, \quad p=2; q=2 \\
\boldsymbol{b}_p(i) &= \int_{T_{pi}} \boldsymbol{f}_{pi}(\boldsymbol{r}) \cdot \boldsymbol{E}^i(\boldsymbol{r}) dS, \quad p=1,2 \\
\tilde{\boldsymbol{Q}}_{p3}(i,j) &= \int_{T_{pi}} \boldsymbol{f}_{pi}(\boldsymbol{r}) \cdot \boldsymbol{L}_{13}(\boldsymbol{f}_{3j}) dS, \quad p=2,3 \\
\tilde{\boldsymbol{Q}}_{p2}(i,j) &= \frac{\eta_1}{\eta_0}\int_{T_{pi}} \boldsymbol{f}_{pi}(\boldsymbol{r}) \cdot \boldsymbol{L}_{12}(\boldsymbol{f}_{2j}) dS, \quad p=2,3 \\
\tilde{\boldsymbol{P}}_{p2}(i,j) &= \int_{T_{pi}} \boldsymbol{f}_{pi}(\boldsymbol{r}) \cdot \boldsymbol{K}_{12}(\boldsymbol{f}_{2j}) dS, \quad p=2,3
\end{aligned} \tag{15}$$

Where, $\{x_1\}$ denotes expansion coefficient of the electric current $\boldsymbol{J}_{1e}$ on the surface S_1, $\{\boldsymbol{c}\}$ denotes expansion coefficient of electric current on the surface S_2, $\{\boldsymbol{d}\}$ denotes the expansion coefficient of magnetic current J_{2m} on the surface S_3, $\{x_3\}$ denotes the expansion coefficient of electric current J_{cd} on the surface S_3.

2.2. Single integral equations for scattering from conductor-dielectric assembly object

To reduce the unknowns on the surface of conductor-dielectric assembly object, the effective electricity $\mathbf{J}_{2e}^{eff}$ is introduced to expressed the field inner the dielectric part instead of the electric $\mathbf{J}_{2e}$ and the magnetic $\mathbf{J}_{2m}$, as well as the electric $\mathbf{J}_{cd}$. Thus, the electric and magnetic field can be expressed as

$$\mathbf{E}_2 = -\eta_1 \boldsymbol{L}_{1d}(\mathbf{J}_{2e}^{eff}) \quad \mathbf{H}_2 = -\mathbf{K}_{1d}(\mathbf{J}_{2e}^{eff}) \tag{16}$$

where, the operator $\boldsymbol{L}_{1d}$ and $\boldsymbol{K}_{1d}$ work inner the whole surface of the dielectric part $S_d = S_2 + S_3$.

The equivalent electromagnetic current on the surface of dielectric part can be expressed by the effective electric $\mathbf{J}_{2e}^{eff}$

$$\mathbf{J}_{2e} = \boldsymbol{n}_2 \times \boldsymbol{H}_2 = -\frac{1}{2}\mathbf{J}_{2e}^{eff} - \boldsymbol{n}_2 \times \tilde{\boldsymbol{K}}_{1d}(\mathbf{J}_{2e}^{eff}) \tag{17}$$

$$-\mathbf{J}_{2m} = \boldsymbol{n}_2 \times \boldsymbol{E}_2 = -\boldsymbol{n}_2 \times \eta_1 \boldsymbol{L}_{1d}(\mathbf{J}_{2e}^{eff}) \tag{18}$$

On the conductor-medium connection surface S_3, the tangential electric field is zero, that is

$$0 = \boldsymbol{n}_3 \times \boldsymbol{E}_2 = -\boldsymbol{n}_3 \times \eta_1 \boldsymbol{L}_{1d}(\mathbf{J}_{2e}^{eff}) \tag{19}$$

The effective electric currents can be approximated with RWG basic function

$$-\eta_0 \mathbf{J}_{2e}^{eff}(\boldsymbol{r}) = \sum_{j=1}^{N_2} x_{2j}\mathbf{f}_{2j}(\boldsymbol{r}) + \sum_{j=1}^{N_3} x_{3j}\mathbf{f}_{3j}(\boldsymbol{r}) \tag{20}$$

$$x_{pj} = \frac{1}{l_{pj}} \int_{l_{pj}} (\hat{l}_{pj} \times \boldsymbol{n}_{pj}) \cdot (-\eta_0 \mathbf{J}_{2e}^{eff}) dl, \; p = 2,3 \tag{21}$$

where, $(\hat{l}_{pj} \times \boldsymbol{n}_{pj})$ denotes the electric direction on the edge that is from T_{2j}^{+} to T_{2j}^{-}.The expansion coefficient $\{c\}$ and $\{d\}$ can be denoted as the average value of the equivalent electric current and magnetic current flowing through the edge, and the equations (17) and (18) can be expresses as the matrix form.

$$\{c\} = (\frac{1}{2}[I_{22}] + [\tilde{P}_{22}])\{x_2\} + [\tilde{P}_{23}]\{x_3\} \tag{22}$$

$$\{d\} = [\tilde{Q}_{22}]\{x_2\} + [\tilde{Q}_{23}]\{x_3\} \tag{23}$$

where, $[I_{22}]$ is the unit matrix with the dimension of $N_2 \times N_2$, the other matrix elements are

$$P_{2q}(i,j) = -\frac{1}{l_{2i}} \int_{l_{2i}} \hat{l}_{2i} \cdot \tilde{\boldsymbol{K}}_{1d}(\boldsymbol{f}_{qj}) dl, \; i = 1,2,\cdots,N_2; \; j = 1,2,\cdots,N_q \tag{24}$$

$$\tilde{Q}_{2q}(i,j) = -\frac{1}{l_{2i}} \int_{l_{2i}} \hat{l}_{2i} \cdot \frac{\eta_1}{\eta_0} L_{1d}(\boldsymbol{f}_{qj}) dl, \; i = 1,2,\cdots,N_2; \; j = 1,2,\cdots,N_q \tag{25}$$

Testing the equation (19) with $\boldsymbol{g}_{3i} = \boldsymbol{n}_{3i} \times \boldsymbol{f}_{3i}$, we can obtain that

$$\{0\} = [\tilde{Q}_{32}]\{x_2\} + [\tilde{Q}_{33}]\{x_3\} \tag{26}$$

$$\tilde{Q}_{3q}(i,j) = \int_{T_{3i}} f_{3i}(r) \cdot \frac{\eta_1}{\eta_0} L_{1d}(f_{qj}) dS,\ i = 1,2,\cdots,N_3;\ j = 1,2,\cdots,N_q \tag{27}$$

By substituting the equations (22) and (23) into equations (9) and (10), and along with the equation (26), we can obtain that

$$\begin{bmatrix} A_{11} & A_{12} & A_{13} \\ A_{21} & A_{22} & A_{23} \\ 0 & A_{32} & A_{33} \end{bmatrix} \cdot \begin{Bmatrix} x_1 \\ x_2 \\ x_3 \end{Bmatrix} = \begin{Bmatrix} b_1 \\ b_2 \\ 0 \end{Bmatrix} \tag{28}$$

Where

$$\begin{aligned}
&[A_{11}] = [Q_{11}] \\
&[A_{12}] = [Q_{12}] \cdot (\frac{1}{2}[I_{22}] + [\tilde{P}_{22}]) + [P_{12}] \cdot [\tilde{Q}_{22}] \\
&[A_{13}] = [Q_{12}] \cdot [\tilde{P}_{23}] + [P_{12}] \cdot [\tilde{Q}_{23}] \\
&[A_{21}] = [Q_{21}] \\
&[A_{22}] = [Q_{22}] \cdot (\frac{1}{2}[I_{22}] + [\tilde{P}_{22}]) + ([U_{22}] + [P_{22}]) \cdot [\tilde{Q}_{22}] \\
&[A_{23}] = [Q_{22}] \cdot [\tilde{P}_{23}] + ([U_{22}] + [P_{22}]) \cdot [\tilde{Q}_{23}] \\
&[A_{32}] = [\tilde{Q}_{32}] \\
&[A_{33}] = [\tilde{Q}_{33}]
\end{aligned} \tag{29}$$

2.3. Scattering from a rough surface

The Kirchhoff approximation (KA) is used to compute the electrical large rough surface. With KA, the rough surface is supposed to consist of numerous little patches, and the mutual interaction inner the rough surface and the edge diffraction as along as the shape point diffraction are neglected, which greatly increase the computational speed. For the isotropic rough surface, the correlation length l_c, the root mean square h_{rms} should satisfy the condition that:

$$k_1 l_c > 6,\ l_c^2 > 2.76 h_{rms}\lambda,\ \frac{l_c^2}{2\sqrt{3}h_{rms}}\left(1 + \frac{2h_{rms}^2}{l_c^2}\right) > \lambda \tag{30}$$

Where, $k_1 = (2\pi/\lambda)\sqrt{\varepsilon_r}$. Supposing that $\hat{n}$ is the unit normal vector in the rough surface, its local orthogonal system is $(\hat{h}, \hat{v}, \hat{k}_i)$, the horizontal and vertical polarization vector are defined as: $\hat{h} = \hat{n} \times \hat{k}_i / |\hat{n} \times \hat{k}|,\ \hat{v} = \hat{h} \times \hat{k}_i$. The local incident angle θ_l is expressed as:

$$\cos\theta_l = -\hat{\boldsymbol{n}} \cdot \hat{\boldsymbol{k}}_{inc} \tag{31}$$

And the local reflection coefficient R_v and R_h on the rough surface are expressed as:

$$R_v(\theta_l) = \frac{\varepsilon_r \cos\theta_l - \sqrt{\varepsilon_r - \sin^2\theta_l}}{\varepsilon_r \cos\theta_l + \sqrt{\varepsilon_r - \sin^2\theta_l}} \quad R_h(\theta_l) = \frac{\cos\theta_l - \sqrt{\varepsilon_r - \sin^2\theta_l}}{\cos\theta_l + \sqrt{\varepsilon_r - \sin^2\theta_l}} \tag{32}$$

Thus, the induced currents on the rough surface can be expressed as:

$$\boldsymbol{J} = \hat{n}\times\boldsymbol{H} = \frac{1}{\eta_0}\left\{-(\hat{e}_i\cdot\hat{h})(\hat{n}\cdot\hat{k}_i)(1-R_H)\hat{h} + (\hat{e}_i\cdot\hat{v})(\hat{n}\times\hat{h})(1+R_V)\right\}E_0 p(x,y)e^{-jk\hat{k}_i\cdot\boldsymbol{r}} \tag{33}$$

$$\boldsymbol{M} = \boldsymbol{E}\times\hat{n} = -\left\{(\hat{e}_i\cdot\hat{h})(\hat{n}\times\hat{h})(1+R_H) + (\hat{e}_i\cdot\hat{v})(\hat{n}\cdot\hat{k}_i)(1-R_V)\hat{h}\right\}E_0 p(x,y)e^{-jk\hat{k}_i\cdot\boldsymbol{r}} \tag{34}$$

Where, the impedance matrix in free space is $\eta_0 = \sqrt{\mu_0/\varepsilon_0} = 120\pi$, the incident electric field is $\boldsymbol{E}^i(\boldsymbol{r}) = \hat{e}_i E_0 p(x,y)e^{-jk\hat{k}_i\cdot\boldsymbol{r}}$, $\boldsymbol{H}^i(\boldsymbol{r}) = \frac{1}{\eta_0}\hat{k}_i\times\boldsymbol{E}^i(\boldsymbol{r})$, $p(x,y)$ is the incident wave modulation function. When the incident wave is plan wave: $p(x,y)=1$.

2.4. *Interaction between object and a rough surface*

The interaction process between object and a rough surface at *n-th* step can be expressed as:

$$\begin{aligned} &\boldsymbol{E}_s^{i(n)}(\boldsymbol{r}) = \boldsymbol{E}^i(\boldsymbol{r}) + \boldsymbol{E}_{so}^{s(n-1)}(\boldsymbol{r}), \boldsymbol{r}\in S_s \\ &\boldsymbol{E}_o^{i(n)}(\boldsymbol{r}) = \boldsymbol{E}^i(\boldsymbol{r}) + \boldsymbol{E}_{os}^{s(n-1)}(\boldsymbol{r}), \boldsymbol{r}\in S_o \quad (n=1,2,.....) \end{aligned} \tag{35}$$

Where, $\boldsymbol{E}^i$ is the incident field, $\boldsymbol{E}_{so}^s$ and $\boldsymbol{E}_{os}^s$ are the interaction field. The interaction field $\boldsymbol{E}_{so}^s$ and $\boldsymbol{E}_{os}^s$ can be expressed as:

$$\begin{aligned} \boldsymbol{E}_{so}^{s(n-1)}(\boldsymbol{r}) = &\iint_{s_1}\{-j\omega\mu_0[\boldsymbol{J}_{1e}^{(n-1)}(\boldsymbol{r}')g(\boldsymbol{r},\boldsymbol{r}')+\frac{1}{k^2}\nabla'\cdot\boldsymbol{J}_{1e}^{(n-1)}(\boldsymbol{r}')\nabla g(\boldsymbol{r},\boldsymbol{r}')]\}dS' \\ &+\iint_{s_2}\{-j\omega\mu_0[\boldsymbol{J}_{2e}^{(n-1)}(\boldsymbol{r}')g(\boldsymbol{r},\boldsymbol{r}')+\frac{1}{k^2}\nabla'\cdot\boldsymbol{J}_{2e}^{(n-1)}(\boldsymbol{r}')\nabla g(\boldsymbol{r},\boldsymbol{r}')]+\boldsymbol{J}_{2m}^{(n-1)}(\boldsymbol{r}')\times\nabla g(\boldsymbol{r},\boldsymbol{r}')\}dS' \end{aligned} \tag{36}$$

$$\boldsymbol{E}_{os}^{s(n)}(\boldsymbol{r}) = \iint_{s_S}\{-j\omega\mu_0[\boldsymbol{J}_{se}^{(n)}(\boldsymbol{r}')g(\boldsymbol{r},\boldsymbol{r}')+\frac{1}{k^2}\nabla'\cdot\boldsymbol{J}_{se}^{(n)}(\boldsymbol{r}')\nabla g(\boldsymbol{r},\boldsymbol{r}')] + \boldsymbol{J}_{sm}^{(n)}(\boldsymbol{r}')\times\nabla g(\boldsymbol{r},\boldsymbol{r}')\}dS' \tag{37}$$

By updating the excitation item at the left part of equations (36) and (37), and then updating the induced electromagnetic current on the surface of a rough surface and object, the electromagnetic current on the surface of a rough surface and object are obtained.

The iteration error at *n-th* step is:

$$\xi_J^{(k)} = \frac{\|\boldsymbol{J}^{(k+1)} - \boldsymbol{J}^{(k)}\|_2}{\|\boldsymbol{J}^{(k)}\|_2}, \xi_M^{(k)} = \frac{\|\boldsymbol{M}^{(k+1)} - \boldsymbol{M}^{(k)}\|_2}{\|\boldsymbol{M}^{(k)}\|_2} \tag{38}$$

Where, $\|\cdot\|_2$ denotes norm-2. The convergence threshold of iteration error is set as 10^{-2}, at which point the induced electric current and magnetic current are supposed to be stable and the iteration process will stop. Finally, the electric current and magnetic current are obtained.

3. Numerical results

In the example, incident frequency is 300MHz, the polarization mode is vertical to vertical polarization (VV polarization), size of the rough surface is $L_x\times L_y$=20×20(m^2); root mean square (rms) is h=0.1m; the correlation of a rough surface is l_x=l_y=1m; size of the sphere is r=2m; height of sphere above underlying a rough surface is H=4m; the properties of the rough surface is medium; the properties of the sphere is perfect electric conductor, conduct-medium combination sphere and medium sphere, respectively. The permittivity and the magnetic conductivity of the medium part of conductor- media combination sphere is ε_r=4+0.3$j$and μ_r=1.2+0.6j, respectively. The permittivity and the magnetic conductivity of the medium sphere is ε_r= 4+0.3*j* and μ_r=1.2+0.6*j*, respectively. The comparing result is shown in figure 3.

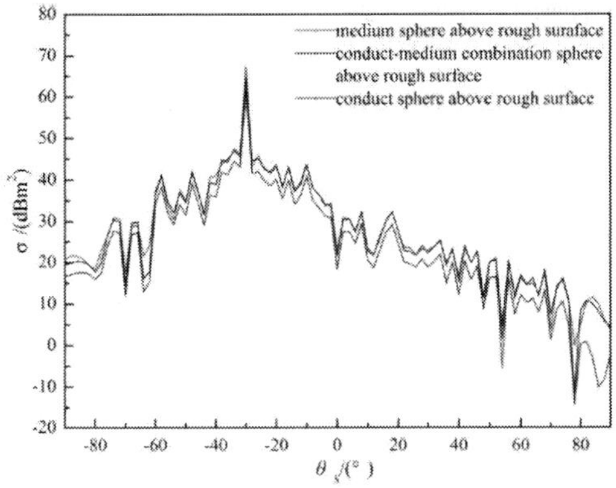

Figure 3. Comparing results of different types of object above a rough surface.

As we see, the scattering from medium object above a rough surface is the strongest and the scattering from conduct sphere above a rough surface is the weakest. What's more, for three examples, the mirror scattering is the strongest. The reason is that the scattering from media is stronger than conductor.

4. Conclusion

In this paper, a hybrid method combining the single integral equation (SIE) and the Kirchhoff approximation (KA)are proposed for computing the conductor- medium assembly object above a rough surface. First, the composite model from conductor-medium assembly object above a rough surface is constructed. Then, the scattering from the object is calculated by SIE and the scattering from a rough surface is calculated by KA. Next, the electromagnetic current on the object and the rough surface is obtained by iteration and the total scattering field is obtained. Finally, the scattering field from conduct- medium assembly object above a rough surface is compared to the scattering field from a conducting object and a medium object above a rough surface, respectively.

Acknowledgment

This work has been supported by the Nature Science Foundation of China (Grant No.61372033).

References

[1] Liang Y, Guo L X, Wu Z S *et al* 2016A study of composite electromagnetic scattering from an object near a rough sea surface using an efficient numerical algorithm *IEEE Antennas & Wireless Propagat. Lett.* **AP-15** 180-90

[2] Wang Q and Guo L X 2017 Composite electromagnetic scattering from a ship located on one-dimensional sea surface with time-domain hybrid method *Acta Phy. Sin.* **66**(18) 180301

[3] Tian G L, Tong C M, Long Z G *et al* 2017 A hybrid method for scattering from 3-D double cubic above a randomly Gaussian a rough surface *Air Force Eng. Univ. J. (Nat. Ed.)* **19** 62-6

[4] Tian G L, Tong M C *et al* 2019 An improved MoM-PO hubrid method for scattering from multiple 3D targets above the 2D random conducting rough surface *Electromag.* **39** 375-92

[5] Rao S M, Cha C C, Cravey R L and Wilkes D L 1991 Electromagnetic scattering from arbitrary shaped conducting bodies coated with lossy materials of arbitrary thickness *IEEE Trans. on Antennas & Propagation* **39** 627-31

[6] Rao S M, Sarkar T K, Midya P and Djordevic A R 1991 Electromagnetic radiation and scattering from finite conducting and dielectric structures: Surface/surface formulation *IEEE*

Trans on Antennas & Propagation **39** 1034-7

[7] Donepudi K C, Jin J M and Chew W C 2003 A higher order multilevel fast multipole algorithm for scattering from mixed conducting/dielectric bodies *IEEE T. Antennas & Propagation* **51** 2814-20

[8] Sarkar T K, Rao S M and Djordevic A R 1990 Electromagnetic scattering and radiation from finite microstrip structures *IEEE T. Microwave Theory and Techniques* **38**1 568-75

[9] Poggio A J and Miller E K 1973 Integral equation solutions of three dimensional scattering problems *Computer Techniques for Electromagnetics* chapter4 (New York: Permagon)

[10] Swatek D R and Ciric I R 1998 Single integral equation for wave scattering by a layered dielectric cylinder *IEEE T. Magnetics* **34** 2724-7

[11] Yeung M S 1999 Single integral equation for electromagnetic scattering by three-dimensional homogeneous dielectric objects *IEEE T. Antennas & Propagation* **47** 1615-22

IOP Conf. Series: Materials Science and Engineering **657** (2019) 012021 doi:10.1088/1757-899X/657/1/012021

Analysis on local force of cable tower in low tower cable-stayed bridge

T S Shen[1], T Z Hao[1,2,3,4], J Z Luo[3] and Q F Chen[3]

[1]School of Civil Engineering and Architecture, Guangxi University, Nanning, 530004, China
[2]Key Lab of Disaster Prevention and Structural Safety of Ministry of Education, Nanning 530004, China
[3]Guangxi Transportation Research & Consulting Co., Ltd., Nanning 530007, China

E-mail: htz0537@163.com

Abstract. Concrete under pylon saddles in extradosed cable-stayed bridges is usually subjected to tremendous cable force, which might easily cause cracking or crushing of concrete and lead to potential structural security problems. At present, the finite element mechanical analysis model for the cable saddle is not accurate enough to analyse the real stress situation. A stress analysis method of concrete under the cable saddle based on the accurate finite element model is presented in this paper. In order to obtain the exact stress results of the concrete under the saddle, this paper modelled the filament dividers in the saddle one by one, and established the loading surface element on the surface of each filament divider layer to apply the equivalent surface force layer by layer. In addition, this paper also studied the stress characteristics of the lower tower column in the tower bridge, and put forward structural reinforcement suggestions for the force of the lower tower column. The results of this paper can provide reference for the design and construction control of pylons in extradosed cable-stayed bridge.

1. Introduction

The external prestressing force of the cable-stayed cable passes through the pylon continuously by installing saddles in the pylon of the extradosed cable-stayed bridge, under the action of enormous cable force, the cable saddle would cause large local compressive stress easily to the concrete on the lower part of the cable saddle which in the pylon due to external prestressing, consequently, it is easy to cause local concrete crushing at the lower part of the cable saddle, as well as large-scale cracks, which buries the hidden danger of structural safety. Domestic and foreign scholars adopted different methods to simulate and analyse extradosed cable-stayed bridges from different perspectives.

Zhang *et al* [1] used ANSYS to analyse the pylon stress of extradosed cable-stayed bridge, they used contact element to simulate the contact relationship between cable and steel pipe, but the saddle structure of sub-steel-pipe is not simplified as a single bundle that ignoring the influence of external steel pipe, and each steel strand interaction of the sub-steel-pipe can't be simulated and analysed. Tan *et al* [2] carried out the model test that main tower saddle segment of prestressed extradosed cable-stayed bridge by using the improved cable saddle structure of the sub-steel-pipe, and a spatial finite element model is established to analyse the stress distribution of concrete under saddle and the stress characteristics of sub-steel-pipe by applying the actual cable force to the corresponding channels according to the surface pressure in the form of spatial paraboloid. Zhang *et al* [3] used three-dimensional finite element software to establish the finite element model of extradosed cable-stayed bridge cable saddle area, selected the most disadvantageous load in the overall calculation for static analysis, the stress distribution of concrete and cable duct were obtained in cable saddle area of bridge pylon. Liu *et al* [4] carried out full-scale model tests of the segment in the cable saddle area combined with a low-tower cable-stayed bridge project, two kinds of double-sleeve cable saddles bond anchorage structures were selected and the transmission degree of unbalanced cable force was tested by low-cycle repeated loading of single-side cable. In addition, the mechanical characteristics of concrete continuum in cable saddle area are analysed and studied, the results show that two kinds of bond anchorage structures have strong and stable anti-slip effect. Zhang [5] used the spatial finite element method to calculate the stress of the saddle. The cable force was converted into parabolic surface pressure and applied to the corresponding channels, the comparison between the sub-steel-pipe and double sleeves shown that the actual effect was closer to the actual situation than the average method and the application method. Zhu [6] took a extradosed cable-stayed bridge as the research object and calculate the stress of cable saddle by space finite element method, the cable force was converted into parabolic surface pressure and applied on the corresponding channels, he considered that the actual effect of the deviation of the tunnel construction on the tower in operation stage was better, which was closer to the actual situation than the previous average method and application method. Wang [7] took an extradosed cable-stayed bridge that the twin-tower single-cable-plane with three-span prestressed concrete as the research background, and used Midas to establish the spatial finite element model of the pylon and saddle area, under the maximum cable force, the spatial force of the whole pylon and the local stress of the concrete in the pylon saddle area were analysed. Liu [8] made use of full-scale model test to study the value and distribution of splitting stress in concrete of main tower, the spatial finite element entity model was established by using ANSYS to analyse the stress distribution under saddle and the stress performance of the filament divider under cable force loading.

From what has been discussed above, although a large number of scholars have carried out a variety of entity analysis on the stress distribution law of saddle and the local concrete stress distribution law of saddle in the model test of extradosed cable-stayed bridge, and get a macroscopic grasp of the stress state about the local concrete with saddle, there are still some problems as follows: the saddle model established mostly uses approximate shape blocks, it fails to describe the concrete stress condition of the structure accurately under the condition that steel strand concentrated action of multiple sub-steel-pipes, the stress distribution of concrete under saddle is less accurate. Therefore, this paper uses ANSYS finite element software to analyse the interaction between the saddle and the concrete of the pylon by establishing the exact model of the cable-stayed saddle pylon and applying the cable force on the corresponding channels by means of the average method, which can provide reference for the design and construction control of the pylon of the extradosed cable-stayed bridge.

2. General situation of the engineering

As shown in figure 1, there is an extradosed cable-stayed bridge with the span arrangement of 90+165+90 m, because of the tower bridge beneath the cable saddle has a large force, it is easy to crush the concrete bridge tower locally. This paper carries out force analysis on the saddle locally stressed area of the bridge. The detailed geometric structure of cable tower shown in figure 2, which including the elevation, side and section. The interaction between the steel cable saddle and the concrete of the cable tower is analysed by accurately establishing the model of the steel cable saddle concrete pylon. In addition, the stress characteristics of the lower tower column which belongs to the tower bridge was studied in this paper, besides, this paper also puts forward structural reinforcement suggestions for the force of the lower tower column.

IOP Conf. Series: Materials Science and Engineering **657** (2019) 012021 doi:10.1088/1757-899X/657/1/012021

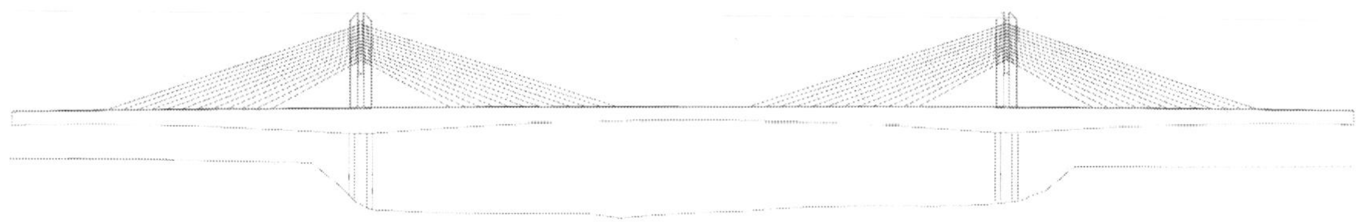

Figure 1. Elevation diagram of low tower cable-stayed bridge.

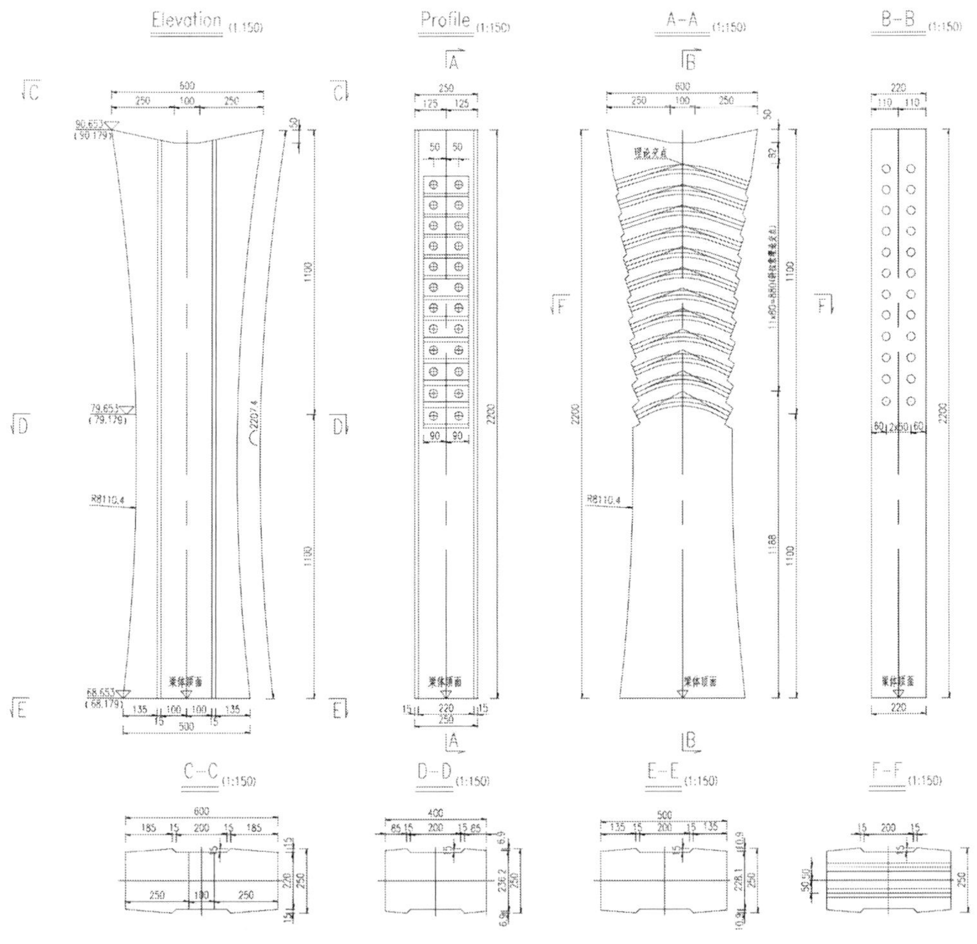

Figure 2. Detailed geometric structure of cable tower.

3. Finite element analysis of cable saddle and tower column

The steel cable saddle adopts solid 45 element, each sub-steel-pipe is represented by solid element, the material properties are defined as steel, and the holes in sub-steel-pipe are ignored, besides, the solid 45 unit is also adopted by the concrete of cable tower, and the joint coupling is adopted for the contact between steel saddle and concrete tower.

Due to the meticulous consideration about the cable saddle sub-steel-pipe, there is a large number of model elements if we establish the sub-steel-pipe accurately by entity elements, it will more difficult to establish the model of cable tower completely. Therefore, the analysis range of the cable tower selected in this paper is C1 and C2 (the number of the cable is C1-C12, which varies from the side of the tower to the middle of the span) corresponding to the cable saddle of the cable tower, the steel wire in the steel cable saddle is not simulated. Besides, the force acting on the cable saddle is simulated by the normal surface force perpendicular to the cable saddle sub-steel-pipe, and the surface force can be calculated by formula (1). The calculation results of normal equivalent forces on the surface of cable saddle sub-steel-pipe corresponding to C1 and C2 cable saddles are shown in table 1, the element type used in the model is shown in table 2.

$$q_x = \frac{F_s}{R} \tag{1}$$

Where, F_s is cable force, R is saddle radius.

Table 1. Normal equivalent forces on the surface of cable saddle sub-steel-pipe corresponding to C1 and C2 cable saddles.

	Surface of sub-steel-pipe with C1	Surface of sub-steel-pipe with C2
The magnitude of cable force (kN)	4170	4237
Radius of cable saddle (m)	3.9	4.0
The magnitude of equivalent force (kN/m)	1069.2	1059.3
The magnitude of equivalent surface load (kN/m^2)	1204.1	1193.0

Table 2. Model / Element table adopted.

Name of component	Types of element	Modulus of elasticity (MPa)
Steel saddle	Solid45	2.1×10e5
Concrete pylon	Solid45	3.45×10e4
Loading element	Shell63(without thickness)	—

The internal forces of the above pylons are simulated by the pylon equivalent weight and the stayed cable vertical force component, and the surface forces are loaded on the finite element model, the calculated values are shown in table 3. Since there are many elements to be divided, a 1/4 model is used to calculate and symillimetersetrical constraint is imposed on the symillimetersetrical surface, the established finite element model is shown in figures 3 and 4. As shown in figure 5, the normal surface force is applied to the surface element through the generating surface element of the sub-steel-pipe.

Table 3. Vertical equivalent forces of cables loaded on the upper surface of cable saddle finite element model.

Cable number	Horizontal inclination angle of cable θ	Sin θ	Cable force kN	Vertical force composition of unilateral cable force kN	Vertical component of cable force kN
C3	22.7	0.385906	4272	1649	3297
C4	21.4	0.364877	4306	1571	3142
C5	20.4	0.348572	4339	1512	3025
C6	19.5	0.333807	4368	1458	2916
C7	18.8	0.322266	4394	1416	2832
C8	18.2	0.312335	4418	1380	2760
C9	17.7	0.304033	4439	1350	2699
C10	17.2	0.295708	4454	1317	2634
C11	16.8	0.289032	4467	1291	2582
C12	16.4	0.282341	4472	1263	2525

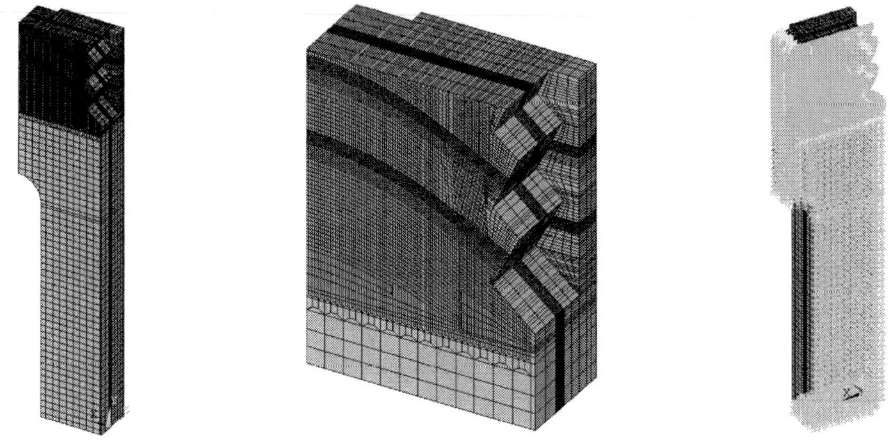

(a) (b) (c)

Figure 3. Finite element model of cable tower. (a) 1/4 model of cable tower (whole), (b) 1/4 model of cable tower (local) and (c) Global boundary conditions of the model.

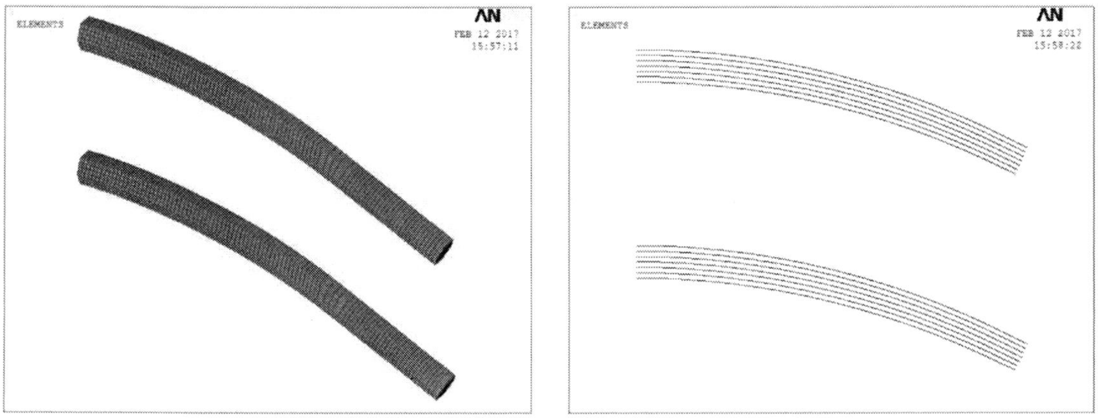

Figure 4. Cable saddle and loading plate element.

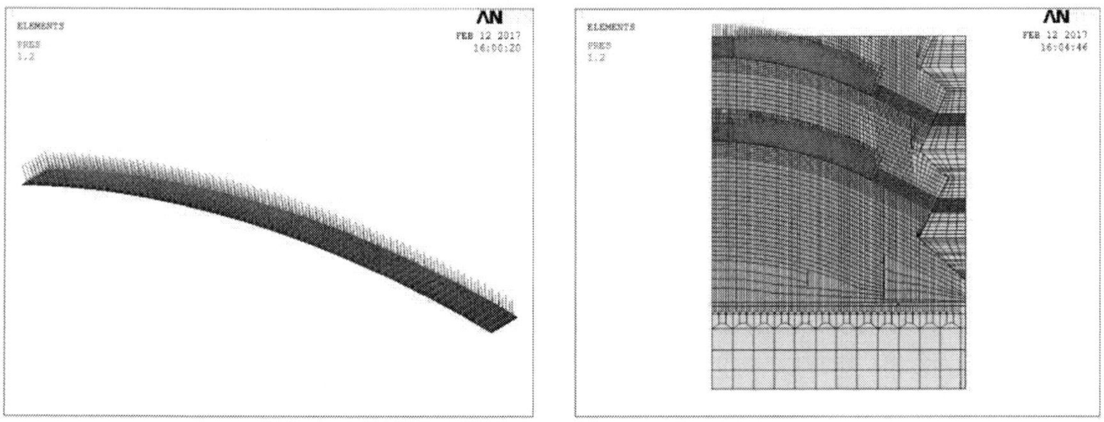

Figure 5. Equivalent force loading of steel wire on cable saddle loading plate element.

The load on the model surface must also include the self-weight of the tower, which is 0.260 N/mm^2, and the vertical force of the cables which causes the pressure load on the model surface is 4.736 N/mm^2. The total of the two loads is 4.996 N/mm^2.

(1) The calculation results of the contact position between the lower part of the cable saddle and the concrete are shown in figures 6-10.

From the calculation results in figures 6-10, it can be seen that the regular vertical (along Y axis) compressive stress distribution appears in the lower part of saddle under the cable saddle force action, and the magnitude of compressive stress is 7.31 MPa-11.41 MPa; The inner side of the saddle arc in the X direction is basically under compression, the tensile stress with a maximum of 1.15 MPa in the X direction will appear at the junction of the arc section and the straight section of the cable saddle.

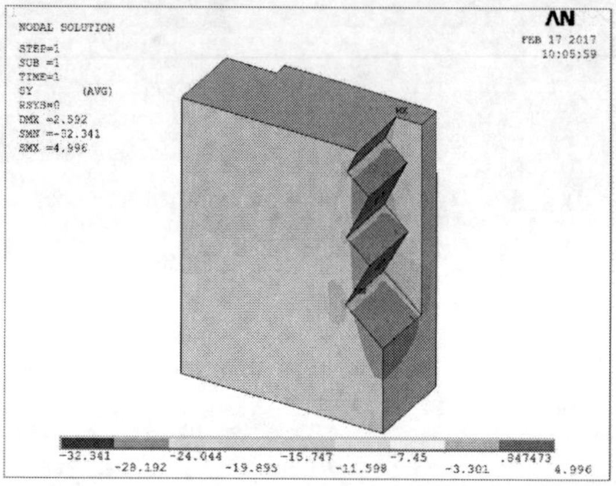

Figure 6. Y-direction stress nephogram on the outer surface of concrete tower column at saddle position.

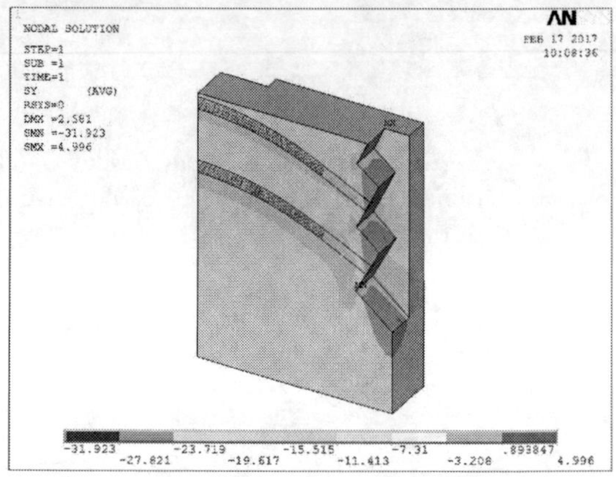

Figure 7. Y-direction stress nephogram of the section of cable saddle of concrete tower column (right side).

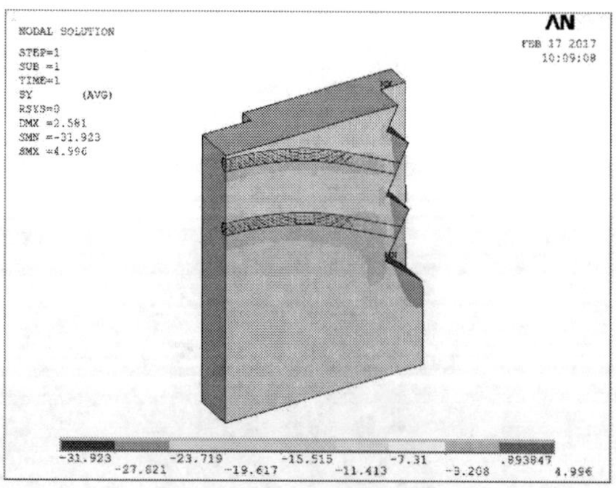

Figure 8. Y-direction stress nephogram of the section of cable saddle of concrete tower column (left side).

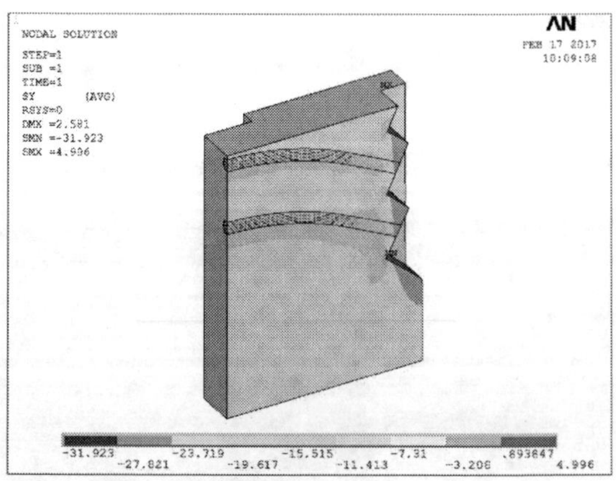

Figure 9. X-directional stress in the section of cable saddle of concrete tower column.

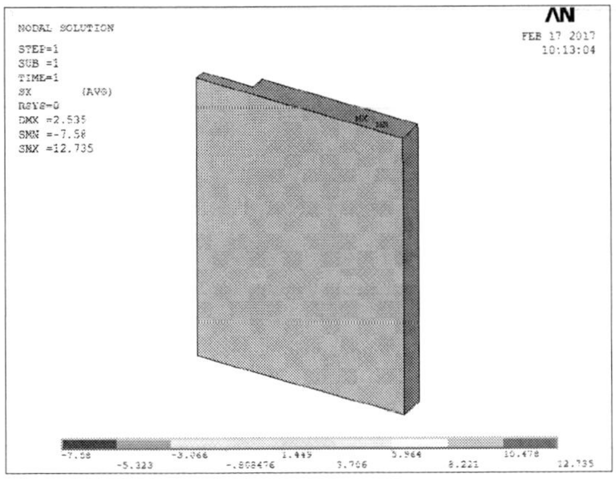

Figure 10. Stress calculation near end plate.

(2) The stress calculation results at the variable section of the main tower are shown in figures 11-16. It can be seen from figures 11 and 12, viewing from the force acting on the whole cable tower in the X-direction, there is a large area of tensile stress along the X-axis direction at the cross section of the cable tower (the junction between the upper tower column and the lower tower column), as shown in figure 13. The range of tensile stress is between 0.41 MPa and 8.10 MPa (mean value is 4.2 MPa), especially in a small area at the top of the arc section, the maximum tensile stress is 8.10 MPa, this is due to the change of section at the upper and lower pylons junction and the eccentric force at the lower pylon section. Figure 15 is the overall Y-direction stress diagram of the pylon, and figure 16 is the overall deformation diagram of the pylon in the X-direction. As can be seen from figure 15, the Y-direction stress in the upper part of the lower pylon in the inner side is larger than the outer side, while the case in the lower part of the lower pylon is contrary, the Y-direction stress in the inner side is smaller than the outer side. From Fig. 16, it can be seen that the bending deformation of the lower pylon in the overall direction of X-direction is the largest with the value of 0.3 mm at a distance about 5-6 meters from the bridge deck. Prestressed tension or transverse reinforcement infill treatment is needed to prevent crack propagation, or a tension member is installed at the lower tower column 5-6 meters away from the bridge deck, which can also reduce the large transverse tension stress at the variable section.

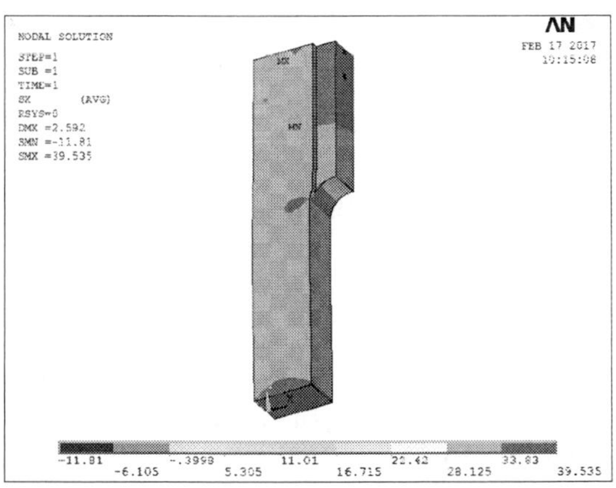

Figure 11. X-axis stress nephogram of the main tower (front).

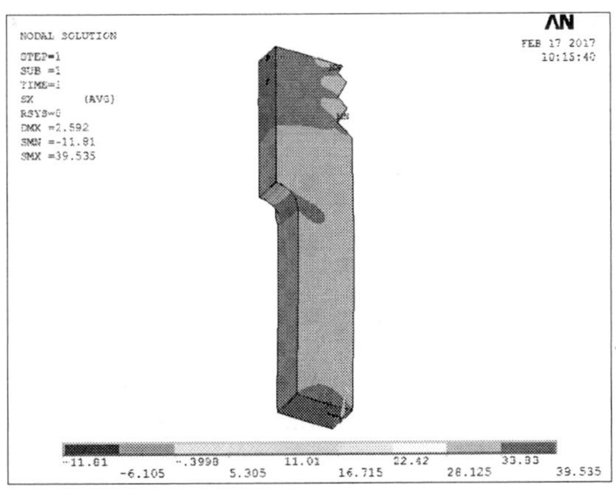

Figure 12. X-axis stress nephogram (interface) of the main tower.

IOP Conf. Series: Materials Science and Engineering **657** (2019) 012021 doi:10.1088/1757-899X/657/1/012021

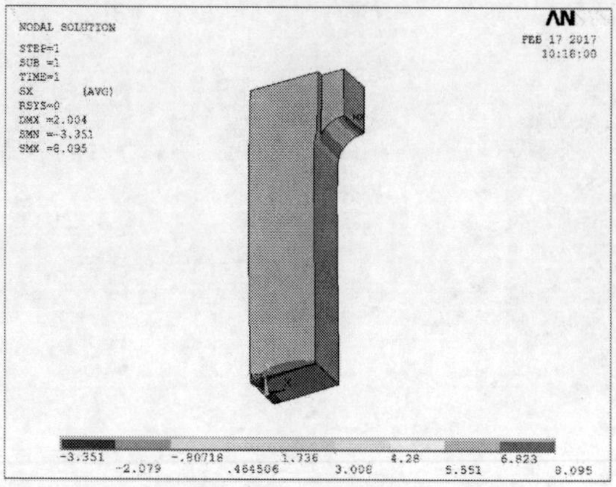

Figure 13. X-axis stress at variable section of main tower.

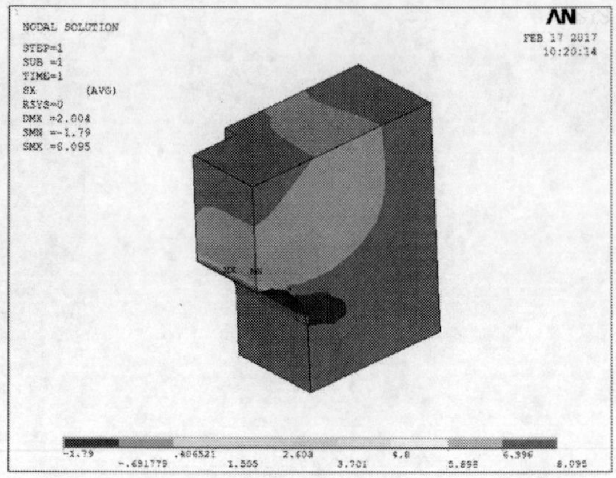

Figure 14. X-axis local stress at variable section of main tower.

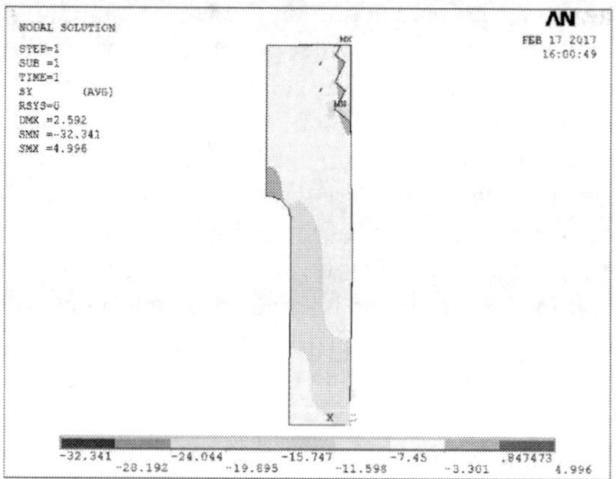

Figure 15. Y-directional stress of cable tower.

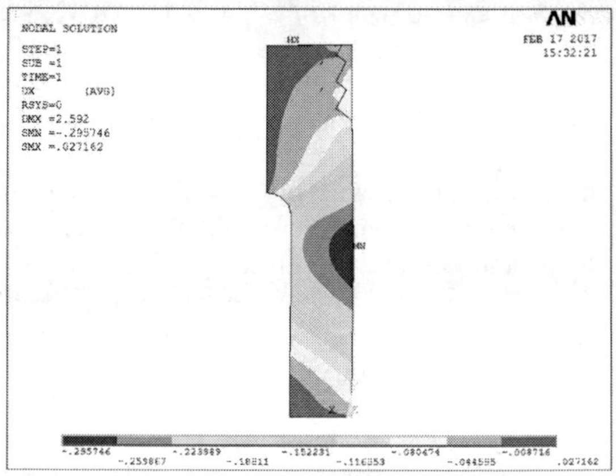

Figure 16. Integral deformation diagram of X-axis cable Tower.

4. Conclusion

In this paper, a stress analysis method of concrete under saddle based on the exact finite element model is presented. The feasibility of this method is verified by the analysis of engineering cases, and the following conclusions are drawn:

- Under the action of local cable saddle force and upper load, regular vertical compressive stress distribution (along Y axis) appears at the lower part of cable saddle with the value of 7.31 MPa to 11.41 MPa. The inner side of the cable saddle arc in the X direction is basically under compression. At the junction of the arc section and the straight section of the saddle, the X direction tensile stress will appear, and the maximum tensile stress is 1.15 MPa, which meets the strength design requirements of C50 concrete.
- From the overall X-direction stress of the pylon, there is a large area of tension stress along the X-axis at the variable section of the pylon (the junction of the upper and lower pylons) , this is due to the change of section at the upper and lower pylons junction and the eccentric force at the lower pylon section. Prestressed tension or transverse reinforcement is needed to prevent crack propagation.

Acknowledgments
This research was supported by Key Laboratory of Disaster Prevention and Structural Safety of Ministry of China (grant no. 2016ZDK004).

References

[1] Zhang H W and Li Y D 2009 Local stress analysis of cable saddle concrete in low tower cable-stayed bridge *Railway Standard Design* **1** 42-4
[2] Tan C J, Zhu B and Ji W H 2008 Model test of cable saddle segment in low tower cable-stayed bridge *J. China & Foreign Highway* **28** 170-2
[3] Zhang S Q and Qu J H 2014 Static analysis of cable saddle in low tower cable-stayed bridge *Transport. Sci. Technol.* **2014** 25-7
[4] Liu Z, Meng S P and Zang H 2007 Model test and design investigation on saddle deviator zone of extra-dosed bridge *J. Southeast Univ.: Nat. Sci.* **37** 291-5
[5] Cai X M, Zhang L M and He H 2009 Analysis of saddle structures in low tower cable-stayed bridge *J. Highway Transport. Res. Dev.* **2009** 89
[6] Zhu J 2006 The analysis of saddle structures in low tower cable-stayed bridge *Shanxi Sci. Technol. Commun.* **4** 44-6
[7] Wang L F, Xiao Z W and Wang Z Q 2012 Spatial stress analysis of pylons and cable saddles in low-pylon cable-stayed bridge Journal of *Journal of China & Foreign Highway* **32** 188-91
[8] Liu Z W 2012 Model test and theoretical study on stress distribution in cable saddle area of railway low tower cable-stayed bridge (Gansu, China: Lanzhou Jiaotong University)

IOP Conf. Series: Materials Science and Engineering **657** (2019) 012022 doi:10.1088/1757-899X/657/1/012022

Lattice Boltzmann simulation of the methane backward flow in coal mine tunnels after methane outburst

Q J Wang[1,2], Y C Zhao[1,3] and Y Nie[1]

[1]College of Emergency Management and Safety Engineering, China University of Mining and Technology, Beijing, China
[2]National Coal Mine Safety Administration, China

E-mail: safety_zyc@student.cumtb.edu.cn

Abstract. Numerical study of the methane backward flow in coal mine tunnels after a methane outburst is necessary to the engineering design and emergency rescue. In this paper, we introduce Lattice Boltzmann Method (LBM) coupled with the Large Eddy Simulation (LES) model to simulate the methane backward flow in the air-intake tunnel of twin-tunnel construction in the coal mine after the methane outburst and analyze the effects of the inlet air velocity, the intensity of the methane outburst and the width of the tunnel to the distance that the methane can flow backward. It demonstrates that with the increase of the inlet air velocity, the distance that the methane can flow backward decreases. With the increase of the intensity of the methane outburst, the distance that the methane can flow backward increases. There exist linear relationships between the inlet air velocity, the intensity of the methane outburst and the distance that the methane can flow backward. The change of the tunnel width has almost no influence on the distance that the methane can flow backward.

1. Introduction

The methane in coal bed emits to the tunnel when tunneling in the underground coal mines. The "methane outburst" is the phenomena that extremely large amounts of coal-bed methane emits to the coal mine tunnel in an extremely brief period. This period can be divided into two stages. In stage one, the coal-bed methane desorbs from the coal body and damages the original coal body structure. In stage two, the coal-bed methane burst into the tunnel and interact with the underground air [1]. Numerous violent methane outbursts have occurred in underground coal mines worldwide, sometimes with fatal consequences for workers. When the methane outburst does not occur, the ventilation system of the coal mine can dilute the methane below the critical concentration which may cause the explosion and asphyxiation accidents [2,3]. However, after methane outburst, the fresh air provided by the ventilation system will not be enough to dilute the methane and even the air flow state will be changed.

Figure 1(a) shows the main network of the ventilation system in a coal mine. Under the action of the main fan, the fresh air flows through the intake-air shafts, the main intake-air tunnels, the intake-air tunnels of the coal face, the coal face, the return-air tunnels of the coal face, the main return-air tunnel and the return-air shafts in sequence. The air that has not yet flowed through the coal face is regarded as fresh air. The air is polluted by the emitted methane, dust and other harmful gas when it flows through the coal face. The region with fresh air is safer than the region with polluted air. Most of the workers and equipment are in the region with fresh air. The blue region in figure 1(a) represents the region with fresh air and the yellow region represents the region with polluted air.

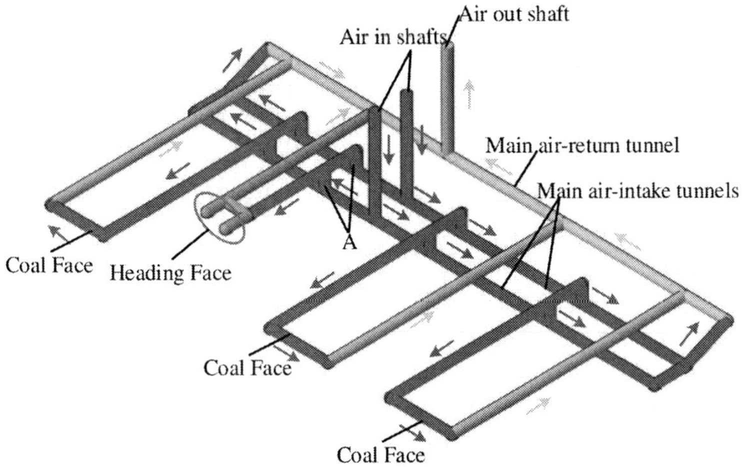

(a)

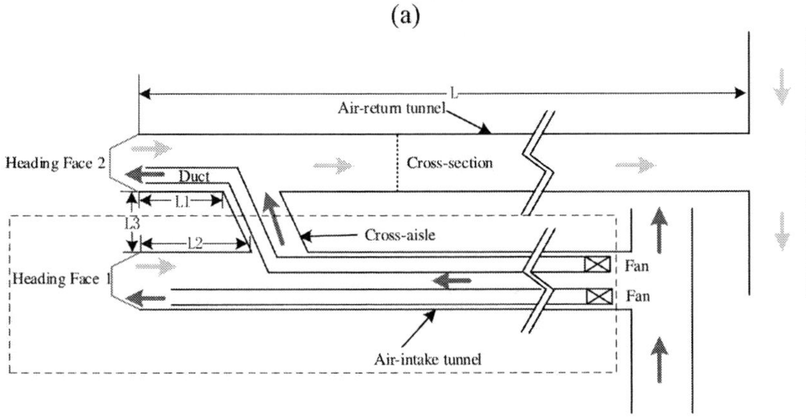

(b)

Figure 1. The ventilation system in a coal mine. (a)The main network of the ventilation system and (b) The local ventilation system of the twin-tunnel.

Methane outburst most likely occurs in the heading face of the coal mine tunnel. Twin-tunnel construction is widely adopted in coal mines when tunneling in the coal bed. Figure 1(b) shows the local ventilation system of the twin tunnel. The heading faces depend on the fans and ducts for forced ventilation. The blue arrows in figure 1(b) represents the flow direction of the fresh air and the yellow arrows represent the flow direction of polluted air. If the methane outburst occurs at the heading face, the huge momentum carried by the emitted methane will change the flow direction of the underground air. The methane will get into the intake-air tunnel due to the interaction between the methane and air. As mentioned above, there are more workers and equipment in this tunnel. Therefore, explosion and asphyxiation accidents will be more easily caused. Furthermore, if the methane can reach the point A as shown in figure 1(a), the methane will enter the main intake-air tunnels and it will spread all over the mine through the ventilation system, then more serious accidents will be caused. This phenomenon is called the "backward flow" of the methane. To study the relationship between the maximum distance that the methane can flow backward and its influencing factors can help conduct engineering design and take rescue measures after methane outburst.

Many authors have investigated the methane behavior in different tunnels of coal mines. Nakayama *et al* [4] developed a Computational Fluid Dynamics (CFD) model for methane distribution in mining face. An important finding of Nakayama's study is that the methane concentration was found higher at location transversely in and along with the corner space where the face end meets the ceiling and the floor as well as the area underneath the ventilation duct. Ichinose *et al* [5] validated Nakayama's model using experimental results. Torano *et al* [6] investigated the airflow behavior and methane dispersion in one and two auxiliary ventilation systems by CFD models and validated them by the data from the underground coal mine Carbonara SA located in Northern Spain. They also analyzed the advantages and drawbacks of risk indexes with a greater impact on the prediction and control of the methane outbursts in a heading driven in the underground coal mines of Hullera Vasco Leonesa SA (HVL). Rodríguez *et al* [7] studied the methane emissions in a tunnel excavated through carboniferous strata based on underground coal mining experience. Fang *et al* [8] investigated the airflow behavior and methane dispersion near the cross-aisle of a twin-tunnel construction by a three-dimensional CFD model and validated the model by field measurements data. They also obtained the methods to eliminate the "dead zone" in the tunnel. Zhou *et al* [9] simulated the methane distribution at a continuous mining face with various curtain setback distance based on the CFD model and validated it by the experimental data collected in a full-scale ventilation test facility. Kurnia *et al* [10] studied the methane dispersion in a mine tunnel with discrete methane sources based on CFD approach and explored the methods to handle the methane in the tunnel. Mishra *et al* [11] conducted a CFD study on the dispersion of methane in the tailgate of a retreating longwall mine and the simulation results agree well with the experimental data obtained from a full-scale coal mine. Qin *et al* [12] investigated the longwall goaf methane flows and borehole drainage performance by CFD simulation. Li *et al* [13] assessed the hazardous zone in the coal mine tunnel after methane outburst based on CFD method. Lu and Wei [14] adopted a kind of multi-component LBM model to study the gas migration and hindering in the underground tunnel but they didn't apply the turbulent model in the simulation. In general, the studies on the methane behavior in the coal mine tunnel mainly focused on the problems that the methane outburst has not occurred, the backward flow of the methane in the tunnel after methane outburst was rarely investigated. The methods used for investigating the methane behavior in the coal mine tunnel are mainly based on the traditional CFD methods which solving the Navier-Stokes (N-S) equation.

The Lattice Boltzmann Method (LBM) as a potential numerical method has several advantages over the conventional CFD methods [15]. The advantages are mainly embodied in its numerical stability [16] and accuracy [17] and the capacity to efficiently handle complex geometries. Besides, the LBM is an explicit numerical scheme with only local operations. It has the advantage of being easy to implement and is especially well suited for massively parallel machines like graphics processing units (GPUs) [18]. As a consequence of the GPU's architecture, computational speed-up of many orders of magnitude can be achieved in comparison to traditional CPU simulation [19]. In addition, the LBM is widely adopted in the simulation of the coal-bed methane migration, adsorption, and desorption process in a coal seam in recent years [20,21]. Adopting the LBM is convenient for making better use of previous research to simulate the whole process of the methane outburst in the future.

The physical essence of the methane backward flow in coal mine tunnels is the interaction between the methane and underground air under turbulent condition. There are mainly three models to simulate the multi-component gas flow using LBM. They are the passive scalar model, the single collision model, and the split collision model respectively [22,23]. The passive scalar model is the easiest to implement but it can only simulate the effect of one of the components to others. This model is just suitable for the situation that the molar fraction of one of the components is extremely higher than others and the effect of other components can be neglected [24]. The single collision model can simulate the interaction among the components but the viscosity of each component cannot be tuned separately [25]. The split collision model can overcome above shortages [26] and only this model has the capability to simulate the methane backward flow in coal mine tunnels after methane outburst. There are mainly three means to simulate the turbulent flow using LBM. They are Direct Numerical Simulation (DNS) [27], Large Eddy Simulation (LES) [28] and Reynolds Average Navier-Stokes (RANS) [29] respectively. The DNS model is well accepted in Physics because of its exceptional accuracy but the DNS model is uneconomical to simulate the engineering problems because it requires enormous computational ability [30]. The RANS models offer the most economical approach for computing complex turbulent industrial flows but the RANS models are the time-averaged turbulence models and it cannot give detailed fluctuation of velocities in the turbulent flow simulation. The LES model can give the detailed fluctuation of velocities in the turbulent flow simulation and it requires less computational ability than the DNS model [31].

The LBM coupled with the LES made great progress in recent years and it has been successfully used in the simulation of many complex engineering problems at the macro level. Ren *et al* [32] designed a GPU-accelerated solver for turbulent flow and scalar transport based on the LBM-LES model. Nathen *et al* [33] studied the adaptive filtering for the simulation of turbulent flows with LBM-LES model. Premnath *et al* [34] investigated the dynamic subgrid-scale modeling of turbulent flows using LBM-LES model. Ahmad *et al* [35] simulated the gust index in an urban area using this method. Cheng *et al* [36] employed this method to simulate the wind gust structure in the atmospheric boundary layer. King *et al* [15] studied urban airflow and natural ventilation using this method. Sajjadi *et al* [37] simulated the indoor airflow and particle dispersion and deposition based on this method. Jacob *et al* [38] assessed the wind comfort in full-scale city area using this method. Mayer and Hazi [39,40] simulated the longitudinal flow along with the triangular array of rods using this method.

The aim of this research is to introduce and develop the multi-component LBM coupled with LES model that can simulate the backward flow of the methane in the tunnel after methane outburst and study the relationship between the maximum distance that the methane can flow backward in the tunnel after methane outburst and the air velocity, the intensity of the methane outburst, and the width of the tunnel.

2. Lattice Boltzmann model

At the macro level, the methane-air flow can be deemed as Convective-Diffusion phenomenon and the Convection-Diffusion equation [41] can describe it.

$$\frac{\partial C}{\partial t} = \nabla \bullet \left(D \nabla C\right) - \nabla \bullet \left(C \vec{u}\right) \tag{1}$$

where C is the volume fraction of methane or air, t is the time, $\vec{u}$ is the velocity of mixed gases, D is the diffusion coefficient between methane and air.

The incompressible Navier-Stokes equation can give the velocity of mixed gases.

$$\frac{D\vec{u}}{Dt} = \vec{F} - \frac{1}{\rho}\nabla P + \upsilon \nabla^2 \vec{u} \tag{2}$$

where ρ is the density of the mixed gases, υ is the kinematic viscosity coefficient, $\vec{F}$ is the body force. The coupled solution of incompressible N-S Equation and Convection-Diffusion Equation can calculate the velocity field of the mixed gases and concentration field of various gases. However, the physical essence of the methane backward flow after methane outburst in the coal mine is the collision between methane molecules and air molecules. The macro control equation is challenging to describe the physical essence of the interaction between methane and underground air. LBM as a numerical simulation method based on Statistical Physics, it has rapidly emerged as a powerful technique with great potential for numerically solving momentum, energy, species transport, and multiphase problem. In this paper, we aim to use LBM to solve the interaction process between the air and methane in the tunnel after methane outburst.

The evolution equation of the split collision Lattice Boltzmann model [42] is

$$f_\alpha^i\left(\vec{r} + e_\alpha^i \Delta t, t + \Delta t\right) - f_\alpha^i\left(\vec{r}, t\right) = \Omega_\alpha^i \tag{3}$$

where $\vec{r}$ is the spatial position, Δt is the time step, f_α^i is the distribution function of component i,

Ω_α^i is the collision term. In two-dimensional case, D2Q9 model [43] is most popular and the e_α^i is

$$e_\alpha^i = \begin{cases} 0 & \alpha = 0 \\ c^i\left(\cos\left((i-1)\pi/2\right), \sin\left((i-1)\pi/2\right)\right) & \alpha = 1,2,3,4 \\ \sqrt{2}c^i\left(\cos\left((i-5)\pi/2+\pi/4\right), \sin\left((i-5)\pi/2+\pi/4\right)\right) & \alpha = 5,6,7,8 \end{cases} \tag{4}$$

c^i represents the lattice speed for component i. We adopt the Different Lattice Speed (DLS) [44,45] scheme herein to simulate the fluids consist of the components with different molecular weights. Assuming component 1 is the lightest component and the definition of c^i is $c^i = c^1\sqrt{M_1/M_i}$. M_i is the molecular weight of component i. The differences of the molecular weight of different components are neglected in this paper and assumes that $c^i=1$.

The definition of the collision term Ω_α^i is

$$\Omega_\alpha^i = \Omega_\alpha^{ii} + \sum_{j=1, j\neq i}^{N} \Omega_\alpha^{ij} \tag{5}$$

where Ω_α^{ii} is the self-collision term and its definition is

$$\Omega_\alpha^{ii} = -\left(f_\alpha^i - f_\alpha^{i,eq}\right)/\tau_i \tag{6}$$

where τ_i is the relaxation time for self-collision which controls the kinematic viscosity of component i via $v_i = c_{s,i}^2\left(\tau_i - 1/2\right)\Delta t$, and $c_{s,i}^2$ is the speed of sound. We can calculate the equilibrium distribution function $f_\alpha^{i,eq}$ through the following equation [26].

$$f_\alpha^{i,eq} = \omega_\alpha \rho_i \left(1 + \frac{1}{c_{s,i}^2}\left(e_\alpha^i - \vec{u}\right)\left(\vec{u}_i^{eq} - \vec{u}\right)\right)\left(1 + \frac{e_\alpha^i \bullet \vec{u}}{c_{s,i}^2} + \frac{\left(e_\alpha^i \bullet \vec{u}\right)^2}{2c_{s,i}^4} - \frac{\vec{u}\bullet\vec{u}}{2c_{s,i}^2}\right) \tag{7}$$

with the weights $\omega_0 = 4/9, \omega_{1\text{-}4} = 1/9, \omega_{5\text{-}8} = 1/36$. The relationships between the mass concentration ρ_i and the equilibrium velocity $\vec{u}_i^{eq}$ of each component and the distribution functions are as follows:

$$\rho_i = \sum_\alpha f_\alpha^i = \sum_\alpha f_\alpha^{i,eq} \tag{8}$$

$$\rho_i \vec{u}_i^{eq} = \sum_\alpha e_\alpha^i f_\alpha^i = \sum_\alpha e_\alpha^i f_\alpha^{i,eq} \tag{9}$$

The expressions for the total concentration and the mixture velocity are

$$\rho = \sum_{i=1}^{N} \rho_i \tag{10}$$

$$\rho\vec{u} = \sum_{i=1}^{N} \rho_i \vec{u}_i^{eq} \tag{11}$$

Ω_α^{ij} represents the effect of cross-collisions between different components. Its definition is [26]

$$\Omega_\alpha^{ij} = -\frac{1}{\tau_{ij}}\left(\frac{\rho_j}{\rho}\right)\frac{f_\alpha^{i,eq}}{c_{s,i}^2}\left(e_\alpha^i - \vec{u}\right)\left(\vec{u}_i^{eq} - \vec{u}_j^{eq}\right) \tag{12}$$

Owing to the large computation domain of the coal mine tunnel, the flow of the methane and air is obvious turbulence and we must introduce the turbulence model. The LBM-LES model means the LBM formation of the standard Smagorinsky model of the Large Eddy Simulation (LES) [46].

Introducing the dimensionless number *Re* and *Sc* to realize the conversion of the physical unit and lattice unit. Both the *Re* and *Sc* are constant no matter in the physical system or the lattice system. The definitions of the dimensionless number *Re* and *Sc* are as follows

$$Re = \frac{ud}{\upsilon}, \quad Sc = \frac{\upsilon}{D} \tag{13}$$

where u is the characteristic velocity of the methane-air mixture, d is the width of the tunnel, υ is the viscosity of the methane-air mixture, D is the diffusion coefficient between the methane and air.

3. Full-scale simulation of the methane backward flow after methane outburst

3.1. Object of the full-scale simulation

In order to explore the mechanism and influencing factors of methane backward flow. We select the region in the dotted line frame in figure 1(b) as the object of the simulation. Figure 2 shows the physical dimensions and boundary conditions of the simulation. The tunnel is 200 m long, 5 m wide and 3 m high. Assuming that the methane outburst will destroy the air ducts. The inlet of the tunnel applies the air velocity boundary, the outlet of the cross-aisle applies the fully developed boundary, the wall of the tunnel applies the bounce back boundary, the heading face applies the methane velocity boundary during methane outburst and the bounce back boundary before and after methane outburst.

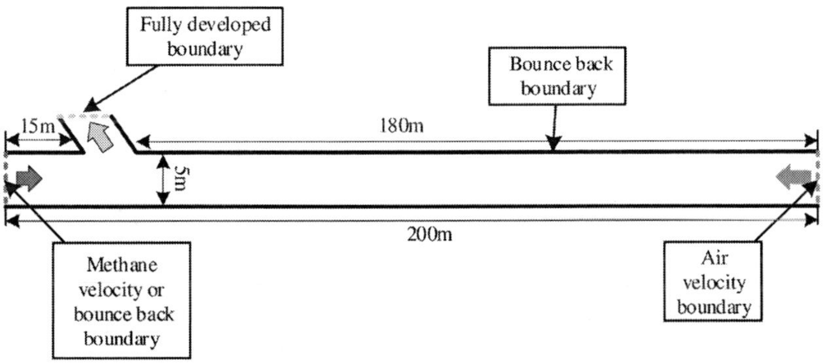

Figure 2. Boundary conditions of the simulation.

We tested three kinds of grid resolution and they were 0.25 m/dx, 0.5 m/dx and 1 m/dx respectively. Three kinds of grid resolution obtained the approximate simulated results. Considering the computing time, this paper adopts the grid resolution of 0.5 m/dx.

3.2. Result and discussion

Figure 3 shows the initial air flow state in the air-intake tunnel before methane outburst (the 50s) when the inlet-velocity is 1 m/s. Figure 4 shows the methane concentration field at different evolution time when the intensity of the methane outburst is 150 m^3/s and the inlet-velocity is 0.5 m/s. As we can see, during the methane outburst (50s-70s), the methane can flow both along and against the initial air flow. The distance that the methane flow backward increases with the evolution time, but the increasing rate slows down gradually. When the methane outburst ends (the 70s), the distance that the methane flow backward reaches the maximum and the fresh air from the inlet of the air-intake tunnel will eliminate the methane which flows back into the tunnel. It is important to note that the elimination rate of methane in the blind tunnel is very low. The methane with high concentration is easily accumulated in the blind tunnel for a long time. It can explain why the workers are strictly forbidden to enter the blind tunnel in the underground coal mine.

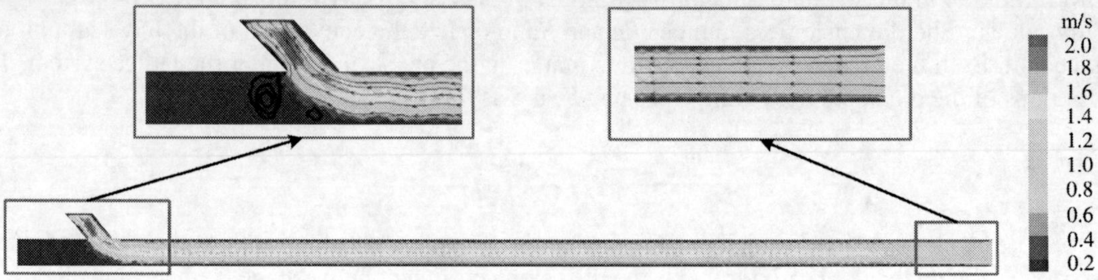

Figure 3. Initial air flow state.

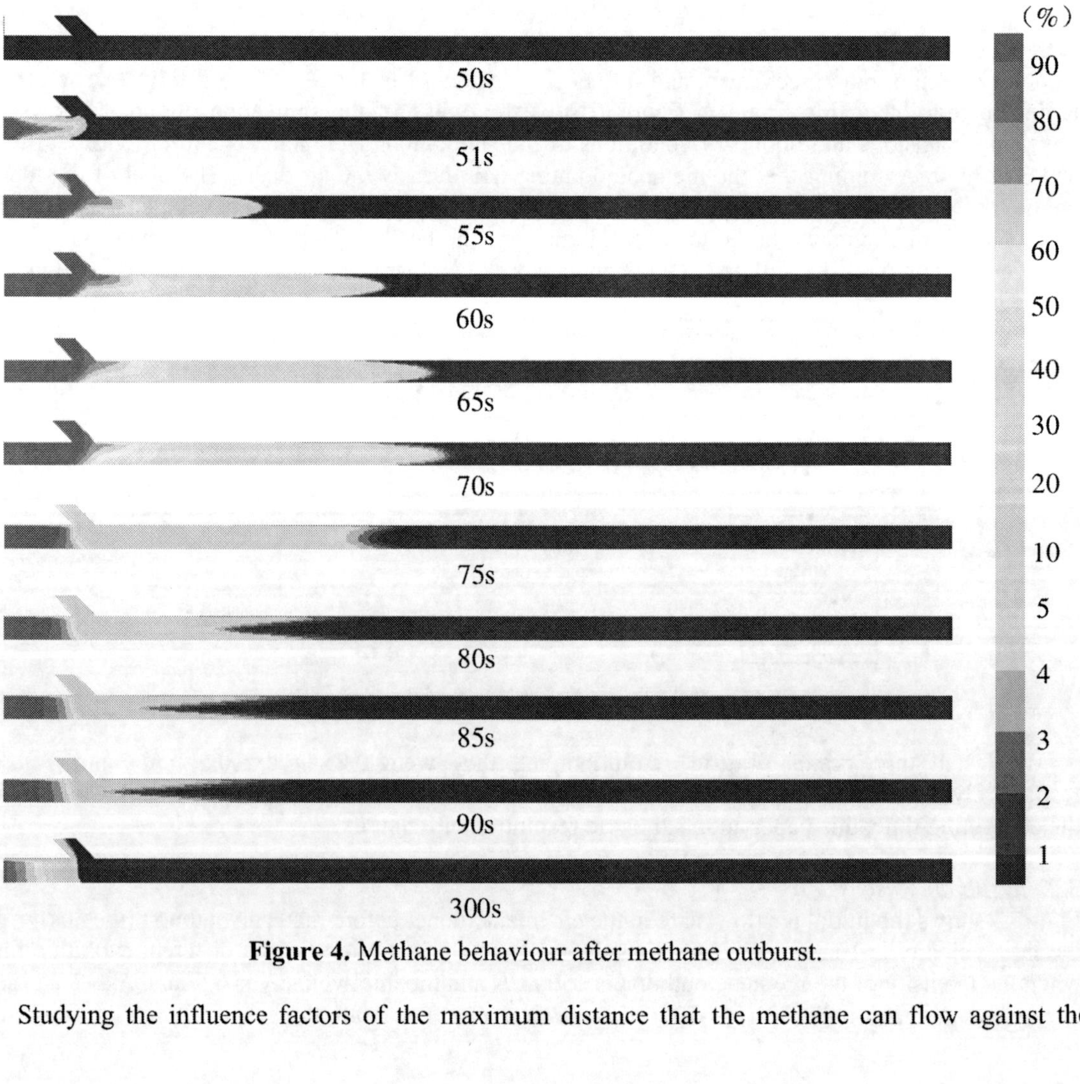

Figure 4. Methane behaviour after methane outburst.

Studying the influence factors of the maximum distance that the methane can flow against the initial air flow is critical to help conduct engineering design and take safety measures. The influences of the intensity of the methane outburst, the initial air velocity and the width of the tunnel to the maximum distance that the methane can flow against the initial air flow are discussed in this part. In order to quantitatively describe the maximum distance that the methane can flow against the initial air flow, we define the distance L_m from the heading face to the farthest cross-section that the average methane concentration is larger than 1% as the maximum distance that the methane can flow backward.

Figure 5 shows the different L_m when the inlet air velocities are 0.25-2 m/s and the intensity of the methane outburst is 150 m^3/s. It demonstrates that the distance L_m decreases with the increase of the inlet air velocities and they show a good linear relationship. The fitting formula and the R-square are shown in figure 5. We can conclude from the fitting formula that increasing the inlet air velocity by 1 m/s can reduce the distance L_m by 26 m when the intensity of the methane outburst is 150 m^3/s. Increasing the inlet air velocity in the air-intake tunnel can reduce the influence range of methane outburst effectively.

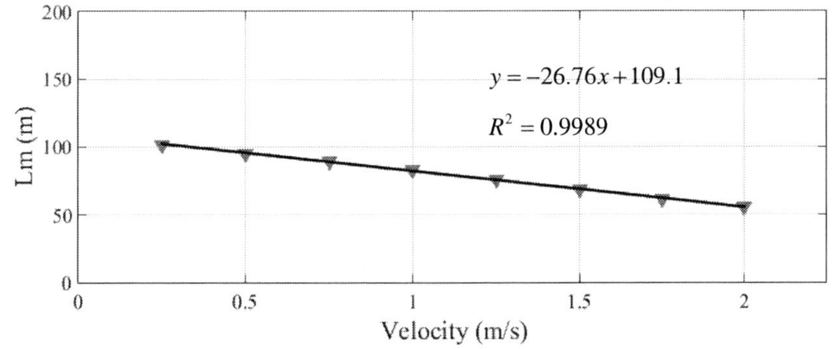

Figure 5. The L_m at different inlet air velocities.

Figure 6 shows the different L_m when the intensity of the methane outburst are 75-180 m^3/s and the inlet air velocity in the air-intake tunnel is 0.5 m/s. It demonstrates that the distance L_m increases with the increase of the intensity of the methane outburst and there exists the linear relationship between them. Figure 6 shows the fitting formula and the R-square.

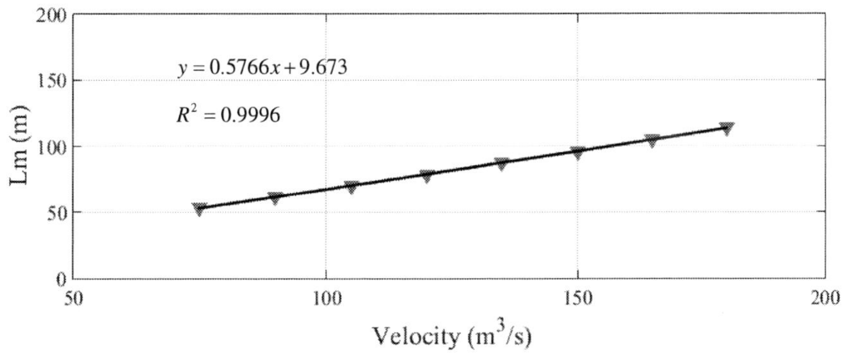

Figure 6. The L_m at different intensities of methane outburst.

Figure 7 shows the different L_m when the width of the tunnel is 3-6 m when the intensity of the methane outburst is 90 m^3/s. It demonstrates that the distance L_m are basically stable at 70 m and it shows a small range of fluctuations with the change of the tunnel width. It demonstrates that the change of the tunnel width has almost no effect on the distance L_m (The proportion of each tunnel remains unchanged).

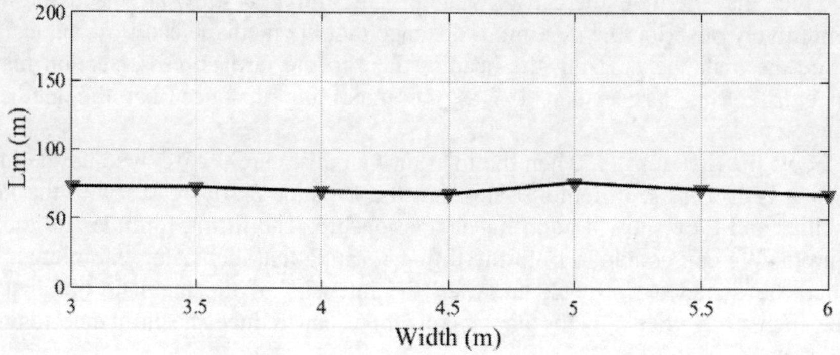

Figure 7. The L_m at different tunnel width.

4. Conclusion

The introduced LBM model coupled with the LES model can simulate the methane backward flow in coal tunnels after methane outburst.

With the increase of the inlet air velocity, the distance that the methane can flow backward decreases. There exists a linear relationship between the inlet air velocity and the distance that the methane can flow backward. Increasing the tunnel air velocity can reduce the maximum distance that the methane can flow backward after methane outburst obviously. The tunnel air velocity should be increased as much as possible when the economic and other conditions are met in practice.

With the increase of the intensity of the methane outburst, the distance that the methane can flow backward increases. There exists a linear relationship between the intensity of the methane outburst and the distance that the methane can flow backward. This relation can help calculate the intensity of the methane outburst which cannot be obtained through the measuring equipment owing to the instantaneity, uncertainty, and destructiveness of the methane outburst in practice.

The change of the tunnel width (The proportion of each tunnel remains unchanged.) has almost no influence on the distance that the methane can flow backward. The maximum distance that the methane can flow backward should not be considered when determining the width of the tunnel in practice. The width of the tunnel should be small as much as possible to reduce the cost.

References

[1] Torano J, Torno S, Alvarez E and Riesgo P 2012 *Int. J. Rock Mech. Min.* **50** 94-101
[2] Bjerketvedt D, Bakke J R and Van Wingerden K 1997 *J. Hazard Mater.* **52** 1-150
[3] Terazawa K, Takatori T, Tomii S and Nakano K 1985 *Am. J. Foren. Med. Path.* **6** 211-4
[4] Nakayama S 1999 *Min. Sci. Amp Tech.* 99 259-62
[5] Ichinose M, Nakayama S, Uchino K and Inoue M 1998 *J. Min. Mater. Proc. Inst. Japan* **114** 769-75
[6] Torano J, Torno S, Menendez M and Gent M 2011 *Tunn. Undergr. Sp. Tech.* **26** 201-10
[7] Rodriguez R and Lombardia C 2010 *Tunn. Undergr. Sp. Tech.* **25** 456-68
[8] Fang Y, Fan J, Kenneally B and Mooney M 2016 *Tunn. Undergr. Sp. Tech.* **58** 30-9
[9] Zhou L, Pritchard C and Zheng Y 2015 *Int. J. Min. Sci. Tech.* **25** 635-40
[10] Kurnia J C, Sasmito A P and Mujumdar A S 2014 *Appl. Math. Model.* **38** 3467-84
[11] Mishra D P, Kumar P and Panigrahi D C 2016 *Environ. Earth. Sci.* **75** 475
[12] Qin Z, Yuan L, Guo H and Qu Q 2015 *Int. J. Coal. Geol.* **150** 51-63
[13] Li C W, Yang W, Xie B J, Cao X and Mei G D 2012 *Appl. Mech. Mater.* 212 (Trans TechPubl) 989-97
[14] Lu Q Q and Wei X G 2009 2nd *Int. Conf. Infor. Comput. Sci.* 4 308-11

[15] King M F, Khan A, Delbosc N, Gough H L, Halios C, Barlow J F and Noakes C J 2017 *Build. Environ.* **125** 273-84
[16] Chen S and Doolen G D 1998 *Annu. Rev. Fluid Mech.* **30** 329-64
[17] Khan M A I, Delbosc N, Noakes C J and Summers J 2015 *Build. Simul-China* 8 (Springer) 405-14
[18] Obrecht C, Kuznik F, Merlier L, Roux J J and Tourancheau B 2015 *Environ. Fluid Mech.* **15** 753-70
[19] Delbosc N, Summers J L, Khan A, Kapur N and Noakes C J 2014 *Comput. Math. Appl.* **67** 462-75
[20] Peng Z, Liu S, Tang S, Zhao Y and Li Y 2018 *Geofluids* **2018** 5169010
[21] Ning Y, Jiang Y, Liu H and Qin G 2015 *J. Nat. Gas. Sci. Eng.* **26** 345-55
[22] Sofonea V and Sekerka R F 2001 *Physica A* **299** 494-520
[23] Inamuro T, Yoshino M, Inoue H, Mizuno R and Ogino F 2002 *J. Comput. Phys.* **179** 201-15
[24] Asinari P and Luo L S 2008 *J. Comput. Phys.* **227** 3878-95
[25] Asinari P 2005 *Phys. Fluids* **17** 067102
[26] Luo L S and Girimaji S S 2002 *Phys. Rev. E.* **66** 035301
[27] Benzi R and Succi S 1990 *J. Phys. A-Math. Genl.* **23** L1
[28] Fernandino M, Beronov K and Ytrehus T 2009 *Math. Comput. Simulat.* **79** 1520-6
[29] Chen X P 2012 *Eng. Appl. Comp. Fluid* **6** 572-80
[30] Jahanshaloo L, Pouryazdanpanah E and Che Sidik N A 2013 *Numer. Heat Tr. A-Appl.* **64** 938-53
[31] Zhao P, Li Q, Kuang S and Zou Z 2017 *Metall. Mater. Trans. B* **48** 456-70
[32] Ren F, Song B, Zhang Y and Hu H 2018 *Comput. Fluids* **173** 29-36
[33] Nathen P, Haussmann M, Krause M and Adams N 2018 *Comput. Fluids* **172** 510-23
[34] Premnath K N, Pattison M J and Banerjee S 2009 *Physica A* **388** 2640-58
[35] Ahmad N H, Inagaki A, Kanda M, Onodera N and Aoki T 2017 *Bound-Lay. Meteorol.* **163** 447-67
[36] Cheng X, Hu F and Zeng Q 2012 *Chinese Sci. Bull.* **57** 1196-203
[37] Sajjadi H, Salmanzadeh M, Ahmadi G and Jafari S 2016 *Build. Environ.* **102** 1-12
[38] Jacob J and Sagaut P 2018 *Build. Environ.* **139** 110-24
[39] Mayer G and Házi G 2006 *Math. Comput. Simulat.* **72** 173-8
[40] Mayer G, Páles J and Házi G 2007 *Ann. Nucl. Energy* **34** 140-9
[41] Mohamad A A 2011 *Lattice Boltzmann Method: Fundamentals and Engineering Applications with Computer Codes* (London, UK: Springer Science & Business Media)
[42] Tong Z X, He Y L, Chen L and Xie T 2014 *Comput. Fluids* **105** 155-65
[43] Qian Y H, d'Humi`eres D and Lallemand P 1992 *EPL (Europhysics Letters)* **17** 479
[44] McCracken M E and Abraham J 2005 *Phys. Rev. E.* **71** 046704
[45] Joshi A S, Peracchio A A, Grew K N and Chiu W K 2007 *J. Phys. D Appl. Phys.* **40** 2961
[46] Krafczyk M, Tölke J and Luo L S 2003 *Int. J. Mod. Phys. B* **17** 33-9

IOP Conf. Series: Materials Science and Engineering **657** (2019) 012023 doi:10.1088/1757-899X/657/1/012023

Study on wind-induced response of a simplified model of transmission tower based on finite particle method

Z Huang[1,3], S Liu[1,2], M Nie[1], Y Yang[1,2], Q S Gao[1,2] and C Zhang[1,2]

[1]Electric Power Research Institute of Guangdong Power Grid Co., Ltd, Guangzhou, Guangdong 510080, P.R. China

[2]Guangdong Diankeyuan Energy Technology Co., Ltd, Guangzhou, Guangdong 510080, P.R. China

E-mail: zhhuang8831@163.com

Abstract. A reduced model of a transmission tower was established based on the principle of equivalent stiffness. The finite particle method was used to analyse the wind-induced response of the reduced model. The analysis of the dynamic characteristics of the reduced model shows that the accuracy of the equivalent formula proposed in this paper was very high. Compared with the first five order natural frequencies of the complete model, the maximum calculation error of the reduced model was only 0.87%. Since the first order shear deformation beam theory was used in the reduced model, the interpolation-independent element shape functions were developed to avoid the shear locking phenomenon. Moreover, the formula for calculating the internal force of Timoshenko beam for finite particle method was derived. By using two successive Euler angles, a new method for separating the rigid motion and pure deformation of the element was proposed. The computational efficiency of this method is significantly improved, and the computational time is less than half of that of the traditional finite particle method. The results of wind-induced response analysis of the transmission tower show that the simplified model proposed in this paper can obtain highly accurate displacement response of the structure, and the maximum error of the RMS of the displacements at the top of the tower is less than 3.2%. This model is suitable for studying the dynamic response of transmission tower line system and evaluating the effect of the wind-induced vibration control of damping device.

1. Introduction

As an important part of the electrical transmission systems, the safety of the transmission tower-line system is of great importance to the normal operation of the power grid. With the increase of the span and height of transmission tower-line system, the influence of strong wind load on the safety of the system becomes more and more obvious. Wind-induced dynamic response of transmission tower-line system has become an important subject in structural wind resistance field.

The transmission tower, consisting of large number of nodes and complicated structure connections, is the main load-bearing structure of the transmission tower-line system. Related researches are mainly developing in two directions, one is the accurate 3D model [1-3], the other is the simplified model [4-8]. The multi-particle model [4-6] is a widely used simplified model, which simplifies the transmission tower to a series of multi-degrees-of-freedom systems with multiple concentrated masses connected by rods, and the stiffness of the tower is usually obtained by inverse of the flexibility matrix using finite element method. Zhang *et al* [7] carried out wind tunnel test based on high frequency force balance and 3D finite element analysis, which proved that the multi-particle model has enough accuracy in calculating the overall response of the tower. Limongelli *et al* [8] obtained a reduced model of the lattice transmission tower without stiffness matrix. In their method, the transmission tower is divided into several sections, and assumed that each section has the same material and stiffness properties. Compared with the multi-particle model, this model established the equivalent model of the cross arms, which can better consider the torsion effect and more accord with the actual situation. Based on this model, Dua *et al* [9] studied the dynamic response of transmission tower-line system under the action of random simulated wind fields, and proved that if the coupling effect of tower line is not taken into account, the displacement of the conductors and insulators will be wrongly estimated. Literature [8] gives a potential way to build the simplified model of transmission tower, but there are still some to be improved. For example, the torsion stiffness formula overestimates the first-order torsional frequency. In addition, the mass of the steel plate, bolt and nut is not considered in the calculation of equivalent density, which will have a great influence on the dynamic characteristics of the transmission tower, especially for a tower with a heavy tower top.

Finite particle method (FPM), developed from vector form finite element method [10,11], is a new structural analysis method which is more suitable for space structure analysis. This method does not need to assemble structural element stiffness matrix and to carry out iterative calculation. Therefore, it has advantages in calculation of the geometric and material nonlinearity, discontinuity, and has been successfully used in the analysis of various types of complex structural behaviors [12-16]. In FPM, Euler-Bernoulli beam model [12], rod model [14] and fine beam model [15,16] have been proposed to compute the internal force of spatial beam. In literature [14], the computational code of FPM was developed with C++ programming language, and it was found that the computational efficient of this method was significantly higher than that of ANSYS.

In this paper, the equivalent simplified model of a lattice transmission tower is established by the method of literature [8], and the formula of torsional stiffness is modified to improve the calculation of torsion natural frequency. In this model, the shear deformation is not negligible, so the formulas of internal forces based on Timoshenko beam model are derived. By using the method based on two successive Euler angles instead of the virtual inverse motion analysis, the calculation formula of rotation matrix is derived. The central difference method is adopted to solve the equations of motion.

2. Reduced model of transmission tower

A transmission tower can be subdivide in four substructures: the tower foot, the tower body, the tower top and the cross arms. The reduced model consists of straight beam elements along the center line of the substructures. The transverse cross arms are rigidly coupled to the vertical tower top [8].

2.1. Equivalent formulas of reduced model

The equivalent properties of tensile, bending, shear, and torsional of the i-th portion of the tower are as follows [8]:

$$k_{A,i}=\frac{EA_{eq,i}}{l_i},k_{F,i}=\frac{EI_{eq,i}}{l_i},k_{S,i}=\frac{GA_{eq,i}}{\chi_{eq,i}},k_{T,i}=\frac{GJ_{eq,i}}{l_i} \tag{1}$$

respectively, and

$$A_{eq,i}=\sum_{j=1}^{n_{ch,i}}A_{ch,j}\cos^3\theta_j \tag{2}$$

$$I_{eq,i}=A_{eq,i}h_{i,\min}h_{i,\max}/4 \tag{3}$$

$$A_{d,i} = \sum_{j=1}^{n_{d,i}} A_{d,j} \cos^2 \delta_j \sin \delta_j \tag{4}$$

$$\chi_{eq,i} = \frac{A_{eq,i}}{2(1+\nu)A_{d,i}} \tag{5}$$

$$J_{eq,i} = 2I_{eq,i} / \chi_{eq,i} \tag{6}$$

where, $n_{ch,i}$ is the number of main bearing components of the i-th portion, θ is the angle between the axial direction of the main bearing component and the axial direction of the equivalent element; $n_{d,i}$ is the number of diagonal components of the i-th portion, δ is the angle between the axial direction of the diagonal component and the shear direction of the equivalent element; $A_{ch,j}$ and $A_{d,j}$ are the transversal area of the j-th main bearing and diagonal components of the i-th portion; $A_{eq,i}$, $I_{eq,i}$, $J_{eq,i}$, $\chi_{eq,i}$ are the equivalent transversal area, moment of inertia, torsional moment of inertia, and shear factor of the i-th portion, respectively; $h_{i,\min}$ and $h_{i,\max}$ are the widths of the two ends of the i-th portion; l_i is the length of the i-th portion; and E, G, ν are Young's modulus, shear modulus, and Poisson's ratio, respectively. See literature [8] for more details. Moreover, the equivalent density of the i-th portion is as follows:

$$\gamma_{eq,i} = \rho \left(\sum_{j=1}^{n_{ch,i}} A_{ch,j} l_j + \sum_{j=1}^{n_{d,i}} A_{d,j} l_j \right) \Big/ \sum_{j=1}^{n_{ch,i}} A_{ch,j} l_j \tag{7}$$

where, ρ is density of steel, l_j is the length of the components of the i-th portion. In order to consider the influence of the mass of the steel plate, bolt and nut, a density correction factor η is introduced. The corrected density is $\eta\gamma_{eq,i}$, and then $\eta\gamma_{eq,i}A_{eq,i}$ equals to the total mass of the i-th portion.

2.2. *Modified formula of torsional stiffness*

Considering a lattice transmission tower shown in figure 1(a), the total height of the tower is 77.5 m. A complete finite element model (refer as complete model) of the tower was established with one element for each component. There were 652 nodes and 1802 elements in total. The element type was chosen as BEAM188, and cubic interpolation shape function was adopted. The reduced model only has 58 nodes and 58 elements, as shown in figure 1(b).

It is found that if the cross-section of the equivalent beam element is assumed to be circular, the torsional stiffness formula is the same as equation (6). In this paper, the cross-section of the equivalent beam element is assumed to be a square ring, then the torsional stiffness can be written as

$$J_{eq,i} = I_{eq,i} / \chi_{eq,i} \tag{8}$$

In table 1, the results of the reduced model using two torsional stiffness formulas are compared with those of the complete model. It can be seen that the error of the first-order torsion natural frequency obtained from equations (1)-(7) is up to 27.27%; while using the modified formulas (equation (8)) proposed in this paper, the errors of the first five natural frequencies are all less than 0.87%. The first five modes of the reduced model are shown in figure 2.

IOP Conf. Series: Materials Science and Engineering **657** (2019) 012023 doi:10.1088/1757-899X/657/1/012023

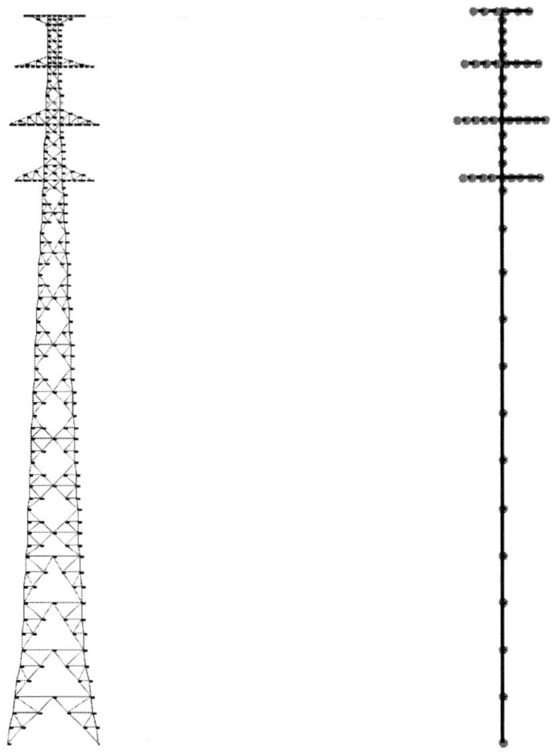

(a) (b)

Figure 1. Models of transmission tower. (a) complete model and (b) reduced model.

Table 1. First five modes and frequencies of the tower.

Mode No.	Mode type	Frequencies (Hz)		
		Reduced model [a]	Reduced model [b]	Complete FE model
1	1st (bending xz plane)	1.0470	1.0470	1.0445
2	1st (bending yz plane)	1.0492	1.0492	1.0468
3	2st (bending xz plane)	3.3590	3.3590	3.3531
4	2st (bending yz plane)	3.4222	3.4231	3.4266
5	1st (torsion about z-axis)	5.4545	3.9616	3.9962

[a] Equation (6) was used to calculate the torsional stiffness;
[b] Equation (8) was used to calculate the torsional stiffness.

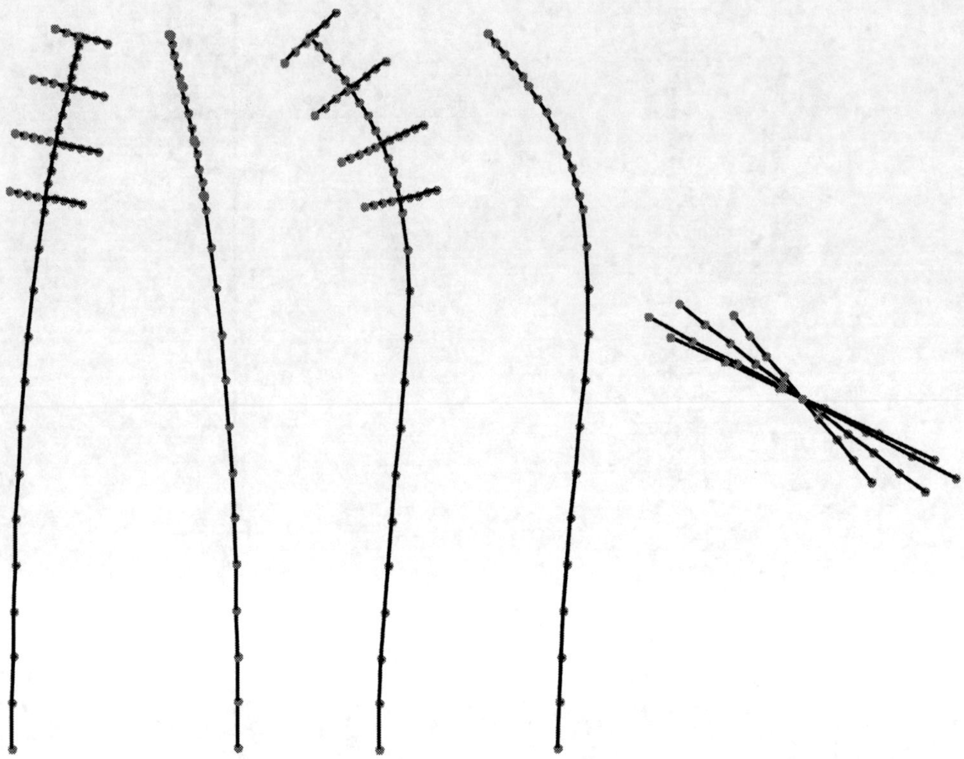

Figure 2. First five modes of the reduced model. (a) mode 1, (b) mode 2, (c) mode 3, (d) mode 4 and (e) mode 5.

3. Finite particle method

The motion of a straight beam element without initial curvatures and torsion can be decomposed into rigid motion and pure deformation by using local coordinate system. Therefore, the motion of an element can be decomposed into two steps: the first step is the rigid body translation and rotation of the element; and the second step is the pure deformation of the element in the local coordinate system. Assuming the length of the element is properly chosen, the pure deformation of the element is always small relative to the local coordinate system. Since the motion of the rigid body does not generate internal force, the internal force of the element is only related to the pure deformation.

3.1. Equation of motion of a particle

According to the characteristics of spatial frames, finite particle method uses a space beam element and two particles connected with it to describe each component. In FPM, it is assumed that each particle is in a state of dynamic equilibrium during the process of rigid motion and deformation. So the motion of all particles follows Newton's second law, which can be expressed as

$$\mathbf{m}_p \frac{\mathrm{d}^2\mathbf{u}_p}{\mathrm{d}t^2} = \mathbf{F}_p^{\mathrm{ext}} - \mathbf{F}_p^{\mathrm{int}} - \mathbf{F}_p^{\mathrm{damp}}, \mathbf{I}_p \frac{\mathrm{d}^2\boldsymbol{\theta}_p}{\mathrm{d}t^2} = \mathbf{M}_p^{\mathrm{ext}} - \mathbf{M}_p^{\mathrm{int}} - \mathbf{M}_p^{\mathrm{damp}} \tag{9}$$

where, $\mathbf{m}_p$ and $\mathbf{I}_p$ are the equivalent mass and rotatory inertia of particle p, respectively; $\mathbf{u}_p$ and $\boldsymbol{\theta}_p$ are the translation and angular displacement vectors of particle p, respectively; $\mathbf{F}_p^{\mathrm{ext}}$ and $\mathbf{M}_p^{\mathrm{ext}}$ are the equivalent external forces and moments acting on particle p, respectively; $\mathbf{F}_p^{\mathrm{int}}$ and $\mathbf{M}_p^{\mathrm{int}}$ are the internal forces and moments of particle p caused by the deformation of connecting elements, respectively; $\mathbf{F}_p^{\text{damp}}$ and $\mathbf{M}_p^{\text{damp}}$ are the damping forces and moments of particle p, respectively, and they are usually assumed to be expressed as $\mathbf{F}_p^{\text{damp}} = \mu \mathbf{m}_p \dot{\mathbf{u}}_p$ and $\mathbf{M}_p^{\text{damp}} = \mu \mathbf{I}_p \dot{\boldsymbol{\theta}}_p$, where μ is the damping factor. Finally, $\mathbf{m}_p$, $\mathbf{I}_p$, $\mathbf{F}_p^{\text{ext}}$, and $\mathbf{M}_p^{\text{ext}}$ are detailed in literature [4].

The explicit central difference method was used to solve Eq. (9), so the displacements of a particle can be written as [12]

$$\mathbf{X}_{p,t+\Delta t} = 2c_1\mathbf{X}_{p,t} - c_2\mathbf{X}_{p,t-\Delta t} + c_1\Delta t^2\mathbf{H}_p^{-1}(\mathbf{F}_p^{\text{ext}} - \mathbf{F}_p^{\text{int}}) \tag{10}$$

where, $c_1 = 1/(1+\mu\Delta t/2)$, $c_2 = c_1(1-\mu\Delta t/2)$, and Δt is the time step; $\mathbf{X}$ is the translation or angular displacement vectors; $\mathbf{H}_p$ is a matrix consisting of $\mathbf{m}_p$ or $\mathbf{I}_p$.

3.2. Description of rigid body rotation and pure deformation

As shown in figure 3, $\mathbf{i}_m$ is a fixed global coordinate system, and the subscript $m = 1,2,3$. Points p_1 and p_2 are the two endpoints of the beam element. At moment t_a, the positions of the two endpoints are $\mathbf{x}_1^a$ and $\mathbf{x}_2^a$, respectively; and the rotation angle vectors about $\mathbf{i}_m$ at the two endpoints are $\boldsymbol{\alpha}_1^a$ and $\boldsymbol{\alpha}_2^a$, respectively. At moment t_b, the positions of the two endpoints are $\mathbf{x}_1^b$ and $\mathbf{x}_2^b$, respectively; and the rotation angle vectors about $\mathbf{i}_m$ at the two endpoints are $\boldsymbol{\alpha}_1^b$ and $\boldsymbol{\alpha}_2^b$, respectively. At moment t_a, the element length denotes l_a; the local coordinate system denotes $\mathbf{a}_m$, and the transformation matrix between $\mathbf{a}_m$ and $\mathbf{i}_m$ is $\mathbf{R}_a$. At moment t_b, the element length denotes l_b; the local coordinate system denotes $\mathbf{b}_m$, and the transformation matrix between $\mathbf{b}_m$ and $\mathbf{i}_m$ is $\mathbf{R}_b$.

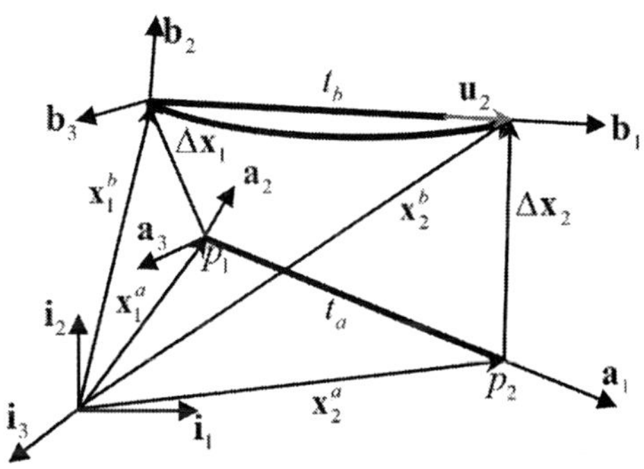

Figure 3. Particle displacement and local coordinate system of spatial beam element.

From moment t_a to t_b, the increments of the rotation angle vectors in $\mathbf{a}_m$ are $\Delta\boldsymbol{\alpha}_1$ and $\Delta\boldsymbol{\alpha}_2$, respectively, i.e.

$$\Delta\boldsymbol{\alpha}_1 = \mathbf{R}_a(\boldsymbol{\alpha}_1^b - \boldsymbol{\alpha}_1^a), \Delta\boldsymbol{\alpha}_2 = \mathbf{R}_a(\boldsymbol{\alpha}_2^b - \boldsymbol{\alpha}_2^a) \tag{11}$$

The displacement of endpoint p_2 in $\mathbf{b}_m$ denotes $\mathbf{u}_2$. According to geometric relations, we have

$$l_a\mathbf{b}_1 + \mathbf{u}_2 = l_b\mathbf{b}_1 \tag{12}$$

$$\mathbf{b}_1 = (\mathbf{x}_2^b - \mathbf{x}_1^b)/l_b = \mathbf{R}_a(\mathbf{x}_2^b - \mathbf{x}_1^b)/l_b = d_m\mathbf{a}_m \tag{13}$$

Apparently, $\mathbf{u}_2 = (l_b - l_a)\mathbf{b}_1$. Notice that the last two equations of equation (13) describe the vector $\mathbf{b}_1$ in local coordinate system $\mathbf{a}_m$.

The finite rotation of the rigid body motion part of the beam element can be described by two Euler angles [17]: firstly, $\mathbf{a}_1$ is translated by $\Delta\mathbf{x}_1$ and rotated by angle $\alpha = \sin^{-1}(\sqrt{d_2^2 + d_3^2})$, where the rotation axis is $\mathbf{n} = (\mathbf{b}_1 \times \mathbf{a}_1)/\|\mathbf{b}_1 \times \mathbf{a}_1\|$; Secondly, the rotated coordinate system is further rotated by angle β about $\mathbf{b}_1$, where β is the first component of $\Delta\boldsymbol{\alpha}_1$. Then, the rotation angle vector of $\mathbf{a}_m$ can be written as

$$\boldsymbol{\beta} = \left[\beta \quad -\alpha d_3/\sqrt{d_2^2 + d_3^2} \quad \alpha d_2/\sqrt{d_2^2 + d_3^2}\right]^{\mathrm{T}} \tag{14}$$

Applying Rodrigues rotation formula twice, the finite rotation tensor from $\mathbf{a}_m$ to $\mathbf{b}_m$ is obtained as

$$\mathbf{R} = \begin{bmatrix} 1 & 0 & 0 \\ 0 & \cos\beta & \sin\beta \\ 0 & -\sin\beta & \cos\beta \end{bmatrix} \begin{bmatrix} d_1 & d_2 & d_3 \\ -d_2 & d_1 + d_3^2/(1+d_1) & -d_2 d_3/(1+d_1) \\ -d_3 & -d_2 d_3/(1+d_1) & d_1 + d_2^2/(1+d_1) \end{bmatrix} \tag{15}$$

Therefore, at moment t_b, the rotation matrix of the local coordinate system with respect to the fixed global coordinate system is

$$\mathbf{R}_b = \mathbf{R}\mathbf{R}_a \tag{16}$$

Usually, it is assumed that the rotation matrix of the local coordinate system with respect to the fixed global coordinate system is known at the initial time. Then, according to equation (16), the rotation matrix at any time can be obtained recursively. In local coordinate system $\mathbf{b}_m$, the pure deformation of the two endpoints of the beam element can be written as

$$\boldsymbol{\theta}_1 = \Delta\boldsymbol{\alpha}_1 - \boldsymbol{\beta}, \boldsymbol{\theta}_2 = \Delta\boldsymbol{\alpha}_2 - \boldsymbol{\beta} \tag{17}$$

3.3. Internal force of the beam element

Consider the two-node space Timoshenko beam element shown in figure 4, each node has six degree-of-freedoms (DOFs), namely three translational DOFs and three rotational DOFs. Assuming that there is no external load acting between the two nodes, and ignoring the torsional warpings, the kinematics relationship of Timoshenko beam can be expressed as

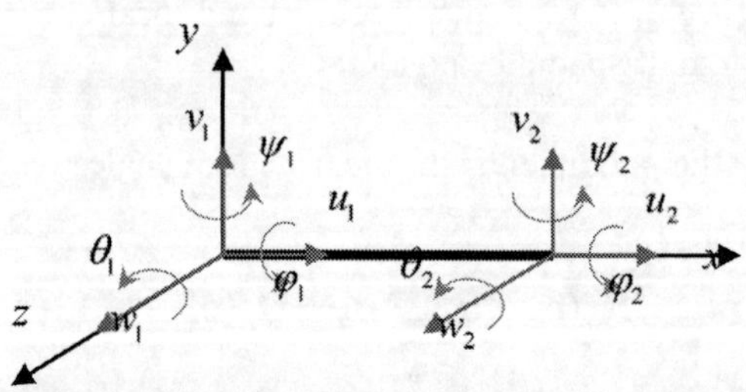

Figure 4. Two node spatial beam element.

$$U = u - y\theta + z\psi, V = v - z\varphi, W = w + y\varphi \tag{18}$$

where, u, v and w are the displacement along the x, y and z axis at a point on the bema reference line, respectively; U, V and W are the displacements in x, y and z directions at any point of the cross-section, respectively; φ, ψ and θ are the small rotation angles of the cross-section with respect to x, y and z, respectively. The non-zeros strains of the beam element can be expressed as

$$\{\varepsilon\}=\{\varepsilon_{xx},\varepsilon_{xy},\varepsilon_{xz}\}^{\mathrm{T}}=[S]\{\chi\} \tag{19}$$

$$\{\chi\}=\{u',\gamma_y,\gamma_z,\varphi',\psi',-\theta'\}^{\mathrm{T}} \tag{20}$$

$$[S]=\begin{bmatrix} 1 & 0 & 0 & 0 & z & y \\ 0 & 1 & 0 & -z & 0 & 0 \\ 0 & 0 & 1 & y & 0 & 0 \end{bmatrix} \tag{21}$$

where, $(\cdot)'$ denotes the derivative with respective to x. γ_y and γ_z are the shear strains in y and z directions, and can written as

$$\gamma_y = v' - \theta, \gamma_z = w' + \psi \tag{22}$$

If the material is isotropic and homogeneous, and the beam reference point coincides with the mass centroid, then the variation of strain energy can be written as

$$\delta U = \int_L \int_A (\sigma_{xx}\delta\varepsilon_{xx} + \sigma_{xy}\delta\varepsilon_{xy} + \sigma_{xz}\delta\varepsilon_{xz})\mathrm{d}A\mathrm{d}x = \int_L \{\delta\chi\}^{\mathrm{T}}\{\hat{F}\}\mathrm{d}x \tag{23}$$

The resultant cross-sectional stress conjugated with the generalized strain are

$$\{\hat{F}\}=\{F_x,Q_y,Q_z,M_x,M_y,M_z\}^{\mathrm{T}}=[D]\{\chi\} \tag{24}$$

$$D_{11}=EA, D_{22}=k_yGA, D_{33}=k_zGA, D_{44}=k_xGJ, D_{55}=EI_y, D_{66}=EI_z \tag{25}$$

where, k_x is the influence factor of torsion stiffness by warping, k_y and k_z are the shear correction factors, A is the cross-section area, I_y and I_z are moment of inertia, and J is polar moment of inertia. Since it is assumed that there is no load acting on the element, the relationship

$$M_z' - Q_y = 0, M_y' - Q_z = 0 \tag{26}$$

The boundary conditions at endpoints p_1 and p_2 in the local coordinate system $\mathbf{b}_m$ are

$$\begin{aligned} &u(0)=v(0)=w(0)=\varphi(0)=0,\psi(0)=\psi_1,\theta(0)=\theta_1, \\ &u(l_a)=u_2,v(l_a)=w(l_a)=0,\varphi(l_a)=\varphi_2,\psi(l_a)=\psi_2,\theta(l_a)=\theta_2 \end{aligned} \tag{27}$$

Linear interpolation was used for axial displacement u and torsion angle φ, and cubic polynomial interpolation was used for transverse displacements v and w. Considering equation (26) and applying boundary conditions (27), we have

$$\{u,v,w,\varphi,\psi,\theta\}^{\mathrm{T}}=[N]\{q\} \tag{28}$$

where, $\{q\}=\{u_2,\varphi_2,\psi_1,\psi_2,\theta_1,\theta_2\}^{\mathrm{T}}$ is the nodal displacement vector, and the non-zero elements of the shape function matrix are

$$
\begin{aligned}
&N_{11} = N_{42} = \xi \\
&N_{25} = l_a\beta_y[(\xi^3 - 2\xi^2 + \xi) + \tfrac{1}{2}\alpha_y(\xi - \xi^2)] \\
&N_{26} = l_a\beta_y[(\xi^3 - \xi^2) + \tfrac{1}{2}\alpha_y(\xi^2 - \xi)] \\
&N_{33} = -l_a\beta_z[(\xi^3 - 2\xi^2 + \xi) + \tfrac{1}{2}\alpha_z(\xi - \xi^2)] \\
&N_{34} = -l_a\beta_z[(\xi^3 - \xi^2) + \tfrac{1}{2}\alpha_z(\xi^2 - \xi)] \\
&N_{53} = -\beta_z[3\xi^2 - 4\xi + 1 + \alpha_z(1 - \xi)] \\
&N_{54} = -\beta_z[3\xi^2 - 2\xi + \alpha_z\xi] \\
&N_{65} = \beta_y[3\xi^2 - 4\xi + 1 + \alpha_y(1 - \xi)] \\
&N_{66} = \beta_y[3\xi^2 - 2\xi + \alpha_y\xi]
\end{aligned} \tag{29}
$$

$$
\alpha_y = \frac{12EI_z}{k_z GAl_a^2}, \beta_y = \frac{1}{1+\alpha_y}, \alpha_z = \frac{12EI_y}{k_y GAl_a^2}, \beta_z = \frac{1}{1+\alpha_z} \tag{30}
$$

here, $\xi = x/l_a$ is dimensionless axial coordinate. The relationship between generalized strains $\{\chi\}$ and nodal displacement vector $\{q\}$ is

$$
\{\chi\} = [\Phi]\{q\} \tag{31}
$$

where the non-zero elements of $[\Phi]$ are

$$
\begin{aligned}
&\Phi_{11} = N'_{11}, \Phi_{25} = N'_{25} - N_{65}, \Phi_{26} = N'_{26} - N_{66} \\
&\Phi_{33} = N'_{33} + N_{53}, \Phi_{34} = N'_{34} + N_{54}, \Phi_{44} = N'_{11} \\
&\Phi_{53} = N'_{53}, \Phi_{54} = N'_{54}, \Phi_{65} = -N'_{65}, \Phi_{66} = -N'_{66}
\end{aligned} \tag{32}
$$

Then, the variation of strain energy can be also written as

$$
\delta U = \int_L \{\delta\chi\}^{\mathrm{T}}\{\hat{F}\}\mathrm{d}x = \int_L \{\delta q\}^{\mathrm{T}}[\Phi]^{\mathrm{T}}\{D\}[\Phi]\{q\}\mathrm{d}x = \{\delta q\}^{\mathrm{T}}\{\bar{F}\} \tag{33}
$$

The nodal force vector $\{\bar{F}\}$ corresponding to the nodal displacement vector $\{q\}$ is

$$
\{\bar{F}\} = \{F_{2x}, M_{2x}, M_{1y}, M_{2y}, M_{1z}, M_{2z}\}^{\mathrm{T}} = \int_L [\Phi]^{\mathrm{T}}\{D\}[\Phi]\{q\}\mathrm{d}x \tag{34}
$$

According to equation (34), the nodal force vector $\{\bar{F}\}$ can be easily obtained by symbolic software Mathematica, and hence it was not given here. Then, according to the static equilibrium conditions, the axial force F_{1x} and the torque M_{1x} of point p_1, as well as the shear forces at the two endpoints can be obtained.

4. Wind induced response

The wind load acting on the transmission tower includes mean wind and turbulent wind [18], where the logarithmic function is used to express the mean wind, and the Kaimal fluctuating wind power spectrum is used to express the turbulent wind [2]. Moreover, the harmony superposition method is used to simulate the turbulent wind. Then, the along wind load is given by

$$
F(t) = \mu_s A_n u^2/1.6 \tag{35}
$$

where, u is the wind speed of simulation point, μ_s and A_n are the shape coefficient and wind area of a portion of the complete transmission tower, respectively.

In the following calculation, the angle of between the wind direction and line direction (y-direction in figure 1(b)) in horizontal direction is defined as wind angle. The modal damping ratio is set to 0.02.

4.1. Validation

At the top of the transmission tower, a concentrated constant load of 10^4N is applied the along line direction. The total simulated time is 50 s. The time history of the displacement responses of the nodes connecting the four cross arms and the tower top are given in figure 5. It can be seen that the results of FPM (reduced model) and FEM (Finite Element Method) (complete model) are very close. Furthermore, the vibration responses caused by the sudden concentrated constant load are approximately zero after 50 s.

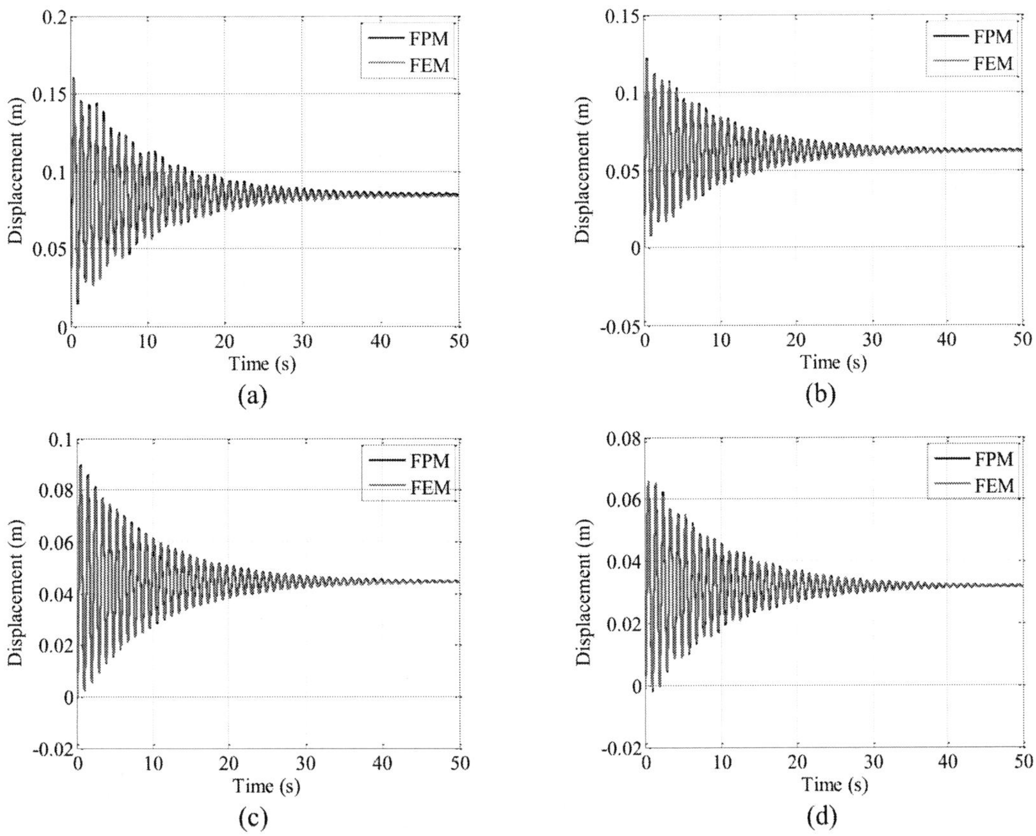

Figure 5. Comparison of displacement responses under concentrated constant load. (a) z = 77.5 m, (b) z = 72.0 m, (c) z = 66.0 m and (d) z = 60.0 m.

Spectrum analysis is carried out on the displacement and acceleration responses of the top of the tower, and the power spectrum curves are shown in figure 6. The response of the reduced model calculated by FPM shows that the first two natural frequencies in the along line direction are 1.04 Hz and 3.38 Hz, which are very close to those of the complete model calculated by FEM. It can be seen that the displacement response is dominated by the first mode, while the acceleration response is dominated by the first two modes. This example shows that the FPM can be used to calculate the dynamic characteristics of the tower, especially if the vibrations are dominated by low-order modes. In addition, the calculation time required by the present method is about 650s, while the time consumed by the traditional FPM is about 1480s, indicating that the proposed method in this paper, which uses two continuous Euler angles to describe the rotation of the local coordinate system of the beam element, is much more efficient.

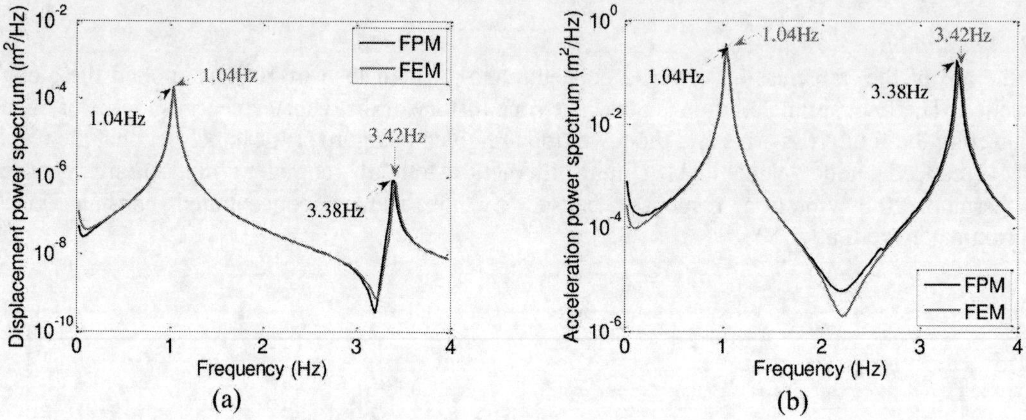

Figure 6. Comparison of the displacement and acceleration power spectrums at the top of the tower. (a) displacement power spectrum and (b) acceleration power spectrum.

4.2. Response various with mean wind speed

In the case of 0° wind angle, the mean wind speeds at height of 10 m are taken 8 values equally spaced between 5 and 40 m/s, and the corresponding fluctuating wind speeds are superimposed to calculate the responses of the top of the tower. Figure 7 shows the time history of the displacement and acceleration of the top of the tower with mean wind speed at 40 m/s. Figure 8 compares the RMS (Root Mean Square) of the along and cross line displacements of the top of the tower at different wind speeds, where only the data after 50 s are taken for calculation. It can be seen that the displacement response obtained by FPM is very consistent with that of FEM, indicating that the reduced model presented in this paper has good accuracy in solving the dynamic response of the transmission tower under random wind loads.

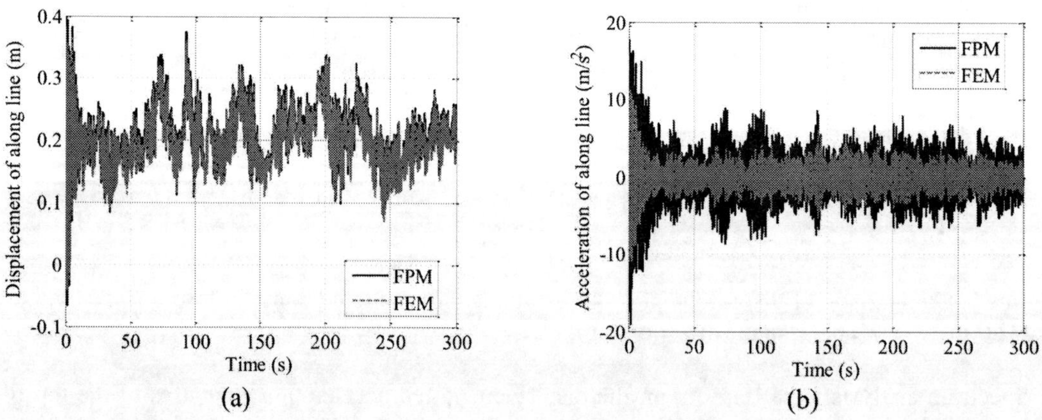

Figure 7. Comparison of the displacement and acceleration responses at the top of the tower. (a) displacement and (b) acceleration.

Taking the results of the complete model as reference values, the error of the RMS of the along line displacements of the reduced model are listed in table 2. The errors are almost constant and relative small with the change of wind speed, which indicates that the reduced model has high reliability for different wind field intensities.

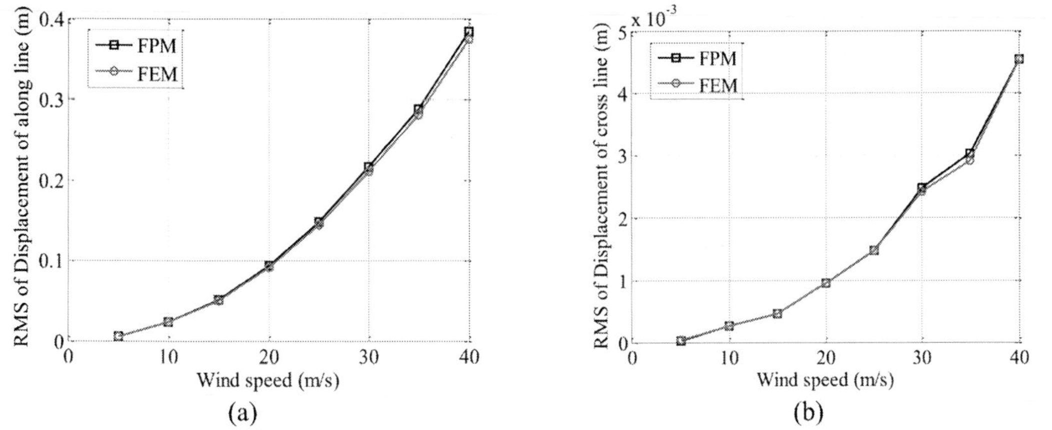

Figure 8. Comparison of the RMS of the displacements at the top of the tower with the wind speed. (a) along line direction and (a) cross line direction.

Table 2. Error of the RMS of the along line displacements at the top of the tower.

Mean wind(m/s)	5	10	15	20	25	30	35	40
Error(%)	2.38	2.43	2.39	2.42	2.39	2.45	2.36	2.47

4.3. Response various with wind angle

In the case of 30 m/s mean wind speed, the wind angles are taken 10 values equally spaced between 0° and 90°, and the corresponding fluctuating wind speed is superimposed to calculate the response of the top of the tower. Figure 9 shows the curves of the RMS of the displacements of the top of the tower various with different wind angles. In figure 9(a), the RMS of the along (cross) line displacement decreases (increases) gradually. In figure 9(b), the RMS of the along wind displacements are gradually smaller, while the RMS of the cross wind displacements first increases and then decreases, and the maximum value is achieved when the wind angle is between 40° to 50°. It can be seen that the maximum RMS errors of the displacements of the top of the tower is 3.2%.

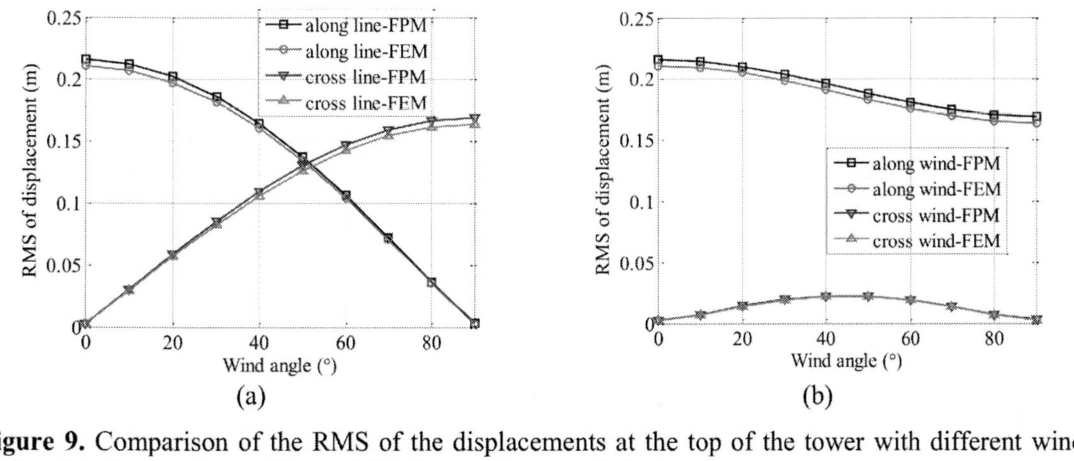

Figure 9. Comparison of the RMS of the displacements at the top of the tower with different wind angles. (a) along and cross line and (b) along and cross wind.

5. Conclusions

Based on the principle of stiffness equivalence, a reduced model of lattice transmission tower with high accuracy can be quickly established according to the geometric characteristics and material properties of the tower. Compared with the multi-particle model, the stiffness matrix of the reduced model is obtained without the need of finite element analysis. Furthermore, since the equivalent model of the cross arms are also established, the wind load acting on the cross arms can be better considered. In order to verify the reduced model, the complete model is also established as reference and solved by finite element method. In the numerical examples, the RMS errors of the displacements of the top of the tower obtained by the reduced model are less than 3.2%, indicating that the calculation accuracy is very high, which also indicates that the correctness of the reduced model and the finite particle method presented in this paper. In this work, two continuous Euler angles are used to describe the rotation of the local coordinate system of the beam element, so that the computational efficiency is significantly improved compared with the traditional finite particle method. Compared with the complete model, the reduced model has fewer degrees of freedom, which provides a potential analysis method for studying the wind-induced response of transmission tower-line system and evaluating the vibration control effect of damping devices.

Acknowledgments

The work described in this paper was financially supported by the Science and Technology Project of China Southern Power Grid co. LTD (No. GDKJQQ20153007). This support for research is greatly appreciated.

References

[1] Shehata A Y, El Damatty A A and Savory E 2005 Finite element modeling of transmission line under downburst wind loading *Finite Eelm. Anal. Des.* **42** 71-89

[2] Tian L, Yu Q Q, Ma R S *et al* 2014 The collapse analysis of a transmission tower under wind excitation *Open Civ. Eng. J.* **8** 136-42

[3] Edgar T-H and Sordo E 2017 Structural behaviour of lattice transmission tower subjected to wind load *Struct. Infrastruct. E.* **13** 1462-75

[4] Ozono S, Maeda J and Makino M 1988 Characteristics of in-plane free vibration of transmission line system *Eng. Struct.* **10** 272-80

[5] Ozono S and Maeda J 1992 In-plane dynamic interaction between a tower and conductors at lower frequencies *Eng. Struct.* **14** 210-6

[6] Li H N and Wang Q X 1997 Dynamic characteristics of long-span transmission lines and their supporting towers *China Civ. Eng. J.* **30** 28-35 (in Chinese)

[7] Zhang Q H, Gu M and Huang P 2012 Multi-degree-freedom of a lattice transmission tower *J. Vib. Shock* **31** 148-52 (in Chinese)

[8] Limongelli M P, Marinelli L and Perotti F 2003 A reduced model for the dynamic analysis of power transmission lines with truss supporting towers *Proceedings of the 5th International Symposium on Cable Dynamics* (Santa Margherita, September) pp 125-32

[9] Dua A, Clobes M, Hobbel T and Matsagar V 2015 Dynamic analysis of overhead transmission lines under turbulent wind loading *Open J. Civ. Eng.* **5** 359-71

[10] Ting E C, Shih C and Wang Y K 2004 Fundaments of a vector form intrinsic finite element: part 1. Basic procedure and a plane frame element *J. Mech.* **20** 113-22

[11] Ting E C, Shih C and Wang Y K 2004 Fundaments of a vector form intrinsic finite element: part 2. Plane solid elements *J. Mech.* **20** 123-32

[12] Yu Y, Glaucio H P and Luo Z Y 2011 Finite particle method for progressive failure simulation of truss structures *J. Struct. Eng.* **137** 1168-81

[13] Luo Z Y, Zheng Y F, Yang C *et al* 2014 Review of the finite particle method for complex behaviors of structures *Eng. Mech.* **31** 1-7+23 (in Chinese)

[14] Yao D, Shen G H, Pan F *et al* 2015 Wind-induced dynamic response of transmission tower using vector-form intrinsic finite element method *Eng. Mech.* **32** 63-70 (in Chinese)

[15] Yuan X F, Chen C, Duan Y F *et al* 2018 Elastoplastic analysis with fine beam model of vector form intrinsic finite element *Adv. Struct. Eng.* **21** 365-79

[16] Long X H, Wang W and Fan J 2018 Collapse analysis of transmission tower subjected to earthquake ground motion *Model Simul. Mater. Sc.* **2018** 2687561
[17] Pai P F 2014 Problems in geometrically exact modeling of highly flexible beams *Thin Wall Struct.* **76** 65-76
[18] Holmes J D 2015 *Wind Loading of Structures* 3rd ed. (Boca Raton, USA: CRC Press)

IOP Conf. Series: Materials Science and Engineering **657** (2019) 012024 doi:10.1088/1757-899X/657/1/012024

A damage constitutive model and its characterization parameters identification for elastic-plastic materials

C H Wang[1,2], W X Jiang[1], K H Xin[1], J Wang[1] and H J Ji[1]

[1]Beijing Institute of Space Long March Vehicle, Beijing 100076, PR China

E-mail: iamwangchh@163.com

Abstract. This paper proposes a simple damage constitutive model for elastic-plastic materials on the basis of the classical elastic-plastic constitutive model and damage mechanics theory. Based on the simple damage constitutive model, ABAQUS built-in Python software was used to develop the virtual experimental platform, and the whole process simulation of the specimens stretching until the fracture was carried out. The correspondence between the damage characterization parameters and the performance indexes such as the fracture strain and the percentage reduction of area was studied. According to the comparison of the virtual/real test results of ZL114A, the situation shows that the simple damage constitutive model can realize the high fidelity simulation of the damage behavior of elastic-plastic materials. The relative accuracy of the important material property parameters such as the ultimate strength, the percentage total elongation at fracture and the percentage reduction of area reaches 5%.

1. Introduction

Historically, the use of structural analysis in aircraft design has been focussed on finite element analysis for the calculation of internal stress distributions and comparing these values with the allowable values of the materials, so as to draw the conclusion that whether the aircraft structure meets the requirement of safety. This stressing approach has no ability to predict damage and failure behavior of structures, and load bearing capacity of the structure can only be obtained through test. Due to the high cost and long cycle of test, and the backward arrangement of many conditions and restrictions, the design defects cannot be exposed in time and the problems are relatively volatile.

The development of progressive damage analysis technology and the improvement of high performance computing capability are expected to fundamentally solve the problem of high fidelity simulation of structural load bearing, so that the designer can master the virtual test evaluation ability in addition to the physical test. However, the theoretical models of classical damage mechanics tend to be ideal [1], and it is difficult to obtain the characterization parameters. In the early stage, there are few applications of theoretical models for complex project, but in a recent period, there are obviously more applications in this area. For example, Lockheed Martin has developed F-35 project based on its special damage analysis tools to effectively reduce test [4]. A user-defined 3D damage model (VUMAT) with solid elements was developed and implemented into the finite element code ABAQUS/Explicit to predict the type and extent of damage through the laminate thickness [7]. In order to assess the seismic vulnerability of historical buildings, the concrete damage plasticity (CDP) material model in ABAQUS was used in numerical analysis [8]. When using ABAQUS for damage analysis, it is necessary to select an available damage model, such as concrete damaged plasticity, concrete smeared cracking etc. If there is no available damage model in ABAQUS, a user-defined damage model is necessary in analysis. There are also applications of damage mechanics for the design of composite cabin section of reentry aircraft [7].

From the traditional check to the high fidelity simulation, the core is to develop a simple and practical material damage constitutive model, and put forward a simple and feasible method to identify the characterization parameters. It is already to implement the monotonic tensile loading test in ABAQUS, but the conventional model in ABAQUS cannot accurately simulate the engineering stress-strain curves and the fracture morphology. The Comparison of the conventional model and the damage model of engineering stress-strain curves and shrinkage of tensile fracture section of specimens is shown in figure 1. The results show that the high fidelity simulation of engineering stress-strain curves, necking phenomenon and fracture morphology is realized based on the damage model.

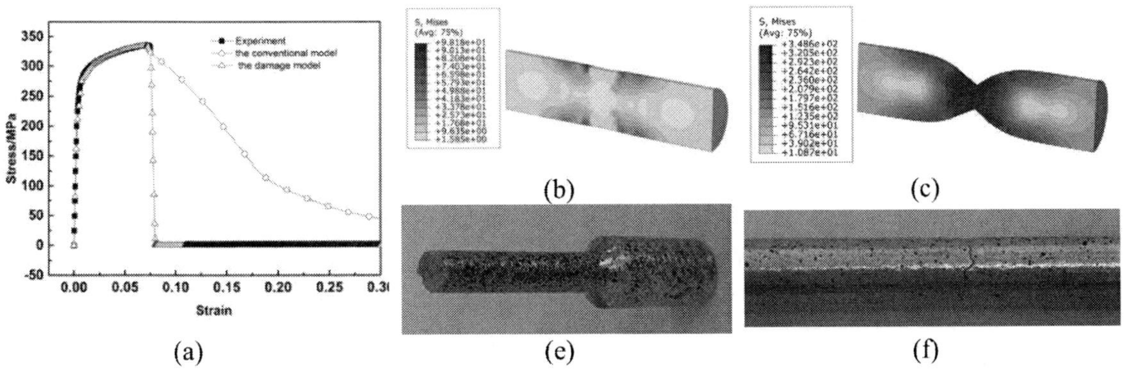

Figure 1. Comparison of the conventional model and the damage model of engineering stress-strain curves and shrinkage of tensile fracture section of specimens. (a) Comparison of engineering stress-strain curves, (b) The damage model, (c) The conventional model, (e) Real test result and (f) Real test result.

In this paper, the whole process of mechanical failure of elastic-plastic materials is divided into two stages: pre-damage and post-damage. The characterization parameters can be identified only by using the percentage reduction of area in conventional performance parameters. Based on the simple damage constitutive model, the digital virtual test capability of structural strength can be developed, which can realize the high fidelity simulation of physical test, and make the traditional finite element analysis level realize technical span.

2. Elastic-plastic simple damage constitutive model

The typical stress-strain curve of elastic-plastic materials (epm) is shown in figure 2. The ultimate stress σ_b is defined as the damage cut-off point. Through the damage cut-off point, the whole process of deformation and failure of elastic-plastic materials is divided into two stages: pre-damage stage and post-damage stage. The constitutive model is established according to the different damage behavior in the pre-damage and post-damage stages.

In figure 2, σ_s is the static yield strength; ε^{ne} is the inelastic strain; σ_b and ε_b^{ne} are the ultimate stress and inelastic strain of the damage cut-off point; σ_f and ε_f^{ne} are the fracture stress and inelastic strain of the fracture point; the damage factor variable d_h reflects the softening degree of the material stiffness, and set the cut-off point $d_h = d_0 = 0$, the fracture point $d_h = d_c$, and after fracture $d_h = 1.0$.

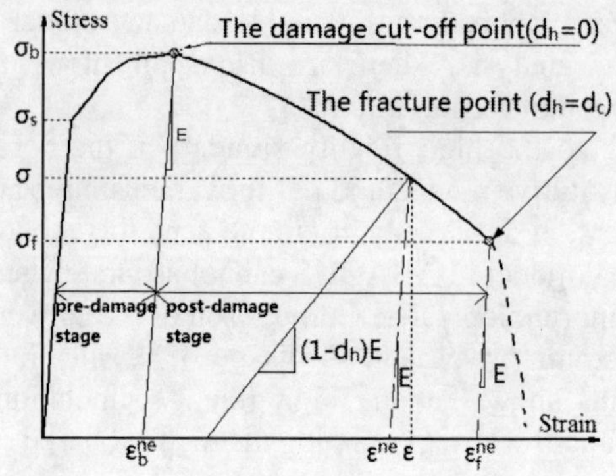

Figure 2. A typical stress-strain curve of elastic-plastic materials.

2.1. Pre-damage constitutive model

The pre-damage stage can be divided into linear elastic section and plastic deformation section according to whether plastic deformation occurs or not. There is no damage in the linear elastic section of materials, After entering the plastic deformation section, materials harden and the stiffness begins to damage According to the classical elastic-plastic theory, the stress-strain relationship should be expressed in an incremental form [11]:

$$\{d\sigma\} = [D]\{d\varepsilon\} \tag{1}$$

where $[D]$ is the stiffness of plastic hardening deformation.

The hardening law and flow law of elastic-plastic materials are the same, which are described by the following yield surface equation [11]:

$$\sigma_{eq} - \left[\sigma_s + R\left(\varepsilon^p\right)\right] = 0 \tag{2}$$

where σ_{eq} is the Von-Mises equivalent stress, σ_s is the static yield strength and ε^p is the equivalent plastic strain. Additionally, the terms in the bracket denotes the hardening yield strength and $R\left(\varepsilon^p\right)$ denotes the hardening function.

According the classical damage mechanics, one can derive the following formula [11]:

$$[D] = [C]\left(1 - d_q\right) \tag{3}$$

where [*C*] is the linear elastic stiffness of materials; d_q is the stiffness damage coefficient caused by plastic hardening.

2.2. Post-damage constitutive model

The continuous deformation of the post-damage stage will inevitably lead to further accumulation of damage, resulting in the reduction of unloading stiffness and the occurrence of stiffness softening and loss of load. The damage factor variable d_h is introduced to reflect the softening degree of the relative ideal plastic state of the material stiffness. According to the hypothesis of Lemaitre equivalent strain, the stress-strain relationship in the post-damage stage can be expressed as follows [2]:

$$\{d\sigma\} = [C](1 - d_h)\{d\varepsilon^e\} \tag{4}$$

where $\{d\varepsilon^e\}$ is the elastic strain increment.

In papers [12] and [13], The integral expression of damage law under the condition of neglecting elastic deformation under proportional loading is given:

$$d_h = d_c - R_v(d_c - d_0)\left(\frac{\varepsilon_c^p - \varepsilon^p}{\varepsilon_c^p - \varepsilon_0^p}\right)^\alpha \tag{5}$$

where d_0 and d_c are the initial and critical damage values, α is the index reflecting the damage accumulation. The R in equation (2) and the R in equation (5) are different. In equation (2), $R(\varepsilon^p)$ is the hardening function and in equation (5), R_v is the stress triaxiality factor [1]:

$$R_v = \frac{2}{3}(1+\nu) + 3(1-2\nu)\left(\frac{\sigma_m}{\sigma_{eq}}\right)^2 \tag{6}$$

where σ_m is the hydrostatic pressure, ν is Poisson's ratio.

The damage factor variable d_h in the equation (5) is directly related to the plastic strain, and the problem is small when the degree of stiffness softening is not obvious, but the proportion of the plastic strain in the whole strain is reduced with the increase of the softening degree of the stiffness, and the applicability of the equation (5) will be limited. The concept of inelastic strain is introduced, and the inelastic strain at any point of stress and strain curve is defined as [1]:

$$\varepsilon^{ne} = \varepsilon - \varepsilon^e \tag{7}$$

where ε^e is the elastic strain.

Based on the division of the material shown in figure 1, the damage law of equation (5) is further developed. Set the cut-off point $d_h = d_0 = 0$, the fracture point $d_h = d_c$, and after fracture $d_h = 1.0$. Assuming that the damage factor varies monotonically and continuously with the inelastic strain in the range of $[0, d_c]$, a general equation of exponential decay type is proposed to describe the variation rule of damage factor with strain in a simpler form:

$$d_h = R_v \cdot \left\{1 - \exp\left[\left(\frac{\varepsilon^{ne} - \varepsilon_b^{ne}}{\varepsilon_f^{ne} - \varepsilon_b^{ne}}\right)^k \ln(1 - d_c)\right]\right\} \tag{8}$$

where ε_b^{ne} is the inelastic strain of damage cut-off point; *k* is the damage shape factor, as shown in figure 3 to control the shape of damage factor curve.

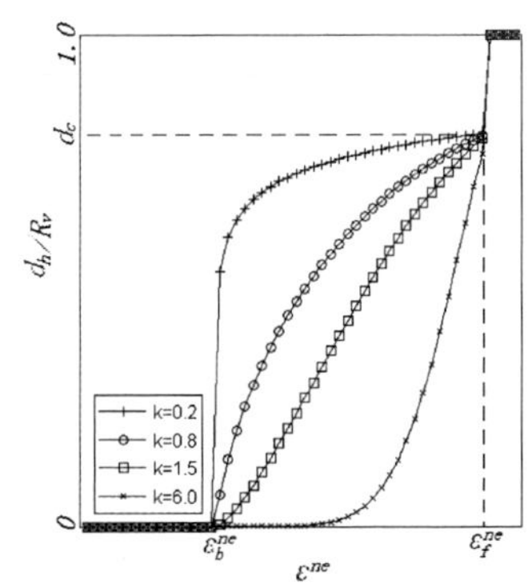

Figure 3. Shape curve of post-damage factor under different *k* value.

In equation (8), R_v, ε^{ne} is the process variable, ε_b^{ne} can be easily calculated from equation

(7),according to the ultimate stress (material strength limit σ_b), which are not characterization parameters. ε_f^{ne} and d_c are related to the stress and strain when the material fracture, together with the damage shape factor, they are the key parameters used to characterize the fracture behavior of the simple damage constitutive model.

3. Algorithm and program realization of simple damage constitutive model

The pre-damage stage of simple damage constitutive model has no difference from that of the ordinary elastic-plastic material. The elastic modulus and stress-plastic strain data of material can be inputted simply. The core of the algorithm is to establish the linkage mechanism of the change of elastic modulus with post-damage factor. In each step of calculation in the post-damage stage, the post-damage factor is calculated in real-time according to the current value of inelastic strain, and the elastic modulus of the material is adjusted in real-time. The material definition in ABAQUS software can associate elastic modulus with field variables, and provide interface of user subroutine to redefine field variables at material point (USDFLD), which provides a convenient condition for the realization of simple damage constitutive program.

Assuming that the field variable FIELD(1) is used to store d_h and is related to the elastic modulus of the material, the state variable STATEV(1) is used to transfer d_h among the calculation steps. Based on the simple damage constitutive model of this paper, the user subroutine to redefine field variables of ABAQUS is developed. The main steps of the program are as follows:

- Call the internal function GETVRM('SINV',......) to get σ_{eq} and σ_m .
- Call GETVRM('E',......), get the strain component, calculate the equivalent strain ε .
- The inelastic strain is calculated from equation (7).
- Compare ε^{ne} and ε_b^{ne} to judge the specific damage of the material.

if$\varepsilon^{ne} \le \varepsilon_b^{ne}$, $d_h = 0$;

if$\varepsilon_b^{ne} < \varepsilon^{ne} < \varepsilon_f^{ne}$, calculate d_h according to equation (8);

if $\varepsilon^{ne} > \varepsilon_f^{ne}$, $d_h = 1$.

- Compare d_h and STATEV(1), if $d_h >$ STATEV(1), re-assign a value to STATEV(1) with d_h .
- Re-assign a value to FIELD(1) with STATEV(1).

Based on the ABAQUS script language python, the relatively simple elastic-plastic constitutive model generation process in the pre-damage stage and the user-defined program generation and compilation process in the post-damage stage are built-in and integrated. Further a special plug-in for damage constitutive creation of elastic-plastic materials is developed. Such a plug-in can help design a damage material model that is as simple and efficient as creating a conventional material model.

4. Identification and verification of characterization parameters for simple damage constitutive model

4.1. Identification method of damage constitutive parameters

If additional performance test requirements are put forward because of the characteristic parameters of damage constitutive model, the engineering application of damage constitutive model will undoubtedly be hindered. According to the classical elastic-plastic and damage mechanics theory, the inelastic strain ε_f^{ne} of the fracture point can be estimated directly from the percentage reduction of area according to the following equation [14]:

$$\varepsilon_f^{ne} \cong -\ln(1-\psi) \tag{9}$$

IOP Conf. Series: Materials Science and Engineering **657** (2019) 012024 doi:10.1088/1757-899X/657/1/012024

where ψ is the percentage reduction of area.

The critical damage factor can also be predicted by equation (10) [1]:

$$d_c \cong 1 - \frac{\sigma_f}{\sigma_b} \tag{10}$$

Fracture stress σ_f is not a conventional performance parameter, so equation (10) is not the same as equation (9), which has obvious guiding significance for engineering selection of characterization parameters. However, the material damage constitutive model is created by taking the ε_f^{ne} estimated by equation (9) and d_c selected by a certain range (usually within 0.2-0.5) as the initial condition. Taking the virtual test plug-in of tensile/compressive strength of metal test rod based on ABAQUS software python language, the virtual test of tensile strength of specimens is carried out, and the virtual/real test results are compared. According to the stress/strain curve, the percentage total elongation at fracture, the percentage reduction of area, the appropriate characterization parameters, and a few rendezvous can identify the specific values of the representation parameters that meet the precision requirements.

4.2. Identification and verification of characterization parameters of ZL114A

Taking ZL114A as an example, the damage constitutive parameters of ZL114A are identified based on the methods and tools presented in chapter 4.1, and the damage constitutive parameters of ZL114A are identified: $k = 0.8$, $d_c = 0.05$, $\varepsilon_f^{ne} = 0.09$. The specimen tensile virtual test was carried out according to the identified material damage constitutive model. The virtual results of the percentage reduction of area and the percentage total elongation at fracture were compared with the experimental results as shown in table 1. The comparison of the virtual/real test results of engineering stress-strain curves in specimen tensile is shown in figure 4. The results show that the high fidelity simulation of engineering stress-strain curves is realized based on simple damage constitutive model, and the relative accuracy of the percentage reduction of area and the percentage total elongation at fracture is up to 5%.

Table 1. Comparison of virtual/real test results.

	Sample	Real test	Virtual test	Error/%
The percentage reduction of area/%	Sample 1	12.94	12.32	3.40
	Sample 2	12.43		
	Sample 3	12.89		
	Average	12.75		
The percentage total elongation at fracture/%	Sample 1	7.52	7.40	1.16
	Sample 2	7.21		
	Sample 3	7.69		
	Average	7.47		

4.3. Analysis of the influence of the characterization parameters on the fracture behavior

On the basis of the simple damage constitutive identification results, the damage parameters are deflected, and the fracture behavior of the material is further studied. Taking ZL114A as an example, the fracture behavior of material is affected by the characterization parameters. Comparison of tensile behavior of materials with different values of damage characterization parameters can be found in figures 5-7.

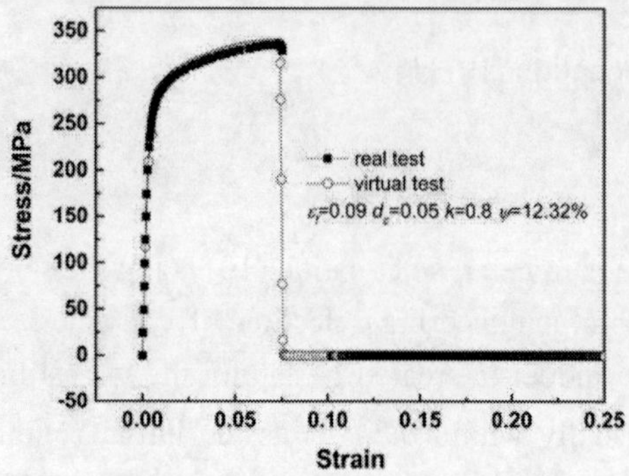

Figure 4. Comparison of virtual/real test results of engineering stress-strain curves.

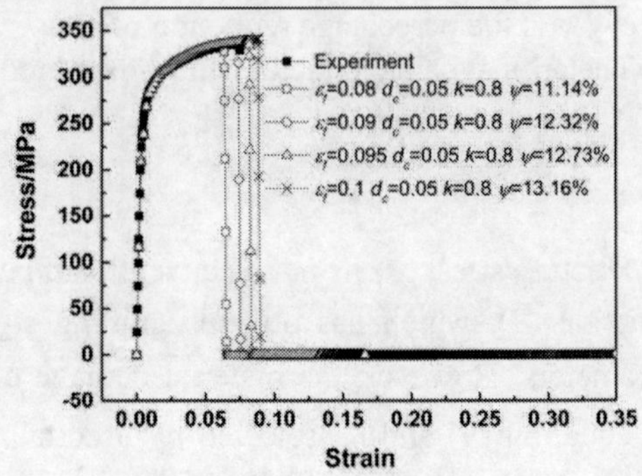

Figure 5. Effect of ε_f^{ne} value change on tensile behavior of material.

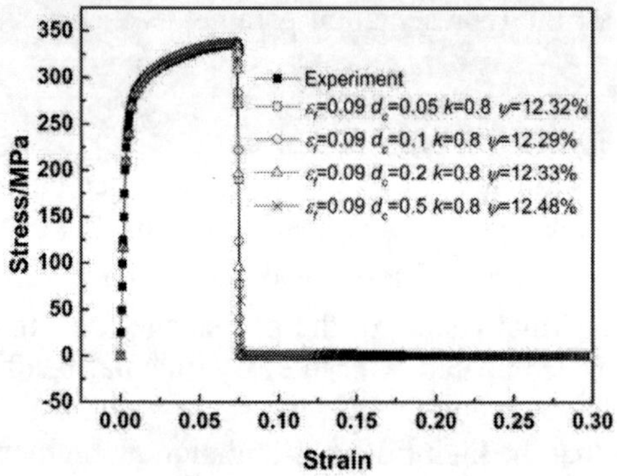

Figure 6. Effect of d_c value change on tensile behavior of material.

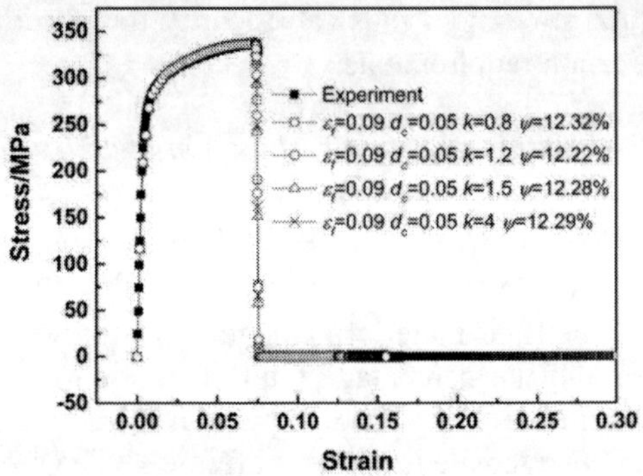

Figure 7. Effect of k value change on tensile behavior of material.

The results show that:

- ε_f^{ne} can affect the macro fracture behavior of ZL114A most significantly, and large ε_f^{ne} means large percentage total elongation at fracture and large percentage reduction of area.
- The change of d_c has no significant effect on the macro fracture behavior of ZL114A, but it can play a fine-tuning role in the percentage total elongation at fracture.
- The difference of k value in the range of 0.8-1.5 does not lead to significant change in fracture behavior of ZL114A, and has sufficient coverage, which is consistent with the conclusion given in reference [6] that $k = 0.8$ can be used as a typical value.

5. Conclusion

In this paper, the classical theory of elastic-plastic and damage mechanics is developed to meet requirements of virtual test construction for high fidelity simulation of structural strength. A simple damage constitutive model of elastic-plastic materials is proposed to solve the core technical problem of high fidelity simulation of structural load bearing failure.

The intrinsic relationship between damage characterization parameters and the percentage total elongation at fracture and the percentage reduction of area is studied. The identification methods and verification tools for characterization parameters is given. The identification verification of typical materials shows: Based on the identification results of simple damage constitutive model and characterization parameters, the mechanical behavior of specimen deformation up to fracture can be simulated with high fidelity. The relative accuracy of the percentage total elongation at fracture and the percentage reduction of area can reach 5%.

The damage parameters were deflected and the influence of the characterization parameters on the fracture mechanical behavior of the materials was studied with ZL114A. The results show that the change of ε_f^{ne} has the most significant effect on the macro-fracture behavior of the material. Large ε_f^{ne} means large percentage total elongation at fracture and large percentage reduction of area. Changing d_c can play a fine-tuning role in the percentage total elongation at fracture, The difference of k value in the range of 0.8-1.5 does not lead to significant change in fracture behavior of materials.

References

[1] Lemaitre J and Desmorat R 2005 *Engineering Damage Mechanics* (New York: Springer-Verlag Berlin Heidelberg)
[2] Lemaitre J 1996 *A Course on Damage Mechanics* (New York: Springer-Verlag Berlin Heidelberg)
[3] Wu H Y 1990 *Damage Mechanics* (Beijing: National Defence Industry Press)
[4] Alex S, Mark W, Steve E and Brandon E 2018 Application of progressive damage failure analysis to large aircraft composite structures *AIAA* **2018-2235**
[5] Norwood D S, Gail L H, Robin J W and Patrick F J 2016 A historical assessment of building block development test programs for modern military aircraft *AIAA* **2016-0231**
[6] Scott R W and Stephen P S 2011 Virtual testing in live fire test and evaluation (LFT & E) *AIAA* **2011-1806**
[7] Shi Y, Swait T and Soutis C 2012 Modelling damage evolution in composite laminates subjected to low velocity impact *Compos. Struct.* **94** 2902-13
[8] Giuseppe F, Marco F F and Paolo L 2017 Survey and seismic vulnerability assessment of the baptistery of san giovanni in tumba (italy) *J. Cult. Herit.* **26** 64-78
[9] Wang C H, Yang Y, Kang Q, Yin D Z and Shui Y T 2016 A stiffness degradation model and the general method for progressive failure analysis of composite structure *China. Eng. Mech.* **33** 17-23
[10] Wang C H, Chen H and Wang J 2016 Ultime strength analysis for the composite material cabin of re-entry vehicle *China. Eng. Mech.* **43** 1-8
[11] Fionn D and Nik P 2005 *Introduction to Computational Plasticity* (New York: Oxford University Press)
[12] Wang T J 1992 Unified CDM model and local criterion for duntile fracture-I. Unified CDM model for ductire fracture *Eng. Fract. Mech.* **42** 177-83
[13] Huang S H and Wang T J 1994 A new local theory for ductile fracture and it's applications to axially and spherically symmetric structures *China. Struct. & Environ. Eng.* **4** 410-4
[14] Yu H Q and Chen J D 1999 *The Principle of Metal Plastic Forming* (Beijing: China Machine Press)

IOP Conf. Series: Materials Science and Engineering **657** (2019) 012025 doi:10.1088/1757-899X/657/1/012025

Constraint conditional finite element method for off-axial interfacial sliding of fiber reinforced composite

R Kitamura[1,2]

[1]Department of mechanical engineering, Faculty of Engineering, Kanagawa University, 3-27-1 Rokkakubashi, Kanagawa, Yokohama, Kanagawa 221-8686, Japan

E-mail: ft102093gj@kanagawa-u.ac.jp

Abstract. The main toughening mechanism of Ceramics Matrix Composite (CMC) is the fiber/matrix interfacial debonding and sliding. To simulate of the contact problem (such as interfacial sliding), penalty method etc have been used in general finite element method. But, these method need the repeated calculations, thus the CPU time is long and the accuracy depends on the number of repetitions. On the other hand, in actual CMCs, the fibers are not necessarily oriented along the loading direction, and the fiber diameter also fluctuates along the axis. Therefore, in this study, Constraint Conditional Finite Element Method (CC-FEM) was formulated to analyze the "off-axis interfacial sliding" with high precision and in a short time without repeated calculations. In CC-FEM, the equality of nodal displacements at the interface and the equilibrium of contact forces are assumed as constraint conditions, in which Coulomb's friction law and equivalent friction coefficient were taken into account. In addition, the validity was verified especially for "off-axis interfacial sliding" by comparing with general-purpose finite element software ANSYS. In the both cases of on- and off-axial interfacial sliding, the resultant stress distributions of the fiber and matrix agreed well with those of ANSYS. As compared to the case of on-axial interfacial sliding, the matrix stress recovered more steeply because of the higher equivalent frictional coefficient.

1. Introduction

Ceramics are excellent for their heat, abrasion and corrosion resistances, but their toughness and strength reliability are insufficient because of their brittle nature. On the other hand, fiber-reinforced ceramic matrix composites (CMCs) such as SiC fiber-reinforced SiC are damage-tolerant materials for which high toughness is achieved through damage mechanisms such as fiber bridging, fiber breakage, pullout, interfacial debonding followed by sliding, and crack deflection deriving from reinforcing materials such as ceramic fibers [1]. From such excellent properties, CMC is anticipated for use as a high-temperature structural material applied for aerospace for which conventional metal are difficult to apply. Accordingly, the formulation for solving mechanical problems in this material is linked to deepened engineering significance.

In general, contact problems are solved through computational algorithms such as penalty method and enlarged Lagrange method, which are included in general finite element method (FEM) software. In these algorithms, stiffness between contact surfaces is expressed by contact spring. Computations are conducted repeatedly until the amounts of sliding and/or overlapping between nodal-points reach to a given allowable value. Therefore, the computation time and precision depend on the allowable value [2].

On the other hand, the constrained conditional finite element method (hereinafter, denoted as CC-FEM) is a mechanical modelling, in which the equivalence of nodal-point displacements and the equilibrium of contact forces are first given as contact conditions on the contact faces. Next, the global stiffness matrix is changed based on the given conditions, and solved numerically. According to this model, the exact numerical solution is obtainable by a single computation with no repetition. If this model is applied for the damage problems in CMCs, then even a complicated damage state including plural matrix cracks and fiber breaks could be solved by only once without divergence. In the previous papers, the formulation of CC-FEM was conducted in the case of on-axial interfacial debonding and sliding problem [1,3-5]. In actual CMCs, however, the fibers are often placed with an inclination along the loading direction; the fiber diameter varies along the fiber axis. Under such circumstances, the applicability of the CC-FEM is not adequate.

The purpose of this study is thus to formulate CC-FEM in the case of an "off-axial interfacial debonding and sliding" problem. The obtained numerical results were compared with the results obtained from ANSYS, a popular general-purpose FEM software. In addition, its validity was discussed.

2. Constrained conditional finite element method (CC-FEM)

2.1. Stiffness equation including contact forces

The present method deals with contact forces occurring at the interface, based on contact analysis. In general, the principle of virtual work including contact forces is given as follows [6,7].

$$\iiint_V \sigma_{ij}\delta\varepsilon_{ij}dV - \left(\iiint_V p_i\delta u_i dV + \iint_{S_{\bar{c}}} T_i\delta u_i dS_{\bar{c}}\right) - \iint_{S_c} R_i\delta u_i dS_c = 0 \quad (1)$$

Where, σ_{ij}, $\delta\varepsilon_{ij}$, δu_{ij} are stress, virtual strain and virtual displacement. As shown in figure 1, p_i, T_i, R_i are boby, surface and contact forces. S_c and $S_{\bar{c}}$ are surface areas of contact surface C (such as fiber/matrix surface) and mechanical boundary except C, while V is volume. Subscripts i and j are the r and z-directions. In equation (1), 1st~3rd terms show internal work, external work (sum of works by external force and body force), and work by contact force, respectively. By discretizing equation (1) and by defining the equivalent nodal force on nodes of each elements, the stiffness equation including contact forces can be derived. The incremental form is given as:

$$[K]\{\Delta u\} = \{\Delta f\} + \{\Delta R\} \quad (2)$$

Where, $[K]$ is the global stiffness matrix, derived from the first term of equation (1). $\{\Delta u\}$ is nodal displacement increments, and $\{\Delta f\}$ is load increments, derived from the 2nd term. $\{\Delta R\}$ is contact force increments, derived from the 3rd term, which are treated as unknown variables.

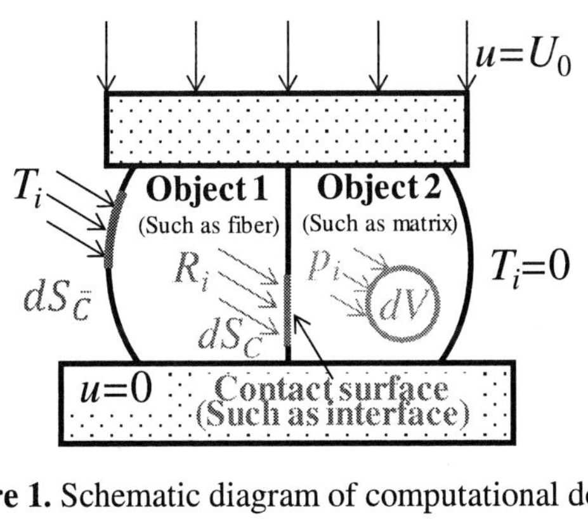

Figure 1. Schematic diagram of computational domain.

2.2. *Definition of interfacial contact states*

As described in the Introduction, reinforcing fibers used for CMCs often have local variation in diameter, resulting in off-axial interfacial debonding and sliding. Therefore, the following interfacial damage process is presumed in the same way as the on-axial case: First, fiber breakage or matrix cracking occurs with subsequent interfacial debonding. Next, off-axial interfacial sliding continues until the fiber-matrix relation reaches the mechanical equilibrium state through interfacial friction. Under such circumstances, the interfacial contact states are defined as shown in figure 2. Every state comprises four nodal points, of which 1 and 3 are placed on the fiber element, and 2 and 4 are on the matrix one. As shown in figure 2(a), bonding state represents a non-damage state. On the other hand, figures 2(b)~2(d) show damage states after relative sliding at the fiber-matrix interface, which occurs at the angle α from the loading axis, where angle α is defined such that counterclockwise direction is positive.

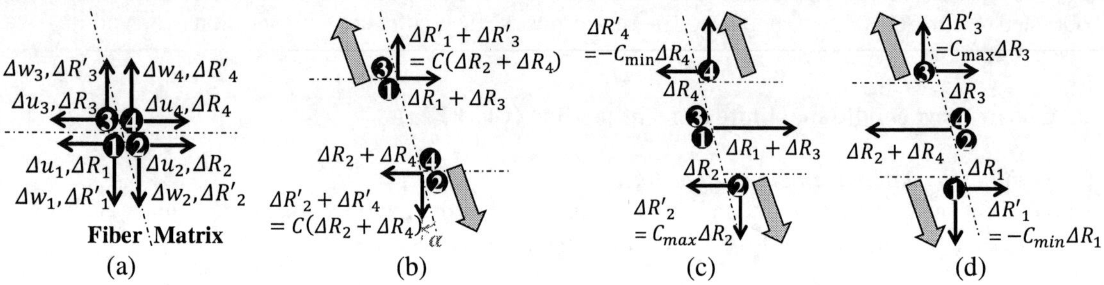

Figure 2. Interfacial contact states for off-axial interface: (a) Bonding, (b) Steady state after interfacial debonding and sliding, (c) with matrix crack, and (d) with fiber breakage.

2.3. *Interfacial contact conditions*

Interfacial contact conditions in the present finite element model, i.e. constrained conditions, are based on the equivalence of displacement increments and the equilibrium of contact forces, as follows.

2.3.1. Bonding. All nodal points are bonded at the same position as shown in figure 2(a). Therefore, all the displacement increments must be identical. Furthermore, the sum of the contact force increments is zero at the stationary state. The conditions are therefore given as below.

$$\begin{aligned} &\Delta u_1 = \Delta u_2 = \Delta u_3 = \Delta u_4 \qquad \Delta w_1 = \Delta w_2 = \Delta w_3 = \Delta w_4 \\ &\Delta R_1 + \Delta R_2 + \Delta R_3 + \Delta R_4 = 0 \quad \Delta R'_1 + \Delta R'_2 + \Delta R'_3 + \Delta R'_4 = 0 \end{aligned} \tag{3}$$

Where, Δu and Δw are displacement increments of nodes 1 to 4 along the r- and z -directions. ΔR and $\Delta R'$ are the contact force increments.

2.3.2. Steady states after interfacial debonding and sliding As shown in figures 2(b)~2(d), in the off-axis case, the direction of frictional forces differs from the direction of contact forces. In figure 2(b), contact forces of r - and z -directions are presented as dissolved to the sliding direction and the normal direction. Under such a situation, Coulomb's law of friction is applicable as presented below.

$$(\Delta R'_1 + \Delta R'_3)\cos\alpha - (\Delta R_1 + \Delta R_3)\sin\alpha = \mu\{(\Delta R_1 + \Delta R_3)\cos\alpha + (\Delta R'_1 + \Delta R'_3)\sin\alpha\} \tag{4}$$

$$\therefore (\Delta R'_1 + \Delta R'_3) = C(\Delta R_1 + \Delta R_3) \tag{5}$$

Therein, μ is the static friction coefficient, and C is the equivalent friction coefficient:

$$C = C_{\max} = (\mu + \tan\alpha)/(1 - \mu\tan\alpha) \tag{6}$$

Equation (6), which means that the friction coefficient apparently varies as a function of α, is established for the case in which the fiber and matrix elements are mutually pushed at the vertical direction of the interface. This equation engenders the relation: $C = C_{max}$. The frictional force increases concomitantly with increasing α. In contrast, when the forces work such that the fiber and matrix elements are mutually separated, the equivalent frictional force is given as;

$$C = C_{\min} = (\mu - \tan\alpha)/(1 + \mu\tan\alpha) \tag{7}$$

Equation (7) is valid when a radial compression works in advance by residual stresses. In equation (7), the frictional force is shown to decrease with increasing α. From the above mention, the interfacial contact conditions of steady states after interfacial debonding and sliding (figure 2(b)) are given as;

$$\begin{gathered} \Delta u_1 = \Delta u_3 \quad \Delta u_2 = \Delta u_4 \quad \Delta w_1 = \Delta w_3 \quad \Delta w_2 = \Delta w_4 \\ \tan\alpha = (\Delta u_2 - \Delta u_1)/(\Delta w_1 - \Delta w_2) \\ (\Delta R'_1 + \Delta R'_3) = C(\Delta R_1 + \Delta R_3) \quad (\Delta R'_2 + \Delta R'_4) = C(\Delta R_2 + \Delta R_4) \\ (\Delta R_1 + \Delta R_3) + (\Delta R_2 + \Delta R_4) = 0 \quad (\Delta R'_1 + \Delta R'_3) + (\Delta R'_2 + \Delta R'_4) = 0 \end{gathered} \tag{8}$$

2.3.3. Interfacial debonding and sliding with fiber breakage.

$$\begin{gathered} \Delta u_1 = \Delta u_3 \quad \Delta w_1 = \Delta w_3 \\ \tan\alpha = (\Delta u_2 - \Delta u_1)/(\Delta w_1 - \Delta w_2) = (\Delta u_3 - \Delta u_1)/(\Delta w_1 - \Delta w_3) \\ \Delta R'_2 = C_{\max}\Delta R_2 \quad R'_4 = -C_{\min}\Delta R_4 \quad \Delta R'_1 + \Delta R'_3 = 0 \\ (\Delta R_1 + \Delta R_3) + \Delta R_2 + \Delta R_4 = 0 \quad (\Delta R'_1 + \Delta R'_3) + \Delta R'_2 + \Delta R'_4 = 0 \end{gathered} \tag{9}$$

2.3.4. Interfacial debonding and sliding with matrix cracking.

$$\begin{gathered} \Delta u_2 = \Delta u_4 \quad \Delta w_2 = \Delta w_4 \\ \tan\alpha = (\Delta u_2 - \Delta u_1)/(\Delta w_1 - \Delta w_2) = (\Delta u_4 - \Delta u_1)/(\Delta w_1 - \Delta w_4) \\ \Delta R'_1 = -C_{\min}\Delta R_1 \quad R'_3 = C_{\max}\Delta R_3 \quad \Delta R'_2 + \Delta R'_4 = 0 \\ \Delta R_1 + \Delta R_3 + (\Delta R_2 + \Delta R_4) = 0 \quad \Delta R'_1 + \Delta R'_3 + (\Delta R'_2 + \Delta R'_4) = 0 \end{gathered} \tag{10}$$

Equations (3) and (8)–(10) agree exactly with the on-axial interfacial contact conditions, when α= 0° [1, 3-5]. In other words, the off-axial conditions include on-axial conditions.

2.4. Insertion of interfacial contact conditions into the global stiffness matrix

From contact conditions (such as equations (3), (8), (9), (10)), several unknown variables (displacements Δu, Δw, and contact forces ΔR, $\Delta R'$) can be reduced. For example, in (b) Steady states after interfacial debonding and sliding (figure 1(b)), from constraint conditions (equation (8)), all variables can be represented by 3 displacements (Δu_1, Δw_1, Δu_2) and 5 contact forces (ΔR_2, ΔR_3, $\Delta R'_3$, ΔR_4, $\Delta R'_4$), and the number of variables to be calculated can be reduced from 16 to 8.

Then, the 1st line of equation (1) is written out as;

$$K_{11}\Delta u_1 + K_{12}\Delta w_1 + K_{13}\Delta u_2 + K_{14}\Delta w_2 + K_{15}\Delta u_3 + K_{16}\Delta w_3 + K_{17}\Delta u_4 + K_{18}\Delta w_4 = \Delta f_{r1} + \Delta R_1 \tag{11}$$

IOP Conf. Series: Materials Science and Engineering **657** (2019) 012025 doi:10.1088/1757-899X/657/1/012025

Using equation (8), equation (11) can be rewritten as:

$$K_{11}\Delta u_1 + K_{12}\Delta w_1 + K_{13}\{(\Delta w_1 - \Delta w_2)\tan\alpha + \Delta u_1\} + K_{14}\Delta w_2 + K_{15}\Delta u_1 + K_{16}\Delta w_1 + K_{17}\{(\Delta w_1 - \Delta w_2)\tan\alpha + \Delta u_1\} + K_{18}\Delta w_4 = \Delta f_{r1} - (\Delta R_2 + \Delta R_3 + \Delta R_4) \quad (12)$$

$$\therefore (K_{11} + K_{13} + K_{15} + K_{17})\Delta u_1 + \{(K_{12} + K_{16}) + (K_{13} + K_{17})\tan\alpha\}\Delta w_1 + \{(K_{14} + K_{18}) - (K_{13} + K_{17})\tan\alpha\}\Delta w_2 - \Delta R_2 - \Delta R_3 - \Delta R_4 = \Delta f_{r1} \quad (13)$$

Equation (13) contains only 8 variables (Δu_1, Δw_1, Δu_2, ΔR_2, ΔR_3, $\Delta R'_3$, ΔR_4, $\Delta R'_4$), and another lines can also be similarly represented by only these 8 variables. Consequently, the following global stiffness equation is ultimately given.

$$[K_c]\{\Delta u_c\} = \{\Delta f\} \quad (14)$$

Where, $[K_c]$ is the global stiffness matrix, which is changed by the interfacial contact conditions. Also, $\{\Delta u_c\}$ consists of nodal displacement increments including contact force increments. For example, in (b) Steady state after interfacial debonding and sliding, the stiffness equations are given as:

$$\begin{bmatrix} K_{11}+K_{13}+K_{15}+K_{17} & (K_{12}+K_{16})+(K_{13}+K_{17})\tan\alpha & 1 & (K_{14}+K_{18})-(K_{13}+K_{17})\tan\alpha & 1 & 0 & 1 & 0 \\ K_{21}+K_{23}+K_{25}+K_{27} & (K_{22}+K_{26})+(K_{23}+K_{27})\tan\alpha & +C & (K_{24}+K_{28})-(K_{23}+K_{27})\tan\alpha & 0 & 1 & +C & 0 \\ K_{31}+K_{33}+K_{35}+K_{37} & (K_{32}+K_{36})+(K_{33}+K_{37})\tan\alpha & -1 & (K_{34}+K_{38})-(K_{33}+K_{37})\tan\alpha & 0 & 0 & 0 & 0 \\ K_{41}+K_{43}+K_{45}+K_{47} & (K_{42}+K_{46})+(K_{43}+K_{47})\tan\alpha & -C & (K_{44}+K_{48})-(K_{43}+K_{47})\tan\alpha & 0 & 0 & -C & 1 \\ K_{51}+K_{53}+K_{55}+K_{57} & (K_{52}+K_{56})+(K_{53}+K_{57})\tan\alpha & 0 & (K_{54}+K_{58})-(K_{53}+K_{57})\tan\alpha & -1 & 0 & 0 & 0 \\ K_{61}+K_{63}+K_{65}+K_{67} & (K_{62}+K_{66})+(K_{63}+K_{67})\tan\alpha & 0 & (K_{64}+K_{68})-(K_{63}+K_{67})\tan\alpha & 0 & -1 & 0 & 0 \\ K_{71}+K_{73}+K_{75}+K_{77} & (K_{72}+K_{76})+(K_{73}+K_{77})\tan\alpha & 0 & (K_{74}+K_{78})-(K_{73}+K_{77})\tan\alpha & 0 & 0 & -1 & 0 \\ K_{81}+K_{83}+K_{85}+K_{87} & (K_{82}+K_{86})+(K_{83}+K_{87})\tan\alpha & 0 & (K_{84}+K_{88})-(K_{83}+K_{87})\tan\alpha & 0 & 0 & 0 & -1 \end{bmatrix} \begin{Bmatrix} \Delta u_1 \\ \Delta w_1 \\ \Delta R_2 \\ \Delta w_2 \\ \Delta R_3 \\ \Delta R'_3 \\ \Delta R_4 \\ \Delta R'_4 \end{Bmatrix} = \begin{Bmatrix} \Delta f_{r1} \\ \Delta f_{z1} \\ \Delta f_{r2} \\ \Delta f_{z2} \\ \Delta f_{r3} \\ \Delta f_{z3} \\ \Delta f_{r4} \\ \Delta f_{z4} \end{Bmatrix} \quad (15)$$

Equation (15) is solvable numerically, and $\{\Delta u_c\}=\{\Delta u_1, \Delta w_1, \Delta R_2, \Delta w_2, \Delta R_3, \Delta R'_3, \Delta R_4, \Delta R'_4\}^T$ can be gotten directly. Subsequently, another unknown displacements (Δu_2, Δu_3, Δw_3, Δu_4, Δw_4) and unknown contact forces (ΔR_1, $\Delta R'_1$, $\Delta R'_2$) can be also calculated using equation (8).

2.5. Finite element mesh and boundary conditions

As reported in the literature [8-12], the interfacial debonding size of CMCs is large-scale, which engenders an extremely small stress concentration around fiber break points or near the matrix crack plane, and which lowers the inter-fiber interaction. Therefore, the present study employed only the analysis of a single fiber composite in which the fiber was embedded in a cylindrical matrix. The finite element models are shown in figure 3. The fiber radius r_f and matrix radius r_m were 0.0055 and 0.011 mm, respectively, and the lengths were also 2.5 mm. The finite element model was composed of axisymmetric linear triangle elements. The respective numbers of nodal points and elements were 1,250 and 1,000. In ANSYS used for comparison, the model was composed of axisymmetric quadratic triangle elements. The respective numbers of nodal points and elements were 752 and 1,000. In these models, the constrained conditional state after interfacial debonding and sliding (figure 2(b)), was applied to area of 1.0 mm long from the matrix crack surface. The remaining area (1.5 mm long) was treated as the bonding state (figure 2(a)). The displacement boundary condition U was given only at the fiber end of the top in compression. To simulate the off-axial interfacial debonding and sliding, a change in fiber diameter Δr and taper length Δz were assumed as shown in figure 3. For example, in α=0.113° model, Δz=0.7mm, $\Delta r=r_f/4$ mm, $\alpha=\tan^{-1}(\Delta r/\Delta z)$=0.113°, and equivalent frictional coefficient C is increased to $C=C_{max}=(\mu+\tan\alpha)/(1-\mu\tan\alpha)$=0.0520°, when μ=0.05.

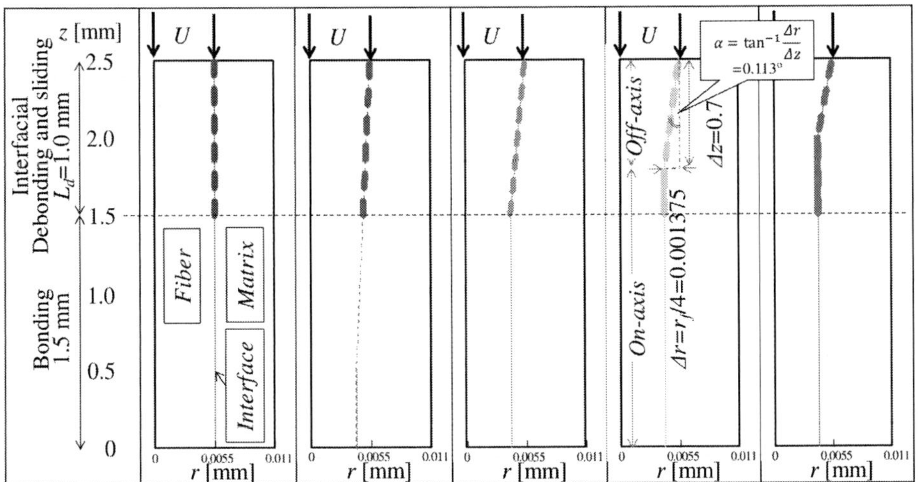

α[°]	0	0.039	0.079	0.113	0.158
Δz[mm]	-	2.0	1.0	0.7	0.5
Δr[mm]	-	0.001375	0.001375	0.001375	0.001375
C	0.05	0.0507	0.0514	0.0520	0.0528

Bonding (off-axis) *Interfacial sliding (off-axis)*
Bonding (on-axis) *Interfacial sliding (on-axis)*

Figure 3. Finite element models.

3. Results and discussion

3.1. Mechanical behavior caused by on-axial interfacial debonding and sliding

Figure 4 presents the axial stress distributions caused by on-axial interfacial debonding and sliding, calculated using CC-FEM and ANSYS. All stresses are normalized such that the axial fiber stress $(\sigma_z)_f$ at the matrix crack surface is applied as −1. The material and physical constants were given for simplicity as follows. Young's modulus: E_f=E_m= 200 GPa, Poisson's ratio: ν_f=ν_m=0.2, static frictional coefficient: μ=0.05. No residual stresses were included in the initial state. Results shown in figure 4(a) demonstrate that the stress distributions $(\sigma_z)_f$ and $(\sigma_z)_m$ mutually agree at the bonding area, i.e. the area more than 1.0 mm in the distance from matrix crack surface, and show a constant. On the other hand, $(\sigma_z)_m$ recovers in compression nonlinearly in area less than 1.0 mm from the matrix crack surface, whereas $(\sigma_z)_f$ increases greatly from the stress level of −1. Such behavior means that, although no load is applied to the matrix, the matrix stress occurs by a frictional effect even at the debonding area. At the debonding end, $(\sigma_z)_f$ and $(\sigma_z)_m$ change intensely because the interfacial contact state is changed here from figure 2(b) to 2(a).

In figure 4(b), the normalized stress distributions along the radial and circumferential directions, σ_r and σ_θ are shown. Both stresses show zero at the bonding area, because the same Poisson's ratios are given to the fiber and matrix. At the sliding area, however, these distributions behave nonlinearly with compression along the r- direction in the fiber and matrix, because radial stresses occur by Poisson's effect as a result of the difference between axial strains in the fiber and matrix. It is also noteworthy that $(\sigma_r)_m$ and $(\sigma_\theta)_m$ decrease in absolute value because distance r is separate from the interface. According to ANSYS the results (broken lines in figure 4), all stress distributions agree well with the CC-FEM results (solid lines). In other words, the CC-FEM validity is demonstrated in the on-axial interfacial state. In addition, the CPU times of ANSYS are about 4~5 sec, while CC-FEM one is about 0.75 sec and analysis can be carried out about 5.3~6.6 times faster.

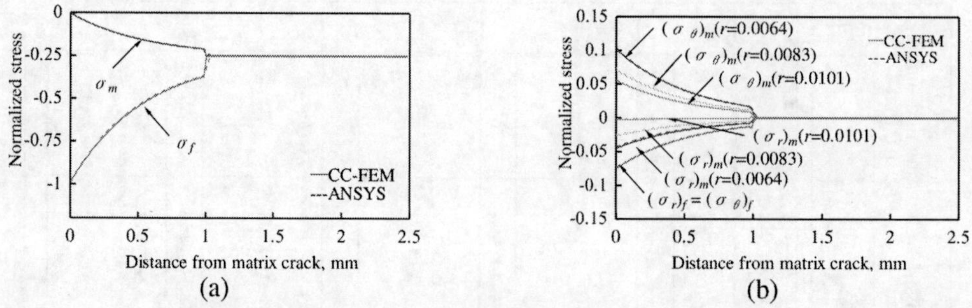

Figure 4. Comparison between stresses by CC-FEM and ANSYS (on-axis sliding). (a) σ_z and (b) σ_r and σ_θ.

3.2. Mechanical behavior caused by off-axial interfacial debonding and sliding

Figures 5(a) and 5(b) show axial stress distributions caused by off-axial interfacial debonding and sliding, calculated using CC-FEM and ANSYS. Results show that both distributions mutually agree well. Such agreement was also confirmed for σ_r and σ_θ distributions. It is concluded that the present CC-FEM formulation is also valid in the case of off-axial interfacial debonding and sliding.

According to the distributions in figure 5, the changes of the fiber and matrix stresses become steeper with increasing off-axial angle α, because the equivalent friction coefficient C is larger, as described earlier, producing more efficient stress recovery in compression. Similar steeper changes in stress were confirmed for σ_r and σ_θ distributions. Such stress recovery means a property improvement caused by a mechanical bond, the so-called 'anchor effect'.

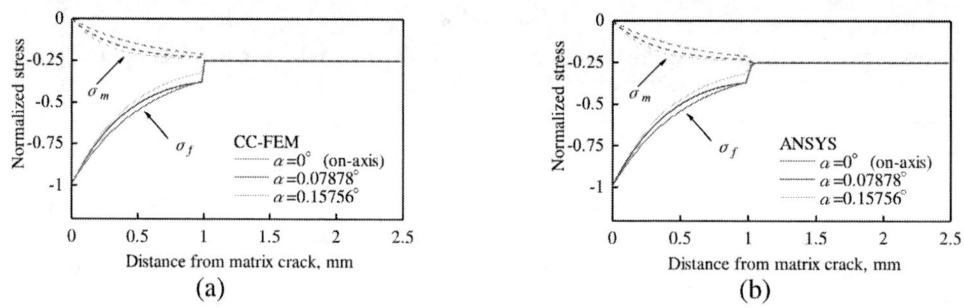

Figure 5. Comparison between axial stresses by CC-FEM and ANSYS (off-axis sliding). (a) CC-FEM (μ=0.05) and (b) ANSYS (μ=0.05).

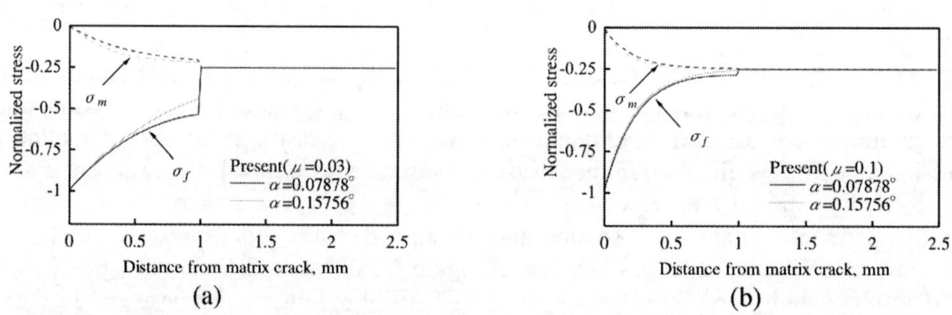

Figure 6. Effect of coefficients of static friction μ on axial stress by CC-FEM. (a) CC-FEM (μ=0.03) and (b) CC-FEM (μ=0.1).

Figure 6 shows the effects of friction coefficient on the longitudinal stress distributions by CC-FEM. In the calculation, the same off-axial angles were used. Comparison between figures 5(a), 6(a) and 6(b) demonstrates that steeper changes of the fiber and matrix stresses occur with increasing α. As presented in figure 6(a), on the other hand, the smaller friction coefficient gives a larger difference in stress between α=0.0788 and 0.158°. Because a smaller α yields a larger ratio, i.e. C/μ, this difference appears more clearly. In other words, for smaller μ, the anchor effect appears more strongly.

4. Conclusions

Formulation of constrained conditional finite element method (CC-FEM) for the case of off-axial interfacial debonding and sliding occurring in ceramic matrix composites was proposed. The calculated stress distributions in longitudinal, radial, and circumferential directions were compared with those of general-purpose software ANSYS. Results showed that these stress distributions agreed well with those of ANSYS at compressive loading to a single fiber composite. Such agreement with ANSYS was also confirmed for the case of on-axial interfacial debonding and sliding.

Calculation results obtained using CC-FEM also demonstrated that the stress recoveries in the fiber and matrix were more efficient with increasing off-axial angle. Such a tendency appeared more strongly, even at the same off-axial angle, when a larger friction coefficient was given.

The proposed CC-FEM can resolve problems of on-axial and off-axial interfacial debonding and sliding by setting the off-axis angle and equivalent frictional coefficient and off-axis angle, appropriately, and establishing equations of displacement equivalence and contact force balance. Results suggest that CC-FEM can be used widely as an effective tool to clarify various physical problems related to any fibrous composite materials.

References

[1] Goda K, Ohgi J, Tamaoki J and Suzuki M 2005 FEM Formulation and strength simulation by interfacial contact model of fiber-reinforced ceramic matrix composites *J. Jpn. Soc. Compos. Mater.* **31** 184-91

[2] Geubelle P H and Baylor J S 1998 Impact-induced delamination of composites: A 2D simulation *Compos. B* **29** 589-602

[3] Kitamura R and Goda K 2014 Verification of damage model of ceramic matrix composites and analysis of cyclic deformation behaviour *Trans. Jpn. Soc. Mech. Eng.* **80** SMM0080

[4] Kitamura R and Goda K 2014 Application of conditional finite element model to ceramics matrix composites with interfacial debonding *J. Adhesion Soc. Jpn* **50** 428-34

[5] Kitamura R, Yamaguchi T and Goda K 2016 Formulation of off-axial interfacial debonding and sliding problem by constrained conditional finite element method *J. Jpn. Soc. Compos. Mater.* **26** 363-73

[6] Kishi M 1988 *Practice Handbook of Finite Element Method for Structural Analysis* (Tokyo, Japan: Morikita Publishing) pp 214-27

[7] Gao Y C, Mai Y W and Cotterell B 1988 Fracture of fiber-reinforced materials *J. Appl. Math. and Phys.* **39** 550-72

[8] Curtin W A 1991 Theory of mechanical properties of ceramic-matrix composites *J. Am. Ceram. Soc.* **74** 2837-45

[9] Phoenix S L and Raj R 1992 Scalings in fracture probabilities for a brittle matrix fiber composite *Acta Metall* **40** 2813-28

[10] Neumeister J M 1993 A constitutive law for continuous fiber reinforced brittle matrix composites with fiber fragmentation and stress recovery *J. Mech. Phys. Solids.* **41** 1383-404

[11] Goda K and Okabe T 2005 A strength reliability model of fiber-reinforced ceramic matrix composites including individual constituent damages given by markov process *Trans. Jpn. Soc. Mech. Eng.* **71A** 919-27

[12] Chiang Y C 2007 On a matrix cracking model using Coulomb's friction law *Eng. Fract. Mech.* **74** 1602-16

Investigation of gas-dynamic processes in a boundary layer on a basis of molecular dynamics simulation

T A Kuryashova[1,4], V O Podryga[1,2] and S V Polyakov[1,3]

[1]Keldysh Institute of Applied Mathematics of RAS, 4 Miusskaya square, 125047, Moscow, Russia
[2]Moscow Automobile and Road Construction State Technical University, 64 Leningradsky prospect, 125319, Moscow, Russia
[3]National Research Nuclear University MEPhI, 31 Kashirskoe shosse, 115409, Moscow, Russia

E-mail: kudryashova@imamod.ru

Abstract. The work is devoted to numerical simulation of the nonlinear gas dynamic processes in technical systems of micron sizes. This problem is relevant for many applications related to the implementation and the use of nanotechnology in various industries. As an example, a gas flow in micro channel with metal walls is considered. Within this problem, we are interested in a calculation of the boundary layer parameters from first principles. The numerical analyses are carried out according to the Newton's equations of classical dynamics. The model under consideration takes into account the molecular composition of the gas, the atomic structure of the metal surfaces, and heat exchange of the gas with the metal. Computer implementation is focused on using high-performance systems with hybrid architecture. The calculations are performed on the example of a nitrogen flow into a nickel micro channel for several micro channel lengths. The flow velocity varied over a wide range. It is considered in the subsonic, transonic, and supersonic ranges. In numerical experiments, macro parameters of gas flow in the boundary layer are obtained and the corresponding near-wall model is formed. It can be used for the calculation of flows in micro channels using continuum mechanics methods.

1. Introduction

Technical (applied) gas dynamics [1] emerged as an independent science in the 30 s of the last century. It received a powerful theoretical and practical impetus in the 40 s-60 s in connection with the development of aviation and rocket technology, nuclear energetics and other industries. The central place in technical gas dynamics is occupied by boundary layer theory [2]. The development of mathematical modeling methods [3] and the use of high-performance computing enable us to study complex gas-dynamic processes without the application of expensive field experiments. It is become necessary to study gas dynamic processes at the atomic-molecular level due to the massive introduction of nanotechnology in industry. Such detailed study is carried out for cases where the similarity theory and/or the continuity hypothesis are violated. The most significant deviations from theoretical estimates are observed on micro and nano scales, taking into account the specific composition of the gaseous medium and structure of real surfaces, streamlined by supersonic gas flows. Up to a certain point in these cases, methods for calculating free-molecular flows have been successfully applied [4]. However, due to increasing the complexity of technical gas dynamics problems, the methods based on the Boltzmann kinetic equation (see, for example, [5]) and the molecular dynamics methods (see, for example, [6,7]) are increasingly used in specific applications.

In this paper, the molecular dynamics (MD) approach [8-18] is chosen. A number of computational experiments on interaction of the nitrogen stream with the nickel surface is carried out. A preliminary model of the boundary layer is formulated. This model can be used at various levels of detail analysis of the flow in the near-wall layer.

2. Mathematical model and numerical technique

The approach based on models and methods of molecular dynamics [8-18] is used to describe the processes in gas-metal micro systems. It is assumed that there are mainly molecules of the gas mixture (N_{gas} total kinds) in the gas part of the micro system. On solid surfaces and in their vicinity, atoms or metal molecules (N_{met} total kinds) are added to them. They constitute surface material and are potentially able to break away from it. The evolution of the micro particle system under study is described by the Newton's equations [15-18]. The system of equations for moving the particles of l kind has the following form. The particles can be both gas and metal.

$$m_l \frac{d\mathbf{v}_{l,i}}{dt} = \mathbf{F}_{l,i}, \quad \mathbf{v}_{l,i} = \frac{d\mathbf{r}_{l,i}}{dt}, \quad i = 1,...,N_l, \quad l = 1,...,N_{tot} \left(N_{tot} \equiv N_{gas} + N_{met} \right), \tag{1}$$

where i is the particle number, l is the particle kind, N_l is the total number of particles of l kind. The particle of l kind with i number has its own mass m_l , $\mathbf{r}_{l,i} = \left(r_{x,l,i}, r_{y,l,i}, r_{z,l,i} \right)$ is particle displacement vector, $\mathbf{v}_{l,i} = \left(v_{x,l,i}, v_{y,l,i}, v_{z,l,i} \right)$ is velocity vector and $\mathbf{F}_{l,i} = \left(F_{x,l,i}, F_{y,l,i}, F_{z,l,i} \right)$ is total force acting on this particle.

The forces are the sum of two terms. The first term is connected with the interaction of the i -th particle with the surrounding particles. It depends on the potential energy. The second term is responsible for external influence. The potential energy of the system is represented as a sum of partial energies. It is calculated according to the formula of the selected interaction potential.

$$\mathbf{F}_{l,i} = -\frac{\partial U}{\partial \mathbf{r}_{l,i}} + \mathbf{F}_{l,i}^{ext}, \quad U = \sum U_{ll'}, \quad i = 1,...,N_l, \quad l = 1,...,N_{tot}, \tag{2}$$

where $U = U\left(\mathbf{r}_{1,1},...,\mathbf{r}_{1,N_1}; \ ...; \mathbf{r}_{N_{tot},1},...,\mathbf{r}_{N_{tot},N_{N_{tot}}} \right)$ is the total potential energy, $U_{ll'}$ is the potential of interaction of l kind particles with l' kind particles, $\mathbf{F}_{l,i}^{ext}$ is the force of interaction with the external environment.

The choice of interaction potentials is based on a comparison of the mechanical properties of the computer model and the real material. At the same time, it is necessary to consider all variants of gas-gas, metal-metal and gas-metal interactions. Only the pair interactions of particles are taking into account. Each type of interaction is described by its potential. For gas mixtures, the Mie potential in the form of "n-6" [9] is usually used. For the interaction of metal atoms, the EAM [13] potential is used, which takes into account the pair interactions of atoms, as well as the effect of collective electron density in the conduction band. To account gas-metal interactions, the Morse [10] or Lennard-Jones [11] potentials are used with parameters calculated on the Lorentz-Berthelot formulas [8].

The initial conditions at the micro level are determined by the equilibrium or quasi equilibrium thermodynamic state of the particle system at a given temperature, pressure, and average momentum. The boundary conditions at the molecular level depend on the modeled situation. To determine the general properties of a medium, it is enough to consider the selected three-dimensional volume with periodic boundary conditions in all coordinates. When studying micro systems of real geometry, such as a micro channel, one or several directions have a finite size. And the preservation of the shape of an object is achieved by selecting a potential or fixing the system. In this case, we consider two variants: either the mirror boundary conditions can be used as boundary conditions, when the particles interact with their mirror reflection and do not consequently leave the boundary; or the condition is applied of the disappearance of a particle at the boundary and its appearance elsewhere in the micro system (so that the equilibrium of the system is not broken). It can also be defined for an input stream of particles on one side of the selected volume and the free exit of the particles on the other side. Monitoring the temperature of the system and its total momentum is carried out using thermostatic methods [14,15].

3. Numerical technique

The system of equations (1) and (2) is solved using the Velocity Verlet integration [12]. At first time step we calculate start forces:

$$\mathbf{F}_{l,i}^{0} = \mathbf{F}_{l,i}\left(\mathbf{r}_{1,1}^{0},...,\mathbf{r}_{1,N_1}^{0};\ ...;\mathbf{r}_{N_{tot},1}^{0},...,\mathbf{r}_{N_{tot},N_{N_{tot}}}^{0}\right),\quad i=1,...,N_l,\quad l=1,...,N_{tot}; \tag{3}$$

Further, for $n = 0,1,2,...,$ we calculate new positions of particles, new forces and velocities:

$$\mathbf{r}_{l,i}^{n+1} = \mathbf{r}_{l,i}^{n} + \mathbf{v}_{l,i}^{n}\Delta t + \frac{\mathbf{F}_{l,i}^{n}}{m_l}\frac{(\Delta t)^2}{2},\quad i=1,...,N_l,\quad l=1,...,N_{tot}; \tag{4}$$

$$\mathbf{F}_{l,i}^{n+1} = \mathbf{F}_{l,i}\left(\mathbf{r}_{1,1}^{n+1},...,\mathbf{r}_{1,N_1}^{n+1};\ ...;\mathbf{r}_{N_{tot},1}^{n+1},...,\mathbf{r}_{N_{tot},N_{N_{tot}}}^{n+1}\right),\quad i=1,...,N_l,\quad l=1,...,N_{tot}; \tag{5}$$

$$\mathbf{v}_{l,i}^{n+1} = \mathbf{v}_{l,i}^{n} + \frac{\mathbf{F}_{l,i}^{n+1} + \mathbf{F}_{l,i}^{n}}{2m_l}\Delta t,\quad i=1,...,N_l,\quad l=1,...,N_{tot}. \tag{6}$$

Here Δt is the integration step, n is the step number, $\mathbf{F}_{l,i}^{n}$ is the force value at the n step, $\mathbf{F}$ is the procedure for calculating the forces. The thermostating procedure [14,15] is added to formulas (3)-(6) to achieve the desired parameters of gas and metal in the state of thermodynamic equilibrium. The Berendsen thermostat [14] is used to achieve a given temperature by the micro system. The Langevin thermostat [15] is used to achieve the given temperature and momentum.

4. Parallel implementation

A parallel implementation of the MD algorithm assumes the use of a cluster (supercomputer) with central or hybrid architecture. Each node has several multicore central processor units (CPU) and several vector or graphic processors units (VPU or GPU). Parallelization of the algorithm is based on the principles of geometric and functional parallelism. The computational domain is divided into sub-domains of the same power. The power of a domain is measured by the number of elementary boxes where all molecules interact with each other. The domain partitioning is performed within the three-dimensional lattice topology. Each calculated domain is directed to a cluster or supercomputer (SC) node. The distribution of domains on SC nodes is implemented using the MPI library.

There is a certain set of elementary boxes grouped into a three-dimensional sublattice inside the computational domain. This set is distributed among the cores and threads of the computational devices of the SC node. Such decision takes place because of the calculations inside the boxes have a higher intensity than among the boxes. As a result, intermolecular interactions in boxes are realized in parallel mode by using OpenMP technology for CPU or VPU. When using graphic accelerators, a copy of the boxes is created in the memory of the GPU. The necessary data to calculate the interactions of the particles are transferred there. The implementation of such mechanism uses CUDA technology.

The basic computational algorithm is as follows. The 1st stage is reading the source data by an MPI application and initializing the data structures on each computational unit. The 2nd stage is basic calculations in a time cycle based on the formulas (4)-(6). The 3rd stage is performing the resulting

IOP Conf. Series: Materials Science and Engineering **657** (2019) 012026 doi:10.1088/1757-899X/657/1/012026

calculations and finalization.

At the first stage, in particular, the number of particles in the computational domain is determined, generation of their placement, generation of uniform on angles and Maxwellian on modulus distribution of their momentum distribution, and the calculation of starting forces is performed.

The following sequence of actions is implemented within the time cycle.

First, new coordinate values are calculated. Then, they are corrected using periodic boundary conditions. After that, particles are exchanged among specific boxes and that is carried out both within the calculated domains and among the calculated domains using MPI functions.

Further, the forces are calculated, on the basis of which the particle velocities are corrected and all the necessary integral characteristics are computed. When the control time points are reached, the necessary data are stored in files.

The difference of the proposed approach from others is the use of data structures based on elementary boxes. The dimensions of the boxes are related to the cutoff radius of the interaction potentials and the lattice parameters of the metal. Such solution is rather costly in terms of the size of RAM needed, but it greatly saves time of calculations. The processing of a small number of particles in several adjacent boxes, located practically in one or several adjacent pages of RAM, is performed as quickly as possible and is well cached. In fact, we implement a RAM localization mechanism. During intensive calculations, the processor does not need to be switched among far-spaced pages of memory.

Also, in the proposed algorithm, we managed to avoid the expensive procedure for determining the belonging of various particles to specific boxes. The other codes perform this procedure at each time step. Another advantage of using the box structure is the transfer of particles among nodes of MVS that associated with the movement of particles over the computational domain, integrated both with the implementation of the periodic boundary conditions and with the exchange of information among MPI processes adjacent in the cubic lattice. This exchange is achieved by using shadow (virtual) boxes with fictitious particles. The interaction of the main particles and the fictitious particles must be taken into account in the potential energy of the system. A more detailed parallel technology is described in [19].

5. Simulation results

As an example, the flow of nitrogen into a nickel micro channel with a length of 1 to 3 microns and a thickness of about 600 nm is considered. The computational geometry contains two nickel plates (top and bottom) with layers of adsorbed nitrogen adhered to them and a free layer of nitrogen between them (see figure 1). In this case, the real structure of the nickel base of the micro channel, the effect of the initial adsorption of nitrogen molecules on the walls of the micro channel, and the energy discharge of the microchannel walls into the environment are taken into account. The calculations are performed with using the hybrid supercomputer K60 of Keldysh Institute of Applied Mathematics of the Russian Academy of Sciences.

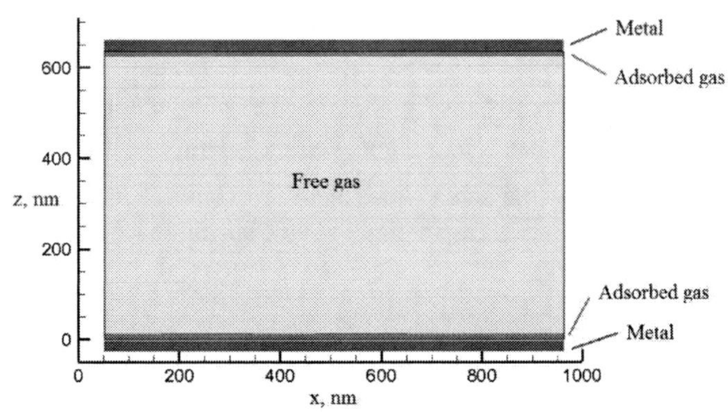

Figure 1. Geometry of simulated micro system.

IOP Conf. Series: Materials Science and Engineering **657** (2019) 012026 doi:10.1088/1757-899X/657/1/012026

The calculations are carried out for several widths of the nitrogen jet and several values of velocity. The velocity is considered to exceed the speed of sound in nitrogen, and for significantly smaller values. To accelerate the calculations on the second direction (*y*), a thin layer and periodic boundary conditions are taken. The general parameters of the gas-metal micro system are proposed in table 1.

Table 1. Physical parameters of the gas-metal micro system.

	Metal plates from Ni	Gas area with N_2
Length (L_x), *nm*	$1017 \cdot n,\ n=1,2,3$	$1017 \cdot n,\ n=1,2,3$
Width (L_y), *nm*	101.7	101.7
Height (L_z), *nm*	8.5	614.5
Temperature, *K*	273.15	273.15 (on start time)
Pressure, *MPa*	-	0.101325 (on start time)
Longitudinal velocity (v_x), *nm/ps*	0	$0 \rightarrow v_{x,\max} = 0.4$ (is varied*)
Transversal velocity (v_y), *nm/ps*	0	0
Vertical velocity (v_z), *nm/ps*	0	0
Particle number, $\times 10^6$	$162.57 \cdot n,\ n=1,2,3$	$42.38 \cdot n,\ n=1,2,3$

* Note that the speed of sound in nitrogen under the normal conditions is 350.6 m/s i.e. 0.3506 nm/ps. Transition from the subsonic range to the supersonic one occurs when a stationary gas is accelerating to a speed of 0.4 nm/ps.

The parameters of the $Ni - Ni$ interaction (it is based on EAM model) were taken from [13]. Our preliminary results show that the equilibrium value of lattice parameter for the normal conditions is $a_{Ni} = 0.35314$ *nm*. The parameters of the $N_2 - N_2$ interaction (it is based on Lennard-Jones "n-m" model [11]) are $n = 11.5$, $m = 6$, $\varepsilon / k_B = 97.78\ K$, $\sigma = 0.3649\ nm$. The parameters of $Ni - N_2$ interaction (it is based on Lennard-Jones "12-6" model [11] with the Lorentz-Berthelot correction [8]) are $\varepsilon / k_B = 343.59\ K$, $\sigma = 0.2795\ nm$. The cutoff radius of the interaction potentials is 0.9 *nm*. Numerical integration by the Verlet scheme are carried out with time step $\Delta t = 0.002\ ps$.

The boundary conditions are periodic along the *x(I)* and *y (II)* coordinates. The first (I) allowed us to escape from the problem of the correct numerical implementation of the incoming gas flow using the molecular dynamics method. The second (II) allowed to reduce the computational capacity of the problem.

Along the z (III) coordinate, the region does not have certain boundary conditions because of external walls of the channel border the vacuum. At the same time, the backs of the metal plates are controlled by a thermostat in order to simulate heat removal to the environment.

Let's consider some calculation results. In the process of evolution, the average (on x and y coordinates) transverse profile of the longitudinal velocity component $\langle v_x \rangle$ gradually turns into a Gaussian profile (see figure 2). At the same time, there are small beats of velocity near the channel walls. They associated with nonlinear gas interaction not so much with the metal, as with the gas adsorbed on its surface.

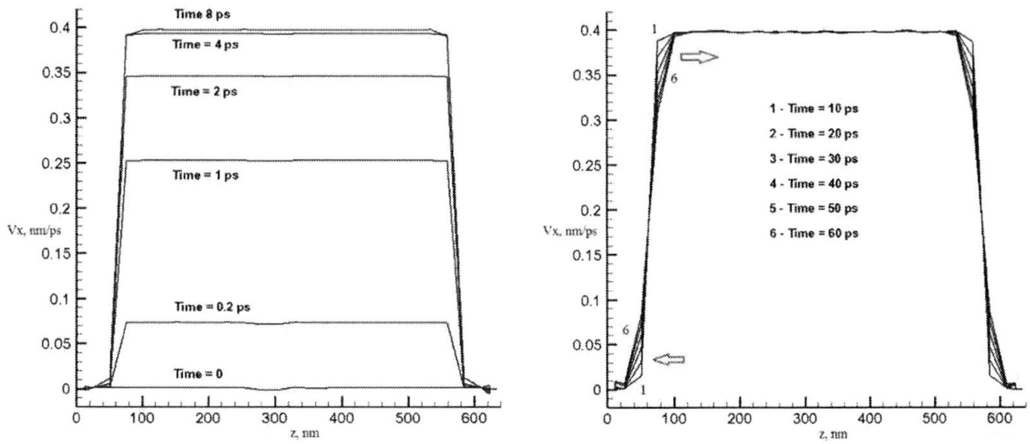

Figure 2. Evolution of the averaged transverse profile of the longitudinal velocity component.

Note that the chaotic behavior of the longitudinal velocity profiles near the walls is associated with two factors. On the one hand, the calculation of gas macro parameters near the wall with a small number of molecules is not very correct. So we suggest adjusting these results below when macro models are using. On the other hand, the presence of a gas layer adsorbed on the metal affects the rest of the gas nonlinear and non-stationary. This is an objective process and it takes into account below when boundary layer model is formulating. The periodic appearance and disappearance of the longitudinal velocity peak near the walls is apparently associated with both of these factors.

Long-term calculations show that the final interaction of the flow with the walls of the micro channel is essentially non-linear and in some cases non-stationary. A particularly complex flow structure is observed at the initial stage of its evolution. The longitudinal velocity profile averaged over the transverse coordinate can have a Gaussian and hyper-Gaussian shape, which changes with time its characteristic features. In some cases, gas beats against the channel walls are observed. With increasing the channel length, the development of a turbulent flow is expected. However, this effect is realized at large times of the order of 20 ns or more (This corresponds to 10 million integration steps).

An analysis of the data obtained suggests that the structure shown in figure 3. It consists of four layers, between which there is an exchange of mass, momentum and energy (specific layer sizes will be obtained as a result of subsequent more accurate calculations). Taking into account this structure, we can say that under the conditions of the adsorption effect, the main gas flow actually interacts not with the metal wall, but with the buffer gas and the gas adsorbed on the surface.

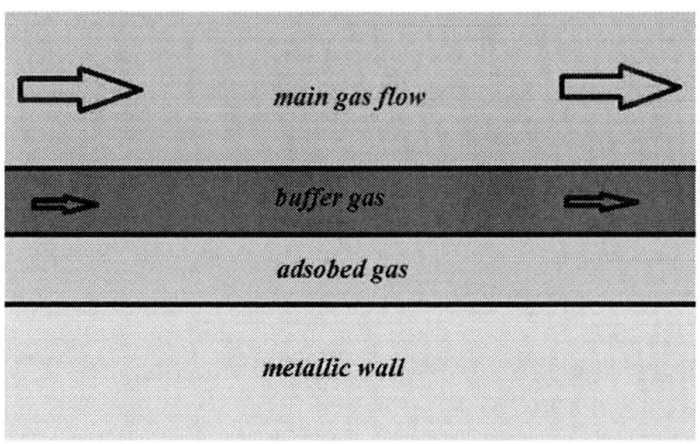

Figure 3. Structure of boundary layers.

Therefore, at the macro level, the following multilayer boundary conditions should be considered:

$$\frac{\partial \rho}{\partial n} = -\eta_1\left(\rho - \rho_{bg}\right), \quad \frac{\partial \rho_{bg}}{\partial n} = -\eta_2\left(\rho_{bg} - \rho_{ag}\right), \quad \frac{\partial \rho_{ag}}{\partial n} = -\eta_3\left(\rho_{ag} - \overset{*}{\rho}_{ag}\right); \tag{7}$$

$$W_n^{(I,-)} = W_{n,bg}^{(I,+)}, \quad W_{n,bg}^{(I,+)} = W_{n,ag}^{(I,-)}, \quad W_{n,ag}^{(I,+)} = W_{n,w}^{(I,-)}; \tag{8}$$

$$W_n^{(E,-)} = W_{n,bg}^{(E,+)}, \quad W_{n,bg}^{(E,+)} = W_{n,ag}^{(E,-)}, \quad W_{n,ag}^{(E,+)} = W_{n,w}^{(E,-)}. \tag{9}$$

Here ρ, ρ_{bg}, ρ_{ag} and $\overset{*}{\rho}_{ag}$ are densities of gas in main flow, buffer layer, adsorbed gas layer and maximal density of gas on metallic surface; η_k are some coefficients; $W_n^{(I,-)}$, $W_{n,bg}^{(I,\pm)}$, $W_{n,ag}^{(I,\pm)}$, $W_{n,w}^{(I,-)}$ are normal components of pulse flows on corresponding boundaries, $W_n^{(E,-)}$, $W_{n,bg}^{(E,\pm)}$, $W_{n,ag}^{(E,\pm)}$, $W_{n,w}^{(E,-)}$ are normal components of energy flows.

In the layer of a buffer gas, it is necessary to solve the equations of multiphase hydrodynamics. In the layer of adsorbed gas, it is necessary to solve the two-dimensional diffusion convection equation. For the metal wall, it is necessary to consider the processes of heat conduction and dissipation of the normal component of the momentum. The formulation of these equations is the subject of a future analysis.

6. Conclusion

The use of the developed numerical technology enable us to calculate, from first principles, the process of nonlinear interaction of a nitrogen flow with the walls of a nickel channel. These calculations allow us to form the macroscopic model of the boundary layer. Within the model, three layers of gas are identified near the channel walls. These layers are clearly indicated when formulating the boundary conditions at the macro level. Firstly, it is the main moving gas layer. Secondly, it is a layer of low-mobility gas molecules adsorbed on the wall. Thirdly, it is a transitional layer in which heat exchange of gas and surface occurs and the flow velocity is lost due to friction against the gas adsorbed on the wall. The selection of specific parameters for this model shows that with certain restrictions on the velocity and density of the gas flow, this model can be used within traditional CFD modeling. However, in the general case, the multi scale modeling method should be used. We discussed it in [20]. This method uses a combination of mesh calculations and molecular dynamics.

Acknowledgments

The work was supported by the Russian Foundation for Basic Research (projects No. 17-01-00973-a, 18-37-20062_mol_a_ved).

References

[1] Abramovich G N 1969 *Applied Gas Dynamics* (Moscow: Nauka)
[2] Schlichting H 1955 *Boundary-Layer Theory* (New York: McGraw-Hill, Inc.)
[3] Samarskii A A and Mikhailov A P 2002 *Principles of Mathematical Modeling. Ideas, Methods, Examples* (London and New York: Taylor and Francis)
[4] Bird G A 1976 *Molecular Gas Dynamics* (Oxford: Clarendon Press)
[5] Kloss Y Y, Ryabchenkov V V, Tcheremissine F G and Shuvalov P V 2011 *J. Math. Models Comput. Simul.* **23** 744-50
[6] Kovalev V L, Sazonova V Yu and Yakunchikov A N 2008 *J. Moscow Univ. Mech. Bull.* **63** 44-6
[7] Norman G E and Stegailov V V 2013 *J. Math. Models Comput. Simul.* **5** 305-33

[8] Berthelot D 1889 *J. Comptes Rendus de L'Academie des Sciences* **126** 1703-6
[9] Von Mie G 1903 *J. Ann. Phys. Leipzig.* **11** 657-97
[10] Morse P M 1929 *J. Phys. Rev.* **34** 57-64
[11] Lennard-Jones J E 1931 *J. Proceedings of the Physical Society* **43** 461-82
[12] Verlet L 1967 *J. Phys. Rev.* **159** 98-103
[13] Daw M S and Baskes M I 1984 *J. Phys. Rev. B.* **29** 6443-53
[14] Berendsen H J C, Postma J P M, Van Gunsteren W F, DiNola A and Haak J R 1984 *J. Chem. Phys.* **81** 3684-90
[15] Heerman D W 1989 *Computer Simulation Methods in Theoretical Physics* (Berlin: Springer-Verlag)
[16] Haile J M 1992 *Molecular Dynamics Simulations. Elementary Methods* (New York: John Wiley & Sons, Inc.)
[17] Frenkel D and Smit B 2002 *Understanding Molecular Simulation. From Algorithm to Applications* (New York: Academic Press)
[18] Rapaport D C 2004 *The Art of Molecular Dynamics Simulation* (Cambridge: Cambridge Univ. Press)
[19] Podryga V O, Polyakov S V and Puzyrkov D V 2015 *J. Vychislitel'nye Metody I Programmirovanie* **16** 123-38
[20] Polyakov S and Podryga V 2018 *Proc. EPJ Web of Conferences* **173** 01010

Forecasting methods in engineering

L F Tratar[1,3] and E Strmčnik[2]

[1]Department of Mathematics, Statistics and Operations Research, Faculty of Economics, University of Ljubljana, Kardeljeva pl 17, 1000 Ljubljana, Slovenia
[2]Laboratory for Fluid Power and Controls LFT, Faculty of Mechanical Engineering University of Ljubljana, Aškerčeva cesta 6, 1000 Ljubljana, Slovenia

[3]liljana.ferbar.tratar@ef.uni-lj.si

Abstract. Forecasting in engineering is one of the most important topics when it comes to optimisation, which is related to energy savings, material savings, increasing efficiency, and appropriate and correct decisions at the level of a company, institution, city, or region. Moreover, forecasting is indirectly related to cost savings and the sustainable development of society and environment. In the energy industry (electricity, natural gas, heat load), there are requirements to balance the supply and demand. Markets are very dynamic, and for this reason forecasting is more challenging. Forecasting errors are usually penalized drastically. However, a well-developed forecasting approach represents a competitive advantage, and so a company may significantly reduce expenditure and increase profits. Many publications on forecasting have been published during recent years. Long-term forecasting methods offer many opportunities for strategic planning and optimal scheduling, whereas a short-term forecasting approach would help attain optimal daily operations and the maximum utilisation of the company's resources. Although different forecasting techniques can be used, the major conclusions are that exponential smoothing methods are the simplest and the least expensive. Distinguished by their simplicity, their forecasts are comparable to forecasts of more complex statistical time series models. In this paper, the forecasting performance of Additive, Multiplicative, and Extended Holt-Winters methods were analysed. We also analyse whether the data format influences the choice of the forecasting method: is the most accurate method for monthly data also the best method for quarterly data?

1. Introduction

Forecasting in engineering is very important topic when it comes to efficient processes, cost reduction, efficient logistic, reduction of energy consumption, sustainable development [1] and key drivers of the transition toward a low carbon society [2]. A main purpose of forecasting is a balancing of the supply and demand. Forecasting is widely used in different branches in the energy industry, for example, there are different methods for prediction of natural gas consumption [3], electricity energy consumption [4,5,6], electric load [7], heat load [8,9,10], cooling [11] and building energy consumption [12,13].

Although there are various forecasting methods, we decided to analyse Additive, Multiplicative and Extended Holt-Winters approach [14,15,16] with and without additional optimisation of initial values. The aim of this article was the identification of the best forecasting methods for monthly and quarterly long- and short-term forecasting. In other words, we focused on the best forecasting methods for monthly forecasting and then explored if the same forecasting methods were also the best choice for quarterly data. A real seasonal time series from the M4-Competition were analysed. The forecasting accuracy was determined taking into account Theil's U-statistics.

IOP Conf. Series: Materials Science and Engineering **657** (2019) 012027 doi:10.1088/1757-899X/657/1/012027

The paper is organised as follows. In Section 2 the forecasting methods are presented, and in Section 3 the data and methodology are described. The results are presented in Section 5 and research findings in Section 6.

2. Forecasting methods

The Holt-Winters method of exponential smoothing involves trend and seasonality and is based on three smoothing equations: an equation for level, for trend, and for seasonality. The decision as to which method to use depends on time series characteristics: the additive method is used when the seasonal component is constant, and the multiplicative method is used when the size of the seasonal component is proportional to the trend level [17].

2.1. Holt-winters' additive procedure (AHW)

The basic equations for the AHW method are:

$$L_t = \alpha(Y_t - S_{t-s}) + (1-\alpha)(L_{t-1} + b_{t-1}) \tag{1}$$

$$b_t = \beta(L_t - L_{t-1}) + (1-\beta)b_{t-1} \tag{2}$$

$$S_t = \gamma(Y_t - L_t) + (1-\gamma)S_{t-s} \tag{3}$$

$$F_{t+m} = L_t + b_t m + S_{t-s+m} \tag{4}$$

where L_t – the estimation of variable in time *t*, Y_t – the observed value, b_t – the trend estimation of time series in time *t*, S_t – the estimation of seasonality in time *t*, F_{t+m} – the forecast in time *t* for *m* period ahead, α, β, γ – smoothing parameters in the interval [0, 1], *m* – the number of forecasted periods, and *s* – the duration of seasonality (for example, number of months or quarters in a year).

For initialisation of the additive method initial values of variable L_t, trend estimation b_t and seasonality estimation S_t are needed. To determine initial estimates we need at least one whole data season (that is, *s* data). Initialisation of variable L_s is calculated with the formula:

$$L_s = \frac{1}{s}(Y_1 + Y_2 + \cdots + Y_s) \tag{5}$$

For trend initialisation, it is more suitable if we use two whole seasons (that is, 2*s* data):

$$b_s = \frac{1}{s}\left(\frac{Y_{s+1} - Y_1}{s} + \frac{Y_{s+2} - Y_2}{s} + \cdots + \frac{Y_{s+s} - Y_s}{s}\right) \tag{6}$$

Seasonal indices are calculated as differences between the observed value and variable estimation L_s:

$$S_1 = Y_1 - L_s,\ S_2 = Y_2 - L_s, \ldots,\ S_s = Y_s - L_s \tag{7}$$

The method is proved to be (regarding costs and calculation itself) comparable with more complex methods (for example Box-Jenkins); in some cases the results gained with the Holt-Winters were even better than more complex methods [17].

2.2. Extended holt-winters' procedure (EHW)

The EHW method differs from AHW only in the equation for level (1); all other equations remain the same as with the AHW (2–7). The equation for level now contains an additional smoothing parameter δ:

$$L_t = \alpha Y_t - \delta S_{t-s} + (1-\alpha)(L_{t-1} + b_{t-1}) \tag{8}$$

This method allows us to smooth the seasonal factors more or less than the AHW method, depending on the value of the parameter δ [17].

2.3. *Holt-winters' multiplicative procedure (MHW)*

The basic equations for the MHW method are as follows:

$$L_t = \alpha(Y_t/S_{t-s}) + (1-\alpha)(L_{t-1} + b_{t-1}) \quad (9)$$

$$b_t = \beta(L_t - L_{t-1}) + (1-\beta)b_{t-1} \quad (10)$$

$$S_t = \gamma(Y_t/L_t) + (1-\gamma)S_{t-s} \quad (11)$$

$$F_{t+m} = (L_t + b_t m) \cdot S_{t-s+m} \quad (12)$$

The second of these equations (10) is identical to the second equation (2) of AHW. The only differences in the other equations are that the seasonal components are now in the form of products and ratios instead of being added and subtracted.

3. Data and methodology

We used a real seasonal time series from the M4-Competition. The M-Competitions are empirical studies that compare the performance of a large number of major time series methods using recognised experts who provide forecasts for their method of expertise [18].

In our study, we analysed five monthly series that refer to industry: M26726, M26727, M26730, M26734, and M26735. For simplicity, we used notation 26 m, 27 m, 30 m, 34 m, and 35 m, respectively (see figure 1). Whereas data for these time series are given for different time periods, we selected consecutive data for a period of 138 months (11 years and 6 months) for unification and easier mutual comparison. The obtained data of each time series were split into initialisation (the first two years, from period 1 to 24), fitting (period from 25 to 120), and testing set (period from 121 to 138). A ratio between initialisation, fitting, and testing set was chosen regarding previous papers [19,20] and experiences. Later on, accumulated quarterly data for each monthly series was prepared and split into initialisation (from period 1 to 8), fitting (period from 9 to 40), and testing set (period from 41 to 46). We used the notation 26q, 27q, 30q, 34q, and 35q, respectively (figure 2). The fitting set was used for method learning. With the testing set we checked a time series learning ability.

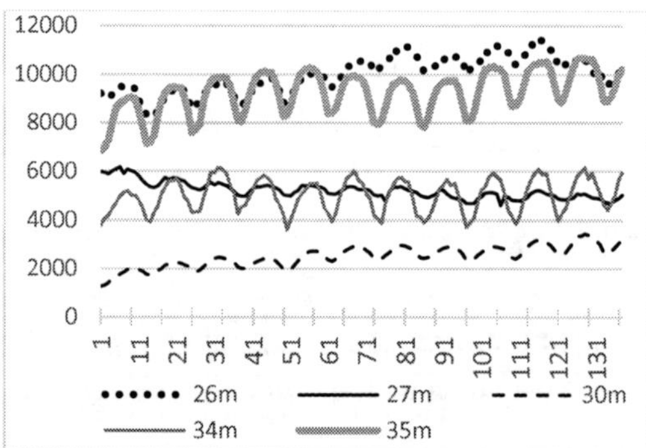

Figure 1. Plots of monthly time series.

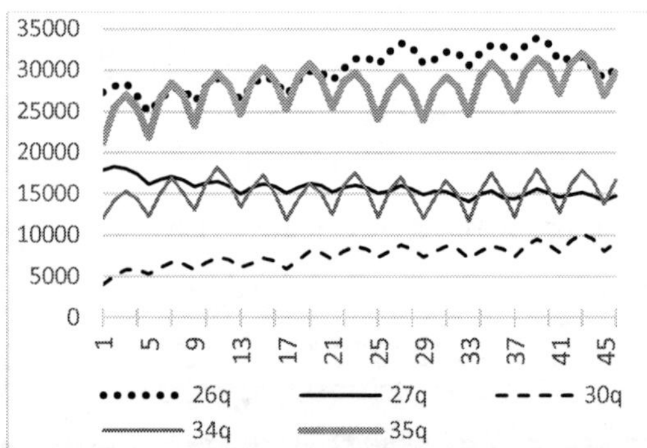

Figure 2. Plots of quarterly time series.

We calculated the forecasting values for the testing subset and then compare these values to independent-real data. In case of the short-term forecasting approach, we calculated the forecasting value F_{t+m} for one month (m=1) ahead (monthly short-term forecasting) and one quarter (m=1) ahead (quarterly short-term forecasting), where t represents an arbitrary time point from the testing set. For the long-term forecasting approach, we calculated the forecasting value F_{t+m} for 18 months (m=18) ahead (monthly long-term forecasting) and six (m=6) quarters ahead (quarterly long-term forecasting), where t represents last month/quarter in the fitting set.

For the evaluation of the forecasting methods we applied three forecasting accuracy measures, Mean

Squared Error (MSE), Mean Absolute Percentage Error (MAPE), and Theil's U-statistics:

$$\text{MSE} = \frac{1}{N}\sum_{t=1}^{N}(Y_t - F_t)^2, \quad t = 1,2,3,\dots N$$

$$\text{MAPE} = \frac{1}{N}\sum_{t=1}^{N}\left|100\frac{Y_t-F_t}{Y_t}\right| \;[\%], \quad t = 1,2,3,\dots N$$

$$U = \left(\sum_{t=1}^{N-1}\left(\frac{F_{t+1}-Y_{t+1}}{Y_t}\right)^2 \Big/ \sum_{t=1}^{N-1}\left(\frac{Y_{t+1}-Y_t}{Y_t}\right)^2\right)^{1/2}$$

where Y_t represents the actual value, F_t the forecasted value, and N the number of samples. As MSE penalizes errors proportional to their squares, minimising MSE leads to smoothing parameters that produce fewer large errors at the expense of tolerating several small errors. Of course, the lower values of MSE, MAPE, and Theil's U-statistics represent a better forecasting performance.

The analysed exponential smoothing methods deal with smoothing parameters that are determined according to the past data. The smoothing parameters were determined by minimising the MSE and MAPE. The starting values of the smoothing parameters were set to 0.5. The minimising problem was solved by using Solver (Microsoft Office Excel 2010). The additional optimisation (notation *method*-init) of the initial values (for level, trend, and seasonality) has been performed.

4. Results

4.1. MSE, MAPE, and Theil's U-statistics

Table 1 shows MSE, MAPE, and Theil's U-statistics results of short-term forecasting for fitting and testing sets for monthly and quarterly data of time series M26726 obtained with six different methods. As expected, MSE values for quarterly data are much higher than for monthly data (since monthly data are smaller than the corresponding accumulated quarterly data), but Theil's U-statistics for quarterly data are lower than for monthly data, which means that forecasts of quarterly data are more accurate. It is further noted that the size of the U-MSE and U-MAPEs, regardless of the fitting or test set, are about the same (see figures 3 and 4). We can see that on average U-MSE values are slightly lower than the U-MAPE value, especially for the test set, which means that the forecasts obtained with minimising MSE are slightly more accurate than those obtained by minimising MAPE. Because we came to similar conclusions in the long-term approach (all results can be obtained upon request), in the continuation of the article, we will present in detail only the results of U-statistics obtained with MSE minimisation.

Table 1. Results for short-term forecasting of time series M26726.

	MSE (fit)	U-MSE (fit)	MSE (test)	U-MSE (test)	MAPE (fit)	U-MAPE (fit)	MAPE (test)	U-MAPE (test)
S_26m-AHW	24,292.67	0.7228	33,332.78	0.7055	1.22	0.7563	1.38	0.6643
S_26m-AHW-init	9,077.67	0.4429	17,062.15	0.5073	0.83	0.5069	1.12	0.5675
S_26m-MHW	25,808.78	0.7415	35,973.53	0.7329	1.22	0.7698	1.46	0.6893
S_26m-MHW-init	9,337.49	0.4515	15,873.97	0.4908	0.74	0.4614	0.97	0.5289
S_26m-EHW	18,459.70	0.6371	25,618.28	0.6191	1.01	0.6225	1.42	0.6545
S_26m-EHW-init	8,690.37	0.4329	15,984.94	0.4922	0.71	0.4507	0.92	0.5289
S_26q-AHW	207,531.70	0.3571	730,794.23	0.6584	0.99	0.3415	2.82	0.6995
S_26q-AHW-init	113,607.43	0.2773	288,901.03	0.4221	0.84	0.2687	2.13	0.5767
S_26q-MHW	221,263.96	0.3676	750,706.93	0.6676	1.08	0.3591	2.74	0.7115
S_26q-MHW-init	149,639.57	0.3201	308,717.39	0.4361	0.90	0.3156	2.31	0.6765
S_26q-EHW	186,568.20	0.3402	673,765.81	0.6287	0.90	0.3242	2.81	0.7062
S_26q-EHW-init	94,678.46	0.2467	237,144.36	0.3801	0.77	0.2543	1.11	0.3957

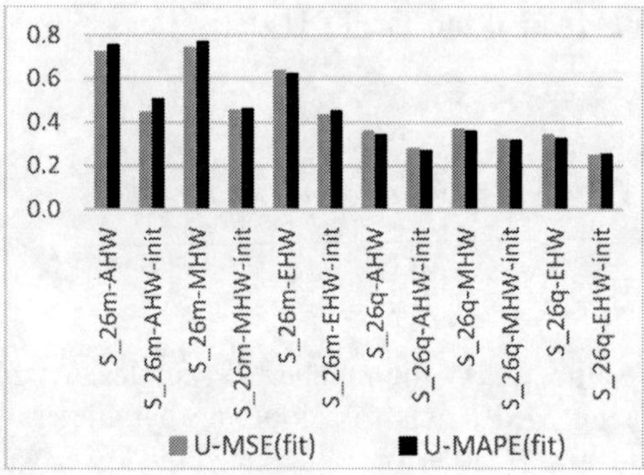

Figure 3. Short-term forecasting, Theil's U-statistics measure accuracy, fitting data.

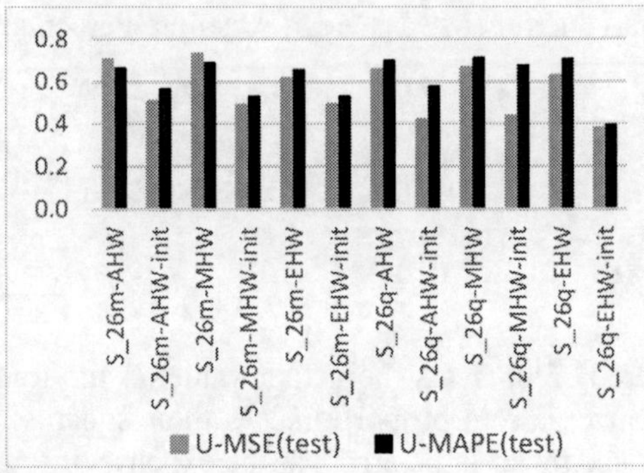

Figure 4. Short-term forecasting, Theil's U-statistics measure accuracy, testing data.

4.2. Forecasts for M26726

The best-obtained fitting result of the long- and short-term approaches was reached by using the EHW-init method (figure 5), minimum fitting Theil's U-statistics value for long-term was 0.23, and for short-term approach was 0.247. If we only consider monthly data, the minimum Theil's U-statistics was 0.421, obtained by the AHW-init method and long-term approach. For the testing set (figure 6), the minimum Theil's U-statistics was 0.38, obtained by EHW-init method for quarterly data and the short-term approach. As the U-statistics for monthly and quarterly data, regardless long or short-term approach, were the smallest for the EHW-init method, this method was selected as the best method.

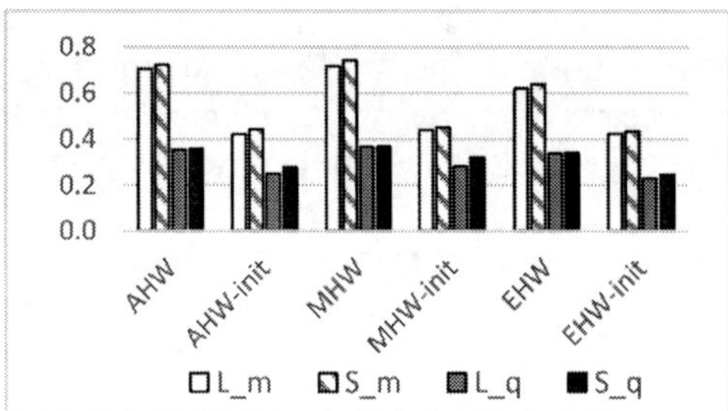

Figure 5. Theil's U-statistics measure accuracy, fitting data, series M26726.

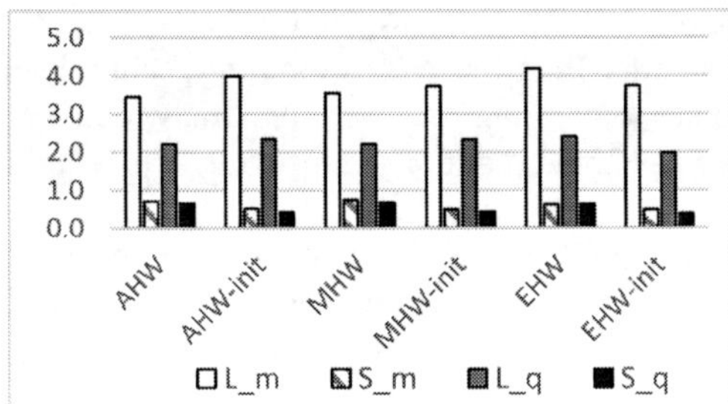

Figure 6. Theil's U-statistics measure accuracy, testing data, series M26726.

4.3. Forecasts for M26727

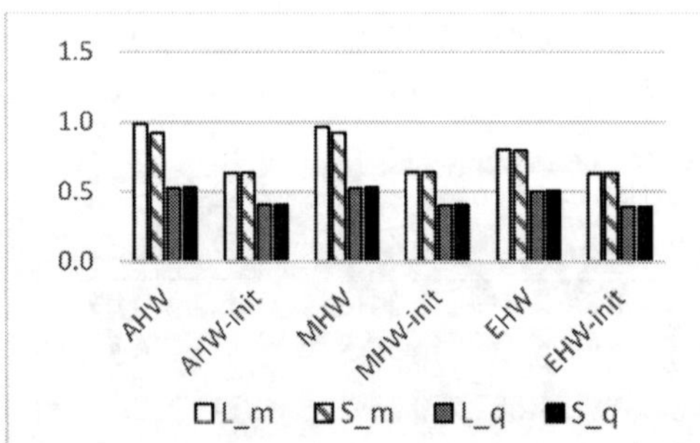

Figure 7. Theil's U-statistics measure accuracy, fitting data, series M26727.

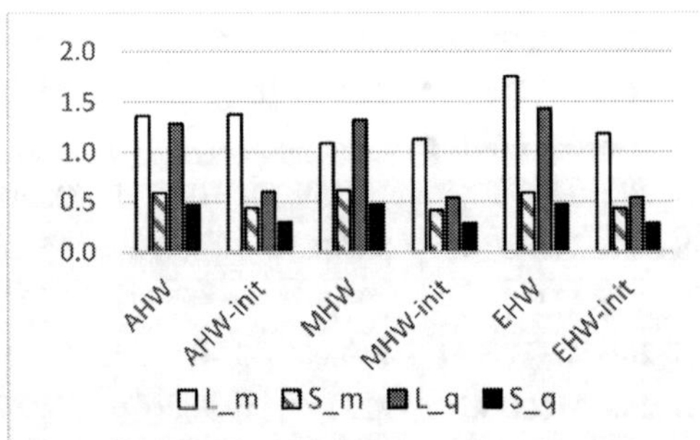

Figure 8. Theil's U-statistics measure accuracy, testing data, series M26727.

According to long and short-term approaches, the EHW-init method was the best fitting method for monthly and quarterly data (see figure 7). The same conclusions could be made for the testing set (figure 8) if the long and short-term approaches are compared, but we noticed that the best method for monthly data is MHW-init.

4.4. Forecasts for M26730

From figure 9 we observed that, for all six methods, obtained values of Theil's U-statistics were almost equal for the long and short-term approaches, so the forecasts on the fitting set are equally precise regardless of whether we use a long or short-term approach. In this case, the EHW-init was chosen as the best method. For the testing set (figure 10), the best method for the long-term approach was MHW, and for the short-term approach MHW-init was best.

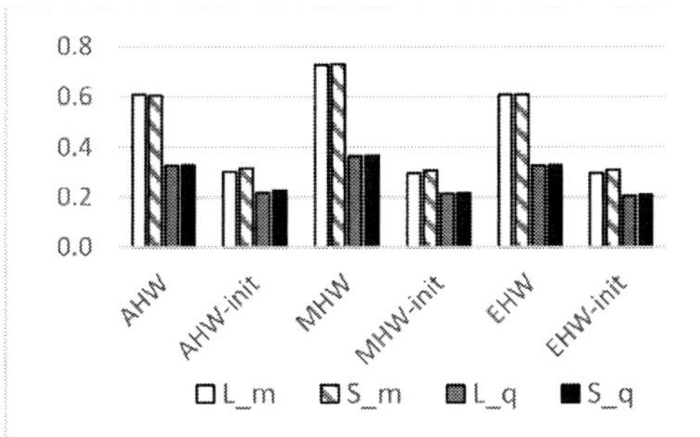

Figure 9. Theil's U-statistics measure accuracy, fitting data, series M26730.

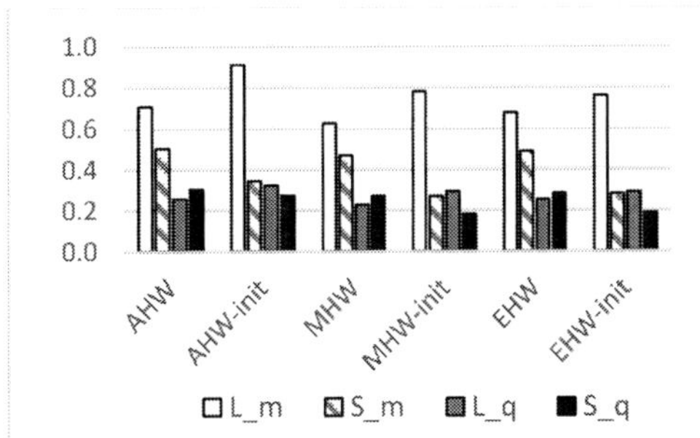

Figure 10. Theil's U-statistics measure accuracy, testing data, series M26730.

4.5. Forecasts for M26734

Similar to the previous time series, the EHW-init method was the best method for the fitting set, regardless of the approach (figure 11). An interesting development for the testing set (figure 12) was that the *init* methods (with the exception of the EHW method) improved forecasts for the short-term approach, and worsened forecasts for the long-term approach. The best method on the test set for the long-term approach was AHW (Theil's U-statistics was 0.43), and for the short-term approach EHW-init (Theil's U-statistics was 0.123).

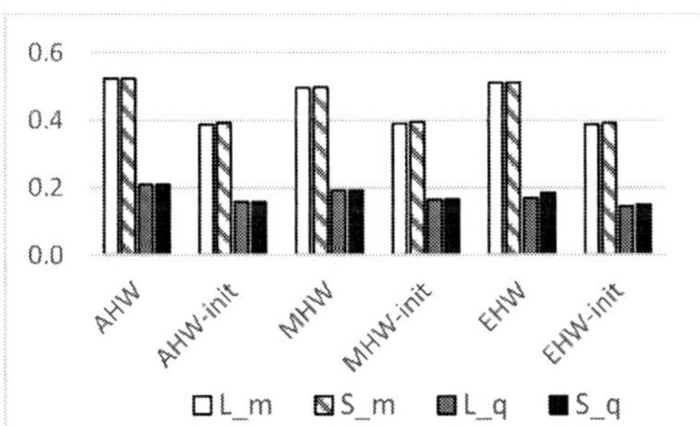

Figure 11. Theil's U-statistics measure accuracy, fitting data, series M26734.

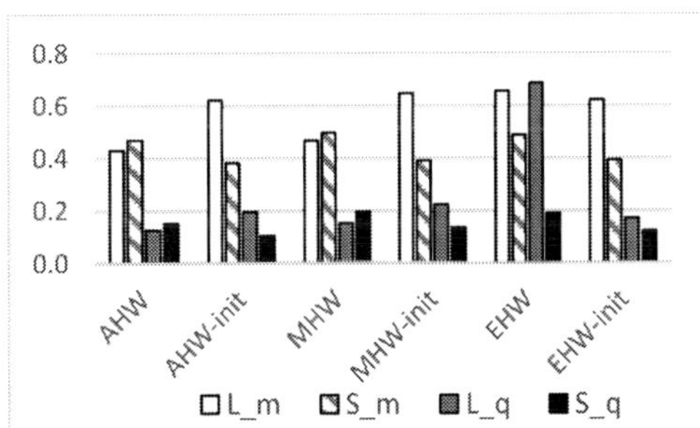

Figure 12. Theil's U-statistics measure accuracy, testing data, series M26734.

4.6. Forecasts for M26735

The best fitting result for the long- and short-term approaches was again attained by the EHW-init method (figure 13). If the results for the long- and short-term approaches for testing set (figure 14) were compared, the best method for the long-term was MHW, and for the short-term approach was EHW. For monthly data, the minimal Theil's U-statistics were reached by AHW-init, and for quarterly data by the EHW method.

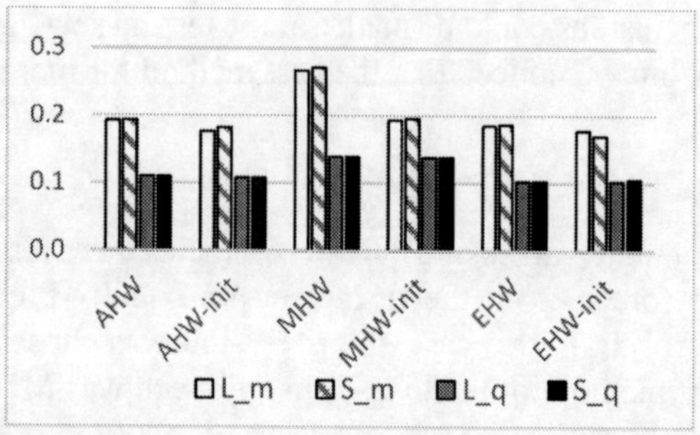

Figure 13. Theil's U-statistics measure accuracy, fitting data, series M26735.

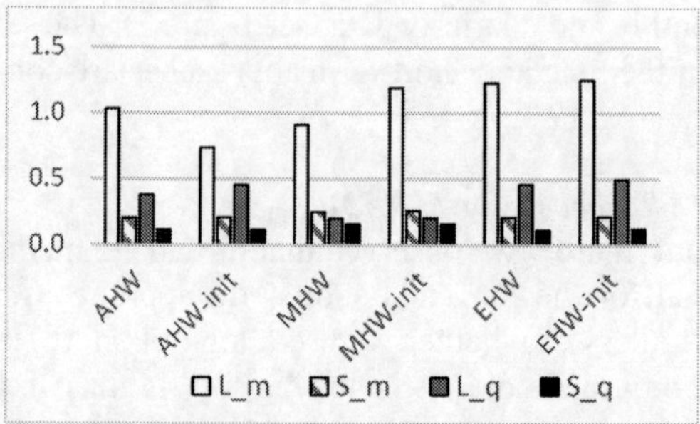

Figure 14. Theil's U-statistics measure accuracy, testing data, series M26735.

5. Research findings

In this section, the testing results of monthly and quarterly, long- and short-term forecasting, obtained by described forecasting methods, are presented.

The best forecasting methods of monthly and quarterly data, which were identified as the most appropriate methods for the industrial forecasting implementation, are collected in table 2. Considering that Theil's U-statistics was selected as the main performance measure, the AHW, MHW, MHW-init, and EHW-init method were recognised as the best method in general (all these methods appeared 4-times as the best), whereas the EHW method was not very applicable for our application. Furthermore, MHW was the best method for long-term, and MHW-init for short-term monthly and quarterly forecasting. AHW and MHW-init were very appropriate for monthly, but EHW-init was the most effective for quarterly long- and short-term forecasting. Moreover, when we focused on a very specific combination regarding the accumulated data (monthly, quarterly) and forecasting approach (long-, short-term), the best forecasting method for every combination was identified. AHW and MHW were very suitable for monthly-long-term forecasting, MHW-init for the monthly-short-term combination, MHW and EHW-init for quarterly-long-term forecasting, and EHW-init for quarterly-short-term forecasting. Theil's U-statistics of quarterly data was very often lower than Theil's U-statistics for a monthly series. A very similar finding may be observed for the short-term forecasting approach, which was much more accurate than a long-term approach.

Table 2. Identification of the best forecasting methods.

	M26726		M26727		M26730		M26734		M26735	
	L	S	L	S	L	S	L	S	L	S
M	AHW	MHW-init	MHW	MHW-init	MHW	MHW-init	AHW	AHW-init	AHW-init	AHW
Q	EHW-init	EHW-init	EHW-init	EHW-init	MHW	MHW-init	AHW	AHW-init	MHW	EHW

6. Conclusions

This study deals with different forecasting methods, taking into account monthly and quarterly data as well as long- and short-term forecasting approaches. The key results can be summarised as follows:

- MHW was recognised as the best method for long-term forecasting,
- MHW-init was the best method for short-term forecasting,
- AHW and MHW-init were the best methods for monthly data,
- EHW-init was the most suitable forecasting method for quarterly data,
- Theil's U-statistics of quarterly data was lower than Theil's U-statistics of a monthly series,
- Short-term forecasting was much more accurate than long-term forecasting.

The result shows that the data format influences the choice of forecasting method. More specifically, if we want to predict quarterly data, we have to use other methods than in cases where we analyse monthly data. As was already emphasized, EHW-init was the most accurate forecasting method for quarterly data. Although the forecasting is much more accurate in the case of quarterly data in comparison to monthly data, different forecasting (daily, weekly, monthly, etc.) are very important for short- and long-term planning, which lead to more efficient operations and better business output. However, long-term forecasting is very important for strategic decision, i.e. production, warehousing, transportation, and logistics in general. In contrast, short-term forecasting help us to manage rapid changes, for instance in consumption choice.

References

[1] Zhao H X and Magoulès F 2012 A review on the prediction of building energy consumption *Renew Sust. Energ. Rev.* **16** 3586-92
[2] European Commission Climate action, roadmap for moving to a low-carbon economy in 2050 available from http://ec.europa.eu/ clima/ policies/roadmap/ (accessed on 2015-02-2)
[3] Soldo B 2012 Forecasting natural gas consumption *Appl. Energ.* **92** 26-37
[4] Tso G K and Yau K K 2007 Predicting electricity energy consumption: A comparison of regression analysis decision tree and neural networks *Energy* **32** 1761-8
[5] Adom P K and Bekoe W 2012 Conditional dynamic forecast of electrical energy consumption requirements in Ghana by 2020: A comparison of ARDL and PAM *Energy* **44** 367-80
[6] Oğcu G, Demirel O F and Zaim S 2012 Forecasting electricity consumption with neural networks and support vector regression *Procedia Soc. Behav. Sci.* **58** 1576-85
[7] Hong T and Fan S 2016 Probabilistic electric load forecasting: A tutorial review *Int. J. Forecast* **32** 914-38
[8] Tratar L F and Strmčnik E 2016 The comparison of holt–winters method and multiple regression method: A case study *Energy* **109** 266-76
[9] Al-Shammari E T, Keivani A, Shamshirband S, Mostafaeipour A, Yee L, Petković D and Ch S 2016 Prediction of heat load in district heating systems by support vector machine with firefly searching algorithm *Energy* **95** 266-73
[10] Potočnik P, Strmčnik E and Govekar E 2015 Linear and neural network-based models for short-term heat load forecasting *Stroj. vestn-J. Mech. E.* **61** 543-50
[11] Vaghefi A, Jafari M A, Bisse E, Lu Y and Brouwer J 2014 Modeling and forecasting of cooling and electricity load demand *Appl. Energ.* **136** 186-96
[12] Zhao H X and Magoulès F 2012 A review on the prediction of building energy consumption *Renew Sust. Energ. Rev.* **16** 3586-92
[13] Kalogirou S A and Bojic M 2000 Artificial neural networks for the prediction of the energy consumption of a passive solar building *Energy* **25** 479-91
[14] Holt C C 2004 Forecasting seasonals and trends by exponentially weighted moving averages *Int. J. Forecast* **20** 5-10
[15] Winters P R 1960 Forecasting sales by exponentially weighted moving averages *Manage Sci* **6** 324-42
[16] Gardner Jr E S 1985 Exponential smoothing: The state of the art *Int. J. Forecast* **4** 1-28
[17] Tratar L F, Mojškerc B and Toman A 2016 Demand forecasting with four-parameter exponential smoothing *Int. J. Prod. Econ.* **181** 162-73
[18] Makridakis S and Hibon M 2000 The M3-competition: Results, conclusions and implications *Int. J. Forecast* **16** 451-76
[19] Tratar L F 2014 Improved holt-winters method: A case of overnight stays of tourists in the Republic of Slovenia *EBR. J.* **16** 5-17
[20] Tratar L F 2015 Forecasting method for noisy demand *Int. J. Prod. Econ.* **161** 64-73

IOP Conf. Series: Materials Science and Engineering **657** (2019) 012028 doi:10.1088/1757-899X/657/1/012028

Analysis of boot rail collision resistance in rocket sled test system

J Xiao[1,2], WW Zhang[1], Q Xue[1], W B Gao[1] and L R Zhang[1]

[1]Huayin Ordence Test Center, Huayin,714200, China

E-mail: 50310644@qq.com

Abstract. The shoe-rail resistance to impact of rocket sled was analyzed. Shoe-rail friction resistance was backward calculated with speed, acceleration, weight and gap of shoe-rail. The influencing factor was gain by the continuous contact force model. The relation of shoe-rail resistance to impact with speed, weight and gap of shoe-rail was gained by means of data fitting. Based on the results, the collision resistance coefficient of a new type of sled is calculated. According to the resistance coefficient, the speed of sled is predicted. The predicted results are consistent with the dynamic experimental results.

1. Introduction

During the rocket sled test, the rocket engine was used as the power to simulate the full ballistic process such as launch and flight of the test item. The ground simulation of flight state of weapons and equipment is realized.

When using rocket sled dynamic test, the parameters such as the full ballistic velocity, acceleration, the quality of the sled vehicle, and the engine thrust obtained by external measurement, the equivalent drag coefficient of the sled vehicle calculated by reverse calculation is greater than the equivalent air drag coefficient of the sled vehicle design. So it predict that in addition to air resistance and friction, there is another resistance in the sled test. The sled is in contact with the track through the sled boots when it is running on the track. Due to the uneven track, the sled boots generate tangential friction when they slide at high speed on the track, which is the main reason why the actual drag coefficient of the sled is greater than the design air resistance coefficient.

The relationship between the boot rail collision resistance and the running speed, quality and boot rail gap of the sled is fitted through the dynamic test data, and a new type of sled's collision resistance is calculated by using the boot rail collision resistance calculation model. Using the calculated collision resistance, the speed of the sled is predicted, and the speed of the sled is consistent with the dynamic test speed of the sled.

2. Boot rail collision resistance

There is a gap between the sliding boots and the track (as shown in figure 1). Due to the uneven track and the influence of the roughness of the track surface, the sliding boots continue to collide with the track when the sled is moving at high speed on the track. The normal contact force is generated, and the collision force is converted into tangent friction resistance in the direction of the sled.

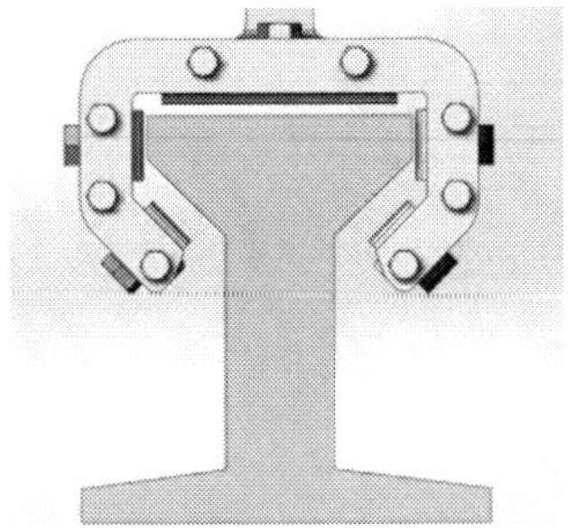

Figure 1. Boot track clearance.

2.1. Foot rail contact force model

Based on Hertz theory[1], Lankarani and Nikravesh proposed a continuous contact model[2,3].The contact force in the direction of the boot rail is expressed as:

$$\begin{cases} F_N = K\delta^m \left[1 + \dfrac{3(1-e^2)}{4} \cdot \dfrac{\dot{\delta}}{\dot{\delta}^{(-)}}\right] \\ K = \dfrac{4}{3\pi(\sigma_i + \sigma_j)} \left(\dfrac{R_i R_j}{R_i - R_j}\right)^{0.5} \end{cases} \tag{1}$$

In above formula, F_N is boot rail normal contact force. K is contact stiffness coefficient. m is power index, which is determined by the characteristics of the contact material. δ is the value of contact deformation. $\dot{\delta}$ is the value of deformation speed after the boot rail contact. $\dot{\delta}^{(-)}$ is boot track collision impact speed. σ_k is contact object yield strength. $\sigma_k = (1-v_k^2)/\pi E_k$. v_k is Poisson's ratio and E_k is Young's modulus corresponding to the material in contact with the object. R_k is the contact radius of the collision object. e is the recovery coefficient after the boot rail collision. The value of pure elastic contact is 1, and the value of pure plastic contact is 0.

The energy dissipation caused by plastic deformation during contact with boot Rails is not considered. Therefore, the boot rail contact force is simplified to:

$$F_N = K\delta^m \tag{2}$$

S P·Timoshenko's theoretical calculation formula for the comprehensive deformation of the two spheres[4]:

$$\delta = (5V^2 m / 4K)^{2/5} \tag{3}$$

In the formula, V is collision point collision speed. m is the comprehensive mass of the two cylinders corresponding to the radius of curvature of the collision point.

The contact stiffness K of the collision point reflects the shape characteristics of the collision surface of the collision object. Different surface shapes correspond to different collision stiffness, and the derivation process of this formula has no limit on the surface shape of the object. Therefore, this formula can be further applied to collision problems of different collision surfaces.

The vibration value caused by the boot track gap is A, and the wavelength caused by the track is λ. When the sled vehicle moves at speed u on the track, the orbital excitation is assumed to be harmonic[5,6]:

$$z(t)=\frac{1}{2}A[1-\cos(\frac{2\pi ut}{\lambda})] \qquad 0\le t\le n\lambda/u \tag{4}$$

In the formula, A is the maximum gap between boot rails. In the formula, λ is orbital wavelength, and u is sled speed.

When the boot rails collide, the vibration speed V of the sled is:

$$V=\dot{z}(t)=\frac{\pi Au}{\lambda}\cdot\sin(\frac{2\pi ut}{\lambda}) \qquad 0\le t\le n\lambda/u \tag{5}$$

From equations (2)-(5), it can be seen that the boot rail contact force F_N is a function related to the quality, speed, and boot rail clearance of the sled:

$$F_N=K'u^{a_1}m^{a_2}A^{a_3} \tag{6}$$

In the formula, K' is the influence coefficient, which is determined by the type, stiffness, strength, contact shape and size of boot rail. a_1,a_2,a_3 are constants which are due to sled running speed, quality and boot rail gap.

2.2. Boot rail collision resistance model

Boot rail shear friction model is[7]:

$$F_t=-u(V_t)\cdot F_n\cdot V_t/|V_t| \tag{7}$$

In the formula, F_t is boot rail shear friction. $u(V_t)$ is boot rail sliding friction coefficient, which is no longer a constant, but a function of the tangent sliding speed.

In the 1980s, a famous expert in tribology from Soviet Union, studied a large number of experiments to study different materials at medium and low speeds (speeds of 0. 004-25m/s, pressure 0. 0008-0.17MPa). Friction coefficient was change under conditions. When sliding at medium and low speeds, friction is mainly caused by local adhesion and shear in the contact area. Friction resistance is manifested as surface heat, but this heating effect will not have a great impact on the overall friction mechanism[8,9]. According to the elastic-viscous contact theory and test data of the friction surface, the relationship between speed and friction coefficient is proposed:

$$u(V_t)=(a+b|V_t|)\cdot e^{-c|V_t|}+d \tag{8}$$

In the formula: a,b,c,d are coefficients, which are determined by the sliding material and positive pressure.

The variation of friction coefficient between metal friction materials under high speed and high pressure with sliding speed is studied, and the high speed and high pressure of friction coefficient still satisfy the variation rule of formula(8). The friction coefficient decreases with the increase of sliding speed under high contact pressure, and the higher the pressure, the smaller the friction coefficient. The friction coefficient changes tend to flatten at high speed. Therefore, it can be considered that the sliding friction coefficient is constant when the boot rail slides at high speed[10,11]. From this, it can be seen that the boot rail collision resistance mainly depends on the boot rail contact force. The influencing factors are the speed, quality, and boot rail gap of the sled. The key of boot rail collision resistance analysis is to determine the index of a_1,a_2,a_3.

3. Data fitting and verification of results

3.1. Data fitting

The purpose of the data fitting is to determine the relationship between boot rail collision resistance and quality, running speed and boot rail clearance of the sled by using the existing boot rail collision resistance[11].

The sled is mainly subjected to rocket thrust T and drag R (including air resistance R_a and boot rail collision resistance F_t), sleigh motion equation expressed as:

$$T - ma = R_a + F_t \tag{9}$$

For the existing test data fitting, it was found tha (as shown in figure2):

$$T - ma = nu^2 \tag{10}$$

In the formula, n is a fixed constant, the value of n is only related to the type of sled. As the same type of sled n is basically the same.

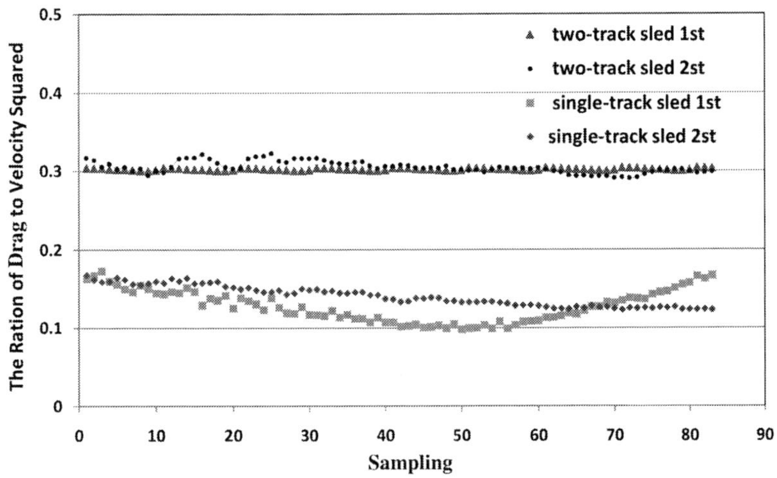

Figure 2. Relation between resistance and velocity square.

In formula (10), $Ra = C_D A\rho u^2/2 = n_1 u^2, n_1$ is the equivalent aerodynamic drag coefficient of the sleigh car. n_1 is only related to the type of sled. As the same type of sled, n_1 is basically the same, as shown in figure 2.

So, from formulas (9) and (10) express of F_t was gained:

$$F_t = (n - n_1)u^2 = n_2 u^2$$

The value of n_2 is only related to the type of sled, and the same type of sled, n_2 is basically the same.

From $\boldsymbol{F_t = [u(V_t)K'm^{a_2}A^{a_3}]u^{a_1} = n_2 u^2}$, $\boldsymbol{a_1 = 2}$ is gained. As the same type of sled, boot rail collision resistance is proportional to the square of speed.

When the sled is running, the gap between the boot and the track is the same. At this time, A^{a_3} is a constant. Express of F_t is:

$$\boldsymbol{F_t = [u(V_t)K'u^2 A^{a_3}]m^{a_2} = n_3 m^{a_2}} \tag{11}$$

In the formula, the gap between the sliding boot and the track is the same, and $\boldsymbol{n_3}$ is a constant.

For the first, second, and third types of sled vehicles, the data fitting of the two-track sled vehicles was found. The Values of $a_2 = 0-10$ are introduced into the above formula to fit the data. It is researched that when the value of a_2 is 0.6 , the value of n_3 was basically the same. The value of the two-track sled vehicle was 0.003, and the value of the single-track sled vehicle was 0.0015. See figure 3 for details.

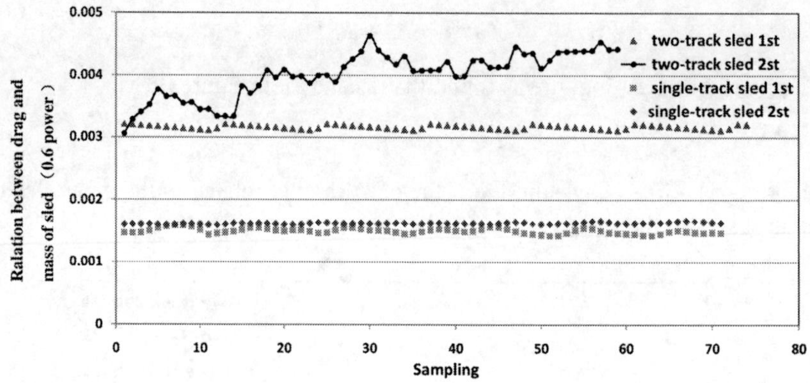

Figure 3. Relation between the collision resistance of boot rails and the power of quality of sled.

The upper and lower clearance of the two-track sled and the track is 3.2mm, and the left and right clearance is 1.6 mm. However, the upper and lower clearance of the skis and rails is 1.3mm and the left and right clearance is 1.2 mm. According to the collision resistance relationship between boots and rails at different gaps, it can be seen that:

$$\frac{3.2^{a_3} + 1.6^{a_3}}{1.3^{a_3} + 1.2^{a_3}} = \frac{0.003}{0.0015} \tag{12}$$

From this fitting: $a_3 = 1.2$

From this, the relationship between boot rail collision resistance and the quality, running speed, and boot track gap of the sled is:

$$F_t = u(V_t)K'u^{a_1}m^{a_2}A^{a_3} = k''u^2m^{0.6}A^{1.2} \tag{13}$$

k'' is a constant.

According to equation (11) and fitting results, the calculation expression of the boot rail resistance of the twin-track sled is:

$$F_t = 1.26\times10^{-3}\times u^2\times m^{0.6}\times A^{1.2}$$

3.2. Validation of the results

According to the relationship of the collision resistance coefficient of the boot rail, the collision resistance coefficient of the boot rail of the sled was calculated as 0.42. According to the engine thrust, the quality of the sled, the air resistance coefficient, and the boot rail collision resistance coefficient, the maximum speed of the sled is predicted to be 300m/s. The predicted results are consistent with the actual dynamic test results(see figure4), and the correctness of the calculation method is verified.

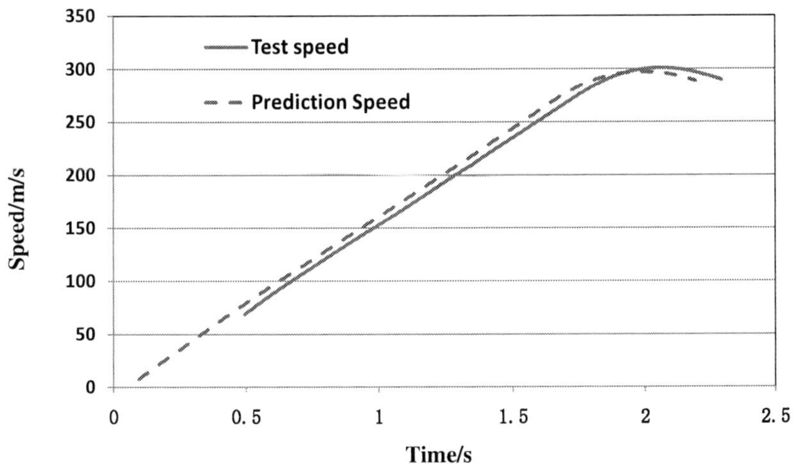

Figure 4. Comparison of forecast speed and dynamic test speed curve of new sled.

4. Conclusion

Due to the objective track roughness and uneven smoothness, the actual boot rail collision resistance generated by the sled is much higher than the static friction force. Through data fitting, the relationship between boot rail collision resistance and the quality, speed and boot rail clearance of skis is obtained. The boot rail collision resistance is proportional to the 0.6 power of the sled mass, the 1.2 power of the boot rail gap, and the square of the speed of the sled. The relationship between the mass, speed and clearance of sled is based on the existing data fitting. Due to the small amount of data, the accuracy of the fitting numerical value needs to be further improved. But this fitting method is correct. The calculation method of boot rail collision resistance was verified by dynamic test, which laid the foundation for the design and ballistic calculation of the subsequent sled.

References

[1] Yu W Y, Ji L H, Yan S Z *et al* 2001 A survey of dynamics model of pairs with clearances in mechanism *Mechanical Science and Technology* **20** 665-70

[2] Su Y W, Zhu A B, Chen W and Xie Y B 2008 Effect of friction and contact in clearance joint on dynamics of multibody system *Lubrication Engineering* **33** 16-9

[3] Ji Z P 2013 Dynamic research on vibro-impact system with clearance (Lanzhou: Lanzhou Jiaotong University)

[4] Timoshenko S·P 2004 *Theory of Elasticity (Photocopy edition)* (Beijing: Tsinghua University Press)

[5] Zhang L Q, Deng Z C, Shi J and Guo F M 2012 Impact of regularity of high-speed slide rail on sled-track dynamic response *Journal of Beijing University of Technology* **38** 544-7

[6] Thompson D 2013 *Railway Noise and Vibration-Mechanism, Model, and Control Methods* (Beijing: Science Press)

[7] Xiao Q, Zhou X J, Wang C G and Che Y X 2012 The finite element analysis of wheel/rail rolling contact with friction coefficient related with sliding speed *Lubrication Engineering* **37** 1-5

[8] Zhang T, Chen X Y, Gu J M and Wang Z L 2018 Influences of preload on the friction and wear properties of high-speed instrument angular contact ball bearings *Chinese Journal of Acronautics* **31** 597-607

[9] Qin L, Wen S H, Zhang X L, Yang B, Chen Y H and Lan G S 2017 A statistical model of static friction coefficient taking into account of adhesion between joint surfaces *Journal of Taiyuan University of Science and Technology* **38** 111-4

[10] Xiao J, Zhang W W, Xue Q, Gao W B and Zhang L R 2018 Modal analysis for single track sled

2018 International Conference on Physics, Mathematics, Statistics Modelling and Simulation (Wu Zhen, China) pp 21-3

[11] Xiao J, Xue Q, Zhang W W, Zhang L R and Zhou X B 2018 Design of single layer vibration dmping system for rocket sled *2018 International Conference on Physics, Mathematics, Statistics Modelling and Simulation* (Wu Zhen, China) pp 16-8

IOP Conf. Series: Materials Science and Engineering **657** (2019) 012029 doi:10.1088/1757-899X/657/1/012029

Derivation of drag calculation model of rocket sled water brake

J Xiao[1,2], WW Zhang[1], Q Xue[1], W B Gao[1] and L R Zhang[1]

[1]Huayin Ordence Test Center, Huayin,714200, China

E-mail: 50310644@qq.com

Abstract. The calculation model of water resistance is derived by using fluid resistance theory. Using the water resistance calculation model, the braking distance calculation model is derived, and the braking distance of the two two-track sled vehicles is estimated. The pre-estimated water resistance, speed, and braking distance of the sled vehicle are in accordance with the actual external measurement results. The correctness of the water resistance calculation model is verified.

1. Introduction

Double-track sled has a wide range of application in scientific research and test. It is a kind of recoverable rocket sled test mode[1,2]. At present, the most widely used is water brake recovery. In the study of hydraulic brake resistance at home and abroad[3], the resistance caused by the change of water momentum is generally considered, while the impact resistance caused by the impact of hydraulic brake on water surface at different altitudes is neglected, resulting in a large deviation between the calculated results and the actual results. At present, there is a lack of accurate and simple drag calculation model of rocket sled water brake.

The drag calculation model of rocket sled water brake directly influences the calculation results of water brake resistance which directly affects the safety and success of test. Based on the fluid resistance theory[4,5], a drag calculation model of rocket sled water brake is derived. Using the water resistance calculation model, the braking distance calculation equation is derived, and the braking distance of two two-track sled vehicles is estimated. The pre-estimated water resistance, speed, and sled braking distance are consistent with the actual external measurement results. The correctness of the drag calculation model of rocket sled water brake is examined.

2. Drag calculation model

When the water brake (as shown in figure1) enters the flume, the water resistance mainly depends on the components immersed in the water. So water brake head-on collision resistance and differential pressure resistance of water brake are the major parts of the water brake resistance.

The vertical section of the immersed fluid of the water brake is perpendicular to the direction of the movement of the water brake. The contribution of the immersed fluid of the water brake to the resistance of the water brake is head-on collision resistance and differential pressure resistance by using the method of hydrodynamic resistance calculation.

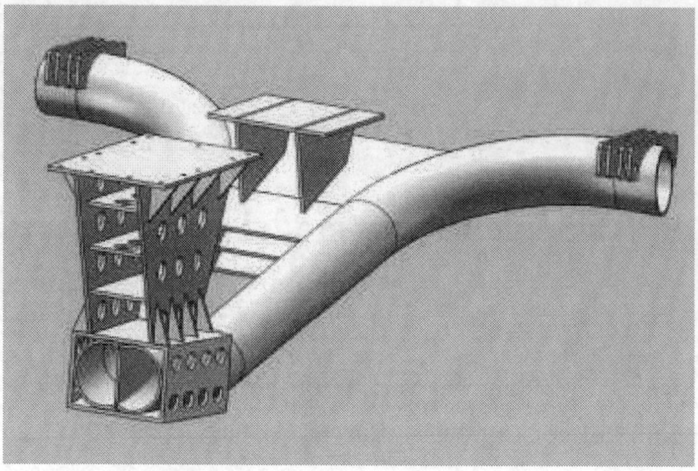

Figure 1. Structure of water brake.

2.1. Water brake head-on collision resistance

The water brake is regarded as arigid body and the deformation of it under water pressure is not considered[6].The water brake is not moving, and the water flows from the opposite direction to the water brake section at the current rate. The movement of the water brake is positive[7].The water flows to the water brake section. The mass of water flowing to the water brake section in a small amount of time:

$$dm = \rho_{water} s_{water} v dt \tag{1}$$

In the formula: dm is the mass of water flowing into the water brake surface during the dt time. It is actually the quality of water encountered by the brake section during the dt time. After the water flow reaches the brake section, the speed is reduced from $-v$ to 0. From the momentum theorem, the water momentum flowing to the brake section decreases in a small amount of time to:

$$dp = dm(-v - 0) = -vdm \tag{2}$$

Surface collision resistance applied by water to the water brake section F_R :

$$F_R = dp / dt = -vdm / dt = -\rho_{water} s_{water} v^2 dt / dt = -\rho_{water} s_{water} v^2 \tag{3}$$

In the type, s is the area where the water brake is immersed in water.

2.2. Differential pressure resistance of water brake

The water brake moves in the water and is also subjected to F_p which is the differential pressure resistance along the water in the direction of movement[8,9]. In front of the water brake section, because the water is stationary, the pressure of the water (p_1) is static pressure. Behind the water brake section, because the water brake moves forward, the water behind the water brake section flows with the water brake, and the flow rate is equal to the speed(v) of the water brake[10]. Therefore, the pressure behind the water brake is the flow pressure (p_2):

$$p_2 = p_1 - \rho_{water} v^2 / 2 \tag{4}$$

So the pressure difference resistance on the front and rear surfaces of the water brake:

$$F_p = p_1 s_{water} - p_2 s_{water} = -\rho_{water} s_{water} v^2 / 2 \tag{5}$$

From this we can see that the water resistance generated by the body of the water brake after entering the water:

$$f_{physical\ water\ resistance} = F_R + F_p = -\rho_{water} s_{water} v^2 - \rho_{water} s_{water} v^2/2 = -3\rho_{water} s_{water} v^2/2 \tag{6}$$

2.3. Hydro momentum resistance

The principle of water brake device is to change the direction of incoming flow speed after the water flow into the water brake device rising to a certain height [10,11]. The water momentum is converted to the water brake momentum, and the water brake resistance is generated through the momentum exchange, thus the sled deceleration braking is realized. When there is a hydrodynamic energy exchange in water brakes, water resistance also includes hydrodynamic energy exchange resistance:

$$f_{Water\ kinetic\ energy\ resistance} = (1 - R_V \cos\theta) A_w \rho_W v^2 \tag{7}$$

In this expression, ρ_W is the density of water (kg/m^3); R_V is the ratio of outlet velocity to inlet velocity for water brakes; A_w is the cross area (m^2) of flow inlet; θ is the angle of entry and exit.

3. Brake distance calculation model

Let the mass of the sled be M, and the initial speed of the sled is v. Water Brake immersion in the tank cross-sectional area is s. When the sled enters the water, the thrust of the rocket engine is zero. The running resistance of sled is air resistance ($C_D S \rho_{air} v^2 / 2$), friction (uMg) and hydraulic braking resistance ($3\rho_{water} s_{water} v^2/2$). In a small period of time, sledding can be regarded as uniformly accelerated motion. Because the change of speed is very small, the water brake resistance related to speed can be regarded as the velocity correlation of the last moment. According to the motion equation of the sled, the sled enters the water at a small time (t_1), then the speed is v_1:

$$v_1 = v - (3\rho_{water} s_{water} v^2/2 + C_D S \rho_{air} v^2 / 2 + uMg) \times (t_1 - 0) / M \tag{8}$$

In the form, u is friction coefficient between slipper and track. g is gravity acceleration. C_D is air resistance coefficient of sled.

When the sled speed is changing from v to v_1, and the sled moves at a uniform acceleration[11,12]. So the braking distance is:

$$l_1 = (v_1 + v) \times (t_1 - 0) / 2 \tag{9}$$

As the sled enters water, the speed of sled is v_1. After a small time, speed is v_2, then:

$$v_2 = v_1 - (-3\rho_{water} s_{water} v_1^2/2 + C_D S \rho_{air} v_1^2 / 2 + uMg) \times (t_2 - t_1) / M \tag{10}$$

The speed of the sled is from $v_1 \rightarrow v_2$, braking distance is:

$$l_2 = (v_1 + v_2) \times (t_2 - t_1) / 2 \tag{11}$$

After many tiny moments, at t_n, the speed of the sled is $v_{n-1} \rightarrow v_n = 0$, and the braking distance is l_n:

$$l_n = (v_{n-1} + v_n) \times (t_n - t_{n-1}) / 2 \tag{12}$$

So when the sled brake ends, the braking distance is l:

IOP Conf. Series: Materials Science and Engineering **657** (2019) 012029 doi:10.1088/1757-899X/657/1/012029

$$l = l_1 + l_2 + \ldots + l_n \tag{13}$$

Through the calculation model of braking distance, we can see that if the calculation model of water resistance is inaccurate, it will be difficult to calculate the braking distance accurately. So we can judge whether the calculation of water resistance is accurate by verifying the accuracy of brake distance calculation.

4. Model applications

The test water resistance and calculation water resistance which is obtained by using the derived water resistance calculation model are opposite as shown as figure 2. The test water resistance is basically in agreement with water resistance by calculation model for two dynamic tests. The maximum error of drag is less than 10%.

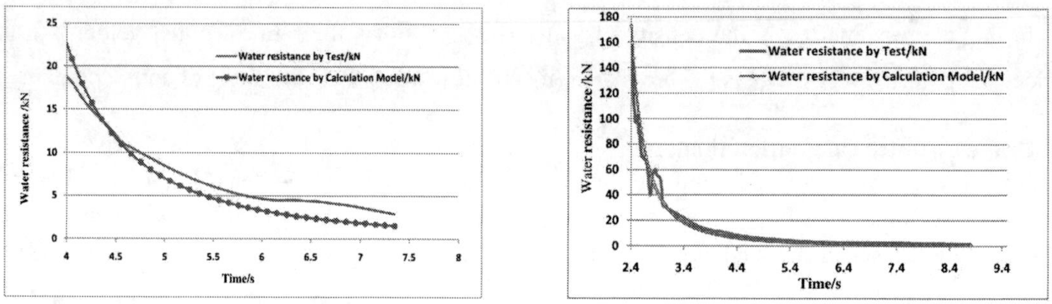

(a) (b)

Figure 2. Comparison of water resistance. (a) Water resistance for the sled for test 1st and (b) Water resistance for the sled for test 2nd.

During the two double-track sled vehicle tests, the speed after sled entering water by test is basically in agreement with speed by calculation model for two dynamic tests (figure 3). The whole speed error of two tests is less than 20m/s.

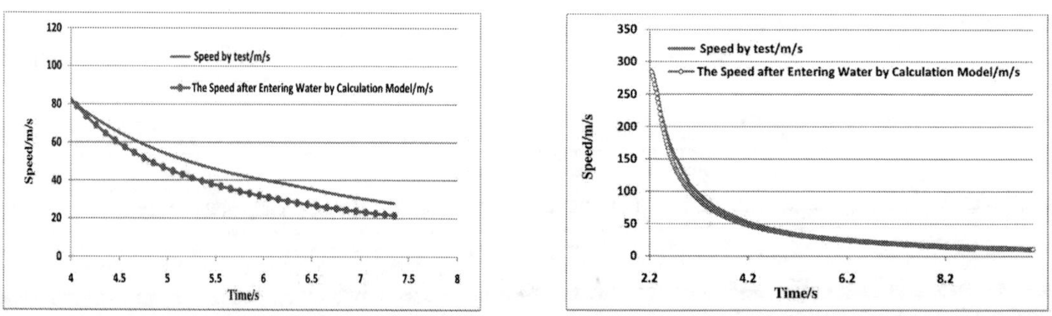

(a) (b)

Figure3. Comparison of speed after entering water. (a) Speed for the sled for test 1st and (b) Speed for test 2nd.

During the two double-track sled vehicle tests, the braking distance by test is basically in agreement with speed by calculation model for two dynamic tests (figure 4). The error of braking distance for two tests is less than 30m.

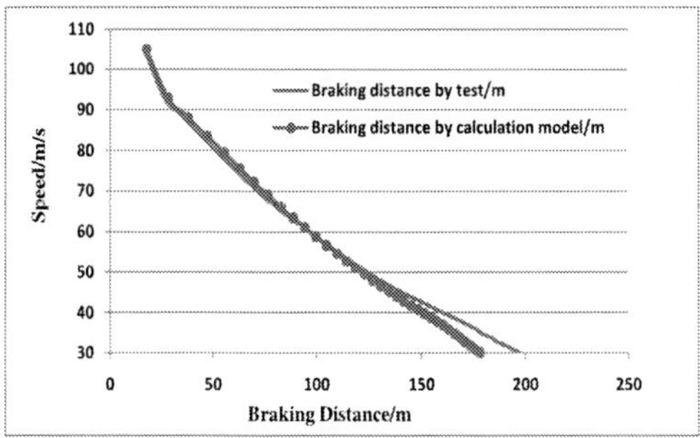

(a)

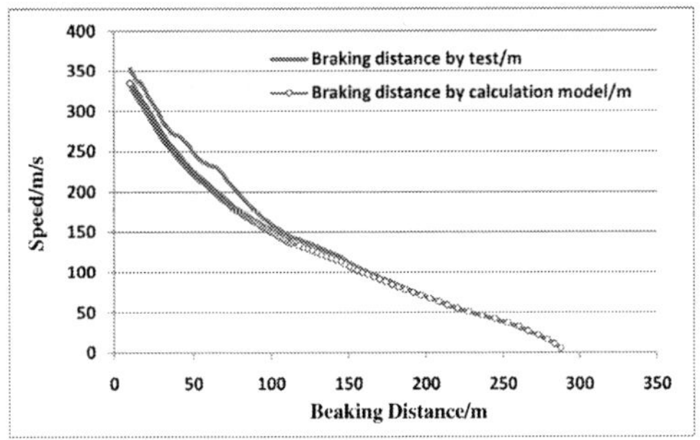

(b)

Figure4. Comparison of speed after entering water. (a) Braking distance for test 1st and (b) Braking distance for test 2nd.

It can be seen from the figure that when the two-track sled vehicle is tested two times. The water resistance which based on the derived water resistance calculation formula was basically consistent with the water resistance obtained by testing. The pre-estimated speed change of sled after entering the water is basically consistent with the actual external measurement speed. For test 1st, it is estimated that the braking distance of the sled is 175m and the actual braking distance is198m. For test 2nd, it is estimated that the braking distance of the sled is 280m, and the actual braking distance is 285m. The braking distance is not much different. The correctness of drag calculation model is verified.

5. Conclusion

Based on the theory of fluid resistance, drag calculation model of rocket sled water brake is derived, and the calculation equation of braking distance is derived. The water resistance, speed and braking distance of the sled were calculated in accordance with the results of external measurement. The specific conclusions are as follows:

- Using the fluid resistance theory, the calculation method of water resistance is derived accurately and simply.
- The calculation equation of braking distance derived by water resistance calculation method is accurate and reliable.

References

[1] Yu Y Y 2008 Design and analysis of high-speed water-brake device in doub1e-track rocket sled test (Nanjing, China: University of Aeronautics and Astronautics) pp 12-4

[2] Liu J, Yin Z P, Ding G Y and Liu J P 2015 Resistance forecast of water bags in braking trays of a rocket sled on the process movement *Adv. Aeronaut. Sci. Eng.* **6** 312-7

[3] Wang J, Zhang X G and Kong W H 2013 Safety analysis on rocket sled water brake high-speed water-entry impact *J. Nanjing Univ. Sci. Technol.* **37** 902-5

[4] Zhao Q B 2009 Numerical analysis about high-speed entry Water of rocket sled water brake machine (Nanjing, China: Nanjing University of Science and Technology) pp 36-7

[5] Xiao J, Li X Y, Zhang L R, Li X P and Zhang W W 2019 Brake force calculation for water-brake device in high speed of doub1etrack rocket sled test *IOP Conf. Ser.-Mat. Sci.* **490** 8-11

[6] Mao Z and Liu G R 2017 A Lagrangian gradient smoothing method for solid-flow problems using simplicial mesh *Int. J. Numer. Method Eng.* **113** 858-90

[7] Wang J, Zhao Q B, Tao G and Wu J J 2010 Numerical simulation on rocket sled water-brake high-speed water-entry impact *Explosion and Shock Waves* **30** 628-31

[8] Zheng H W 2013 *Movement of Objects under the Action of Ideal Fluid* (Beijing: Guangming Daily Press)

[9] Xiao J, Zhang W W, Xue Q, Gao W B and Zhang L R 2018 Modal analysis for single track sled 2018 International Conference on Physics, Mathematics, Statistics Modelling and Simulation

IOP Conf. Series: Materials Science and Engineering **657** (2019) 012029 doi:10.1088/1757-899X/657/1/012029

(Wu Zhen, China) pp 21-3
[10] Xiao J, Xue Q, Zhang W W, Zhang L R and Zhou X B 2018 Design of single layer vibration dmping system for rocket sled *2018 International Conference on Physics, Mathematics, Statistics Modelling and Simulation* (Wu Zhen, China) pp 16-8
[11] Xiao J, Zhang W W, Xue Q, Zhang L R and Zhou X B 2019 Reliability analysis of engine shell strength under rocket sledge vibration environment *IOP Conf. Ser.-Mat. Sci.* **1176** 304-8
[12] Xiao J, Zhang W W, Xue Q, Gao W B and Deng B 2019 Analysis of rocket sled vibration signal transmission based on zigbee application *J. Phys. Conf. Ser.* **1176** 28-32

IOP Conf. Series: Materials Science and Engineering **657** (2019) 012030 doi:10.1088/1757-899X/657/1/012030

Numerical study on flow around vane in a channel using immersed boundary method

J F Huang[1,3], S Yang[1], J Yao[2] and L X Zhang[2]

[1]College of Hydraulic Engineering, Yunnan Agricultural University, Kunming, Yunnan 650201, China
[2]Department of Engineering Mechanics, Kunming University of Science and Technology, Kunming, Yunnan 650051, China

E-mail: hjf30@126.com

Abstract. The immersed boundary method is applied to simulate the flow around a guide vane of hydro-turbine. The complex dynamic boundary effect caused by the moving guide vanes is modelled as the forcing term in the Navier-Stokes equation. Numerical simulation of three-dimensional transient turbulence is conducted to flow around one moving guide vane of a benchmark hydro-turbine unit in a channel. The dynamical Smagorinsky-Lilly model of large eddy simulation is also used. The evolution process of the dynamic vortex structure is obtained by the interaction between the guide vane and the flow around vane to reveal the physical mechanism of wake oscillation flow around vane. This method serves as a reference for the fine simulation and study of the flow around the multi-stage cascade flow in hydro-turbine. The research on the transient effects of the coupling upstream pipeline system and hydro-turbine has theoretical guiding significance.

1. Introduction

The fluid-structure interaction problem in hydro-turbine has been under the attention of engineering and academia in recent 20 years. With the characteristics of simple and fast power regulation, the hydro-turbine has a very important role in the regulation of the power system of the hydropower station. The guide vane of hydro-turbine is controlled by the wicket gates to regulate the flow rate of hydro-turbine, so as to change the power output of hydro-turbine. In the transition process of the hydropower station, the regulating movement of the moving guide vane of hydro- turbine will cause the pressure oscillation on the upstream pressure piping system, and the transient characteristics of the downstream hydro-turbine will be the main reason for the vibration and even stop of the machine. The regulation motion of the moving guide vane of hydro-turbine is a complex transient dynamic boundary fluid dynamics problem. Its accurate simulation is restricted by the computational fluid-structure coupling dynamic method and the computer hardware condition.

Influenced by free flow, the guide vane trailing edge will be dragged out of the wake when rotating the guide vane. The wake is merged into intense spiral Karman vortex, whose azimuth is changed instantaneously. The wake further changed the guide vane hydrodynamic and vibration response, thus produced important influence on performance and vibration amplitude of hydro-turbine. Therefore, the development of appropriate and accurate calculation method about the wake and its distribution is the foundation of the fluid-structure coupling dynamics. It is an important and difficult part of the hydro-turbine technology field. Numerical simulation of flow around airfoil with moving boundary is one of the hot issues in recent years, because of its wide range of flow background and complex flow characteristics. The different numerical and experimental methods were used to study the flow characteristics of various airfoils by many scholars [1-3]. This paper attempts to develop a way to consider the interaction between the fluid and the guide vanes, and the interaction between the stay vanes and the guide vanes. However, because the fluid motion in the rotating machinery is quite complex, the simplified moving guide vane is classified as a guide vane in a channel. Firstly the flow around a guide vane in a channel is simulated, and then the dynamic flow around the double linear cascade in the channel is simulated. These works are the basis for the simulation of the whole hydro-turbine double cascade in the future.

The traditional grid generation technology can be divided into structured grid and unstructured grid. The characteristics of structured grid is directly in the near the object surface to generate orthogonal the body fitted grid (Body-fitted grid), then on the grid nodes solving control equations through coordinate transformation. The structure grid has the advantages of standard structured grid, and easy programming, which provides high precision simulation in solving simple regions. The drawback is that it is difficult for complex geometry to generate the body fitted grid. With a different structured grid, unstructured grid node with the adjacent grid has the same unit, so the finite control nodes of the body (finite control volume) can be used in arbitrary shape. Unstructured grid can deal with the complex geometry boundary problems. Due to the poor nodes correlation, the computation is complex, and its efficiency is low. In fact, only the 2D unstructured grid generation technology is relatively mature, but the 3D unstructured grid still faces many problems.

Regarding the moving boundary problems, the use of adaptive technology in structured and unstructured grid is a way to solve this kind of problem, but the grid needs to constantly generate new grid and refine the grid programming complexity, and thus the workload is very large and the loss of accuracy too [4]. Because each time-step needs recalculating by grid generation especially in the fluid-structure interaction problem, and the variables of the time layer is calculated by the interpolation of the flow field value at last time step. The force of the fluid and the solid is inconsistent with the real time layer, which results in a deviation from the actual physical. In recent years, a series of fixed Descartes Cartesian grid for solving complex boundary technology obtained the full attention and rapid development, especially in the so-called "immersed boundary method (IBM)". The immersed boundary method first was proposed by professor C. S. Peskin of New York University mathematics in 1977 [5], and used to simulate the coupling mechanism between the blood flow and heart muscle. The biggest characteristic of this method is that the calculation is performed in a fixed orthogonal Descartes Cartesian grid, and Peskin presented a new method to deal with the influence on flow field of immersed boundary. Until recently, the immersed boundary method has developed many different modifications and improvements [6-19], and now the range of applications from the initial biomechanics extends to almost all areas of computational fluid mechanics [20-22]. In addition, the immersed boundary method is not only used for flow field calculation of fluid-solid boundary, also for flow of fluid-fluid and fluid-gas boundary calculation [23,24].

The immersed boundary method is the most critical of how the forcing term added to the governing equations. It is a sign of the distinction between different types of immersed boundary methods. In the governing equations, the forcing term can be applied in two ways: one is adding forcing term in the momentum equations, and another is adding forcing term in the continuity equation. According to the different processing methods of the forcing term, the immersed boundary method can be divided into two categories [25]: "continuous forcing approach" and "discrete forcing approach". The forcing term for continuous forcing approach is incorporated into the continuous equations before discretization. An attractive feature of the continuous forcing approach is that it is formulated being independent of the underlying spatial discretization. The discretization of forcing source generally has analytical expression, which meets the requirements of some specific mechanical relationship type (such as Hooke's law). It is mainly used to deal with elastic boundary problems. This approach has reliable physical foundation, and its implementation is very simple, which has been successfully applied to biological flow and multiphase flow problems. The forcing term for discrete forcing approach is introduced after the equations are discretized. The discrete forcing approach depends on the discretization method. It generally cannot obtain the analytical expressions. It is mainly used for the treatment of solid interface problems (such as flow around a rigid body). Although it is not as good as continuous forcing approach, but it can be very good to describe the immersed boundary with high Reynolds number flows. This allows direct control over the numerical accuracy, stability, and discrete conservation properties of the solver.

In this paper, the immersed boundary method (IBM) is applied to simulate flow around vane, and cascade in hydro-turbine, combined with dynamic Smagorinsky-Lilly model of large eddy simulation (LES), numerical simulation of flow around one moving vane passage of the wicket gate in a benchmark hydro-turbine used in our laboratory at Kunming University of Science and Technology. The numerical results show that the propagation characteristics of blade airfoil wake, analysis of the unsteady flow around a vane, vortex structure development process and interaction.

2. The combined LES-IBM approach

Using the Gauss filter function, the original N-S equation through the filter processing to achieve the separation of large and small scale structures, and the introduction of the force source of the solid structure immersed boundary motion effect, flow control equations are:

$$\frac{\partial \bar{u}_i}{\partial t} + \frac{\partial}{\partial x_j}\left(\bar{u}_i \bar{u}_j\right) = -\frac{1}{\rho}\frac{\partial \bar{p}}{\partial x_i} + \nu \frac{\partial^2 \bar{u}_i}{\partial x_j \partial x_j} - \frac{\partial \tau_{ij}}{\partial x_j} + f_i \tag{1}$$

$$\frac{\partial \bar{u}_i}{\partial x_i} = 0 \tag{2}$$

In the above equations, $\bar{u}_i$ and $\bar{u}_j$ is component of velocity. t is time. ρ is fluid density. $\bar{p}$ is the pressure. ν is the kinematic viscosity coefficient of fluid. f_i is the force source of the characterization of immersed boundary effect. The horizontal bar on the variable head indicates the variable after filtering. τ_{ij} considering the effect of small-scale subgrid stress can be expressed as

$$\tau_{ij} = \overline{u_i u_j} - \bar{u}_i \bar{u}_j \tag{3}$$

$$-\left(\tau_{ij} - \frac{\delta_{ij}}{3}\tau_{kk}\right) = 2\nu_{sgs}\bar{S}_{ij} \tag{4}$$

$$\bar{S}_{ij} = \frac{1}{2}\left(\frac{\partial \bar{u}_i}{\partial x_j} + \frac{\partial \bar{u}_j}{\partial x_i}\right) \tag{5}$$

δ_{ij} is Kronecker constant. τ_{kk} is viscous stress. ν_{sgs} is the subgrid eddy viscosity coefficient. $\bar{S}_{ij}$ is fluid strain rate tensor. In order to reflect the differences of flow characteristics in the guide vane closing movement process at different times, and conveniently combined with the existing solver, this paper uses dynamic subgrid stress on Smagorinsky-Lilly model. This model overcomes the shortcomings of the conventional Smagorinsky model. The subgrid scale eddy viscosity coefficient is $\nu_{sgs} = C_d \Delta^2 \left|\bar{S}_{ij}\right|$, and C_d is the coefficients of dynamic Smagorinsky-Lilly model with temporal and spatial variations. It is no longer a constant but with changes in flow coefficient model. Δ is the filtering characteristics of grid scale.

Discrete force method by handling regulating vane motion has been used to stimulate the effect of immersed boundary. By adding a force body similar momentum source through the interface program, the flow field and the IBM problem of moving guide vane combined with the solver of ANSYS-CFX reached a complete solution. The solver will be updated automatically each time step the immersed boundary grid points from the beginning of the position, and then establish a series of internal nodes in the interior of immersed solid. In order to make the fluid velocity trend immersed boundary velocity solver will use force source to deal with fluid node in the interior of immersed solid.

Specifically, because there is no coupling relation between the flow field and the surface of the immersed boundary nodes, the information transfer between them requires an interpolation process. Peskin first proposed to use the Dirac function to construct the interpolation scheme, but proved its computational efficiency is low in the three-dimensional case. We use bilinear interpolation format to replace the Dirac function. In the current study, the fluid velocities $\vec{u}(\vec{X},t)$ are interpolated to a immersed boundary point, $\vec{X}=(X,Y)$ from the four surrounding grid points denoted by the indices (i,j), $(i+1,j)$, $(i,j+1)$ and $(i+1,j+1)$, using bilinear interpolation,

$$\vec{u}(\vec{X},t)=\sum_{i,j}^{i+1,j+1} D_{i,j}(\vec{X})\vec{u}_{i,j} \tag{6}$$

$$D_{i,j}(\vec{X})=d(X-x_i)d(Y-y_j) \tag{7}$$

d in the above equation is elaborated in the following:

$$\begin{cases} d(X-x_i)=\dfrac{(X-x_{i+1})}{(x_i-x_{i+1})} & if \quad x_i<X \\ d(X-x_i)=1 & if \quad x_i=X \\ d(X-x_i)=\dfrac{(X-x_{i-1})}{(x_i-x_{i-1})} & if \quad x_i>X \end{cases} \tag{8}$$

For the 3D case, eight nodes on the grid flow field surrounded by $\vec{X}$ interpolation, using the same format.

Obviously, this interpolation method is more simple and accurate than the Dirac function method. At the same time, bilinear interpolation is only needed to calculate the adjacent grid points without the need for full or partial numerical integration. With the increase of the dimension, the amount of computation is not increased significantly. Therefore, its computational approach is more efficient than that of the Dirac function method.

3. The IBM-LES simulation of dynamical flow around a moving guide vane in a channel

3.1. The establishment of computational model

For the calculation of one guide vane of the wicket gate in a benchmark hydro-turbine in our laboratory at Kunming University of Science and Technology and its closing movement of the representative, the type of hydro-turbine rated head 10 m, rated flow flux of 0.7 m^3/s, the rated speed of 600 r/min. The hydro-turbine and guide vane are shown in figures 1 and 2. In order to reveal formation and evolution mechanism of wake because of guide vane regulation, vane-channel model corresponding to one guide vane passage is built. One passage is expanded into a channel, and formed flow around one moving vane in a channel for guide vane regulation. The vane-channel model was shown in figure 3. The Reynolds number of calculation condition corresponding to one guide vane based on chord is 120421. The computational domain is 12L×4L×1.1L, where L is chord length of one guide vane. In order to fully capture wake, the trailing edge of vane at streamwise direction is 8L.

IOP Conf. Series: Materials Science and Engineering **657** (2019) 012030 doi:10.1088/1757-899X/657/1/012030

After grid independence verification, the total grids numbers are 1498572. The vane region is divided by a hybrid grid, and the grid size is 3 mm. The nodes numbers are 10870, and the grids numbers are 49113. The fluid region is divided by hexahedron grid with eight nodes, and the grid size is 4 mm. The nodes numbers are 1509872, and the grids numbers are 1449459. The computational grid is verified to be grid independently. For the interpolation of IBM, non-uniform mesh interpolation technique is used at the interface of fluid-solid. The dynamic Smagorinsky-Lilly model of large eddy simulation is used for the subgrid stress. The immersed boundary method is applied to deal with the moving boundary of vane. The fully implicit multi-grid coupled solution technology has achieved in accelerating the convergence.

Figure 1. The hydro generator unit.

Figure 2. Hydrofoil of the wicket gate.

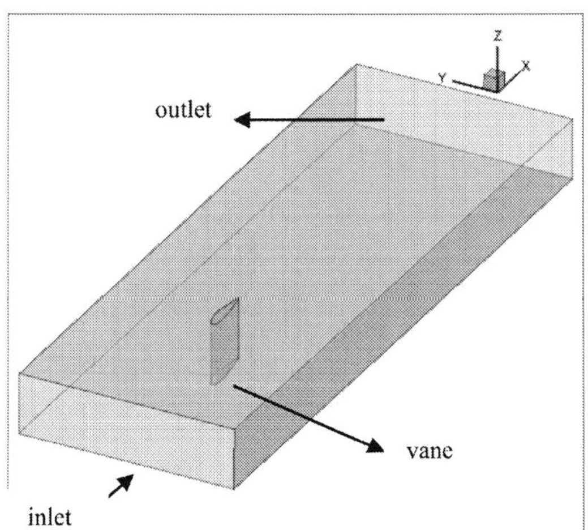

Figure 3. Vane-channel model used in computation.

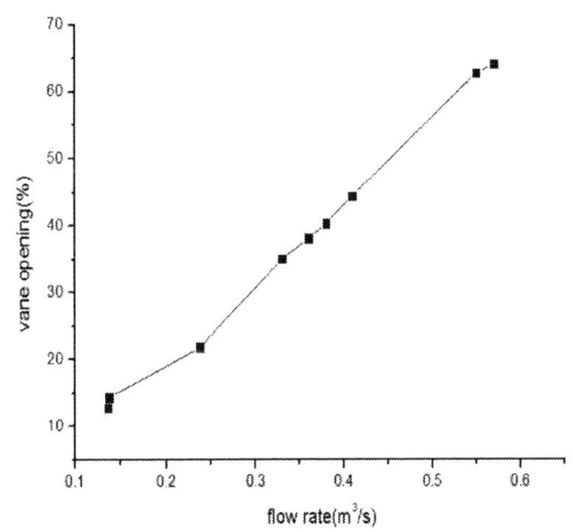

Figure 4. Curve of vane opening via flow rate.

3.2. Boundary conditions

Inlet: the actual unit vane opening corresponding to the change flow rate (show in figure 4), converted into one vane corresponding to the velocity, imposed on inlet as a known velocity boundary. Outlet:

IOP Conf. Series: Materials Science and Engineering **657** (2019) 012030 doi:10.1088/1757-899X/657/1/012030

the pressure outlet boundary condition. Assuming normalwise periodic boundary conditions and spanwise rigid non-slip wall, no-slip conditions are applied to the surface of vane.

The closing law of vane in hydro-turbine is divided into two straight line. The first of time is 5s, the second time is 10s, and total is 15s. Calculation time step is 0.001s. Using the computer as the 8 core CPU, and 24G memory workstation, one calculation takes about 360 hours.

3.3. Analysis of computational results

3.3.1. The changes of pressure in the process of moving vane. In the guide vane two segment closing process, the moving vane in a channel was simulated and recorded to the spanwise section (z=0.5 L) and normalwise section (y=2 L) on the change of pressure with time. Figure 5 shows the pressure distribution for several typical times in the spanwise section (z=0.5 L) and normalwise section (y=2 L). As can be seen in close action guide vane rotation process, the whole flow field pressure distribution becomes non-uniform and asymmetric, the upstream region variable for positive pressure area and downstream region variable for negative pressure zone, pressure distribution in a channel with variation of time showed a complex unsteady evolution process. At t=1s (vane rotates 9°), the small range high pressure region is found in leading edge of vane on spanwise section due to flow impact, and the negative pressure center appears at the bottom of vane. The negative pressure zone emerges near the upper and lower surface of the vane trailing edge. There are two negative pressure zones around the negative force surface of the vane on normalwise section. At t=3s (vane rotates 27°), the negative pressure center of vane is merged into a circular area, and a small range of negative pressure center surrounding the zones. The flow separation points near the negative force surface of vane formation gradually developed into vortex. With the development of the flow, the vortices begin to move away from the negative force surface of the guide vane, which influence the flow field in the downstream zones. At the t=5s (vane rotates 45°), a long narrow negative pressure zone is formed behind the vane on spanwise section. The distance between a strong negative pressure center and the vane trailing edge is very close, a strong negative pressure center is also appear to the negative force surface of vane on normalwise section. At t=7s (vane rotates 54°), strong low pressure vortex moves to downstream zones, strip low pressure zone appeared on the normalwise section. At t=9s (vane rotates 63°), the negative pressure center transfer to near the vane trailing edge, a zone of negative pressure on normalwise section decomposition for multiple small negative pressure center transfer to the downstream region. At t=11s (vane rotates 72°), a large area of low pressure is reformed near the negative pressure center of the vane, the negative pressure center on normalwise section changes complex. At t=13s (vane rotates 81°), the downstream low pressure zones separate into a large range of low pressure zones and a strip of low pressure near negative force surface of vane. At t=15s (vane rotates 90°), the low pressure zone continues to develop, in the distance by three times of chord lengths position dissociate a negative pressure center. The negative pressure center range gradually expanded on normalwise section, the obvious oscillation flow around vane is formed.

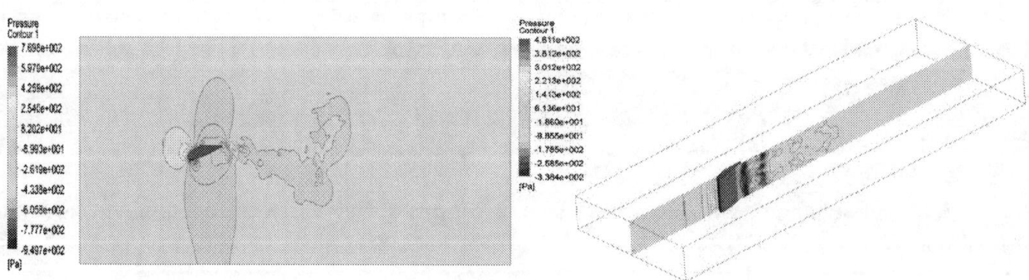

(a)

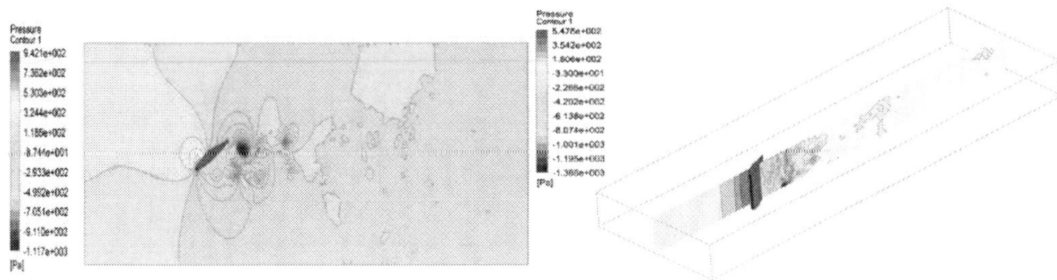

(b)

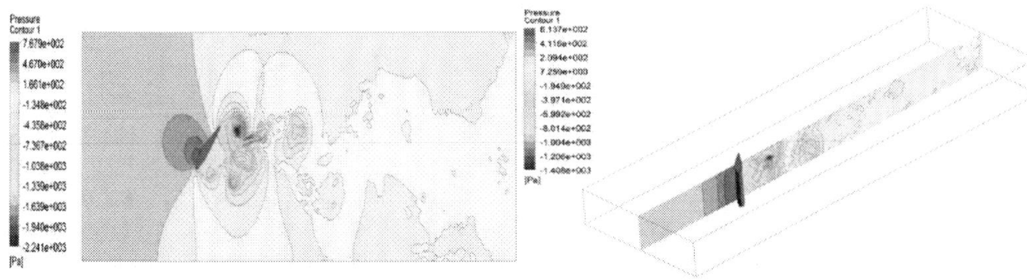

(c)

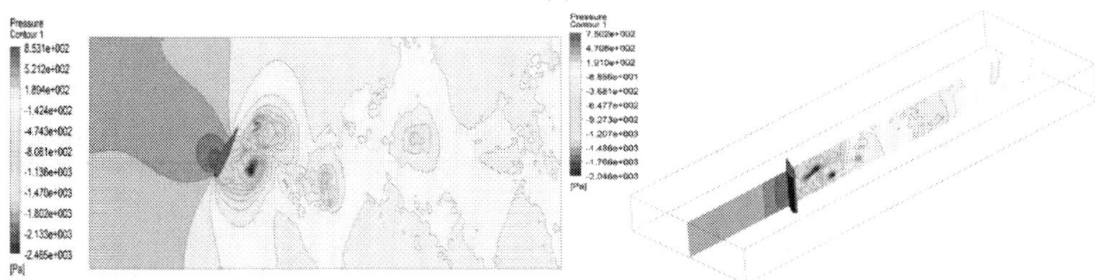

(c)

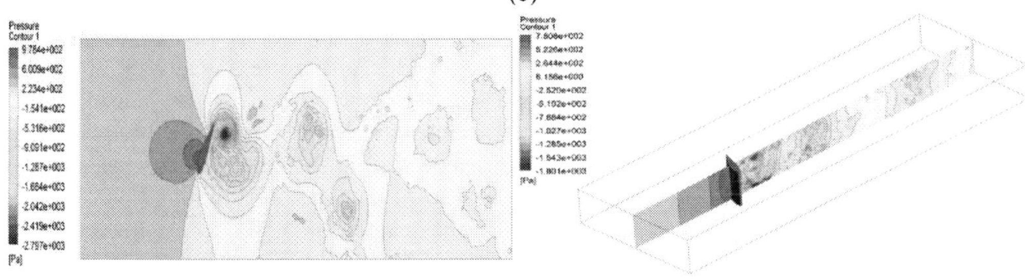

(d)

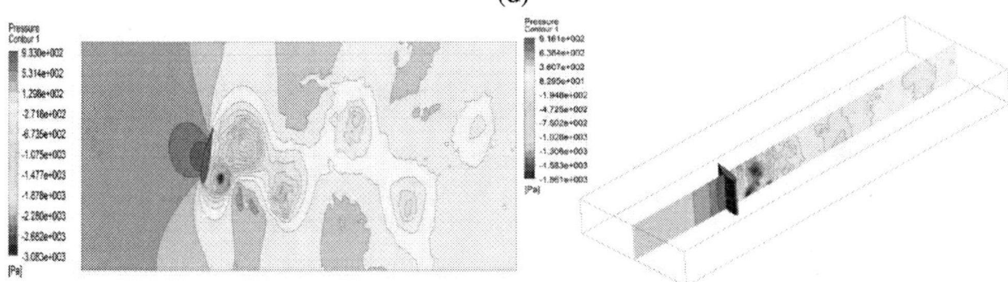

(e)

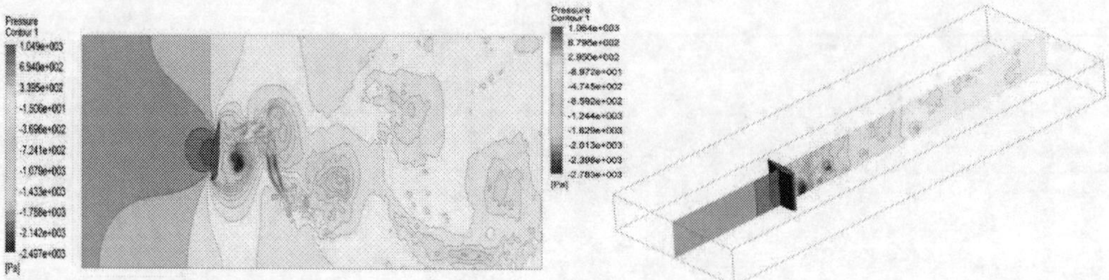

(f)

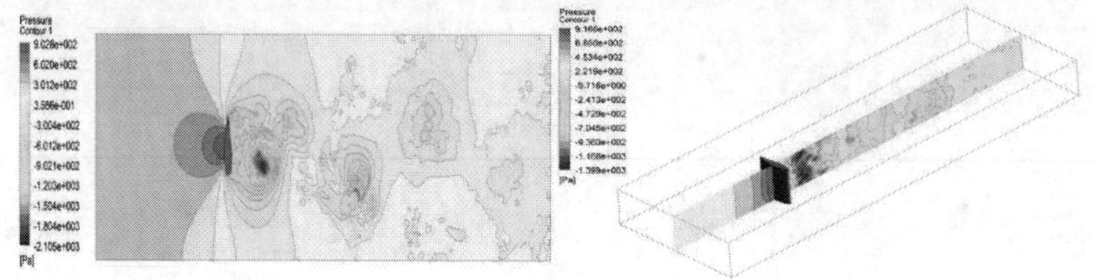

(g)

Figure 5. Pressure distributions at the spanwise section z=0.5L and pitchwise section y=2 L. (a) t=1s, (b) t=3s, (c) t=7s, (d) t=9s, (e) t=11s, (f) t=13s and (g) t=15s.

3.3.2. The changes of wake structure in the process of moving vane. In order to analyze 3D topological structure of the wake, including temporal and spatial distribution characteristics, the non-dimensional λ_2-criterion was used to identify the wake. The wake distribution at typical moment of moving vane was shown in figure 6. As indicated, with the moving vane the development of wake and vortex structures in flow field demonstrated obvious strong three-dimensional, nonlinear and asymmetric distribution characteristics. The evolution of wake is from the order to the disorder. At t=1s (vane rotates 9°), with the vane begin moving the large scale of "starting vortex" was reeled up in the flow field, the near wake is formed by the trailing vortex and vortex shedding. At t=3s (vane rotates 27°), on both sides of vane have many large scale vortex structure one after another to roll up, continuously to promote the downstream direction, in the development process of the vortex size increases, and the distance between the vortex structure also constantly stretched. At t=5s (vane rotates 45°), large eddies break into small eddies in downstream development zone, no longer appearing to the same magnitude size of vortex structure. At t=7s (vane rotates 54°), the continuous roll up and development mechanism of vortex reflect the space quasi periodic of the coherent structure. At t=9s (vane rotates 63°), the shear layer separation and roll up vortex in the wake around the spiral spread out formation to the complex three-dimensional structure. At t=11s (vane rotates 72°), the energy gradually transmit from large scale eddy to small scale eddy in downstream propagation of shedding vortex, because of fluid viscosity, until a certain level of small scale eddies to transfer energy by viscous dissipation, and shedding process is caused by the asymmetry of the vortex development. At t=13s (vane rotates 81°), vortex topology like a long rope appear in the region near the outlet, a braided vortices is formed. At t=15s (vane rotates 90°), through the interaction between the vortex structure, the energy transfer process of different scale eddy is completed. The energy dissipation and transfer part always occur in the place of vortex violently. In the guide vane turn off process, various scales of vortex structures appear, with the nonlinear effect of the increase in different scales of the eddies in the evolution process between continuously exchange energy, induce the formation of complex nonlinear large eddy coherent structure and unsteady dynamic interactions.

In summary, in the guide vane closing process, the downstream area of vane continued to appear the phenomenon of alternating vortex structures of different scales. This is due to the different scales of the vortex structure in the evolution process because of the energy transfer mode alternation, thus inducing a flow around unsteady flow oscillation phenomenon.

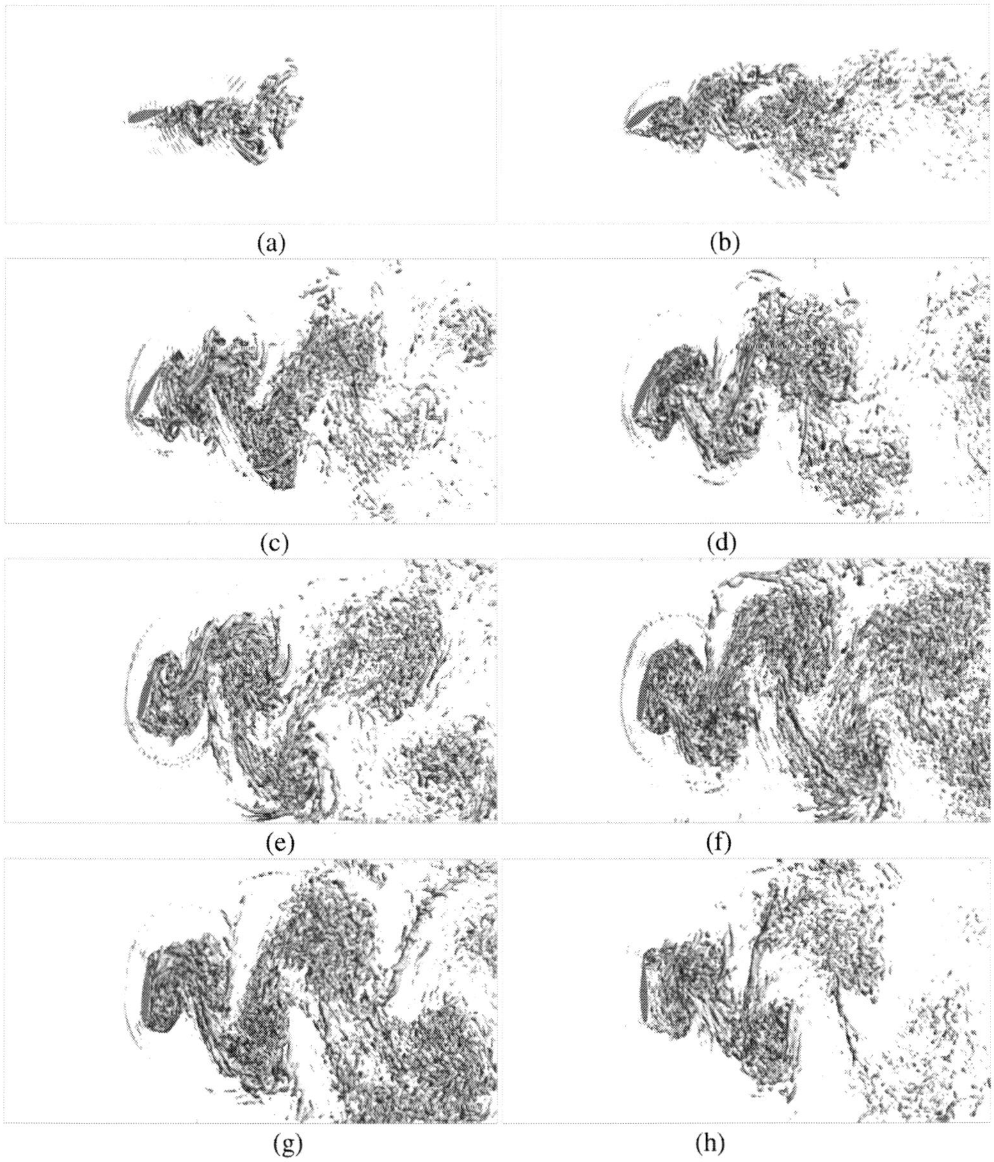

Figure 6. Wake structure distributions flow around kinetic guide vane. (a) t=1s, (b) t=3s, (c) t=5s, (d) t=7s, (e) t=9s, (f) t=11s, (g) t=13s and (h) t=15s.

In 8 core CPU, 24G memory workstations, we simulate all the numerical results produced animation display, realistically reproduce the large eddy rolled up, vortex merger and multiple large eddy interaction process.

4. Conclusions

The numerical results indicated that, after full development zone, the rich coherent vortex and interaction in flow around the vane were improved, and the distribution characteristics of temporal evolution wake was obtained. The immersed boundary method can better simulate complex shape boundary surface of structure and deal with flow around moving boundary problems, and can accurately calculate the wake flow around airfoil vane. The immersed boundary method by adding the corresponding source terms represents fluid-structure interaction in the Navier-Stokes equations. The biggest advantage of this method lies in that the Cartesian grid can achieve almost any complex geometry boundary, simple and efficient without loss of accuracy. In addition, the immersed boundary method can be easily combined with advanced turbulence model (LES and DES), and wake structure to simulate the fine, reasonable and intuitive. The calculation is also superior.

The objective of the calculation is to reveal the guide vane dynamic flow around oscillating wake physical and mechanical mechanism. In order to make the process of the evolution of the wake structure not influenced by the channel geometries, and to reveal the internal mechanism of oscillating wake, the vane passage is simplified to the channel. Based on the simulation results of the vane-channel model, despite the large differences between the channel geometry and the true vane passage, the flow characteristics of flow around a vane in the near wall region of the guide vane are similar to that of the real vane passage. Therefore, the results based on the flow characteristics obtained by the vane-channel model can be used to illustrate the dynamic flow around the real vane passage.

Acknowledgments

The project is financially supported by National Natural Science Foundation of China (NSFC) (Grant No. 51541913 and 51279071) and Yunnan Agricultural University Natural Science Foundation for Young Scientists (No. 2015ZR17).

References

[1] Aparna P, Pothanna N, Ramana Murthy J V and Sreelatha K 2017 *Alexandria Eng. J.* **56** 679-85
[2] Aparna P, Pothanna N and Ramana Murthy J V 2019 *Numerical Heat Transfer and Fluid Flow* (Singapore: Springer) pp 285-93
[3] Aparna P and Ramana Murthy J V 2007 *Bull. Calcutta. Math. S.* **99** 293-304
[4] Hay A, Etienne S, Garon A and Pelletier D 2015 *Comput. Methods Appl. Mech. Eng.* **295** 172-95
[5] Peskin C S 1972 *J. Comput. Phys.* **10** 252-71
[6] Gronskisa A and Artanab G A 2016 *Comput. Fluids* **124** 86-104
[7] Chiu P H, Lin R K and Tony W H S 2010 *J. Comput. Phys.* **229** 4476-500
[8] Yang F C and Chen X P 2015 *J. Hydrodynam.* **27** 658-67
[9] Liao C C and Lin C A 2012 *Comput. Methods Appl Mech. Eng.* **213-216** 58-70
[10] Mori Y and Peskin C S 2008 *Comput. Methods Appl Mech. Eng.* **197** 2049-67
[11] Zhang H, Yu A B, Zhong W Q and Tan Y Q 2015 *Int. J. Heat Mass Transf.* **91** 178-89
[12] Ilinca F and Hétu J F 2010 *Comput. Fluids* **39** 1656-71
[13] Seo J H and Mittal R 2011 *J. Comput. Phys.* **230** 1000-19
[14] Nicolaou L, Jung S Y and Zaki T A 2015 *Comput. Fluids* **119** 101-14
[15] Sui Y, Chew Y T, Roy P and Low H T 2007 *Int. J. Numer. Methods Fluids* **53** 1727-54
[16] Yang J, Preidikman S and Balaras E 2008 *J. Fluids Struct.* **24** 167-82
[17] Sun X S and Sakai M 2016 *Chem. Eng. Sci.* **139** 221-40
[18] Ji C, Munjiza A and Williams J J R 2012 *J. Comput. Phys.* **231** 1797-821
[19] Ilinca F and Hétu J F 2012 *Comput. Fluids* **59** 31-43
[20] Zhang H, Trias F X, Oliva A, Yang D M, Tan Y Q and Shu S 2015 *Powder Techol.* **272** 1-13
[21] Weymouth G D and Yue D K P 2011 *J. Comput. Phys.* **230** 6233-47
[22] Khosronejad A, Kang S, Borazjani I and Sotiropoulos F 2011 *Adv. Water Resour.* **34** 829-43
[23] Anderson D M, McFadden G B and Wheeler A A 1998 *Annu. Rev. Fluid Mech.* **30** 139-65
[24] Scardovelli R and Zaleski S 1999 *Annu. Rev. Fluid Mech.* **31** 567-603
[25] Mittal R and Iaccarino G 2005 *Annu. Rev. Fluid Mech.* **37** 239-61

IOP Conf. Series: Materials Science and Engineering **657** (2019) 012031 doi:10.1088/1757-899X/657/1/012031

Piezo-actuated device for a bio-structural monitoring application through vibration-based condition and electromechanical impedance measurements

D R Barco [1], H A Tinoco[1,2,3], C I Cardona [1] and F M Peña [1]

[1]Experimental and Computational Mechanics Laboratory, Universidad Autónoma de Manizales, Antigua Estación del Ferrocarril, Edificio Fundadores, 170001 Manizales, Colombia
[2]Institute of Physics of Materials, Czech Academy of Sciences, Zizkova 22, 616 62 Brno, Czech Republic
[3]Central European Institute of Technology-Brno University of Technology, Technická 3058/10, 61600 Brno, Czech Republic

E-mail: htinoco@autonoma.edu.co

Abstract. This study presents a numerical and experimental development of a piezo-actuated device used for monitoring the stiffness variations of its support through electromechanical impedance measurements. The piezo-device and its components define a clamped beam system activated dynamically by two piezo-transducers that transmit vibrations to the support (monitoring substrate). An harmonic finite element analysis was carried out to understand the effects of the substrate properties on the dynamics of the piezo-device. Experimental tests corroborated the simulations with the correspondence of modal shapes and frequency response functions (FRFs) when the substrate varies its stiffness. A biomedical application was conducted in a bone specimen with three embedded teeth to monitor the stiffness variations induced by drillings in the bone. Results showed that the bone stiffness monitoring could be possible through the teeth due to that the drillings effect were quantified by electrical impedance signals.

1. Introduction

Structural health monitoring (SHM) techniques have been applied in several practical approaches during the last few years; the majority focused on evaluating structural damages [1-4]. The understanding of structural behaviour is a requirement for the application of SHM methodologies in operating conditions [5,6]. Methods based on vibrations deal with the determination of changes in the modal properties, these changes are caused by damages that modify the structural condition. Non-destructive identification techniques can be classified into two categories: those focused on local damage and others in the global dynamic behaviour [7]. Local identification methods are applied on the basis of non-destructive evaluation tools such as visual inspections, acoustic emission, ultrasound, magnetic particle inspection, radiography and Foucault currents, among others [8-10]. The majority of tools requires the previous location of the diagnosis area to proceed to an inspection. However, methods based on the global vibration show to be an alternative that overcomes those limitations as evidenced by different studies [11-16]. The application principle is supported on the basis of the structural dynamics which is governed by the following physical properties; mass, damping, and stiffness. Variations of these properties will cause modifications in the dynamics; it means that modal properties are altered. Recent studies have explored the application of SHM in the biomedical field through the EMI technique; for example, biomedical sensors to monitor the bone condition [17,18] or the assessment of dental implants [19,20].

This paper presents a numerical and experimental analysis of a piezo-actuated device used to monitor the structural changes of its support by means of mechanical vibrations. The study shows how the elastic variations of the support modify the vibration conditions and how these modifications can be quantified with the electrical impedance. An application is shown with teeth embedded in a bone sample, in which rigidity variation due to drillings done on the sample are quantified with the electrical impedance measurements taken in the piezo-device.

2. Materials and methods

2.1. Finite element analysis for the piezo-actuated system

This section describes a harmonic finite element analysis (FEA) in order to obtain the frequency response functions that characterize the dynamic behavior of a piezo-actuated system. Figure 1(a) illustrates a piezo-device attached to a coupling interface and a supporting substrate. On this system, a parametric analysis is carried out to observe the effects on the vibratory states (kinematic changes) when the elastic properties of the substrate are varied as well as the coupling interface type (cone or tooth).

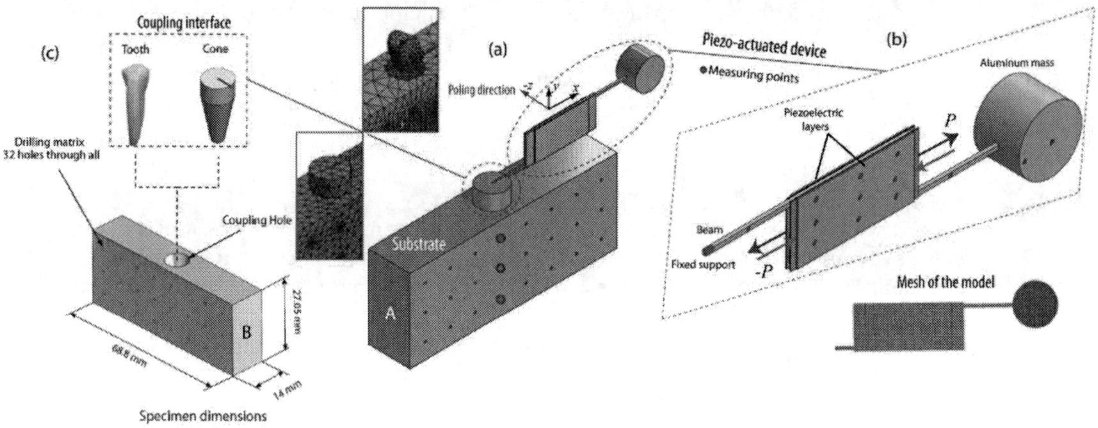

Figure 1. Finite element models and boundary conditions.

For the FEA simulations, the physical and mechanical properties are listed in table 1.

Table 1. Physical and mechanical properties of the materials assigned in ANSYS®.

Part	Material	Young's modulus [GPa]	Poisson ratio	Density [kg/m^3]
Orthodontic Wire		177	0.31	7750
Mass	Aluminum	70	0.33	2700
Piezo-sheets	PZT-5H	Anisotropic (equation(1))	-	7800
Substrate	Copper alloy	127	0.34	8670
Tooth		100	0.3	2500

All materials were considered isotropic linear elastic except for the piezo-sheets that was assumed anisotropic according to the following material properties; C_{11}=126 GPa, C_{12}=84.1 GPa, C_{13} = 84.1GPa, C_{21} = 84.1GPa, C_{22} = 126GPa, C_{23} = 79.5 GPa, C_{31} = 84.1 GPa, C_{32} = 79.5 GPa, C_{33} = 126GPa, C_{44} = 23GPa, C_{55} = 23GPa, C_{66} = 23GPa. Poling direction was defined at *z*-direction. For the activation of the piezo-device, a harmonic distributed force *P* was applied in the frequency range of 1-50 kHz as described in figure 1(b). It is noted that both forces are applied synchronously but in opposite directions so that one sheet remains in compression and the another in tension. The forces were applied with the same magnitude, and it was estimated with the electromechanical models reported by [21]. Boundary conditions were defined in the following way: fixed support in the cable (beam) in the first part of it; rigid contact surfaces were established between the interfaces that couple the substrate and the device. This condition was created in order to guarantee the movements between the bodies. Therefore, there are not permitted relative displacements between interfaces that interact in the model. Additionally, fixed support conditions were applied on the faces A and B of the substrate. Rectangular elements (piezoelectric sheets, brass plates, cable) and triangular elements (concentrated mass) were chosen for meshing, as seen in figure 1.

2.2. *Experimental setup for the velocity measurements*

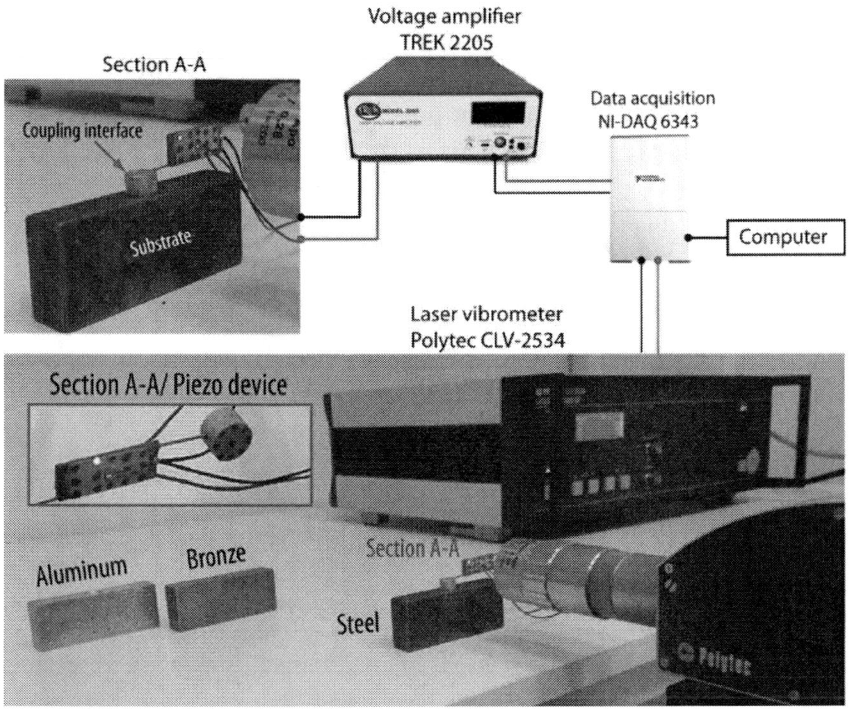

Figure 2. Velocity measurements of the piezo-actuated system with different substrates (Aluminum, Bronze and Steel).

An experimental setup was designed to velocity measures in the piezo-system evaluated through simulations. Therefore, a computer, a data acquisition card (NI DAQ 6211), a high voltage amplifier 2205 (TREK Inc., Lockport, NY, USA) and a laser vibrometer POLYTEC CLV-2534 (Polytec Inc., Auburn, MA, USA) are necessary elements for this purpose. The configuration of this experiment can be seen in figure 2. Velocity measurements were taken in the piezo-device for different metallic substrates (aluminum 70 GPa, Bronze 120 GPa and Steel 200 GPa, according to the local material supplier) coupled to a connection interface (aluminum cone). The measurement points were defined in the following way; 8 points on the electric piezo-sheet, 2 points on the wire and 2 points on the mass; locations are marked in figure 1. An epoxy adhesive was used to bond all interacting parts of the system. The piezo sheets were connected to the output of the voltage amplifier, in which a 35 V excitation signal was applied. A frequency sweep was imposed from 0 to 20 kHz in 1s with a sampling frequency established in 100 kHz. The substrates were bonded at the bottom surface in a rigid table for avoiding non-desired motions.

2.3. Electromechanical impedance measurements in metallic substrates and a bone sample

In figure 3(a) is shown a first experimental setup settled with the aim to measure the electrical impedance of the piezo-device coupled to the same metallic substrates described in section 3.2. The substrate dimensions are depicted in figure 1. The aim of this test is to simulate Young´s modulus variation in the substrate. An impedance analyzer (E4990A, Agilent, Palo Alto, CA) was configured with a sweep signal setup between 0-20 kHz with 1601 resolution points. For the context of a bio-structural application, a second experimental setup was set to perform impedance measurements in cancellous bone samples of cleaned cow hip bone obtained from a local butchery. The sample consists of a cancellous bone prismatic specimen with three embedded teeth that emulate the human maxillary bone-teeth system. Maxillary bone cavities were manufactured by mechanical drillings. All teeth were integrated with a commercial silicone layer that represents an artificial periodontal ligament as depicted in figure 3(b).

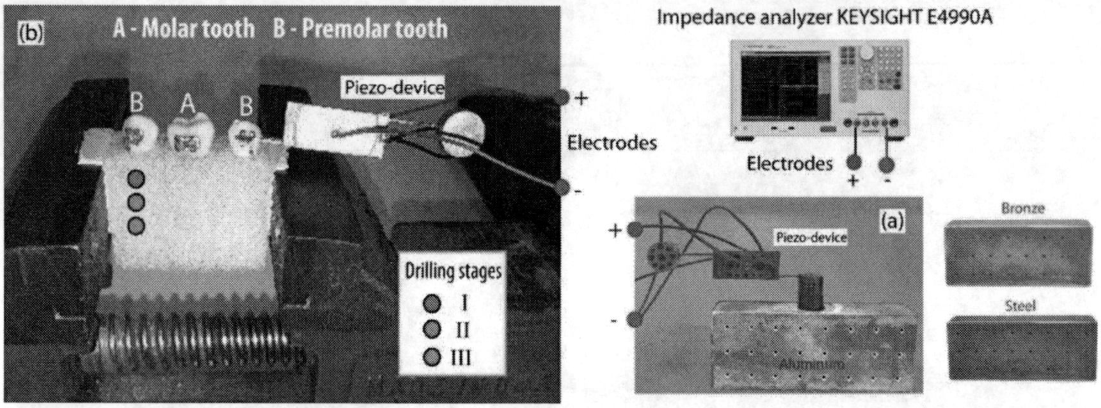

Figure 3. Electromechanical impedance measurements. (a) Variations of the metallic substrates. (b) Stiffness variation in a bone sample.

Dimensions of the sample are $36.9mm \times 27mm \times 18.4mm$ that correspond to length $\times$ height $\times$ width. In order to study the bio-monitoring, damage conditions were induced by a sequence of mechanical drillings of $2.5mm$ (diameter). The drilling order is described by colors in figure 3(b). Electrical resistance measurements were acquired with the same configuration of the first experiment. EMI technique consists of exciting both piezo-transducers by a harmonic voltage and obtaining the current that passes through these simultaneously; the details of the method were reported by [16].

3. Results

3.1. Finite element analysis results: Effects of the elastic variations of the substrate on the kinematic of the piezo-device

In this section, the results of the finite element analysis (FEM) are described in order to analyze the dynamic behavior of the complete system shown in figure 1. The mean velocities for the points corresponding to the piezoelectric sheets (demarked in figure 1) and the mass are shown in figures 4(a) and 4(b). The blue highlighted line corresponds to the velocities when the canine tooth acts as the interface of the piezo-device, and the red line, when it is coupled to the aluminum cone. In both figures, the black line represents the velocities of the clamped piezo-device; it means that there is no substrate and no coupling interface. In figure 4(a), in the interval of 1-10 kHz is analyzed that the coupling interface has no effect on the kinematics of the piezo sheet since basically the signals present the same trending. But from 10 kHz and beyond, it is corroborated that the piezo sheets vibrate in different ways when the interface is changed. The resonance peaks are numbered to point out the effects of each interface (tooth and cone). It is observed that the resonances 1 and 2 present the same values of frequency for both interfaces. The relationships between resonances 3, 4, 5, and 6 are affected by the changes in

IOP Conf. Series: Materials Science and Engineering **657** (2019) 012031 doi:10.1088/1757-899X/657/1/012031

mass and stiffness introduced by the interfaces in the kinematics of the device. In figure 4(b) is seen that the velocities of the mass are higher when the cone acts as a coupling interface; it can be associated with the mobility that permits the tooth inside the substrate. This can be easily verified with the amplitudes of the velocity shown in figure 4(b).

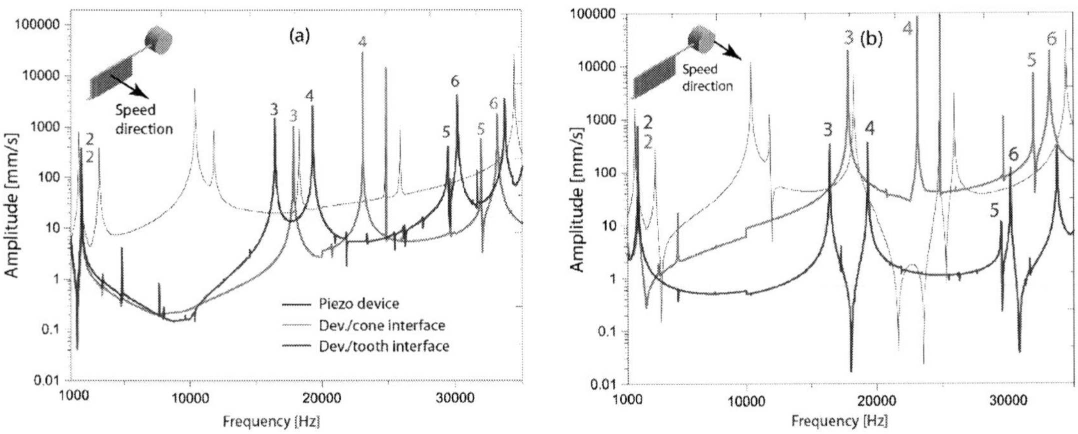

Figure 4. Mean velocities in the piezo-device with a substrate of Young's modulus 5 GPa. (a) Piezoelectric sheet. (b) Mass.

Young's modulus variations (substrate) were considered as a correlation parameter to observe the effects generated in the dynamics of the system (piezoelectric sheets, interface, and substrate). The following elastic moduli were assigned to the substrate, 5 GPa, 10 GPa, 15 GPa, and 20 GPa. The results of the mean velocities for each interface are illustrated in figure 5. There is observed that the piezo-device can detect the elastic variations of the substrate in the velocity magnitudes of the piezo-sheets, which reflect a higher sensitivity to the variations in 33180 Hz (cone) and 30402 Hz (tooth).

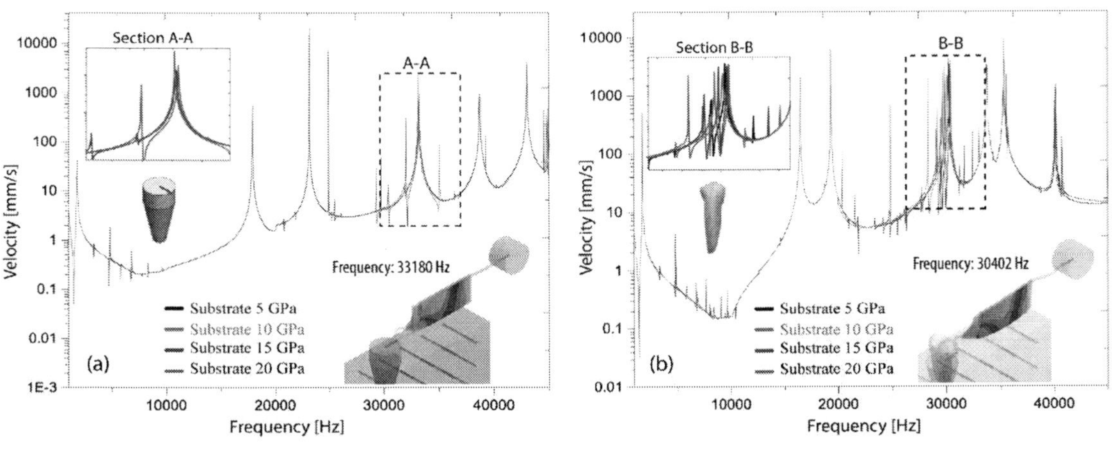

Figure 5. Mean velocities in the piezo-sheet. (a) Cone comparison. (b) Canine tooth comparison.

The characteristic peaks show that the signal conserves the structure; however, there is a shift of the signal to the left when the modulus of elasticity is increased. It means that the vibration condition is modified when the substrate change its elastic properties. With the vibration mode is demonstrated that the vibratory condition is the same independently of coupling interface as illustrated in figure 5; but it is altered by the elastic variations of the substrate. The compared modes were determined for a substrate of 10 GPa.

3.2. Experimental results: Effects of the elastic variations of the substrate on electro-mechanical conditions of the piezo-device

In figure 6(a) are observed the mean velocities (time-domain) taken in the piezo-sheet (red points marked) and transformed to the frequency spectrum with Fast Fourier Transform (FFT). These velocities correspond to three different substrates (Aluminum 70 GPa, Bronze 120 GPa, and steel 200 GPa) used with the aim to verify the vibration condition when the substrates are changed. Two regions of interest that show sensitivity to the elasticity changes of the substrate were observed; these are delimited in between 0-100 Hz and 3-11 kHz. In these zones, it is verifiable as the slopes increased its value when Young's modulus augmented. Particularly, the section A-A is pointed out since a sensitive resonance peak at 9406 Hz is evidenced. This region was extended in figure 6(b) and the vibration mode of the piezo-sheet was reconstructed (using nine measurement points) for the test with the aluminum substrate. The determined vibration shape can be compared with the vibration mode obtained with FEA, which is shown in figure 5. The comparison illustrates that the piezo-sheet is deformed by the vibratory condition del piezo-device which depends on the properties of the substrate. According to the results, the effects of the velocity changes are reflected in the electrical properties of the piezo-sheets, as shown in figure 6(b).

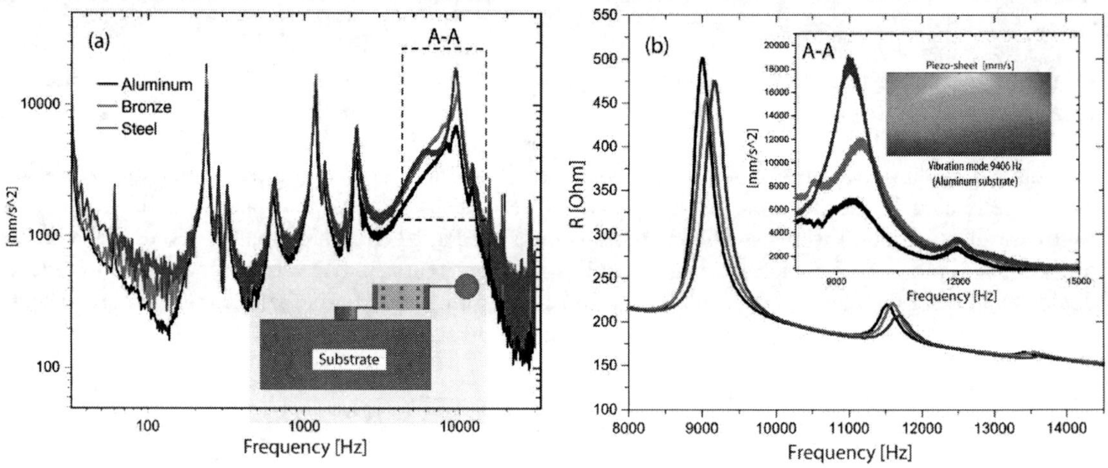

Figure 6. (a) Mean velocity of the piezoelectric sheet in the frequency spectrum. (b) Electrical resistance of the piezoelectric sheet.

In order to determine a correlation between the electromechanical behavior of the piezo-device, the electrical resistance was measured through an impedance analyzer with the experiment described in section 3.3. The frequency range in between 8 to 15 kHz (figure 6(b)) is compared with the obtained velocity measurements in the same interval. There are identified two main peaks in the electrical resistance; first peak at 9 kHz and a second peak at 11.5 kHz, it is evident that the electrical peaks present a correlation with the kinematic ones since the resonances identified in the velocity functions describe the same values in the frequency. The results demonstrate that electrical resonances come strictly from the vibration mode of the piezo-device.

3.3. Bone monitoring through electromechanical impedance measurements

In this section the results of an application of bio-monitoring are presented according to the proposed experiment in section 3.3, detailed in figure 3(b). Damage locations were defined in the bone sample with the objective to evaluate the damage from the neighbour teeth through electromechanical impedance measurements. In this test, it is considered that a human tooth can be used as a probe since it is naturally coupled to biological structures such as the case of maxillary bone. Bone changes its structural properties by factors like; force application, natural remodelling process and metabolic deceases. The aim is to demonstrate the usefulness of the piezo-device for detecting variations in mechanical parameters of bio-structures.

Figures 7(a) and 7(b) show the electrical resistances measured for the bone sample from a molar and premolar tooth, both located in the same specimen. Analyzing the curves, one observability window was chosen between 9-10.5 kHz (first peak at 9.6 kHz) where all resistance signals presented sensitivity to the induced damages showing differences in the maximum amplitudes. For the molar tooth, it was observed that the damage 1 shifts downwards from the pristine signal and then these shift upwards until the third damage. For the premolar tooth is evidenced the same trend. These pertubartions evidenced that the damage can be monitored from the tooth and these changes can be quantified trhough statistical indexes.

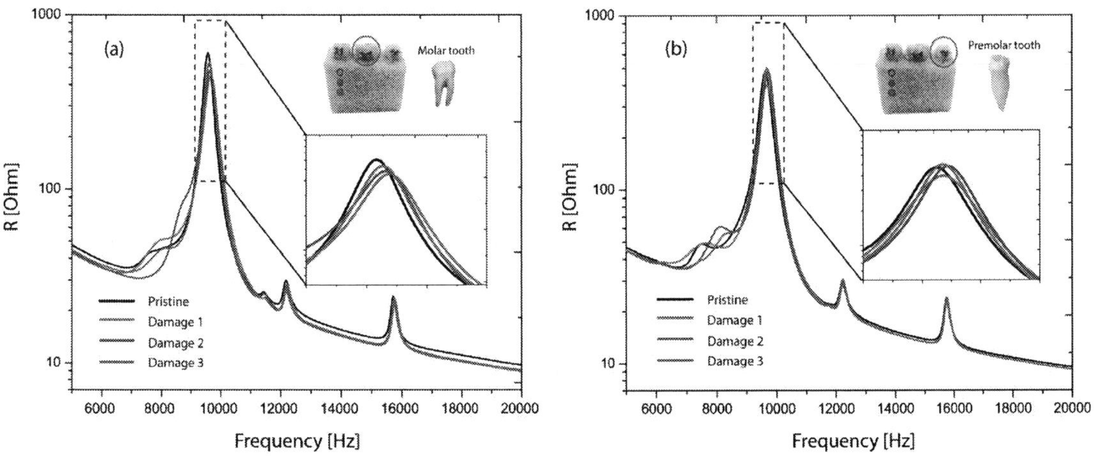

Figure 7. Electrical resistance signals measured in the bone sample. (a) Acquired resistance signals with the molar tooth. (b) Acquired resistance signals with the premolar tooth.

4. Conclusions

Numerical and experimental evaluation of the system showed the effects of the elastic variations in the substrate on the vibration condition of the piezo-device. The application of the electromechanical impedance technique (EMI) verified that the resonances in the frequency corresponded with the mechanical resonances of the system. Two intervals were identified between 0-100 Hz, and 3-11 kHz where the device presented a higher sensitivity to the elastic changes of the substrate. This result leads to conclude that the designed device can be used as a structural monitoring instrument since it presents the ability to capture the elastic changes in its vibratory condition.

Acknowledgments

This research was funded by grant from Departamento Administrativo de Ciencia Tecnología e Innovación (COLCIENCIAS) (Grant code 121974455599) legalized contract 739-2016 and obtained in the announcement 744, year 2016.

References

[1] Doebling S W, Farrar C R, Prime M B and Shevitz D W 1996 Damage identification and health monitoring of structural and mechanical systems from changes in their vibration characteristics: A literature review (No. LA-13070-MS) (United States: Los Alamos National Lab.)

[2] Das S, Saha P and Patro S K 2016 Vibration-based damage detection techniques used for health monitoring of structures: A review *J. Civil Struct. Health Monit.* **6** 477-507

[3] Hu W H, Thöns S, Rohrmann R G, Said S and Rücker W 2015 Vibration-based structural health

monitoring of a wind turbine system, Part I: Resonance phenomenon *Eng. Struct.* **89** 260-72

[4] Tinoco H A, Robledo-Callejas L, Marulanda D J and Serpa A L 2016 Damage detection in plates using the electromechanical impedance technique based on decoupled measurements of piezoelectric transducers *J. Sound Vib.* **384** 146-62

[5] Soman R, Kyriakides M, Onoufriou T and Ostachowicz W 2018 Numerical evaluation of multi-metric data fusion based structural health monitoring of long span bridge structures *Struct. and Infrastruc. Eng.* **14** 673-84

[6] Ochieng F X, Hancock C M, Roberts G W and Le Kernec J 2018 A review of ground-based radar as a noncontact sensor for structural health monitoring of in-field wind turbines blades *Wind Energy* **21** 1435-49

[7] Fan W and Qiao P 2011 Vibration-based damage identification methods: a review and comparative study *Struct. Health Monit.* **10** 83-111

[8] Willberg C, Duczek S, Vivar-Perez J M and Ahmad Z A B 2015 Simulation methods for guided wave-based structural health monitoring: A review *Appl. Mech. Rev.* **67** 010803

[9] Strantza M, Aggelis D, de Baere D, Guillaume P and van Hemelrijck D 2015 Evaluation of SHM system produced by additive manufacturing via acoustic emission and other NDT methods *Sensors* **15** 26709-25

[10] Yadav, S. K., Chung, H., Kopsaftopoulos, F., & Chang, F. K. Damage quantification of active sensing acousto-ultrasound-based SHM based on a multi-path unit-cell approach. In *11th Int. Workshop on Struct. Health Monit.* September 12-14, 2017, Palo Alto, CA, USA, pp 3093-3103, DEStech Publications, Inc

[11] Doebling S W, Farrar C R and Prime M B 1998 A summary review of vibration-based damage identification methods *Shock Vib. Dig.* **30** 91-105

[12] Salawu O S 1997 Detection of structural damage through changes in frequency: a review *Eng. Struct.* **19** 718-23

[13] Doebling S W, Farrar C R, Prime M B and Shevitz D W 1998 A review of damage identification methods that examine changes in dynamic properties *Shock Vib. Dig.* **30** 91-105

[14] Sohn H, Farrar C R, Hemez F M, Shunk D D, Stinemates D W, Nadler B R and Czarnecki J J 2003 A review of structural health monitoring literature: 1996–2001 (USA: Los Alamos National Laboratory)

[15] Farrar C R and Worden K 2006 An introduction to structural health monitoring Philosophical Transactions of the Royal Society A: Mathematical, Physical and Engineering Sciences **365** 303-15

[16] Tinoco H A, Cardona C I, Peña F M, Gomez J P, Roldan-Restrepo S I, Velasco-Mejia M A and Barco D 2019 Evaluation of a piezo-actuated sensor for monitoring elastic variations of its support with impedance-based measurements *Sensors* **19** 184

[17] Zou Y, Tong L P S G and Steven G P 2000 Vibration-based model-dependent damage (delamination) identification and health monitoring for composite structures—a review *J. Sound Vib.* **230** 357-78

[18] Moll J, Arnold P, Mälzer M, Krozer V, Pozdniakov D, Salman R and Nuber A 2018 Radar-based structural health monitoring of wind turbine blades: The case of damage detection *Struct. Health Monit.* **17** 815-22

[19] Boemio G, Rizzo P and De Nardo L 2011 Assessment of dental implant stability by means of the electromechanical impedance method *Smart Mater. Struct.* **20** 045008

[20] Ribolla E L M, Rizzo P and Gulizzi V 2015 On the use of the electromechanical impedance technique for the assessment of dental implant stability: Modeling and experimentation *J. Intell. Mater. Syst. Struct.* **26** 2266-80

[21] Sirohi J and Chopra I 2000 Fundamental behavior of piezoceramic sheet actuators *J. Intell. Mater. Syst. Struct.* **11** 47-61

IOP Conf. Series: Materials Science and Engineering **657** (2019) 012032 doi:10.1088/1757-899X/657/1/012032

Influence of different oil-paper configurations on the electric field distribution simulation of the internal insulation of converter transformer

Z D Cheng[1], L Cheng[1,3], J H Wang[1] and L Wang[2]

[1]Country State Key Laboratory of Power Equipment & System Security and New Technology, Chongqing University, China

[2]China Electric Power Research Institute, Haidian Xiaoyingdong Road 15, Beijing, China

E-mail: chengl16@cqu.edu.cn

Abstract. Converter transformer is an important equipment in HVDC system and it is the key to the stable operation of the whole power transmission and distribution system. Unlike the traditional AC transformer, the oil-paper insulation of converter transformer will receive both AC and DC voltage. So the requirement for the structure design of oil-paper insulation system will be more stringent. It is of great engineering significance to study the electric field distribution of converter transformer and to study the influence of different oil-paper insulation configuration on the electric field distribution in the internal insulation of converter transformer. In this paper, first, an oil-paper insulation structure of converter transformer is established and electric field distribution under AC/DC composite voltage is studied. Then change the thickness and layers of pressboard, the distance of oil channel and the oil-paper ratio to get different oil-paper configurations. Use the simulation software to calculate the electric field distribution under the different oil-paper configurations. Finally, the influence of different oil-paper configuration on the electric field distribution of internal insulation structure of converter transformer is analysed which will provides significant reference for the internal insulation design of converter transformer.

1. Introduction

When the society is growing rapidly, the demand for electricity is increasing year by year. In order to meet the requirements of long distance transmission and large capacity, the HVDC transmission system is born, which plays a significantly critical role in energy transmission and distribution system [1]. Converter transformer, as a key role in HVDC system, will have a serious impact on the whole transmission and distribution system in case of failure, bringing incalculable economic losses to social production and life [2]. In addition, it is different from the ordinary ac transformer. During the operation, the valve side winding of converter transformer will receive the combined ac and dc voltages, which has more severe requirements on the insulation system [3]. Oil-paper insulation, which is the main form of insulation structure in converter transformer, will receive the action of composite voltage. The study on the of electric field distribution in oil-paper insulation inside converter transformer under the action of composite voltage will provide important reference for the design of typical insulation structure inside converter transformer [4].

Multi-physics field simulation software is getting more popular as a simulation platform [5]. Based on the finite element calculation of multiple physical field simulation software, it can be quick, easy for the analysis of multi-physics field distribution in electric equipment, such as transformer. Now researches generally use multi-physics simulation software based on the finite element to calculate and analyze the internal field distribution for converter transformer.

There have been many calculation and analysis of electric field distribution in oil-paper insulation structure of transformer done by researches and equipment manufacturers, including: 1) calculation of electric potential distribution of transformer internal insulation structure [6]; 2) calculation and analysis of characteristics of electric field distribution in internal insulation under the action of dc and ac voltages and compound voltage [7]; 3) study the transient change rule of field-distribution of internal main insulating structure in the case of voltage polarity reversal [8]; 4) calculation and analysis of temperature rise inside transformer under oil flow state [9]. However, the analysis and designing of the oil-paper insulation structure in converter transformer with large capacity and high voltage are still in the stage of calculation and analysis from the empirical formula, and there are few simulations of the influence of different oil-paper configuration on typical insulation structure inside the converter transformer.

To figure out the influence of oil-paper configuration on the field-distribution in insulation structure of the converter transformer, this paper establishes a two-dimensional model of main insulation structure of converter transformer. Including the end winding insulation and main insulation of valve side and line side. Then using the multi-physics simulation software to get the distribution of field distribution. After that, the layers and thickness of pressboard are changed to get different oil-paper configuration. And influence of layers and thickness of pressboard on the electric field distribution inside the internal insulation structure of converter transformer are studied. So that it can provide some references for the design, manufacture, operation and maintenance of converter transformer.

2. Establishment of typical internal insulation structure of converter transformer

Because of the huge size and complex inside construction of converter transformer, it will be very difficult to test the typical insulation structure of large converter transformer [10]. But the multi-physics field simulation software can provide a significant approach to calculate and analysis the field distribution of insulation structure inside converter transformer [11]. In order to calculate the field distribution in insulation structure, this paper first builds the typical inside insulation structure of converter transformer which can be shown in figure 1.

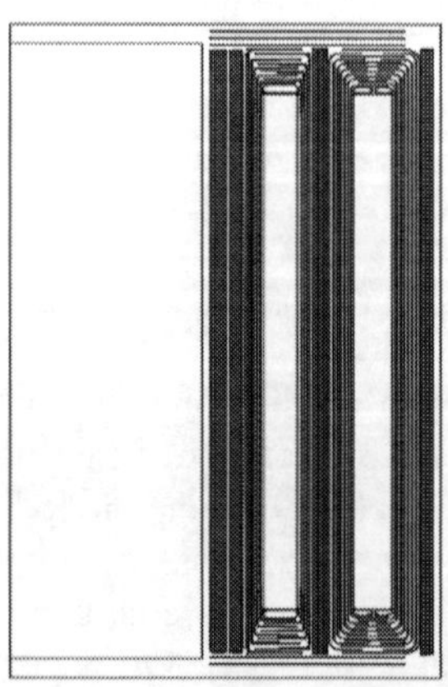

Figure 1. A typical internal structure of converter transformer.

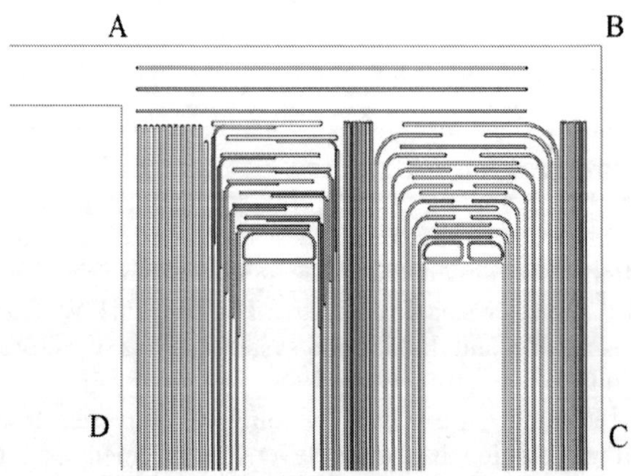

Figure 2. Main insulation and end winding insulation structure of converter transformer.

In order to facilitate the calculation and minimize the calculation deviation, this paper has appropriately simplified the internal structure of converter transformer. And neglected the influence of the structure such as lead wire, connector and casing on the distribution of electric field. The internal insulation structure of a typical converter transformer shown in figure 1 is mainly composed of iron core, iron yoke, line side winding, valve side winding, oil-paper insulation and transformer metal wall [12]. Due to the actual distribution of electric field in practical engineering, the insulation material at the end winding is often subjected to higher voltage and temperature, which is the key part to be examined in design and manufacture [13]. The simplified model of typical winding end insulation and main insulation of converter transformer is shown in figure 2.

As shown in figure 2, the AB is the iron yoke and BC is the metal wall of transformer shell. AD is the iron core. The end winding and main insulation of converter transformer are mainly composed of pressboard. And forming angle ring is also made of pressboard. The line side winding and the valve side winding are covered with several layers of paper. Several layers of pressboard and forming angle ring are arranged with certain distance, and finally form a number of insulation oil channels. Thus the whole internal main insulation and end winding insulation model of converter transformer is formed.

3. Simulation of field distribution of internal insulation structure of converter transformer

3.1. Mathematical model for electric field distribution simulation

With the development of computer software technology, there has been more and more simulation software. The Multi-physics simulation software that this paper will use has been widely used in many industry areas and scientific research. It has powerful functions and is widely used by researchers and scholars in various industries [14].

In converter transformer, since the valve windings of converter receives ac and dc composite voltage, this paper selects the current module given in simulation software to conduct transient analysis of field distribution in insulation structure. The transient calculating model of electric field distribution is shown in equations (1) to (3).

$$\nabla J = Q_{j,\mathrm{V}} \tag{1}$$

$$J = \sigma E + \partial D/\partial t + J_e \tag{2}$$

$$E = -\nabla V \tag{3}$$

As shown in these equation, J is current density, Q stands for the charge density while E means the field strength. V is potential while the D stands for the electric displacement. By further combining and simplifying the above equations, the partial differential equation of electric field transient distribution can be obtained. If three-dimensional spatial coordinate system is established, the three-dimensional transient electric field equation can be obtained as shown in equation (4) [15].

$$\left(\gamma + \varepsilon \frac{\partial}{\partial t}\right)\left(\frac{\partial^2 \varphi}{\partial x^2} + \frac{\partial^2 \varphi}{\partial y^2} + \frac{\partial^2 \varphi}{\partial z^2}\right) = 0 \tag{4}$$

As shown equation (4), γ is conductivity and ε is permittivity. In the setting of boundary conditions, as shown in figure 2, AB section of iron yoke, BC section of converter transformer wall and transformer core section are all grounded, and 0 potential is maintained [16]. The valve side winding receives dc voltage and line side winding receives ac voltage.

3.2. Simulation of electric field distribution under applied composite voltage

When converter transformer is operating in actual condition, the network side windings of converter transformer only receives ac voltage, while the valve side windings receives both of ac and dc voltages. To simulate the field-distribution of converter transformer and the appropriate to simplify the calculation, this paper set up a working condition of the converter transformer: in this situation, the line side winding of converter transformer receives 900 kV ac voltage with 50 Hz, and the valve side winding of converter transformer receives 1250 kV dc voltage. The electric field distribution of typical insulation structure of converter transformer is shown in figure 3.

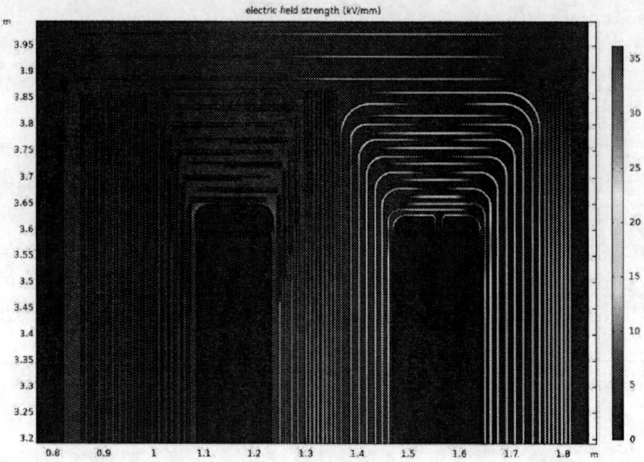

Figure 3. Electric field distribution under applying ac and dc voltages.

From the result, the electric field closely around the valve side winding mainly is composed by the dc component while the ac electric field is mainly accumulated around the line side winding. The maximum value of electric field strength in pressboard is about 35 kV/mm and about 10 kV/mm in oil.

4. Influence of different oil-paper configuration on field-distribution of insulation structure

4.1. The setting of different oil-paper configuration

With the increasing voltage grade of transformer, the configuration of oil-paper insulation structure will also change in the design of transformer. Including the number of pressboard layers, pressboard thickness, oil-paper ratio and so on [17]. The ratio of oil-paper in transformer is complementary to the number and thickness of pressboard. Different thickness and layer of pressboard will result in different oil gap space. That is, different oil-paper configuration change the ratio of oil and paper. Thus forming different oil-paper configuration. Therefore, it is of great engineering significance to study the influence of different oil-paper configuration on field distribution.

In order to simplify the research method, this paper sets 4 different thickness and 4 different numbers of layers of pressboard to form different oil-paper configuration. Then calculate electric field distribution with these different oil-paper insulation configuration under the composite voltages. These 4 different thickness of insulating board refers to the thickness of 3, 4, 5 and 6cm. Number of layers of pressboard that form the Angle ring around the valve side are set as 5, 6, 7, 8. Thus under these settings, there forms different number of oil-paper ratio and oil-paper insulation configuration mode. Next step is to study the influence of different oil-paper configuration on the field distribution under composite voltage.

4.2. The simulation about influence of oil-paper configuration on the electric field distribution

Different oil-paper configurations can be formed by changing the number of layers and thickness of the pressboard. In each oil-paper structure configuration, the field distribution is simulated under the composite voltage. Maximum electric field strength in insulating oil or in insulating paper of each oil-paper configuration are observe and record. Finally the variation rule of field distribution in insulation structure under different oil-paper configuration is obtained.

4.2.1. Change layers of pressboard. Change the number of layers of pressboard. The relationship between the field distribution and different number of layers of pressboard in typical insulation

IOP Conf. Series: Materials Science and Engineering **657** (2019) 012032 doi:10.1088/1757-899X/657/1/012032

structure of converter transformer under composite voltage are shown in figures 4(a), 4(b), 4(c) and 4(d).

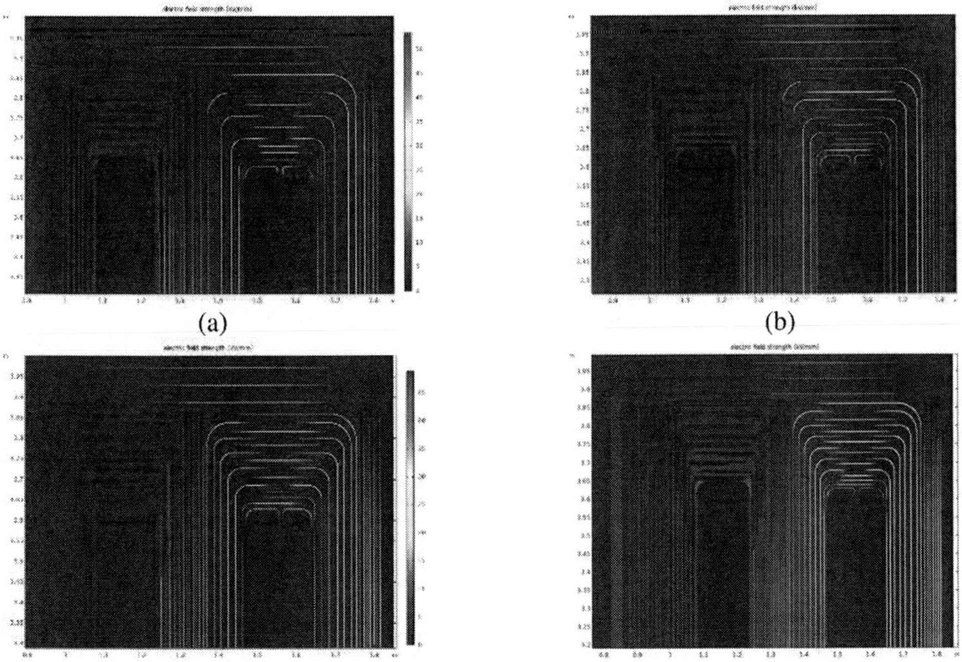

Figure 4. Influence of layers of pressboard on electric field distribution. (a) 5, (b) 6, (c) 7 and (d) 8.

The maximum value of electric field strength in insulating oil and insulating paper are extracted and plotted in figure 5.

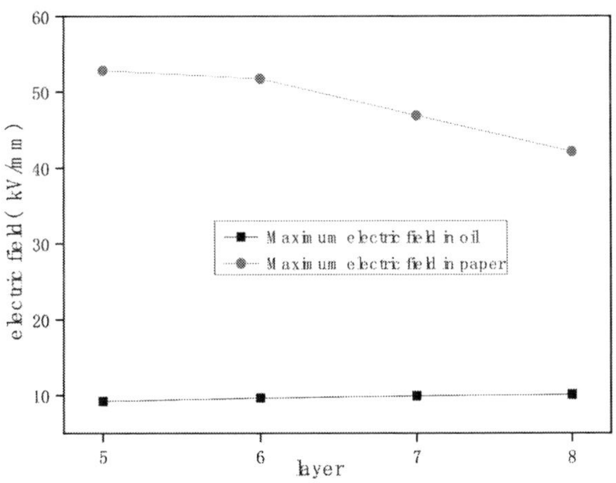

Figure 5. Influence of layers of pressboard on the maximum value of electric field strength.

As the number of layers of pressboard increases, the electric field strength in the pressboard decreases significantly, with a drop range up to 10 kV/mm. At the same time, the electric field strength in oil only rises slightly, with a range of about 0.5 kV/mm, and the rise is basically negligible.

4.2.2. Change the thickness of pressboard. Change the thickness of pressboard. The relationship between the field distribution and the thickness of pressboard in the typical structure of converter transformer under composite voltage is obtained, as shown in figures 6(a), 6(b), 6(c) and 6(d).

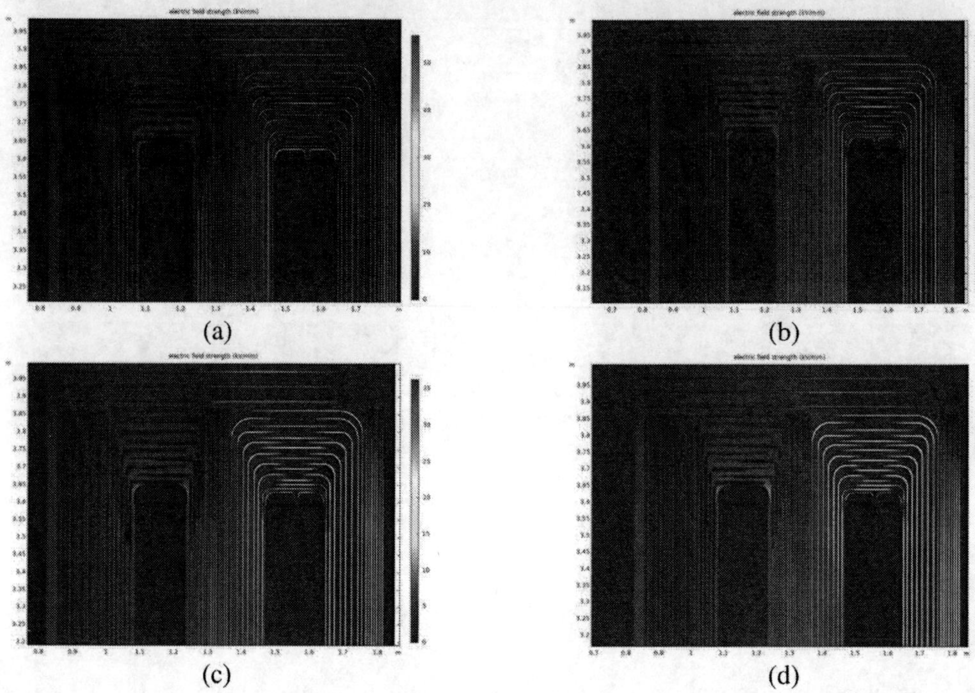

Figure 6. Influence of thickness of pressboard on electric field distribution. (a) 3 cm, (b) 4 cm, (c) 5 cm and (d) 6 cm.

The maximum value of electric field strength in oil and paper are extracted and plotted in figure 7.

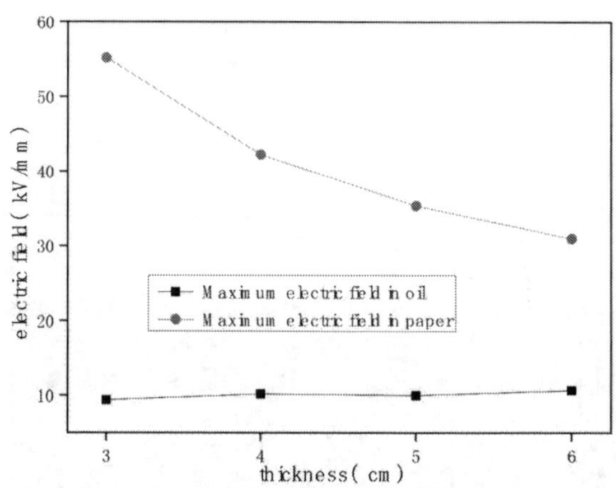

Figure 7. Influence of thickness of pressboard on the maximum value of electric field strength.

It can be seen that when the thickness of pressboard increases, the field strength distributed in pressboard will drop substantially, with a drop range of about 20 kV/mm. Due to the distribution characteristics of ac voltage, the proportion of the pressboard increases, and the electric field strength in oil will rise slightly, but also with a small increase.

4.3. Discussion about the influence of oil-paper configuration on electric field distribution

From the simulation results, it can be seen that when the composite voltage is applied, the field strength in the oil is low that the maximum value of field strength is no more than 12 kV/mm. The minimum electric field strength in pressboard is more than 30 kV/mm, and the maximum value is close to 60 kV/mm. And the region where the maximum filed strength exists on the pressboard is located close to the valve side end winding, which belongs to the dc component. The maximum electric field strength in pressboard is much higher than that in oil. And the influence result of different oil-paper configuration on the field distribution of typical insulation structure of converter transformer under composite voltage is shown in table 1.

Table 1. Variation of electric field strength in oil-paper under different configuration.

	Electric filed in oil	Electric filed in paper
Increase of layers	increase	decrease
Increase of thickness	increase	decrease

From the simulation results, it can be found out that the increase in the number of layers of pressboard can significantly reduce the electric field strength in paper. However, because of the distribution characteristics of ac voltage [18], the increasing of layers of pressboard increases the proportion of paper in the ratio of oil-paper. It will increase the electric field strength in oil slightly. Increasing of thickness of pressboard and the number of layers of pressboard both have a similar mechanism of action. Increasing of thickness will greatly reduce the concentration of voltage in pressboard while the electric strength in oil will rise slightly.

In conclusion, the change in the number of layers and thickness of pressboard reflects the influence of different oil-paper ratio on the characteristic of electric field distribution. When applying the composite voltage, if the proportion of paper increases, the electric field of dc component in the insulation structure decreases while the electric field of ac component increases slightly. When the proportion of paper decreases, the dc electric field strength in insulation structure increases while the ac electric field strength decreases slightly. Therefore, for a large converter transformers, the proportion of dc and ac components in the voltage waveform must be carefully considered in the designing of internal insulation structure to determine the ratio of oil and paper. When setting the oil-paper configuration, the numbers of layer and thickness of pressboard should be increased appropriately. On the one hand, it can decrease the maximum value of field strength in pressboard, on the other hand, the maximum value of field strength in oil should be well controlled to let it only rise only slightly and far above its critical breakdown field strength. Under this circumstance, the electric field distribution will be in a stable and safe situation.

5. Conclusion

- In converter transformer, when line side winding receives the ac voltage while valve side winding receives composite voltage, electric field strength in pressboard is much higher than that in insulating oil, and the maximum electric field strength in oil is located near the line side while the maximum field strength in paper is located close to the valve side.
- Applying the composite voltage, when the number of layers of pressboard increases, the electric field strength in pressboard decreases. While the electric field in insulating oil slightly increases. When the thickness of pressboard increases, the electric field strength of pressboard decreases, while the electric field strength in insulating oil increases slightly.
- In the actual operating situation, when the proportion of dc component dominants, it should be considered to decrease the ratio of oil-paper appropriately in the designing and manufacturing of converter transformer. But when the proportion of ac component dominants, more ratio of

oil-paper should be set appropriately.

Acknowledgments
This research work is supported by National Key R & D Program of China (2016YFB0900804).

References

[1] Li Y *et al* 2008 Transient response characteristics of new HVDC transmission system based on new converter transformer *2008 3rd Int Conf Electric Utility Deregulation and Restructuring and Power Technologies* (Nanjing) pp 1873-7

[2] Li Y, Zhang Z, Rehtanz C, Luo L, Rüberg S and Liu F 2011 Study on steady- and transient-state characteristics of a new HVDC transmission system based on an inductive filtering method *IEEE Trans. Power Electron.* **26** 1976-86

[3] Ji L I *et al* 2006 Study on electric field characteristics at ends of valve-side winding in converter transformer *High Voltage Eng.* **32** 121-4

[4] Wang Y, Wei X, Chen Q, Huang Y and Nie H 2009 Breakdown characteristics of converter transformer insulation under composite AC and DC voltage *2009 IEEE 9th Int Conf the Properties and Applications of Dielectric Materials* (Harbin) pp 634-7

[5] Christen T 2015 HVDC insulation boundary conditions for modeling and simulation *IEEE Trns. Dielectr. Electr. Insul.* **22** 35-44

[6] Zhang H, Gong W, Cao L and Huang P 2013 Simulation of distribution of space charge in ±500kV converter transformer in DC tests *2013 Annual Report Conf Electrical Insulation and Dielectric Phenomena* (Shenzhen) pp 242-5

[7] Hou S, Tian Y, Fu M and Zhuge X 2015 Influential factor analysis and computation on electric field of converter transformer barrier *2015 IEEE 11th Int Conf the Properties and Applications of Dielectric Materials (ICPADM)* (Sydney)

[8] Liu G *et al* 2012 Analysis of transient electric field and charge density of converter transformer under polarity reversal voltage *IEEE Trans. Magn.* **48** 275-8

[9] Wang Q, Li X and Yin Y 2009 Computer simulation and analysis of electric and temperature fields of HVDC cables *2009 IEEE 9th Int Conf the Properties and Applications of Dielectric Materials* (Harbin, China) pp 101-4

[10] Hagiwara M, Pham P V and Akagi H 2008 Calculation of DC magnetic flux deviation in the converter-transformer of a self-commutated BTB system during single-line-to-ground faults *IEEE Trans. Power Electron.* **23** 698-706

[11] Singh A K, Das P and Panda S K 2014 High voltage high frequency resonant DC-DC converter for electric propulsion for micro and nanosatellites *2014 IEEE 36th Int Telecommunications Energy Conference (INTELEC)* (Vancouver, BC) pp 1-5

[12] Shuai Y, Han X, Zhang L, Yang C, Hu X and Wu H 2016 Major insulation design consideration of converter transformer *2016 Int Conf Condition Monitoring and Diagnosis (CMD)* (Xi'an, China) pp 1004-7

[13] Grant D H and McDermid W 2004 Assessment of thermal aging of HVDC converter transformer insulation *Conference Record of the 2004 IEEE International Symposium on Electrical Insulation* (Indianapolis, IN, USA) pp 230-2

[14] Liu H, Hao Y P, Fu M L, Wang D B and Yang L 2017 Study on ventilation of indoor substation main transformer room based on COMSOL software *2017 1st Int Conf Electrical Materials and Power Equipment (ICEMPE)* (Xi'an, China) pp 296-300

[15] Wu H, Li C, Qi B, Zhao X, Lv J and Zhao L 2012 The electric field distribution in oil-paper insulation under combined AC-DC voltage *2012 IEEE Int. Conf. Condition Monitoring and Diagnosis* (Bali, Indonesia) pp 1097-101

[16] Liu G, Li L, Zhao X, Li B and Sun Y 2012 An effective method of solving anisotropic nonlinear periodic electric field in oil-paper insulation under the AC-DC hybrid voltage *2012 Sixth Intl Conf Electromagnetic Field Problems and Applications* (Dalian, China) pp 1-4

[17] Cheng J, Peng Z and Liu P 2012 Research on the electric field distribution of multilayer dielectric under AC-DC composite voltage *2012 IEEE 10th Int Conf the Properties and Applications of Dielectric Materials* (Bangalore, India) pp 1-4
[18] Qi B, Zhao X, Li C and Wu H 2016 Electric field distribution in oil-pressboard insulation under AC-DC combined voltages *IEEE Trns. Dielectr. Electr. Insul.* **23** 1935-41

IOP Conf. Series: Materials Science and Engineering **657** (2019) 012033 doi:10.1088/1757-899X/657/1/012033

Parameter optimization of centrifugal pump impeller

X Z Zhang[1,2], Y W Wang[1] and J S Hu[1]

[1]School of Mechanical & Electrical Engineering, Wuhan Institute of Technology, Wuhan 430205, China

E-mail: 940107865@qq.com

Abstract. After the hydraulic calculation of the centrifugal pump, some design parameters of the impeller can be obtained. The axial projection of the impeller can then be drawn by conventional methods, but in the process, the determination of some parameters in the axial projection map ignores whether it is an optimal solution. In this paper, orthogonal test method is used to determine the optimal solution and optimal combination of some design parameters in the axial projection map, to achieve multi-objective optimization, and to optimize the axial projection of the impeller, improve the flow curve of the impeller and improve Work efficiency.

1. Introduction

At present, most of the centrifugal pump design methods are based on the one-way theory and the similarity theory based on the model-changing algorithm and the velocity coefficient method [1]. The core component of the centrifugal pump is the impeller, and the efficiency of the pump depends mainly on the shape of the impeller. The shape of the impeller is mainly determined by the axial projection of the impeller [2]. When the conventional method is used to draw the axle projection map, the main parameters of the impeller are not optimized and further screened. Of course, for the main design parameters of optimization, Xu and Yuan introduced the application of new optimization techniques such as multi-objective optimization design, genetic algorithm and neural network in the optimization design of impeller, and pointed out the problems that should be paid attention to in the optimization design. [3-4], but genetic algorithms and neural network methods need to build a huge database, which also results in a large workload. For the three-dimensional model of the impeller, Zhang, Huang, Yuan *et al* [5], The performance of small flow centrifugal pump was simulated by computational fluid dynamics method. Taking low specific speed centrifugal pump as an example, the impeller and volute runners of model centrifugal pump were numerically calculated by ANSYS CFX 14.5 software. [6-8]. In addition, Wang, Shi, Jiang *et al* selected four parameters, such as blade exit width, impeller exit diameter, number of blades, and blade exit angle, each factor took 3 levels, based on orthogonal test and numerical calculation. Optimize the impeller [9]. In this paper, the orthogonal projection experiment is used to optimize the axial projection of the impeller, and more parameter factors and horizontal factors are selected. And foreign A. Goto [10-11] combines CFD technology with reverse design to optimize the flow structure within the pump cavity through three-dimensional flow field analysis. Jose, Shulka and others studied the relationship between turbulence model, meshing technology and internal flow characteristics and external characteristics, and obtained the results of the experiment through experimental verification [12]. For the first time, Wahba *et al* introduced the genetic algorithm into the optimization design of centrifugal pump blades and achieved the desired effect [13].

This paper takes the Q5H26 electro-hydraulic circulation pump as the research object, and shows the electro-hydraulic circulation pump with the flow rate of $5m^3 / h$ and the head size of 26 m. Under this research object, the orthogonal design experiment method is used to establish an orthogonal experimental table to select the theoretical optimal parameter matching.

At the same time, after using the orthogonal experimental method to optimize the impeller parameters, the optimized impeller axial projection and the cross-sectional area of the cross section and the flow center line (F-L) can be drawn by MATLAB, and the axial plane drawn by the traditional method. The projected image and its cross-sectional area of the cross-section are compared with the flow-line midline (F-L) diagram to reflect the rationality of using this method.

The advantage of orthogonal test method is that when three or more impeller variation factors are analyzed, there may be interaction between the factors. Orthogonal experiment can eliminate the interference between factors, and can achieve a large number of comprehensive tests with the least number of tests, effectively and quickly select the best combination of impeller parameters.

2. Drawing of traditional axial projection

After the dimensions of the various parts of the impeller are determined, the axial projection of the impeller can be drawn. The shape of the axial projection map should be modified repeatedly to achieve smooth and smooth. General method steps:

① Add a circle in a straight line, respectively as the front and rear cover profiles

② Make a set of inscribed circles that are tangent to the front and rear covers, and make the center line of the inner circle as the center line of the inscribed circle, as shown in figure 1.

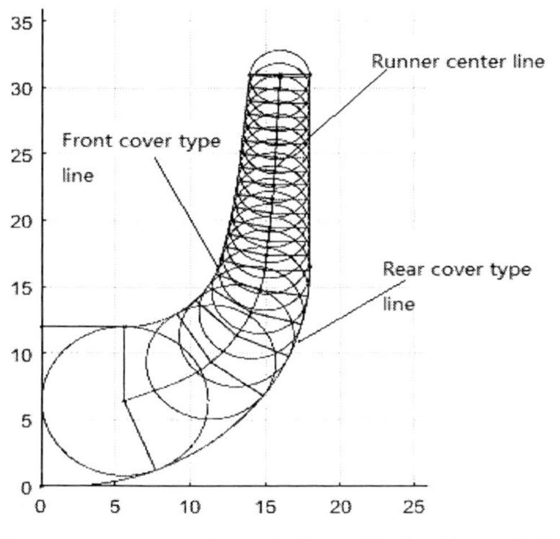

Figure 1. Formation of the middle line of the flow channel.

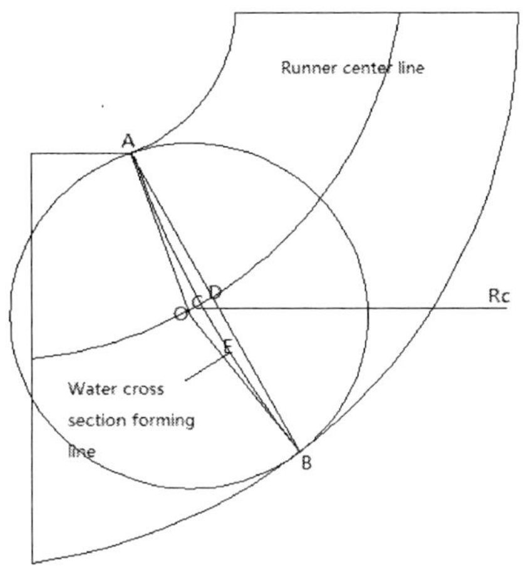

Figure 2. Shaft surface flow through the water section.

③ After drawing the axial projection, check the cross-section change of the flow passage, as shown in figure 2. In the figure, make the inscribed circle of the two flow lines, and the cutting point is A, B. Connect AB to the center of the circle O into a triangle AOB. The high OD of the triangle is equally divided into OE, EC, and CD. A curve AEB passing through point E and perpendicular to the axial flow line is a line forming a cross section of the water. Its length is b. The center of gravity of the line forming the cross section of water is approximately the same as the center of gravity of the triangle AOB (point C), and the radius of the center of gravity is R_c. Since the cross-section of the axial flow through the water must be perpendicular to the flow line of the axial surface, the liquid flows out from the periphery of the impeller, so the cross-section of the flow of the axial flow is a paraboloid formed by the formation of the cross-section of the cross-section as a busbar around the axis. The calculation formula for the area of the cross section of the axial flow can be obtained as follows:

IOP Conf. Series: Materials Science and Engineering **657** (2019) 012033 doi:10.1088/1757-899X/657/1/012033

$$b = \frac{2}{3}(s + p) \tag{1}$$

$$F = 2\pi R_c b \tag{2}$$

Where s is the length of the inscribed circular chord AB and p is the radius of the inscribed circle.

④ By judging the smoothness of the cross-sectional area of the cross-section and the flow-line midline (F-L), the curve should be a straight or smooth line.

The known design parameters of the centrifugal pump research object selected in this paper are shown in table 1 below:

Table 1. Design parameters of Q5H26 electro-hydraulic circulation pump.

Import diameter $D_{in}(mm)$	Outlet diameter $D_{out}(mm)$	flow $Q(m^3/h)$	Head $H(m)$	Rotating speed $n(r/\min)$	Medium density $\rho(Kg/m^3)$
24	22	5	26	7000	1200

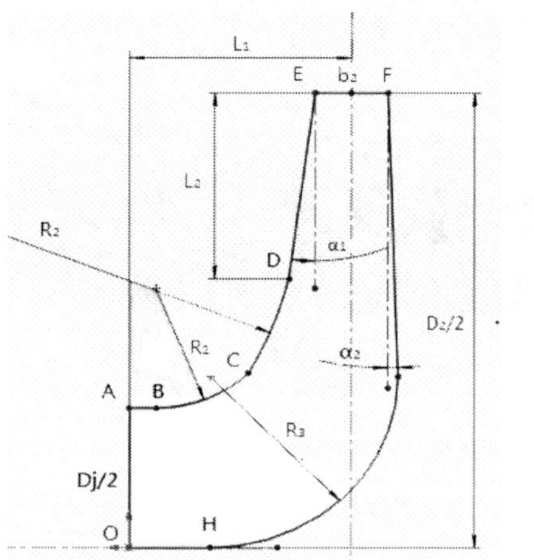

Figure 3. Basic structure of the axial projection of the impeller.

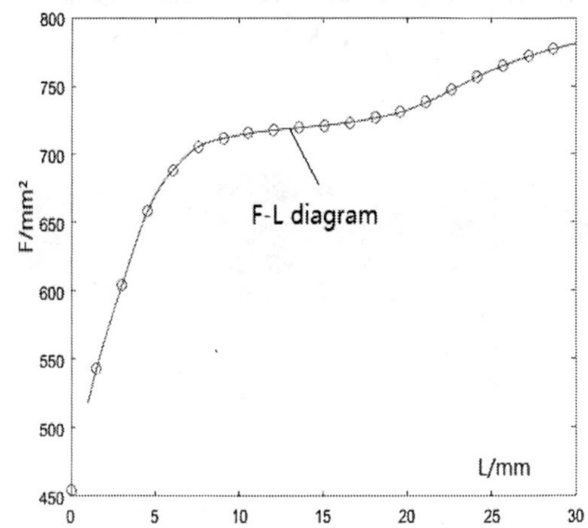

Figure 4. Cross-sectional area of water and flow line (F-L).

Through the design variables of the above Q5H26 electro-hydraulic circulation pump, the theoretical formula and the empirical coefficient can be used to calculate the impeller outlet diameter D_2=62 mm, the impeller outlet width b_2=4 mm, the impeller inlet diameter D_j=24 mm, and the impeller structure size L_1=16 mm. The radius R1 of the impeller front cover profile arc segment BC is R_1=5 mm, the radius of the impeller front cover profile arc segment CD R_2 =40mm, and the impeller front cover profile line exit straight segment DE and the vertical angle $\alpha_1 = 6^\circ$, the radius of the arc of the impeller rear cover profile line GH is R_3=16 mm, the angle between the straight line segment GF of the impeller rear cover profile line and the vertical angle $\alpha_2 = 0°$, the vertical distance from the diameter of the impeller exit to the end point D of the outlet straight line L_2=4 mm. After determining the values of all the above design variables, the structure diagram of the impeller axis projection map can be determined. The basic structure of the axial projection of the impeller is shown in figure 3; At this time, the change of the cross-surface flow path over the water surface should be checked, and the relationship between

IOP Conf. Series: Materials Science and Engineering **657** (2019) 012033 doi:10.1088/1757-899X/657/1/012033

the cross-sectional area of the cross-section and the center line of the flow path (F-L) and the axial projection view at this time can be programmed (figure 4).

As can be seen from the above figure, the image between the area of the cross section and the midline of the channel (F-L) is basically a smooth curve. The results show that the inspection of flow passage section is qualified and meets the basic working conditions of impeller. By the traditional drawing method of axial projection graph, the impeller parameters obtained above are regarded as the final target value, while the optimal combination of these factors is neglected.

The advantage of the orthogonal experimental method is that when the main variation factors of the impeller are analyzed in three or more, and there may be interaction between the factors, the orthogonal experimental method can be used to optimize the impeller parameters and find the impeller parameters. The optimal solution is to optimize the overall axial projection of the impeller.

3. Orthogonal experimental design

3.1. Level factors

The radius R_1 of the arc section BC of the impeller front cover type line (value range: $0.8D_2 \leq R_1 \leq 0.14D_2$), and because $D_2 = 62$ mm, that is, 4.96 mm $\leq R_1 \leq$ 8.86 mm. The radius R_2 of the circular arc segment CD of the front plate of the impeller (value range: $0.6D_2 \leq R_2 \leq 1.2\,D_2$), the same reason, 37.2 mm $\leq R_2 \leq$ 74.4 mm. The range of the angle between the DE and the vertical of the straight section of the impeller front cover profile line is $5° \leq \alpha_1 \leq 8°$. The radius R_3 of the arc segment GH of the impeller rear cover profile line (value range: $L_1 - 0.5b_2 \leq R_3 \leq L_1 + 0.5b_2$), since the values of L_1 and b_2 are known, that is, 14≤R2≤18 mm. The impeller rear cover profile line exit straight section GF and the vertical angle α_2 (value range: -3° $\leq \alpha_2 \leq 5°$). The vertical distance L_2 of the impeller exit diameter to the end point D of the outlet straight line ($0.06D_2 \leq L_2 \leq 0.12D_2$), that is, 3.72 mm $\leq L_2 \leq$ 7.44.

The above-mentioned several parameters are paintings that directly affect the axial surface diagram of the impeller, but in order to make the established orthogonal experimental table not too cumbersome and increase the difficulty of calculation, the impeller rear cover type line exit straight line segment can be made. The angle between GF and the vertical $\alpha_2 = 0$. Thus, the variation factor is determined as five factors of R_1, R_2, α_1, R_3 and L_2.

3.2. factor level

The state in which the factor is located is called the factor level. In short, it is the constraint condition corresponding to the design variable in the nonlinear programming. When selecting the factor level, the factor level is generally 2~5. To ensure optimization the accuracy of the results and the ease of calculation. The number of factors determining the factor is 4.

3.3. Evaluation indicators

The evaluation index is also used to evaluate the merits of the orthogonal experiment results. The evaluation index can be multiple or one. In this experiment, an evaluation index is selected. The evaluation method is as follows:

According to the known design parameters of the centrifugal pump described above, the flow line center length L and the over flow area corresponding to the three points a, c near the inlet and outlet of the impeller and the point b of about half the length of the middle line of the flow path can be used. F, and establish a quadratic curve equation for F-L, which is used as a fitting target curve. Define the curve equation as: $F_{pi} = f(L_i) = al_i^2 + bl_i + c$ .Through MATLAB programming calculation, calculate the length of the centerline of the flow path at the three points a, b, c above, L_a, L_b, L_c and the value of the over-current cross-sectional area of these three points F_a, F_b, F_c, will The F and L values calculated at these three points serve as three coordinate points for establishing the above curve equation.

When L_a=1.5 mm, the overflow area at a can be calculated by MATLAB: $F_a = 530\ mm^2$

When L_b=15 mm, the overflow area at b can be calculated by MATLAB: $F_b = 690\ mm^2$

When L_c=28 mm, the c-flow cross-sectional area can also be calculated as $778mm^2$.

Therefore, the relationship between the three functions of a, b, and c can be established as (1.5, 530), (15, 690) and (28, 778), and the curve equation is solved by the three-point coordinates, so the established the curve equation is as follows:

$$F_{pi} = f(L_i) = -0.2L_i^2 + 15L_i + 509 \tag{3}$$

And because the area of the cross section of the flow of the axial flow of the impeller is calculated as:

$$F_i = 2\pi R_{ci} b_i \tag{4}$$

Then the values solved by the above two companies are obtained to obtain ΔF, and it can be calculated:

$$\Delta F = \sqrt{\sum_{i=1}^{n} (F_i - F_{pi})^2} \quad (i=1,\ 2,\ \cdots,\ n) \tag{5}$$

In the above formula, it is the evaluation index of the orthogonal experimental results of the impeller flow path profile.

When continuously adjusting the above five horizontal change factors of R_1, R_2, α_1, R_3 and L_2, draw a plurality of random point images and curve equation F_{pi} images of the overflow area F_i and L through MATLAB. The degree of fit on the two images can be visually seen in the image. When the value of ΔF is small, the random point image composed between the over-current cross-sectional area F_i and L will be closer to the image of the curve equation. The more obvious the degree of fitting, the more ideal the impeller runner profile will be. In the process of calculation, in order to simplify the calculation process, a lambda value can generally be preset. In this test, λ=65. That is, when the calculated ΔF value is less than λ, the entire calculation can be terminated without the need to establish an orthogonal experiment again. That is, R_1, R_2, α_1, R_3 and L_2 corresponding to the time can be used as the final impeller design parameters.

3.4. Orthogonal experiment table type

Since the above variables have been determined as R_1, R_2, α_1, R_3 and L_2. For the calculation of the simple and reasonable selection of the horizontal factor of 4, the orthogonal experimental table can be selected as $L_{16}\left(4^5\right)$.

4. Orthogonal experimental calculation analysis

4.1. Establishment of the factor level table

In the establishment of the orthogonal experimental table, since the range of the various variation factors and the number of horizontal numbers have been discussed above in this paper, a factor and its factor level table can be preset to perform the initial value of the first orthogonal experiment. The preset factors and their factor levels are shown in table 2 below:

4.2. Orthogonal experiment visual analysis table establishment

By writing a calculation program, calculate the value of the corresponding evaluation index ΔF under each group of different change factors, select the item with the smallest ΔF value, and subdivide the five change factors corresponding to the minimum ΔF, and re-establish The new horizontal factor table and the orthogonal experimental visual analysis table, the first orthogonal experiment visual analysis of the calculation table, as shown in table 3 below:

Table 2. Factors and their factors level table.

Factor	First	The second	The third	The fourth	The fifth
content	R_1	R_2	R3	α_1	L_2
Level	1,2,3,4	1,2,3,4	1,2,3,4	1,2,3,4	1,2,3,4
Numerical value	5,6,7,8	40,50,60,70	14,15,16,17	5,6,7,8	4,5,6,7

Table 3. Orthogonal experiment visual analysis calculation table.

Experiment number	Factor 1 R_1	Factor 2 R_2	Factor 3 R_3	Factor4 α_1	Factor 5 L_2	Evaluation index ΔF
1	5	40	14	5	4	ΔF_1=187.95
2	5	50	15	6	5	ΔF_2=128.13
3	5	60	16	7	6	ΔF_3=126.01
4	5	70	17	8	7	ΔF_4=160.67
5	6	40	15	7	7	ΔF_5=141.99
6	6	50	14	8	6	ΔF_6=229.88
7	6	60	17	5	5	ΔF_7=78.77
8	6	70	16	6	4	ΔF_8=99.95
9	7	40	16	8	5	ΔF_9=287.21
10	7	50	17	7	4	ΔF_{10}=206.17
11	7	60	14	6	7	ΔF_{11}=97.36
12	7	70	15	5	6	ΔF_{12}=96.99
13	8	40	17	6	6	ΔF_{13}=104.18
14	8	50	16	5	7	ΔF_{14}=107.70
15	8	60	15	8	4	ΔF_{15}=276.59
16	8	70	14	7	5	ΔF_{16}=183.99

It can be seen intuitively from the above table, when R_1=6 mm, R_2=60 mm, R_3=17 mm, $\alpha_1=5^{\circ}$, L_2=5 mm, at this time, it is the minimum value of ΔF, Indicated that the five variables at this time are optimal, and ΔF =78.77, at which point this value is greater than λ=65.

Therefore, the five variables will be further refined on the basis of this interval, and a second level factor table will be established to continue the calculation. As shown in table 4 below:

Table 4. Factor and its factor level table.

Factor	First	The second	The third	The fourth	The fifth
content	R_1	R_2	R_3	α_1	L_2
Level	1,2,3,4	1,2,3,4	1,2,3,4	1,2,3,4	1,2,3,4
Numerical value	6.2,6.4,6.6,6.8	61,63,66,69	15.3,15.7,16.5,16.8	5.1,5.3,5.5,5.7	5.4,5.8,6.4,6.8

Through the same method as above, it is found in the second orthogonal experiment that when R_1 =6.6 mm, R_2=61 mm, R_3=16.5 mm, α_1=5.7°, L_2=5.8 mm, This time ΔF takes the minimum value, ie ΔF= 63.54, that is, the value of ΔF at this time is smaller than λ=65. That is, the calculation can be stopped, and the horizontal change factor corresponding to this time is taken as the optimal solution.

5. Case comparison analysis

The design parameters of the pump are known, so the centrifugal pump design calculation formula is used to obtain each structural size in the axial projection structure, and then the axial projection is drawn and the axial flow is checked through the water surface. . This is to use the traditional design calculation method to design the axial projection of the impeller. The design parameters of the axial projection structure calculated by the calculation formula are shown in Table 5 below.

Table 5. The main parameters of the impeller under the traditional axial projection drawing.

D_2 (mm)	b_2 (mm)	D_j (mm)	L_1 (mm)	R_1 (mm)	R_2 (mm)	R_3 (mm)	α_1 (°)	α_2 (°)	L_2 (mm)
62	4	24	16	5	40	16	6	0	4

While the above has been discussed, the optimal solution for the five varying factors of the impeller has been obtained by orthogonal experiment. Also by sorting out the following table 6.

Table 6. Main parameters of the impeller after optimization by orthogonal experiment.

D_2 (mm)	b_2 (mm)	D_j (mm)	L_1 (mm)	R_1 (mm)	R_2 (mm)	R_3 (mm)	α_1(°)	α_2 (°)	L_2 (mm)
62	4	24	16	6.6	61	16.5	5.7	0	4

Thus, when different impeller parameters are obtained under two different methods, an image of the degree of fitting of the plurality of discrete points composed of the overflow area F_i and the length of the center line of the flow path to the curve equation F_{pi} can be drawn separately.

That is, ΔF = 225.98 can be calculated, which is the main parameters of the impeller determined by the conventional method of drawing the axial projection. The image of the F-L can be plotted as figure 5; the optimal solution for the main parameters of the impeller after two orthogonal experiments, at this time ΔF = 63.54. Can draw the image as figure 6.

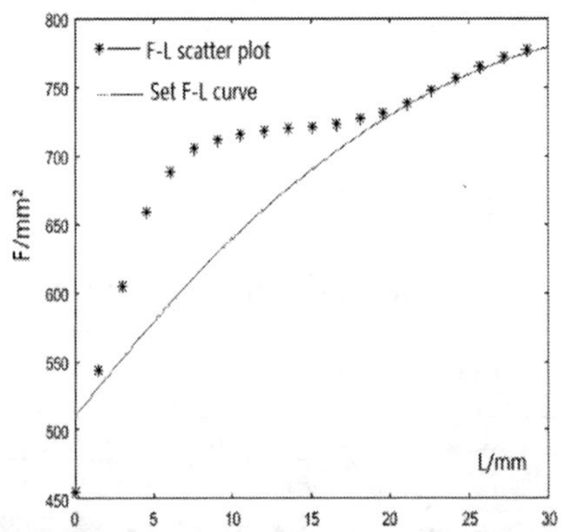

Figure 5. Traditional Axial Projection Drawing Method (F-L).

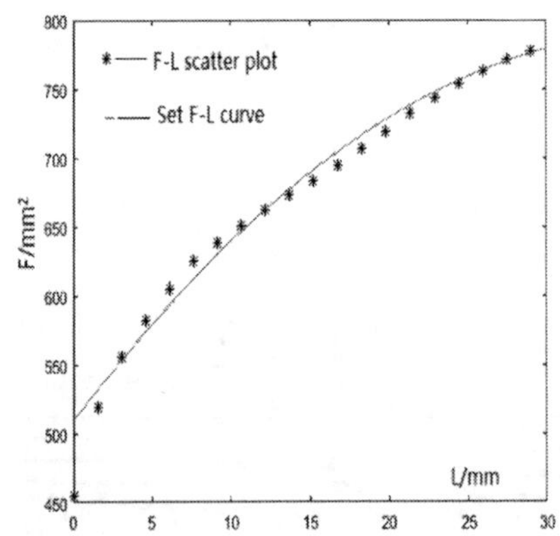

Figure 6. Under orthogonal experiment (F-L).

It can be clearly seen from the comparison between figures 5 and 6 above that figure 6 is better than the figure 5 in the fitting of the plurality of random points composed of the over-current area F and the length of the flow line center line L. Many, and when the main parameters of the impeller determined by the traditional method of drawing the axial plane, ΔF = 225.98, which is much larger than the ΔF calculated by the main parameters of the impeller determined by the orthogonal experimental method. ie ΔF = 225.98>>63.54. It is further proved that through the orthogonal experimental design method, while selecting reasonable level change factors, the purpose of optimizing the change factors is also achieved. Each of the five level change factors has an accuracy of 0.1. At the same time, the preset requirement of $\Delta F = 225.98 < \lambda = 65$ is also satisfied. The corresponding horizontal factor at this time can be taken as the last optimal solution of the impeller design variation parameter. At the same time, under the condition of the optimal solution, the optimized cross-section of the axial flow passage can also be drawn. As shown in figure 7 below:

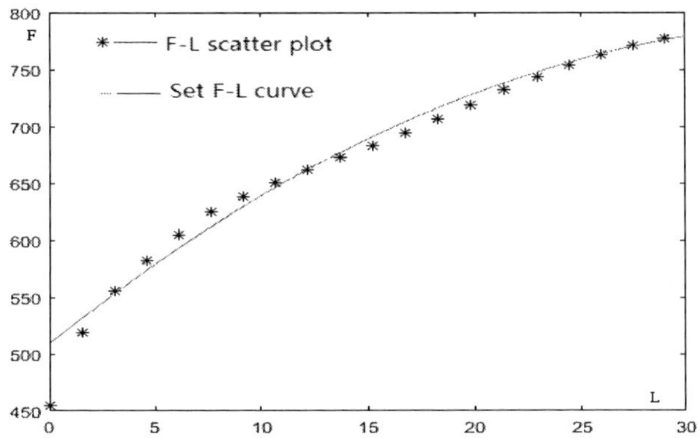

Figure 7. Cross-sectional area of the water and the flow line (F-L).

At this time, the relationship between the cross-sectional area of the water passage and the center line (F-L) of the flow passage can be compared with the cross-sectional area of the flow passage and the center line of the flow passage when the main parameters of the impeller are determined by the conventional axial projection drawing method. F-L) The relationship diagram 4 is contrasted. It can be seen that the F-L curve in figure 7 is smoother than that in figure 4, and there is no large undulation, which can be approximated to a straight line. Once again, it can be reflected that the orthogonal experimental method is used to correct the variation of the cross section of the flow passage by adjusting and optimizing the main parameters of the impeller to optimize the axial projection of the impeller of the cross section.

6. Summary

In this paper, through the orthogonal experiment method, the reasonable combination of the five horizontal factors is selected, and the flow path profile inside the impeller is also optimized, At the same time, when the main parameters of the impeller are determined by the conventional method of drawing the axial plane projection, Can calculate ΔF=225.98 , After the optimal solution of the main parameters of the impeller determined by the orthogonal experimental method, the above calculation has been calculated ΔF=63.54 , which is ΔF=225.98 >> 63.54 .

It can be shown that the flow path profile inside the impeller is also optimized.so that the axial projection shape of the impeller is improved very well, to the greatest extent. The main design parameters of the impeller and the axial projection of the impeller are optimized.

References

[1] Guan X F 1986 *Theory and Design of Pump* (Beijing: Mechanical Industry Press) pp 207-30
[2] Hu J S, Dong C, Xiong B and Zhang X H 2014 Optimization of drawing of the axial surface of centrifugal pump impeller *Chem. Eng. Eqpt.* **2014** 24-9
[3] Tahsin E and Akif K 2003 Prediction of centrifugal slurry pump head reduction: An artificial neural networks approach *J. Fluids Eng. T. ASME* **1** 199-202
[4] Xu W X and Yuan S Q 2006 Progress in optimization design of low specific speed centrifugal pump impeller *Fluid Machinery* **2006** 39-42
[5] Zhang J F, Huang W, Yuan S Q *et al* Grid-independent of low specific speed centrifugal pump based on PIV *J. Mech. Eng.* **2016** 567-572, 583

[6] Shojaeefard M H, Tahani M, Ehghaghi M B *et al* 2012 Numerical study of the effects of some geometric characteristics of a centrifugal pump impeller that pumps a viscous fluid *Comput. Fluids* **2012** 61-70
[7] Choi Y D, Kurokawa J and Matsui J 2006 Performance and internal flow characteristics of a very low specific speed centrifugal pump *J. Fluids Eng.* **2** 341-9
[8] Shen Y, Yuan S, Lu W, Zhang J and Yuan J 2010 The orthogonal test design method for numerical simulation of compound impeller centrifugal pump *T. Chinese Soc. Agric. Mach.* **41** 22-6
[9] Wang W, Shi W D and Jiang X P 2016 Optimization design of multistage centrifugal pump impeller based on orthogonal experiment and CFD *J. Drainage and Irrigation Mach. Eng.* **34** 191-7
[10] Wang W 2017 Optimization design of tangent pump based on orthogonal test (Gansu, China: Lanzhou University of Technology)
[11] Goto A and Zangenneh M 2002 Hydrodynmaie design of pump diffuserusing inverse design method and CFD *ASME J. Fluid Eng.* **124** 319-28
[12] Jose-Gonzalez Eduardo-Bianco Carlos—Santolaria 2002 Numerical simulation of the dynamic effects due to impeller volute interaction in a centrifugal pump *ASME J. Fluids Eng.* **124** 348-55
[13] Wahba W and Tourlidakis A 2001 A genetic algorithm applied to the design of blade profiles for centrifugal pump impeller *CFDC 15th* (June 1-14)

IOP Conf. Series: Materials Science and Engineering **657** (2019) 012034 doi:10.1088/1757-899X/657/1/012034

Construction and analysis of a district heating / cooling network system based on thermal bus

T Xu[1], J Yan[1], H Wang[2], X L Wang[1], T Zhu[2] and H Y Wang[2,3]

[1]Beijing Huajian Power Design and Research Institute, Beijing 102400, China
[2]School of Mechanical Engineering, Tongji University, Shanghai, 200092, China

E-mail: haiyingw@tongji.edu.cn

Abstract. A district energy system based on "thermal bus" was proposed, in which the heating and cooling water transmission processes shared one set of the distribution pipe network. Combined with the gas energy stations, ground source heat pumps, small-scale photovoltaic power stations and rooftop solar collectors and other facilities, it provided the users in the region with heating in winter, cooling in summer and power supply all year round. From the perspective of the bidirectional (supply/return) flow of heat energy along the thermal bus, the mathematical model of the main equipment applied for the source, pipe network, load and storage was established. The optimization model with the lowest system operation cost was proposed considering the fuel cost, the external electricity purchase fee and the satisfaction degree of all users' energy demand. The operation cost and energy consumption of the regional energy system scheme and the traditional distributed energy system scheme are studied by a practical case study. The results showed that the new scheme reduces the operating cost by 21.5% and 16.8% respectively under the typical weekly scenarios of cooling in summer and heating in winter. When no storage equipment was used, the new scheme reduced the amount of purchased electricity in summer and winter by 4.1% and 5.2% respectively.

1. Introduction

The distributed energy systems have been increasingly applied in the regional comprehensive energy supply system by providing combined cold, heat and power supply (CCHP) [1,2]. However, a CCHP system with a single energy station can adapt to several buildings' energy demand changes. When expanded to the district scale, a district distributed energy system (DDES) must be applied. To achieve high power generation, high thermal efficiency and high utilization rate, the DDES must have several interconnected energy stations and an adaptive energy distribution network to make full use of regional renewable energy, waste heat and other resources and to transfer the corresponding cooling, heating and electric energy from all energy stations to geographically scattered different kinds of buildings.

Several authors have investigated the role of transmission networks in optimizing the distributed energy systems [3-10]. Mehleri *et al* [3] proposed a mixed integer optimization model for the design and operation of a distributed energy system. And a district energy supply system was optimized combined with the heating network. Bracco *et al* [3] developed an optimized model for solving the energy supply of towns based on the cogeneration system. In the model, the heating network and the distributed energy were combined and used for the urban buildings, which reduced the annual cost and carbon dioxide emissions of the system. Yang *et al* [5] proposed an optimization model with the minimum cost of regional energy supply as the optimization target. The energy supply networks of heating, cooling and power were integrated in the distributed energy system. And the strong coupling relationship between the supply side and the users' side is partly eliminated by integrating the energy supply networks. Li *et al* [6] studied the economic and environmental optimization for distributed energy resource systems coupled with district energy networks, which also considered the equipment capacity of the energy supply network and the optimization of the network path. Jeddi *et al* [7] proposed a new modified harmony search algorithm to solve the proposed robust dynamic DER planning model. Van der Meer *et al* [8] and Mohamed *et al* [9,10] used the finite element method to model geothermal systems.

In district scale, however, to achieve the decoupling of the distributed energy systems and the energy distribution networks, some difficulties should be overcome: A variety of the power supply, heat and cold sources need to be merged in the energy system with their own special functions and features; Since multiple sources are connected to multiple users, the network may have the ability to provide energy from all sources to all users in an optimized way; It is necessary to consider how to reduce the temperature difference between the working fluid (water) and the environment to make full use of the pipe network and further reduce the loss of heating or cooling network.

Based on the configuration and operation mode of thermal bus, a new district comprehensive energy system is proposed in this paper. A mathematical model for the proposed system was established including the source, network, user's load and energy storage. The combination of gas energy station, low-grade renewable energy and waste heat resources and various types of buildings were considered. The distributed variable-speed water pumps and the distributed heat pumps were used respectively to solve the hydraulic and thermal balance problems caused by the multiple heat/ cold sources. Then, the performance of heating/cooling water distribution through the thermal bus, the utilization of low-grade energy and the reduction of network heat loss were studied. Furthermore, an optimization operation model was proposed to achieve optimal operation cost for the new comprehensive energy system, which was analyzed by a practical case.

2. Construction of the network system for cooling and heating

The configuration principle of the new regional thermal bus system in summer and winter is shown in figures 1(a) and 1(b) respectively. Water is used as the heating or cooling medium in the heating or cooling network, in which cascade utilization of energy should be the basic principle.

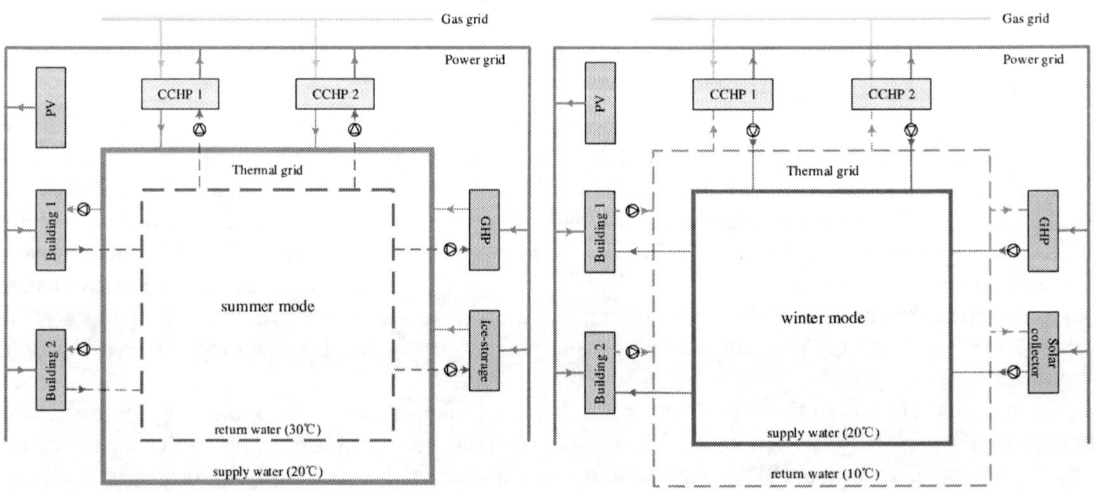

(a) (b)

Figure 1. Configuration of a district energy system based on thermal bus. (a) In summer and (b) In winter.

For a district scale, the proposed "thermal bus" configuration must be based on a ring-shaped topological network (including the supply and return water networks). The equalizing pipes should be arranged at each heat source [11]. The throttle valves and other equipment should be placed on the branch pipelines, which can be used only for local branch flow regulation. The water chillers or the heat pumps should be arranged in each building to further meet the cooling and heating load of each user in the building. However, unlike the traditional air conditioning units, the chillers inside each building in this energy system would exchange heat with the water supplied from the main network. The distributed heat pump can be also used to get the decoupling of the thermal transmission and distribution to achieve the thermal balance of the system. To achieve the hydraulic balance of the system, the distributed variable-speed water pumps should be adopted to get the decoupling of the hydraulic transmission and distribution.

In summer, the main network (as the bold line shown in figure 1(a)) provides cooling water to the buildings. The temperature of the cooling water is close to the temperature of the soil around the pipelines to reduce the heat loss, for instance 20°C. Inside each building, water chillers are used to provide chilled water to the users in each building. After passing through the water chillers, the cooling water will be heated to about 30°C and flow through the return water pipelines back into the energy station, the ground-source heat pump or the ice storage system to be cooled down before returning to the water supply pipelines.

In winter, the main network (as the bold line shown in figure 1(b)) provides low temperature heating water to the buildings, for instance 20°C. Inside each building, the heat pumps are used to provide the space heating and the domestic hot water (higher than 55°C). After passing through the heat pumps, the low temperature heating water will be cooled down to about 10°C and flow through the return water pipelines back into the energy station, the ground source heat pump or the solar collector to be heated before returning to the supply pipelines.

3. Models

3.1. Mathematical models

The model of a district integrated energy system mainly includes the sources, the network and the users. The sources are supposed to generate cold, heat and electricity energy, which might be the combined cooling heating and power station (CCHP), the gas-fired boilers, the renewable energy and the industrial waste heat. The network includes a power network, a gas pipe network and a cooling/heating pipe network. The users mainly refer to the buildings. In addition, the integrated energy system includes a variety of electrical, thermal or cold storage devices.

In the district integrated energy system, the source might be the gas turbines, the fuel cells, the gas boilers, the renewable energy and/or the waste heat resources etc. The thermodynamic model of each source may be complex and characteristic. While, for the energy network, each source can be simplified into a node model with only lumped parameters in order to conveniently fit the sources into the model of the network. When establishing the node model for a source, only its energy balance, hydraulic and thermal characteristics should be considered. The buildings can also be simplified as the node of energy using for the energy network, where the curve of heat load over time of the buildings can be obtained through simulation or measurement.

The district energy network includes the electric power network and the heating/cooling network. Through the power network, the energy stations, small solar photovoltaic and wind power stations can power the buildings in the area. The equivalent circuit of the cable transmission process in the power network is shown in figure 2. According to Zheng *et al*, the power loss in the transmission process can be calculated by the parameters such as voltage and phase angle [12], as shown in equation (1).

$$W_{\text{grid}} = \sum_{k}^{K_{\text{E}}} g_k \left[v_i^2 - v_j^2 - 2 v_i v_j \cos\left(\theta_i - \theta_j\right) \right] \tag{1}$$

Where, q_k is the equivalent conductance of the transmission cable *k,* which connects the power transmission line i and line j. V_i and θ_i are the voltage and phase angle of the power transmission line i respectively. V_j and θ_j are the voltage and phase Angle of the power transmission line j respectively.

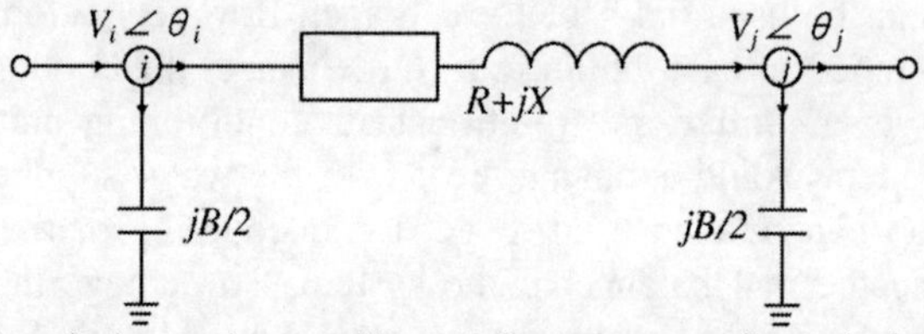

R-equivalent resistance; *X*- equivalent reactance; *B*- equivalent capacitance; *V* - voltage; *θ* - phase Angle; *i, j*- Power transmission lines

Figure 2. Equivalent circuit diagram of cable transmission network.

Modeling of the district heating and/or cooling system includes several kinds of the networks, such as the hot water heat-supply system, the steam heat-supply system and the cold supply system. Generally, the district heating and/or cooling system can be simplified into a network similar to a circuit according to the graph theory, and then based on the kirchhoff's first and second laws, the node flow equilibrium equations and loop equations can be established [13-17].

3.2. Operational optimization model

The objective of operation optimization is to select the scheme with the lowest operation cost among all the feasible energy supply schemes for a certain energy system according to the predicted energy demands of the users. In the model, the proposed objective function considers the operating cost of all devices and meet the energy requirements of all users. The controllable variables in the energy system are the decision variables, such as the energy supply parameters of the power supply and the heat source, the degree of valve opening and the frequency control of the water pump. The constraint condition is the operating parameter ranges of all kinds of the energy supply/use equipment and the relation between the parameters.

The absolute value of the difference between the energy supplied and required is adopted as the indicator of the operation optimization in this paper. It can not only reflect the heat/cold loss of the network indirectly, but also show whether the system reaching its thermal equilibrium. The new indicator, GAP_H as shown in equation (2), is named as the satisfaction degree of the user's heat/cold demand.

$$Gap_H = \gamma \cdot \left[\sum_{h=0}^{168} \sum_{n=1}^{N} \left| H_n - H_n^{\text{need}} \right| \right] \tag{2}$$

Where, H_n is the actual heating/cooling energy supplied to the user n; H_n^{need} is the required heating/cooling loads for the user n; γ is the penalty factor, CNY/GJ. Larger γ means that the optimization model pays more attention to the user's satisfaction degree of heating/cooling; h is the serial number of the hours.

The objective function of operation optimization is defined as shown in equation (3),

$$C_{\text{tot}} = \min \left\{ C_{\text{fuel}} + Gap_H \right\} \tag{3}$$

Where, C_{fuel} is the total cost of the natural gas and electricity consumed by the sources; C_{tot} is the total operation cost.

For the nonlinear and strongly coupled energy system, intelligent optimization algorithm for global search, such as genetic algorithm (GA), should be adopted to do the optimization.

4. Case analysis

4.1. System planning

The planning Area of the Eco-Park is shown in figure 3. The planning area is in the central area, covering an area of about 0.57 km^2. The planned floor-space of the buildings in the central area is 1.83 km^2. According to the landform of the planning area and the location of the main users, three main heat sources are set. S1 is the gas-fired energy station with the gas-turbine generator set, the heat recovery boilers and the flue-gas absorption refrigeration units. S2 is the water source heat pump station with the double-pipe heat exchangers. S3 is the ground source heat pump station with air-cooled cooling tower. As shown in figure 4, in the district energy system, a ring-shaped pipe network and several branch pipes are arranged to connect the heat sources and the main users numbered from C1 to C13. For the power system planning, the gas-fired energy station is used as the main power source to supply electricity to the users in the region. In addition, PV solar panels are installed on the roof of the public venue buildings in the central area, serving as the auxiliary power supply to all buildings in the region through the local grid.

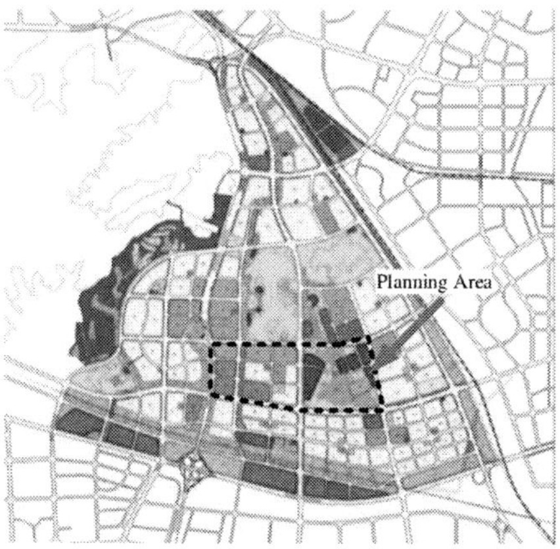

Figure 3. Planning area of the eco-park, Zhenjiang, China.

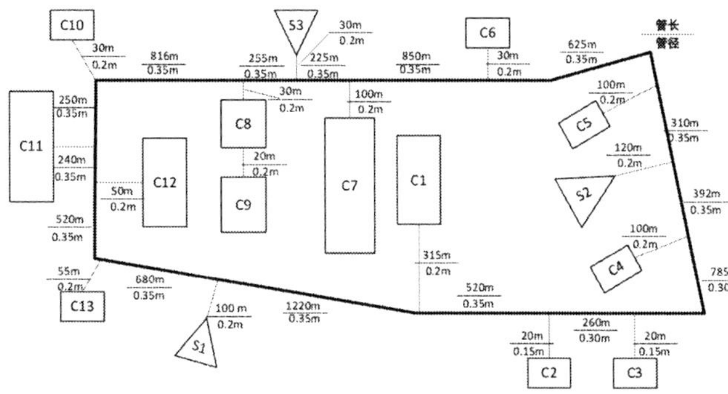

Figure 4. Topology of pipeline network.

The contrast of the integrated energy system based on thermal bus proposed in this paper with the traditional distributed energy system are shown in table 1.

Table 1. Contrast in energy planning of the integrated energy systems with the traditional distributed energy systems.

Season	Item	Traditional distributed energy system	Integrated energy system
Year-round	Power supply mode	Gas-fired energy station/PV	Gas-fired energy station /PV
Summer	Supply/return water temperature	Chilled water (7/12°C)	Cooling water (20/30°C)
	Cooling equipment in the building	Air conditioner/coils	Chiller/air conditioner/coils
	Domestic hot water equipment in the building	Solar collector/electric water heater	Waste heat recovery heat pump/solar collector
	Cold storage	Water from a fire pool	Water in the main pipe network
Winter	Supply/return water temperature	High temperature hot water (70/40°C)	Preheat the hot water (30/20°C)
	Heating equipment in the building	Wall /floor radiator	Heat pump unit/radiator

Domestic hot water equipment in the building	Plate heat exchanger	Heat pump units
Heat storage	None	Water in the main pipe network

4.2. Heating/cooling load

The users are numbered from Cl to C13 including a hotel, a hospital, a school, a railway station, a business area and some office/residential buildings etc. The typical weekly (168 h) heating, cooling and electricity loads of the users were calculated, as shown in figure 5. For the power system planning, the gas-fired energy station is used as the main power source to supply electricity to the users in the region. In addition, PV solar panels are installed on the roof of the public venue buildings in the central area, serving as the auxiliary power supply to all buildings in the region through the local grid.

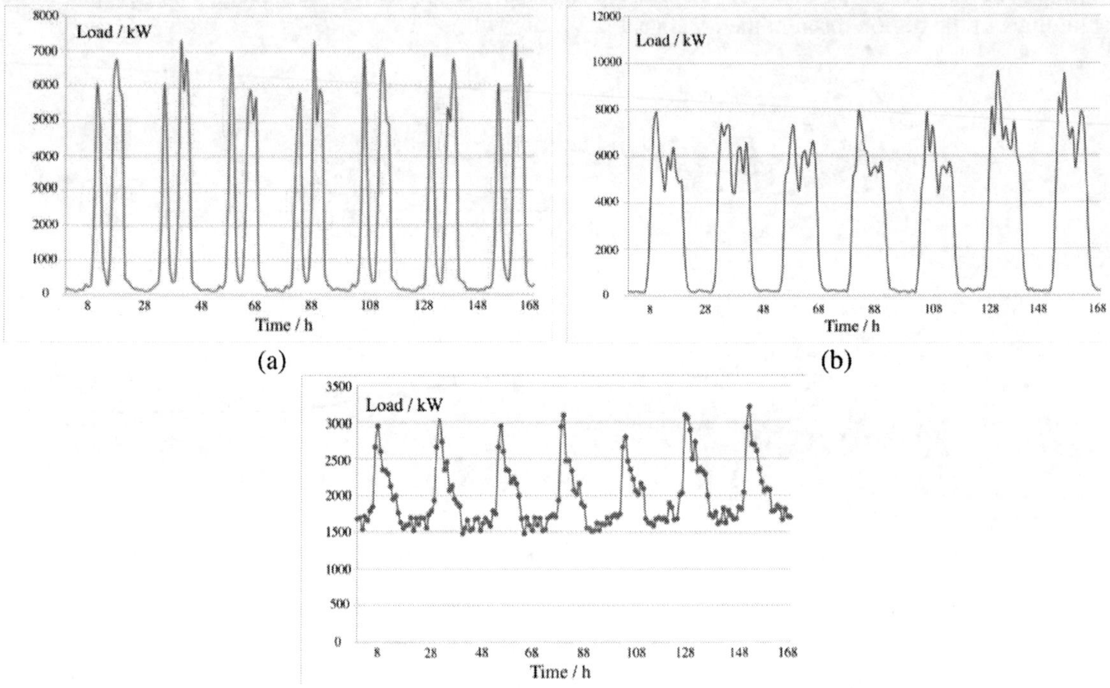

Figure 5. Typical weekly (168 hours) heating, cooling and electricity loads of the user C1. (a) Heating load in winter, (b) Cooling load in summer and (c) Electricity Load.

4.3. Optimization calculation

The gas price was 2.26 CNY/Nm^3. The electricity price was 0.5003 CNY/kWh. The penalty factor was 100.0 (CNY/GJ). Genetic algorithm (GA) was used to solve the model numerically. The solution flow of the model was shown in figure 6.

The difference in the calculation process of the traditional distributed energy planning approach and the approach presented were in step 2 and 4 in figure 6.

4.4. Results and analysis

The optimized parameters of the equipment in each energy station (S1, S2 and S3), the PV battery capacity and the parameters of the user's heating/cooling equipment was calculated after the above solution process. Then according to the calculated optimal rated parameters, the equipment for three source and PV battery were selected as shown in tables 2 and 3. PV cells were set according to the building roof and open space area of the site.

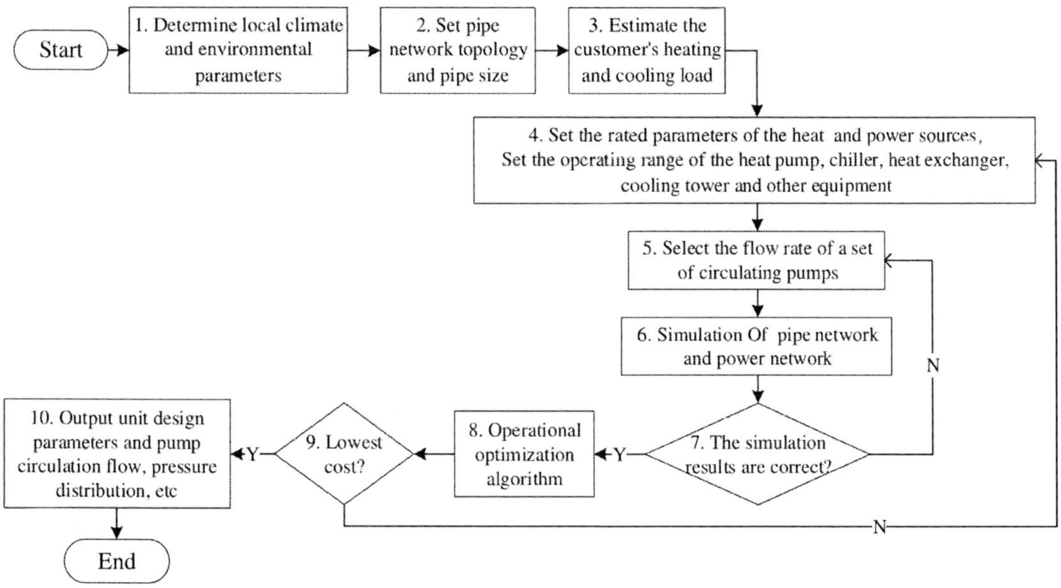

Figure 6. The solution flow of the model of the integrated energy system planning based on thermal bus.

Table 2. Parameters of the energy station equipment and photovoltaic cell.

Station	**Summer**				
	Rated parameter for electricity supply		**Rated parameters for cooling supply**		
	Power (kW)	**Sets of equipment**	**Power (kW)**	**EER**	**Quantity**
S1	635	3	1116	1.3	3
S2	-	-	428.5	3.5	2
S3	-	-	756.5	3.5	2
PV	5.0	1200	-	-	-
Station	**Winter**				
	Rated parameter for electricity supply		Rated parameters for heating supply		
	Power (kW)	Sets of equipment	Power (kW)	EER	Quantity
S1	635	2	1200	0.92	2
S2	-	-	300	4.8	2
S3	-	-	600	4.8	2
PV	5.0	1200	-	-	-

Table 3. Parameters of the user's equipment.

SN	Summer						Winter					
	Water chiller			Heat pump for domestic hot water			Heat pump for heating			Heat pump for domestic hot water		
	Power (kW)	COP	Sets of equipment	Power (kW)	EER	Sets of equipment	Power (kW)	EER	Sets of equipment	Power (kW)	EER	Sets of equipment
C1	1407	3.6	2	-	-	-	556.2	4.8	2	-	-	-
C2	176	3.5	2	-	-	-	85.5	4.5	2	-	-	-
C3	1055	3.6	2	-	-	-	502.2	4.8	2	-	-	-
C4	105.5	3.5	2	-	-	-	85.5	4.2	2	-	-	-
C5	281.4	3.5	2	-	-	-	156.5	4.5	2	-	-	-
C6	422.0	3.5	2	19.6	4.6	2	212.8	4.5	2	22.6	4.3	2
C7	1055	3.6	2	47.3	4.6	2	502.2	4.8	2	43.6	4.3	2
C8	562.7	3.5	2	24.0	4.6	2	303.6	4.5	2	22.6	4.3	2
C9	492.4	3.5	2	21.2	4.6	2	303.6	4.5	2	22.6	4.3	2

C10	703.4	3.5	2	26.4	4.6	2	428.5	4.5	2	22.6	4.3	2
C11	562.7	3.5	2	29.4	4.6	2	428.5	4.5	2	22.6	4.3	2
C12	562.7	3.5	2	29.4	4.6	2	428.5	4.5	2	22.6	4.3	2
C13	3517	3.8	2	99.6	4.8	2	1506	4.8	2	86.5	4.5	2

Comparing the energy system proposed in this paper with the traditional distributed energy system, the total operating cost in a typical week was reduced by 21.5% in summer and 16.8% in winter. If using no storage equipment, the amount of purchased electricity of the new scheme was reduced by 4.1% in summer and 5.2% in winter. The purchased electricity required for the energy system proposed in this paper and the traditional distributed energy system, as shown in figure 7.

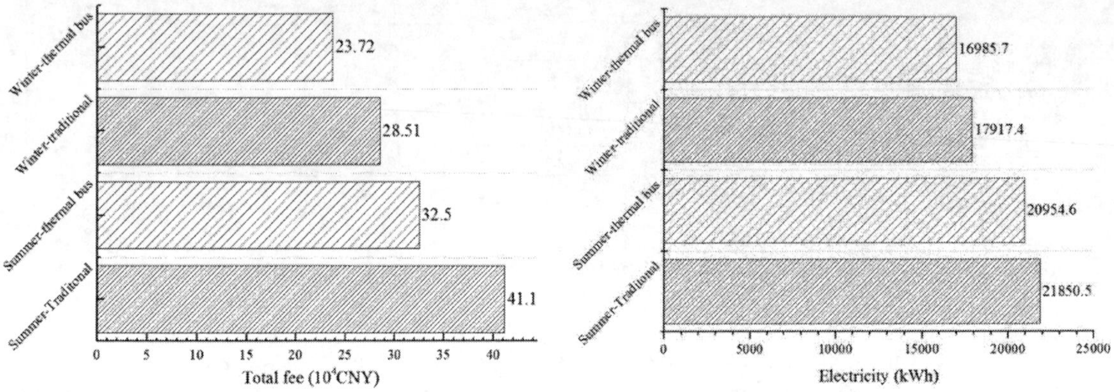

(a) (b)

Figure 7. Typical weekly costs for the thermal bus system and the traditional distributed energy system. (a) weekly operating costs and (b) weekly electricity purchased.

5. Conclusions

An energy planning method based on thermal bus mode was proposed in this paper. In the district integrated energy system, different types and qualities of energy can be utilized in a more flexible way. The system can be used to improve the overall system energy efficiency, reduce investment and operating costs and fully meet the user's energy needs. The proposed pipe network can be used to distribute heating water in winter and cooling water in summer efficiently. The supply pipelines for heating are also the return pipelines for cooling. The low temperature of water in the supply/return pipes is beneficial for reducing the heat loss of the network and easily connecting the ground/water source heat pumps to the heating network. The operating costs can be saved and the electricity purchased from the grid can be reduced when using the proposed system.

References

[1] Cho H J, Smith A D and Mago P 2014 Combined cooling, heating and power: A review of performance improvement and optimization *Appl. Energ.* **136** 168-85

[2] Liu M X, Shi Y and Fang F 2014 Combined cooling, heating and power systems: A survey *Renew Sust. Energ. Rev.* **35** 1-22

[3] Mehleri E D, Sarimveis H, Markatos N C and Papageorgiou L G 2012 A mathematical programming approach for optimal design of distributed energy systems at the neighbourhood level *Energy* **44** 96-104

[4] Bracco S, Dentici G and Siri S 2013 Economic and environmental optimization model for the design and the operation of a combined heat and power distributed generation system in an urban area *Energy* **55** 1014-24

[5] Yang Y, Zhang S and Xiao Y 2015 Optimal design of distributed energy resource systems coupled with energy distribution networks *Energy* **85** 433-48

[6] Li L X, Mu H L, Li N and Li M 2016 Economic and environmental optimization for distributed energy resource systems coupled with district energy networks *Energy* **109** 947-60

[7] Jeddi B, Vahidinasab V, Ramezanpour P, Aghaei J, Ahafie-khah M and Catalao J P S 2019 Robust optimization framework for dynamic distributed energy resources planning in distribution networks *Int. J. Elec. Power* **110** 419-33

[8] Van der Meer F P, Al-Khoury R and Sluys L J 2009 Time-dependent shape functions for modeling highly transient geothermal systems *Int. J. Numer. Meth. Eng.* **77** 240-60

[9] Mohamed M S, Seaid M, Trevelyan J and Laghrouche O 2013 Time-independent hybrid enrichment for finite element solution of transient conduction–radiation in diffusive grey media *J. Comput. Phys.* **251** 81-101

[10] Diwan G C, Mohamed M S, Seaid M, Trevelyan J and Laghrouche O 2015 Mixed enrichment for the finite element method in heterogeneous media *Int. J. Numer. Meth. Eng.* **101** 54-78

[11] Meng H, Wang H and Long W D 2017 District-energy planning method based on thermal bus system in hot summer and cold winter areas *J. Refrig.* **38** 50-8 (In Chinese)

[12] Wang H, Wang H Y and Zhou H Z 2012 Analysis of multi-sources looped-pipe network based on object-oriented methodology *J. Zhejiang Univ. (Eng. Sci.)* **46** 1900-9 (in Chinese)

[13] Wang H, Wang H Y and Zhu T 2017 A novel model for steam transportation considering drainage loss in pipeline networks *Appl. Energ.* **188** 178-89

[14] Oppelt T, Urbaneck T, Gross U and Platzer B 2016 Dynamic thermo-hydraulic model of district cooling networks *Appl. Therm. Eng.* **102** 336-45

[15] Shaikh M M, Massan S R and Wagan A I 2015 A new explicit approximation to Colebrook's friction factor in rough pipes under highly turbulent cases *Int. J. Heat Mass. Tran.* **88** 538-43

[16] Lopes A M G 2004 Implementation of the hardy-cross method for the solution of piping networks *Comput. Appl. Eng. Educ.* **12** 117-25

[17] Duquette J, Rowe A and Wild P 2016 Thermal performance of a steady state physical pipe model for simulating district heating grids with variable flow *Appl. Energ.* **178** 383-93

IOP Conf. Series: Materials Science and Engineering **657** (2019) 012035 doi:10.1088/1757-899X/657/1/012035

Numerical simulation of gravity instabilities in gas flows by use of the quasi gas dynamic equation system

E Shilnikov[1,2,3] and T Elizarova[1]

[1]Keldysh Institute of Applied Mathematics of RAS, 4, Miusskaya sq., 125047, Moscow, Russia
[2]Moscow Automobile and Road Construction Technical University, 64, Leningradsky av., 125319, Moscow, Russia

[3]E-mail: shilnikov@imamod.ru

Abstract. Paper presents two models for gravity instability simulations using compressible gas formulations, and especially, using regularised gas dynamic equations, or, in other words, the quasi-gas dynamic (QGD) equation system. The first part of the paper describes the layer flow for a single-component gas and the second one is one-fluid approximation for a gas mixture. Here a density equation for a mixture is split in two equations for each gas species. Momentum and energy equations are presented for a mixture with unified velocity and temperature and aggregated density and pressure. Examples of numerical calculations of Rayleigh-Taylor instability for different Atwood numbers are shown. The obtained results correspond to the well-known behaviour features of the Rayleigh-Taylor instability.

1. Introduction

The problems related with gravity, or Rayleigh-Taylor instabilities in layer gas flows arise in aerodynamic, engineering and environmental applications. However, for the numerical modelling incompressible flow models are commonly applied, beginning from the very first papers e.g., [1]. Detailed description of Rayleigh-Taylor instability simulation in compressible flow may be found in [2]. The review [3] contains a significant amount of information about both Rayleigh–Taylor and Richtmyer–Meshkov instabilities, including analytical investigations, experimental as well as computational results.

In this paper we show two models for gravity instability simulations using compressible gas formulations, and especially, using regularized gas dynamic equations, or, in other words, the quasi-gas dynamic (QGD) equation system. One of the advantages of this system is the nice possibilities for numerical modelling of strongly non stationary flows. The QGD system can be interpreted as the Navier-Stokes equation system averaged or smoothed over some small time and space intervals. This smoothing leads to the appearance of the strongly non-linear additional dissipation terms proportional to the small parameter τ which has a dimension of time. The first variants of the system have been published more than three decades ago by Chetverushkin, Elizarova and Sheretov. Over these years the QGD system was used to construct numerical algorithms for simulations of wide variety of gas dynamic problems. The obtained results may be found in, e.g., [4, 5] and in the recent papers, such as [6]. A number of theoretical results such as the close relations between QGD and Navier-Stokes equation systems and exact solutions of QGD system have been established in [5]. Theoretical investigations presented in [7] permit to establish Petrovskii parabolicity and linearized stability of equilibrium solutions of the QGD equation system. In the present paper the QGD system is examined in simulation of the Relay-Taylor instabilities in single gas and in a mixture of two different non-reacting gases.

2. Single gas convection in QGD formulation

For gravity flows of idcal polytropic gas QGD equations in Cartesian coordinates have the form

$$\frac{\partial \rho}{\partial t} + \frac{\partial}{\partial x_i} J_i = 0, \tag{1}$$

$$\frac{\partial}{\partial t} \rho u_i + \frac{\partial}{\partial x_j} J_j u_i + \frac{\partial}{\partial x_i} p = \rho_* F_i + \frac{\partial}{\partial x_j} \Pi_{ji}, \tag{2}$$

$$\frac{\partial}{\partial t} E + \frac{\partial}{\partial x_j} J_j \left(\frac{E+p}{\rho} \right) + \frac{\partial}{\partial x_j} q_j = J_j F_j + \frac{\partial}{\partial x_j} \Pi_{ji} u_i. \tag{3}$$

Here $E = \rho\left(\mathbf{u}^2/2 + \varepsilon\right)$ is the total energy per unit volume, $F_i = -g\delta^{i3}$ is the gravity force with g is the gravity acceleration, δ^{ij} is the Kronecker delta. Here we use the notations:

$$J_i = \rho(u_i - w_i), \quad w_i = \frac{\tau}{\rho}\left(\frac{\partial}{\partial x_j} \rho u_j u_i + \frac{\partial}{\partial x_i} p - \rho F_i \right), \quad \rho_* = \rho - \tau \frac{\partial}{\partial x_i} \rho u_i. \tag{4}$$

The viscous stress tensor $\mathbf{\Pi}$ and the vector of heat flux $\mathbf{q}$ are given by

$$\Pi_{ij} = \Pi_{ij}^{NS} + \tau \rho u_i \left(u_k \frac{\partial}{\partial x_k} u_j + \frac{1}{\rho} \frac{\partial}{\partial x_j} p - F_i \right) + \tau \delta_{ij} \left(u_k \frac{\partial}{\partial x_k} p + \gamma p \frac{\partial}{\partial x_k} u_k \right). \tag{5}$$

$$q_i = q_i^{NS} - \tau \rho u_i u_j \left(\frac{\partial}{\partial x_j} \varepsilon + p \frac{\partial}{\partial x_j} \frac{1}{\rho} \right). \tag{6}$$

Navier-Stokes expressions for the viscous stress tensor $\mathbf{\Pi}^{NS}$ and the heat flux vector $\mathbf{q}^{NS}$ are

$$\Pi_{ij}^{NS} = \mu \left(\frac{\partial}{\partial x_i} u_j + \frac{\partial}{\partial x_j} u_i - \frac{2}{3} \delta^{ij} \frac{\partial}{\partial x_k} u_k \right), \quad q_i^{NS} = -\kappa \frac{\partial}{\partial x_i} T. \tag{7}$$

Here the indices i, j, k, taking values 1,2 and 3, denote Cartesian coordinates. The closing relations and the heat conductivity coefficient are given by

$$p = \rho R T, \quad \varepsilon = \frac{RT}{\gamma - 1}, \quad \kappa = \mu \frac{\gamma R}{(\gamma - 1)\Pr}, \tag{8}$$

where Pr is the Prandtl number, R is a gas constant and $\gamma = c_p/c_v$ is the adiabatic index. The gas viscosity coefficient μ, appearing in expressions for the viscous stress tensor and the heat flux, is defined by a temperature power-law [8]

$$\mu = \mu_0 \left(T / T_0 \right)^{\omega}. \tag{9}$$

Here ω value is related to the intermolecular potential for gas molecules. The bulk viscosity is supposed to be zero. The relaxation parameter τ appearing in formulae (4) – (6) is defined as

$$\tau = \alpha \cdot h / c, \quad c = \sqrt{\gamma RT}, \tag{10}$$

where c is the sound speed and h is the grid resolution, estimated locally. The value of τ defines the time required for a perturbation to travel across a grid cell and the coefficient α, which is adjusted experimentally, defines the level of subgrid dissipation. According to numerical practice its value is restricted in the interval 0.1-1.0. The coefficient α is the only tuning parameter of the QGD computational model.

3. Gas mixture convection. One-fluid model

In one-fluid model for a binary gas mixture simulation (gas a and gas b) we suppose that velocities and temperatures of gas components are in equilibrium with each other and are equal, so $\mathbf{u}^a = \mathbf{u}^b = \mathbf{u}$ and $T^a = T^b = T$. One-fluid approximation for a binary gas mixture in QGD formulation was proposed and tested for rarefied flows in e. g., [4, 9]. However this approach loses the applicability for dense gas simulations. One of the reasons is the large density and pressure gradients of the components ρ^{a} and ρ^{b} near the interface of domains, occupied by separated gas species. It means that the density of a gas component may change significantly, while the aggregated density $\rho^{\mathrm{a}} + \rho^{\mathrm{b}}$ conserves near a constant value.

Here we propose a new QGD formulation for a dense binary gas mixture. In contrast to the equation (1) the density equation is split into two equations for each species (a and b)

$$\frac{\partial \rho^{\mathrm{a}}}{\partial t} + \frac{\partial}{\partial x_i} J_i^{\mathrm{a}} = 0, \quad \frac{\partial \rho^{\mathrm{b}}}{\partial t} + \frac{\partial}{\partial x_i} J_i^{\mathrm{b}} = 0, \tag{11}$$

where mass flux for each gas component is presented in the form

$$J_i^{\alpha} = \rho^{\alpha} (u_i - w_i), \quad \alpha = \mathrm{a}, \mathrm{b}. \tag{12}$$

The additional velocity components w_i (4) and momentum and energy equations for gas mixture are similar to equations (2) - (6), but include gas mixture parameters with aggregate values of density, pressure and mass fluxes. Note, that here $\mathbf{J} = \mathbf{J}^{\mathrm{a}} + \mathbf{J}^{\mathrm{b}}$. In mixture gas equations gas parameters γ and R depend on the mass fractions of the species and are calculated as averaged values by:

$$\gamma = \frac{\gamma^{\mathrm{a}} \rho^{\mathrm{a}} + \gamma^{\mathrm{b}} \rho^{\mathrm{b}}}{\rho}, \quad R = \frac{R^{\mathrm{a}} \rho^{\mathrm{a}} + R^{\mathrm{b}} \rho^{\mathrm{b}}}{\rho}, \quad \rho = \rho^{\mathrm{a}} + \rho^{\mathrm{b}}. \tag{13}$$

Viscosity and heat conductivity coefficients μ and k are calculated for the mixture as the sum of corresponding species values.

4. Numerical algorithm and computational results

For numerical implementation of QGD system we use the explicit-in-time finite-volume scheme with an approximation of all space derivatives by second-order central differences on a rectangular grid.

The calculation process is organized as follows: we calculate densities using (11). Than we found the mixture parameters and calculate the common velocity components u_i and total energy E. Using first two equations (8) and supposing that the temperatures of gases are equal $T^{\mathrm{a}} = T^{\mathrm{b}} = T$ we found the temperature and the mixture pressure p.

The first test is a flow simulation of the behavior of two cold and hot single gas layers in a cavity. Such flows strongly depend on the value of Atwood number, that characterizes a density difference ρ_1 and ρ_2 in the layers

$$\mathrm{A} = (\rho_1 - \rho_2)/(\rho_1 + \rho_2). \tag{14}$$

We consider a following problem. A rectangular container 14cm × 14cm × 20cm is filled with argon under atmospheric pressure p = 101325 Pa. The gas in the lower half of the container is hot with the density ρ_1 = 1.62 kg/m^3 and the upper half is filled with cold gas with the density ρ_2. Two numerical experiments were held for different values of ρ_2. This density was taken equal to ρ_2 = 3 kg/m^3, which corresponds to the Atwood number A= 0.3 (variant 1) and ρ_2 = 5 kg/m^3 with A = 0.51 (variant 2). Gas viscosity was taken equal to $\mu_0 = 2.26 \cdot 10^{-5}$ N$\cdot$s/m^2 for T_0 = 273 K, viscosity index ω = 0.81, γ = 5/3, molecular mass 39.9166 and Prandtl number Pr = 0.67 [8]. At the beginning of the simulations the velocities in the whole computational region are equal to zero, except the boundary surface between the hot and cold gases. In this surface the vertical velocity has a sinusoidal profile with the amplitude of 0.2 m/s. The no slip boundary conditions are set at all borders. Figure 1 shows the computational domain with the density distributions along the walls (left) and the temperature distribution in the middle section (right) for variant 1. Figure 2 demonstrates the density lines together with the stream function lines in the middle section of the domain for variant 1 (left) and 2 (right).

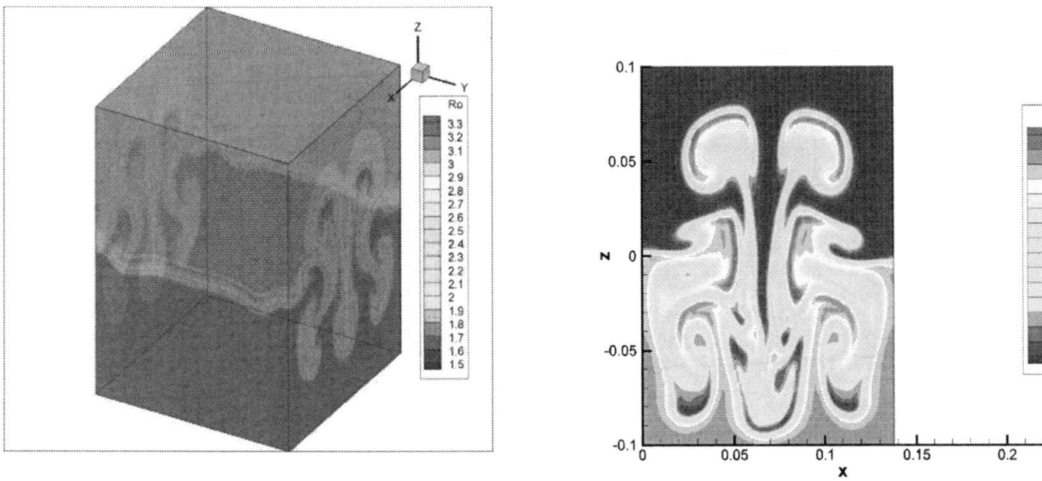

Figure 1. Density distributions along the walls (left picture) and the temperature distribution in the middle section (right picture) for variant 1, A=0.3, t = 0.7 s.

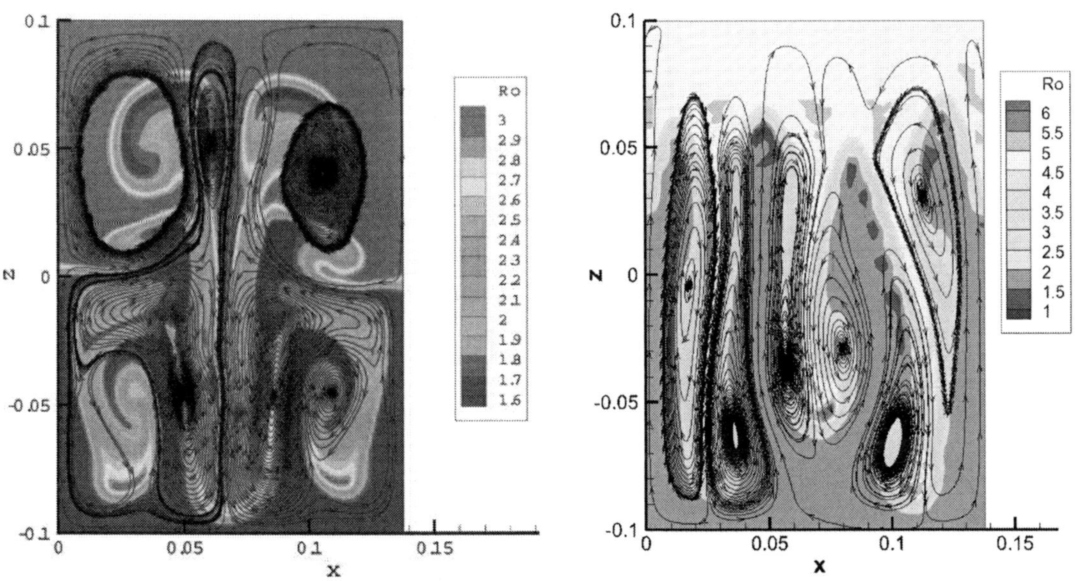

Figure 2. Density distributions together with the streamfunction lines in the middle section of the domain for variant 1, A=0.3 (left picture) and variant 2, A=0.51 (right picture), t = 0.7 s.

It is seen that the increasing of Atwood number leads in changing the flow picture with "mushroom" form to the picture with "knife" form of falling down streams of heavy gas.

Both calculations were carried out until the time moment t_{fin} = 0.7 s with the coefficient in regularization parameter in (10) $\alpha = 0.2$ and the Courant number $\beta = 0.4$. It corresponds to a time step $\Delta t = 2 \cdot 10^{-6}$s. The calculation grid contains 80×80×120 cells.

The second test problem is similar to previous one but two gases in the container are different. For the sake of simplicity we solved this problem in 2D formulation. A rectangular container 12.5 cm × 50 cm is considered. Its upper half is filled with heavy gas sulfur hexafluoride (SF_6, $\gamma = 1.094$, Pr = 0.736, molecular mass 146.0554). The gas in the lower half is light. We considered two variants of the light gas: argon (Ar, it's parameters are mentioned above) and helium (He, $\gamma = 5/3$, Pr = 0.67, molecular mass 4.003). The corresponding Atwood numbers are equal to A = 0.57 for Ar – SF_6 and A = 0.95 for He – SF_6. The initial state of the gases at time $t = 0$ is determined by hydrostatic equilibrium assuming adiabaticity and zero gas velocities. The normal atmospheric pressure and temperature (as in the first problem) are set at the interface surface. In this surface the vertical velocity has a sinusoidal profile with larger amplitude of 10 m/s. The boundary conditions are: no slip conditions are set at the bottom, slip conditions at the vertical walls and constant pressure and zero vertical derivatives for the rest variables are imposed at the upper boundary. Such conditions allow perturbations of the solution to leave the calculation region through the upper boundary.

Figure 3 demonstrates the distributions of the flow parameters at the advanced stage of flow development in the Ar – SF_6 mixture. The mushroom shape of the emerging bubbles of light gas and the downward bubbles of heavy gas is clearly visible, while the edges of the bubbles twist inward and become thinner over time. Acoustic disturbances forming in the gas-dynamic fields of light and heavy gases in the process of their evolution are also clearly visible.

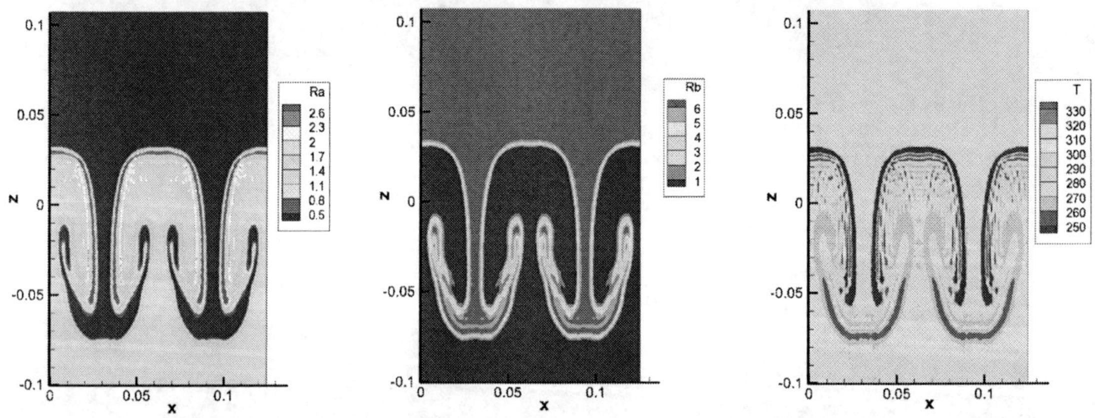

Figure 3. Distributions of the light gas density ρ^a (left picture), heavy gas density ρ^b (center picture) and the temperature T (right picture) for the Ar – SF_6 mixture at the time moment $t = 0.2$ s.

An increase in temperature is observed on the temperature distribution, near the boundaries of the bubbles, which is associated with the light gas heating during its compression in the structures formed. Over time, the picture of gravitational mixing of gases becomes more and more complex, and more and more small details and vortex formations appear in it. Ultimately, these new details become comparable in size with the spatial grid, which leads to the termination of the calculations.

The results of calculations for a mixture of He-SF_6 are presented in figures 4 and 5. Comparison with the previous calculation shows that, in general, the overall picture of the instability development persists. However, with an increase in Atwood number all processes proceed with a significantly greater speeds, the formation of bubbles-fungi occurs more intensely, the magnitude of acoustic disturbances increases. One can see a sharp lowering of heavy gas into the zone of very light helium gas, which forms a thin stream of heavy gas. At the ends of SF_6 jets mushroom-like formations occur, while the jets themselves tend to split and form drops. In addition, a rapid heating of helium due to its compression is observed under the action of heavy gas, which plays the role of a piston.

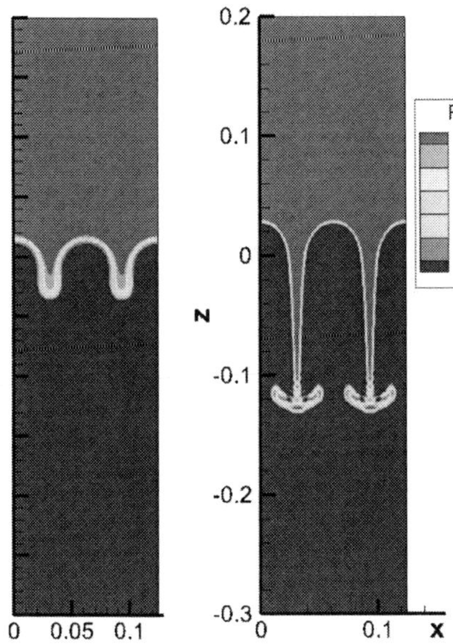

Figure 4. Density distributions of SF_6 at time moments $t = 0.05$ s (left) and $t = 0.15$ s (right).

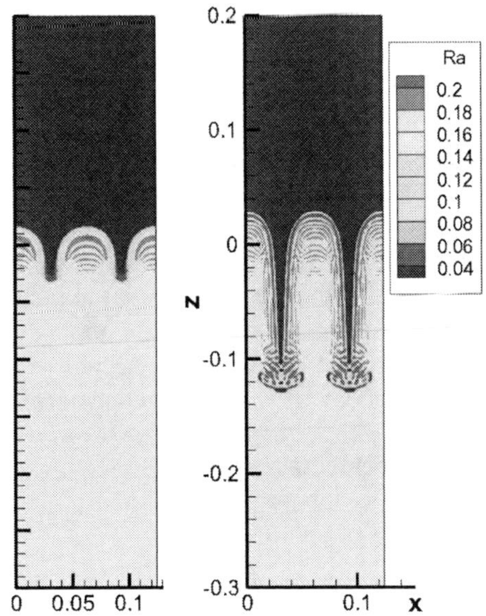

Figure 5. Density distributions of He at time moments $t = 0.05$ s (left) and $t = 0.15$ s (right).

The calculations were held on the uniform spatial grid containing 125×400 cells. To analyze the dependence of the numerical solution on the grid size the calculations were held for the Ar – SF_6 mixture on the grids 60×200 and 250×800 nodes. Figure 6 shows the results obtained on these three spatial grids for the same time moment $t = 0.12$ s.

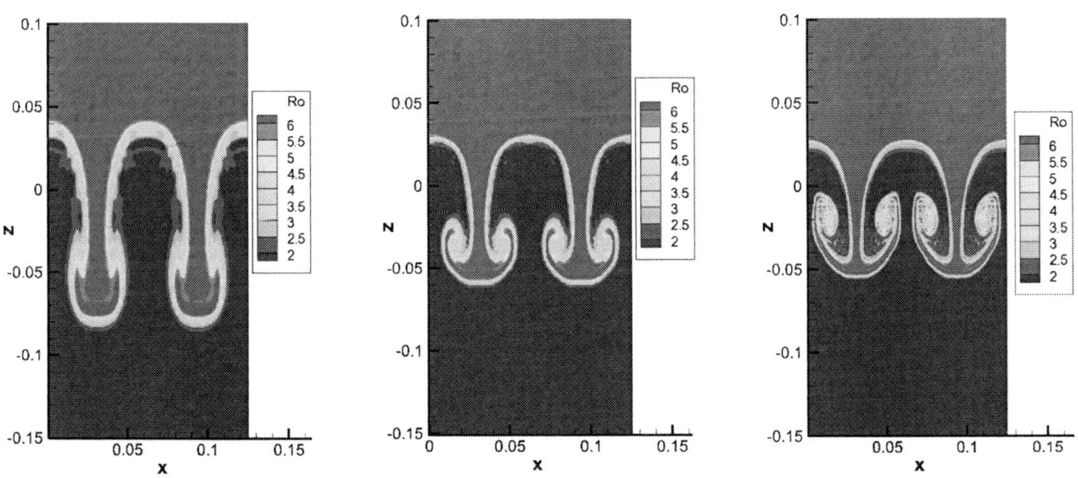

Figure 6. Aggregate density distributions obtained on different spatial grids: 60×200 (left picture), 125×400 (center picture) and 250×800 (right picture), At=0.57.

These results clearly show that even on rough mesh the overall picture flow is obtained, with the exception of a specific curving shape of the boundaries between the gases, since this boundary contains only one or several grid cells. When the mesh is refined, additional finer effects are revealed.

IOP Conf. Series: Materials Science and Engineering **657** (2019) 012035 doi:10.1088/1757-899X/657/1/012035

It is known that numerical modeling of such problems with high Atwood numbers present serious difficulties due to the huge difference in gas densities and high speed processes. As an advantage of the algorithm used here it is worth noting that the symmetry properties of the numerical solution with respect to x - axis are preserved with high accuracy.

5. Conclusion

Presented here a new numerical method based on the QGD equations, allows to perform numerical simulation of Rayleigh-Taylor type instability problems in a wide range of Atwood numbers. The results obtained are qualitatively consistent with those known in the literature. Numerical solution has proper symmetry. The improvement of its quality is shown with the spatial grid refinement. The numerical algorithm is constructed to simulate the 3D flows. 2D calculations described in the paper are performed by use of 3D code with the minimum number of nodes in the third direction.

In the future we plan to work on the inclusion of the developed algorithm for the gas mixtures flow simulation into the open software complex OpenFOAM [10]. In addition, the developed technique can be applied to other problems connected with gas mixtures flows, including jets, both in subsonic and supersonic modes, as well as to geophysical flows at atmospheric scales.

Acknowledgements

This work was supported by Russian Foundation for Basic Research, grants No 19-01-00262, 18-01-00587 and 18-01-00405.

References

[1] Taylor G I 1950 *Proc. of Roy. Soc.* A **201** 192
[2] Reckinger S J, Livescu D and Vasilyev O V 2016 *J. Comput. Phys.* **313** 181–208
[3] Zhou Y 2017 *Phys. Reports* **723–725** 1–160
[4] Elizarova T G 2009 *Quasi-Gas Dynamic Equations* (Berlin: Springer-Verlag)
[5] Sheretov Yu 2009 *Continuum dynamics under spatiotemporal averaging* (Moscow-Izhevsk: R&C Dynamics) in Russian
[6] Shirokov I A and Elizarova T G 2014 *Journ. of Turb.* **15** 707–30
[7] Zlotnik A A 2010 *Dokl. Math.* **82** 811–5
[8] Bird G A 1994 *Molecular Gas Dynamic and the Direct Simulation of Gas Flows* (Oxford: Clarendon Press)
[9] Elizarova T G, Graur I A and Lengrand J C 2001 *European Journal of Mechanics* B **3** 351–69
[10] Kraposhin M V, Smirnova E V, Elizarova T G and Istomina M A 2018 *Computers & Fluids* **166** 163–75

IOP Conf. Series: Materials Science and Engineering **657** (2019) 012036 doi:10.1088/1757-899X/657/1/012036

Simulation of thermal characteristics of lithium batteries for electric vehicles

L W Zhang[1,2] and Y S Yu[1,2]

[1]School of Mechanical and Power engineering, Dalian Ocean University, 52 Heishijiao Street, Dalian, Liaoning Province, China, 116023

E-mail: zlw_1982@163.com; yingshui_y@foxmail.com

Abstract. To study the thermal characteristics of lithium batteries in electric vehicles, a single lithium ion battery under natural convection, forcedair cooling and water cooling conditions were simulated. On this basis, a parallel water-cooling model for battery pack is proposed, and CFD method was used to simulate and analyze the heat characteristics and temperature field distribution of the model at different flow rates, and an optimum flow rate of the cooling water was determined at which the battery pack has the best cooling result and thermal homogeneity corresponding to the given discharge rate.

1. Introduction

Lithium ion batteries are widely used in the energy system of electric vehicles for their advantages of high energy density, low self-discharge, long life and good stability. However, with the widely application of lithium batteries in electric vehicles, the volume is getting smaller and the power is getting larger, and the heat is increasing correspondingly, which affects the battery life and safety [1].

Bandhauer *et al* [2] researched the capacity of lithium battery under different temperature, and pointed out that the capacity of the battery will be reducedwhen the working temperature exceeds 50°C, and suggested the working temperature be controlled below 50°C. PESARAN [3] compared air cooling with liquid cooling and concluded that although the structure of liquid cooling is more complex than air cooling, the heat exchange efficiency and temperature controlling effect of liquid cooling are better. Chen [4] made a comparative analysis of serial ventilation and parallel ventilation by using CFD method, and improved the air inlet and outlet to optimize the temperature field of the battery pack. An *et al* [5] compared the effects of flow pipe Numbers, section shape of flow passage, and length-width ratios of flow passage on the maximum temperature rise and thermal homogeneity of battery pack.

In this paper, first of all, the heat dissipation of single lithium battery under different cooling modes and different ambient temperatures is simulated by virtue of FLUENT, and the problems of air cooling are pointed out. On this basis, a new type of parallel water cooling model of the battery pack including heterogeneous components is established, and detailed simulations are carried out on the heat dissipation and temperature field of the battery pack.

2. Thermophysical model

The actual thermo genesis of lithium ion battery is quite complex. When solving the internal temperature field of the battery, a simplified three-dimensional thermal mathematical model can be adopted, as shown in equation (1) [6]:

$$\rho C_P \frac{\partial T}{\partial t} = \lambda_x \frac{\partial^2 T}{\partial x^2} + \lambda_y \frac{\partial^2 T}{\partial y^2} + \lambda_z \frac{\partial^2 T}{\partial z^2} + q \tag{1}$$

Where, t stands for time, T stands for battery temperature, ρ stands for the average density of the battery, q stands for the heat generation rate per unit volume of the battery, C_P stands for the specific heat capacity of a single battery, λ_x, λ_y, λ_z respectively stands for the thermal conductivity coefficient of the battery along the x, y, z axis.

D Bernard's heat generation model is commonly used to calculate the heat generation rate of batteries [7], expressed as:

$$q = \frac{I}{V_d}\left[(U_0 - U) - T\frac{dU_0}{dT}\right] = \frac{I}{V_d}\left[I \times R_r - T\frac{dU_0}{dT}\right] \tag{2}$$

Where, V_d represents the volume of battery; I represents the charge or discharge current; U_0 represents the open circuit voltage of the battery; U represents the load voltage; dU_0/dT represents the temperature coefficient, which is negligible because of its small value. R_r is the internal resistance of the battery;

The heat generation rate of the positive and negative poles of the battery is:

$$q_{\mathrm{Al}} = \frac{Q_{\mathrm{Al}}}{V_{\mathrm{Al}}} = \frac{I^2 R_{\mathrm{Al}}}{V_{\mathrm{Al}}} \tag{3}$$

$$q_{\mathrm{Cu}} = \frac{Q_{\mathrm{Cu}}}{V_{\mathrm{Cu}}} = \frac{I^2 R_{\mathrm{Cu}}}{V_{\mathrm{Cu}}} \tag{4}$$

Where, q_{Al}, q_{Cu} represents the heat generation rate of positive and negative pole column; Q_{Al}, Q_{Cu} represent the calorific value of the positive and negative pole column; V_{Al}, V_{Cu} represent the volume of positive and negative pole column; R_{Al}, R_{Cu} represents the resistance of positive and negative

In general, the heat capacity of lithium ion batteries can be obtained by using the weighted average method based on the specific heat capacity of components. The expression is:

$$C_P = \frac{1}{m}\sum C_i m_i \tag{5}$$

Where, m is the mass of lithium battery; $C_P$is the specific heat capacity of lithium ion battery. m_i and C_i representrespectively the mass and specific heat capacity of each component in the battery.

For the analysis of heat generation and conduction of heterogeneous materials, some valid methods such as partition of unity finite element method are presented [8-10].

3. Heat dissipation simulation of single battery

3.1. Numerical model of battery

Figure 1. Model mesh of single battery.

The research object here is a certain type of cuboid lithium iron phosphate battery for electric vehicle, which size is 70 mm × 27 mm ×88 mm. Hypermesh was used to conduct the modeling and meshing for the battery. The model finished is shown in figure 1.

3.2. Physical parameters of battery

The physical parameters of copper and aluminum materials were determined by query, and the thermal conductivity andSpecific heat capacityof the core of the battery obtained by calculation. The specific parameters are shown in table 1.

Table 1. Battery physical parameters.

Component	Material	Density/(kg/m^3)	Specific heat capacity/ (J/(kg·k))	thermal conductivity coefficient/ (W/(m·k))
Battery body	Mixed material	2329	291	K_x=1.1, K_y=1.4, K_z=1
Positive poles	Aluminum alloy	2730	963	201
Negative poles	Copper	8450	390	116

The heat generation rates of the battery core , positive and negative poles under different discharge rates are shown in table 2.

Table 2. The heat generation rates of different components of the battery.

discharge rates/C	Current /A	q	q_{Al}	q_{Cu}
1	11	6 329.5	8 799.2	22 012.5
2	22	19 789.6	34 551.8	89 218.1
3	33	43 336.3	77 741.5	200 732.2

The heating condition of lithium battery in electric vehicle will change dynamically with the working condition and power demand of the vehicle. For different working conditions, it can be expressed by the charge ordischarge rate of the battery (1 C, 2 C, 3 C).The higher the discharge rate, the higher the heat generation. Limited to the paper space, here only give analysis of the battery under 3C discharge rate, environmental temperature 25°C and 40°C, under three different cooling conditions (natural convection, forced air cooling and water cooling (water temperature is 25°C)). In order to save the calculation time, the simulation of single battery does not include the fluid part, and the fluid effect is realized by setting the ambient temperature and the heat transfer coefficient of the battery surface. The heat transfer coefficients under the three cooling conditions are: natural convection α= 10 W /(m^2·K), forced air cooling α=25 W /(m^2·K), and water cooling α= 390 W /(m^2·K).

3.3. Results and analysis

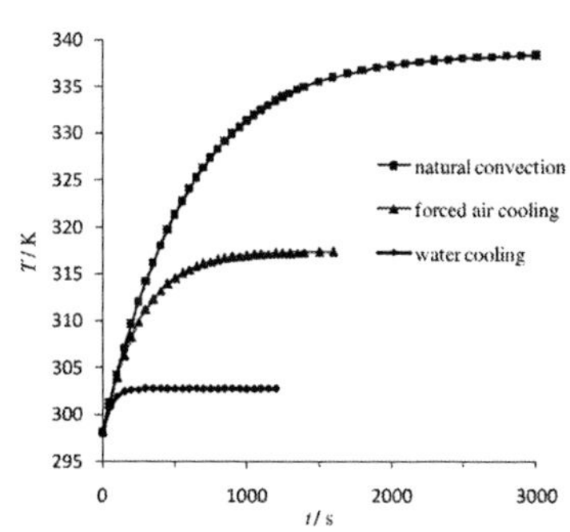

Figure 2. Temperature curves.

Figure 2 shows the changes of highest temperature of the battery with the time under environmental temperature 25°C, three different cooling conditions. It is obvious that with the increase of time, the highest temperature rises nonlinearly. At the beginning the curves rise fast, and then after a certain time, the heating and cooling gradually come into balance, temperature curves become flat. In the meanwhile, among the three cooling conditions, the effect of water-cooling is the best, as it can achieve the heat balance in a short time, and the final temperature is the lowest.

Figure 3 shows the comparison of battery temperature distribution after the battery reaches thermal equilibrium under two ambient temperatures and three cooling conditions. When the environment temperature is 25°C, the highest center temperature under natural convection exceeds 50°C, which does not meet the requirement of the heat dissipation. When the environment temperature is 40°C, the highest center temperatures under natural convection and forced air cooling both exceed 50°C. However, the overall temperature of the battery is quite low with water cooling, and the heat dissipation effect is good.

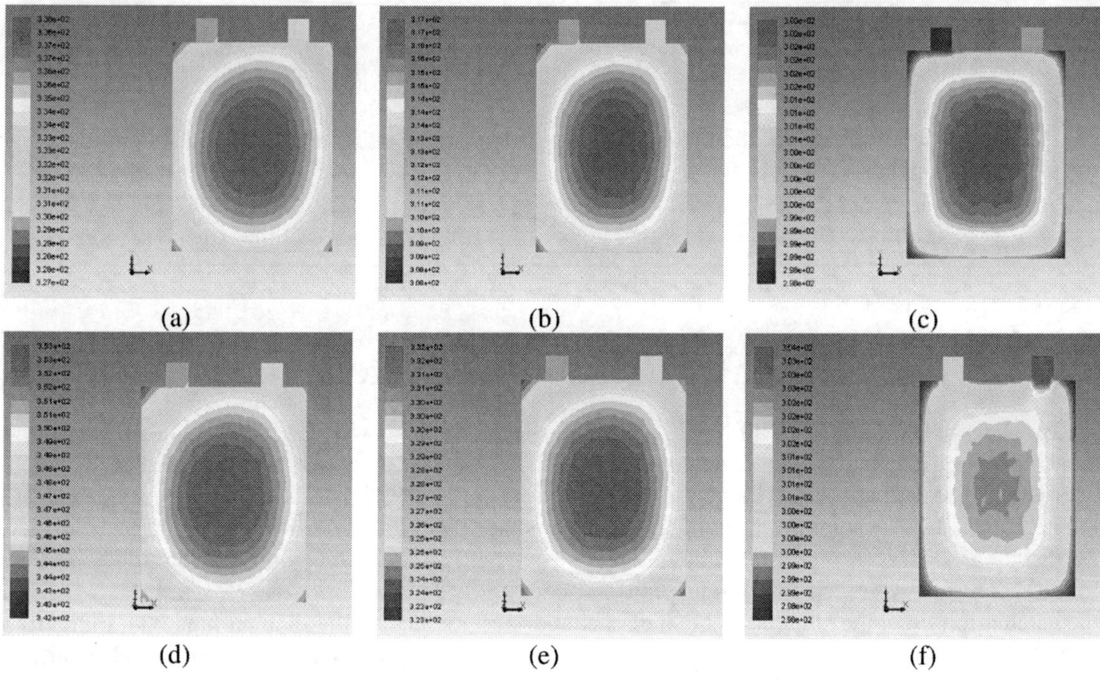

Figure 3. Final temperature distribution under different ambient temperatures and cooling conditions. (a) 25°C, natural convection, (b) 25°C, forced air cooling, (c) 25°C, water cooling, (d) 40°C, natural convection, (e) 40°C, forced air cooling and (f) 40°C, water cooling.

Meanwhile, under the conditions of natural convection and forced air cooling, the maximum temperature of the battery rises with the change of the ambient temperature. Under the same cooling condition, the change of the maximum temperature of the battery is nearly the same as that of the ambient temperature. Therefore, according to the calculation results, the maximum temperature of the battery can be inferred from the ambient temperature under natural convection and forced air cooling conditions. But water cooling effect is less affected by environmental temperature, when the environmental temperature change from 25°C to 40°C, the highest temperature of battery only increased by 1°C. Therefore, water cooling is more suitable for accurate control of battery temperature.

4. Battery pack with water cooling

4.1. Model and settings
In general, there are two modes of heat dissipation for battery pack: serial heat dissipation and parallel heat dissipation. For serial heat dissipation, as the coolant flows in the pipe, the farther it is from the inlet, the higher the temperature of the coolant is, which leads to poor temperature uniformity of the battery pack. In addition, the longer the pipe is, the greater the flow resistance is, and the more energy is needed. Therefore, a new parallel water-cooling model is designed here, as shown in figure 4. This model is composed of 12 single batteries, cooling pipes and aluminum alloy frames, and the batteries and pipes are embedded together, and they are in close contact with each other. The cooling water flow passage has 4 entrances and 4 exits, both of which are 10 mm in diameter. Considering the symmetry of the model, quarter model is used for calculation, as shown in figure 5. The model is composed of 418,487 tetrahedral elements and 64,271 nodes.

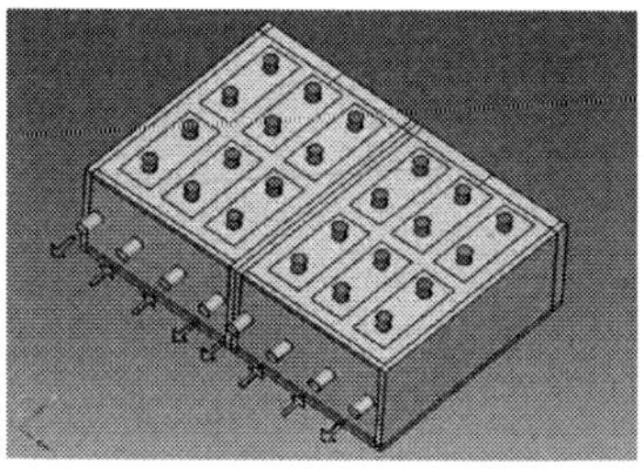

Figure 4. Parallel cooling model.

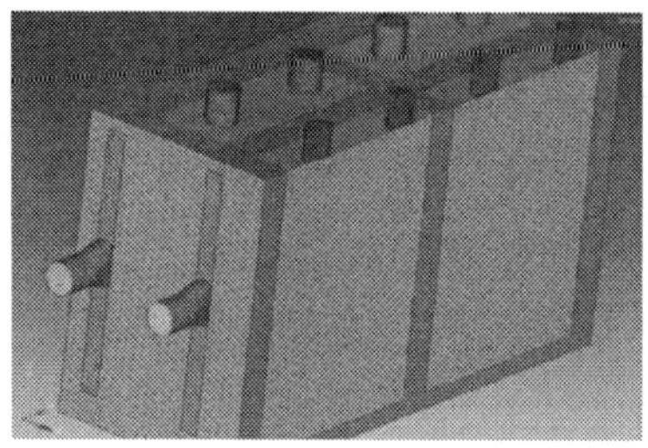

Figure 5. Quarter model.

Since the cooling water flow passage was plate-shaped, the laminar flow model was selected according to the Reynolds number. The inlet is set to speed entry, the flow rates were 0.01 m/s, 0.03 m/s and 0.05 m/s, and inlet water temperature of 25°C; The outlet is set as the pressure outlet;The heat exchange between different parts of the packadopts the coupling model, and the software calculates the heat transfer automatically. Radiation is used for the heat exchange between the frame, battery body,electric polesand the external environment; Setting the environment temperature at 40°C.

4.2. Results and analysis
Figure 6 shows the final temperature distribution of the battery pack after reaching the heat balance under different flow rates. It can be seen that the highest temperature appears in the negative pole, and the internal temperature of the battery body is relatively low. In flow rate of 0.01 m/s, top temperature of the battery pole is 42°C, and the two batteries adjacent has certain temperature difference. The temperature near the entrance is relatively low, far away from the entrance, the temperature is higher, which indicates that the flow rate is too small. In flow rate of 0.03 m/s, top temperature of the batteries is 37°C, the battery body temperature distribution is quite uniform, indicating that heat dissipation of the battery and thermal homogeneity are good. In 0.05 m/s, the highest temperature is still 37°C, and battery temperature distribution is similar to that of 0.03 m/s, only a slight decline in local.

Figure 7 shows the variation of the battery's temperature uniformity at different flow rates.It tells that when the flow rate is 0.03 m/s, the temperature difference between the center and the edge of a single battery and the difference between the two adjacent battery centers are the smallest. It is meaningless to increase the flow rate to improve the heat dissipation and temperature uniformity of the battery after 0.03 m/s. Therefore, the flow rate of 0.03 m/s can be used as the optimumrate to control the temperature under 3C discharge rate.

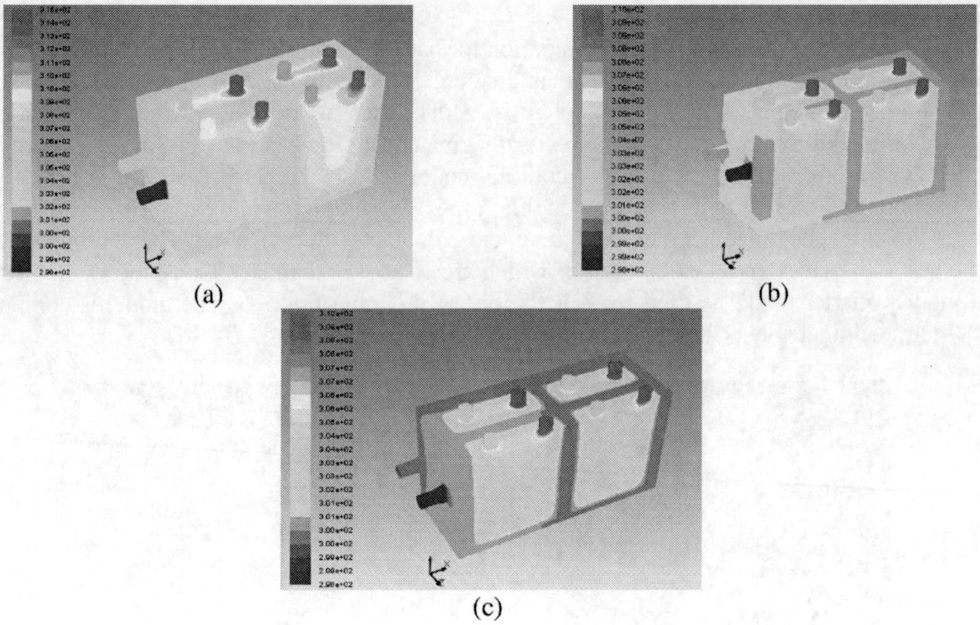

Figure 6. Battery pack temperature distribution in different flow rates. (a) 0.01 m/s, (b) 0.03 m/s and (c) 0.05 m/s.

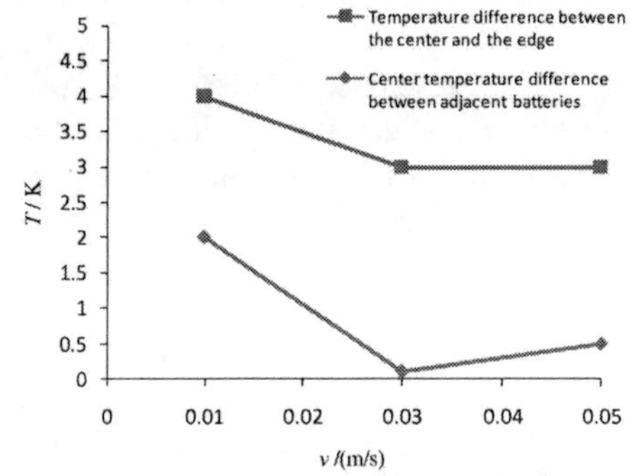

Figure 7. Uniformity of battery temperature at different flow rates.

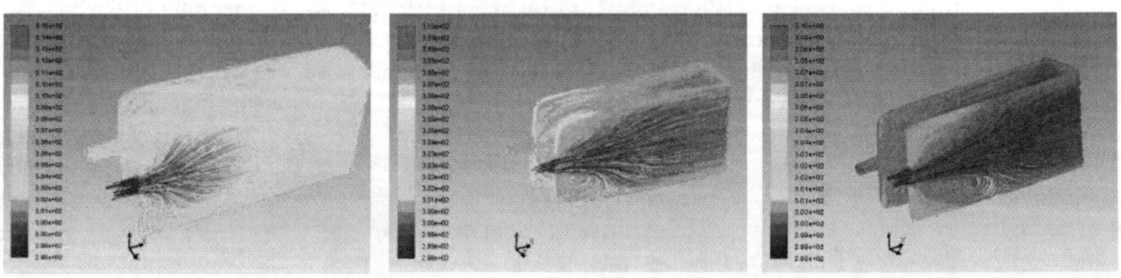

(a) (b) (c)

Figure 8. Movement and temperature distribution of cooling water in flow passage. (a) 0.01 m/s, (b) 0.03 m/s and (c) 0.05 m/s.

Figure 8 shows the movement and temperature distribution of cooling water in the flow passagein different flow rates. It can be seen that, near the inlet, vortexes generate along the height of the flow passage, and with the increase of the flow rate, the range of vortexes gradually expand, which is to increase the flow loss of the fluid, and increase the energy consumption of the system. From the perspective of energy conservation and temperature control, the inlet flow rate of cooling water in this model should be set at 0.03 m/s under 3C discharge rate.

5. Conclusions

In this paper, the thermal characteristics of lithium battery under natural convection, forced air cooling and water cooling are simulated and analyzed by changing the ambient temperature and heat transfer coefficient, and the results of natural convection and forced air cooling and the effect of water cooling are obtained.On this basis, a parallel water-cooling heat dissipation model of the battery pack was proposed, and the thermal characteristics, temperature field and water movement of the model at different flow rates were simulated and analyzed using CFD method, and the optimum flow rate was determined through analysis.The research method and calculation results can provide reference for the design and optimization of the thermal management system of power lithium battery.

References

[1] Lu L, Han X, Li J *et al* 2013A review on the key issues for lithium-ion battery management in electric vehicles *J. Power Sources* **226** 272-88
[2] Bandhauer T M, Garimella S and Fuller T F 2011 A critical reviewof thermal issues in lithium-ion batteries *J. Electro-Chemical Soc.* **158** 1-25
[3] Pesaran A A 2001 Battery thermal management in EVs and HEVs: issues and solutions *Proceedings of Advanced Automotive Battery Conference* (Las Vegas, USA)
[4] Chen Z 2018 Thermal analysis and temperature control of lithium ion battery for pure electric vehicle (Beijing: Beijing Jianzhu University) (in Chinese)
[5] An Z G, Ding Y Z, Liu Q *et al* 2018 Investigation on temperature field performance for lithium-ion batteriesby liquid-cooled tubes *Chinese J. Power Sources* **42** 795-8 (in Chinese)
[6] Li Z and Xing D H 2016 Thermal performance characteristics of lithium-ion power batterybased on finite element analysis *J.Hunan University of Science &Technology (Natural Science Edition)* **31** 42-8 (in Chinese)
[7] Bernardi D, Pawlikowski E and Newman J 1985 A general energy balance for battery systems *J. Electrochem.l Soc.* **132** 5-12
[8] Diwan G C, Mohamed M S, Seaid M *et al* 2015 Mixed enrichment for the finite element method in heterogeneous media *Int. J. Numer. Meth. Eng.* **101** 54-78
[9] Mohamed M S, Seaid M, Trevelyan J *et al* 2013 Time-independent hybrid enrichment for finite element solution of transient conduction–radiation in diffusive grey media *J. Comput. Phys.* 251 81-101
[10] Malek M, Izem N, Seaid M *et al* 2019 A partition of unity finite element method for nonlinear transient diffusion problems in heterogeneous materials *Comput. Appl. Math.* **38** 31

Numerical modelling and vibration analysis of human middle ear for application in sound conduction reconstruction

Y Liu[1,3], T Higashimachi[1] and R Toriya[2]
[1]Department of Mechanical Engineering, Sojo University, 4-22-1 Ikeda, Nishi-ku, Kumamoto-shi, Kumamoto 860-0082, Japan
[2]Toriya ENT Clinic, 5-7-12 Kuhonji, Chuo-ku, Kumamoto-shi, Kumamoto 862-0976, Japan
E-mail: liu@mec.sojo-u.ac.jp

Abstract. In this paper, we introduce a new numerical approach for estimating the hearing restoration eect prior to the tympanoplasty operation using the nite element harmonic vibration analysis. This kind of approach makes it possible to propose a new medical treatment for the recovery of conductive or cochlear hearing loss. First of all, precise geometric models of the middle ear, including healthy type model and tympanoplasty type models, are constructed on the basis of the computerized tomography (CT) scanning data. Then, frequency response characteristics of the stapes displacement in sound conduction are claried using the three-dimensional nite element harmonic vibration analysis. Based on the investigation results, we propose that the hearing restoration eect can be estimated by a comparison of the dierences in the stapes displacement between the reconstruction model and the healthy subject.

1. Introduction

The conductive hearing loss occurs when the middle ear is damaged by various ear diseases, and sounds hardly transfer through the ear canal to the eardrum and three ossicles. The chronic otitis media or middle ear cholesteatoma in which the ossicular chain in middle ear is damag causes severe conductive hearing loss. The tympanoplasty operation is often carried out to reconstruct the damaged ossicular chain, and to improve the sound conduction efficiency. In the ossicular chain reconstruction, the column article called columella is produced and used instead of incus which is generally broken. In this operation, the sound conduction efficiency changes by the variations in the shape, material and mounting position of the columella.

Several experimental researches have been carried out to investigate the dynamics characteristics of the middle ear [1-3]. However, it is extremely difficult for the experimental researches to deal with the individual variations of the dynamic characteristics, because experiments on live humans and cadavers are very limited. On the other hand, many researchers use the finite element analysis (FEA) in their study. The FEA is a powerful tool to analyse the vibration of a middle ear because the middle ear has a complicated shape [4-7]. Nevertheless, there are few studies regarding clinical applications and aiming at the improvement of the sound conduction efficiency in tympanoplasty.

In this paper, we introduce a numerical approach for estimating the hearing restoration effect prior to the tympanoplasty operation using the harmonic vibration analysis. By the approach, it is possible to find an optimal operative method for the sound conduction reconstruction. Fist of all, precise geometric models of the middle ear, including healthy type model and tympanoplasty type models, were constructed through CT data. Then, the harmonic vibration analysis was carried out to investigate the frequency response of the stapes footplate. Moreover, III type and IV type operation models in which mounting positions of the columella are umbilical and intermediate parts of the malleus were analyzed as an application.

2. The human middle ear and its function

The ear is the organ of controlling hearing and balance, and has three parts: the outer ear, the middle ear and the inner ear as shown in figure 1. The middle ear connects the eardrum of the outer ear and the oval window of the inner ear. The main function of the middle ear is to efficiently transfer the vibrations of the eardrum to fluid waves within the cochlea. As shown in figure 2, the middle ear contains three small ossicles: malleus, incus and stapes, which are connected to each other by the joints and supported by ligaments and muscles. By the rotary motion of the auditory ossicles, the vibration on the tympanic membrane can be amplified and transmitted to the stapes footplate, which moves the labyrinthine fluid in the internal ear to generate electrical signals. Then, the electrical signals are sent directly to the brain to be recognized as a sound.

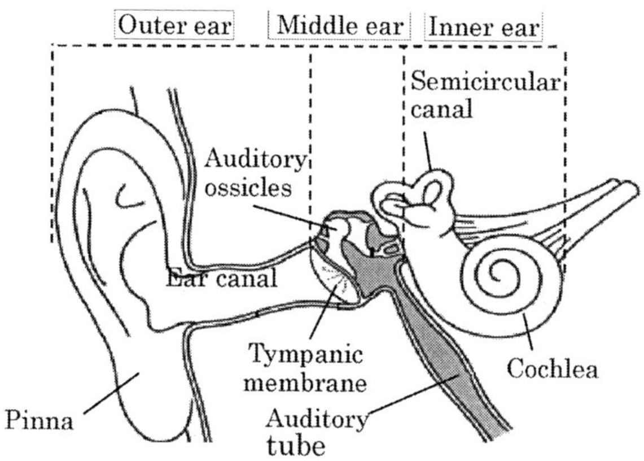

Figure 1. Structure of the human ear.

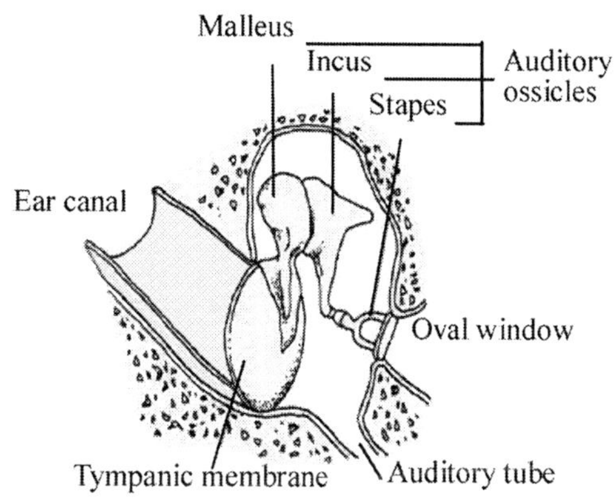

Figure 2. Structure of the middle ear.

3. Geometric modelling of a middle ear

A precise geometric model of ossicles can be created through CT scanning of the human head. As shown in figures 3-5, the two-dimensional slice images of the CT scanning data were converted into three-dimensional solid geometries using a post-processing software and the region which contains the three ossicles can be extracted. Then, the 3D geometries were refined and transformed to the DICOM data, and further transformed into the STL data, which were imported into a general-purpose structure analysis software to create a FEA model. The FEA model of a healthy subjec is shown in figure 6 which contains the tympanic membrane, auditory ossicles chain(malleus, incus and stapes), ligaments, joints, stapedial muscle and others. The anatomical name of each part in figure 6 is given in table 1.

IOP Conf. Series: Materials Science and Engineering **657** (2019) 012037 doi:10.1088/1757-899X/657/1/012037

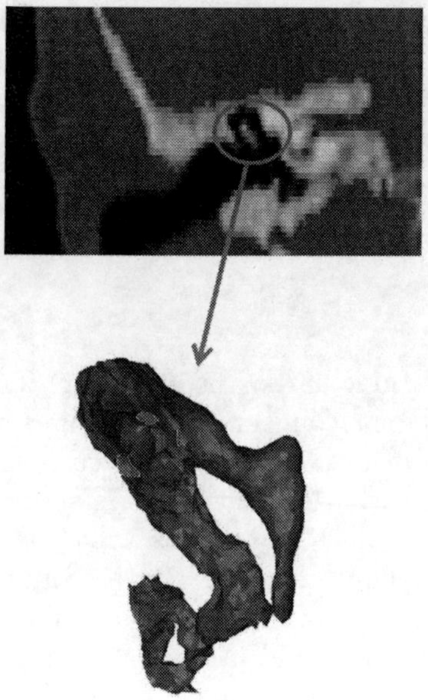

Figure 3. Extracting 3D model from CT.

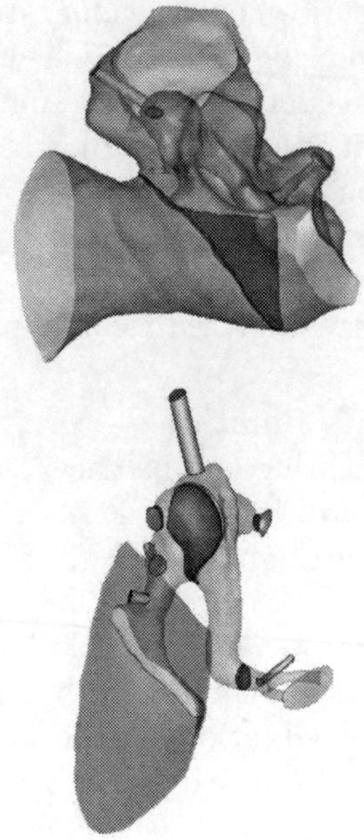

Figure 4. Refine, design and prepare model for FEM.

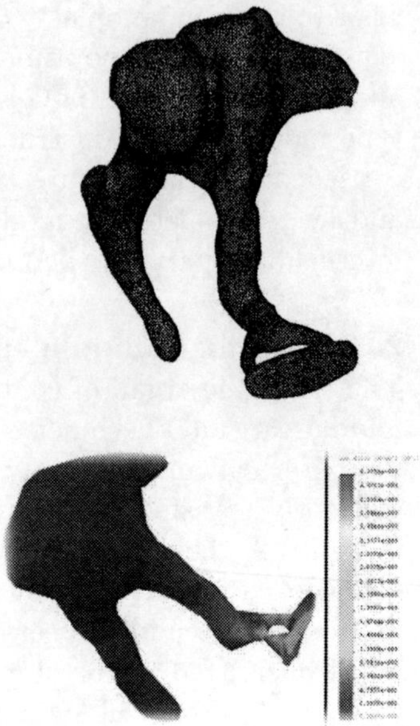

Figure 5. Vibration analysis by FEM.

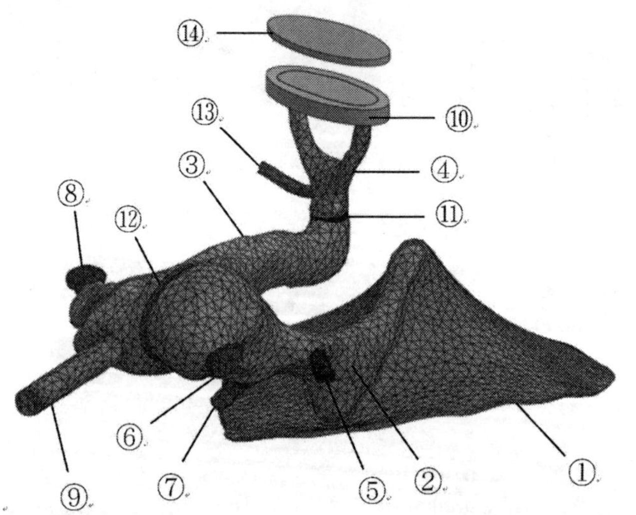

Figure 6. Geometric model of the healthy middle ear

4. Dynamic characteristics of a healthy middle ear

Firstly, dynamic characteristics of a healthy model was investigated by the harmonic vibration analysis using a general-purpose finite element method code.

4.1. Material properties and boundary conditions

Material properties used in the analysis model were determined by referring to other researches [6-8], and were summarized in table 1. The interaction between the stapes footplate

Table 1. Material properties used in the analysis model.

Anatomical parts name	Young's Modulus [MPa]	Mass density [kg/m^3]	Poisson's ratio
1 Tympanic membrane			
Pars tensa	33.4	1200	0.3
Pars flaccida	11.1	1200	0.3
Annulus tympanicus	0.6	1200	0.3
2 Malleus	13436	4350	0.3
3 Incus	13436	4350	0.3
4 Stapes	13436	4350	0.3
5 Lateral mallear ligament	21	2500	0.3
6 Superior mallear ligament	21	2500	0.3
7 Anterior mallear ligament	21	2500	0.3
8 Posterior incudal ligament	0.65	2500	0.3
9 Superior incudal ligament	0.65	2500	0.3
10 Stapedial annular ligament	0.65	2500	0.3
11 Incudostapedial joint	6	1200	0.3
12 Incudo mallear joint	6	1200	0.3
13 Stapedial muscle	0.52	2500	0.3
14 Base plate	1×10^{10}	-	0.3

and the cochlea labyrinthine fluid was modelized with a set of translational springs installed between the stapes footplate and a virtual base plate (a rigid body). The spring constant was given as 40N/m by considering the research of Gan et al [5]. In addition, the damping matrix $[C]$ of the solid elements was expressed in equation (1).

$$[C] = \zeta[I] + \alpha[K] + \beta[M] \tag{1}$$

where, $[I]$ was the unit matrix, $[M]$ was the mass matrix, and $[K]$ was the stiffness matrix. ζ, α, β express the damping coefficients. Rayleigh damping was only considered in this study, ie., $\zeta = 0$, and $\alpha = 0, \beta = 7.5 \times 10^{-5}$ were used for the coefficients of Rayleigh damping.

As the boundary conditions, outer circumferences of the tympanic membrane and the stapedial muscle, ends of the six ligaments, and the inner side of the base plate were clamped. In the harmonic vibration analysis, a uniform pressure p was applied at the whole surface of the tympanic membrane, and the applied pressure p=0.632 Pa can be calculated from the definition equation of the sound pressure level (SPL) expressed in equation (2).

$$L_p = 20log_{10}(p/p_0) \tag{2}$$

where, $L_p = 90$dB was set as the sound pressure level. $p_0 = 20 \times 10^{-6}$Pa denotes the reference value or hearing threshold of the sound pressure.

4.2. Analysis results of the healthy model

As the analysis results, the frequency responses curve of the healthy model is shown in figure 7, where the vertical axis represents the average displacement of the stapes footplate, and the horizontal axis shows the input frequency. Figure 7 shows that a resonance region appears at 0.5~2 kHz in frequency, witch is called "conversation range". The peak value of stapes footplate average displacement reaches 5.09 nm at about 1.3 kHz, and the responses falls off steadily above

2 kHz. In this study, the frequency response curve of the healthy model was used as a standard curve to be compared with that of a operation model, where the damaged auditory ossicles was substituted by the columella (a medical device). Then, it possible to find an optimal operative method for the sound conduction reconstruction.

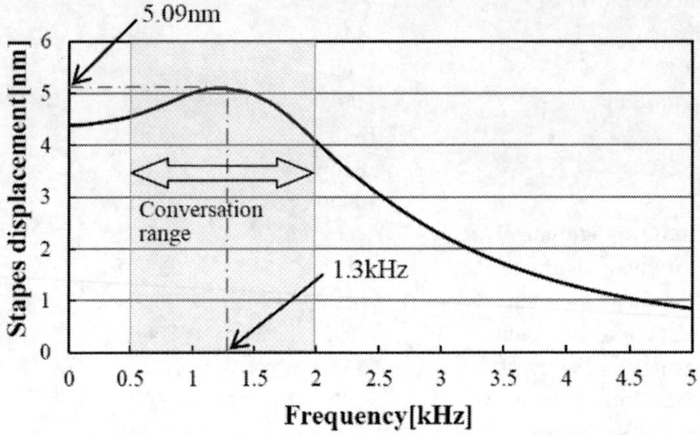

Figure 7. Harmonic vibration analysis results of a healthy model

5. Application for clinics in tympanoplasty

As mentioned in Section 1, the tympanoplasty operation is often carried out to reconstruct the damaged ossicular chain to improve the sound conduction efficiency. In the ossicular chain reconstruction, the column article called columella is produced and used instead of incus which is generally broken. Currently, the tympanoplasty operation can be classified into two types, that is Auditory ossicles formation III type and IV type. The main difference is that the stapes is kept in almost normal in the III type, but only the stapes footplate is left in the IV type.

Figure 8 shows a III type tympanoplasty model, where the columella was installed between the umbilical region (the tip of the malleus) and the stapes. Its analysis result of the frequency response curve shows that the maximum displacement of the stapes footplate is 4.7 nm near 1.5 kHz. Comparing with the displacement 5.09nm of the healthy model, the value of this tympanoplasty model reaches about 94% of the healthy one. Figure 9 shows another model, where the columella was installed at the intermediate region, that is, between the tip of the

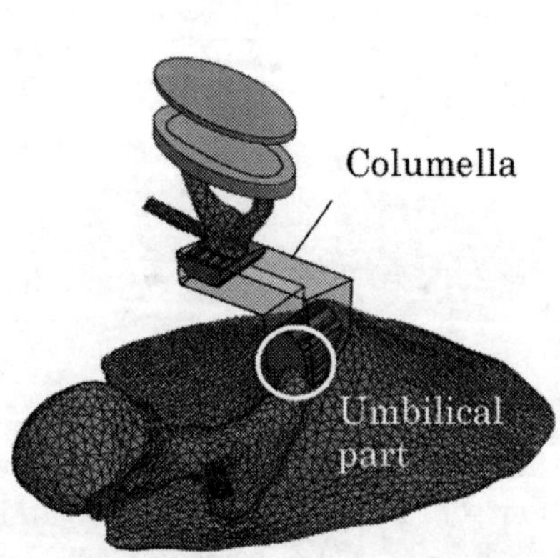

Figure 8. Umbilical connection model in the III type tympanoplasty.

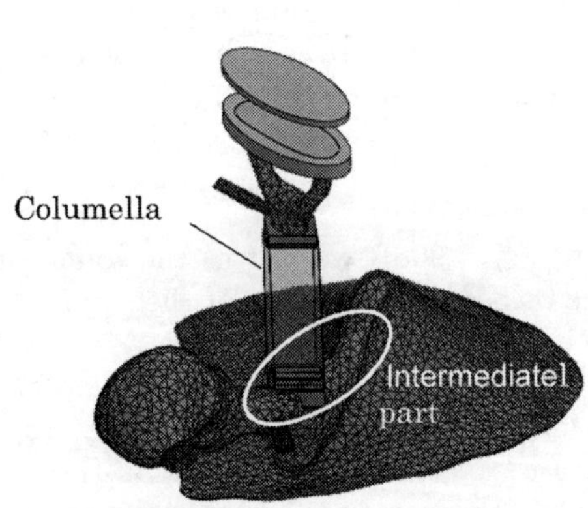

Figure 9. Intermediate connection model in the III type tympanoplasty.

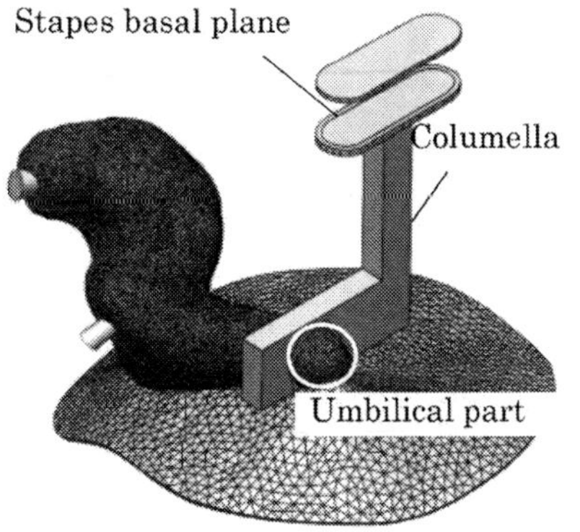

Figure 10. Umbilical connection model in the IV type tympanoplasty.

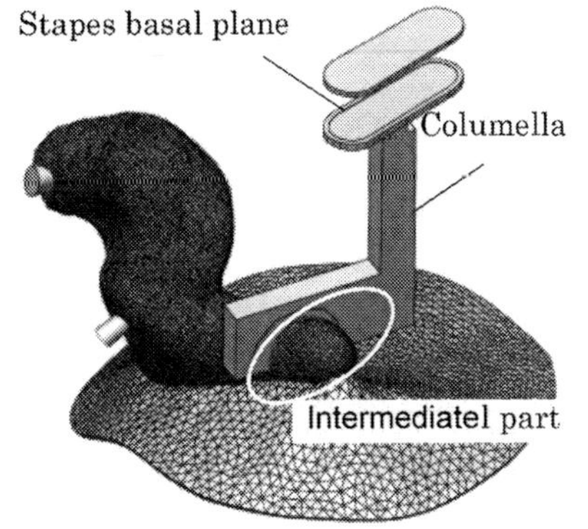

Figure 11. Intermediate connection model in the IV type tympanoplasty.

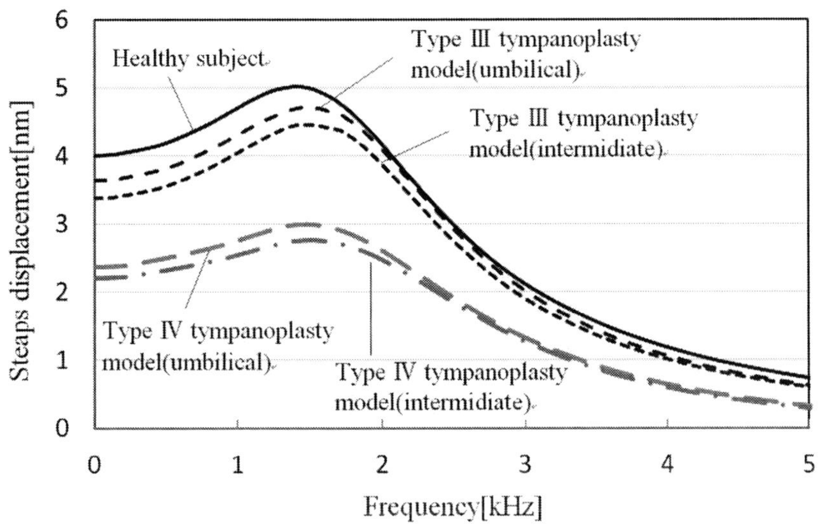

Figure 12. Frequency response graphs of tympanoplasty models

malleus and the short process of the malleus. In this case, analysis results show that the maximum displacement decreases to 4.4 nm which is approximately 88% of the healthy model.

A IV type tympanoplasty model is shown in figure 10, where a columella was installed at the umbilical region. The analysis result of the frequency response curve shows that the maximum response of the stapes footplate is 3.01 nm near 1.5 kHz. This displacement of 3.01 nm is about 60% of the healthy model. As another example, a model in which the columella is installed between the intermediate region is shown in figure 11. Analysis result is shown in figure 12, in which the maximum displacement decreases to 2.79 nm which is approximately 56% of the healthy model.

The results of two tympanoplasty models, both the a III type and the IV type, shows that the connection position of the columella at the umbilical region is more suitable for obtaining a larger displacement, that contributes to a higher hearing restoration ratio. It is clear that the comparison between the stapes footplate displacement of the tympanoplasty model and the healthy one can estimate the restoration effect, not only in the III type operation but also in the IV type operation, and then find an optimal operative method in the sound conduction reconstruction.

6. Conclusions

This paper proposed a new numerical approach for estimating the hearing restoration effect using the harmonic vibration analysis. Numerical modelling and harmonic vibration analysis of the human middle ear were carried out to investigate the dynamic characteristics. Then, the sound conduction efficiency in tympanoplasty was improved by comparison of the differences in the stapes footplate displacement between the reconstruction model and the healthy one. The validity of the proposed method was confirmed by the application of two types of operation models to find suitable connection positions of the columella.

Acknowledgement

This work was supported by JSPS KAKENHI Grant Number JP17K14581.

References

[1] H, Kobayashi T, Nagamura H and Tashizaki H 1990 *Trans. Jpn. Soc. Mech. Eng. Ser. C* **56** 1431-4
[2] Voss S E, Rosowski J J, Merchant S N and Peake W T 2000 *Hear. Res.* **150** 43-69
[3] Aibara R, Welsh J T, Puria S and Goode R L 2001 *Hear. Res.* **152** 100-9
[4] Wada H, Metoki T and Kobayashi T 1993 *J. Acoust. Soc. Am. JASA* **92** 3157-68
[5] Gan R Z, Feng B and Sun Q 2004 *Ann. Biomed. Eng.* **32** 847-59
[6] Sun Q, Gan R Z, Chang K H and Dormer K J 2002 *Biomech. Model. Mechanobiol.* **1** 109-22
[7] Koike T and Wada H 2002 *J. Acoust. Soc. Am. JASA* **111** 1306-17
[8] Higashimachi T, Suga K and Sasahara H 2009 *J. Japan Soc. Prec. Eng.* **75** 428-33

IOP Conf. Series: Materials Science and Engineering **657** (2019) 012038 doi:10.1088/1757-899X/657/1/012038

Development and engineering application of high-precision and multi-function simulation system for double reheat ultra-supercritical units

B L Cai[1,3] and H C Wang[2]

[1]Xi'An Thermal Power Research Institute Co., Ltd, Xi'an 710054, Shaanxi Province, China
[2]Department of Electrical Engineering, University of Liverpool, Suzhou 215123, Jiangsu Province, China

E-mail: caibaoling@tpri.com.cn

Abstract. Laiwu ultra-super critical double reheat unit is the highest ultra-super critical double reheat unit, with the highest parameters in China, from the safe and economical to run level put forward higher requirements. Taking ultra-super critical double reheat unit as the research object, a high-precision and multi-functional simulation system for ultra-super critical double reheat unit was developed before the unit was generated, to realize the dynamic analysis of ultra-super critical double reheat unit, control strategy verification and operation technique validation; and conduct operator operation and accident management training. provides technical support for the smooth construction, and safe and stable operation of the ultra-super critical double reheat unit.

1. Introduction

The resource characteristics of "rich coal, poor oil and little gas" determine the important position of coal-fired power generation in China's energy structure. "700°C" Ultra-super critical power generation technology, as new high temperature resistant material is still in the development phase, the double reheat super critical technology can effectively improve the thermal efficiency of thermal power unit and unit capacity of CO^2 emission reduction, so the double reheat application of the ultra-super critical technology (USCT) is the practical choice, to solve the efficiency and environmental protection. China double reheat ultra-super critical (DRUC) unit design, manufacture and operation maintenance technology are at the exploratory stage, Anyuan and laiwu ultra-super critical double reheat units are the first batch of DRUC units in China,, The lack of the DRUC unit actual operation experience related to the higher economic and safe requirement which make the solution is required. This paper describes the development of a high precision and multi-functional simulation system for the DRUC unit, which can analyze the dynamic characteristics of the unit, verify the control strategy, and verify the operation technology before the unit generates electricity. In addition, training on operation personnel and accident handling was conducted by providing technical support for the smooth construction, and stable operation of the double reheat ultra-super critical unit [1-3].

IOP Conf. Series: Materials Science and Engineering **657** (2019) 012038 doi:10.1088/1757-899X/657/1/012038

2. Double reheat ultra-supercritical unit model

2.1. Simulation support platform software

The APROS (Advanced Process Simulator) simulation supporting is used with the 3-equation, 5-equation and 6-equation calculation modules for one-dimensional two-phase flow calculation, which can be used in various stages of preliminary feasibility study and system design. Training of operators, operation guidance, safety analysis and technical transformation of power plants. Using the process model of thermal power units developed by APROS, physical phenomena such as false water level, pump vaporization, fan scavenging, working medium expansion, working medium pulsation and furnace deflagration can be reflected naturally by improper operation model calculation, without any special treatment or fault setting [4].

2.2. The double reheat ultra-supercritical unit

HuaNeng LaiWu 1000 MW boiler has highest parameter in the world: the double reheat boiler main steam pressure is 32.87 MPa, the main steam temperature/high pressure reheat steam temperature /low pressure reheat steam temperature of 605/623/623°C respectively. The boiler type is dc Benson furnace, tower layout, two intermediates reheat, square chamber, the bottom chamber of the furnace coil water cooling wall, low NOx thick-light burner with SOFA wind, Quadrangle tangential combustion mode. Tail double flue duct layout, low temperature reheater inlet layout flue gas regulation baffle. The smoke re-circulation extraction position is at the induced draft fan outlet. Tail flue design flue gas waste heat utilization system (high- and low-pressure economizer), heater system. The steam turbine is a super critical condensing steam turbine with double reheat, which adopts a series arrangement of ultra-high-pressure cylinder, high pressure cylinder, medium pressure cylinder and two low pressure cylinders, 10-stage reheat system, and no. 2 and no. 4 high pre-installed steam cooler (See figure 1).

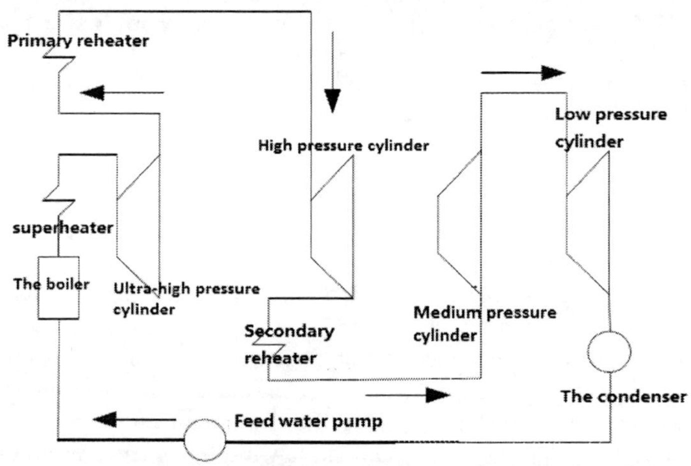

Figure 1. A simplified diagram of the thermal system of double reheat Ultra-supercritical units.

2.3. Thermodynamic process model

The model is divided into several subsystems. The divide reason is thermal system function and process flow of 1000 MW ultra-supercritical double reheat unit. Each subsystem model is composed of several calculation modules. Each module corresponds to an equipment unit or a physical process of the thermal system, and has clear physical significance and mathematical independence. Nodes with little fluctuation of temperature, pressure and other parameters were selected as the connection points between subsystem models. The boiler turbine generator body model describes detailed steam turbine

IOP Conf. Series: Materials Science and Engineering **657** (2019) 012038 doi:10.1088/1757-899X/657/1/012038

system, and auxiliary system of the double reheat ultra-supercritical boiler, which meets the requirements of real-time calculation while meeting the calculation precision of mode [5, 6].

Boiler model is strictly divided into several parts. They are the boiler economizer, water wall, steam separator, two-stage super heater, high and low pressure reheater, bypass economizer, flue gas deep waste heat utilization system, combustion system, pulverization system, air preheater, three-stage bypass and other subsystems according to boiler steam flow and flue gas flow. Taking the water wall as an example, the model calculation reflects the change rule of working medium reserves, evaporation pressure and outlet enthalpy of the water wall in the dynamic and static process of boiler.

Key technical problems solved in the model development and debugging process include 1. The ultra-supercritical double reheat unit boiler is tower furnace, the convection heating surface series increase, bilateral arrangement and left and right cross, superheater and reheater system is more complex than the primary reheat unit. It is difficult to calculate the heat balance of boiler model and the resistance of flue. 2. Steam turbine part is to increase the ultra-high-pressure cylinder, the process extension, extraction and heat recovery series increase, three-stage bypass layout, increase the high and low temperature economizer system. Therefore, the thermal system of steam turbine is more complex, the computational nodes of fluid network increase, and the difficulty of model development and debugging increases. Unit based on the design data and parameters (including the system P & ID diagram, operating procedures, equipment and system specifications, the I/O point table, etc.), isolators to boiler thermodynamic calculation, heat balance diagram and other structural design parameters (such as heat transfer surface size, elevation, pipe diameter, material, and pump, fan and adjustable valve characteristic curve data, etc.) as input data, ensure the unit model calculation value of the main parameters and design values, to ensure the simulation calculation results true, credible, and conform to the laws of physics [7-12].

The calculated data of this model were compared with the design value of the unit in the conditions of 100%, 75% and 50%. The deviation between the calculated data of the main parameter model and the design value met the requirements of engineering analysis. The data of 100%THA conditions are shown in table 1:

Table 1. Model calculation data and the design parameters (100% THA condition).

project	Design value	Model values
1 Power/MW	1000.0	993.9
2 Main steam flow/(t/h)	2562.8	2718.0
3 Main steam pressure/MPa	30.79	29.85
4 Main steam temperature/°C	600.0	599.1
5 Primary reheat steam pressure /MPa	10.32	10.39
6 Primary reheat steam temperature /°C	620.0	621.4
7 Secondary reheat steam pressure /MPa	3.06	3.15
8 Secondary reheat steam temperature /°C	620.0	621.5
9 Feed water pressure /MPa	34.43	33.96
10Feed temperature /°C	324.6	316.5
11The total amount of coal /(t/h)	351.6	365.0
12The total air volume /(t/h)	3209	3169
13Proportion of recycled flue gas /%	12.12	10.62
14Exhaust temperature /°C	112.0	137.8

2.4. Control model

The control model adopts virtual DCS technology, that is, "VDCS" replaces the operation process of control logic in the actual DPU and outputs the operation results to realize all control functions of the unit. The control system of the double reheat ultra-supercritical unit is Emerson DCS system, and the DEH control system is Siemens T3000 system, so the system adopts Emerson virtual DCS system + DEH Siemens T3000 virtual system. The actual unit configuration engineering files are translated , separately to achieve a high consistency between the control algorithm block and the actual control system. The operator interface is simulated at 1:1 [13,14].

3. Analysis and research on dynamic characteristics

Take the DRUC process models + virtual DCS way, firstly the actual unit construction structures, close to the actual project the DRUC running environment, dynamic response analysis under different load condition, get the detailed test data for research and development. The control strategy of the unit operation is to provide theoretical basis for the establishment of procedures and technical support. In addition, the whole start up and shutdown simulation test of the double reheat ultra-supercritical unit is carried out, and the full range real-time dynamic simulation verification of the unit control strategy is carried out [15].

The reheater of double reheat ultra-supercritical unit is divided into high and low pressure reheater. Therefore, the dynamic characteristics of primary and secondary reheat steam temperature are the focus of research.

Through the research and analysis of the dynamic response characteristics of the ultra-supercritical double reheat unit, the influence law of various means of temperature regulation on the key operating parameters of the boiler is obtained. Examples are as follows:

3.1. Recirculation smoke volume dynamic characteristics test

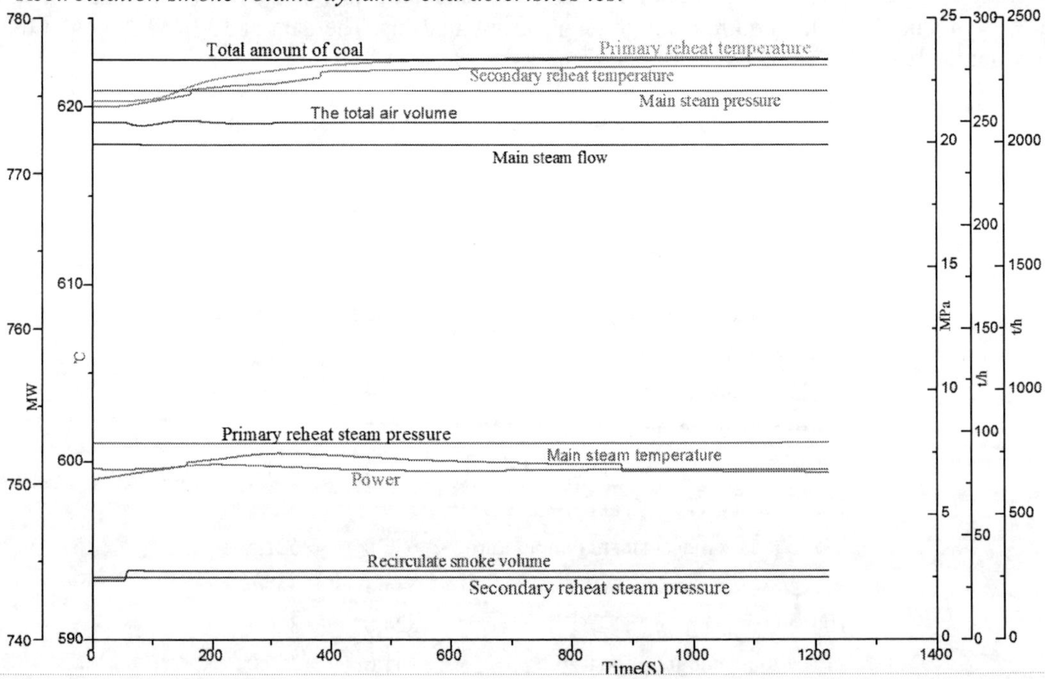

Figure 2. Recurrent flue gas dynamic characteristics test data curve.

After the step of the recirculation flue gas increased by 10% (accounting for about 1% of the total flue gas), the temperature of the main steam decreased slightly, and the temperature of the first and second reheated steam gradually increased, from the initial steady state value until reaching a new stable value. The actual unit operation shows that the conclusion is correct (As shown in figure 2)

3.2. Dynamic characteristics test of burner swing angle

Burner angular swing up 10%, furnace flame centre position up, exports in final stage super heater of superheated steam temperature is gradually reduced, the primary and secondary reheat steam temperature increased 1.9°C and 3.06°C. The dynamic characteristics of primary and secondary reheat steam temperature are similar, to that of flue gas recirculation when the burner swing Angle changes. as shown in figure 3. The actual unit operation shows that the conclusion is correct.

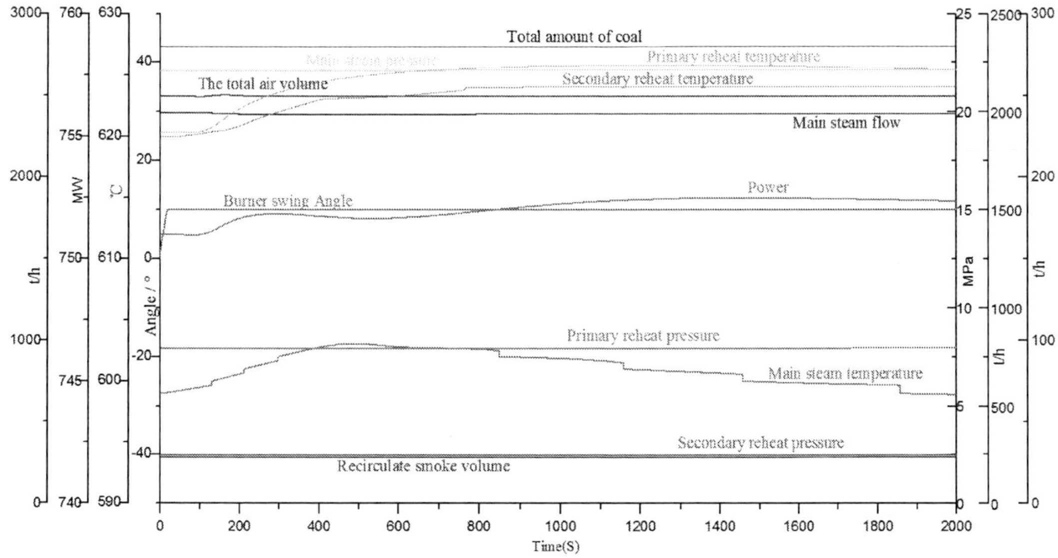

Figure 3. Test data curve of dynamic characteristics of burner swing angle.

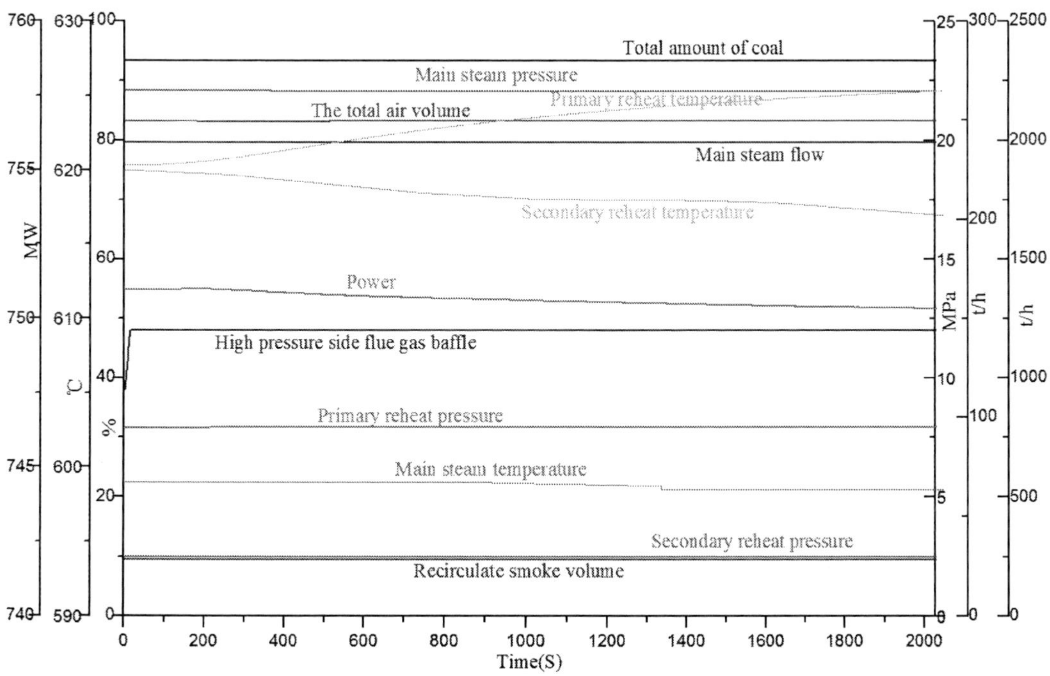

Figure 4. Flue gas baffle dynamic characteristic test data curve.

3.3. Dynamic test of flue gas baffle

After the flue gas baffle on the high-pressure side was opened up by 10% (while the flue gas baffle on the low-pressure side was closed down by 10%), the pressure and temperature of the main steam remained basically unchanged, and the temperature of the primary reheat steam gradually increased , and that of the secondary reheat steam gradually decreased. Flue gas baffle represents the means to adjust the deviation of primary and secondary reheat steam temperature, as shown in figure 4. The actual unit operation shows that the conclusion is correct [16,17].

4. Verification of control strategy and operation technology

Through the DRUC advocate complementary equipment configuration, process flow, operation characteristics and control requirements, combined with dynamic response study conclusion, develop a complete set of large double reheat ultra-supercritical coal-fired power generating units of the automatic control system scheme, to ensure that the type of double reheat the safe and stable operation of the unit. It mainly includes coordinated control, water-coal ratio control, superheated steam temperature control, primary and secondary reheat steam temperature control scheme, etc.

The control strategy of superheated steam temperature adopts the way of coal-water ratio + two-stage water spraying to reduce the temperature, and considers the influence factors of flue gas recirculation on the superheated steam temperature. According to the analysis results of the dynamic response characteristics of flue gas recirculation, flue gas baffle, burner swing Angle, and reheater accident water spray step disturbance, the first and second reheat steam temperature control is developed to adopt the combined regulation mode, of flue gas recirculation + flue gas baffle + burner swing Angle + reheater accident water spray (particularity of double reheat unit). The flue gas recirculation control input is the sum of the temperature deviation of primary reheat steam and that of secondary reheat steam. The flue gas baffle control input is from the difference between the temperature deviation of primary reheat steam and secondary reheat steam.

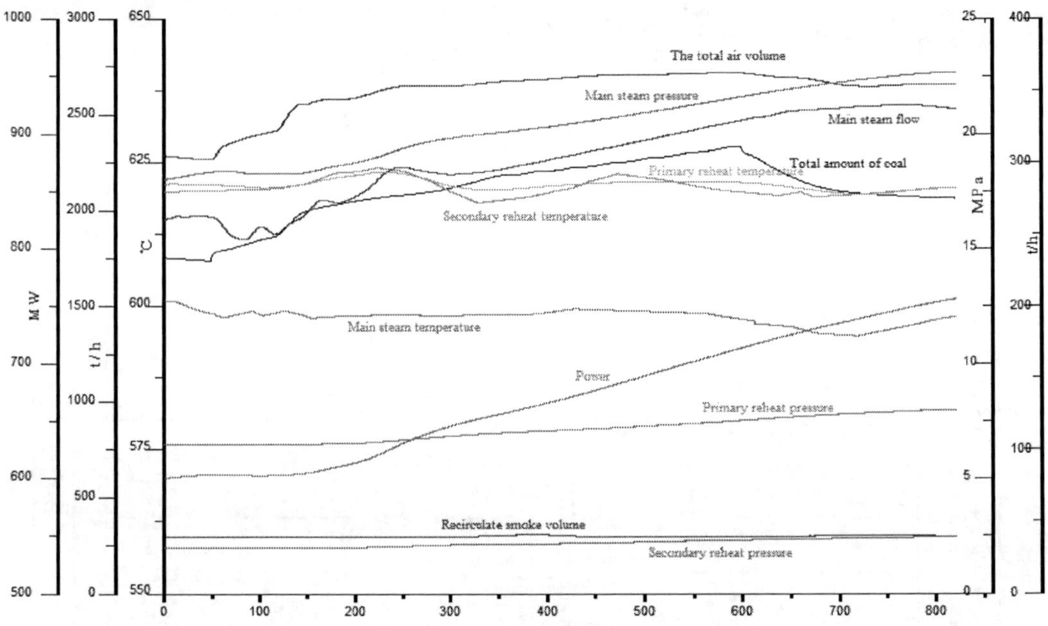

Figure 5. Real-time data curve of increase load simulation test at 1.5% rated load /min.

To engineering analysis of simulation system test platform verifies the control strategy. The investment unit coordinated control mode. The another main control loop is put into automatic, continuous lift load test, the load change rate of 1.5% of the rated load according to the unit/min, 3% /

min, 2% / min verification experiment was carried out respectively, analysis of the unit under different rate changes of the main parameters. Figure 5 shows the real-time simulation dynamic process test of load change, and the load change rate is 1.5% of the rated load /min.

The test results show that the key to the control of the double reheat ultra-supercritical unit is still to control the coal-water ratio and the wind-coal ratio. The superheated steam temperature is controlled by the coal-water ratio + the secondary water-jet temperature reduction method. The primary and secondary reheat steam temperature is controlled by the flue gas recirculation + burner swing Angle + flue gas baffle + reheater accident water-jet temperature reduction method. The variable load rate below 1.5% can be controlled within a reasonable range. Laiwu double reheat ultra-supercritical unit was officially put into commercial operation in 2016 after 168 hours of trial operation. During the trial operation, thermal protection and automatic input rate of the unit were both 100%, which verified the correctness of the overall control strategy of the unit. Figure 6 shows the load variation process at 1.5% rated load /min after the unit was put into commercial operation. According to the trend of the two figures and comparative data analysis, the simulated load variation test conducted on the engineering analysis test platform is basically consistent with the actual load variation test process after the unit is put into operation.

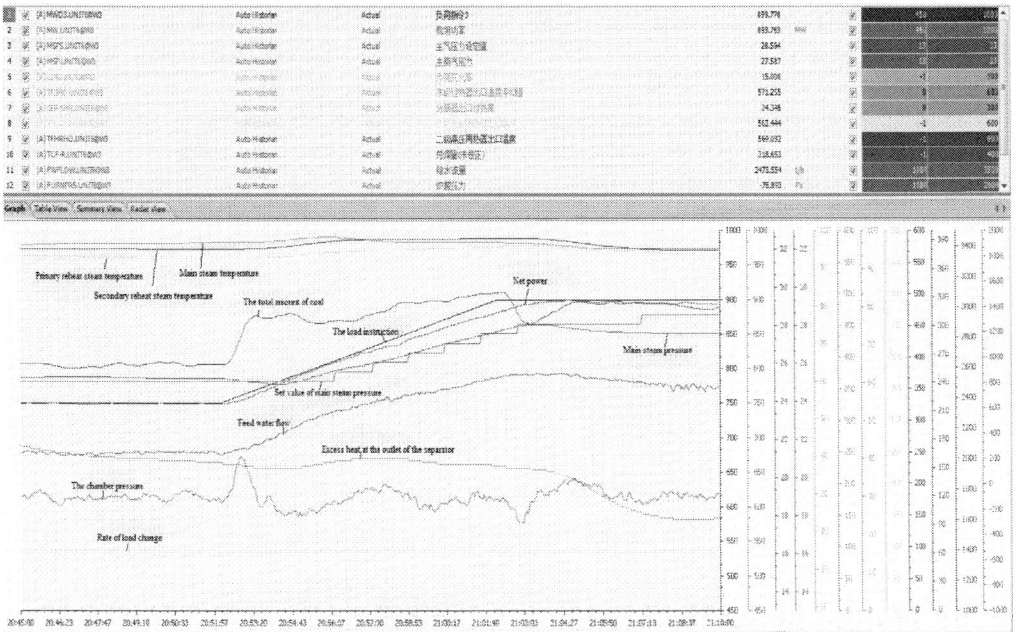

Figure 6. Unit actual operation 1.5% load rate increase load test.

The operating regulations of units are compiled in the design stage according to the design parameters of units. The operating regulations of similar units in operation. The running technology verification is carried out through the real-time dynamic test of the starting and stopping of the simulation unit and the fault process, which provides the basis for the modification of the unit's operating regulations, and makes the performance of the actual unit's operating operation reach the optimal level.

5. Conclusion

The high-precision and multi-function simulation system of the ultra-supercritical unit for double reheat is developed, to realize the dynamic characteristic analysis, control strategy verification and operation technology verification of the ultra-supercritical unit for double reheat before generating, so as to shorten the start-up and commissioning time of the actual unit.

Operating personnel training and operation accidents exercise, which make them to the new unit running characteristics of various equipment and systems, logic, the function of the control circuit, man-machine interface of monitoring content and way of operation and all kinds of accidents and the unit start-up condition of correct operation and treatment, etc.

Sum it up, The new unit was put into operation. Skilfully Control operation of the unit reduces the occurrence of wrong operation. It solved DRUC lack the practical operation experience. The problem of the operation with the lack of professional and technical personnel, for the smooth construction double reheat ultra-super critical unit provides the technical support with the safe and stable operation.

Acknowledgments
I would like to thank the other members of the project team, such as Gao H D and Wu H Y, for their kind assistance.

References

[1] Gao J Q and Zhang C 2017 Numerical simulation on dynamic performance of combustion and flue gas system in a 300 MW oxy-coal fired boiler *Thermal Power Generation* **2017** 94-9
[2] Chen J H, Qiao Q, Zhang B *et al* 2014 Critical pressure ratio characteristics of steam turbine governing stage *J. Zhejiang Univ.: Eng. Sci.* **48** 2072-9
[3] Liang Q J, Zeng D L, Liu J Z *et al* 2011 Modeling and simulation analysis for 1000 MW ultra-supercritical unit *East China Electric Power* **39** 1889-92 (in Chinese)
[4] Paananen M and Henttonen T 2009 Investigations of a long-distance 1000 MW heat transport system with APROS simulation software *Proc. 20th Int. Conf. Structural Mechanics in Reactor Technology* (Espoo, Finland)
[5] Sun W W and Duan X H 2015 Research on SimuPlat effect on simulation accuracy of thermal power boiler system *Comput. Simul.* **32** 161-4
[6] Wang W S and Wang X 2017 Dynamic simulation for absorbed heat of super-heater in supercritical unit boilers *Thermal Power Generation* **2017** 64-9
[7] Liu Q J, Liu D Y, Zhu T Y *et al* 2015 Analysis of thermodynamic performance of cavity absorber in the parabolic trough solar concentrators *Proc. CSEE* **35** 126-32
[8] Lu X L and Ma P 2015 Model of super critical unit's water supply system based on standard particle swarm optimization *Comput. Simul.* **32** 446-8
[9] Wu P, Wang Q Y, Yu Z H *et al* 2014 Research on cold aerodynamic field of tangentially fired boiler *Ind. Control Comput.* **2014** 92-4
[10] Wang F J, Huang Y, Liu H Y *et al* 2013 Study and preliminary design of double-reheat ultra-supercritical boiler *Power Equipment* **27** 73-7
[11] Xu H and Xu E S 2013 Dynal simulation model for ultra supercritical 1000 MW unit boilers *Thermal Power Generation* **42** 26-31 (in Chinese)
[12] Yang X M, Xiao Y and Chen F 2014 Present situation and development prospects of coal-fired power plant monitoring and control system *Thermal Power Generation* **43** 15-8
[13] Cui N, Wang B S, Deng Y *et al* 2008 Dynamic simulation model for the heavy duty gas turbine thermodynamic system *Proc. CSEE* **28** 110-7 (in Chinese)
[14] Gu J J and Wang P 2015 Modeling and simulation for ultra supercritical units load regulation by condensate throttling *J. Syst. Simul.* **27** 2920-6
[15] Näveri J, Tahvonen T and Hakasaari P 2010 Testing and utilization of loviisa full scope apros model in engineering and development simulator *International Youth Nuclear Congress of South Africa* (Cape Town, South Africa)
[16] Yin Y N 2013 Application status and development of USC unit with double reheat cycles *Power Syst. Eng.* **29** 37-8
[17] Yu Y, Liu X L and Fan S W 2013 Analysis and discussion for key technologies of USC double reheat turbines *Thermal* **42** 69-72

IOP Conf. Series: Materials Science and Engineering **657** (2019) 012039 doi:10.1088/1757-899X/657/1/012039

Numerical simulation of cryosurgery in biological tissue with developed circulatory system

K E Shilnikov[1,2] **and M B Kochanov**[1]

[1]Department of applied mathematics, National Research Nuclear University "MEPhI", Russia

E-mail: shilnikov.k@yandex.ru

Abstract. This paper is devoted to the modeling of the cryogenic impact on biological tissue with a developed circulatory system: large blood vessels, dense distributed small vessels and capillaries penetrating the tissue. The case of cryogenic destruction of a cancer tumor localized near a large artery is considered. The paper presents a direct numerical simulation of the temperature distribution evolution in biological tissue including the calculation of blood flow parameters in a vessel. The mathematical model is based on the Pennes bioheat transfer model with enthalpy modification and the quasi-one-dimensional hemodynamic model of blood flow in arteries, taking into account local changes in the elastic properties of the vessel walls due to strong cooling of adjacent tissue. A numerical algorithm based on explicit finite-volume approximation of the heat transfer model and the hybrid characteristic scheme (after splitting the equation with respect to physical processes) for numerical solution of hyperbolic systems is developed in order to simulate cryosurgery for the tumor near the artery. The main focus of the paper is a numerical study of the circulatory system parameters' influence on the shape of the final distribution of the cellular necrosis zone. We study the dependence of the necrosis front penetration depth on the perfusion power which characterizes the density of capillary distribution, the peak and background rates of blood flow in the artery, the diameter of the blood vessel and the thickness of its walls.

1. Introduction

During the last decades cryosurgery has become a competitive approach for human cancer treatment [1–3].Taking into account the risk of relapse and undesired destruction of healthy tissue the prediction of the cryosurgery results is very important [4]. The cryosurgical treatment is based on the tissue necrosis effect caused by the impact of extremely low temperatures on tumor tissue [5–8]. In case of the cryotip based cryosurgery necrosis effect is caused by the freezing front propagation from the tips of cryoneedles in the tissue. Current paper is concerned with the direct numerical simulation of the temperature distribution evaluation in biological tissue near the large artery including the calculation of the blood flow parameters in a vessel. The mathematical model is based on the Pennes bioheat transfer model with enthalpy modification [9, 10] and the quasi-one-dimensional hemodynamic model of blood flow in arteries [11, 12]. The influence of circulatory system parameters such as vessel wall thickness, blood flow speed and capillary blood perfusion rate is numerically studied.

2. Mathematical model

For the numerical modeling of the cryosurgery on the tumor near the artery a cylindrical computational region D_1 with height $Z = z_2 - z_1$ was considered:

$$D_1 = \left(\mathbf{r} \in \mathbb{R}^3 \middle| R_1 < r < R_2, 0 < \phi \leq 2\pi, z_1 < z < z_2, \mathbf{r}\overline{\in} D_3\right).$$

Here R_2 is an outer radius of the computational region. R_1 is a radius of the artery $\widehat{D}_{00}$. A middle section of the computational region orthogonal to the cylinder axis is shown in figure 1. $D_2 \subset D_1$ is a tumor zone and D_3 marks the position of the cryotip. The enthalpy modified

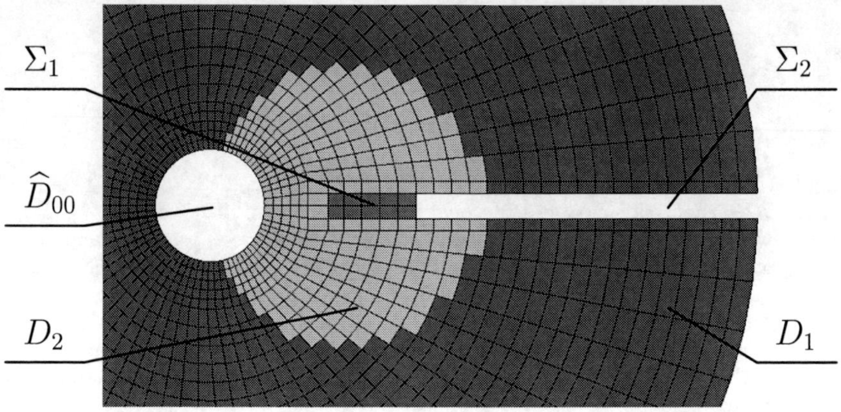

Figure 1. Computation area slice at $z = 0$.

Pennes model [9] was used in order to describe the cryogenic heat transfer in the target tissue:

$$\widehat{C}\frac{\partial T}{\partial t} = -\operatorname{div}\mathbf{Q}_F + \omega_b C_b (T_c - T), \tag{1}$$
$$\mathbf{Q}_F = -k\nabla T, \quad \mathbf{r} \in D_1,$$
$$\widehat{C}(T) = \frac{dH(T)}{dT}, \quad H(T) = \begin{cases} C_1 T, & T \leq T_1 \\ F(T), & T_1 < T < T_2 \\ H_2 + C_2 T, & T \geq T_2 \end{cases},$$
$$H_1 = C_1 T_1, \quad H_2 = H_1 + (C_1 + C_2)\frac{T_2 - T_1}{2} + L,$$
$$F(T) = [C_2(T_2 - T_1) + H_1 - H_2]\frac{(T - T_1)^2 (T - T_2)}{(T_1 - T_2)^3} + H_2 \frac{T - T_1}{T_2 - T_1}$$
$$+ [C_1(T_2 - T_1) + H_1 - H_2]\frac{(T - T_1)(T - T_2)^2}{(T_1 - T_2)^3} - H_1 \frac{T - T_2}{T_2 - T_1}$$

Here $C, T, k, \omega_b, C_b, T_c, L$ are the effective heat capacity, temperature, thermal conductivity of biological tissue, perfusion rate, heat capacity, temperature of blood in capillaries penetrating the biological tissues and latent heat of phase transition, correspondingly. Numerical values of the parameters used in the model are presented in table 1. The thermal conductivity coefficient k is linearly interpolated within an interval of $T \in (T_1, T_2)$, where T_1 and T_2 are the corresponding temperature values at the beginning and at the end of the phase transition process.

Graphs of temperature dependent regularized thermophysical quantities for healthy biological tissue and tumor tissue are shown in figure 2.

	Healthy tissue		Tumor tissue	
	$T > 272\,\mathrm{K}$	$T < 265\,\mathrm{K}$	$T > 272\,\mathrm{K}$	$T < 265\,\mathrm{K}$
C, $\mathrm{erg\,K^{-1}\,cm^{-3}}$	3.16×10^7	1.80×10^7	3.89×10^7	2.01×10^7
k, $\mathrm{erg\,K^{-1}\,cm^{-1}\,s^{-1}}$	0.48×10^5	1.68×10^5	0.56×10^5	2.22×10^5
$\omega_b C_b$, $\mathrm{erg\,K^{-1}\,cm^{-3}\,s^{-1}}$	24.2×10^4	0	48.5×10^4	0

Table 1. Thermophysical parameters of healthy tissue and tumor tissue.

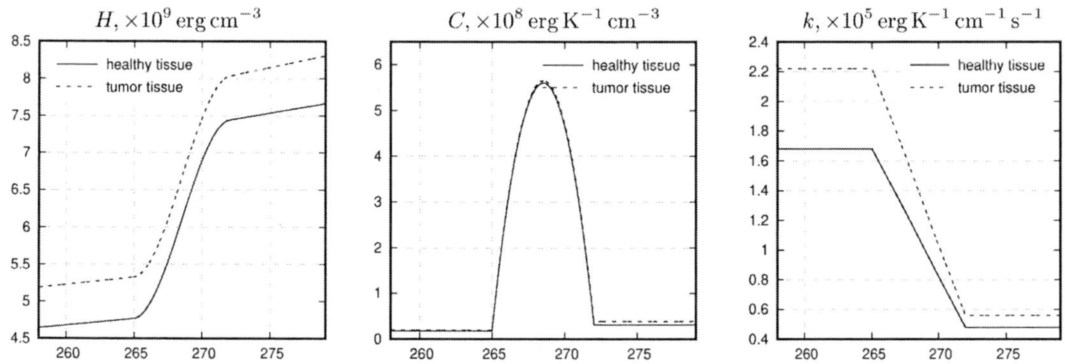

Figure 2. Plots of thermodynamic characteristics of a healthy (—— line) biological tissue and angioma (- - - - line).

For the numerical modeling of the blood flow in the artery the quasi-one-dimensional hemodynamic model was applied:

$$\begin{aligned}
&\frac{\partial S}{\partial t} + \frac{\partial}{\partial z}(Su) = 0, \quad \frac{\partial u}{\partial t} + \frac{\partial}{\partial z}\left(\frac{u^2}{2} + \frac{1}{\rho_b}P(S)\right) = 0, \\
&\frac{\partial T_b}{\partial t} + u\frac{\partial T_b}{\partial z} - \frac{k_b}{C_b}\frac{\partial^2 T_b}{\partial z^2} = q, \quad z \in \widehat{D}_0, \quad 0 < t, \\
&q(z,t) = \frac{2\sqrt{\pi}}{C_b\sqrt{S}} \oint_{\partial D_{S(z)}} \overline{k}\frac{T|_{r=R_1} - T_b(z,t)}{h} dl, \\
&\overline{k} = \frac{k_b\left(T_b(z,t)\right) + k\left(T|_{r=R_1}\right)}{2}, \quad P(S) = P_0 + \alpha\left(1 - \sqrt{\frac{S_0}{S}}\right).
\end{aligned} \tag{2}$$

Here S, u, P are the cross-sectional area of the vessel, the average cross-sectional velocity of blood flow and pressure, respectively. T_b is the average temperature of blood in the vessel, $k_b = 0.56 \times 10^5\,\mathrm{erg\,K^{-1}\,cm^{-1}\,s^{-1}}$, $C_b = 4.212 \times 10^7\,\mathrm{erg\,K^{-1}\,cm^{-3}}$ are thermal conductivity and heat capacity of blood. $\widehat{D}_0$ is the interval of height variation along the vessel $\widehat{D}_{00}$. $q(z,t)$ is the Newton's type heat exchange term describing averaged heat transfer between biological tissue and artery through the vessel wall with thickness of h. S_0, η are the cross-sectional area of the vessel in the equilibrium state and the relative change in the cross-sectional area correspondingly. $\partial D_{S(z)}$ is the boundary of the vessel cross-section.

IOP Conf. Series: Materials Science and Engineering **657** (2019) 012039 doi:10.1088/1757-899X/657/1/012039

The initial and boundary conditions used in this paper are:

$$T(\mathbf{r},t)|_{t=0} = T_0, \quad T_b(z,t)|_{t=0} = T_0, \quad u(z,t)|_{t=0} = u_0, \quad S(z,t)|_{t=0} = s_0,$$
$$T(\mathbf{r},t)|_{\mathbf{r}\in\partial D_1\setminus(\partial\hat{D}_{00}\cup\partial D_3)} = T_0, \quad k\frac{\partial T}{\partial n}\bigg|_{\mathbf{r}\in\partial D_0} = -\theta(\mathbf{r};T,T_b)\left(T|_{r=R_1} - T_b(z,t)\right),$$
$$T_b(z,t)|_{z=z_1} = T_0, \quad S(z,t)u(z,t)|_{z=z_1} = Q(t).$$

On the active zone of the cryoprobe surface $\Sigma_1 \subset \partial D_3$ the Dirichlet boundary condition $T(\mathbf{r} \in \Sigma_1) = T_s(t)$ was used. We use the homogeneous Neuman boundary condition on the surface separating domain $\Sigma_2 \subset \partial D_3$ from D_1.

3. Numerical algorithm

A finite volume explicit scheme for indexed convex grids [13, 14] is used for the numerical modeling of the tissue freezing problem:

$$C^n_{ijk}\frac{T^{n+1}_{ijk} - T^n_{ijk}}{\tau}V_{ijk} = -\left\{\sum_{l=1}^{6}(Q_F)_l\right\}^n_{ijk} + \omega_b C_b\left(T_c - T^n_{ijk}\right)V_{ijk}. \tag{3}$$

Here $\{(Q_F)_l\}^n_{ijk}$ is the heat flux through the l-th face of the cell v_{ijk} (of volume V_{ijk}) on the n-th time layer and $F^n_{ijk} = F(\mathbf{r}_{ijk}, t^n)$, where $\mathbf{r}_{ijk}$ is the center of the cell v_{ijk}. Heat fluxes through faces are calculated by means of the estimations for the space derivatives of the temperature field. For the improvement of applied scheme stability the flux relaxation approach [15, 16] is used. (1) is modified in following manner:

$$\begin{aligned} \widehat{C}\frac{\partial T}{\partial t} &= -\operatorname{div}\mathbf{Q} + \omega_b C_b\left(T_c - T\right), \\ \mathbf{Q} + \tau_r\frac{\partial}{\partial t}\mathbf{Q} &= \mathbf{Q}_F. \end{aligned} \tag{4}$$

Here $\mathbf{Q}_F = -\kappa\nabla T$ is Fourier flux, $\tau_r \sim \tau$ is the stabilizing parameter [17,18]. Thus the following estimation for the heat flux can be obtained:

$$\mathbf{Q}^n = D\mathbf{Q}^{n-1} + (1-D)\mathbf{Q}^n_F, \quad D = e^{-\frac{\tau}{\tau_r}}.$$

Taking into account the heat equilibrium state of the tissue before beginning of the freezing process, the trivial initial conditions for the fluxes are used.

For the numerical computation of the hemodynamical part of the model the process decomposition method is applied. Finite difference approximation of the hyperbolical part is given by the hybrid characteristic scheme [19] with Fedorenko type switching analyzer [20]. Reaction-diffusion part is approximated directly.

4. Numerical results

The angioma is present in model as a sphere with the radius of 0.75 cm. The distance between the center of the sphere and the axis of the vessel is 0.75 cm. The center of the sphere is located on a straight line passing through the origin and the cryotip that is located on a radial beam in the $z = 0$ plane. The active zone of the cryoprobe tip has a fixed square cross-section of 0.02 cm^2 and length of $l_c = 0.5$ cm. The frontal part of the tip is located 0.35 cm away from the vessel wall.

Note that the cryoprobe parameters are corresponded to the parameters of industrial cryotips already used during the cryosurgery procedures. In our model the parameters are based on the parameters of Endocare cryoprobes. The protocol of the cryosurgery procedure was $T_s(t) = \max(310 - t, 77.4)\,\mathrm{K}$, corresponding to the constant cooling rate 1 K per second.

Preliminary numerical experiments showed that in order to obtain the results that correspond to the error of the order of 1 °C, it is sufficient to use a rather coarse grid (grid step ≈ 1 mm).

The amount of blood entering the artery is

$$Q(t) = Q_1 + Q_0 e^{-10^3(t \mod 1 - 0.5)^2},$$

where Q_1 is the background flow, Q_0 is the flow associated with heartbeats.

The basic values of the parameters used in the process of numerical modeling are presented by the following set of values: $h = 1$ mm, $R_1 = 3$ mm, $Q_1 = 3\,\mathrm{ml\,s^{-1}}$ and $Q_0 = 15\,\mathrm{ml\,s^{-1}}$.

Figure 4(b) presents a graph of the phase transition boundary at $z = 0$ through $t = 300$ s after the beginning of cryogenic exposure. For convenience of comparison, all the results of numerical simulation obtained below are given at the same time $t = 300$ s at $z = 0$. Figure 3 shows two orthogonal sections ($z = 0$ and $y = 0$) of phase transition boundary. Also it shows two iso-surfaces of temperature at $t = 265$ K and $t = 272$ K that correspond to the temperature interval of phase transition.

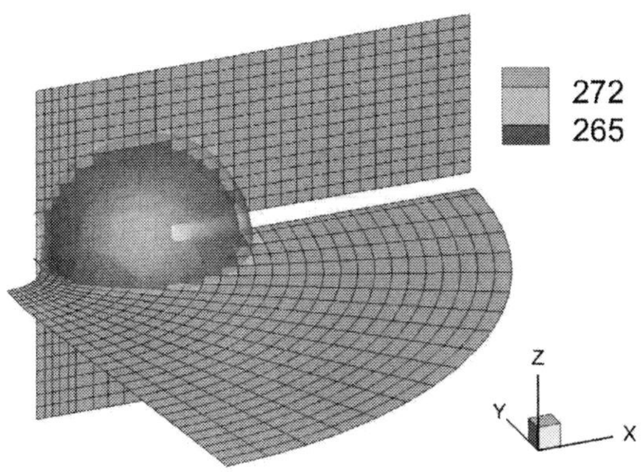

Figure 3. The two orthogonal sections ($z = 0$ and $y = 0$) of phase transition boundary at $t = 300$ s.

Firstly, the effect of blood perfusion on the propagation of the freezing front was investigated. From figures 4(c) and 4(d) it can be seen that an increase of the perfusion coefficient $\omega_b C_b$ leads to a decrease of the volume of the area that has been frozen. The shape of the area is preserved.

A decrease of wall thickness (see figure 4(e)) leads to an increase of heat flow through the vessel wall. Thus, the leading edge of the phase transition is located at a greater distance from the vessel wall. When the vessel wall thickness is doubled (see figure 4(f)) the vessel wall is cooled to lower temperatures.

The change of cross-sectional area (figures 4(g) and 4(h)) has little effect on the volume of the frozen part of the tissue. The shape of the frozen area adjacent to the vessel slightly varies due to a small change in the curvature radius of the vessel wall.

A change in the flow parameters (an increase of the background flow rate and amplitude of the peak velocity) does not lead to noticeable changes, the phase transition front is not distinguishable in the graphs.

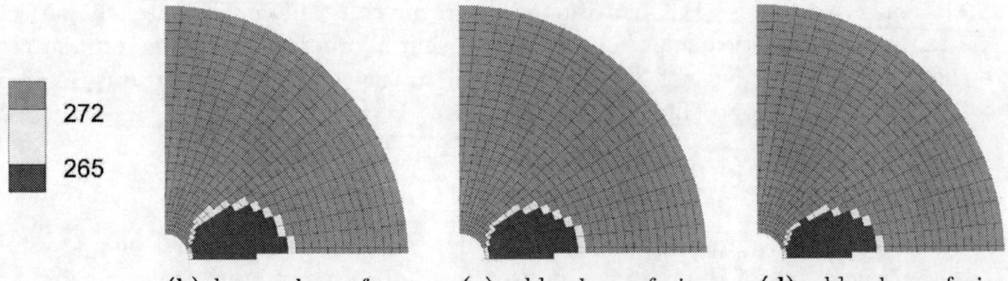

(a) legend **(b)** base values of parameters **(c)** blood perfusion $2\omega_b C_b$ **(d)** blood perfusion $3\omega_b C_b$

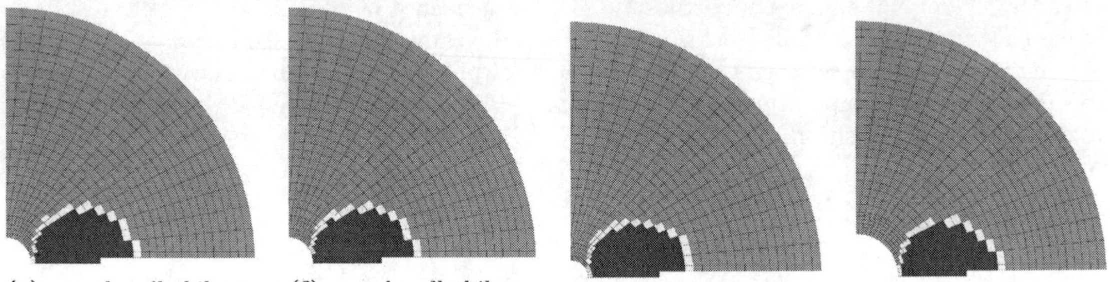

(e) vessel wall thikness $\frac{1}{2}h$ **(f)** vessel wall thikness $2h$ **(g)** vessel radius $\frac{2}{3}R_1$ **(h)** vessel radius $\frac{5}{3}R_1$

Figure 4. Boundary of phase transition in section $z = 0$ at time $t = 300\,\text{s}$.

5. Conclusion

The direct numerical simulation of the temperature distribution evolution in biological tissue near a large artery and the calculation of the blood flow in the artery were made.

The mathematical model of heat transfer process in tissue is based on the Pennes model with enthalpy modification. To describe the blood flow in the artery the quasi-one-dimensional hemodynamic model is used. The model takes into account local changes in the elastic properties of the vessel wall during strong cooling of the adjacent biological tissue.

The proposed model was simulated via numerical algorithm based on explicit finite-volume method and the hybrid characteristic scheme (after splitting the equation with respect to physical processes) for numerical solution of hyperbolic system.

The influence of circulatory system parameters such as vessel wall thickness, blood flow speed and capillary blood perfusion rate is numerically studied. It is shown that capillary blood perfusion, artery size and the thickness of its walls can cause valuable changes in tissue necrosis propagation when the influence of blood flow speed is negligible.

Acknowledgments

This work is supported by the Russian Science Foundation under grant No.18-71-00108.

References

[1] Gdal-On M and Gelfand Y A 1998 *Ophthal. Surg. Lasers* **29** 969–73
[2] Grunder W, Goldammer A, Schober R and Vitzthum H E 2003 *Z. Med. Phys.* **13** 203–7
[3] Baust J, Gage A A, Ma H and Zhang C 1997 *Cryobiology* **34** 373–84
[4] Stanczyk M and Telega J J 2003 *Acta Bioeng. Biomech.* **5** 3–22
[5] Mazur P 1963 *J. Gen. Physiol.* **47** 347–69
[6] Toner M, Cravalho E G and Karel M 1990 *J. Appl. Phys.* **67** 1582–93

[7] Zhao G, Takamatsu H and He X 2014 *J. Appl. Phys.* **115** 144701
[8] Devireddy R V, Smith D J and Bischof J C 2002 *J. Heat. Transf.* **124** 365–74
[9] Pennes H H 1948 *J. Appl. Physiol.* **1** 93–122
[10] Shilnikov K E, Kudryashov N A and Gaiur I Y 2017 *J. Phys. Conf. Ser.* **937** 012048
[11] Vassilevski Y V, Salamatova V Y and Simakov S S 2015 *Comp. Math. Math. Phys.* **55** 1567–78
[12] Olufsen M S, Peskin C S, Kim W Y, Pedersen E M, Nadim A and Larsen J 2000 *Ann. Biomed. Eng.* **28** 1281–99
[13] Shilnikov E V 2014 Viscous gas flow simulation based on QGD equations system on non-orthogonal index grids *Preprints of Keldysh Institute of Applied Mathematics RAS, 33*
[14] Kudryashov N A and Shilnikov K E 2016 *Math. Models Comput. Simul.* **8** 680–8
[15] Chetverushkin B N, Shilnikov E V and Davydov A A 2013 *Adv. Eng. Softw.* **60** 42–7
[16] Chetverushkin B N and Shilnikov E V 2010 *Proceedings of the II International Conference on Engineering Optimization, EngOpt-2010 (Lisbon)*
[17] Myshetskaya E E and Tishkin V F 2015 *Comp. Math. Math. Phys.* **55** 1270–5
[18] Chetverushkin B N 2013 *Math. Mod. and Comp. Simul.* **5** 266 – 79
[19] Magomedov K M and Kholodov A S 1969 *USSR Comp. Math. Math. Phys.* **9** 158–76
[20] Fedorenko R P 1963 *USSR Comp. Math. Math. Phys.* **2** 1355–65

IOP Conf. Series: Materials Science and Engineering **657** (2019) 012040 doi:10.1088/1757-899X/657/1/012040

Thermal and mechanical characteristics of contact friction pair based on 3-D wheel/rail-foundation contact vertical system

Y P Wei[1,2,3] and Y P Wu[1]

[1]School of Civil Engineering, Lanzhou Jiaotong University, Lanzhou 730070, China
[2]School of Civil Engineering, Qinghai University, Xining 810016, China

E-mail: ypweiinchina2@126.com

Abstract. The main goal of this article is to analyze the influence of wheel/rail friction heat on the wheel/rail contact friction pair. The nonlinear constitutive relation and heat conduction equation are established firstly. In addition, a direct coupling model of 3-D wheel/rail-foundation contact vertical system is established. The mechanical, thermal and geometrical characteristics of wheel/rail friction pair are researched. The results show that the temperature rise on the wheel/rail contact surface is significant, and the influence of friction heat on the contact patch, contact pressure and stress is obvious. However, the influence focuses on the surface and subsurface of the wheel/rail contact. Once the depth reaches certain distance below the contact surface, the influence can be ignored. The temperature does not change when the distance to the rail surface surpasses 3.28 mm; also the influence of temperature rise on the stress becomes very weak when the distance exceeds 8.05 mm.

1. Introduction

Comparing with the motor and airplane, railway transport has incomparable advantages. Railway has great influence on the communication, economy and environment, etc. The railway transport has increasingly developed in recent years. A lot of researchers widely carry out the studies about material strength, heavy haul railway, high-speed railway and wheel/rail relationship, etc. Wheel and rail are the important components of railway, playing vital roles in ensuring safety and transferring load. The research of wheel/rail relationship includes the analysis of fatigue damage, wear, creep and friction, etc. The friction is one of the common phenomena when a train is braking or passing through a small radius curved track. The friction heat can cause the peeling of the wheel/rail surface [1], and it is tightly linked with the surface scratch of wheel and rail. Recently, the friction and stress of wheel/rail contact has been widely researched. Li *et al* [2] established a 2-D rail model to investigate the influence of creep on the temperature and residual stress. The research results show that friction heat has a significant effect on the residual stress when the creepage is large. Kuminek *et al* [3], Wu [4] and Pun *et al* [5] widely researched the temperature and stress of wheel/rail under condition of sliding or rolling contact. Esmaeili *et al* [6] predicted that the growing of thermo-mechanical crack on the wheel treads when the tread bears mechanical and thermal load because of the rolling contact and braking. A 3-D model is used to analyze the temperature and strain feature of wheel/rail when a vehicle wheel slides on the rail surface in [7]. The calculation results show that the data of 3-D model are closer to the actual situation and more reasonable than the results gotten from plane model. Liu *et al* [8], Athukorala *et al* [9] and Nikas *et al* [10] studied the damage of wheel/rail contact.

Since there are many difficulties in accurately measuring the temperature and stress of wheel/rail contact region through experiment, the numerical simulation becomes a good solution. Many researchers studied the wheel/rail contact and friction heat by means of finite element methods. A large number of vital research results have been obtained [1-3,11-13]. Those research results have positive effect on the understanding of wheel/rail contact. In the previous studies, some of the models are 2-D models, the calculation parameter and the friction coefficient are not temperature-dependent. Besides, the friction heat of model sometimes is replaced by a moving heat source. Meanwhile, the tread profiles of wheel/rail are always simplified. In this article, the nonlinear constitutive relation and heat conduction equation are firstly established. The mechanical, thermal and geometrical features of wheel/rail friction pair are researched. The node temperature and the influences of temperature on node stress are deeply analyzed. Of course, the thermal and mechanical parameters of wheel/rail material are seen as the function of temperature. The geometry of wheel/rail are real. Moreover, the coupling of wheel/rail is direct coupling.

2. Coupling equation

2.1. Equation of friction heat transfer

Wheel/rail contact links the vehicle with the track. The force which acts on the contact surface of wheel/rail is especially complex. Large amount of friction heat will generate on contact surface during the sliding of wheel on rail surface. The schematic diagram of heat conduction on rail surface can be illustrated in figure 1 [7].

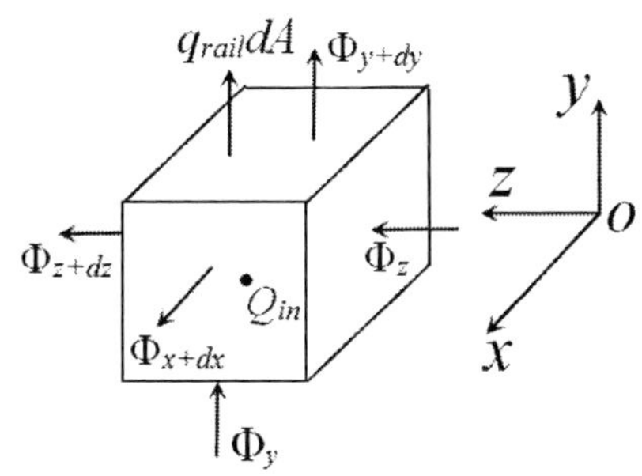

Figure 1. Micro element of heat conduction.

where Φ_x, Φ_y and Φ_z is respectively the heat flow in three coordinate axes. Q_{in} is internal heat generation per unit time and volume. q_{rail} and q_{wheel} are the heat flux transferred into the rail and wheel contact surface. They can be written as follows [7],

$$\begin{cases} q_{rail} = \xi \gamma q_t = \xi \gamma \mu(T) v p(x,z) \\ q_{wheel} = \xi(1-\gamma) q_t = \xi(1-\gamma)\mu(T) v p(x,z) \end{cases} \tag{1}$$

where q_t is the total braking mechanical energy between wheel and rail in unit time and area. ξ is heat generation rate. γ is the heat partition coefficient between the wheel and rail, and its value is 0.5. $\mu(T)$ is temperature-dependent friction coefficient. v is the sliding speed of wheel. $p(x, z)$ is contact pressure, it is [14]:

$$p(x,z) = \begin{cases} 0 & \text{if,} \quad u > 0, \text{separation} \\ K_n \cdot u_n + \eta_{i+1} & \text{if,} \quad u \le 0, \text{contact} \end{cases} \tag{2}$$

where K_n is wheel/rail contact stiffness, u_n is the distance between contact interfaces. η_{i+1} is

$$\eta_{i+1} = \begin{cases} \eta_i + K_n \cdot u_n, & \text{if, } |u_n| > \Delta \\ \eta_i, & \text{if, } |u_n| < \Delta \end{cases} \tag{3}$$

where Δ is calculation tolerance of wheel/rail, η_i is Lagrange multiplier.

Based on the Fourier law and the law of energy conservation [7,8], the equation of heat conduction in the wheel/rail contact region can be written in integral form as follows,

$$\int_V \rho C(T)\frac{\partial T}{\partial t}dV + \int_V \nabla \bullet [-k(T)\nabla T]dV = \int_V Q_{in}dV + \int_A [\xi\gamma\mu(T)p(x,z)v]dxdzdV \tag{4}$$

where ρ is steel density, its value is 7790 kgm^{-3}, and it does not change with temperature. $C(T)$ is specific heat capacity. $k(T)$ is thermal conductivity. $C(T)$ and $k(T)$ are function of temperature (table 1). Q_{in} is internal heat source of wheel/rail, its value is 0. ∇ is Hamiltonian Operator.

Table 1. Mechanical and thermal parameter.

T (°C)	Elastic modulus (GPa)	Poisson's ratio	Yield Stress (MPa)	Tensile Strength (MPa)	$C(T)$ (J·kg^{-1}·°C^{-1})	$k(T)$ (W·m^{-1}·°C^{-1})	Coefficient of thermal expansion (°C^{-1})	$\mu(T)$
25	209	0.30	608	1000	490.1	47.7	11.0×10^{-6}	0.334
100	207	0.30	608	998.9	499.9	48.9	11.6×10^{-6}	0.301
650	105	0.36	502	985.7	571.5	57.8	14.8×10^{-6}	0.139
1000	50	0.39	237.9	740.9	617.1	63.4	15.7×10^{-6}	0.085
1450	2.1	0.40	7	42	671.8	76.4	16.1×10^{-6}	0.045

2.2. Equation of thermal stress

The thermal and mechanical parameters are seen as function of temperature in equation (4). At the same time, the above temperature-dependent parameters should be considered in the constitutive relation [15]. Based on the free energy of isotropic material, the constitutive relation considering the temperature-dependent parameters can be obtained as follows,

$$\sigma_{ij} = 2G(T)\varepsilon_{ij} + \lambda(T)\varepsilon_{kk}\delta_{ij} - \beta(T)\Delta T\delta_{ij} \tag{5}$$

where ε_{kk} is body strain, δ_{ij} is Kronecker symbol. ΔT is temperature difference, and the temperature is controlled by the equation (4). $G(T)$, $\lambda(T)$ and $\beta(T)$ is respectively shear modulus, Lame constant and thermal stress coefficient. They can be described as follows,

$$G(T) = G_0 + G_1T^*,\ \lambda(T) = \lambda_0 + \lambda_1T^*,\ \beta(T) = \beta_0 + \beta_1T^* \tag{6}$$

where G_0, λ_0 and β_0 is respectively shear modulus, Lame constant and thermal stress coefficient at initial time. G_1, λ_1 and β_1 is respectively coefficient of influence of the unit relative temperature difference on the shear modulus, Lame constant and thermal stress coefficient. T^* is relative temperature difference, it is equal to

$$T^* = \Delta T / T_0 \tag{7}$$

where T_0 is initial temperature.

2.3. Thermal boundary condition

The thermal boundary conditions of wheel/rail friction can be written as follows [16]:

$$q_c = h(T - T_s),\ q_s = \varepsilon\sigma(T^4 - T_s^4) \tag{8}$$

where h is heat transfer coefficient, ε is emissivity, σ is Stefan-Boltzmann constant. T is temperature of wheel/rail. T_s is surrounding temperature.

IOP Conf. Series: Materials Science and Engineering **657** (2019) 012040 doi:10.1088/1757-899X/657/1/012040

3. Finite element model

The finite element model of wheel/rail-foundation contact vertical system is built by the ANSYS software [14]. Solid226 is selected to simulate the wheel/rail component. Conta174 and Targe170 are adopted to simulate the contact behavior of wheel/rail. Fasteners and foundation are simulated by Combin14. The Solid226, Conta174 and Targe170 have structural-thermal capability. In order to reduce the element and node numbers, the multiple point constraint (MPC) method is used. The model has 48672 nodes and 42101 elements. The smallest element size is 0.5 mm in this article [7,8]. At the same time, the coupling of both heat and force is strong coupling, and the full integration method is adopted. At last, the Solid186 is selected to establish temperature-independent model, in which the temperature is not considered. And calculation model is shown in figure 2.

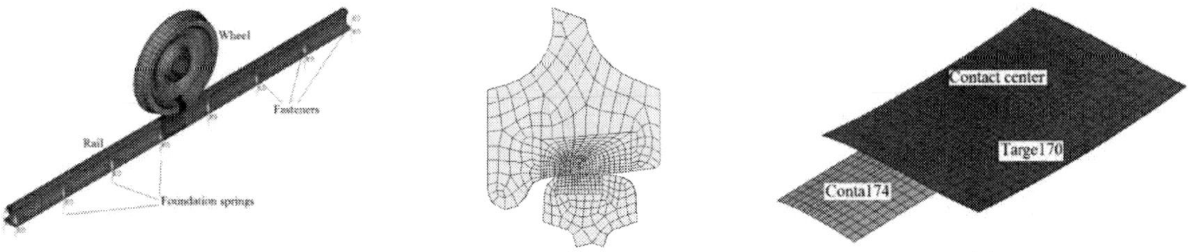

(a) (b) (c)

Figure 2. 3-D finite element model. (a) Wheel/rail-foundation contact system, (b) Detail of contact and (c) Contact pair.

When the rail length exceeds 4.2 m, the rail length has little effect on the result. Therefore, the length of rail in FEM is 4.2 m [8]. The longitudinal and transverse displacement on the two sides of rail are constrained. The vertical displacement at the bottom of foundation springs is constrained. For the fastener springs, the vertical displacement at the top is constrained. The wheel can slide on the rail surface along longitudinal direction, but it can not move in the transverse direction.

The wheel load is 100 kN. The initial temperature is 20°C, the heat transfer coefficient is 25 $W \cdot m^{-2} \cdot K^{-1}$. Stefan-Boltzmann constant is 5.68×10^{-8} $W \cdot m^{-2} \cdot K^{-4}$. The stiffness of single fastener is 2 $MN \cdot m^{-1}$, and the foundation stiffness is 50 $MN \cdot m^{-1}$ [17]. The wheel sliding speed is equal to 1.0 $m \cdot s^{-1}$ [7]. The sliding distance is 0.1 m. The material properties used for the FEA are listed in table 1 [18].

4. Result

4.1. Stress of wheel/rail friction pair

In this section, the basic features of stress on wheel/rail friction pair are illustrated. The contact characteristics is very fundamental for wheel/rail friction pair. The maps of von Mises stress on the wheel/rail surface at the beginning of sliding are presented in figure 3. The von Mises stress fields obtained by sectioning the wheel/rail through the contact center in two directions are illustrated in figure 4. The contact patch of wheel/rail is shown in figure 6.

Figure 3 shows the von Mises stress of wheel/rail surface is very large in the contact center. On wheel and rail surface, the maximum von Mises stress is respectively 600.3 MPa and 585.6 MPa at the initial time (t=0 s, sliding distance (s) is equal to 0 m). The maximum stress is close to the yield stress of wheel/rail.

Figure 4 shows the phenomenon of stress concentration is very obvious near the wheel/rail contact region. The von Mises stress fields of wheel/rail contact region are similar to figure 4 in two directions during each iteration. The calculation result shows that the maximum von Mises stress occurs at the subsurface of wheel/rail. In the range of 3.93 mm below the wheel/rail surface, von Mises stress is close to the maximum value. The stress of the wheel/rail contact area is very large, which is one of the reasons that leads to the damage of wheel and rail (figure 5 [19,20]). The result calculated in this paper is in agreement with the cases of wheel/rail damage.

IOP Conf. Series: Materials Science and Engineering **657** (2019) 012040 doi:10.1088/1757-899X/657/1/012040

(a) (b)

Figure 3. Von Mises stress (t=0 s, s=0 m, Pa). (a) Wheel surface and (b) Rail surface.

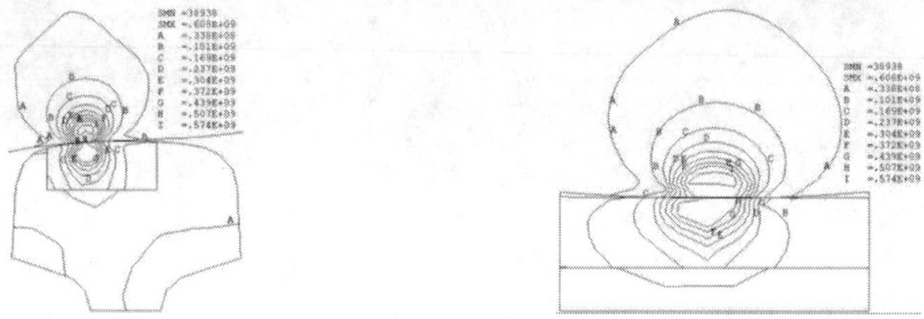

(a) (b)

Figure 4. Von Mises stress fields (t=0 s, s=0 m, Pa). (a)Transverse direction and (b) Longitudinal direction.

Figure 5. Damage of wheel/rail.

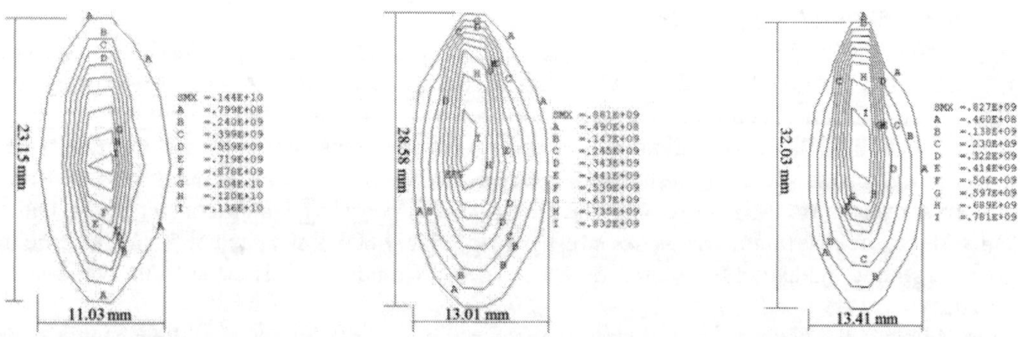

(a) (b) (c)

Figure 6. Contact patch. (a) t=0 s, s=0 m (P_{max}=1440 MPa,A=153.1 mm^2), (b) t=0.05 s, s=0.05 m (P_{max}=881 MPa A=245.4 mm^2) and (c) t=0.1 s, s=0.1 m (P_{max}=827 MPa, A=272.6 mm^2).

Figure 6 shows the shape of the wheel/rail contact patch on the rail surface is similar to ellipse. The shape does not change so much in the process of sliding. The minor axis and major axis of contact patch at the initial contact is respectively 11.03 mm and 23.15 mm. However, because of temperature rise, the contact area increases by 78.1% and the maximum pressure decreases by 42.6% at the end of sliding time (figure 6).

4.2. *Temperature of wheel/rail friction pair*

The stripping and abrasion of wheel/rail surface are closely related to friction heat. Huge friction heat generates during the wheel sliding, and the heat will influence the stress of wheel/rail contact region (equations (4) and (5)). Figure 7(a) is temperature sectioning maps in longitudinal and transverse direction through the contact point. Figures 7(b) and 7(c) is temperature field of the wheel and rail surface respectively.

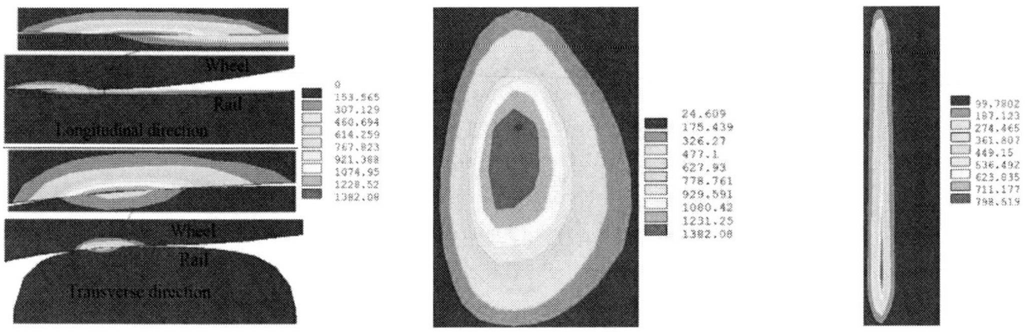

(a) (b) (c)

Figure 7. Temperature field (t=0.1 s, s=0.1 m, °C). (a) Slicing map, (b) Wheel surface and (c) Rail surface.

The temperature field is similar to the figure 7(a) at every moment during wheel sliding. The contact location of wheel does not change, so the temperature on the wheel surface is much higher than that on the rail surface. The high temperature region locates on the wheel/rail surface. However, the influence area and depth of temperature are limited. For wheel and rail surfaces, figure 7 shows the highest temperature is respectively 1390°C and 798.5°C. On the rail, the influence of temperature on the depth is no more than 3.28 mm. However, the temperature influence depth on wheel is two times as large as that on rail.

4.3. *Node temperature and stress of rail*

In order to research the thermo-mechanical coupling effect, seven nodes are selected from rail surface and subsurface (figure 8(a)). The nodes locate at the depth of 0~14.92 mm below the rail surface. The relationship between node temperature and wheel sliding distance is presented in figure 8(b). The von Mises stress curve obtained from temperature-dependent model is shown in figure 8(c). Figure 8(d) illustrates the von Mises stress curve obtained from the temperature-independent model. Moreover, the distribution of maximum von Mises stress of nodes along depth is plotted in figure 8(e).

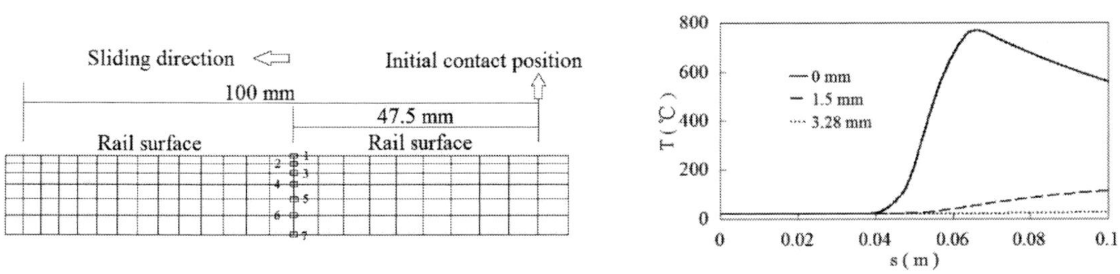

(a) (b)

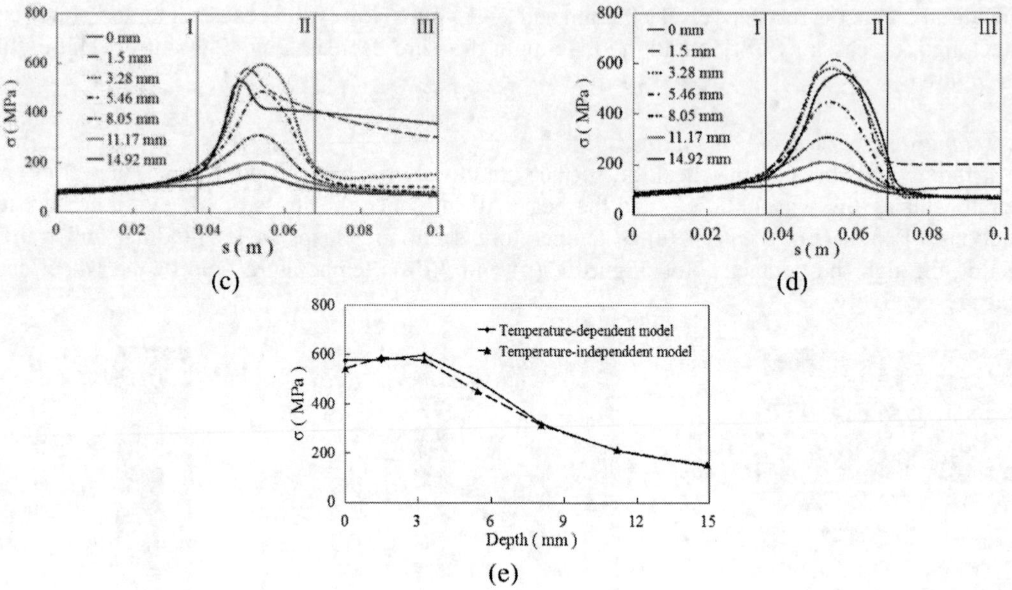

Figure 8. Temperature and von Mises stress curve. (a) Node position, (b) Node temperature (°C), (c) Node stress with friction heat (MPa), (d) Node stress without friction heat (MPa) and (e) Maximum von Mises stress vs depth (MPa).

Figure 8(b) shows the temperature curve of node 1 is kept at room temperature before wheel/rail contact. The temperature rises rapidly during wheel/rail contact and then decreases gradually. The temperature time history curve of node 1 corresponds to three processes: the wheel approaching the node 1 (stage I), sliding over the node 1 (stage II) and departing away the node 1 (stage III). However, the temperature of nodes 2 ascends slowly with the sliding distance increasing. And the temperature change of node 3 is not obvious.

Both figure 8(c) and (d) indicate that there are three stages in von Mises stress curves. However, the curves of node 1 and 2 in figure 8(c) are obviously different from the curves at the same locations in figure 8(d). Figure 8(c) shows that the curves of both node 1 and 2 decrease gradually after reaching the maximum values. It is mainly because the temperature rise of node 1 and 2 is obvious in stage II and III (figure 8(b)). The temperature has obvious influence on the von Mises stress of node 1 and 2. In the stage III, the von Mises stress of both node 1 and 2 in figure 8(c) is respectively 70.9 MPa and 170.2 MPa higher than that in figure 8(d).

Figure 8(e) presents the change law of two curves is same. The maximum von Mises stress of nodes increases slowly first and then decreases gradually with the increasing of depth. The maximum value occurs at the rail subsurface. Because the influence of temperature on the von Mises stress descends gradually with the growth of the depth, the maximum von Mises stress calculated through two models are almost the same when the depth exceeds 8.05 mm.

5. Conclusions

In this paper, the stress and temperature of wheel/rail sliding contact are studied. The calculation results indicate that the friction heat has a significant effect on the stress and the contact patch. Main conclusions are as follows:

- The maximum von Mises stress of wheel and rail surface is respectively 600.3 MPa and 561.7 MPa at initial time. In the range of 3.93 mm below the wheel/rail surface, the von Mises stress is close to the maximum value. At the initial contact moment, the minor axis and major axis of contact patch is 11.03 mm and 23.2 mm respectively.
- The temperature of wheel/rail surface increases with the sliding distance increasing. Comparing

with the initial contact, the contact area ascends by 78.1% and the maximum pressure decreases by 42.6% because of temperature rise at the end of sliding time. And the highest temperature of wheel and rail is respectively 1390°C and 798.5°C. However, the influence of temperature on depth is limited, and the value is no more than 3.28 mm on the rail.

- The stress curves of rail surface and subsurface (node 1 and 2) obtained from the temperature-dependent model are obviously different from the curves in temperature-independent model. Especially in the stage III of the curves, the von Mises stress obtained from the temperature-dependent model is significantly higher than the stress calculated by the temperature-independent model.
- The maximum von Mises stress of nodes ascends slowly first and then descends gradually with the growth of depth. The maximum value occurs at the rail subsurface. However, when the depth exceeds 8.05 mm, the difference of von Mises stress in two models can be ignored.
- Based on the actual geometric dimension and the thermal property of material, the mechanical, thermal and geometrical characteristics of wheel/rail friction pair are truly reflected.

Acknowledgments
This study is a key project supported by the National Natural Science Foundation of China (grant number 51236003).

References

[1] Ahlstom J and Karlsson B 2002 Modeling of heat conduction and phase transformations during wheel sliding-theoretical predictions and comparison with results of full-scale experiments *Wear* **253** 291-300
[2] Li W, Wen Z F, Jin X S and Wu L 2014 Numerical analysis of rolling-sliding contact with the frictional heat in rail *Chinese J. Mech. Eng-En.* **27** 41-9
[3] Tomasz K and Krzysztof A 2014 Methodology and verification of calculations for contact stresses in a wheel-rail system *Vehicle Syst. Dyn.* **52** 111-24
[4] Wu B, Wen Z F, Wu T and Jin X S 2016 Analysis on thermal effect on high-speed wheel/rail adhesion under interfacial contamination using a three-dimensional model with surface roughness *Wear* **366-367** 95-104
[5] Pun C L, Kan Q, Mutton P J, Kang G and Yan W 2014 A single parameter to evaluate stress state in rail head for rolling contact fatigue analysis *Fatigue Fract. Eng. M.* **37** 909-19
[6] Esmaeili A, Walia M S, Handa K, Ikeuchi K, Ekh M, Vernersson T and Ahlström J 2017 A methodology to predict thermomechanical cracking of railway wheel treads: From experiments to numerical predictions *Int. J. Fatigue* **105** 71-85
[7] Wu Y P, Wei Y P, Liu Y, Duan Z D and Wang L B 2017 3-D analysis of thermal-mechanical behavior of wheel/rail sliding contact considering temperature characteristics of materials *Appl. Therm. Eng.* **115** 455-62
[8] Liu Y, Jiang S, Wu Y P, Duan Z D and Wang L B 2016 Effects of spalling on rail thermo-elasto-plasticity in wheel-rail sliding contacts *J. Traffic Transp. Eng.* **16** 47-54 (in Chinese)
[9] Athukorala A C, De Pellegrin D V and Kourousis K I 2017 A unified material model to predict ratcheting response in head-hardened rail steel due to non-uniform hardness distributions *Tribol. Int.* **111** 26-38
[10] Nikas D, Ahlström J and Malakizadi A 2016 Mechanical properties and fatigue behaviour of railway wheel steels as influenced by mechanical and thermal loadings *Wear* **366-367** 407-15
[11] Ahlström J, Kabo E and Ekberg A 2016 Temperature-dependent evolution of the cyclic yield stress of railway wheel steels *Wear* **366-367** 378-82
[12] Vo K D, Tieu A K, Zhu H T and Kosasih P B 2015 The influence of high temperature due to high adhesion condition on rail damage *Wear* **330-331** 571-80
[13] Meysam N, Li S G, Li Z L, Wu R 2017 Thermomechanical analysis of the wheel-rail contact

IOP Conf. Series: Materials Science and Engineering **657** (2019) 012040 doi:10.1088/1757-899X/657/1/012040

using a coupled modelling procedure *Tribol. Int.* **117** 250-60
[14] Alawadhi and Esam M 2015 *Finite Element Simulations Using Ansys* 2nd Ed. (Boca Raton: CRC Press)
[15] Huang H G 1989 *Introduction to Thermoelasticity* (Beijing: Tsinghua University Press) (in Chinese)
[16] Li Z L 2002 *Wheel-Rail Rolling Contact and Its Application to Wear Simulation* (Delft: Delft University Press)
[17] Wang K Y, Cai C B and Zhu S Y 2013 Vertical dynamic model and vibration characteristic of rail fastening system *Engineering Mechanics* **30** 146-9 (in Chinese)
[18] Chen Y C and Lee S Y 2009 Elastic-plastic wheel-rail thermal contact on corrugated rails during wheel braking *J. Tribol.* **131** 1-9
[19] Nan X 2011 Method of thermo-mecanical fatigue strength assessment on railway heavy-haul freight car wheel plate (Beijing: Beijing Jiaotong University) p 2 (in Chinese)
[20] Xing L X 2008 Research on defect characteristics and classification of higher speed rails (Beijing: China Academy of Railway Sciences) p 66 (in Chinese)

IOP Conf. Series: Materials Science and Engineering **657** (2019) 012041 doi:10.1088/1757-899X/657/1/012041

Relation between stress intensity factor of circumferential crack and adhesive thickness in bonded round bar

K Oda[1,3], Y Kasamura[2], M Nishi[2] and N Tsutsumi[1]

[1]Division of Mechanical Engineering, Faculty of Science and Technology, Oita University, 700 Dannoharu, Oita 870-1192, Japan
[2]Mechanical and Energy Systems Engineering Course, Graduate School of Engineering, Oita University, 700 Dannoharu, Oita 870-1192, Japan

E-mail: oda-kazuhiro@oita-u.ac.jp

Abstract. The fracture of the bonded dissimilar materials generally occurs near the interface or at the interface edge. In this study, the stress intensity factors of the circumferential crack near the interface and the circumferential interface crack are analysed by the crack tip stress method. The relation between the dimensionless stress intensity factor of the circumferential crack and the adhesive thickness is investigated. The dimensionless stress intensity factors are defined by using the singular stress value at the crack tip point in bonded round bar without crack. The dimensionless stress intensity factors based on the singular stress value become constant irrespective of the adhesive thickness. In addition, the dimensionless stress intensity factor of the circumferential crack in bonded round bar is compared with that of the edge crack in bonded plate. Both factors coincide with each other when the geometrical condition and material combinations are the same.

1. Introduction

Adhesive bonding has advantages such as bonding of different kind of materials, reduction of weight, smoothing of parts surfaces, reduction of equipment cost and workload. Therefore, it is used in various fields such as industrial and medical fields such as automobiles, airplanes and dental adhesives. With the spread of adhesive structures, the strength evaluation of adhesive bonding has become an important problem [1-11].

Numerous studies have been conducted on the adhesive strength of materials. Noda *et al* [5-8] have clarified that the intensity of the singular stress field (ISSF) at the interface edge is useful as a criterion for the interface strength. Furthermore, Noda *et al* [9,10] reported an efficient numerical method to calculate the intensity of singular stress field (ISSF) at the interface edge by using the ratio of the stress values along the interface analysed by the finite element method (FEM). Naito *et al* [11] have studied the effects of adhesive thickness on static joint strength of polyimide adhesive for butt jointed round bar and single lap joints from the experiments and FEM stress analysis.

The authors have studied the interface strength evaluation based on the stress intensity factors of the interface crack and the edge crack near the interface in butt joint plate [12-15]. The previous studies have shown that the stress intensity factors of small interface crack and small edge crack near the interface in butt jointed plate regularized by the corner singular stress at the crack tip position when there is no crack is constant regardless of the adhesive thickness. However, the method of describing the dimensionless stress intensity factors for three-dimensional problems has not been studied.

IOP Conf. Series: Materials Science and Engineering **657** (2019) 012041 doi:10.1088/1757-899X/657/1/012041

In this study, problems of circumferential crack close to the interface and circumferential interface crack in a butt jointed round bar will be treated. The stress intensity factors (SIFs) for circumferential cracks will be evaluated by varying the crack depth and the adhesive thickness, and the useful expression of the circumferential crack solution will be examined.

2. Analysis method and conditions

The aim of this study is to examine the strength evaluation method of bonded round bar shown in figure 1(a). In order to evaluate the intensity of the singular stress field at the interface edge, the circumferential crack near the interface and the circumferential interface crack as shown in figures 1(b) and 1(c) are assumed in this study. This is because the stress intensity factor of the small crack near the interface is strongly affected by the intensity of the singular stress field at interface edge. In figure 1, the notations '*a*', '*c*', '*h*' and '*d*' indicate the crack depth, the distance from the interface to the crack, the adhesive layer thickness and the diameter of round bar, respectively, and the red point indicates the crack tip position.

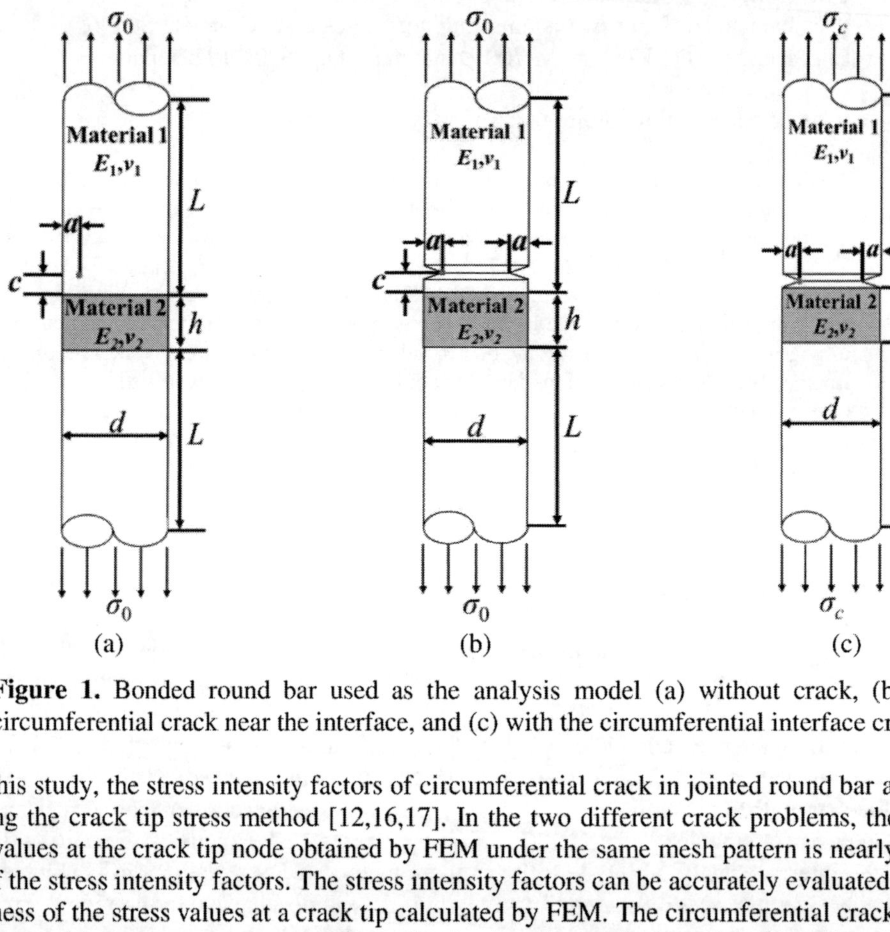

Figure 1. Bonded round bar used as the analysis model (a) without crack, (b) with circumferential crack near the interface, and (c) with the circumferential interface crack.

In this study, the stress intensity factors of circumferential crack in jointed round bar are calculated by using the crack tip stress method [12,16,17]. In the two different crack problems, the ratio of the stress values at the crack tip node obtained by FEM under the same mesh pattern is nearly equal to the ratio of the stress intensity factors. The stress intensity factors can be accurately evaluated based on the usefulness of the stress values at a crack tip calculated by FEM. The circumferential crack problems in bonded round bar are assumed as a perfect adhesive bonding and are modeled by using the four-node-quadrilateral-axisymmetric element. A very refined mesh pattern is used in the vicinity of the crack tip.

Table 1 shows the elastic constants of the adhesive and the adherend used in this analysis [11]. In this study, the aluminum alloy (material 1) is used as the adherend and the polyimide (material 2) is used as the adhesive material as shown in figure 1. Dundurs' composite parameters α, β and the index of corner stress singularity λ are also indicated in table 1. Here, the Dundurs' composite parameters α, β and the singularity index λ at the interface edge can be obtained from equations (1) and (2) [10].

$$\alpha = \frac{G_1(\kappa_2+1)-G_2(\kappa_1+1)}{G_1(\kappa_2+1)+G_2(\kappa_1+1)}, \quad \beta = \frac{G_1(\kappa_2-1)-G_2(\kappa_1-1)}{G_1(\kappa_2+1)+G_2(\kappa_1+1)}, \quad \kappa_i = 3-4\nu_i \ (i=1,2) \tag{1}$$

$$\left[\sin^2\left(\frac{\pi}{2}\lambda\right)-\lambda^2\right]^2\beta^2 + 2\lambda^2\left[\sin^2\left(\frac{\pi}{2}\lambda\right)-\lambda^2\right]\alpha\beta + \lambda^2(\lambda^2-1)\alpha^2 + \frac{\sin^2(\lambda\pi)}{4} = 0 \tag{2}$$

Table 1. Elastic constants and Dundurs' parameters used in the analyses [11].

	E[GPa]	G[GPa]	ν	α	β	λ
Material 1(Aluminum)	69.6	26.2	0.33	0.896	0.214	0.7397
Material 2(Polyimide)	3.77	1.40	0.342			

In the analysis, we set the diameter of the round bar d=12.7 mm, the length of the bar L=38.1 mm, the bond line thickness h=0.218 mm to 0.59 mm, and the tensile load σ_0=1 MPa. The two geometrical patterns of the circumferential crack near the interface in figure 1(b) are assumed a=0.001 mm, c=0.1 mm (a/d=7.87×10^{-5}, a/c=0.01) and a=0.0005 mm, c=0.05 mm (a/d=3.94×10^{-5}, a/c=0.01). In both patterns, the ratio between the crack depth and the distance a/c is 0.01.

3. Numerical results and discussion

3.1. Stress intensity factor of circumferential crack near the interface

In this study, the stress intensity factor of the circumferential crack near the interface in the bonded round bar with the adhesive layer shown in figure 1(b) is analysed. Figure 2 illustrates the relationship between the dimensionless stress intensity factor $F_I^{\sigma 0}$ and the bond line thickness when the bond line thickness h is changed. Here, the stress intensity factor is normalized by the remote stress σ_0 as shown in equation (3). From figure 2, it can be seen that the value of the dimensionless stress intensity factor $F_I^{\sigma 0}$ increases as the bond line thickness h increases in both of the two crack patterns. This means that the stress intensity factor K_I of the circumferential crack near the interface is controlled by the intensity of the singular stress field of the interface end.

$$F_I^{\sigma 0} = K_I/\sigma_0\sqrt{\pi a} \tag{3}$$

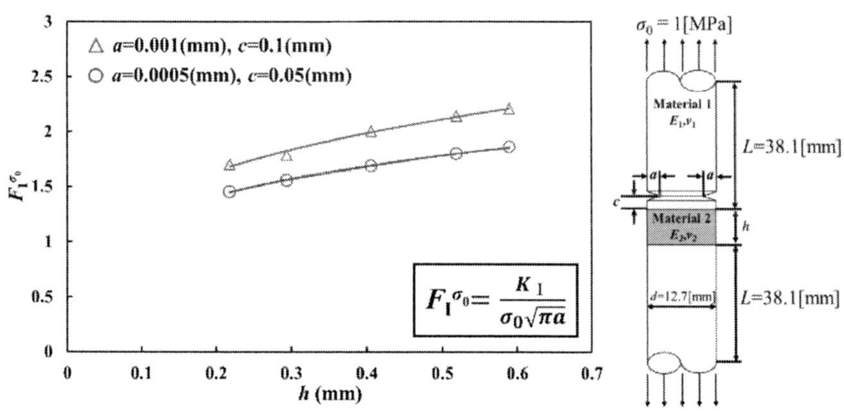

Figure 2. Relation between $F_I^{\sigma 0}$ defined by equation (3) and adhesive thickness.

Next, we verified the dimensionless stress intensity factor defined in equation (4). In equation (4), $\sigma_y(a,c)$ is the singular stress value at the red point in round bar without the crack as shown in figure 1(a) calculated by FEM. Figure 3 shows the relationship with the dimensionless stress intensity factor F_I defined by equation (4) and the adhesive thickness h.

$$F_{\rm I} = K_{\rm I}/\sigma_y(a,c)\sqrt{\pi a} \tag{4}$$

As shown in figure 3, it is found that the dimensionless stress intensity factor $F_{\rm I}$ becomes constant at $F_{\rm I}$=1.12 regardless of the adhesive layer thickness h. For readers convenience, $\sigma_y(a,c)$ analysed directly by FEM directly are also shown in figure 3. The values of $\sigma_y(a,c)$ increase as the adhesive thickness h increases. Therefore, the stress intensity factor of the circumferential small crack near interface can be expressed by the following equation (5).

$$K_{\rm I} = 1.12\sigma_y(a,c)\sqrt{\pi a} \tag{5}$$

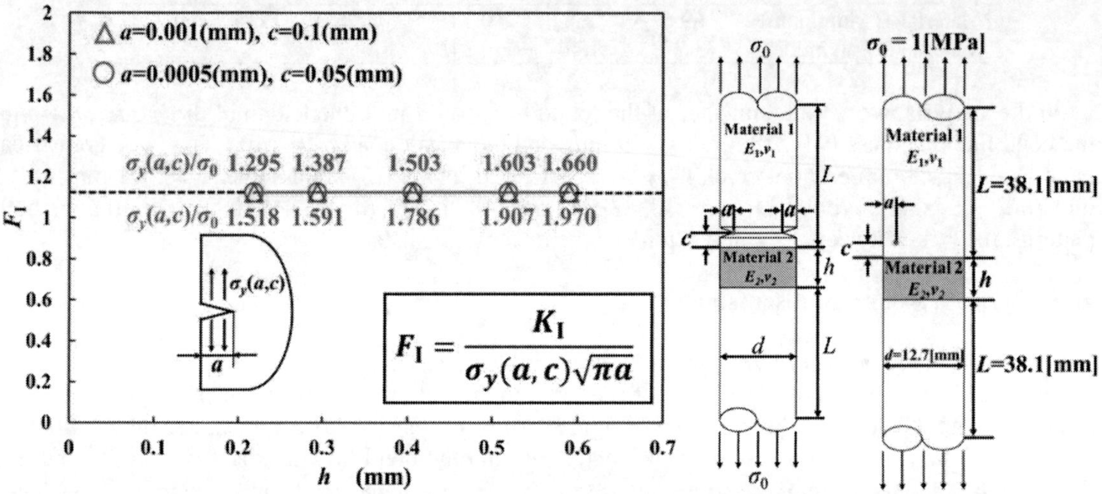

Figure 3. Relation between $F_{\rm I}$ defined by equation (4) and adhesive thickness.

In this crack model, the crack depth and position is important. The previous studies [14,15] have reported that the dimensionless stress intensity factor $F_{\rm I}$ of edge crack in butt joint plate becomes constant when the ratio between the crack length and the crack distance a/c<0.01. In this study, the circumferential cracks are set within this range.

3.2. Stress intensity factor of circumferential interface crack

Next, the problem of circumferential interface crack shown in figure 1(c) is analysed, and the same examination is conducted on the dimensionless stress intensity factor. The stress intensity factor of the interface crack is defined as the equation (6). In the definition (6), 'ε' expresses the oscillation singularity index of interface tip. As shown in equation (7), by using the singular stress $\sigma_y(a)$ at the crack tip position in the bonded round bar without the crack, the dimensionless stress intensity factor in the three-dimensional bonded round bar is calculated. Figures 4 and 5 show the dimensionless stress intensity factors $F_{\rm I}$ and $F_{\rm II}$ when the adhesive layer thickness is changed. In figures 4 and 5, red triangles indicate the results of a=0.001 mm and the blue circles show the results of a=0.0005 mm. The numerical values above and below the marks are stress values $\sigma_y(a)$ analysed directly by FEM when there is no crack. Since the values of $\sigma_y(a)$ are changed as the adhesive thickness increases, $F_{\rm I}$ and $F_{\rm II}$ for the circumferential interface crack in bonded round bar are almost constant, $F_{\rm I}$ = 1.25 and $F_{\rm II}$ = -0.41, regardless of the adhesive layer thickness h. In this analysis, the sign of $F_{\rm II}$ means the crack tip displacement in tangential direction of crack surface.

$$\sigma_y + i\tau_{xy} = \frac{K_{\rm I} + iK_{\rm II}}{\sqrt{2\pi r}}\left(\frac{r}{2a}\right)^{i\varepsilon} \tag{6}$$

$$F_{\mathrm{I}} = \frac{K_{\mathrm{I}}}{\sigma_y(a)\sqrt{\pi a}}, \quad F_{\mathrm{II}} = \frac{K_{\mathrm{II}}}{\sigma_y(a)\sqrt{\pi a}} \tag{7}$$

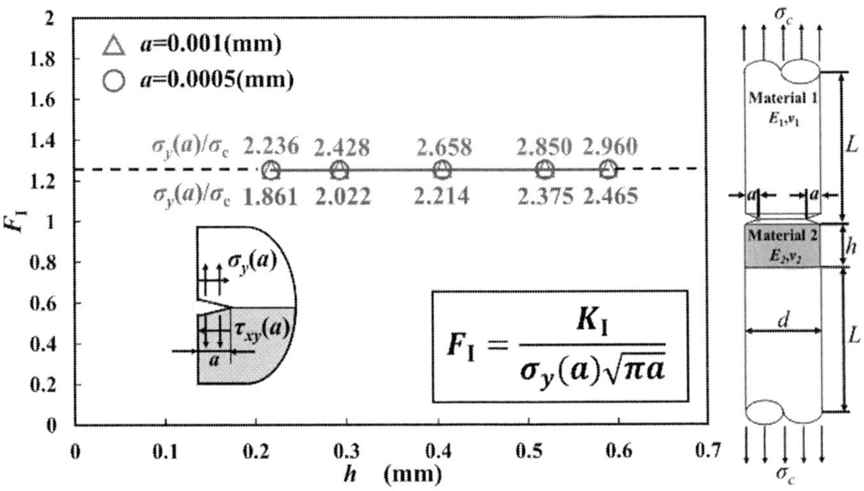

Figure 4. Relation between F_{I} defined by equation (7) and adhesive thickness.

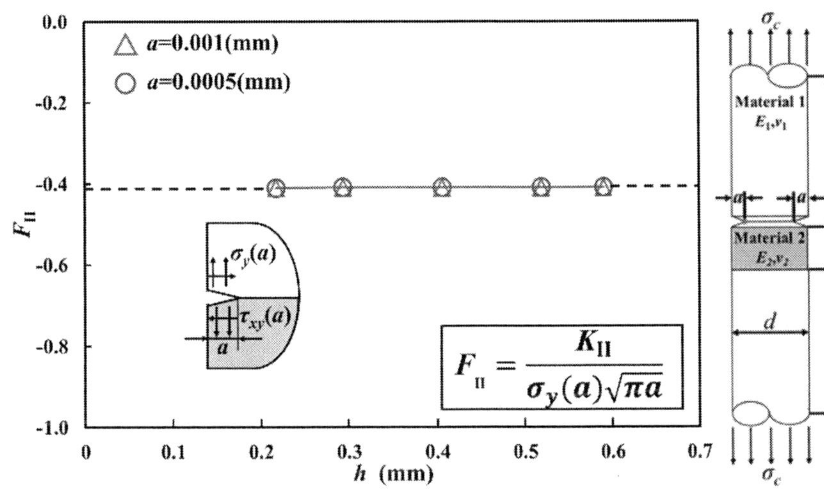

Figure 5. Relation between F_{II} defined by equation (7) and adhesive thickness.

3.3. Comparison of dimensionless stress intensity factor between 2D and 3D models

Finally, we compare the dimensionless stress intensity factors calculated by two dimensional and three dimensional models as shown in figure 6. Table 2 shows the dimensionless stress intensity factors defined by equation (4) of the circumferential crack and the double edge crack close to the interface. Table 3 also indicates the dimensionless stress intensity factors defined by equation (7) of the circumferential interface crack and the double edge interface crack. In table 2 and 3, $F^{3\mathrm{D}}$ stands for the dimensionless stress intensity factor of circumferential crack in bonded round bar and $F^{2\mathrm{D}}$ stands for that of edge crack in bonded plate. Two dimensional model is analysed by using the plane strain finite element. As shown in tables 2 and 3, both factors coincide with each other when the geometrical condition and material combinations are the same. Since the singular stresses $\sigma_y(a,c)$ and $\sigma_y(a)$ in the equations (4) and (7) differ in the two and three dimensional models, the stress intensity factors are naturally deferent.

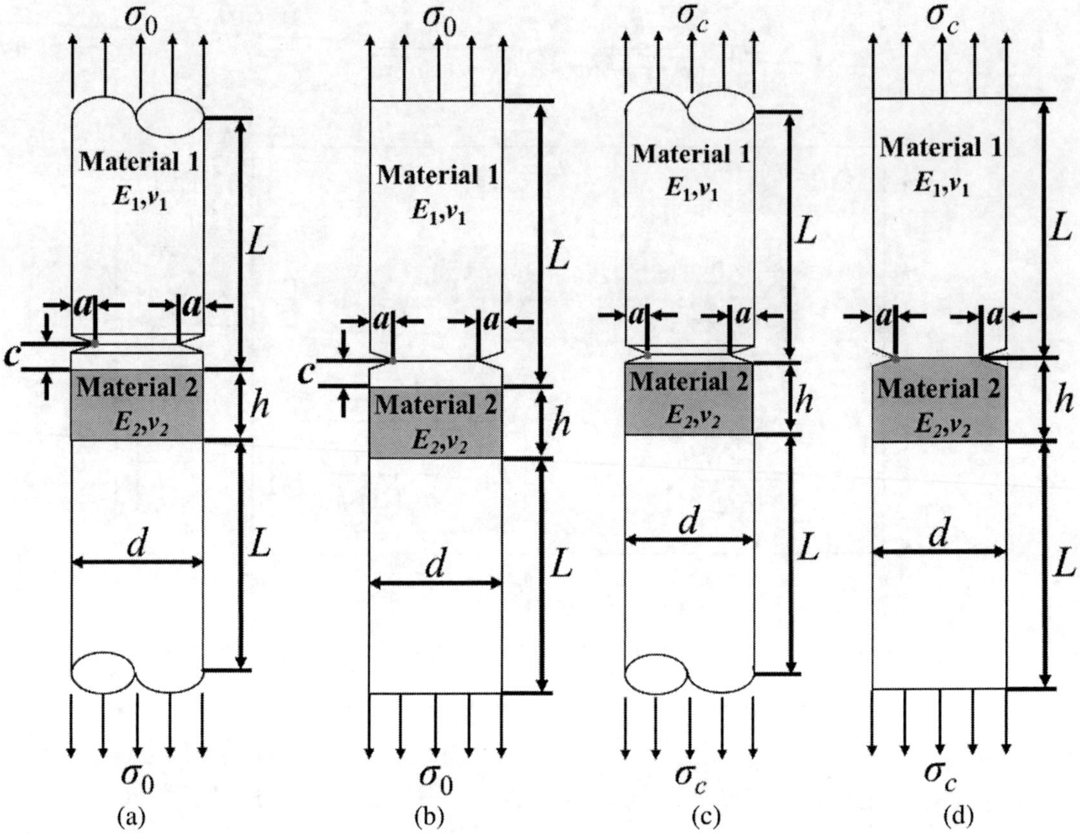

Figure 6. Comparison of circumference crack model and edge crack model. (a), (b) the crack near the interface, (c), (d) the interface crack.

Table 2. Comparison of F_I between two and three dimensional models (Bonded round bar and bonded plate with crack near the interface).

a=0.001[mm], c=0.1[mm]					
h [mm]	0.218	0.294	0.406	0.519	0.59
F_I^{3D} (*Bonded Round Bar*)	1.1235	1.1235	1.1237	1.1238	1.1238
F_I^{2D} (*Bonded Plate*)	1.1232	1.1234	1.1236	1.1237	1.1237
a=0.0005[mm], c=0.05[mm]					
h [mm]	0.218	0.294	0.406	0.519	0.59
F_I^{3D} (*Bonded Round Bar*)	1.1237	1.1238	1.1239	1.1239	1.1239
F_I^{2D} (*Bonded Plate*)	1.1236	1.1237	1.1238	1.1215	1.1239

Table 3. Comparison of F_I, F_{II} between two and three dimensional models (Bonded round bar and bonded plate with interface crack).

		a=0.001 [mm]			
h [mm]	0.218	0.294	0.406	0.519	0.59
F_1^{3D} (*Bonded Round Bar*)	1.2559	1.2552	1.2586	1.2581	1.2575
F_1^{2D} (*Bonded Plate*)	1.2557	1.2550	1.2586	1.2581	1.2574
F_2^{3D} (*Bonded Round Bar*)	-0.4103	-0.4094	-0.4102	-0.4098	-0.4096
F_2^{2D} (*Bonded Plate*)	-0.4103	-0.4095	-0.4103	-0.4099	-0.4097
		a=0.0005 [mm]			
h [mm]	0.218	0.294	0.406	0.519	0.59
F_1^{3D} (*Bonded Round Bar*)	1.2526	1.2524	1.2520	1.2518	1.2517
F_1^{2D} (*Bonded Plate*)	1.2526	1.2540	1.2519	1.2517	1.2516
F_2^{3D} (*Bonded Round Bar*)	-0.4102	-0.4087	-0.4085	-0.4084	-0.4084
F_2^{2D} (*Bonded Plate*)	-0.4103	-0.4066	-0.4086	-0.4085	-0.4084

By using the normalized factors, the stress intensity factors of circumferential crack can be easily obtained. Then, the stress intensity factor of the crack can be applied to strength evaluation of the butt joint material instead of the intensity of singular stress field. In [6,15], the strength evaluation of the butt joint plate has been studied by using the stress intensity factor of fictitious crack.

4. Conclusion

In this study, the stress intensity factors of the circumferential crack near the interface and the circumferential interface crack were analysed by the crack tip stress method. The relation between the dimensionless stress intensity factor of the circumferential crack and the adhesive thickness was investigated. In addition, the dimensionless stress intensity factor of the circumferential crack in bonded round bar was compared with that of the edge crack in bonded plate. The conclusions can be summarized as follows.

- The stress intensity factor of a circumferential crack near the interface in bonded round bar with adhesive layer can be expressed as $K_I = 1.12\sigma_y(a,c)\sqrt{\pi a}$ when the crack depth is small enough. In this expression, $\sigma_y(a, c)$ is the singular stress in the y-direction at the crack tip position in bonded round bar without crack.
- The dimensionless stress intensity factors F_I and F_{II} of the circumferential interface crack in bonded round bar become constant value regardless of the adhesive layer thickness h. In this problem, the dimensionless factors are based on the singular stress $\sigma_y(a)$ at the crack tip point in the bonded round bar without the interface crack.
- The dimensionless stress intensity factors of the circumferential crack model in bonded round bar were compared with that of the edge crack model in bonded plate. Both factors coincide with each other when the geometrical condition and material combinations are the same.

References

[1] Charalambides P G, Lund J, Evans A G and McMeeking R M 1989 a test specimen for determining the fracture resistance of bimaterial interfaces *Trans. of ASME Series E, J. Appl. Mech.* **56** 77-82

[2] Akisanya A R and Fleck N A 1997 Interfacial cracking from the free-edge of long bi-material strip *Int. J. Solids Struct.* **34** 1645-65

[3] Asada B, Shinya A and Yokozuka S 1990 Effect of dental adhesive film thickness on bond strength *Japan Soc. for Adhesive Dentistry* **8** 201-26 (in Japanese)

[4] Ashida T 2001 Application of adhesive joint for automobiles *J. Japan Welding Society* **70** 248-52 (in Japanese)

[5] Noda N-A, Miyazaki T, Uchikoba T, Li R, Sano Y and Takase Y 2014 Convenient debonding strength evaluation based on the intensity of singular stress for adhesive joints *J. Japan Institute of Electronics Packaging* **17** 132-42 (in Japanese)

[6] Noda N-A, Miyazaki T, Li R, Uchikoba T, Sano Y and Takase Y 2015 Debonding strength evaluation in terms of the intensity of singular stress at the interface corner with and without fictitious crack International *Int. J. Adhes. Adhes.* **61** 46-64

[7] Li R, Noda N-A, Takaki R, Sano Y, Takase Y and Miyazaki T 2018 Most suitable evaluation method for adhesive strength to minimize bend effect in lap joints in terms of the intensity of singular stress field *Int. J. Adhes. Adhes.* **86** 45-58

[8] Noda N-A, Ren F, Takaki R, Wang Z F, Oda K, Miyazaki T and Sano Y 2018 Intensity of singular stress field over the entire bond line thickness range useful for evaluating the adhesive strength for plate and cylinder butt joints *Int. J. Adhes. Adhes.* **85** 234-50

[9] Wang Z, Noda N-A, Miyazaki T, Uchikoba T, Wang G and Sano Y 2014 Analysis on intensity of singular stress for bonded pipe and cylinder in comparison with bonded plate *Int. J. Fracture Fatigue and Wear* **2** 50-6

[10] Miyazaki T, Noda N-A, Ren F, Wang Z, Sano Y and Iida K 2017 Analysis of intensity of singular stress field for bonded cylinder and bonded pipe in comparison with bonded plate *Int. J. Adhes. Adhes.* **77** 118-37

[11] Naito K, Onta M and Kogo Y 2012 The effect of adhesive thickness on tensile and strength of polyimide adhesive *Int. J. Adhes. Adhes.* **36** 77-85

[12] Oda K, Lan X, Noda N-A and Michinaka K 2012 Effect of arbitrary bi-material combination and bending loading conditions on stress intensity factors of an edge interface crack *Int. J. Struct. Int.* **3** 457-75

[13] Oda K, Takahata Y, Noda N-A and Tsutsumi N 2016 Relation between intensity of singular stress at interface edge and stress intensity factor of small edge interface crack in butt joint *Trans. of the JSME* **82** 15-00557 (in Japanese)

[14] Oda K, Kawakita M and Tsutsumi N 2017 Stress intensity factor for small edge crack in interfacial singularity region *J. Physics: Conf. Series* **843** 012010

[15] Sato T, Oda K and Tsutsumi N 2018 Strength evaluation of butt joint by stress intensity factor of small edge crack near interface edge *IOP Conf. Ser.-Mat. Sci.* **372** 012014

[16] Nisitani H, Kawamura T, Fujisaki W and Fukuda T 1999 Determination of highly accurate values of stress intensity factor or stress concentration factor of plate specimen by FEM *Trans. of the JSME* **65A** 26-31 (in Japanese)

[17] Nisitani H and Teranishi T 2004 KI of a circumferential crack emanating from an ellipsoidal cavity obtained by the crack tip stress method in FEM *Eng. Fract. Mech.* **71** 579-85

IOP Conf. Series: Materials Science and Engineering **657** (2019) 012042 doi:10.1088/1757-899X/657/1/012042

The estimation method of time of on-orbit spacecraft cabin decompression based on effective circulation area

Y Zhang[1], P Sun[1,2], Z W Liang[1] and W P Dong[1]

[1]National Key Laboratory of Human Factors Engineering, China Astronaut Research and Training Center, Beijing 100094, China

E-mail: sunpeng177287377@163.com

Abstract. This paper presented a method to estimate the pressure relief time of spacecraft cabin by testing a single component in the airway. Comparison between the theoretical prediction and experimental measurement indicates that there exits two methods which can effectively predict the pressure relief time of spacecrafts: measuring the effective flow area each single component or direct measurement of the effective flow area of the pressure relief system composed of single component. It also should be noted that the latter one can bring about a smaller error. Therefore, when conditions were met, the measurement and calculation of the system composed of components could more accurately predict the time required for the spacecraft pressure to drop to the target point.

1. Introduction

Pressure relief function is an important function of pressure control system of spacecraft with sealed cabin, and pressure relief time was one of the key indicators of spacecraft cabin pressure relief [1-4]. Subject to design boundary conditions such as installation location, safety, reliability, life and operability, the cabin pressure relief system was composed of a number of manual, electronic control valves and pipelines through series and parallel. The pressure relief path was zigzag, crisscross, the cross - sectional area was variable [5]. The estimation error of flow resistance was large because the common empirical formula was difficult to estimate the complex and changeable flow passage, and the pressure relief parameters calculated were difficult to be used in engineering design. Therefore, the effective pressure relief parameters were obtained through pressure relief tests, and the spacecraft body was usually placed in a large vacuum container in engineering [6-9]. But the method requires all piping and products were installed in place before the test, which needed a longer time. If the estimated test results were wrong before the test, a lot of resources will be consumed after several iterations of trial-design-retest.

Therefore, this paper presents a method to estimate the pressure relief time of spacecraft cabin by test, combining the measured flow capacity with the empirical formula. The method could shorten test cycle, reduced the number of trial iterations, saved test resources.

2. Test principle

2.1. System equivalent testing principle

The external cabin pressure of a spacecraft in orbit is $P_{out} \leq 1.33\times10^{-6}$ kPa, when the pressure is released from inside ($P_{in} \approx 100$ kPa) to outside meet $P_{out}/P_{in} < 0.528$, so the process of the pressure relief from inside to outside through the spacecraft pressure relief system could be regarded as the supercritical state. If the pressure relief system of the spacecraft was equivalent to a pressure relief hole with area and no thickness installed in the cabin, as long as the equivalent pressure relief hole area was obtained, the pressure relief time and other parameters of the supercritical state of the cabin could be calculated.

The complex gas circuit of the pressure relief system could be simplified into a series and parallel composition of several pipelines and valve products. By setting up the test platform, the equivalent cross-sectional area of single pipeline and valve could be measured in the supercritical condition. The equivalent cross-sectional area of the whole system can be obtained by calculating the equivalent area in series and parallel. In this way, the pressure relief time of spacecraft cabin could be obtained through test and calculation of single product.

2.2. Thermodynamic analysis of pressure relief process

Spacecraft pressure relief was a variable process of vacuum exhaust from pressure-tight cabin. In engineering, it was usually treated by isothermal process or adiabatic process. When a gas was discharged into a vacuum, it expanded to do work, and the temperature dropped sharply and caused the temperature of the bulkhead near the pressure relief port to drop. The metal used in the cabin had better thermal conductivity, as the pressure relief progresses, the heat transferred from the pressure relief port was gradually balanced with the heat transferred from other parts of the bulkhead to the pressure relief port. Spacecraft shells and loads were mostly metal components. Large space stations, in particular, could weigh tens of tons. But that air was usually in the hundreds of kilograms, the heat capacity of spacecraft was much larger than that of air. After the local temperature of the pressure relief port was balanced with the whole cabin, the temperature had little effect on the whole cabin temperature, so the pressure relief of spacecraft in orbit could be analyzed by isothermal process. Figure 1 was the temperature curves of the pressure relief test process of a certain type of large cabin, the figure 1 illustrated the change of pressure relief port and whole tank temperature. Therefore, when the test design and result analysis were carried out in this paper, they were simplified according to the temperature process.

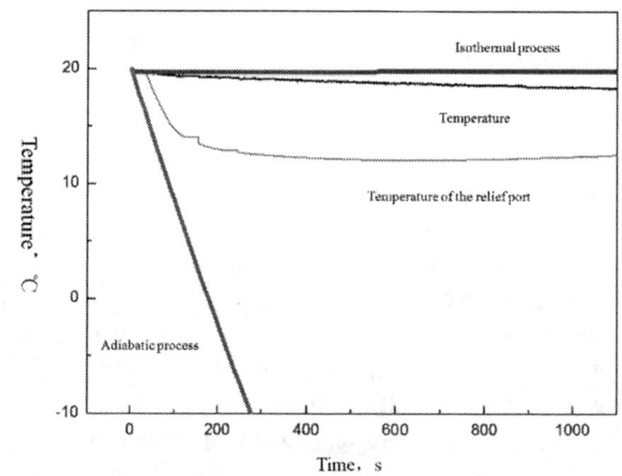

Figure 1. The temperature changes in large cabin vacuum pressure relief process.

2.3. Principle of equivalent cross section test

The single machine of equivalent test was worked in the supercritical flow. According to the method shown in figure 1, the pressure-tight cabin, products and vacuum tank needed to be set up. But it required higher quality equipment and the general laboratory was difficult to meet the demands [5,6]. So, the equivalent method in figure 2 was used in this paper. In the early stage of the re-pressure of the vacuum tank, the test was carried out in the supercritical state with the ratio of internal and external pressure less than 0.528.

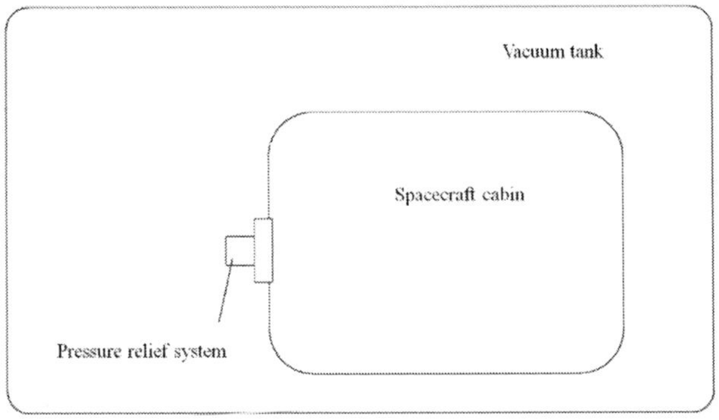

Figure 2. Schematic diagram of testing methods of common pressure relief time.

The complex pressure of cabin could be simplified to isothermal and constant volume process under supercritical condition [10].

$$t = \frac{V}{\mu ACR\sqrt{T}}\left(\frac{P_2 - P_1}{P_s}\right) \tag{1}$$

Where t is repressing time, A is repressing the equivalent cross-sectional area of the product which was tested, V is repressing vessel volume, P_1 is repressing the starting, P_2 is repressing the stopping pressure, P_s is repressing air supply pressure and T is repressing temperature. μ, C and R are constants. When testing different products on the same platform, the product of the equivalent area A and the repressing time t is a definite value. So an approximate standard leak could be made by measuring the repressing time, used

$$A_t t_t = A_s t_s \tag{2}$$

The equivalent flow area of the tested machine could be got by calculating. This method could avoid the influence of empirical coefficients in the formula.

3. Test equipment and results

3.1. Test equipment

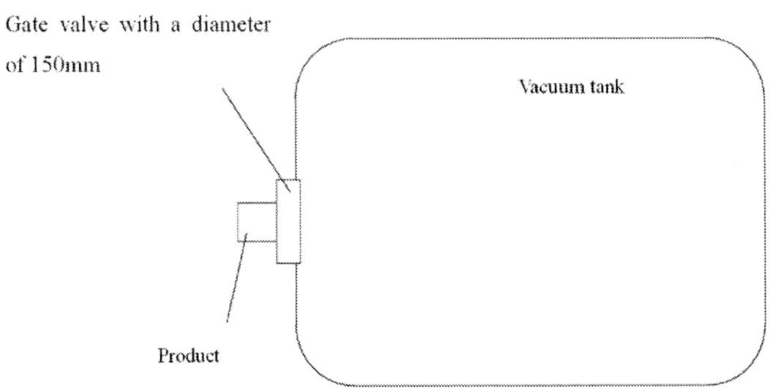

Figure 3. Schematic diagram of test platform.

Figure 3 is a schematic diagram of the test platform. The pressure outside the vacuum tank was normal pressure. Therefore, when the pressure in the tank did not exceed 50 kPaA, the repressing airflow was supercritical state. In this paper, the repressing process from 1 kPaA to 20 kPaA was analyzed and calculated.

Electronic pressure relief valves, manual pressure relief valves and vacuum pipelines of aerospace products were selected to build the test platform. These products were widely used in real spacecraft. During the test, the entrances and exits were also set according to the gas flow direction equipped in real spacecraft. The standard hole diameter for testing was selected to be 50 mm, which was directly installed on the bulkhead of vacuum tank. In order to avoid the influence of valve on product testing, the gate valve with a diameter of 150 mm was selected as the control valve of pressure relief path.

3.2. Experimental procedure

The vacuum tank recharge processes from 1 kPaA to 20 kPaA were timed using only the standard hole, the valves and the pipe. Then the valves and the pipe were connected as one pipe system, the same recharge processes from 1 kPaA to 20 kPaA was timed using the pipe system. The test results are shown in table 1.

Table 1. Test and calculation results.

No.	Name	Recharge time	Equivalent circulation area
1	Standard hole	58s	1962.5mm^2
2	the electronically controlled relief valve	118	964.4 mm^2
3	the manual relief valve	185	618.4 mm^2
4	the pipeline	184	615.1 mm^2
5	the pipe system	289	393.8 mm^2

4. Analysis of the results

4.1. Calculation of equivalent circulation area

The equivalent circulation areas calculated by formula 2 was shown in table 1 according to recharge time of the vacuum tank. The equivalent circulation area of the pipe system S_{sc}, which was established by connecting the electronically controlled relief valve S_1, the manual relief valve S_2 and the pipeline in series S_3, was 397.5 mm^2. It is in accordance with the experimental measurement value of 393.8 mm^2.

$$\frac{1}{S_{sc}} = \frac{1}{S_1} + \frac{1}{S_2} + \frac{1}{S_3} \tag{3}$$

4.2. Pressure relief time calculation

Compared with the surface of the spacecraft cabin, the surface area of the pressure relief system is usually very small. The spacecraft cabin is usually made of metallic material with good thermal conductivity. Therefore, the spacecraft cabin could be treated as an isothermal process when it is decompressing on orbit. According to the pressure relief data of some Chinese spacecrafts, the temperature change of the cabin before and after pressure relief was less than 1 K, so the pressure relief of spacecrafts could be calculated approximately according to the isothermal process.

The on-orbit spacecrafts operate in a vacuum environment and the pressure relief process was in a supercritical state. The volumes of certain on-orbit spacecrafts are constant. According to Figure 1, temperature does not change much and the pressure relief process can be regarded as isothermal process. Therefore the pressure change of a cabin in a supercritical case in this paper is simplified to an isothermal constant volume process and the following formulas were applicable.

Volume flow rate Q [11,12] can be expressed as

$$Q=-\frac{1}{\rho}\cdot\frac{\mathrm{d}m}{\mathrm{d}t}=-\frac{V}{kRT\rho}\cdot\frac{\mathrm{d}P}{\mathrm{d}t} \tag{4}$$

In isentropic steady flow volume flow rate Q [11,12] can also be expressed as

$$Q=7.3\times10^{-3}SP_s\sqrt{\left(\frac{P_e}{P_s}\right)^{\frac{2}{k}}-\left(\frac{P_e}{P_s}\right)^{\frac{k+1}{k}}}\cdot\sqrt{\frac{273}{T}} \tag{5}$$

According to formulas (4) and (5), dt is expressed as formula (6).

$$\mathrm{d}t=-5.217\frac{V}{kS}\sqrt{\frac{273}{T}}\cdot\frac{\mathrm{d}P}{P\left(\frac{P}{P_s}\right)^{k-\frac{1}{2k}}} \tag{6}$$

Integral operation is used to obtain cabin pressure relief time t [11,12], which is expressed as

$$t=\frac{2k}{k-1}\left[\left(\frac{P_s}{P_e}\right)^{\frac{k-1}{2k}}-1\right]\left(5.217\frac{V}{kS}\sqrt{\frac{273}{T}}\right) \tag{7}$$

Temperature in ShenZhou-7 with 2 astronauts was controlled at 22.7°C in thermal tests [13], accordingly the air temperature was controlled at 23°C±0.5°C in this experiment. Adiabatic index k is considered as a fixed value at 23°C and under normal pressure. Therefore it takes 1.4 in this paper. V is the volume of the spacecraft cabin, S is the equivalent flow area of the relief channel, T is the absolute temperature, P_s is the initial pressure of the relief, and P_e is the pressure at the end of the relief.

The time of the cabin with a volume 90 m^3 decompressing from 97 kPaA to 27 kPaA is 1136s, which is got by substituting the test data of table 2 (line 2,line 3 and line 4) into formulas (3) and (4). The time of the same process by substituting the test data of table 2 (line 5) into formula (3) is 1147 s.

In the course of Chinese spacecraft test, similar electronic pressure relief valves, manual pressure relief valves and pipelines are connected in series as pressure relief channels. After conversion, it takes about 1200s for the pressure relief system to complete the same pressure relief process.

Table 2. Time to complete the same pressure relief process.

No	Test conditions	Equivalent flow area	Time	Remarks
1	Measuring the Equivalent Area of a Single Machine and Calculating the Equivalent Area of the System	397.5 mm^2	1136s	The product is not ready
2	Measuring Equivalent Area of the pipe System	393.8 mm^2	1149s	The product is ready
3	Measured data	-	1200s	-

The calculation methods and the results were compared in table 2. The results show that both methods can accurately predict the pressure relief time, but method 2 is more accurate. The measurement and calculation errors are smaller by using the pipe system for calculation, when the valves and pipes are ready. If they are not ready, the relief time can be effectively predicted by measuring the equivalent flow area of valves and pipes at different time or location.

5. Conclusion

The method of estimating the decompression time of spacecraft cabin by valves and pipes test was presented in this paper. The decompression time of a spacecraft can be effectively predicted by measuring both the equivalent area of the unit (valves and pipes) and the pipe-system. By comparing the calculated results with the measured results, the error of the latter method is smaller. Further research will be carried out on relationship between equivalent area and pipe-system. And then the decompression time of spacecraft cabin will be estimated more quickly and precisely.

Acknowledgments

This project is supported by the Foundation of National Key Laboratory of Human Factors Engineering, Grant NO: SYFD160051809.

References

[1] Fan Y F, Liu B Q, Huang J R and Man G L 2013 *J Astronautics* **34** 293-8
[2] Shang Z, Zhang B N and Pan T 2009 *M. Spaceflight* **2** 16-21
[3] Cao M L 2007 *Hydraulics P. S.* **27** 51-5
[4] Steven B and Zachary B 2019 *Int Conf Environmental Systems* (USA: Boston) p 373
[5] Pang H W, Huang B C,Qian B X and Jia Y 2002 *Chinese Space Sci. Tech.* **22** 6-12
[6] Wu G X 2004 *Space Int.* **2** 12-7
[7] Xu S H, Yan L, Ma A J, Zhao W, Zhang L, Lu Z G, Deng J H and Bi J Z 2019 *Spacecraft Env. Eng.* **36** 1-6
[8] Su X M, Wang J, Qie D F, Ji X Y, Xie J H, Xu Z W, Xu D Y and Li R H 2018 *Spacecraft Env. Eng.* **35** 86-91
[9] Zhu G C and Jia S J 2009 *M. Spaceflight* **3** 48-53
[10] Huang Z D, Gao F, Li J, Song X Y and Wu B 2009 *M. Spaceflight* **4** 32-9
[11] Kashima K, Ishii Y and Funaki T 2004 *ASME Transaction: J. Fluid Eng.* **126** 273-9
[12] Yang G, Xu X W, Gao L L and Li B R 2010 *J. Lanzhou Uni. Tech.* **36** 42-7
[13] Huang J R, Fan Y F, Liu B Q and Jin Y 2009 *J. Astronautics* **30** 2086-91

IOP Conf. Series: Materials Science and Engineering **657** (2019) 012043 doi:10.1088/1757-899X/657/1/012043

Weak-form differential quadrature element method for dynamic analysis of fluid-saturated soil

G H He[1,2], Z H Kuang[1] and X K Li[1]

[1]Shanghai Construction Group Co., Ltd., Shanghai 201114, China

E-mail: flamehe@163.com

Abstract. It is formulated the weak-form of differential equations that govern the two-dimensional dynamical behavior of fluid-saturated soil, then the weak-form equations are discretized by the differential quadrature technique, and finally solved by the implicit Euler method. The proposed weak-form equations and numerical programs developed are verified through comparisons with benchmark solutions, and the convergence performance of the presented method is investigated. Numerical results show that the proposed weak-form quadrature element method not only possesses significantly higher computational efficiency, for the dynamic analysis of saturated soil, than the conventional finite element method, but it is also significantly alleviated the problem of numerical smoothness in the stress analysis.

1. Introduction

Safety assessments of soils and foundations are of great importance in most of the infrastructure engineering, especially the offshore engineering, bridge engineering and underground engineering. To carry out the mechanical analyses and engineering design, the Biot [1,2] theory and the theory of porous media (TPM) [3] have matured into the most widespread theories depicting the behaviors of fluid-saturated soils. Based on the Biot theory, several aspects of analyses have been performed, including the consolidation of foundation [4], the dynamic analysis of saturated soil [5] and the seepage prediction of underground water [6], etc. Compared with the Biot theory, the TPM complies with more rigorous continuous media mixture laws and the concept of volume fraction, and is capable of dealing with nonlinear problems [7] more naturally. As a result, the TPM is getting popular, and there had been researchers investigated the saturated soil in the framework of TPM, e.g. Breuer [8] had analyzed the dynamical response of two-dimensional foundation using the standard Galerkin finite element method (FEM); Hu *et al* [9] had developed the strong-form differential quadrature (DQ) method for the dynamical response analysis of axisymmetric saturated porous media [10].

The FEM has nowadays undoubtedly evolved into one of the most powerful and practical analyzing techniques in the engineering practice, and commercial finite element analysis software equipped with Biot porous medium are common. Despite the success of FEM, there still exists shortcomings of the conventional FEM, e.g. the unsatisfactory discretization efficiency, that is, the acceptable analyzing results are usually based on large number of degrees of freedom (DOFs), memory and CPU/GPU time. To improve the discretization efficiency, the DQ technique [10] has recently been known as a good candidate. The weak-form differential quadrature element method (WQEM) [11,12], which incorporates DQ technique into the conventional FEM, inherits merits of both methods. And the CPU time and memory requirements may be reduced significantly by WQEM.

In this paper, the TPM is used for dynamic analysis of two-dimensional fluid-saturated soils. The

IOP Conf. Series: Materials Science and Engineering **657** (2019) 012043 doi:10.1088/1757-899X/657/1/012043

weak-form equations governing the motions of soil skeleton, pore water together with pore pressure are obtained. With the help of DQ technique, the initial value problem is formulated and the computing programs are implemented. The implicit Euler method is used to solve the initial value problem. The computer programs are verified through the numerical comparisons, and the computational efficiency of the proposed method versus the conventional FEM is examined.

2. Formulations

2.1. Governing differential equations

Let us consider a fluid-saturated porous medium with immiscible and incompressible constituents in domain Ω. The behaviors of water and solid skeleton can be formulated by TPM [7,8] as

$$\begin{cases} \operatorname{div}\left(\boldsymbol{\sigma}^{\mathrm{SE}} - p\mathbf{I}\right) + \rho\left(\mathbf{b} - \dot{\mathbf{v}}^{\mathrm{S}}\right) - \rho^{\mathrm{F}}\dot{\mathbf{w}}^{\mathrm{F}} = \mathbf{0} \\ \rho^{\mathrm{F}}\left(\mathbf{b} - \dot{\mathbf{v}}^{\mathrm{S}} - \dot{\mathbf{w}}^{\mathrm{F}}\right) - S_v\mathbf{w}^{\mathrm{F}} - n^F \operatorname{grad} p = \mathbf{0} \\ \operatorname{div}\left(\mathbf{v}^{\mathrm{S}} + n^{\mathrm{F}}\mathbf{w}^{\mathrm{F}}\right) = 0 \\ \mathbf{v}^{\mathrm{S}} - \dot{\mathbf{u}}^{\mathrm{S}} = \mathbf{0} \end{cases} \tag{1}$$

where, $\boldsymbol{\sigma}^{\mathrm{SE}}$, $\mathbf{u}^{\mathrm{S}}$ and $\mathbf{v}^{\mathrm{S}}$ are the tensors of effective stress, displacement and solid skeleton velocity, respectively; p, $\mathbf{w}^{\mathrm{F}}$ and n^{F} are the pore pressure, fluid-solid relative velocity and volume fraction of water, respectively; ρ^{F} denotes the partial density of fluid, that is, $\rho^{\mathrm{F}} = n^{\mathrm{F}}\rho^{\mathrm{FR}}$, in which ρ^{FR} is the real density of water; ρ is the sum of ρ^{F} and ρ^{S}, and herein $\rho^{\mathrm{S}} = n^{\mathrm{S}}\rho^{\mathrm{SR}}$, with n^{S} and ρ^{SR} the volume fraction and real density of solid skeleton, respectively; $\mathbf{b}$ is the body force density acting upon both constituents; S_v is the coupling coefficient, determined by $S_v = (n^{\mathrm{F}})^2\gamma^{\mathrm{FR}} / k^{\mathrm{F}}$, with $\gamma^{\mathrm{FR}} = \rho^{\mathrm{FR}}|\mathbf{b}|$ the real specific weight of water, and k^{F} the Darcy permeability coefficient; I is the Kronecker tensor; div is the operator for divergence, and grad denotes the operator for gradient with respect to Cartesian coordinate $\mathbf{x}$.

On the boundary S, the boundary conditions for the fluid-solid mixture and fluid can be described as follows

$$\begin{cases} \mathbf{u}^{\mathrm{S}} = \bar{\mathbf{u}}^{\mathrm{S}},\ \mathbf{x} \in \mathrm{S}_{\mathrm{u}}, & \mathbf{n}\cdot(\boldsymbol{\sigma}^{\mathrm{SE}} - p\mathbf{I}) = \bar{\mathbf{T}},\ \mathbf{x} \in \mathrm{S}_{\mathrm{T}} \\ p = \bar{p},\ \mathbf{x} \in \mathrm{S}_{\mathrm{p}}, & \mathbf{n}\cdot(n^{\mathrm{F}}\mathbf{w}^{\mathrm{F}}) = \bar{Q},\ \mathbf{x} \in \mathrm{S}_{\mathrm{Q}} \end{cases} \tag{2}$$

where, $\mathrm{S} = \mathrm{S}_{\mathrm{p}} \cup \mathrm{S}_{\mathrm{Q}} = \mathrm{S}_{\mathrm{u}} \cup \mathrm{S}_{\mathrm{T}}$, $\mathrm{S}_{\mathrm{u}} \cap \mathrm{S}_{\mathrm{T}} = \mathrm{S}_{\mathrm{p}} \cap \mathrm{S}_{\mathrm{Q}} = \phi$; $\mathbf{n}$ is the unit vector normal to the boundary S; $\bar{\mathbf{u}}^{\mathrm{S}}$ and $\bar{p}$ are the prescribed solid displacement on S_{u} and surface pore pressure of water on S_{p}, respectively; $\bar{\mathbf{T}}$ and $\bar{Q}$ are the known surface traction on S_{T} and water flux on S_{Q}, respectively.

In the initial time t_0, the state of all the unknowns is known as

$$\mathbf{u}^{\mathrm{S}} = \mathbf{u}_0^{\mathrm{S}},\quad \mathbf{v}^{\mathrm{S}} = \mathbf{v}_0^{\mathrm{S}},\quad \mathbf{w}^{\mathrm{F}} = \mathbf{w}_0^{\mathrm{F}} \qquad t = t_0 \tag{3}$$

For the isotropic elastic soil skeleton with small deformation, and in the case of plane strain hypothesis,

$$\sigma_{\mathrm{x}} = (2\mu^{\mathrm{S}} + \lambda^{\mathrm{S}})\frac{\partial u_{\mathrm{x}}}{\partial x} + \lambda^{\mathrm{S}}\frac{\partial u_{\mathrm{y}}}{\partial y},\ \sigma_{\mathrm{y}} = (2\mu^{\mathrm{S}} + \lambda^{\mathrm{S}})\frac{\partial u_{\mathrm{y}}}{\partial y} + \lambda^{\mathrm{S}}\frac{\partial u_{\mathrm{x}}}{\partial x},\ \tau_{\mathrm{xy}} = \mu^{\mathrm{S}}\left(\frac{\partial u_{\mathrm{x}}}{\partial y} + \frac{\partial u_{\mathrm{y}}}{\partial x}\right) \tag{4}$$

where, μ^S and λ^S are the Lamé constants of soil skeleton,

$$\boldsymbol{\sigma}^{\mathrm{SE}} = \begin{bmatrix} \sigma_{\mathrm{x}} & \tau_{\mathrm{xy}} \\ \tau_{\mathrm{xy}} & \sigma_{\mathrm{y}} \end{bmatrix}, \ \mathbf{u}^{\mathrm{S}} = \begin{bmatrix} u_{\mathrm{x}} \\ u_{\mathrm{y}} \end{bmatrix} \tag{5}$$

2.2. Weak-form equations

Equations (1) and (2) can be transformed into an integral form, which can be summarized as

$$\begin{aligned} &\int_{\Omega} \left[\tilde{\mathbf{u}}^{\mathrm{T}} \left(\rho^{\mathrm{F}} \dot{\mathbf{w}}^{\mathrm{F}} + \rho \dot{\mathbf{v}}^{\mathrm{S}} \right) + \left(\rho^{\mathrm{F}} \tilde{\mathbf{w}}^{\mathrm{T}} + \beta \rho^{\mathrm{FR}} \nabla^{\mathrm{T}} \tilde{p} \right) \left(\dot{\mathbf{v}}^{\mathrm{S}} + \dot{\mathbf{w}}^{\mathrm{F}} \right) + \tilde{\mathbf{v}}^{\mathrm{T}} \dot{\mathbf{u}}^{\mathrm{S}} \right] \mathrm{d}\Omega \\ &+ \int_{\Omega} \left[\tilde{\boldsymbol{\varepsilon}}^{\mathrm{T}} \boldsymbol{\sigma} - \tilde{\varepsilon}_{\mathrm{v}} p + \tilde{\mathbf{w}}^{\mathrm{T}} \left(s_{\mathrm{v}} \mathbf{w}^{\mathrm{F}} + n^{\mathrm{F}} \nabla p \right) + \tilde{p} \nabla^{\mathrm{T}} \mathbf{v}^{\mathrm{S}} + \nabla^{\mathrm{T}} \tilde{p} \beta \nabla p - \tilde{\mathbf{v}}^{\mathrm{T}} \mathbf{v}^{\mathrm{S}} \right] \mathrm{d}\Omega \\ &= \int_{\Omega} \left[\tilde{\mathbf{u}}^{\mathrm{T}} \rho \mathbf{b} + \tilde{\mathbf{w}}^{\mathrm{T}} \rho^{\mathrm{F}} \mathbf{b} + \rho^{\mathrm{FR}} \beta \nabla^{\mathrm{T}} \tilde{p} \mathbf{b} \right] \mathrm{d}\Omega + \int_{S_{\mathrm{T}}} \tilde{\mathbf{u}}^{\mathrm{T}} \bar{\mathbf{T}} \mathrm{d}S - \int_{S_{\mathrm{Q}}} \tilde{p} \bar{Q} \mathrm{d}S \end{aligned} \tag{6}$$

in which, $\beta = k^{\mathrm{F}} / \gamma^{\mathrm{FR}}$; $\boldsymbol{\sigma} = \begin{bmatrix} \sigma_{\mathrm{x}} & \sigma_{\mathrm{y}} & \tau_{\mathrm{xy}} \end{bmatrix}^{\mathrm{T}}$ is the stress vector; $\tilde{\boldsymbol{\varepsilon}}$ and $\tilde{\varepsilon}_{\mathrm{v}}$ are composed of arbitrary function in Ω as

$$\tilde{\boldsymbol{\varepsilon}} = \begin{bmatrix} \dfrac{\partial \tilde{u}_{\mathrm{x}}}{\partial x} & \dfrac{\partial \tilde{u}_{\mathrm{y}}}{\partial y} & \dfrac{\partial \tilde{u}_{\mathrm{x}}}{\partial y} + \dfrac{\partial \tilde{u}_{\mathrm{y}}}{\partial x} \end{bmatrix}^{\mathrm{T}}, \ \tilde{\varepsilon}_{\mathrm{v}} = \frac{\partial \tilde{u}_{\mathrm{x}}}{\partial x} + \frac{\partial \tilde{u}_{\mathrm{y}}}{\partial y} \tag{7}$$

$\nabla = \begin{bmatrix} \partial/\partial x & \partial/\partial y \end{bmatrix}^{\mathrm{T}}$; $\tilde{\mathbf{w}}$, $\tilde{\mathbf{u}}$ and $\tilde{\mathbf{v}}$ are composed of arbitrary functions in Ω, which have expressions

$$\tilde{\mathbf{w}} = \begin{bmatrix} \tilde{w}_{\mathrm{x}} & \tilde{w}_{\mathrm{y}} \end{bmatrix}^{\mathrm{T}}, \ \tilde{\mathbf{u}} = \begin{bmatrix} \tilde{u}_{\mathrm{x}} & \tilde{u}_{\mathrm{y}} \end{bmatrix}^{\mathrm{T}}, \ \tilde{\mathbf{v}} = \begin{bmatrix} \tilde{v}_{\mathrm{x}} & \tilde{v}_{\mathrm{y}} \end{bmatrix}^{\mathrm{T}} \tag{8}$$

As the test functions, $\tilde{\mathbf{u}}$, $\tilde{\mathbf{v}}$ and $\tilde{p}$ are arbitrary except that $\tilde{\mathbf{v}} = \tilde{\mathbf{u}} = \mathbf{0}$ on $\mathbf{x} \in \mathrm{S}_{\mathrm{u}}$ and $\tilde{p} = 0$ on $\mathbf{x} \in \mathrm{S}_{\mathrm{p}}$.

2.3. DQ approximation

The core idea of the DQ technique is that the mth-order ($m \in \mathrm{Z}$, $m \geq 1$) derivative with respect to ξ of function $F(\xi, t)$ with C^m continuity, can be approximated through combination of function values at N given nodes: $-1 = \xi_1 < \xi_2 < \cdots < \xi_N = 1$, that is

$$\frac{\partial^m F(\xi_i, t)}{\partial \xi^m} = \sum_{j=1}^{N} C_{ij}^{(m)} F(\xi_j, t) \tag{9}$$

where $C_{ij}^{(m)}$ is the combination coefficient, whose generation procedure was initially proposed by Bellman *et al* [13,14] and improved by Shu and Richards [10]. The algorithms for generation of $C_{ij}^{(m)}$ have been reported by many researchers, e.g. [15]. The distribution of grid points has major influences on the approximation precision of equation (9). Among all types of grid point distribution, the expanded-Chebyshev distribution

$$\xi_i = -\cos\left[(2i-1)\pi/(2N) \right] \sec\left[\pi/(2N) \right], \quad i = 1, 2, \cdots, N. \tag{10}$$

is an excellent approximation to the Lebesgue-optimal grid distribution. Although the present differential quadrature is of Newton-Cotes type, which is less efficient than the Gaussian type or generalized Gaussian rules [16], there is an excellent property that the combination coefficients $C_{ij}^{(m)}$ are constant and whose generation cost is very low, even without solving a linear system.

When dealing with two-dimensional problems, equation (9) can be extended into a form of

$$\frac{\partial^{m+n}F(\xi_i,\eta_j,t)}{\partial\xi^m\partial\eta^n}=\sum_{l=1}^{M}\sum_{k=1}^{N}C_{ik}^{(m)}\tilde{C}_{jl}^{(n)}F(\xi_k,\eta_l,t)=\mathbf{N}_{i,j}^{(m,n)}\mathbf{F}(t) \tag{11}$$

where, M and N are the numbers of points projecting to η and ξ axes, respectively; $\mathbf{N}_{i,j}^{(m,n)}$ may be called the shape function vector at coordinate (ξ_i,η_j), and $\mathbf{F}(t)$ is the vector of nodal value or the DOF. For simplicity, M is set with same value as N (see in figure 1) in the present paper.

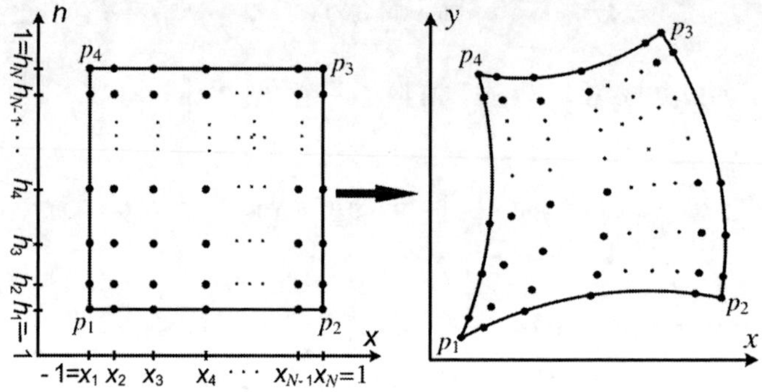

Figure 1. Two-dimensional expanded-Chebyshev distribution of grid points.

With the help of equation (9), all the basic unknowns $\mathbf{u}^{\mathrm{S}}$, $\mathbf{v}^{\mathrm{S}}$, $\mathbf{w}^{\mathrm{F}}$, p and their partial derivatives with respect to $\mathbf{x}$ can be approximated in terms of $\mathbf{N}_{i,j}^{(m,n)}$ and vector of DOF.

$$\frac{\partial^{m+n}\mathbf{u}^{\mathrm{S}}(\xi_i,\eta_j,t)}{\partial\xi^m\partial\eta^n}=\begin{bmatrix}\frac{\partial^{m+n}u_{\mathrm{x}}^{\mathrm{S}}(\xi_i,\eta_j,t)}{\partial\xi^m\partial\eta^n} & \frac{\partial^{m+n}u_{\mathrm{y}}^{\mathrm{S}}(\xi_i,\eta_j,t)}{\partial\xi^m\partial\eta^n}\end{bmatrix}^{\mathrm{T}}=\begin{bmatrix}\mathbf{N}_{i,j}^{(m,n)}\boldsymbol{\delta}_{ux}\\ \mathbf{N}_{i,j}^{(m,n)}\boldsymbol{\delta}_{uy}\end{bmatrix} \tag{12}$$

where, $\boldsymbol{\delta}_{ux}$ and $\boldsymbol{\delta}_{uy}$ are the nodal value vectors of $u_{\mathrm{x}}^{\mathrm{S}}$ and $u_{\mathrm{y}}^{\mathrm{S}}$ over all the points (ξ_i,η_j), respectively; the component arrangements of $\boldsymbol{\delta}_{ux}$ and $\boldsymbol{\delta}_{uy}$ are consistent with $\mathbf{N}_{i,j}^{(m,n)}$. Similarly, the following can also be obtained

$$\frac{\partial^{m+n}\mathbf{v}^{\mathrm{S}}(\xi_i,\eta_j,t)}{\partial\xi^m\partial\eta^n}=\begin{bmatrix}\mathbf{N}_{i,j}^{(m,n)}\boldsymbol{\delta}_{vx}\\ \mathbf{N}_{i,j}^{(m,n)}\boldsymbol{\delta}_{vy}\end{bmatrix},\ \frac{\partial^{m+n}\mathbf{w}^{\mathrm{F}}(\xi_i,\eta_j,t)}{\partial\xi^m\partial\eta^n}=\begin{bmatrix}\mathbf{N}_{i,j}^{(m,n)}\boldsymbol{\delta}_{wx}\\ \mathbf{N}_{i,j}^{(m,n)}\boldsymbol{\delta}_{wy}\end{bmatrix},\ \frac{\partial^{m+n}p(\xi_i,\eta_j,t)}{\partial\xi^m\partial\eta^n}=\mathbf{N}_{i,j}^{(m,n)}\boldsymbol{\delta}_p \tag{13}$$

For the convenience of formulation, the above DOF vectors can be augmented,

$$\frac{\partial^{m+n}\mathbf{u}^{\mathrm{S}}(\xi_i,\eta_j,t)}{\partial\xi^m\partial\eta^n}=\begin{bmatrix}\mathbf{N}_{i,j}^{(m,n)} & \mathbf{0}\,\mathbf{0}\,\mathbf{0}\,\mathbf{0}\,\mathbf{0}\,\mathbf{0}\\ \mathbf{0}\ \mathbf{N}_{i,j}^{(m,n)} & \mathbf{0}\,\mathbf{0}\,\mathbf{0}\,\mathbf{0}\,\mathbf{0}\end{bmatrix}\boldsymbol{\delta}_{\mathrm{e}},\ \frac{\partial^{m+n}\mathbf{v}^{\mathrm{S}}(\xi_i,\eta_j,t)}{\partial\xi^m\partial\eta^n}=\begin{bmatrix}\mathbf{0}\,\mathbf{0}\ \mathbf{N}_{i,j}^{(m,n)} & \mathbf{0}\,\mathbf{0}\,\mathbf{0}\,\mathbf{0}\\ \mathbf{0}\,\mathbf{0}\,\mathbf{0}\ \mathbf{N}_{i,j}^{(m,n)} & \mathbf{0}\,\mathbf{0}\,\mathbf{0}\end{bmatrix}\boldsymbol{\delta}_{\mathrm{e}} \tag{14}$$

$$\frac{\partial^{m+n}\mathbf{w}^{\mathrm{F}}(\xi_i,\eta_j,t)}{\partial\xi^m\partial\eta^n}=\begin{bmatrix}\mathbf{0}\,\mathbf{0}\,\mathbf{0}\,\mathbf{0}\ \mathbf{N}_{i,j}^{(m,n)} & \mathbf{0}\,\mathbf{0}\\ \mathbf{0}\,\mathbf{0}\,\mathbf{0}\,\mathbf{0}\,\mathbf{0}\ \mathbf{N}_{i,j}^{(m,n)} & \mathbf{0}\end{bmatrix}\boldsymbol{\delta}_{\mathrm{e}},\ \frac{\partial^{m+n}p(\xi_i,\eta_j,t)}{\partial\xi^m\partial\eta^n}=\begin{bmatrix}\mathbf{0}\,\mathbf{0}\,\mathbf{0}\,\mathbf{0}\,\mathbf{0}\,\mathbf{0}\ \mathbf{N}_{i,j}^{(m,n)}\end{bmatrix}\boldsymbol{\delta}_{\mathrm{e}} \tag{15}$$

in which, $\boldsymbol{\delta}_{\mathrm{e}}=\begin{bmatrix}\boldsymbol{\delta}_{ux}^{\mathrm{T}} & \boldsymbol{\delta}_{uy}^{\mathrm{T}} & \boldsymbol{\delta}_{vx}^{\mathrm{T}} & \boldsymbol{\delta}_{vy}^{\mathrm{T}} & \boldsymbol{\delta}_{wx}^{\mathrm{T}} & \boldsymbol{\delta}_{wy}^{\mathrm{T}} & \boldsymbol{\delta}_{p}^{\mathrm{T}}\end{bmatrix}^{\mathrm{T}}$; $\mathbf{0}$ is a zero row vector with length N^2.

Since all the prime unknowns and their partial derivatives are approximated by the DQ technique in (ξ,η) coordinate system, the isoparametric coordinate transformation may be applied to transform coordinate (ξ,η) into physical coordinate (x,y) (see in figure 1), and finally it can be obtained that

$$\frac{\partial^{m+n}\mathbf{u}^{\mathrm{S}}(x_i,y_j,t)}{\partial x^m \partial y^n}=\mathbf{N}_{\mathrm{u}}\boldsymbol{\delta}_{\mathrm{e}},\quad \frac{\partial^{m+n}\mathbf{v}^{\mathrm{S}}(x_i,y_j,t)}{\partial x^m \partial y^n}=\mathbf{N}_{\mathrm{v}}\boldsymbol{\delta}_{\mathrm{e}} \tag{16}$$

$$\frac{\partial^{m+n}\mathbf{w}^{\mathrm{F}}(x_i,y_j,t)}{\partial x^m \partial y^n}=\mathbf{N}_{\mathrm{w}}\boldsymbol{\delta}_{\mathrm{e}},\quad \frac{\partial^{m+n}p(x_i,y_j,t)}{\partial x^m \partial y^n}=\mathbf{N}_{p}\boldsymbol{\delta}_{\mathrm{e}} \tag{17}$$

According to the theory of Galerkin weighted residual value method, all the test functions are approximated with the same shape functions as the basic unknowns,

$$\begin{aligned}&\frac{\partial^{m+n}\tilde{\mathbf{u}}^{\mathrm{S}}(x_i,y_j,t)}{\partial x^m \partial y^n}=\mathbf{N}_{\mathrm{u}}\tilde{\boldsymbol{\delta}}_{\mathrm{e}},\quad \frac{\partial^{m+n}\tilde{\mathbf{v}}^{\mathrm{S}}(x_i,y_j,t)}{\partial x^m \partial y^n}=\mathbf{N}_{\mathrm{v}}\tilde{\boldsymbol{\delta}}_{\mathrm{e}},\\ &\frac{\partial^{m+n}\tilde{\mathbf{w}}^{\mathrm{F}}(x_i,y_j,t)}{\partial x^m \partial y^n}=\mathbf{N}_{\mathrm{w}}\tilde{\boldsymbol{\delta}}_{\mathrm{e}},\quad \frac{\partial^{m+n}\tilde{p}(x_i,y_j,t)}{\partial x^m \partial y^n}=\mathbf{N}_{p}\tilde{\boldsymbol{\delta}}_{\mathrm{e}}\end{aligned} \tag{18}$$

where $\tilde{\boldsymbol{\delta}}_{\mathrm{e}}$ is the nodal values of all the test functions in the same arrangement as $\boldsymbol{\delta}_{\mathrm{e}}$.

$$\tilde{\boldsymbol{\delta}}_{\mathrm{e}}=\left[\tilde{\boldsymbol{\delta}}_{\tilde{u}\mathrm{x}}^{\mathrm{T}}\ \tilde{\boldsymbol{\delta}}_{\tilde{u}\mathrm{y}}^{\mathrm{T}}\ \tilde{\boldsymbol{\delta}}_{\tilde{v}\mathrm{x}}^{\mathrm{T}}\ \tilde{\boldsymbol{\delta}}_{\tilde{v}\mathrm{y}}^{\mathrm{T}}\ \tilde{\boldsymbol{\delta}}_{\tilde{w}\mathrm{x}}^{\mathrm{T}}\ \tilde{\boldsymbol{\delta}}_{\tilde{w}\mathrm{y}}^{\mathrm{T}}\ \tilde{\boldsymbol{\delta}}_{\tilde{p}}^{\mathrm{T}}\right]^{\mathrm{T}} \tag{19}$$

2.4. Weak-form DQ discretization

Applying equations (16) and (18) to equation (6), and using the condition that $\tilde{\boldsymbol{\delta}}_{\mathrm{e}}$ is arbitrary in each sub-domain Ω_{e}, it can be finally obtained the weak-form differential quadrature element (WQE) equations

$$\mathbf{C}_{\mathrm{e}}\dot{\boldsymbol{\delta}}_{\mathrm{e}}+\mathbf{K}_{\mathrm{e}}\boldsymbol{\delta}_{\mathrm{e}}=\mathbf{R}_{\mathrm{e}} \tag{20}$$

where,

$$\mathbf{C}_{\mathrm{e}}=\int_{\Omega}\left[\mathbf{N}_{\mathrm{u}}^{\mathrm{T}}\left(\rho^{\mathrm{F}}\mathbf{N}_{\mathrm{w}}+\rho\mathbf{N}_{\mathrm{v}}\right)+\left(\rho^{\mathrm{F}}\mathbf{N}_{\mathrm{w}}+\beta\rho^{\mathrm{FR}}\nabla\mathbf{N}_{p}\right)^{\mathrm{T}}\left(\mathbf{N}_{\mathrm{v}}+\mathbf{N}_{\mathrm{w}}\right)+\mathbf{N}_{\mathrm{v}}^{\mathrm{T}}\mathbf{N}_{\mathrm{u}}\right]\mathrm{d}\Omega \tag{21}$$

$$\begin{aligned}\mathbf{K}_{\mathrm{e}}=\int_{\Omega}\Bigg[&(\mathbf{G}\mathbf{N}_{\mathrm{u}})^{\mathrm{T}}\mathbf{D}\mathbf{G}\mathbf{N}_{\mathrm{u}}-\left(\mathbf{r}_1\frac{\partial\mathbf{N}_{\mathrm{u}}}{\partial x}+\mathbf{r}_2\frac{\partial\mathbf{N}_{\mathrm{u}}}{\partial y}\right)^{\mathrm{T}}\mathbf{N}_{p}+\mathbf{N}_{\mathrm{w}}^{\mathrm{T}}\left(s_{\mathrm{v}}\mathbf{N}_{\mathrm{w}}+n^{\mathrm{F}}\nabla\mathbf{N}_{p}\right)\\ &+\mathbf{N}_{p}^{\mathrm{T}}\left(\mathbf{r}_1\frac{\partial\mathbf{N}_{\mathrm{v}}}{\partial x}+\mathbf{r}_2\frac{\partial\mathbf{N}_{\mathrm{v}}}{\partial y}\right)+\beta\left(\frac{\partial\mathbf{N}_{p}^{\mathrm{T}}}{\partial x}\frac{\partial\mathbf{N}_{p}}{\partial x}+\frac{\partial\mathbf{N}_{p}^{\mathrm{T}}}{\partial y}\frac{\partial\mathbf{N}_{p}}{\partial y}\right)-\mathbf{N}_{\mathrm{v}}^{\mathrm{T}}\mathbf{N}_{\mathrm{v}}\Bigg]\mathrm{d}\Omega\end{aligned} \tag{22}$$

$$\mathbf{R}_{\mathrm{e}}=\int_{\Omega}\left[\mathbf{N}_{\mathrm{u}}^{\mathrm{T}}\rho\mathbf{b}+\mathbf{N}_{\mathrm{w}}^{\mathrm{T}}\rho^{\mathrm{F}}\mathbf{b}+\left(\frac{\partial\mathbf{N}_{p}}{\partial x}+\frac{\partial\mathbf{N}_{p}}{\partial y}\right)^{\mathrm{T}}\rho^{\mathrm{FR}}\beta\mathbf{b}\right]\mathrm{d}\Omega+\int_{S_{\mathrm{T}}}\mathbf{N}_{\mathrm{u}}^{\mathrm{T}}\bar{\mathbf{T}}\mathrm{d}S-\int_{S_{\mathrm{Q}}}\mathbf{N}_{p}^{\mathrm{T}}\bar{Q}\mathrm{d}S \tag{23}$$

Herein, $\mathbf{r}_1=[1\ \ 0]$, $\mathbf{r}_2=[0\ \ 1]$; $\mathbf{G}$ is a differential operator for strain; $\mathbf{D}$ is a elasticity constants matrix. $\mathbf{G}$ and $\mathbf{D}$ have expressions as follows

$$\mathbf{G}=\begin{bmatrix}\partial/\partial x & 0\\ 0 & \partial/\partial y\\ \partial/\partial y & \partial/\partial x\end{bmatrix},\ \mathbf{D}=\begin{bmatrix}2\mu^{S}+\lambda^{S} & \lambda^{S} & 0\\ \lambda^{S} & 2\mu^{S}+\lambda^{S} & 0\\ 0 & 0 & \mu^{S}\end{bmatrix} \tag{24}$$

It can be observed that equations (21), (22) and (23) cannot be evaluated through numerical quadrature such as the Gauss-Legendre quadrature for the moment, because the integrand there can only be acquired in the two-dimensional expanded-Chebyshev points. To cope with this, [17] and [15] have already presented solutions to this problem by using the Lagrange interpolation. As a result, the Gauss-Legendre quadrature can be used in the present numerical program, where only the inner and constant sum of points of each element may be used. This is different from the high efficient quadrature rules [18,19] in the spline FEM, whose quadrature point sampling spans many elements, and the quadrature points number in each element varies. After assembly of equation (20) and other procedures of the conventional FEM the first-order initial value problem is solve by the implicit Euler method step by step.

3. Numerical verification

According to the above formulations, the numerical programs are implemented using Julia language on a PC, which configures: OS Windows 10 64 bit, RAM 16 GB and CPU intel i7 8700k. A full-band sparse matrix storage is used for the assembly of element matrices, and the gaussian factorisation method is used for solution of linear equation system.

3.1. Numerical example 1

There is a half-space foundation consisting of saturated-soil. And its top surface is subjected to a uniform presseure $q(t) = 3000[1 - \cos(75\,t)]$. In this case, the three-dimensional problem degrades into a one-dimensional one. And de Boer *et al* [20] had contributed the analytical solutions to this problem. In this numerical example, the uniform mesh shown in figure 2 is used to represent the one-dimensional case approximately, where 50 four-node elements with side length 0.2 m are used; the time step length is fixed 0.1 ms; all the physical parameters are also listed in figure 2.

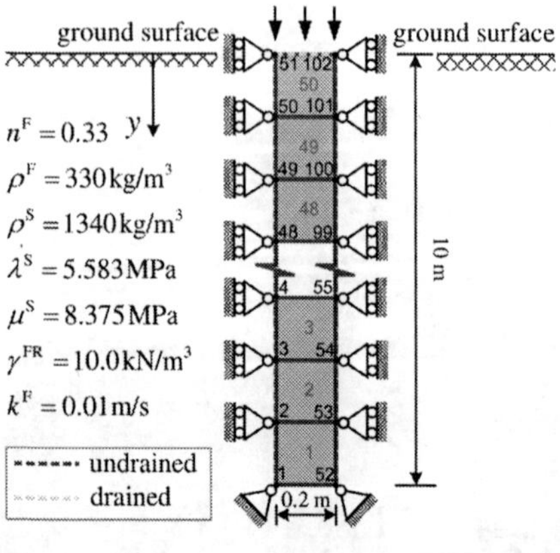

Figure 2. Physical, geometric and meshing parameters together with boundary conditions.

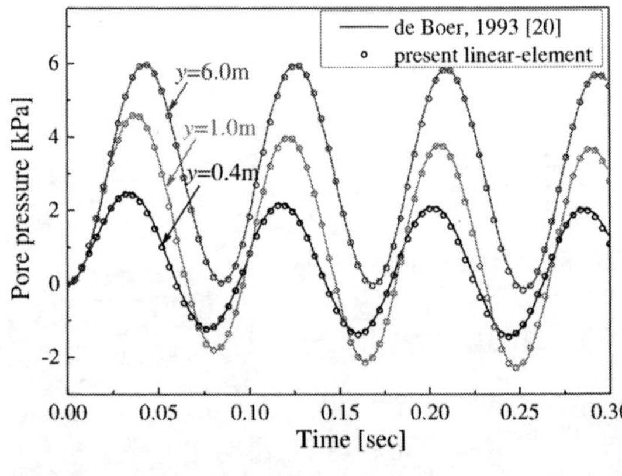

Figure 3. Verification for solutions of pore pressure.

Figure 3 illustrates the pore pressure time-history at depth from ground top 0.4 m, 1.0 m and 6.0 m. There an excellent agreement of pore pressure results is observed, which to some extent verifies the present formulations and numerical program.

IOP Conf. Series: Materials Science and Engineering **657** (2019) 012043 doi:10.1088/1757-899X/657/1/012043

3.2. Numerical example 2

The same physical parameters as numerical example 1 are used in this example, except that the Darcy permeability coefficient herein is modified as $k^F = 0.0001$ m/s. A square fluid-saturated soil block with side length 10 m, shown in figure 4, is analyzed, where four uniform five-order WQEs are used to discretized the whole block, and the detailed displacement, water flux and stress boundary conditions are marked in the figure. A fixed time length 0.01 s is used in the implicit Euler method.

Figure 5 shows the vertical displacement at the two top-surface corners, p_1 and p_2 (see in figure 4). It can be seen in figure 5 that in the very beginning of the loading the rise of p_1 and the sink of p_2 almost coincide with each other due to the low-permeability and incompressibility of the saturated soil, while the rise and sink diverge when more time elapses, due to the seepage from the drainage surface. The good agreement of results generated by the WQEM and FEM of [7] again verifies the work of the present paper.

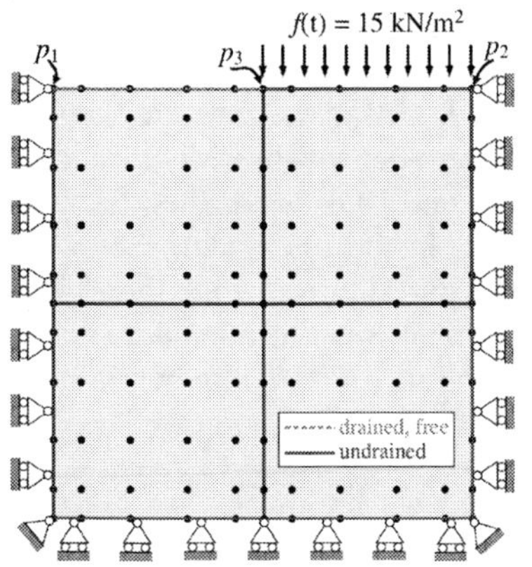

Figure 4. A square saturated soil with half top surface subjected to uniform load.

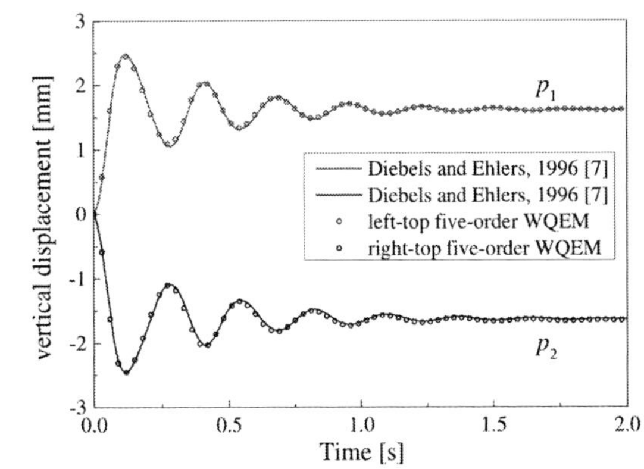

Figure 5. Time history results of vertical displacements at two top-corners.

4. Numerical performances

4.1. Numerical convergence

Keeping all the physical and geometric parameters unchanged except the Darcy permeability coefficient $k^F = 0.01$ m/s, the saturated-soil block analyzed by section 3.2 is re-analyzed. There are two factors, the time step-length and the mesh size, controlling the convergence of analysis results. To investigate these factors, the vertical displacement of the central point p_3 (see in figure 4) in time t = 0.1 s is computed. Considering three types of uniform quadrilateral mesh 64x64, 16x16 and 6x6, which respectively denote element sizes 10/64 m, 10/16 m and 10/6 m, figure 6 shows the convergence course of vertical displacement at p_3 as the number of fixed time-length increases. It can be seen, from figure 6, clearly that the vertical displacement converges stably for soil block with different meshing densities.

IOP Conf. Series: Materials Science and Engineering **657** (2019) 012043 doi:10.1088/1757-899X/657/1/012043

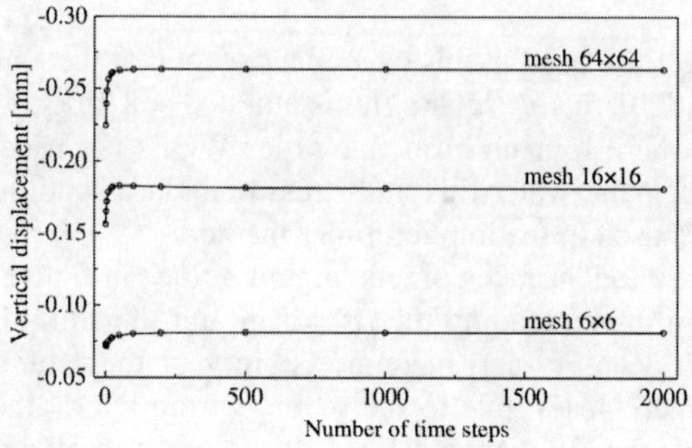

Figure 6. Influence of time-integration on convergence of vertical displacement at p_3.

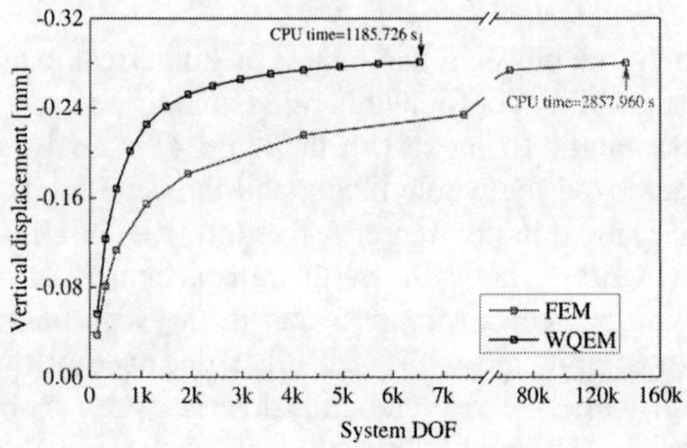

Figure 7. Influence of system DOF on convergence of vertical displacement at p_3.

From figure 6, it can be judged that a fixed time step-length 0.2 ms may provide time integration with enough precision. As a result, figure 7 is further presented, setting the time-step length fixed 0.2 ms, to investigate the convergence efficiency of WQEM. Herein, the soil-block is meshed uniformly. The WQE model is discretized by only four WQEs, whose system DOF is adjusted through changing the number of expanded-Chebyshev grid points each element. In the FE model, the system DOF is adjusted by varying the meshing density. It is observed from figure 7 that there is much less requirement of system DOF for the WQEM than the conventional FEM to pursue a certain accuracy goal. The CPU time spent by the FEM is 2857.960 s for an accuracy of 0.01 mm, while it takes 58.5% less CPU time in WQEM for the same accuracy.

4.2. Stress smoothness

A foundation pit shown in figure 8 is analyzed, whose physical parameters are the same as those used in section 3.1. The geometry information is also illustrated in figure 8. As the structure is vertically symmetric, half structure is considered and meshed with two types of mesh grids (see in figure 9). In mesh 0 and mesh 1, there are 16 six-order WQEs and 736 conventional FEs used, respectively; the half structure is discretized into a system with DOFs 4209 in WQE model and 5349 in FE model, respectively. A fixed time step-length 0.5 ms is used in the implicit time integration.

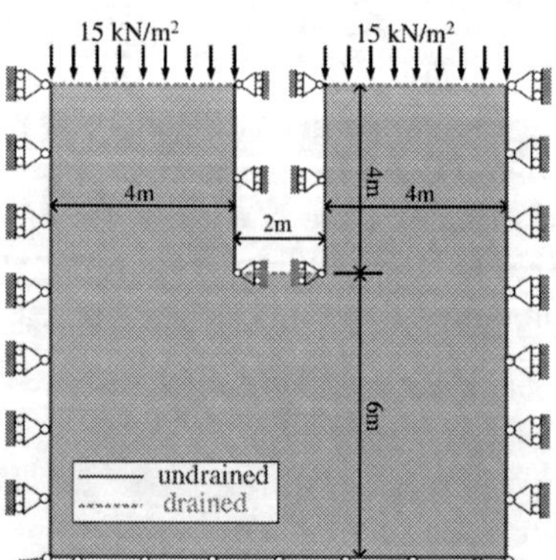

Figure 8. Dimensions and boundary conditions of a foundation pit.

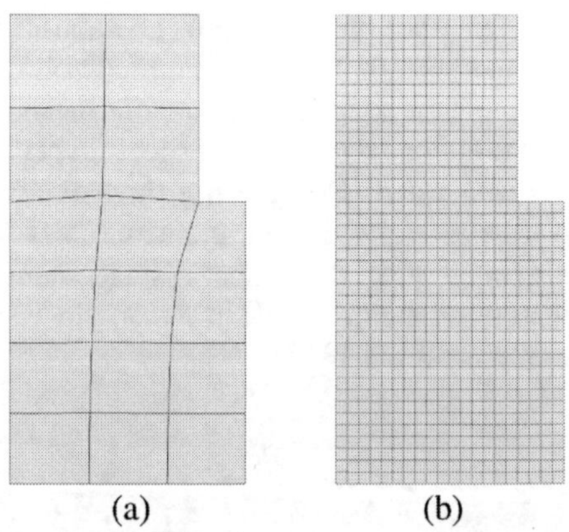

Figure 9. Mesh grid for (a) mehs0: higher order QEM model, (b) mesh 1: conventional FEM model.

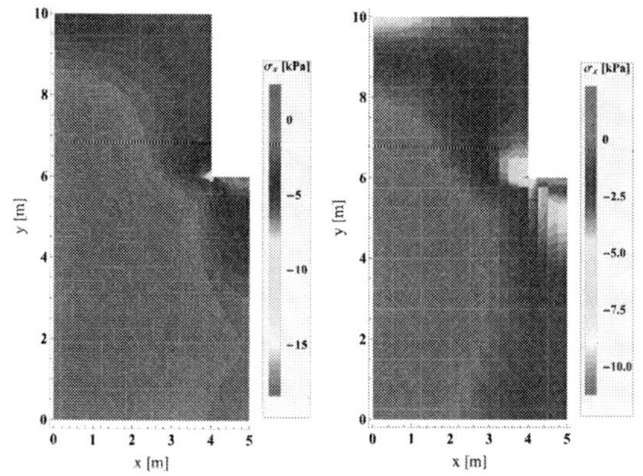

(a) (b)

Figure 10. Effective stress σ_x, using (a) six-order QEM and (b) conventional FEM.

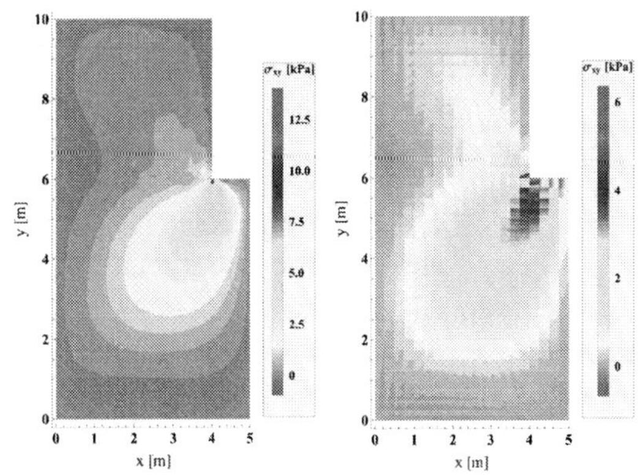

(a) (b)

Figure 11. Effective stress τ_{xy}, using (a) six-order QEM and (b) conventional FEM.

The effective stresses responses, σ_x and τ_{xy}, are captured by the two models in the moment when time reaches 0.1 s, and are showed in figures 10 and 11. In the figures, no numerical smoothness technique is applied to the nodes connecting neighboring elements. And it can be observed that a serious discontinuity of stress occurs in the FEM model, while it is not a problem in the QEM model. This also demonstrates why it is usually required of numerical smoothness in some of the commercial FE software.

5. Conclusions

Taking advantage of the DQ technique, a WQE for the dynamic analyses of two-dimensional fluid-saturated soil is developed. The formulations and computer programs are verified through two comparison examples with the previous results. Convergence analyses for both time and space discretization are performed, from which it is found that the proposed WQEM is able to converge at much less cost of system DOFs than the conventional FEM; in the stress analyses it is demonstrated that the presented WQEM dramatically alleviates the smoothness problem of the conventional displacement-based FEM.

Acknowledgments

The authors wish to acknowledge the support from National Key R & D Program of China (Grant No. 2017YFC0805500).

References

[1] Biot M A 1941 General theory of three-dimensional consolidation *J. Appl. Phys.* **12** 155-64
[2] Biot M A 1956 Theory of propagation of elastic waves in a fluid-saturated porous media, I. low frequency range *J. Acoust. Soc. Am.* **28** 168-78
[3] de Boer R 2011 *Theory of Porous Media* (Berlin Heidelberg: Springer)
[4] Yuan S and Zhong H 2014 Consolidation analysis of non-homogeneous soil by the weak form quadrature element method *Comput. Geotech.* **62** 1-10
[5] Zou D, Teng X, Chen K and Yu X 2018 An extended polygon scaled boundary finite element method for the nonlinear dynamic analysis of saturated soil *Eng. Anal. Bound. Elem.* **91** 150-61
[6] Bui H H and Nguyen G D 2017 A coupled fluid-solid SPH approach to modelling flow through deformable porous media *Int. J. Solids Struct.* **125** 244-64
[7] Diebels S and Ehlers W 1996 Dynamic analysis of fully saturated porous medium accounting for geometric and material non-linearities *Int. J. Numer. Meth. Eng.* **39** 81-97

IOP Conf. Series: Materials Science and Engineering **657** (2019) 012043 doi:10.1088/1757-899X/657/1/012043

[8] Breuer S 1999 Quasi-static and dynamic behavior of saturated porous media with incompressible constituents *Transport Porous. Med.* **34** 285-303

[9] Hu Y, Zhu Y and Cheng C 2009 DQM for dynamic response of fluid-saturated visco-elastic porous media *Int. J. Solids Struct.* **46** 1667-75

[10] Shu C and Richards B E 1992 Application of generalized differential quadrature to solve two-dimensional incompressible Navier-Stokes equations *Int. J. Numer. Meth. Fl.* **15** 791-8

[11] Chen C N 1997 The two-dimensional frame model of the differential quadrature element method *Comput. Struct.* **62** 555-71

[12] Chen C N 2000 A generalized differential quadrature element method *Comput. Method. Appl. M.* **188** 553-66

[13] Bellman R and Casti J 1971 Differential quadrature and long-term integration *J. Math. Anal. Appl.* **34** 235-8

[14] Bellman R, Kashef B G and Casti J 1972 Differential quadrature: A technique for the rapid solution of nonlinear partial differential equations *J. Comput. Phys.* **10** 40-52

[15] Fu C and Yang X 2018 Dynamic analysis of partial-interaction Kant composite beams by weak-form quadrature element method *Arch. Appl. Mech.* **88** 2179-98

[16] Hiemstra R R, Calabrò F, Schillinger D and Hughes T J R 2017 Optimal and reduced quadrature rules for tensor product and hierarchically refined splines in isogeometric analysis *Comput. Method Appl. M.* **316** 966-1004

[17] Jin C, Wang X and Ge L 2014 Novel weak form quadrature element method with expanded Chebyshev nodes *Appl. Math. Lett.* **34** 51-9

[18] Barton M and Calo V M 2016 Optimal quadrature rules for odd-degree spline spaces and their application to tensor-product-based isogeometric analysis *Comput. Method Appl. M.* **305** 217-40

[19] Barton M and Calo V M 2017 Gauss-Galerkin quadrature rules for quadratic and cubic spline spaces and their application to isogeometric analysis *Comput. Aided Design* **82** 57-67

[20] de Boer R, Ehlers W and Liu Z 1993 One-dimensional transient wave propagation in fluid-saturated incompressible porous media *Arch. Appl. Mech.* **63** 59-72

IOP Conf. Series: Materials Science and Engineering **657** (2019) 012044 doi:10.1088/1757-899X/657/1/012044

Influences of various orifice layouts on the properties of aerostatic journal bearings based on finite element method

S Y Gao[1,2,5], Y G Shi[2], S Yang[2,3], P Sun[2] and T J Li[4,5]

[1]The State Key Laboratory of Mechanical Transmissions, Chongqing University, 400044, Chongqing, PR China
[2]Institute of Advanced Manufacturing Technology, Hefei Institutes of Physical Science, Chinese Academy of Sciences, Changzhou, 213164, Jiangsu, PR China
[3]State Key Lab of Digital Manufacturing Equipment and Technology, Wuhan, 430074, Hubei, PR China
[4]School of Mechanical Engineering, Shanghai University of Technology, Shanghai 200093, PR China

E-mail: sygao@iamt.ac.cn/litianjian99@163.com

Abstract. It is essential to thoroughly investigate influences of orifice layouts on properties of aerostatic journal bearings employed in ultra-high speed spindles. In this paper, four orifice layouts of aerostatic journal bearings which have a large length-diameter ratio are designed and investigated, particularly in light of the nonlinear compressible Reynolds equation and the associated computational analysis and algorithms using the FEM-based Galerkin weighted residual method. A series of pressure distributions as well as the load capacity, attitude angle, volume flow rate and air film stiffness of the aerostatic journal bearings for the cases of four orifice layouts are analysed with respect to different rotational speeds and eccentricity ratios. It can be concluded that increasing orifice sets is not beneficial to enhance the performance of the aerostatic journal bearing. The aerostatic journal bearing divided into two identical independent aerostatic journal bearings has relative optimal performance at rotational speeds from 0 r/min to 160,000 r/min, such as the highest load capacity, the lowest attitude angle and modest air film stiffness, however, with consumption of the largest quantities of pressurized gas. The designers can use the analysis results to optimize the orifice layout for improving the performance of the aerostatic journal bearing.

1. Introduction

Air bearings have outstanding features of almost zero friction, clean and high supporting accuracy with respect to ball bearings [1,2]. Hence, air bearings become the essential components of high speed motorized spindles which are commonly used in precision micro-machining fields with the top speed of greater than 100,000 r/min [3,4]. Aerostatic bearings, one type of air bearings, have many types of restrictors, such as porous restrictor, slot restrictor, orifice restrictor and so on. However, orifice restrictors play a dominant role in the academic and industrial communities. It is of great significance to study the varied properties of aerostatic bearings influenced by orifice layouts for developing higher accuracy bearings which are employed to the ultra-precision aerostatic spindles.

Large quantity of research papers had been reported to investigate properties of the aerostatic bearing by numerically solving the Reynolds equation with finite difference method (FDM) or finite element method (FEM). Gas film pressure and load capacity of the aerostatic bearing with a length-diameter ratio of 1 were analysed based on the modified Reynolds equation [5]. Rotation effects of hybrid air journal bearings were studied by numerically solving the Reynolds equation with FDM [6]. Effects of the geometrical and running parameters on the characteristics of the micro air journal bearing were obtained by applying the Newton iteration and the FDM to analyse the Reynolds equation [7]. Aerodynamic and aerostatic effects of aerostatic bearings were investigated through numerically calculating the Reynolds equation associated with the Newton's method [8]. The Reynolds equation was numerically analysed by FDM and successive over-relaxation (SOR) method to calculate the load capacity and stiffness of air bearings [9]. Air film pressures of aerostatic bearings were analysed with Euler explicit approach and central FDM [10,11]. Besides, air film pressures and stiffnesses of aerostatic journal and thrust bearings were studied based on the FDM and the successive relaxation method [12]. Furthermore, the Reynolds equation was numerically calculated by FEM to study the radial load capacity and air film stiffness of the air spherical bearing [13], and the effects of pressure-equalizing grooves of aerostatic journal bearings were also investigated [14]. The Reynolds equation with the velocity item was further discretized by FEM to study the performance characteristics of the aerostatic journal bearing under conditions of ultra-high speeds and eccentricities [15]. Based on the aforementioned analyses, the performance investigations of the aerostatic bearings were conducted in many papers, however, most of the length-diameter ratios of the aforementioned aerostatic journal bearings were less than 2, and few research literatures were published to discuss the influences of orifice layouts on the properties of aerostatic journal bearings which have a large length-diameter ratio. Therefore, in this study, influences of various orifice layouts on the properties of the aerostatic journal bearings with a large length-diameter ratio are investigated by numerically solving the nonlinear compressible Reynolds equation based on the FEM.

2. Aerostatic journal bearings with various orifice layouts

Figure 1 shows the designed four orifice layouts of the aerostatic journal bearing. Different orifice sets are designed in each bearing and twelve equispaced orifices are distributed in each orifice set. The bearing length, L, is 60 mm, the bearing bore diameter, D, is 20 mm, the orifice diameter, d, is 0.2 mm, the gas film thickness without eccentricity, h_n, is 20 μm, the supply gas pressure, p_s, is 0.7 MPa. Section a-a is the axial middle plane of the aerostatic journal bearing.

Case 1: two orifice sets are symmetrical with respect to the section a-a, and l/L=1/5.

Case 2: four orifice sets are symmetrical with respect to the section a-a, and $l_1/(L/2)$=1/5, $l_2/(L/2)$=4/5.

Case 3: four orifice sets are symmetrical with respect to the section a-a. There is an exhaust vent in the axial middle position of the bearing. Thus, the whole bearing is divided into two identical independent small aerostatic journal bearings and the section b-b is the axial middle plane of the small aerostatic journal bearing. Two orifice sets are symmetrical with respect to the section b-b. The axial length of the small aerostatic journal bearing is $L/2$, and $l_3/(L/2)$=1/5.

Case 4: six orifice sets are symmetrical with respect to the section a-a, and l_4/L=1/10, l_5/L=1/4, l_6/L=2/5.

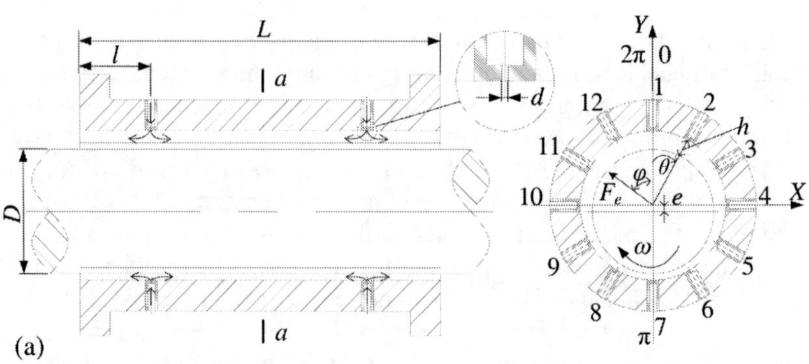

(a)

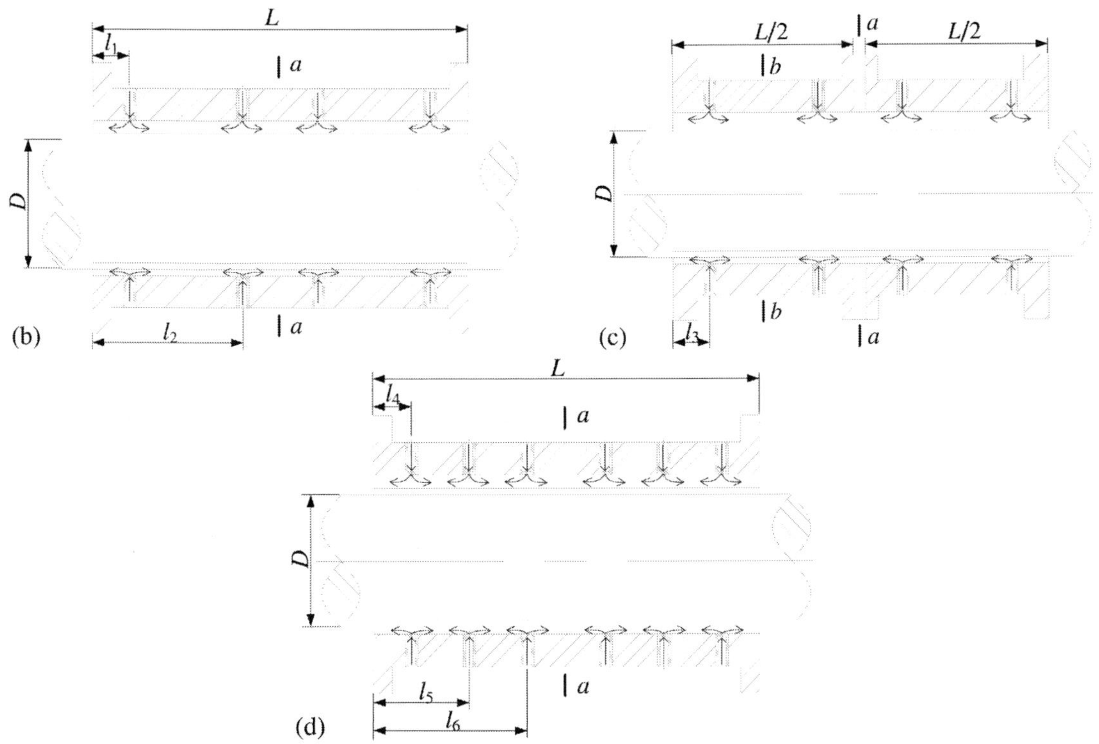

Figure 1. Four orifice layouts of the aerostatic journal bearing. (a) Case 1. (b) Case 2. (c) Case 3. (d) Case 4.

3. Computational model

3.1. Governing equations

The dimensionless Reynolds equation is depicted in equation (1) [15,16]

$$\frac{\partial}{\partial \bar{x}}\left(\bar{h}^3 \frac{\partial \bar{p}^2}{\partial \bar{x}}\right)+\frac{\partial}{\partial \bar{z}}\left(\bar{h}^3 \frac{\partial \bar{p}^2}{\partial \bar{z}}\right)=\Lambda_x \frac{\partial\left(\bar{h}\bar{p}\right)}{\partial \bar{x}}+\Lambda_z \frac{\partial\left(\bar{h}\bar{p}\right)}{\partial \bar{z}} \tag{1}$$

where Λ_x and Λ_z are dimensionless bearing numbers, $\bar{x}$ and $\bar{z}$ are dimensionless Cartesian coordinates, $\bar{h}$ is the dimensionless thickness, $\bar{p}$ is the dimensionless pressure.

In this study, the dimensionless Reynolds equation has the following boundary conditions.

- Pressure at the outer border of the bearing: $p=p_a$, p_a is ambient pressure;
- Pressure at the symmetric boundary: $\frac{\partial p}{\partial n}=0$, n is the normal direction of the symmetric boundary;
- Pressure at the orifice boundary: $p=p_r$, p_r is the pressure at the orifice outlet.

The numerical solution of the Reynolds equation is calculated by FEM with the triangular element, and the finite element formula of the Reynolds equation is deduced as follows

$$\sum_{e\in\Delta i}\int_{\Delta e}\left(\mathbf{N}^{eT}\mathbf{h}^{e}\right)^{3}d\bar{x}d\bar{z}\left(c_{i}\mathbf{c}^{eT}+b_{i}\mathbf{b}^{eT}\right)\mathbf{f}^{e}\frac{1}{\left(2\Delta e\right)^{2}}-\bar{Q}\zeta_{i}$$
$$-\sum_{e\in\Delta i}\lambda\left(c_{i}\bar{u}+b_{i}\bar{\omega}\right)\mathbf{h}^{eT}\int_{\Delta e}\mathbf{N}^{e}\mathbf{N}^{eT}d\bar{x}d\bar{z}\frac{1}{2\Delta e}\left(\mathbf{f}^{1/2}\right)^{e}=0 \tag{2}$$
$$\left(i=1,2,\ldots,q\right)$$

where $\mathbf{N}^e$ is the shape function, $\mathbf{h}^e$ is the column matrix of the node thickness, c_i and b_i are the coefficients of the interpolation function, $\mathbf{c}^e$ and $\mathbf{b}^e$ are the coefficient matrixes of the interpolation function, $\mathbf{f}^e$ is the column matrix of dimensionless node pressure square, $\bar{Q}$ is the mass flow factor, ζ_i is the Kronecker function, $\bar{u}$ and $\bar{\omega}$ are the dimensionless velocities, λ is the bearing coefficient.

Iteration method is applied to solve the finite element formula and the unknown nodes' pressures are obtained. Furthermore, properties of the aerostatic journal bearing, such as load capacity, attitude angle, volume flow rate and stiffness, can be calculated based on the obtained air film pressure (to study the calculation theory in detail, please refer to Liu *et al* [16]).

3.2. Computational domain

In this study, the computational domain of each aerostatic journal bearing is determined by half of the fluid zone which is symmetrical with respect to section a-a (section b-b for case 3). The meshed computational domain with right triangle elements and boundary conditions are illustrated in figure 2. Each orifice is regarded as a point in the domain and only one node is distributed to an orifice zone.

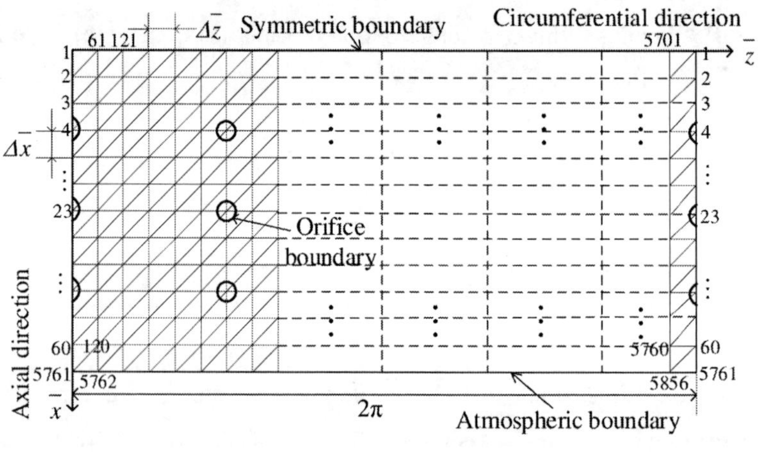

Figure 2. Meshes and boundary conditions of the computational domain (case 4).

Grid independence test is conducted to determine the computation grid size. Case 3 is chosen as the testing object. Two size schemes are scheduled as follows, scheme A is that the circumferential length of the bearing is equally cut into 96 parts in Z direction and the axial length is equally cut into 60 parts in X direction, producing 5856 nodes and 11520 triangular elements; Scheme B is that the circumferential length of the bearing is equally cut into 120 parts in Z direction and the axial length is equally cut into 60 parts in X direction, producing 7320 nodes and 14400 triangular elements. Rotational speed (n) of the bearing is 0 r/min, and the eccentricity ratios (ε) of the bearing is 0.3. The calculated load capacities and volume flow rates of the case 3 bearing are presented in table 1.

It can be found from table 1 that there are very small differences of the performance parameters between scheme A and scheme B. Besides, the grid quantity of scheme A is less than that of scheme B. Hence, scheme A is employed to mesh the computation domain.

Table 1. Load capacities and volume flow rates of two size schemes.

Parameters	Scheme A	Scheme B	$\frac{\text{Scheme A} - \text{Scheme B}}{\text{Scheme A}} \times 100\%$
Load capacity (N)	126.16	124.98	0.94%
Volume flow rate (L/min)	85.94	85.4	0.63%

3.3. *Experimental verification*

In order to verify the aforementioned analysing method, experimental results tested by Chen *et al* [9] are employed to this study. Air film stiffnesses of the bearings reported by Chen *et al* [9] are numerically analysed under condition of various supply pressures based on the proposed computational method. Table 2 presents the experimental and calculated results, indicating that the maximum difference between the experimental and calculated values is 10.02% which is acceptable for engineering purposes. Therefore, the proposed computational model is validated.

Table 2. Experimental [9] and calculated stiffnesses of aerostatic journal bearings.

Supply pressure (MPa)	Experimental (N/μm)	Calculated (N/μm)	Error (%)
Front bearing (L/D=1.5)			
0.6	10.1	11.038	-9.29
0.7	12.5	13.752	-10.02
0.8	16.2	16.903	-4.34
0.9	19.2	20.629	-7.44
Rear bearing (L/D=1.0)			
0.6	8.50	8.8733	-4.39
0.7	10.5	11.140	-6.10
0.8	13.6	13.836	-1.74
0.9	16.5	17.086	-3.55

4. Results and discussions

An analysing procedure is developed by MATLAB. The nonlinear equations are numerical solved by the 'fsolve' function with the termination tolerance of 1e-20. Pressure distribution, load capacity, attitude angle, volume flow rate and air film stiffness of four orifice layouts are thoroughly studied at various speeds and eccentricities.

Dimensionless air film pressure distributions of four orifice layouts under two running conditions (n=0 r/min, ε=0; n=200,000 r/min, ε=0.3) are shown in figure 3. It should be noted that the whole aerostatic journal bearing of case 3 is divided into two identical independent small aerostatic journal bearings and the dimensionless pressure distributions of each small aerostatic journal bearing are shown in figures 3(e) and 3(f). It can be found from figures 3(a), 3(c), 3(e) and 3(g) that an identical pressure value appears among twelve orifices of each orifice set along the circumferential direction when there is no eccentricity. Similar pressure profiles happen between case 1 and case 3, and the pressure distributions of case 2 and case 4 are also similar. What's more, the pressure uniformity decreases as in the sequence of cases of 4, 2, 1 and 3 under the same operating condition.

The load capacities of aerostatic journal bearings with four orifice layouts under working conditions of diverse rotational speeds and ε of 0.3 are described in figure 4. Figure 4 illustrates that load capacities of four orifice layouts increase as rotational speeds increased at ε of 0.3, and the load capacity of case 1 has the maximum growth rate. The load capacity of case 3 is the biggest one among four orifice layouts at rotational speeds from 0 r/min to 160,000 r/min, and the load capacity of case 1 becomes the largest one when the rotational speed is more than 170,000 r/min. Case 4 has the lowest load capacity at rotational speeds from 0 r/min to 200,000 r/min, and case 2 has the second lowest load

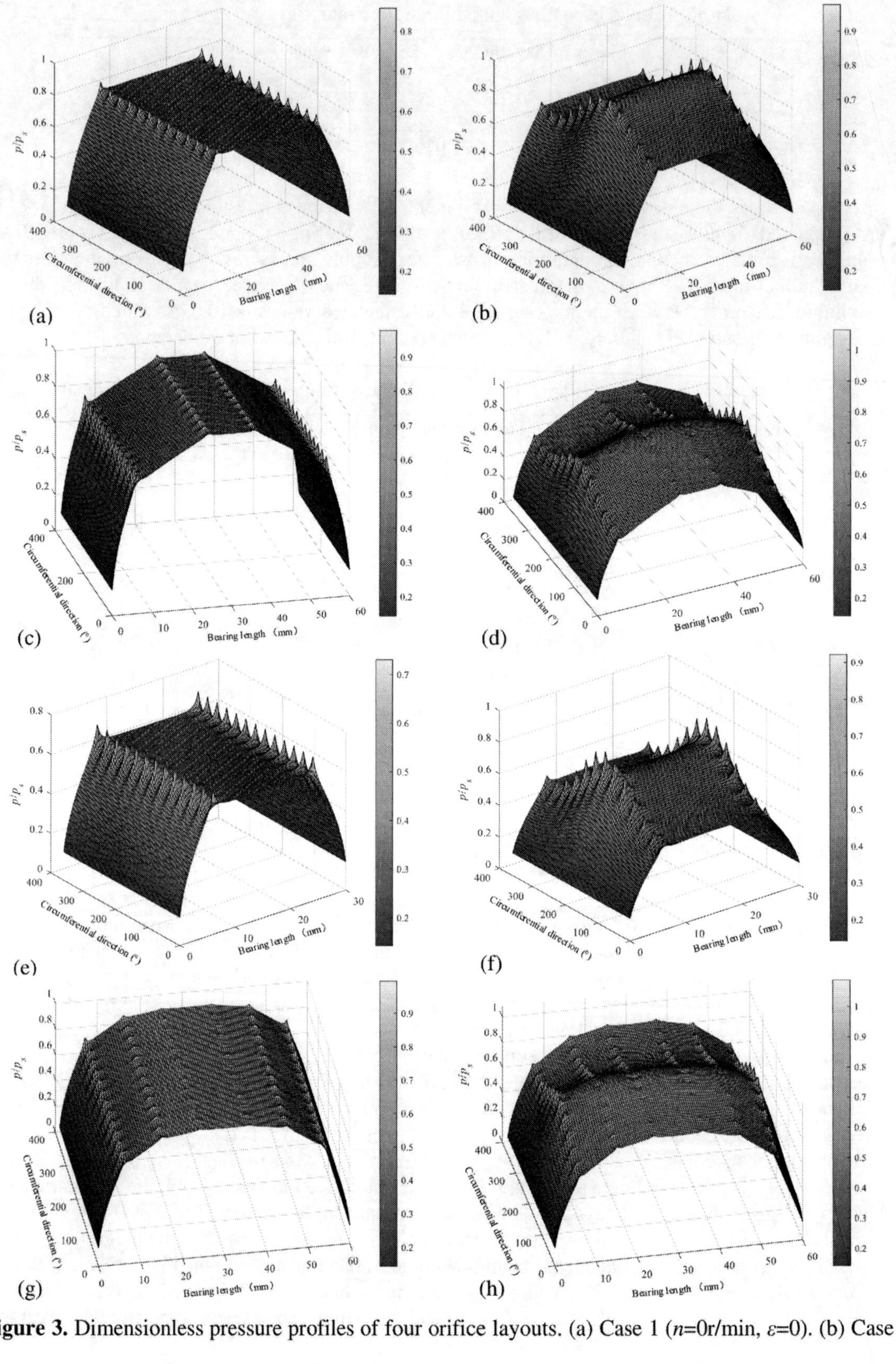

Figure 3. Dimensionless pressure profiles of four orifice layouts. (a) Case 1 (n=0r/min, ε=0). (b) Case

1 (n=200,000 r/min, ε=0.3). (c) Case 2 (n=0 r/min, ε=0). (d) Case 2 (n=200,000r/min, ε=0.3). (e) Case 3 (n=0 r/min, ε=0). (f) Case 3 (n=200,000 r/min, ε=0.3). (g) Case 4 (n=0r/min, ε=0). (h) Case 4 (n=200,000 r/min, ε=0.3).

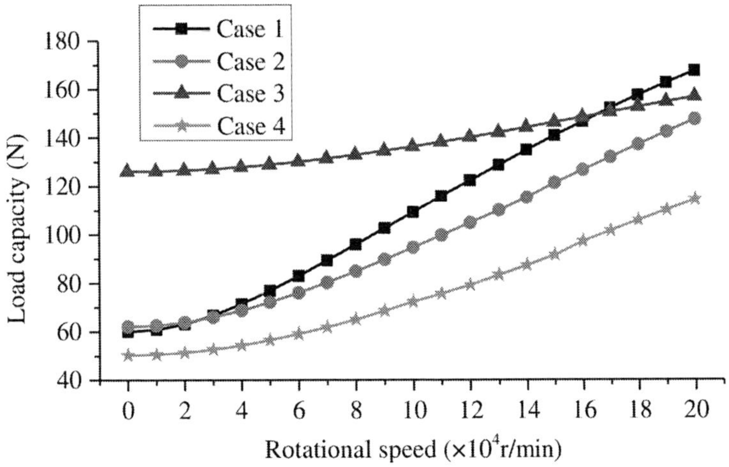

Figure 4. Load capacities of four orifice layouts with various rotational speeds.

capacity at rotational speeds from 30,000 r/min to 200,000 r/min. For cases 1, 2 and 4, the load capacity decreases with increasing orifice sets at speed of more than 20,000 r/min. Based on the aforementioned analysis, increasing orifice sets is not beneficial to increase the load capacity of the aerostatic journal bearing. However, the aerostatic journal bearing which is divided into two identical independent aerostatic journal bearings can greatly improve the load capacity at rotational speeds from 0 r/min to 160,000 r/min.

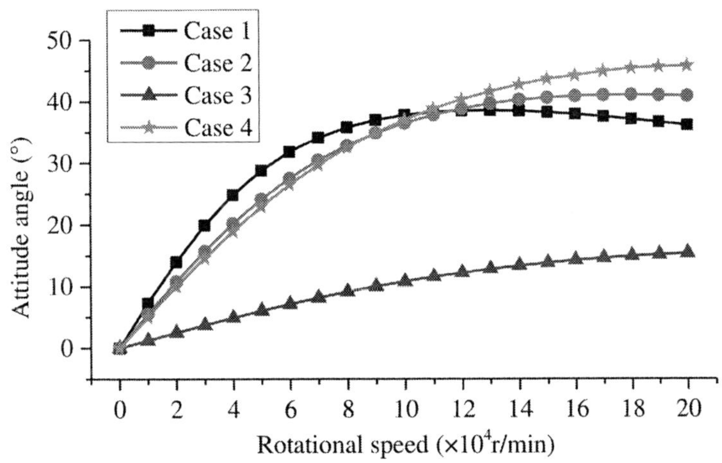

Figure 5. Attitude angles of four orifice layouts with various rotational speeds.

The attitude angles of aerostatic journal bearings with four orifice layouts under running conditions of diverse speeds and ε of 0.3 are shown in figure 5. It can be found from figure 5 that the attitude angles of case 1 and case 2 go up first and then they will go down with advancing speeds, and the attitude angles of case 3 and case 4 increase with increasing rotational speeds but the increasing rate is gradually slowed down. For cases 1, 2 and 4, the attitude angle increases with increasing orifice sets at speed of more than 110,000 r/min. However, case 3 has the lowest attitude angle at rotational speeds from 0 r/min to 200,000 r/min, which means that case 3 has the lowest whirl force, presenting that case 3 has the best stability among four orifice layouts.

Volume flow rates of four orifice layouts at disparate speeds and ε of 0.3 are illustrated in figure 6. Figure 6 describes that the volume flow rates of four cases slowly decrease with increasing rotational speeds in general. For cases 1, 2 and 4, the volume flow rate increases with advancing orifice sets under identical operating conditions. The volume flow rate of case 3 is larger than those of other cases under the same operating conditions. The volume flow rates of four orifice layouts are ranked with sort descending under identical conditions, i.e. cases of 3, 4, 2 and 1. Based on the aforementioned analysis, the aerostatic journal bearing which is divided into two identical independent aerostatic journal bearings consumes the largest quantities of pressurized gas.

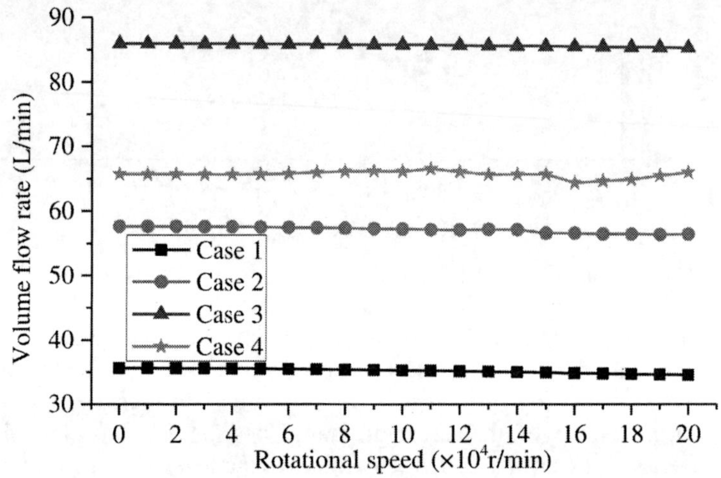

Figure 6. Volume flow rates of four orifice layouts with various rotational speeds.

Figure 7 depicts air film stiffnesses of four orifice layouts under conditions of various eccentricities and n of 200,000 r/min.

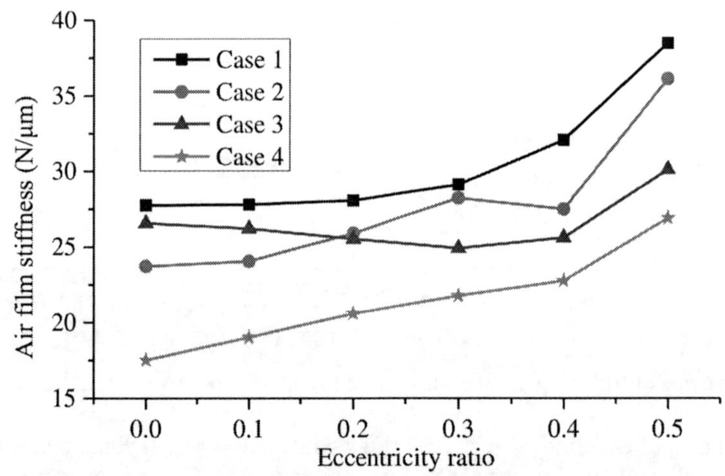

Figure 7. Air film stiffnesses of four orifice layouts with various eccentricity ratios.

As depicted in figure 7, it can be found that the air film stiffnesses of cases 1, 2 and 4 increase with increasing eccentricity ratios in general. Nevertheless, the air film stiffness of cases 3 decrease first and then increase with increasing eccentricity ratios. For cases 1, 2 and 4, the air film stiffness decreases with increasing orifice sets under the same eccentricity ratios, presenting that increasing orifice sets cannot enhance the air film stiffness of the bearing. Eccentricity ratio of 0 is the point at which the bearing is firstly subjected to an external force. Hence, the air film stiffness at ε of 0 is the greatest important to the bearing among all air film stiffnesses. The difference is 4.27% between the air film stiffnesses of case 1 and case 3 at ε of 0, which can be ignored, presenting that the air film stiffness of case 3 is acceptable for engineering purposes with respect to that of case 1.

5. Conclusions

This paper investigated influences of various orifice layouts on properties of the aerostatic journal bearing by numerically solving the Reynolds equation based on FEM, and air film pressure, bearing load capacity, bearing attitude angle, bearing volume flow rate and air film stiffness of four different orifice layouts are analyzed under different operating conditions. Based on the aforementioned analysis results of four orifice layouts, it can be concluded that increasing orifice sets is not beneficial to strengthen abilities of aerostatic journal bearings. Aerostatic journal bearing which is divided into two identical independent aerostatic journal bearings (case 3) has the relative optimal performance, the highest load capacity, the lowest attitude angle and modest air film stiffness, at rotational speeds from 0 r/min to 160,000 r/min. However, the case 3 bearing consumes the largest quantities of pressurized gas. Based on the obtained analysis results, engineers can better design the orifice layout for higher precision aerostatic journal bearing with a large length-diameter ratio.

Acknowledgments

The authors thank for the funding support for this research by the State Key Laboratory of Mechanical Transmissions (Grant No. SKLMT-KFKT-201801), the National Natural Science Foundation of China (Grant No. 51705501, 51705500), the CASHIPS Director's Fund (Grant No. YZJJ201622) and the State Key Lab of Digital Manufacturing Equipment and Technology in China (Grant No. DMETKF2017017).

References

[1] Stanev P T, Wardle F and Corbett J 2004 Investigation of grooved hybrid air bearing performance *P. I. Mech. Eng. K.-J. Mul.* **218** 95-106
[2] Gao Q, Chen W, Lu L, Huo D and Cheng K 2019 Aerostatic bearings design and analysis with the application to precision engineering: State-of-the-art and future perspectives *Tribol. Int.* **135** 1-17
[3] Abele E, Altintas Y and Brecher C 2010 Machine tool spindle units *CIRP. Ann.-Manuf. Techn.* **59** 781-802
[4] Cheng K and Huo D 2013 *Micro Cutting: Fundamentals and Applications* (West Sussex: Wiley)
[5] Lo C Y, Wang C C and Lee Y H 2005 Performance analysis of high-speed spindle aerostatic bearings *Tribol. Int.* **38** 5-14
[6] Su J C T and Lie K N 2003 Rotation effects on hybrid air journal bearings *Tribol. Int.* **36** 717-26
[7] Xiao H, Li W, Zhou Z, Huang X and Ren Y 2018 Performance analysis of aerostatic journal micro-bearing and its application to high-speed precision micro-spindles *Tribol. Int.* **120** 476-90
[8] Liu Z S, Zhang G H and Xu H J 2009 Performance analysis of rotating externally pressurized air bearings *P. I. Mech. Eng. J.-J. Eng.* **223** 653-63
[9] Chen Y S, Chiu C C and Cheng Y D 2010 Influences of operational conditions and geometric parameters on the stiffness of aerostatic journal bearings *Precis. Eng.* **34** 722-34
[10] Belforte G, Colombo F, Raparelli T, Trivella A and Viktorov V 2011 Comparison between grooved and plane aerostatic thrust bearings: static performance *Meccanica* **46** 547-55
[11] Belforte G, Colombo F, Raparelli T, Trivella A and Viktorov V 2008 High-speed electrospindle

running on air bearings: design and experimental verification *Meccanica* **43** 591-600

[12] Zhang G and Ehmann K F 2015 Dynamic design methodology of high speed micro-spindles for micro/meso-scale machine tools *Int. J. Adv. Manuf. Tech.* **76** 229-46

[13] Yao Y X, Qin D L and Zhang H B 2006 Static performance analysis of orifice compensated externally pressurized gas spherical bearing based on FEM *Key. Eng. Mater.* **315-316** 860-3

[14] Du J, Zhang G, Liu T and To S 2014 Improvement on load performance of externally pressurized gas journal bearings by opening pressure-equalizing grooves *Tribol. Int.* **73** 156-66

[15] Gao S, Cheng K, Chen S, Ding H and Fu H 2017 Computational design and analysis of aerostatic journal bearings with application to ultra-high speed spindles *P. I. Mech. Eng. C.-J. Mec.* **231** 1205-20

[16] Liu T, Liu Y H and Chen S J 1990 *Aerostatic Lubrication* (Harbin: Press of Harbin Institute of Technology)

IOP Conf. Series: Materials Science and Engineering **657** (2019) 012045 doi:10.1088/1757-899X/657/1/012045

Influence of bias voltage on structure and mechanical properties of TiCrN-Mo_2N-Ni coatings

A V Chernogor[1,2], I V Blinkov[1], D S Belov[1], A O Volkhonskiy[1] and V S Sergevnin[1]

[1]National University of Science and Technology "MISiS", Moscow, Russia

[2]E-mail: avchernogor@gmail.com

Abstract. In this work, the influence of bias voltage on the structure, composition and physical properties of TiCrN-Mo_2N-Ni arc Physical Vapor Deposition (arc-PVD) coatings were studied. Multilayer structure formed at all applied bias voltage due to planetary rotation of substrates. Thickness of sublayers range from 1 to 32 nm and defined by the plasma flow intensity from evaporators. Monte-Carlo calculation was used to study structure of prepared coatings. According to calculation and analysis of TEM images, sublayers composed by nitrides mixture. All coating characterized by high hardness, which correlated with residual stress of coatings.

1. Introduction

Nowadays, surface engineering methods are widely used in many industries [1-8]. In particular, they are used to increase the service life of the cutting tool by applying wear-resistant coatings. These coatings should be characterized by high hardness, increased adhesion strength with the substrate, resistance to oxidation, low friction coefficient. To create such coatings, methods of magnetron deposition [7] and arc-PVD deposition [8] are widely used. In contrast to magnetron, arc Physical Vapor Deposition (arc-PVD) is characterized by a higher deposition rate and increased adhesive strength of the coating with the substrate.

Refractory and transition metals frequently used as material for coatings, due to their high hardness (titanium nitrides with hardness more than 25 GPa [9]), low the coefficient of friction (coatings base on molybdenum described as self-lubricant [10]), and oxidation resistance (chromium nitrides can be applied up to 700°C [11]). At present time, studies of multicomponent coatings based on these elements are being actively conducted [12-15]. It was shown that the coatings of the Ti-Cr-N and Cr-Mo-N systems are characterized by high wear resistance, and due to Cr, the oxidation resistance of such coatings reaches 700°C [12]. Additionally, Cr prevents the process of evaporation of the molybdenum oxides up to 900°C [13]. Nickel addition into coatings based on nitrides of these metals leads to the refinement of nitride grains and an increase in the wear resistance [6,16]. To achieve the high mechanical and tribological properties, multicomponent coatings are used. In this paper, system TiCrMoN-Ni were proposed, in which a high concentration of chromium is designed to provide high oxidation resistance, molybdenum to reduce the coefficient of friction, and nickel to increase hardness, due to the grinding of nitride grains.

2. Methods

The arc-PVD method was used to prepare three series of coatings with a bias potential of 80, 120 and 140 V. All depositions were carried out for 90 min and electric currents on cathodes are listed in table

1. The values of electric current applied to the cathodes were chosen so that the plasma flow density corresponded to the maximum values, at which the processes of the droplet formation are insignificant. TiNi evaporated through the droplet magnetic separator, since titanium is tend to intensive formation of large droplets, up to several micrometers in diameter. The inclusion of such droplets in the structure of the coating, whose thickness is about 2.5 microns, leads to a dramatically decrease of tribological properties of the coatings. Also, a high content of molybdenum in the coating can reduce the oxidation resistant due to the avalanche-like oxidation of molybdenum-based compounds, known as molybdenum pest [17]. Therefore, it was decided to evaporate molybdenum through a magnetic separator, which reduces the density of the plasma flow [18]. Thus, it was possible to reduce the concentration of molybdenum to 5 atomic percent.

Table 1. Deposition parameters.

Bias	Cathode current, A		
(V)	TiNi	Cr	Mo
-80	140	120	130
-120			
-140			

The morphology and elemental composition were studied using a field emission scanning electron microscope (SEM) JSM-6700 equipped with an energy dispersive spectrometer (EDS, JEOL, Japan) at acceleration voltage of 20 kV. Structural transformations and the phase composition were studied by means of transmission electron microscopy (TEM). Structure of the coatings was studied by Monte-Carlo calculation, as reported in [19,20]. Thin lamellas of the coatings were prepared by ion beam etching. Mechanical properties were studied using a Micro-Hardness Tester (CSM, Switzerland) equipped with a diamond Berkovich indenter in the continuous multi-cycle (CMC) indentation mode. The unloading was limited to 5% of the corresponding maximum load (50 μN). The duration of a single loading–unloading cycle was 60 s. From the analysis of the resulted load–displacement curves, hardness and elastic modulus (as well as the degree of elastic and plastic strain were determined. Phase analysis was performed on a D8 Discover apparatus by means of glancing angle X-ray diffraction utilizing copper radiation. The macrostresses were measured by means of sin2ψ using an angle of incidence of 2°. The wear resistance and friction coefficient were measured with Pin-On-Disk tribometer (Nanovea, USA). The wear tests were performed under 25 and 550°C, Al_2O_3 ⌀6 mm, sliding speed of 20 m/s and an applied load of 5 N. The wear track after tests was studied using optical profilometer and X-ray Photoelectron Spectroscopy (XPS).

3. Results

3.1. Elemental composition and surface morphology

The EDS results are presented in table 2. It can be seen that the main element of the coating is chromium, whose high concentration is provided primarily by sputtering the chromium cathode through a direct-flow evaporator. While the deposition of titanium and molybdenum leads to a decrease in plasma flow by more than two times. Is should be noted, that the concentration of nickel in the coating is directly proportional to the bias voltage applied to the substrates. This is because the presence of elements with a high nitrogen affinity, such as chromium, molybdenum and titanium, completely suppresses the formation of nickel nitrides. As a result, nickel presents in the coating is either in the form of a solution in nitrides (with a solubility limit of less than 0.5%) or in the metallic state [21]. The cohesive energy of metallic nickel is significantly lower than the cohesive energy of transition metal nitrides; therefore, in the deposition process, nickel is re evaporated by plasma ions with an energy of more than 150 eV [22].

Table 2. Chemical composition of coatings.

Bias (V)	Element, (at. %)				
	Ti	Cr	Mo	Ni	N
-80	9.6	35.2	5.7	5.4	44
-120	8	42	7	1.4	41.6
-140	7.5	42	7.5	2.5	40

According to SEM (figure 1), it can be seen a droplet fraction on the coatings surface. Evaporating titanium through a droplet phase separator reduces the fraction of the micro-droplet fraction in the coating structure, as well as their average size. However, as a result of elastic collisions of drops with the walls of the chamber [23], some of them still reach the surface of the coating. In general, the surface is characterized by continuity and cellular structure, which is a replica of the surface of the substrate.

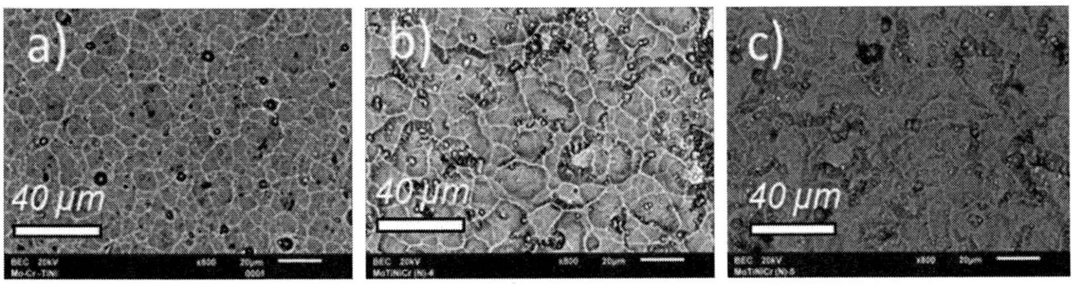

Figure 1. SEM images of coating morphology, prepared at – 80 V (a), - 120 V (b) and -140 V (c).

3.2. Phase composition and coating structure

Table 3. Binding energies of elements.

Bias, (V)	Cr/Ti ratio	Binding Energy, (eV)			
		Cr-$2p_{3/2}$	Ti-$2p_{3/2}$	Mo-$3d$	Ni-$2_{p/2}$
-80	22.4	575.5	454.8	228.2	852.3
	2.9	574.3	455.3	228.1	852.3
-120	21.5	575.4	454.9	228.2	852.1
	2.5	574.2	455.3	228.1	852.2
-140	18.8	575.6	454.8	228.1	Below the limit of detection
	4.1	574.4	455.2	228.1	

The study of the phase composition of multicomponent coatings entails a number of difficulties. In the case of the TiCrMo-N systems, the only presented phase is the *fcc* solid solution of nitrides of the corresponding metals. And taking into account the similarity of the lattice parameters, as well as the unlimited solubility of the components in each other, it is not possible to take an accurate conclusion about the phase composition based the diffraction patterns. In addition, the high value of microstrain, leads to the broadening of diffraction lines, which also complicates the analysis. Therefore, to study the phase composition, an X-ray photoelectron spectroscopy (XPS) analysis was also carried out, the results of which are presented in table 3. In all coatings, a change in the binding energy of titanium and chromium depending on the concentration of these elements is detected. At the same time, taking into account X-ray diffraction analysis (XRD), the formation of nitrides or metals with a structure other than fcc is excluded. Consequently, this behavior is caused primarily by the redistribution of ionic-metal-covalent bonds, which is observed in the case of the formation of solid solutions $Ti_{1-x}Cr_xN$ with different concentrations x [24]. Taking into account that the planetary rotation of the substrates in the plasma flow propagating in space according to a law close to sinusoidal [19], it is possible to form a number of solid solutions with different concentrations. Interestingly that the binding energy of molybdenum

IOP Conf. Series: Materials Science and Engineering **657** (2019) 012045 doi:10.1088/1757-899X/657/1/012045

remains constant for all coatings and corresponds to γ-Mo_2N and does not depend on the elemental composition of the studied areas of the coatings. Despite the fact that molybdenum is also characterized by unlimited solubility in TiN and CrN, its dissolution does not occur. Nickel in the coatings is present in a metallic state.

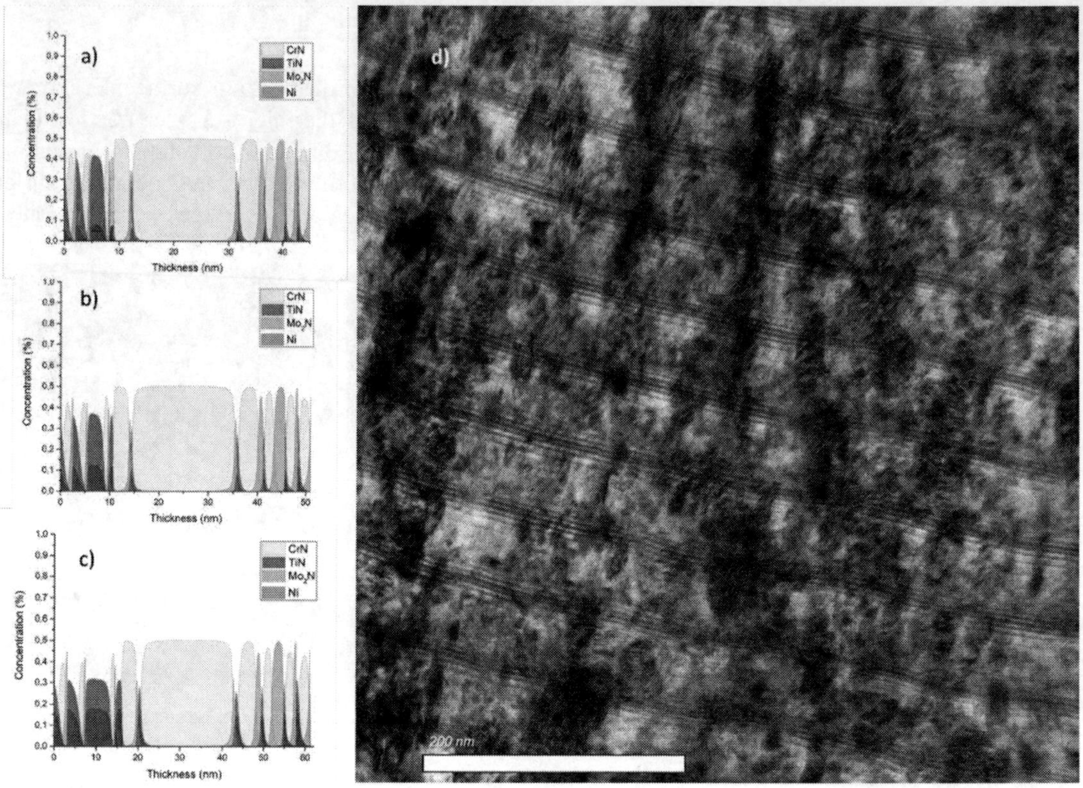

Figure 2. Calculated structure of coatings prepared at -140 V (a), -120 V (b) and – 80 V (c) and TEM image of coating, prepared at -140 V(d).

The planetary rotation of the substrates leads to the formation of a layered structure, the model of which is shown in figure 2. According to the TEM, the planetary rotation of the substrates leads to the formation of layers with a thickness equals to 45, 51 and 62 nm for coatings prepared at -140, -120 and -80 V respectively. Each such layer consists of sublayers formed as a result of the rotation of the substrate around own axis. The thickness of the sublayer ranges from ~1 to 32 nm. At the same time, as the bias potential increases, the average sublayer thickness decreases. Examination of these sublayers by TEM and XPS methods with layer-by-layer etching does not give an idea of their composition due to the insufficient accuracy of these methods [20]. To investigate the distribution of elements in the coating, numerical calculations were performed, the results of which are presented in figures 2(a)-2(c). As can be seen from the calculated composition of coating layer, one sublayer consists of a mixture of nitrides CrN/TiN/γ-Mo_2N, while, as mentioned above, CrN/TiN form a complex solid solution TiCrN. Also, all coatings are characterized by the presence of a single-phase CrN sublayer with a thickness of up to 25 nm, the formation of which is caused by the evaporation of the chromium cathode through a direct-flow evaporator. As the concentration of nickel in the coating increases, the nitride grains are refined, which is associated with Ni segregation on the surface of the growing grains.

It should be noted that with an increase in the bias voltage, the level of compressive residual stress decreases. As a rule, the dependence of the bias voltage and the level of compressive residual stress is directly proportional [25]. But in this case, nickel leads to refinement of nitride grains [26], which

IOP Conf. Series: Materials Science and Engineering **657** (2019) 012045 doi:10.1088/1757-899X/657/1/012045

reducing the probability of relaxation processes of residual stress by plastic deformation. This is also confirmed by the results of microindentation presented in table 4. It can be seen that coatings with a high level of compressive stresses correspond to greater resistance to plastic deformation and its share in the total work of deformation, despite the plasticity of nickel.

Table 4. Mechanical properties of coatings.

Bias (V)	Stress (GPa)	Hardness (GPa)	Elastic Module (GPa)	Plastic Deformation (%)	Temperature, (°C)			
					Wear, (m^2/N)		CoF	
					25	550	25	550
-80	-4.4	42	440	43	0.9	0.8	0.59	0.43
-120	-0.6	32	450	57	0.9	1.3	0.61	0.45
-140	-1.2	32	400	63	0.5	1.5	0.58	0.45

3.3. Mechanical and tribological properties

All coatings are characterized by high hardness, at the same time, the influence of residual stresses on the coating's hardness is pronounced, as can be seen from table 4.

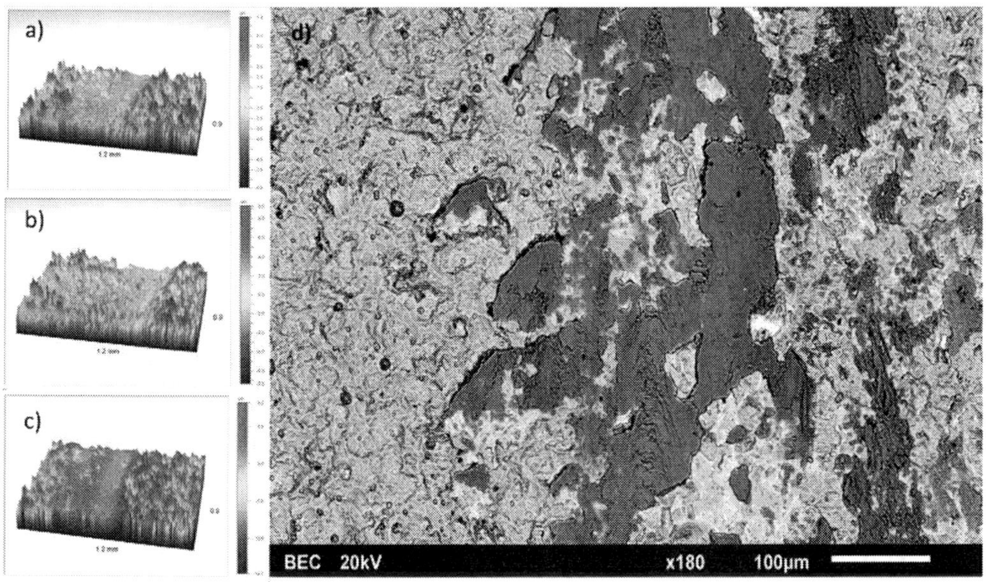

Figure 3. Wear track of coating prepared at -80 V (a), -120 V (b) and -140 V (c) and SEM images of wear track of coating, prepared at -140 V (d).

It can be seen that despite the increase of nickel concentration, which should contribute to an increase in plasticity and a decrease in hardness, residual stress plays a dominant role in mechanical properties. The coating prepared at - 140 V is characterized by greater hardness and wear resistance at room temperature. However, as the temperature of tribological tests increases up to 550°C, the wear resistance of highly stressed coatings begins to be significantly inferior to coatings with low values of residual stress. Taking into account the fact that after annealing there is a complete relaxation of macrostresses (according to XRD analysis the level of residual macrostresses reduce to -0.1 GPa), a possible reason for the reduced wear resistance of such coatings is more active diffusion processes, leading to degradation of the multilayer coating structure and growth nitride grains. However, the wear of all coatings after friction consists only in the abrasion of the protruding irregularities of the coating without opening the substrate, as evidenced by the optical profilometer of the friction track and the SEM image presented in figure 3. According to XPS data, molybdenum oxides (Mo$3d$231.6 eV) are present in the wear track at 550°C, as opposed to tribological tests at 25 C. The formation of Mo containing oxide film leads to a decrease in the friction coefficient to values about 0.45 at high temperatures.

4. Conclusions

TiCrMoN-Ni coatings were prepared at a bias potential of 80, 120 and 140 V by the arc-PVD method. All coatings are characterized by a multilayer structure with sublayer thicknesses from 1 to 32 nm, consisting of a mixture of complex solutions $Ti_{1-x}Cr_xN$ and γ-Mo_2N. As the bias increases, the thickness of the coatings decreases from 2.7 to 2.4 μm as a result of an increase in the re-evaporation of adatoms of the growing coating. An increase in nickel in coatings prepared at a bias potential of 80 V leads to the grinding of nitride grains and an increase in residual stress. This contributes to the growth of mechanical properties, including hardness up to 42 GPa, but in the process of heating there is a complete relaxation of compressive stress, and highly stressed coatings are characterized by lower wear resistance.

Acknowledgments
This work was supported by the Russian Foundation for Basic Research, project N 18-32-00372 (mol_a).

References

[1] Hoseinzadeh S, Ghasemiasl R, Bahari A and Ramezani H 2018 *J. of Electronic Materials* **47** 3552
[2] Faure C, Hänni W, Schmutz C and Gervanoni M 1999 *Diam. Relat. Mater.* **8** 830
[3] Ramezani H, Hoseinzadeh S and Sari H 2019 *J. of Nanoelectronics and Optoelectronics* **14** 425
[4] Vipindas K and Mathew J 2019 *Wear* **424–425** 165
[5] Hoseinzadeh S and Ramezani H 2018 *J. of the Chinese Society of Mech Eng* **39** 501
[6] Akbari A, Templier C, Beaufort M, Eyidi D and Riviere J 2011 *Surf. Coatings. Technol* **206** 972
[7] Zhang X, Tian X-B, Zhao Z-W, Gao J-B, Zhou Y-W, Gao P, Guo Y-Y and Lv Z *Surf. Coatings. Technol* **364** 135
[8] Malvajerdi S, Malvajerdi A and Ghanaatshoar M 2019 *Ceramics Int.* **45** 3816
[9] Bobzin K 2017 *CIRP J. of Manufacturing Science and Technology* **18** 1
[10] Liu A, Deng J, Cui H, Chen Y and Zhao J 2012 *Int. J. Refract Met Hard Mater* **31** 82
[11] Gassner G, Mayrhofer P, Kutschej K, Mitterer C and Kathrein M 2006 *Surf. Coatings Technol.* **201** 3335
[12] Lee D, Kim M, Lee Y and Kwon S 2001 *Surf. Coatings Technol.* **201** 232
[13] Koshy R, Graham M and Marks L 2010 *Surf. Coatings. Technol* **204** 1359
[14] Hong E, Kim T, Lee S-H, Lee J and Lee H 2019 *Ceramics Int.* **45** 9192
[15] Thampi A, Bendavid A and Subramanian B 2016 *Ceramics Int.* **42** 9940
[16] Zhang Z, Rapaud O, Mercs D, Brien V, Dong C and Coddet C 2009 *Thin Solid Films* **517** 3304
[17] Solak N, Ustel F, Urgen M, Aydin S and Cakir A F 2003 *Surf. Coatings Technol.* **174** 713
[18] Dodonov A and Bashkov V 2003 Producing electric arc plasma in a curvilinear plasmaguide and substrate coating *US6635156B1*
[19] Chernogor A, Blinkov I and Volkhonskiy A 2018 *Int. J. Comput. Phys. Ser.* **1** 30
[20] Chernogor A, Blinkov I, Belov D, Sergevnin V and Volkhonskii A 2019 Tech Phys Letter **45** 75
[21] 2015 *Thermodynamic Properties of Inorganic Materials Compiled by SGTE*, ed Dwatson A and Markus T V. (Springer-Verlag Berlin Heidelberg) p 250
[22] Hanson D, Stephens B, Saravanan C and Kress J 2001 *J. Vac. Sci. Technol. A* **19** 820
[23] Anders A 2008 *Cathodic Arcs: From Fractal Spots to Energetic Condensation* (New York: Springer) p 555
[24] Kunchenko Y, Kunchenko V, Neklyudov I, Kartmazov G and Andreyev A 2007 *Probl. At Sci.* **90** 203
[25] Oettel H and Wiedemann R 1995 *Surf. Coatings Technol.* **76** 265
[26] Belov D, Blinkov I and Volkhonskii A 2014 *Surf. Coatings Technol.* **260** 186

IOP Conf. Series: Materials Science and Engineering **657** (2019) 012046 doi:10.1088/1757-899X/657/1/012046

Optimizing basket transplanting parameters using the mechanical properties of tomato plug seedlings

Y Liu[1,2], Y C He[3,5], Y X Li[1], H P Mao[2], B Li[1], T Wang[1], S G Wang[1] and H Li[4]

[1]Mechanical Equipment Research Institute, Xinjiang Academy of Agricultural and Reclamation Science, Shihezi, China
[2]Key Laboratory of Modern Agriculture Equipment and Technology, Ministry of Education, Jiangsu University, Zhenjiang, China
[3]College of Mechanical and Electronic Engineering, Tarim University, Alar, China
[4]College of Machine and Electrical Engineering, Shihezi University, Shihezi, China

E-mail: hyc2003@163.com

Abstract. The collision between tomato plug seedlings and the planting device during basket transplanter function causes the seedlings' root lump to lose some mass. To resolve this problem, we establish a contact mechanics model using Hertzian contact theoretical and Newtonian methods, enabling us to formally derive the maximum deformation of the root lump. The equation of motion describing the collision between the tomato plug seedling and planting device is constructed using the principles of impulse and momentum. Then, we identify the optimization function of contact angle θ—which consists of the maximum deformation of the root lump, the angular velocity of the tomato plug seedling, and the absolute velocity of the root lump—via multi-objective methods. When the angular velocity of the planting device is 1.05 rad/s or 1.15 rad/s, the optimal contact angle is 67°. Further, we examine the velocity of the tomato plug seedling after collision. A small plant height reduces the root lump's mass loss; we found that 12-cm-tall tomato plug seedlings lose less root lump mass than taller seedlings. Finally, for angular velocities of 1.05 rad/s and 1.15 rad/s, we carry out of tomato plug seedling crush tests with the optimized parameters, achieving minimum root lump mass losses of 8.76% and 7.43%, respectively.

1. Introduction

The basket transplanters can form holes and cast seedlings at one time and thus are applied into the transplanting of tomato plug seedlings [1]. To improve their efficiency, basket transplanters generally have auxiliary seedling transplanting machines [2]. Workers place tomato plug seedlings into the cast seedling cup of the feed device. When a tomato plug seedling falls freely from the cast seedling cup, its root lump collides with the inner wall of the planting device. This motion results in loss of the root lump's mass and decreases the plug's survival rate.

Under a theoretical framework, Song *et al* [3] revealed the relationship between rice seedlings' movement and distance when the plugs are turned into a guide. Zhang *et al* [4] adopted a high-speed video camera to observe the collision between the rice plug seedling and the guide device; concluding that the rice plug seedling slides along the guide device. Wang *et al* [5] studied the breakage rate of broccoli seedlings during mechanized planting. Orthogonal experiments based on agronomic requirements have been carried out to obtain optimal parameters. Chen *et al* [6] developed a transplanting mechanism with planetary deformed elliptic gears for plug seedling transplant and built differential equations for the collision of broccoli plug seedlings and the planting derive. In all the studies discussed above, the collision between the root lump and planting device is plastic, the plug seedling slides along the wall of the planting device, and the velocity of the plug seedlings was neglected during the collision between the stem and planting device

In this paper, we focus on the collision between tomato plug seedlings and the planting device of a basket transplanter. The contact model and collision model of tomato plug seedlings are established, the relationships of the working parameters of transplanters with the deformation of root lumps and the movement speed of tomato plug seedlings are studied, and the working parameters are optimized to reduce the quality loss of root lumps.

2. Materials and method

2.1. Collision analysis

We assume uniform rotation of the planting device about a fixed axis in a plane perpendicular to this axis. When cast seedlings cup put tomato seedling plug, the attitude of the seedling (its verticality or slope) is a stochastic process. Two common sets of contacts are collisions between the root lump and planting device, and between the stem and the edge of the planting device.

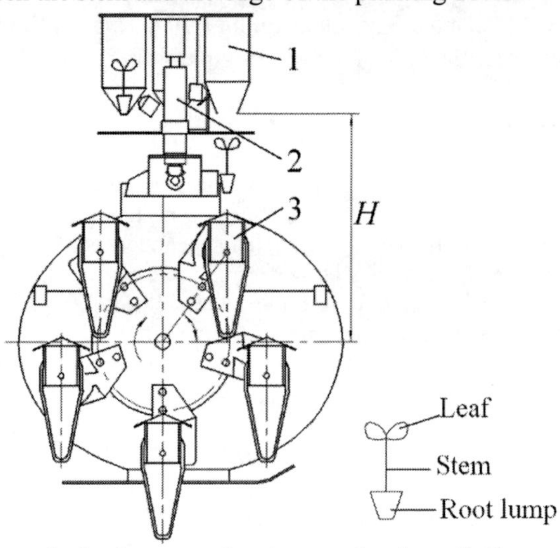

Figure 1. Structural diagram of a basket transplanting mechanism: 1) Cast seedling cup; 2) Feeding device; 3) Planting device.

2.2. Mechanics model

We divide our analysis into two branches: a contact mechanics analysis and a collision analysis. Contact deals with the instant impact between a root lump and the planting device—that is, those that deform the surface—whereas collision addresses the velocity from just before to just after the collision.

2.3. Contact mechanics analysis

When the root lump collides with the planting device, deformation of the root lump occur. These deformations are classified as contact mechanics due to the instantaneous impact [7]. The colliding parts of the root lump can be compared to the face of sphere, and those of the planting device can be considered to be a surface. Thus, the instantaneous collision model can be simplified as describing contact between a flexible sphere and a rigid plane.

According to Hertzian contact theoretical analysis [7], the normal force F can be described as

$$F = \frac{4}{3} E^* r^{\frac{1}{2}} \delta^{\frac{3}{2}} \tag{1}$$

Whcre

$$\frac{1}{E^*} = \frac{1-\gamma_1^2}{E_1} + \frac{1-\gamma_2^2}{E_2} \tag{2}$$

When the equation of motion is applied to each body of the system (the root lump and planting device), similar equations are derived by Newtonian methods.

$$F = m\frac{du_1}{dt} = -M\frac{du_2}{dt} \tag{3}$$

Then, equation (3) can be solved:

$$\frac{M+m}{Mm}F = \frac{d}{dt}(u_2 - u_1) = \frac{d^2\delta}{dt^2} \tag{4}$$

Substituting equation (4) into equation (1) gives

$$\left(\frac{Mm}{M+m}\right)\frac{d^2\delta}{dt^2} = \frac{4}{3}E^* r^{\frac{1}{2}} \delta^{\frac{3}{2}} \tag{5}$$

Thus, integrating the resulting equation (5) yields

$$\frac{1}{2}\left(\frac{Mm}{M+m}\right)\left[u_N^2 - \left(\frac{d\delta}{dt}\right)^2\right] = \frac{8}{15}E^* r^{\frac{1}{2}} \delta^{\frac{5}{2}} \tag{6}$$

When the slope of deformation of the root lump is zero, the maximum compression of the root lump $\delta_{\max}$ is

$$\delta_{\max} = \left[\frac{15Mmu_N^2}{16(M+m)E^* r^{\frac{1}{2}}}\right]^{\frac{2}{5}} \tag{7}$$

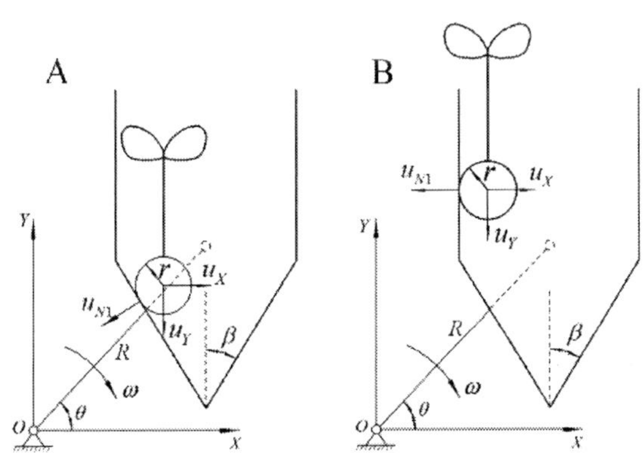

Figure 2. The configuration of the collision of the rootlump and planting device: A, The collision between the root lump and the sloped wall of the planting device; B, The collision between the root

lump and the vertical wall of the planting device.

In general, there are two types of impact. The motion in figure 2A is caused by the collision between the root lump and the sloped wall of the planting device. The motion in figure 2B occurs due to the collision between the root lump and the vertical wall of the planting device. Global inertial coordinates *XOY* are considered. The symbols in figure 2 are listed in table 1.

Table 1. Specification of analysis symbols.

Symbol	Implication	Symbol	Implication
ω	Angular velocity of the planting device	u'_C	Velocity of the center of mass of the tomato plug seedling after collision between the root lump and planting device
R	Radius of gyration of the planting device	$u'_{CX'}$	Velocity of the center of mass of the tomato plug seedling in the X′ direction after collision between the root lump and planting device
H	Vertical distance between the cast seedling cup nozzle and the axis of rotation of the planting device	$u'_{CY'}$	Velocity of the center of mass of the tomato plug seedling in the Y′ direction after collision between the root lump and planting device
θ	Angle between the direction of the planting device and the horizontal plane after collision of the tomato plug seedling and planting device	ω_1	Angular velocity of the tomato plug seedling before collision between the root lump and planting device
E_1	Modulus of elasticity of the planting device	ω_2	Angular velocity of the tomato plug seedling after collision between the root lump and planting device
E_2	Modulus of elasticity of the root lump	J_C	Moment of inertia of the collision model to the center of mass
γ_1	Poisson's ratio of the planting device	m_1	Mass of ball A in the collision model
γ_2	Poisson's ratio of the root lump	m_2	Mass of ball B in the collision model
r	Radius of the root lump of in the contact mechanics model	l	Distance between the center of mass of balls A and B in the collision model
M	Mass of the planting device	d	Distance between the center of mass of the tomato plug seedling and the center of mass of the root lump
m	Mass of the tomato plug seedling	r_1	Radius of ball A in the collision model
u_1	Velocity of the tomato plug seedling in the collision direction	r_2	Radius of ball B in the collision model
u_2	Velocity of the planting device in the collision direction	u'_A	Velocity of the center of mass of the root lump after collision between the root lump and planting device
δ	Compression of the root lump	$u_{AX'}$	Velocity of the center of mass of the root lump in the X′ direction before collision between the root lump and planting device
u_N	Relative velocity of the root lump to the planting device before collision	$u'_{AX'}$	Velocity of the center of mass of the root lump in the X′ direction after collision between the root lump and planting device
δ_{max}	Maximum compression of the root lump	$u'_{AY'}$	Velocity of the center of mass of the root

			lump in the Y′ direction after collision between the root lump and planting device
u_E	Velocity of the planting device	u'_{AC}	Relative velocity of the center of mass of the root lump to the center of mass of the tomato plug seedling after collision
u_X	Velocity of the tomato plug seedling in the X direction before collision between the root lump and planting device	k_1	Coefficient of restitution after collision between the root lump and planting device
u_Y	Velocity of the tomato plug seedling in the Y direction before collision between the root lump and planting device	k_2	Coefficient of restitution after the collision between the stem and planting device
u_{N1}	Relative velocity of the root lump to the planting device on the collision direction before collision between the root lump and inclined wall of the planting device	α	Angle between the stem and vertical plane before collision between the stem and planting device
u_{N2}	Relative velocity of the root lump to the planting device on the collision direction before the collision between the root lump and the vertical wall of the planting device	ω_3	Angular velocity of the tomato plug seedling after collision between the stem and planting device
β	Angle between the inclined wall of planting device and the vertical plane	b	Distance between the collision point and the center mass of the root lump
$I_{X'}$	Impulse in the X′ direction	v'_A	Velocity of the center of mass of the root lump after collision between the stem and planting device
$I_{Y'}$	Impulse in the Y′ direction	$v_{AX'}$	Velocity of center of mass of the root lump in the X′ direction before collision between the stem and planting device
$M_C(I^{(e)})$	Geometric sum of the impulse moment to the center of mass	$v'_{AX'}$	Velocity of the center of mass of the root lump in the X′ direction after collision between the stem and planting device
$u_{CX'}$	Velocity of the center of mass of the tomato plug seedling in the X′ direction before collision between the root lump and planting device	$v'_{AY'}$	Velocity of the center of mass of the root lump in the Y′ direction after collision between the stem and planting device
$u_{CY'}$	Velocity of the center of mass of the tomato plug seedling in the Y′ direction before collision between the root lump and planting device	v'_C	Velocity of the center of mass of the tomato plug seedling after collision between the stem and planting device
v'_{DC}	Relative velocity of the collision point to the center of mass of the tomato plug seedling after collision between the stem and planting device	$v_{CY'}$	Velocity of the center mass of the tomato plug seedling in the Y′ direction before collision between the stem and planting device
v'_D	Velocity of the collision point after collision between the stem and planting device	$v'_{CX'}$	Velocity of the center of mass of the tomato plug seedling in the X′ direction after collision between the stem and planting device
$v'_{DY'}$	Velocity of the collision point in the Y′	$v'_{CY'}$	Velocity of the center of mass of the

direction after collision between the stem and planting device	tomato plug seedling in the Y′ direction after collision between the stem and planting device

Just before the root lump strikes the planting device, its angular velocity ω_1 and velocity u_X in the X direction is zero. Thus, the Y direction with a velocity of u_Y is

$$u_Y = \sqrt{2g(H - R\sin\theta)} \tag{8}$$

Neglecting air resistance, the planting device is subjected to a constant angular velocity about the axis of rotation. Then, the velocity u_E of the planting device is

$$u_E = \omega R \tag{9}$$

Also neglecting the lump-device frictional force in the collision and applying the relative velocity equation to the collision points [8], the relative velocity u_{N1} of the root lump to the planting device in the collision direction is

$$u_{N1} = \left(\omega R \sin\theta \cos\beta - \omega R \cos\theta \sin\beta\right) + \sqrt{2g(H - R\sin\theta)}\sin\beta \tag{10}$$

If the root lump crashes into the vertical wall of the planting device, angle β between the inclined wall of planting device and the vertical plane is set to 0. Thus, equation (10) is reduced to

$$u_{N2} = \omega R \sin\theta \tag{11}$$

The relative velocity u_{N1} and u_{N2} are shown as functions of the planting device's angular velocity ω and contact angle θ. The structural parameters of the planting device and tomato plug seedling are given in table 2 [9]. Then, three-dimensional representations of the solutions given by both equations (10) and (11) can be obtained. As shown in figure 3, velocity u_{N1} is higher than velocity u_{N2}. Thus, attention is focused on the collision between the root lump and the sloped wall of the planting device.

Table 2. The relative parameters of the contact mechanics.

Parameters	Values	Parameters	Values
H (mm)	490	γ_1	0.31
R (mm)	240	γ_2	0.44
β (°)	20	M (g)	1000
E_1 (Pa)	2.1×10^{11}	m (g)	14.7
E_2 (Pa)	3×10^{6}	r (mm)	16.7

From equation (7), the maximum deformation δ_{max} of the root lump depends on the relative velocity before collision. The influence of the interaction between angular velocity ω of the planting device and contact angle θ on the maximum deformation δ_{max} of the root lump is plotted in figure 4. From the curved surface diagram, when angular velocity ω is fixed at a low speed, maximum deformation δ_{max} gradually decreases with increasing contact angle θ. When angular velocity ω is fixed at a high velocity, the trend in maximum deformation δ_{max} has the opposite effect on contact angle θ. When contact angle θ is fixed at a low value, the change in maximum deformation δ_{max} with angular velocity ω is at approximately one level. However, when the contact angle θ was fixed at a large angle, the maximum deformation δ_{max} gradually increased with the angular velocity ω. Therefore, contact angle θ is the major factor influencing the deformation of the root lump.

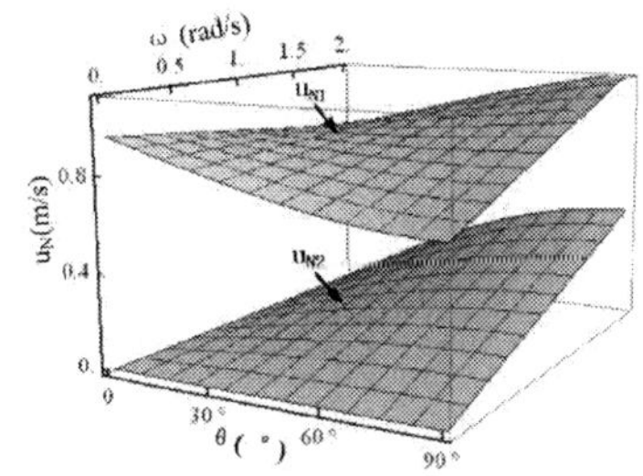

Figure 3. The effect of angular velocity ω and contact angle θ on relative velocity u_N.

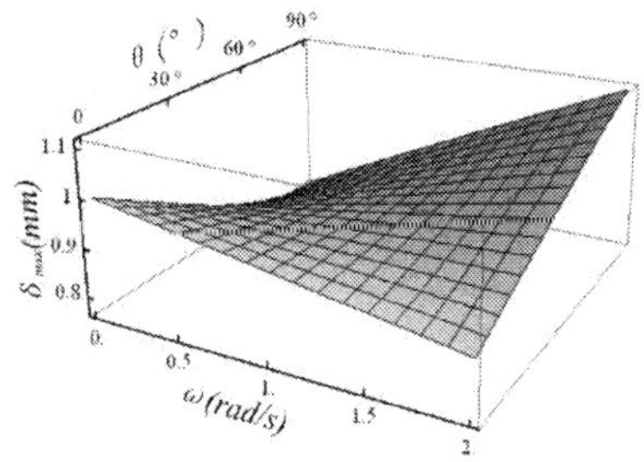

Figure 4. The effect of angular velocity ω and contact angle θ on the maximum deformation δ_{max} of the root lump.

Manual planting frequencies of 50 plants/minute and 55 plants/minute were set as a design parameter, based on the worker productivity [9]. There are five planting devices in the transplanting mechanism in figure 1, resulting in rotation velocities of ω = 1.05 rad/s and ω = 1.15 rad/s, respectively. According to figure 4, when angular velocity ω = 1.05 rad/s, the maximum deformation of the root lump δ_{max} gradually increases from 0.94 mm to 0.96 mm when the contact angle θ increases from 0° to 90°. When angular velocity ω = 1.15, the maximum compression δ_{max} changed from 0.92 mm to 0.97 mm when increasing θ.

2.4. *Collision motion analysis*

2.4.1. Collision between the root lump and planting device. To describe the characteristics of this collision, we assume a dumbbell shape for the distribution of the tomato seedling plug's structure [10]. The link mass between balls A and B is negligible. Figure 5 shows an example of such a body. The origin of the X and Y coordinates is located at the ground O. The origin of the relative X′ and Y′ coordinates has been placed at collision point O′. The symbols in figure 5 are listed in table 1.

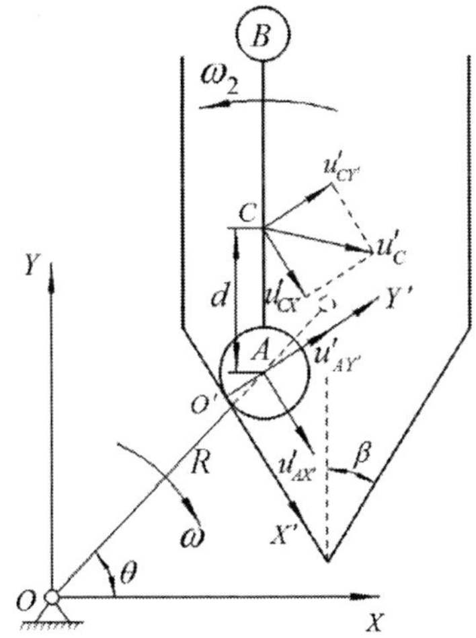

Figure 5. The collision diagram between the root lump and planting device.

Using impulse and momentum [11], it is possible to write three equations that define the particle's motion:

$$mu'_{CX'} - mu_{CX'} = \sum I_{X'} \tag{12}$$

$$mu'_{CY'} - mu_{CY'} = \sum I_{Y'} \tag{13}$$

$$J_C\omega_2 - J_C\omega_1 = \sum M_C(I^{(e)}) \tag{14}$$

The first two of these equations represent the linear impulse in the X' and Y' directions, and the third equation represents the impulse momentum about point C of the center of mass. Neglecting fractional terms, impulse $I_{X'}$ in the X' directions is zero. The X' and Y' components of the velocity are indicated in figure 5; then,

$$u_{CX'} = \sqrt{2g(H - R\sin\theta)}\cos\beta,\ u'_{CX'} = \sqrt{2g(H - R\sin\theta)}\cos\beta \tag{15}$$

$$u_{CY'} = \sqrt{2g(H - R\sin\theta)}\sin\beta \tag{16}$$

Substituting equation (16) into equation (13) leads to

$$mu'_{CY'} - m\sqrt{2g(H - R\sin\theta)}\sin\beta = I_{Y'} \tag{17}$$

Since the tomato plug seedling is in translation just before collision, angular velocity $\omega_1 = 0$. equation (14) can be written as

$$J_C\omega_2 = I_{Y'}d\cos\beta \tag{18}$$

Where

$$J_C = \frac{2(m_1r_1^2 + m_2r_2^2)}{5} + \frac{2m_1m_2l^2}{m_1 + m_2} \tag{19}$$

$$d = \frac{m_2 l}{m_1 + m_2} \tag{20}$$

To determine the relation between the velocities of points A and C, the absolute velocity can be determined:

$$u'_A = u'_C + u'_{AC} \tag{21}$$

Then, the X' and Y' components of absolute velocity $u'_{AX'}$ and $u'_{AY'}$ can be written as

$$u'_{AX'} = u'_{CX'} + \omega_2 d\sin\beta \tag{22}$$

$$u'_{AY'} = u'_{CY'} + \omega_2 d\cos\beta \tag{23}$$

The ratio of the restitution impulse to the deformation impulse is called the coefficient of restitution k_1 [11]. From the above equations, this value for ball A can be expressed in terms of the particles' initial and final velocities as

$$k_1 = \frac{u'_{AY'} - (\omega R\sin\theta\cos\beta - \omega R\cos\theta\sin\beta)}{(\omega R\sin\theta\cos\beta - \omega R\cos\theta\sin\beta) + \sqrt{2g(H - R\sin\theta)}\sin\beta} \tag{24}$$

Equations (17), (18), (22), (23), and (24) can be solved simultaneously to obtain the X' and Y' components of velocities $u'_{AX'}$ and $u'_{AY'}$ of ball A and the angular velocity ω_2 of the tomato plug seedling just after collision.

$$u'_{AX'} = \frac{\cos\beta\left(d^2(1+k_1)mR\omega\cos(2\beta-\theta)-d^2(1+k_1)mR\omega\cos\theta\right)}{2\left(J_c+d^2m\cos^2\beta\right)} + \frac{\left(2J_c+d^2k_1m-d^2(-2+k_1)m\cos2\beta\right)\sqrt{2g(H-R\sin\theta)}}{2\left(J_c+d^2m\cos^2\beta\right)} \tag{25}$$

$$u'_{AY'} = \sqrt{2g(H - R\sin\theta)}k_1\sin\beta - (1 + k_1)R\omega\sin(\beta - \theta) \tag{26}$$

$$\omega_2 = \frac{md(-1+k_1)\sin\beta\sqrt{2g(H-R\sin\theta)} - md(1+k_1)\omega R\sin(\beta-\theta)}{J_C + md^2\cos\beta} \tag{27}$$

The absolute velocity u'_A of ball A has a magnitude that is found from

$$u'_A = \sqrt{u'^2_{AX'} + u'^2_{AY'}} \tag{28}$$

In the studies discussed above, the planar movement equation of the tomato plug seedling was obtained. The relative parameters in the above equations are given in tables 2 and 3 [6].

Table 3. The relative parameters of collision motion.

Parameters	Values	Parameters	Values
m_1 (g)	11.2	r_2 (mm)	5.6
m_2 (g)	3.5	k_1	0.42
r_1 (mm)	16.7	k_2	0.75

Given equation (27), the effects of the angular velocity ω of the planting device and contact angle θ on the angular velocity ω_2 of the tomato plug seedling can be examined in figure 6. When the angular velocity ω of the planting device is fixed, the angular velocity ω_2 of the tomato plug seedling decreases as contact angle θ increases. When contact angle θ is fixed at a low value, angular velocity ω_2 increases with the rotation velocity ω; however, when the contact angle θ is fixed at a high value, there is a single decreasing curve. Thus, the angular velocity ω_2 of the tomato seedling decreases from -5.78 rad/s to -2.04 rad/s or from -5.89 rad/s to -1.93rad/s when the angular velocity ω of the planting device is 1.05 rad/s and 1.15 rad/s, respectively.

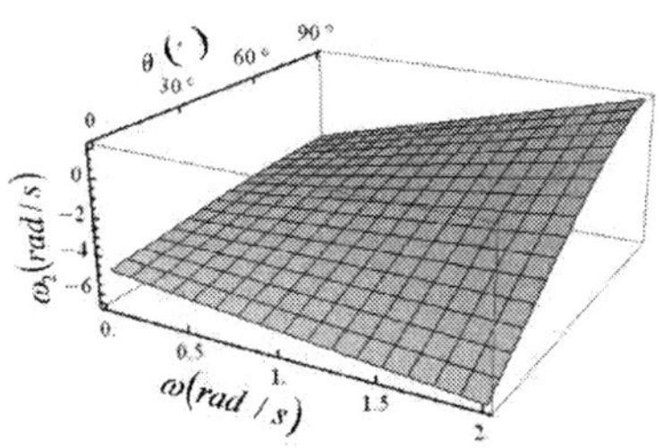

Figure 6. The effect of angular velocity ω and contact angle θ on angular velocity ω_2 of the tomato seedling plug.

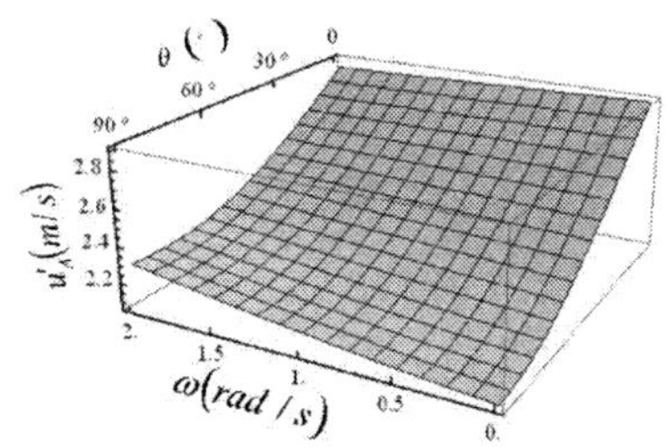

Figure 7. The effect of angular velocity ω and contact angle θ on the absolute velocity u'_A of the root lump.

From equation (28), the influence of the interaction between angular velocity ω of the planting device and contact angle θ on the absolute velocity u'_A of the root lump is graphed in figure 7. When the angular velocity ω of the planting device is fixed, the absolute velocity u'_A decreases as contact angle θ increases. If the contact angle θ is a small constant, absolute velocity u'_A is on the order of angular velocity ω. However, when contact angle θ is fixed at a high value, absolute velocity u'_A increases asymptotically with angular velocity ω. When the angular velocity ω is 1.05 rad/s, the absolute velocity u'_A decreases from 2.87 m/s to 2.19 m/s, and when the angular velocity ω is 1.15 rad/s, the absolute velocity u'_A decreases from 2.88 m/s to 2.18 m/s.

2.4.2. Collision between the stem and planting device. The free-body diagram of the inclined tomato plug seedling shown in figure 8 indicates that during the brief time from just before to just after the collision, the weights of the stem and resistance force have been ignored. The symbols in figure 8 are listed in table 1.

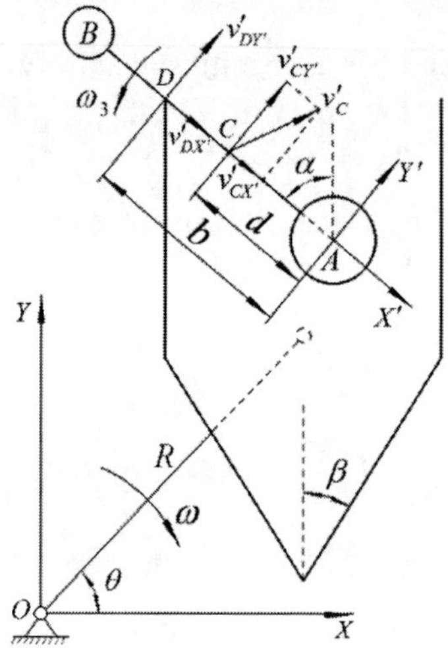

Figure 8. Collision diagram between the stem and planting device.

According to both impulse and momentum theories, impulse $I_{X'}$ in the X' direction is zero because fractional terms are ignored. Then

$$v_{CX'} = \sqrt{2g(H - R\sin\theta)}\cos\alpha, \quad v'_{CX'} = \sqrt{2g(H - R\sin\theta)}\cos\alpha \tag{29}$$

$$v_{CY'} = \sqrt{2g(H - R\sin\theta)}\sin\alpha \tag{30}$$

Substitution of equation (30) into equation (13) leads to

$$mv'_{CY'} - m\sqrt{2g(H - R\sin\theta)}\sin\alpha = I_{Y'} \tag{31}$$

Because the tomato plug seedling is not rotating just before collision, angular velocity $\omega_1 = 0$. Equation (14) can be rewritten as

$$J_C\omega_3 = I_{Y'}(b - d) \tag{32}$$

To study the relation between the velocities of points D and C, the absolute velocity after the collision can be determined:

$$v'_D = v'_C + v'_{DC} \tag{33}$$

From absolute motion analysis, the Y' projections of absolute velocity $v'_{DY'}$ can be written as

$$v'_{DY'} = v'_{CY'} + (b-d)\omega_3 \tag{34}$$

From the definition of the coefficient of restitution, k_2 for particle D can be expressed in terms of the particles' initial and final velocities as

$$k_2 = \frac{v'_{DY'} - (\omega R\sin\theta\cos\alpha - \omega R\cos\theta\sin\alpha)}{(\omega R\sin\theta\cos\alpha - \omega R\cos\theta\sin\alpha) + \sqrt{2g(H - R\sin\theta)}\sin\alpha} \tag{35}$$

Equations (31), (32), (34), and (35) can be solved simultaneously to obtain the relative velocity $v'_{CY'}$ of the center of mass and angular velocity ω_3 of the tomato plug seedling just after collision:

$$v'_{CY'} = \frac{\left(J_C k_2 + m(b-d)^2\right)\sin\alpha\sqrt{2g(H - R\sin\theta)} - J_C(1+k_2)\omega R\sin(\alpha-\theta)}{J_C + m(b-d)^2} \tag{36}$$

$$\omega_3 = -\frac{m(b-d)\left[(1+k_2)\omega R\sin(\alpha-\theta) + (1-k_2)\sin\alpha\sqrt{2g(H - R\sin\theta)}\right]}{J_C + m(b-d)^2} \tag{37}$$

According to planar kinematics analysis, the X' and Y' components of the velocity of ball A are

$$v'_{AX'} = \sqrt{2g(H - R\sin\theta)}\cos\alpha \tag{38}$$

$$v'_{AY'} = \frac{\left(J_C k_2 + m(b-d)(b-2d+dk_2)\right)\sin\alpha\sqrt{2g(H - R\sin\theta)}}{J_C + m(b-d)^2} - \frac{(1+k_2)\left(J_C + md(b-d)\right)\omega R\sin(\alpha-\theta)}{J_C + m(b-d)^2} \tag{39}$$

The absolute velocity v'_A of ball A has a magnitude that is found from

$$v'_A = \sqrt{v'^2_{AX'} + v'^2_{AY'}} \tag{40}$$

We now discuss the collision of the stem, whose physical structure is defined in tables 2 and 3. From equation (37), we can plot the data as shown in figure 9, allowing us to examine the effects of the angular velocity ω of the planting device and contact angle θ on the rotation velocity ω_3 of the tomato seedling plug. When angular velocity ω is fixed, angular velocity ω_3 decreases as contact angle θ increases. When contact angle θ is fixed at a low value, rotation velocity ω_3 increases with angular velocity ω; however, when contact angle θ is fixed at a high value, the reverse occurs. Thus, the angular velocity of a tomato seedling ω_3 decreases from -8.27 rad/s to -1.96 rad/s or from -8.56 rad/s to -1.89 rad/s when the angular velocity ω of the planting device is 1.05 rad/s or 1.15 rad/s, respectively.

Using equation (40), we can show the influence of the interaction between angular velocity ω of the planting device and contact angle θ on the absolute velocity v'_A of the root lump in figure 10. When angular velocity ω of the planting device is fixed, absolute velocity v'_A decreases as contact angle θ increases; when contact angle θ is fixed, absolute velocity v'_A decreases in roughly inverse proportion with angular velocity ω of the planting device. These results reveal that when angular velocity ω is 1.05 rad/s, absolute velocity v'_A decreases from 2.67 m/s to 2.17 m/s, and when angular velocity ω is 1.15 rad/s, absolute velocity v'_A decreases from 2.65 m/s to 2.19 m/s.

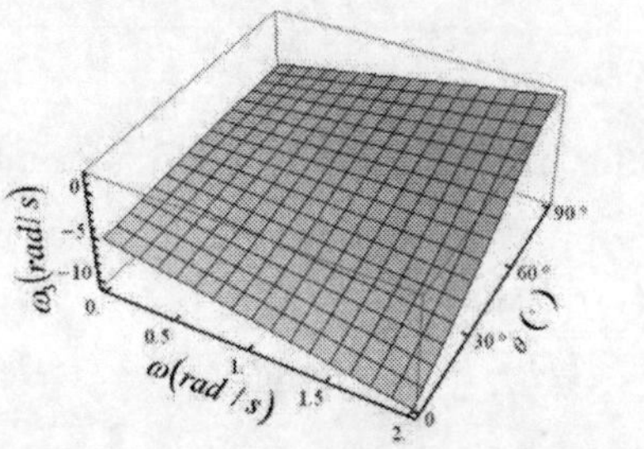

Figure 9. The effect of angular velocity ω and contact angle θ on the angular velocity ω_3 of the tomato seedling plug.

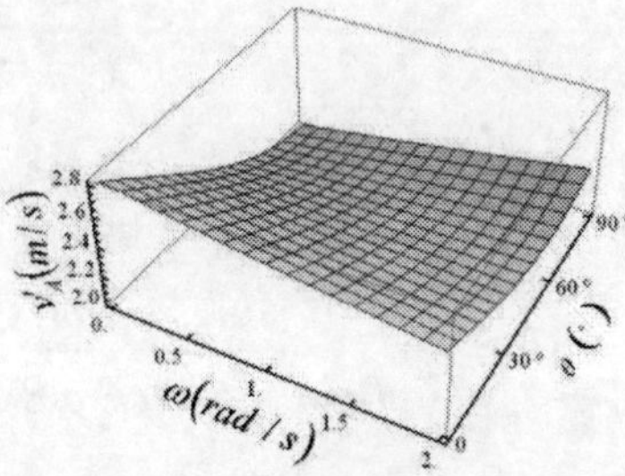

Figure 10. The effect of angular velocity ω and contact angle θ on the absolute velocity v'_A of the root lump.

Using equation (37), we examine the effects of distance b from collision point D to the center A of the root lump mass and incline angle α on the angular velocity ω_3 of the tomato plug seedling in figure 11. When distance b is equal to d, collision point D coincides with point C—the center of mass. Angular velocity ω_3 equals zero and the tomato plug seedling is in translation. When collision point D is far from point C, rotation velocity ω_3 increases with incline angle α. When the distribution of collision point D is subjected to either the upward or downward sides of point C, note that the directions of rotation of the tomato plug seedling are reversed.

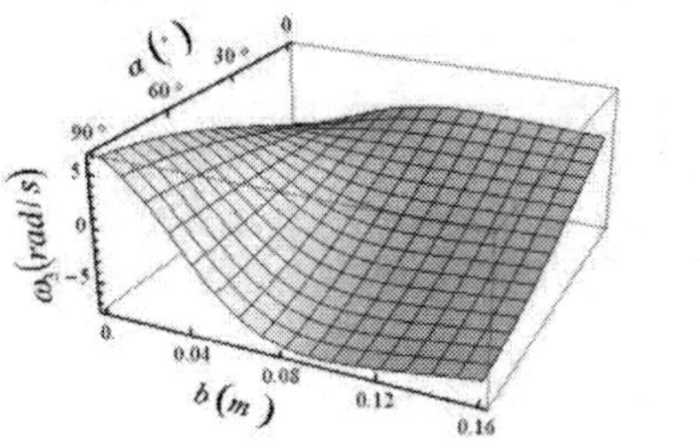

Figure 11. The effect of distance b from collision point D to the center A of the root lump and incline angle α of the tomato plug seedling on angular velocity ω_3.

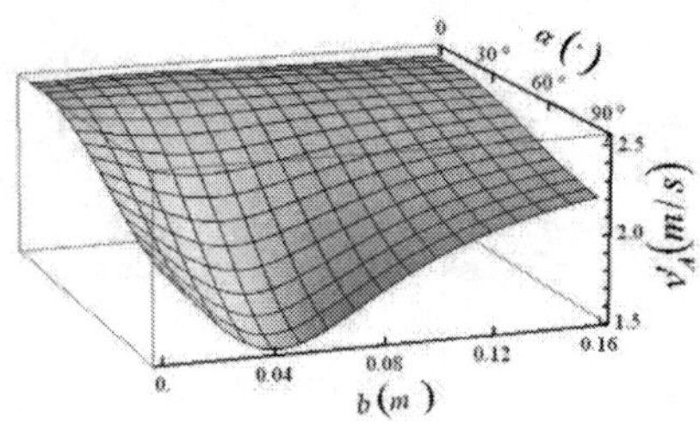

Figure 12. The effect of the distance b from collision point D to the center A of the root lump and the incline angle α of the tomato plug seedling on the absolute velocity v'_A of the root lump.

We next discuss the effects of distance b from collision point D to the center A of the root lump mass and incline angle α on the absolute velocity v'_A of ball A in figure 12. When b is fixed, absolute velocity v'_A decreases as incline angle α increases. When incline angle α is fixed at a low value, absolute velocity v'_A remains the same. However, if incline angle α is fixed at a large angle, absolute velocity v'_A is concave. Thus, when distance b is equal to d, absolute velocity v'_A is effectively at its minimum.

3. Results and discussion

3.1. Optimization of working parameters

Decreasing the velocity of the tomato plug seedling during the collision and the deformation of the root lump can reduce the root lump mass loss [12]. Thus, the maximum deformation δ_{max}, angular velocities ω_2 and ω_3 of the tomato seedling plug, and absolute velocities u'_A and v'_A of the root lump should be decreased. However, when the angular velocity ω of the planting device equal 1.05 rad/s or 1.15 rad/s in previous research of contact and collision analysis, the contact angle θ is decreased to minimize the maximum deformation δ_{max}, and increased for other parameters. Therefore, a multi-objective method is utilized to obtain the optimizing equations for contact angle θ [13]. A series of sub-objective functions—equations (7), (27), (28), (37), and (40)—are combined to construct objective function $f(\theta)$. Thus, the optimization function is:

$$f(\theta) = W_1\left(\frac{\delta_{\max}-\delta^*_{\max}}{F^*_{\max}}\right)^2 + W_2\left(\frac{V'_A-V^*_A}{V^*_A}\right)^2 + W_3\left(\frac{\omega_2-\omega^*_2}{\omega^*_2}\right)^2 + W_2\left(\frac{u'_A-u^*_A}{u^*_A}\right) + W_3\left(\frac{\omega_3-\omega^*_3}{\omega^*_3}\right)^2 \quad (41)$$

where W_1, W_2, and W_3 are weighting factors, and δ^*_{max}, v^*_A, ω^*_2, u^*_A, and ω^*_3 represent the most reasonable values arising from the minimum sub-objective function. These factors are listed in table 4.

Table 4. The relative parameters of the optimization function.

Parameters	Values	
	ω = 1.05 rad/s	ω = 1.15 rad/s
W_1	0.6	
W_2	0.2	
W_3	0.2	
δ^*_{max} (mm)	0.94	0.92
ω^*_2 (rad/s)	-2.04	-1.93
u^*_A (m/s)	2.19	2.18
ω^*_3 (rad/s)	-1.96	-1.89
v^*_A (m/s)	2.17	2.19

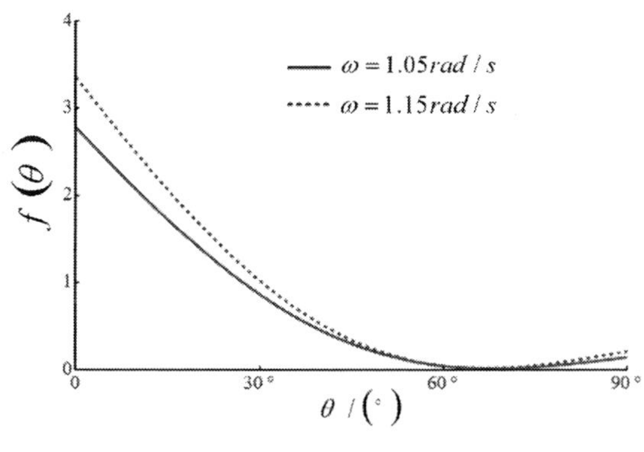

Figure 13. The optimizing function curves of contact angle θ.

The curve in figure 13 is established by solving equation (41), specific parameters of which are listed in tables 2 and 3. Figure 13 reveals the different optimizing function curves of contact angle θ. When the angular velocity ω of the planting device is 1.05 rad/s or 1.15 rad/s, $f(\theta)$ is minimized and the contact angles are 68.1° and 66.5°, respectively. Hence, the optimal contact angle is 67° because the manual planting frequency was changed from 50 to 55 plants/minute.

According to figures 11 and 12, as the distance between collision point D and the center of mass C decrease, the angular velocity ω_3 of the tomato plug seedling and the velocity v'_A of the root lump also decreases. If collision point D is located below center of mass C, angular velocity ω_3 is positive and is identical to the direction of ω_3 in figure 8. The results indicated that the root lump failed to enter the planting device due to its rotation. Reducing the plant height of tomato plug seedlings decreases the distance between the center of mass and the root lump, which can reduce the occurrence of collision phenomena between the planting device and the bottom of the center of mass of the tomato plug seedling. According to agronomic requirement, the general plant height range is 12–19 cm [14]. Then, based on the studies discussed above, 12-cm-tall tomato plug seedlings have an advantage.

3.2. Experimental verification test

The experiments described here were designed to confirm the optimized working parameters in crush resistance experiments. In this experiment, the plug seedling tray has 128 cells, the tomato seedlings were 45 days old and 12–19 cm in height. The transplant mechanism experimental bed is shown in figure 14, and includes the planting device, feeding device, cast seedling cup.

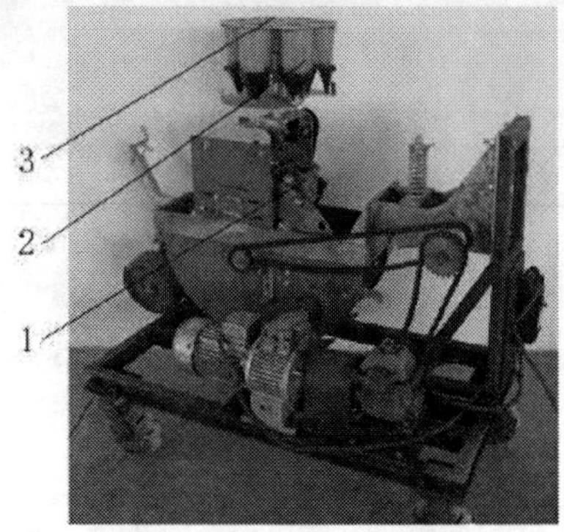

Figure 14. The transplanting mechanism test bench. 1) Planting device; 2) Cast seedling cup; 3) Feeding device.

Table 5 provides a set of direct test parameters. Conditions 1 and 5 is the optimum work parameters through optimizing analysis. The conditions were chosen by varying the contact angle and plant height. In this experiment, 200 tomato plug seedlings were fed in succession, and each condition was repeated three times.

Table 5. Test parameters and results.

Test No.	Planting frequency (plants min^{-1})	Contact angle (°)	Plant height (cm)	Percent of weight loss (%)
1	55	67	12	8.76
2	55	67	19	9.52
3	55	80	12	11.93
4	55	60	12	13.84
5	50	67	12	7.43
6	50	67	19	9.17
7	50	80	12	10.75
8	50	60	12	12.41

The weight of the tomato plug seedling was measured before and after testing. The percent of weight loss during collision is

$$p = \frac{M_B - M_A}{M_B} \times 100\% \tag{42}$$

Here, P is the percent of weight loss, M_B is the weight of the tomato plug seedling before collision, and M_A is the weight of the tomato plug seedling after collision.

The crush test results are shown in table 5. The percent of weight loss under Conditions 1 and 5 were 8.76% and 7.43%, respectively. Those experimental values are generally smaller than other test values. We therefore have experimentally validated the optimal working parameters. Wang *et al* [5] transplanted broccoli plug seedlings with a planetary elliptic gear transplanting machine. Their results showed that the percentage of damage to the root lump is 12.2%. Although the types of vegetables to be transplanted were different, it could still prove that the optimized working parameters in this paper have good working performance.

4. Conclusions

- We employed Hertzian contact theory and the principles of impulse and momentum to derive equations to maximize root lump compression and formalize the motion of tomato seedling plugs, respectively. The trends in maximum compression and velocity were discussed in detail. Under angular velocity conditions $\omega = 1.05$ rad/s and 1.15 rad/s of the planting device, reducing the contact angle decreases the maximum deformation, whereas the contact angle is increased for other parameters (the angular velocity of the tomato plug seedling and the absolute velocity of the root lump).
- We derived optimizing equations for contact angle θ by a multi-objective method. Using angular velocities of $\omega = 1.05$ rad/s or 1.15 rad/s for the planting device, the optimum contact angle was 67°. We then studied the effects of the location of the collision point and the incline angle on the velocity of the tomato seedling plug, revealing that 12-cm-tall tomato plug seedlings experience minimal root lump mass loss.

Acknowledgments
This work was supported by the National Scientists Foundation of China (No. 51475216, 51765059); The Open Fund of the Ministry of Education Key Laboratory of Modern Agricultural Equipment and Technology & High-tech Key Laboratory of Agricultural Equipment and Intelligentization of Jiangsu Province (No. NZ201607); Innovation Project of Science and Technology for College Graduates of Jiangsu Province (No. KYZZ16_0329); The corresponding author of this study is Y C He.

References

[1] Kumar G V P and Raheman H 2008 Vegetable transplanters for use in developing countries—A review *Int. J. Veg. Sci.* **14** 232-55
[2] Nandede B and Raheman H 2015 Multi-stage metering mechanism for transplanting of vegetable seedlings in paper pots *J. Inst. Eng. India Ser. A* **96** 295-300
[3] Song J N, Wang P and Wang Q X 2004 Movement analysis of seedlings and parameter optimization of the set of distributing and transplanting seedlings with channel pipe *J. China Agric. Univ.* **9** 71-4
[4] Zhang G F, Zhao Y and Chen J N 2009 Characteristic analysis of rice plotted-seedlings motion in air and on turbination-type guide-canal *J. Zhejiang Univ. Sci.* **43** 529-34
[5] Wang Y, Chen J N, Wu J W and Zhao Y 2014 Mechanics property experiment of broccoli seedling oriented to mechanized planting *Trans. CSAE* **30** 1-10
[6] Chen J N, Xia X D, Wang Y and Yan J J 2015 Motion differential equations of seedling in duckbilled planting nozzle and its application experiment *Trans. CSAE* **31** 31-9
[7] Valentin L P 2010 *Contact Mechanics and Friction Physical Principles and Application* (New York: Springer-Verlag Berlin Heidelberg)
[8] Eugster S R and Glocker C H 2013 Constraints in structural and rigid body mechanics: A frictional contact problem *Ann. Solid Struct. Mech.* **2013** 1-13
[9] Liu J D, Cao W B, Tian D Y and Tang H Y 2016 Kinematic analysis and experiment of planetary five-bar planting mechanism for zero-speed transplanting on mulch film *Int. J. Agric. & Biol. En.* **9** 84-91
[10] Lee S and Marghitu D B 2009 Analysis of a rigid body obliquely impacting granular matter *Nonlinear Dyn.* **2009** 89-301
[11] Hibbeler R C 2016 *Statics and Dynamics* (New Jersey, America: Pearson Prentice Hall)
[12] Han L H, Mao H P and Hu J P 2015 Development of a doorframe-typed swinging seedling pick-up device for automatic field transplantation *Span. J. Agric. Res.* **13** 1-14

[13] Hwang C L and Yoon K 1981 *Multiple Objective Decision Making: Methods and Applications* (Berlin, Heidelberg: Springer)

[14] Kumi F, Mao H and Li Q 2016 Assessment of tomato seedling substrate-root quality using x-ray computed tomography and scanning electron microscopy *Trans. ASABE* **32** 1-11

IOP Conf. Series: Materials Science and Engineering **657** (2019) 012047 doi:10.1088/1757-899X/657/1/012047

Experimental investigation on micro-drilling properties of Mg-based MMCs reinforced with HAP nanoparticles

F Z Sun[1,4], G Y Fu[2] and J R Shen[3]

[1]School of Mechanical and Power Engineering, Nanjing TECH University, Nanjing, Jiangsu 211816, China
[2]Mechanical Engineering School of Engineering, Newcastle University, Newcastle upon Tyne, NE1 7RU, UK
[3]School of Material and Science, Shenyang Ligong University, Shenyang, Liaoning, 110159, China

E-mail: sunfuz@njtech.edu.cn

Abstract. Mg-based metal matrix composites (MMCs) reinforced with Hydroxyapatite (Ca10(PO4)6(OH)2, HAP) nanoparticles have been studied as bone substitute materials because of their excellent biocompatibility and biological activity. In this paper, Mg/HAP MMCs with different HAP contents (0.5, 1.0 wt%) were selected and the micro-drilling properties were revealed. Firstly, the chip of MMCs was analyzed. Secondly, the surface roughness of Mg/HAP MMCs under different cutting parameters was studied. The results showed that the surface roughness decreases with the increasing of feedrate and decreases with the increasing of spindle speed, which are contrary to pure Mg. Furthermore, the shapes of burr are all classified as uniform burr, and the burr height increased with the increase of spindle speed and feed. Finally, the size effect of micro-drilling was studied. The minimum cutting thickness per tooth of Mg/HAP MMCs was determined to be 1.0 μm.

1. Introduction

Magnesium (Mg), the sixth richest element in nature, has been studied for nearly 200 years [1]. At the same time, Mg is also the fourth abundant cation in the human body (~1 mol/adult), which is mainly found in bone tissue to stabilize DNA and RNA structures [2]. Mg has been extensively studied as a human implant material due to its excellent biocompatibility and indications of local or systemic toxicity free [3,4]. However, the disadvantages of low strength, low Young's modulus, poor ductility and corrosion resistance at room temperature of pure Mg, which limits its wide application [5]. Matrix metal composites (MMCs) can solve the shortcoming of low ductility and low mechanical resistance of pure Mg by adding other nanoparticles [6]. Hydroxyapatite (Ca10(PO4)6(OH)2, HAP) has chemical and structural similarities to bone and dental minerals and is used as a biomedical material because of its excellent biocompatibility and biological activity [7]. However, HAP is difficult to form in the specific form required for bone repair and implantation due to its inherent hardness and brittleness, which limits its use as a biomedical material. Therefore, Mg-based MMCs reinforced with HAP nanoparticles, combining the advantages of pure Mg and HAP, are considered to be a promising biodegradable implant. Campo *et al* [8] produced a Mg/HAP MMCs containing 5, 10 and 15% by weight of hydroxyapatite using a dynamic metallurgical method. It was found that the addition of HAP can increase the microhardness of the composite and improve the corrosion resistance of the composite.

Researchers have done a lot of research on MMCs, mainly focusing on tool wear and cutting parameters. Li and Seah [9] studied the tool wear mechanism of Al/SiC MMCs using tungsten carbide tools, and the results showed that tool wear accelerated due to particle interference when the particle weight fraction reached the critical value. Ciftci *et al* [10] studied the effects of cutting speed, feedrate and depth of cutting by turning machining methods and found that the tool side wear which cutting speed is 150m/min was less than that in 100 and 200 m/min. Research on micromachining of Mg-based composites has rarely been published in all MMCs material cutting studies. Teng *et al* [11] studied the influence of cutting parameters on the cutting force and surface morphology in micromachining of magnesium-based MMCs reinforced with two types of nanoparticles, namely, titanium (Ti) and titanium diboride (TiB2), as well as studied the size effect and the minimum chip thickness according to the specific cutting energy and surface morphology at different feed per tooth.

In order to reveal the effect of HAP nanoparticles on the machinability of Mg-based MMCs, pure Mg and two MMCs with different contents (0.5, 1.0 wt.%) of HAP nanoparticles were selected for micro-drilling experiments. Firstly, the formation characteristics of burrs and chips in MMCs with different contents were analyzed. Then, the influence of different cutting parameters on the surface morphology was obtained. Finally, the size effect of Mg/HAP was analyzed. The experiment results presented an attempt to fill the gap and provide a comprehensive understanding on the micro-drilling machinability of Mg-based MMCs reinforced with different HAP nanoparticles.

2. Experimental setup

2.1. Workpiece materials

The micromachining experiment was performed on an ultra-precision desktop micro-machine tool (MES5R) consisting of three axes (X, Y, Z) driven by a DC servo motor, shown in figure 1 (to the left). The processing equipment has a 100W (240 V) continuous power driven high-speed spindle with a speed range of 20,000-800,000 rpm. It also has the ability to use high feedrates and small diameter tools with a minimum feed of 0.1 μm. Uncoated tungsten carbide micro twist drill with tool diameter of 1.0 mm and shank diameter of 3 mm were utilized. The geometry of all the new tools used in the experiment was examined by an electron scanning microscope (SEM, Hitachi TM3030) before drilling. The cutting edge radius was estimated to be 1.3 μm by SEM. One new drill was used for each workpiece to ensure the accuracy of the measurement. The characteristics and parameters of micro drill is stated in figure 1 (to the right). Before the drilling experiment, the billet was processed into a rectangular parallelepiped of 60 mm x 7 mm x 4 mm.

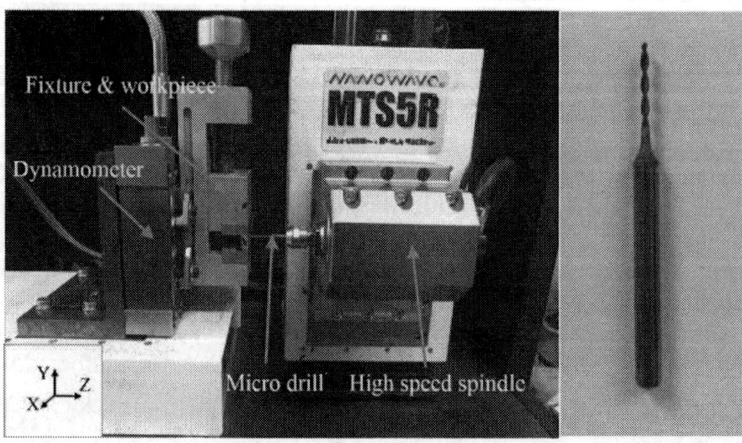

Properties	Value
Drill Type	Twist Drill
Tool diameter	1mm
Shank diameter	3mm
Drill Bit Length	38mm
Flute Length	9.5mm
Point Angle	140 degree
Rake Angle	34 degree
Number of Flutes	2

Figure 1. Machine setup with dynamometer (to the left) and micro twist drill properties (to the right).

2.2. *Experimental procedure*

In order to study the influence of micro-drilling parameters on the hole quality in the micro-drilling experiment, the first experiment used a full-factor design, in which two controlled quantitative factors, that are spindle speed and feedrate. Three spindle speeds were selected to be 20 000, 40 000 and 60 000 rpm, respectively. Three feedrate were selected to be 0.005, 0.01 and 0.015 mm/rev, respectively, and nine combinations were finally formed for micro drilling, as shown in table 1. In order to study the size effect during the drilling process, a second experiment was designed in which 16 different feedrate were selected with constant spindle speed (20,000 rpm), as shown in table 2.

Table 1. Cutting conditions for the first experiment.

No.	Spindle speed(rpm)	Feedrate (mm/rev)
1	20000	0.005
2	20000	0.01
3	20000	0.015
4	40000	0.005
5	40000	0.01
6	40000	0.015
7	60000	0.005
8	60000	0.01
9	60000	0.015

Table 2. Cutting conditions for the second experiment.

No.	Spindle speed (rpm)	Feedrate (mm/rev)	Feedrate (mm/min)
1	20000	0.0001	4
2	20000	0.0002	8
3	20000	0.0003	12
4	20000	0.0004	16
5	20000	0.0006	24
6	20000	0.0008	32
7	20000	0.001	40
8	20000	0.0016	64
9	20000	0.0022	88
10	20000	0.0028	112
11	20000	0.0034	136
12	20000	0.004	160
13	20000	0.006	240
14	20000	0.008	320
15	20000	0.01	400
16	20000	0.012	480

3. Results and discussions

3.1. *Chips formation*

The chip was first studied because the formation of the chips is closely related to the surface quality, which can reveal the effect of HAP nanoparticles on the quality of the drilled surface. The chips were collected after the end of the processing, and SEM observed the shape and size of the chips. It is necessary to analyze the specific surface morphology of the chip while exploring the shape of the chip due to the chip surface and the machined surface of the workpiece will exhibit similar surface features.

Figure 2 shows the microscopic morphology of the chip when the spindle speed is 40,000 rpm and the feedrate is 0.01 mm/rev. The chip morphology of pure Mg, seen in figure 2(a), shows that the chip shape is regular, the edge is smooth and grooves exist on the chip surface, which is mainly due to the high ductility and low hardness of pure Mg that increases the blurred material between tool and chip and results in grooves on the inner surface of the chip. Figure 2(b) illustrates the chip morphology of Mg/HAP at 0.5 wt%, in which the chip shape is irregular, the edge of the chip appears serrated, and cracks and fragments exist on the chip. This is because the nanoparticle size of HAP is 1-5 micron and agglomeration inevitably occurs in Mg matrix material, which leads to the discontinuous chip formation [8]. Chip deformation and stress concentration occur along the shear zone due to the existence of hard nanoparticles HAP, so some nanoparticles will debond from the matrix during drilling, which will lead to the formation of cracks and some small voids. The cracks on the chips of Mg/HAP MMCs 1.0wt% become larger, which is the main reason for the chip to become powder and also explains the phenomenon of chip reduction with the increase of HAP.

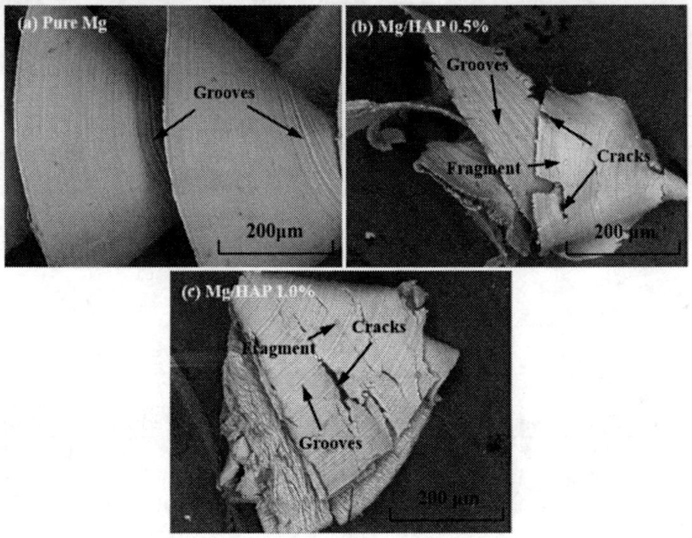

Figure 2. SEM micrographs of chips morphology when spindle speed is 40000 rpm, feedrate is 0.01 mm/rev.

3.2. Surface morphology

The SEM was used to obtain the microscopic morphology of the drilled hole in order to further explain the variation of roughness with cutting parameters. Figure 3 show the surface morphology of pure Mg and Mg/HAP MMCs 1.0 wt% when spindle speed is and 60,000 rpm, respectively. Significant material flow can be seen on the pure Mg surface when the spindle speed is 20000 rpm and the feedrates are 0.005 mm/rev and 0.01 mm/rev, which proves that the heat generated during the cutting process causes the pure magnesium to soften and flow during the rotation of the tool. Irregular grooves are generated on the surface when the feedrate increases to 0.015 mm/rev, which is a tool mark generated in the cutting process and causes an increase in surface roughness. It is apparent that the flow of the material increases as the spindle speed increases to 60,000 rpm due to the higher spindle speed produces more heat and the softening of the material increases with increasing speed.

Grooves appear on the surface of the chip when the feedrate is 0.01 mm/rev and 0.015 mm/rev, and the deformation is brittle, resulting in better surface quality than plastic deformation, which also explains the phenomenon of roughness variation. Grooves, fragments and micro-pits appear, and the surface is severely damaged when feedrate is 0.005 mm/rev and spindle speed is 60,000 rpm. Grooves on the chip surface become more obvious, and a large number of cracks are produced due to the spalling of the material when the feedrate increases, which results in deterioration of the surface finish. Therefore, in order to obtain high surface quality, high spindle speed and small feed should be selected in pure Mg drilling, while low spindle speed and large feed should be selected in Mg/HAP MMCs drilling.

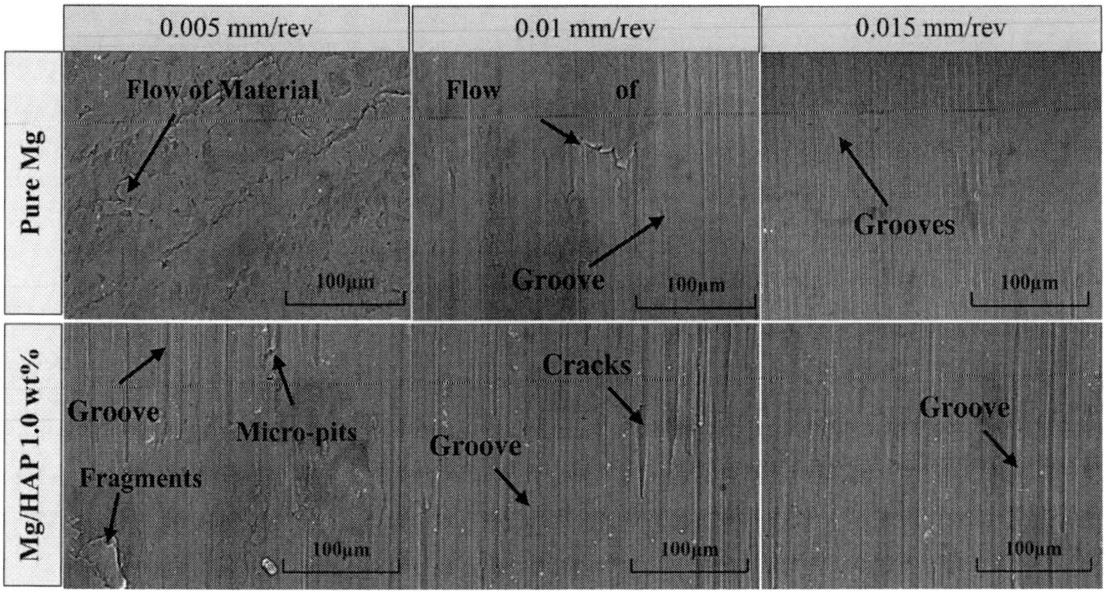

Figure 3. Surface morphology when spindle speed is 60 000 rpm.

3.3. Burr formation
The formation of burrs is inevitable in the process of drilling, so the study of burr formation and burr reduction is the focus of micro-machining. Three types were classified according to the shape and size of burrs in the previous drilling research [12]. Burrs with relatively small and uniform height and thickness around the hole are classified as Uniform burrs, burrs with larger size and irregular height distribution are classified as crown burrs, and burrs formed during the transition between uniform burrs and crown burrs are classified as transient burrs, which is used to analyze the shape and height of burrs in this study.

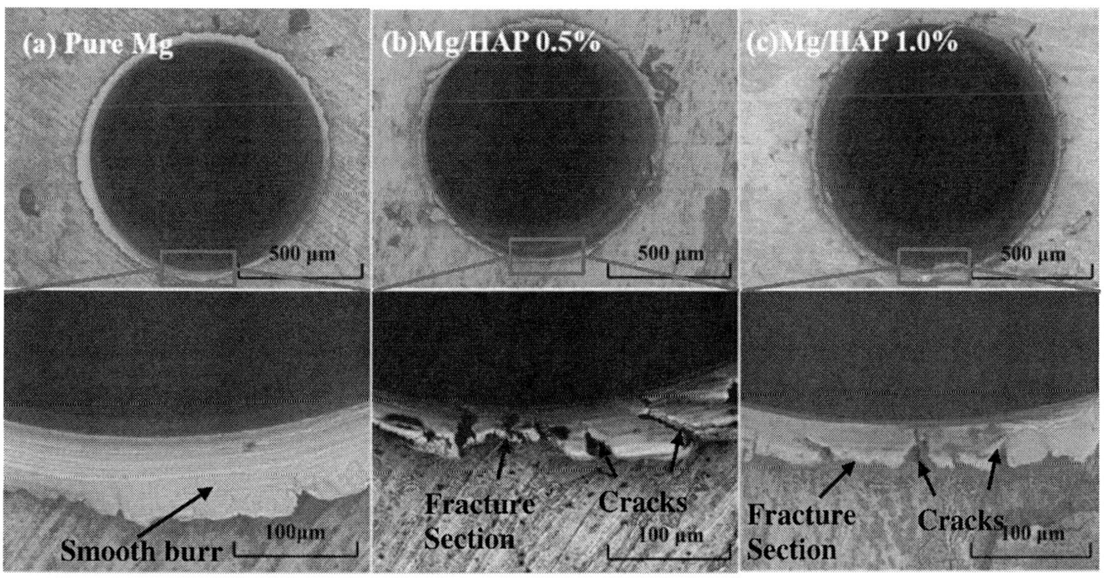

Figure 4. Burr morphology of pure Mg and Mg/HAP MMCs.

Figure 4 shows the burr types of pure Mg and Mg/HAP MMCs when the spindle speed is 60 000 rpm and the feedrate is 0.01 mm/rev. The size of all burrs is small and the burr height is uniformly distributed, so all can be classified as uniform burr. The lower part of figure 4 is a local enlargement of burrs. It can be seen that the burr size of pure Mg is larger and the surface is smooth, while the burr height of Mg/HAP MMCs decreases with the increase of HAP nanoparticle content, and obvious cracks and fracture sections appear on the burrs. Pure Mg has good ductility and large plastic deformation, so the burr surface is smooth. However, the brittleness of Mg/HAP MMCs increases with the addition of HAP nanoparticles, and the brittleness deformation in drilling is much greater than plastic deformation, so chips are prone to fracture and fracture sections are produced.

3.4. Size effect

Macro machining and micromachining are similar in kinematics, but there are many basic differences as well. When the uncut chips thickness becomes comparable to the radius of cutting edge, the size effects becomes the dominant factor affecting the cutting process. When the thickness of the uncut chip is below the critical value, the material is mainly elastic and will be ploughed out.

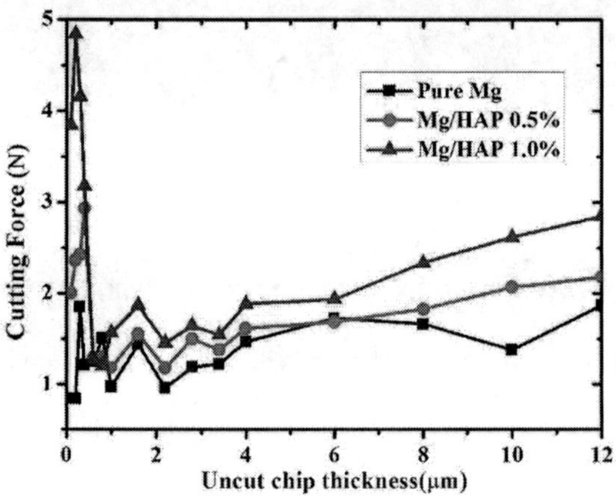

Figure 5. Cutting force versus uncut chip thickness.

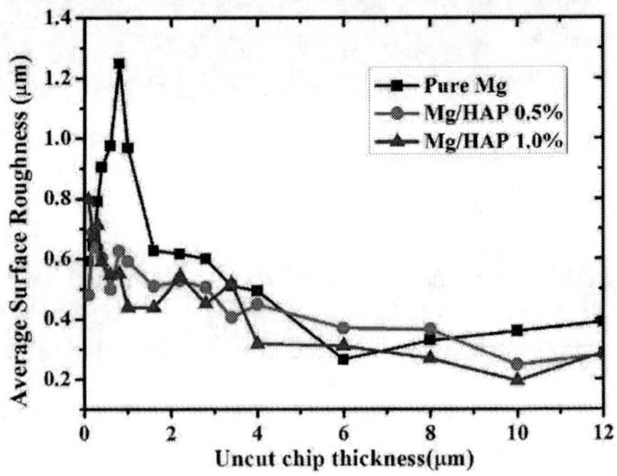

Figure 6. Surface morphology versus uncut chip thickness.

Figure 5 indicates the variation of the cutting force and the thickness of the uncut chip. The cutting force of pure Mg and Mg/HAP MMCs has undergone three stages. When the thickness of uncut chips ranges from 0.1 to 0.3 μm, the cutting force increases greatly with the increase of the thickness of uncut chips. In this region, the material has elastic deformation and no chips are produced because the thickness of uncut chips is much smaller than the radius of the tool edge. Therefore, the material will return to its original state after cutting, resulting in the increase of the cutting force between the workpiece and the tool. The cutting force of this three materials decreases dramatically when the thickness of the uncut chip increased to 0.8 μm. In this region, the thickness of uncut chips is close to the critical cutting thickness, and the plastic deformation and elastic deformation occur during the cutting process. The proportion of plastic deformation increases with the increase of the uncut chip thickness, causing the interaction force between workpiece and tool decreases. When the thickness of uncut chips is greater than 0.8 μm, the cutting force increases with the increase of the thickness of uncut chips. Figure 6 shows the variation of the uncut chip thickness and the surface roughness. The surface roughness of pure Mg and Mg/HAP MMCs, shown in figure 6, goes through three stages with the increase of the thickness of uncut chips. The surface roughness of pure Mg increases with the increase of the thickness of uncut chips which ranges from 0.1 to 0.8 μm. The surface roughness decreases when the thickness of uncut chips increases from 0.8 to 6.0 μm, and the reduction rate of surface roughness from 0.8 to 1.7 μm is greater than that from 1.7 to 6.0 μm. Higher cutting force and worse machined surface quality are obtained at the small feed per tooth ranging from 0.05 to 1.7 μm/tooth in pure Mg and from 0.05 to 1.0 μm/tooth in Mg/HAP MMCs indicating a strong size effect. Considering the influence of cutting force and surface roughness, the minimum cutting thickness of pure Mg and Mg/HAP MMCs should be 1.7 μm and 1.0 μm respectively.

4. Conclusion

The micro-machinability of pure Mg and Mg/HAP MMCs was comprehensively studied by micro-drilling with uncoated cutting tools. The conclusions of the study are as follows:

- The chip shape of pure Mg and Mg/HAP MMCs increases with the increase of feedrate and spindle speed. With the increase of HAP content, the larger the crack produced on chips, and the addition of HAP is beneficial to cutting process.
- The surface roughness of pure Mg increases with the increase of feedrate, but decreases with the increase of feedrate in Mg/HAP MMCs. The surface roughness of pure Mg decreases with the increase of spindle speed, but the change trend of surface roughness of Mg/HAP MMCs is opposite.
- The burrs of pure Mg and Mg/HAP MMCs are all classified as uniform burrs. The burr height increases with the increase of spindle speed and feedrate, in addition, the burr height decreases with the increase of HAP nanoparticles.
- In the study of size effect, the cutting force and surface roughness of pure Mg and Mg/HAP MMCs have undergone three stages with the increase of the thickness of uncut chips. The minimum cutting thickness of pure Mg and Mg/HAP MMCs should be 1.7 μm and 1.0 μm respectively. It is not recommended to micro-drill Mg/HAP MMCs under the minimum cutting thickness per tooth.

Acknowledgment

This project is supported by National Natural Science Foundation of China (Grant No. 51805246) and Natural Science Foundation of Jiangsu Province (Grant No. BK20180704).

References

[1] Davy H 1808 Electro-chemical researches, on the decomposition of the earth; with observations on the metals obtained from the alkaline earths, and on the amalgam procured from ammonia *Philos. Trans. R. Soc. London* **98** 333-70

[2] Asit K K, Hwa C H, Seung H Y, Kug S H and Kwang S S 2010 Microstructure and mechanical properties of Mg-HAP composites *Bull. Mater.* **58** 145-51

[3] Toda-Caraballo I, Galindo-Nava E I and Rivera-Diaz-del-Castillo P E 2014 Understanding the factors influencing yield strength on Mg alloys *Acta Mater* **75** 287-96

[4] Johnson S, Shi Z and Atrens A 2015 The influence of pH on the corrosion rate of high-purity Mg, AZ91 and ZE41 in bicarbonate buffered Hanks' solution *Corros. Sci.* **101** 182-207

[5] Parande G, Manakari V, Meenashisundaram G K and Gupta M 2016 Enhanced the hardness/Compression/damping response of magnesium by reinforcing with biocompatible silica nanoparticulates *Int. J. Mater. Res.* **107** 1-9

[6] Seetharaman S, Subramanian J, Tun K, Hamouda A and Gupta M 2013 Synthesis and characterization of nano boron nitride reinforced magnesium composites produced by the microwave sintering method *Materials* **6** 1940-55

[7] Denissen H, De Groot K, Makkes P C, Van den Hoogg A and Klopper P 1980 Tissue response to dense apatite implants in rats *J. Biomed. Mater. Res.* **14** 713-721

[8] del Campo R, Savoini B, Munoz A, Monge M A and Pareja R 2017 Processing and mechanical characteristics of magnesium-hydroxyapatite metal matrix biocomposites *J. Mech. Behav. Biomed.* **69** 135-43

[9] Li X and Seah W H K 2001 Tool wear acceleration in relation to workpiece reinforcement percentage in cutting of metal matrix composites *Wear* **247** 161-71

[10] Ciftci I, Turker M and Seker U 2004 CBN cutting tool wear during machining of particulate reinforced MMCs *Wear* **257** 1041-6
[11] Teng X, Huo D, Wong E *et al* Micro-machinability of nanonanoparticle-reinforced Mg-based MMCs: An experimental investigation *Int J Adv Manuf Technol* **87** 1-14
[12] Zheng L J, Wang C Y, Yang L P, Song Y X and Fu L Y 2012 Characteristics of chip formation in the micro-drilling of multi-material sheets *Int. J. Mach. Tool Manu.* **52** 40-9

IOP Conf. Series: Materials Science and Engineering **657** (2019) 012048 doi:10.1088/1757-899X/657/1/012048

Structural optimization of second-stage intertank composite X-shaped truss

Y Liu[1], H Wang[1], J Su[1] and C P Yin[1,2]

[1]College of Aerospace Science and Engineering, National University of Defense Technology, Changsha,, Hunan, China

E-mail: ychangping@163.com

Abstract. An optimization procedure for composite second-stage intertank truss assembly used for connecting the fuel (Liquid Hydrogen, LH_2) tank and the oxidizer (Liquid Oxygen, LO_2) tank is proposed. The typical working conditions of X-shaped truss are 30 tons tensile load and 5.55 tons compressive load in sequence. Parametric geometrical model and finite element model are created in Catia R21 and Abaqus 6.14, respectively. An optimization procedure integrating geometrical model and finite element model has been constructed in Isight 2018, the algorithm of which is multi-islands genetic algorithm. The result shows that the weight of the optimal structure decreased by 26.50% comparing with the initial design.

1. Introduction

The connection adapter is one of the important part in rocket, which used to connect the different parts of rocket such as stages, payloads, fuel tanks and so on. At present, the connection adapter can be divided into two classes by the structural form, shell adapter and truss adapter. The shell adapter including cylindrical and conical adapter which depends on whether the parts have the same diameter or not. The shell adapter has been used in many rockets [1]. A cylindrical interstage adapter was used to connect the common core booster (CCB) and payload in Atlas V heavy-lift vehicle (HLV)series [1]. A conical interstage adapter transition, consists of hand laid-up carbon fiber reinforced polymer (CFRP) skin over an aluminum honeycomb core, was used to connect the CCB and upper stage [1]. Du *et al* [2] analyzed and manufactured a circular bearing structure subjected compressive load. The circular bearing structure can be integrated molded through flexible-mold assisted resin transfer molding (RTM). The composite lattice structure, one of the shell-like adapter, can also be used as a load-bearing part [3,4]. Vasilie *et al* [5] analyzed the relationship between buckling loading and geometrical parameters of the composite lattice structure. Zheng *et al* [6] studied the mechanical properties of the grid lattice cylinders subjected axial compression. The shell adapter can be manufactured by integrated molding. However, the defects during the manufacturing process are difficult to control, which can make such a significant impact on the reliability [7]. The thrust truss adapter assembled by robs and joints can avoid this problem because robs and joints can be manufactured respectively. What's more, a payload truss adapter consists of composite tubes was adopted to connect the payload and the end of the Centaur's forward adapter in Atlas V [1]. For composite truss adapter, connection is one of the main problem. The stress of the connection region between load-bearing robs and the frame of parts to be connected is quite complex. Generally, metallic end fitting was used to increase the reliability. However, the weight of the connection adapter increases because of the metallic end fitting.

Refer to the intertank truss assembly used in Delta IV to connect the fuel (Liquid Hydrogen, LH_2) tank and oxidizer (Liquid Oxygen, LO_2) tank assembled in second-stage rocket, a new composite truss adapter is designed to carry 30 tones tensile load and 5.55 tons compressive load. Unlike the general truss adapter which have metallic joint [8], the body and joint of the X-shaped truss are integral. The joint of the X-shaped truss connects the frame of fuel/oxidizer tank by bolts while the straight section gradually changes to the joint through transition section. This paper is structured as follows. Section 2 introduces the material and its mechanical properties used for X-shaped truss, the geometrical and optimization models. Section 3 presents the analysis of the results while Section 4 is the conclusion of this study.

2. Materials and methods

2.1. Materials and mechanical properties

The material of X-shaped truss is T800/602 epoxy plain-weave composite. The X-shape truss consists of two hat stringers (inner/outer hat stringer) and one flat plate, which can be manufactured in the following steps: 1) Two rigid female molds were manufactured for two hat stringers respectively. Rigid female and male molds were prepared for flat plate. 2) Layups of the plain-weave prepreg layers. The hat stingers were sealed by vacuum bag. 3) Curing in the autoclave at 80°C for 1 hour and 125°C for 6 hours subsequently. The curing pressure is 0.6 MPa. 4) These three parts were demoulded and then glued together.

We simplified the mechanical properties of T800/602 epoxy plain-weave composite as in-plane isotropic mechanical properties to decrease the difficulty of the finite element model and increase the optimization efficiency. The mechanical properties are listed in Table 1.

Table 1. Mechanical properties of T800/602 epoxy plain-weave composites.

Engineering constants	Value	Strength properties	Value
E_{11}/GPa	70.75	X_t/MPa	900
E_{22}/GPa	70.75	Y_t/MPa	900
G_{12}/GPa	3.787	X_c/MPa	700
$G_{13}=G_{23}$/GPa	2.931	Y_c/MPa	700
v_{12}	0.08	S/MPa	90

2.2. Geometrical model

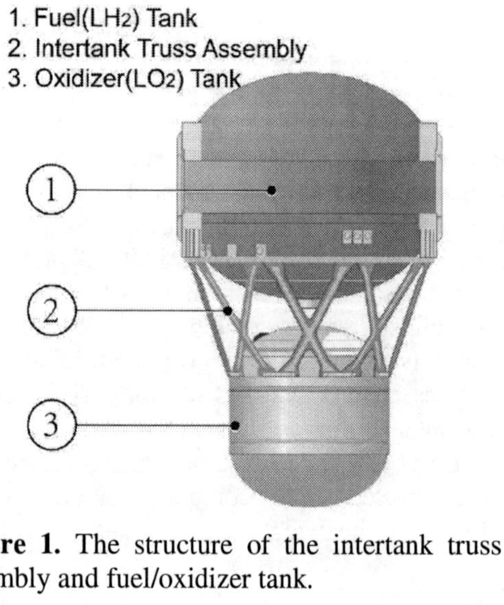

Figure 1. The structure of the intertank truss assembly and fuel/oxidizer tank.

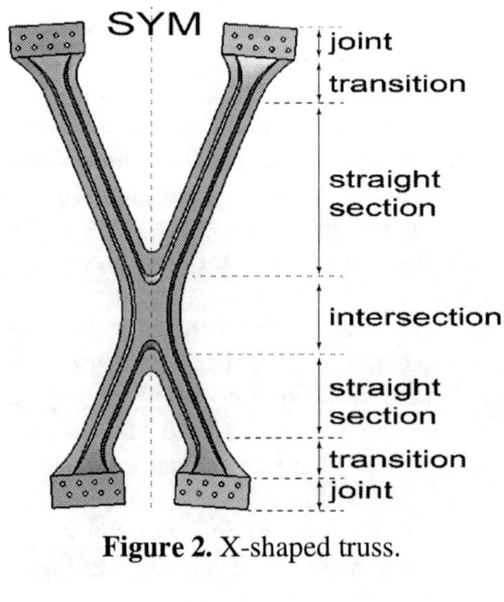

Figure 2. X-shaped truss.

2.2.1. Structural features. The fuel (LH_2) tank and oxidizer (LO_2) tank assembled in a certain type of second-stage rocket are connected by an intertank truss assembly (Figure 1), which consists of 12 identical X-shaped trusses in circle arrangement. X-shaped truss is symmetrical with SYM plane, which can be divided into four parts: joint, transition, straight section and intersection, according to structural features (figure 2).

The X-shaped truss connect the frame of the fuel/oxidizer tank by bolts in the joint. The joint has eight bolt holes (figure 3). The cross section of the straight section is shown in figure 4. The main role of the flat plate is to carry the tensile and compress load while the hat stringers can improve the stability of X-shaped truss under compressive load. The hat stringer include one flange portion and two legs. The transition section (figure 5) was used to implement the gentle transition between the joint and straight section which are not in the same plane. The hat stringer in straight section gradually reduces the protruding height and increases the width to the joint section through the transition section. The arms of the X-shaped truss converged and intersected in the intersection (figure 6). To decrease the stress concentration, the intersection should be as smooth as possible.

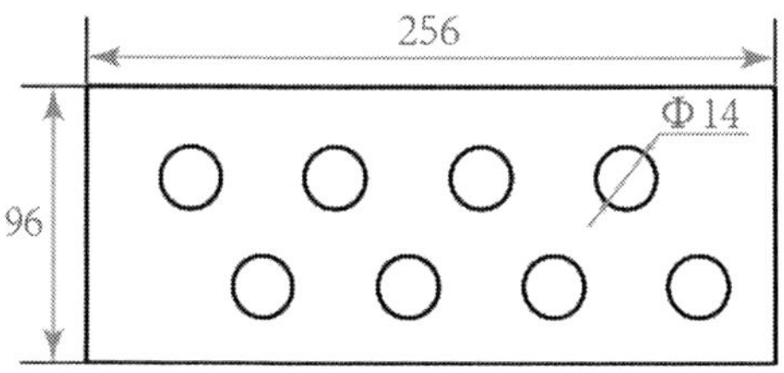

Figure 3. The dimensions of the joint.

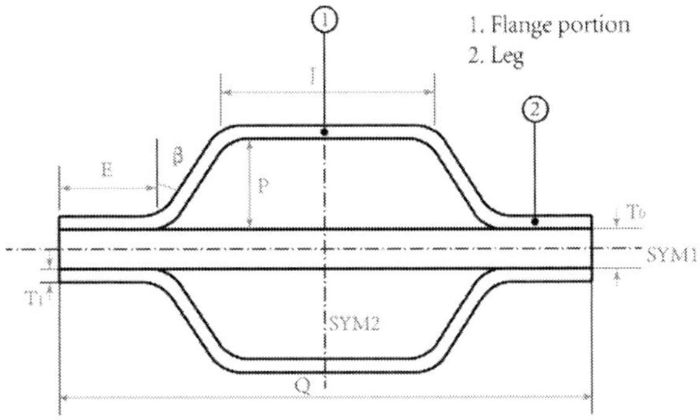

Figure 4. Structure and geometrical parameters of the cross-section of the straight section.

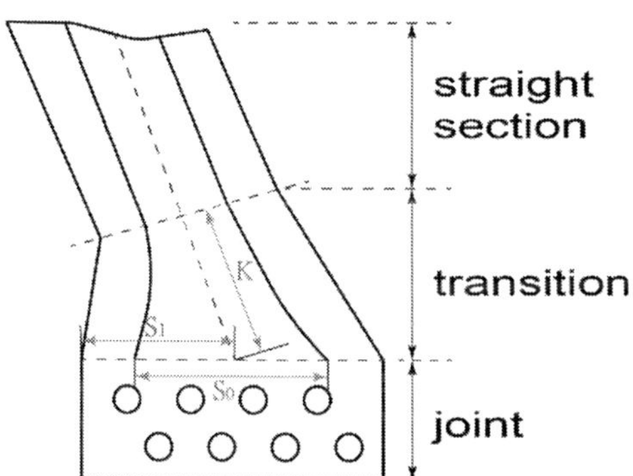

Figure 5. Structure and geometrical parameters of the transition section.

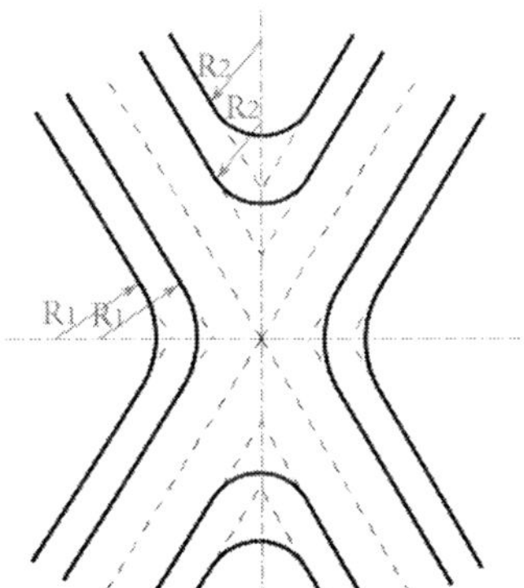

Figure 6. Structure and geometrical parameters of the intersection.

2.2.2. Geometrical parameters. The geometrical model with high-precision is the key phases in numerical analysis [8]. We developed the parametric geometrical model in Catia R21. To describe X-shaped truss completely, a set of geometrical parameters needed to be defined.

- Joint. The height and width of the joint are 96 mm and 256 mm respectively. The bolt holes whose radius is 7 mm have been designed in a staggered arrangement in two rows. The relative position of these bolt holes is shown in figure 3. The dimensions of joint section are provided by design department, which are predefined dimensions.

- Straight section. The cross section of the straight section (figure 4) is symmetrical with SYM1 plane and SYM2 plane. Table 2 lists all of the geometrical parameters of straight section. The cross section of the straight section has seven parameters, six of them are independent. Five variables are needed to define the cross-section's dimension: the thickness T_1, the draft angle β, the width of the leg E, the height and width of the flange portion, P and J respectively. In addition, the flat plate includes two variables: the thickness and the width, T_0 and Q respectively. The relationship between the parameters Q, E, J, P and β:

$$Q = J + 2(E + P \cdot \tan(\beta)) \tag{1}$$

Table 2. Parameters of the geometrical model.

Region	Symbol	Explanation
Straight section	E	The width of the leg
	J	The width of the flange portion
	P	The height of the hat stringer
	Q	The width of the flat plate
	β	The draft angle of the hat stringer
	T_1	The thickness of the hat stringer
	T_0	The thickness of the flat plate
Transition	K_u	The length of the upper transition section
	K_d	The length of the lower transition section
Intersection	R_1	The radius of the fillet on the left and right sides
	R_2	The radius of the fillet on the upper and lower sides

- Transition. As shown in figure 5, the length of transition section is measured by the length of central axis K. The central axis of the transition section intersects with the joint at the midpoint of the joint's upper edge which means the width of the joint W is two times of S_1.

$$W = 2S_1 \tag{2}$$

The width of flange portion S_0 in the connection part closed to the joint is half of the width of the joint W.

$$W = 2S_0 \tag{3}$$

- Intersection. The fillet radius of intersection in the same side are equal. R_1 indicate the radius of fillet on the left and right sides while R_2 indicate the radius of fillet on the upper and lower sides (figure 6).

2.3. Finite element model

Abaqus is used in many fields, such as statics [9] and dynamics [10,11]. The elements' types used in the finite element model are S4R and S3R. The element size is 5 mm. The finite element model has about 40,000 S4R elements and 1000 S3R elements. The certain number of the elements depend on the geometrical parameters. The two hat stringers were bound to flat plate by tie constraints. We developed the finite element model using the half of the X-shaped truss to reduce the amount of computation. The CPU time of each iteration is about 400 seconds. Symmetry constraints was applied to the symmetrical plane. The edges of each bolt hole are coupled to the center of the hole to simulate the effect of the bolt. All of the node on the edges of each bolt hole have the same kinematic behavior in all directions as the center of the hole. The center of the bolts in the lower joint are full clamped and a fixed displacement in normal direction of the joint surface was applied to the upper joint. The concentrate force was applied to the center of the hole (figure 7). According to the rocket launching process, the tensile and compressive load of the X-shaped truss are 300 kN and 55.5 kN respectively. We assume that the loads on each bolt are the same. The load on each bolt can evaluated as

$$\text{The load of the single bolt} = \frac{\text{Total load of X} - \text{shaped truss}}{\text{The number of bolts } \times 2} \tag{4}$$

Therefore, the forces on each bolt are 18.75 kN tensile load and 3.47 kN compress load, separately.

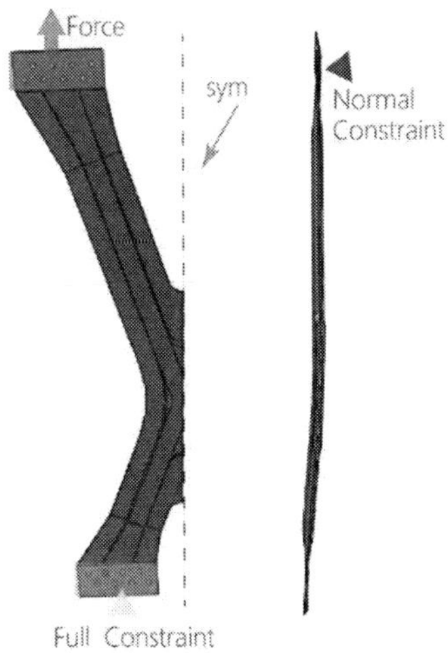

Figure 7. Boundary conditions of the finite element model.

2.4. *Optimization model*

2.4.1. Definition. Ten parameters including geometrical parameters and thickness of the shell are chosen as the optimization variables: E, J, P, β, T_0, T_1, R_1, R_2, K_u and K_d. We set the initial values and ranges as shown in Table 3. The objective of the optimization is the mass of the X-shaped truss. The constraints can be summarized as follows: 1) The maximum of the principal strain (TENSILE_E_MAX) under the tensile load. 2) The damage factors calculated by Hashin's criterion: HSNFTCRT (Hashin's fiber tensile damage initiation criterion), HSNFCCRT (Hashin's fiber compressive damage initiation criterion), HSNMTCRT (Hashin's matrix tensile damage initiation criterion), HSNMCCRT (Hashin's matrix compressive damage initiation criterion). 3) The external load (PRESS_F_MAX) under compressive load. Table 4 lists the initial value and the boundary value of each constraint.

Table 3. The initial values and ranges and optimal solutions of the parameters.

Region	Parameters	Lower bound	Initial value	Upper bound	Optimal solution
Straight section	E(mm)	25	30	60	53.61
	J(mm)	55	60	90	64.14
	P(mm)	10	20	45	12.30
	β(deg)	5	10	20	18.66
	T_0(mm)	5	8	12	7.55
	T_1(mm)	1	5	6	2.69
Transition	K_u(mm)	130	150	250	205.0
	K_d(mm)	130	150	210	143.0
Intersection	R_1(mm)	250	400	600	295.24
	R_2(mm)	40	50	80	45.99

IOP Conf. Series: Materials Science and Engineering **657** (2019) 012048 doi:10.1088/1757-899X/657/1/012048

Table 4. The constraints of optimization.

Parameters	Lower bound	Initial value	Upper bound	Optimal solution
TENSILE _E_MAX	\	0.00787	0.006	0.00598
HSNFTCRT	\	0.382	1	0.098
HSNFCCRT	\	0.014	1	0.053
HSNMTCRT	\	0.182	1	0.166
HSNMCCRT	\	0.038	1	0.119
PRESS_F_MAX	27750.0N	134008.0N	\	43263.2N

2.4.2. Optimization flow. The optimization flow was constructed in Isight (figure 8). Parameters of geometrical model are correlated with the written parameter file in Isight using data exchanger component GEO_DATA. The simcode component UPDATA_GEO updates the geometrical model based on the parameters in GEO_DATA. The finite element model developed in Abaqus using python script is written in data exchanger component FEM_DATA. The finite element model will be submitted after the update of geometrical model and finite element model. The component of READ_RES will extract the results that read by RES_DATA. The optimization algorithm, objective and constraints can be set in optimization component Optimization1. Optimization1 will point out the new direction of the optimization based on the result obtained in last step.

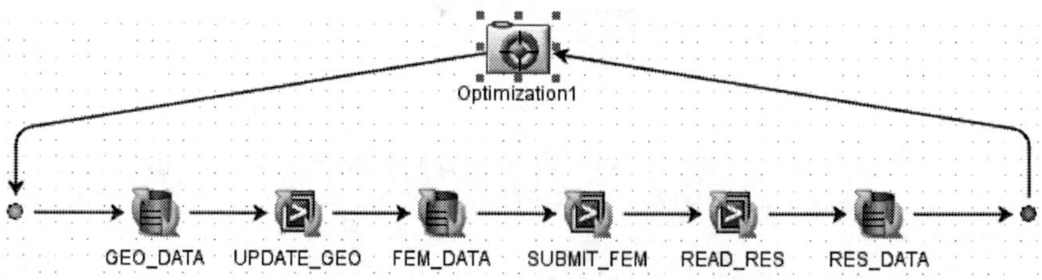

Figure 8. The optimization flow.

2.4.3. Optimization algorithm. The problem can be summarized as an optimization problem with multiple constraints, multiple variables and single objective. The common algorithms used in such problems include response surface method (RSM) [12,13], genetic algorithm (GA) [14] and multi-island genetic algorithm (MIGA) [13,15]. The MIGA is an improved version of the GA. The population divided into several islands. Then the same operations as GA was performed separately. The migration of population from one island to another island will improve the species diversity, which can promote the global convergence. The migration between the islands depend on the rate of migration and the interval of migration. Thus, we use MIGA to search the optimal design. Table 5 shows the parameters of MIGA.

Table 5. The parameters of MIGA.

Parameters	Value
Sub-population size	10
Number of islands	5
Number of generations	20
Rate of crossover	0.01
Rate of mutation	0.01
Interval of migration	5

3. Results and discussions

3.1. Design variables

The optimal design is obtained after 1000 iterations through MIGA (figure 9). The blue spots and orange triangle respectively indicate whether the design satisfies the constraints or not. The green square indicate the optimal solution. The optimal results of design variables were listed in Table 3. None of the values of the optimal design is equal to the boundary value, which mean the ranges of the variables are reasonable. For the initial design, the tensile strain does not satisfy the constraint. The buckling loading, however, is about 5 times as much as the constraint boundary value. The optimal design increases the load bearing capacity under tensile load by increasing the value of the E, J, β. As can be seen in the figure 10, these three variables are negatively correlated with the maximum of the principal strain (TENSILE_E_MAX). What's more, the P value was decreased, which is positively correlated with mass and the buckling loading (PRESS_F_MAX). Although the decrease in P value will induce the decrease in buckling loading, it can save the mass. Furthermore, the optimal design decreases the thickness of the hat stringers and flat plate to reduce the mass.

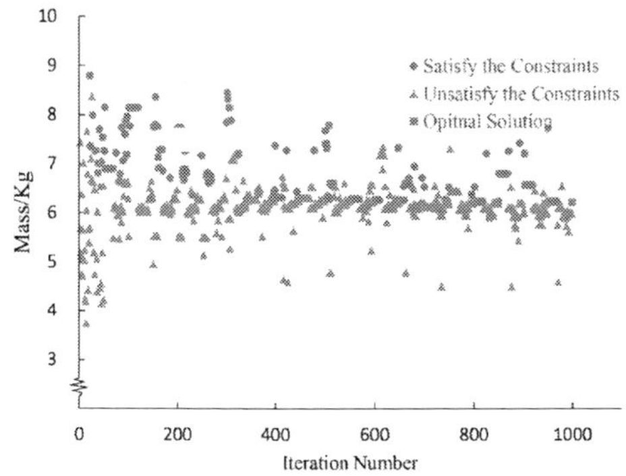

Figure 9. The iteration process.

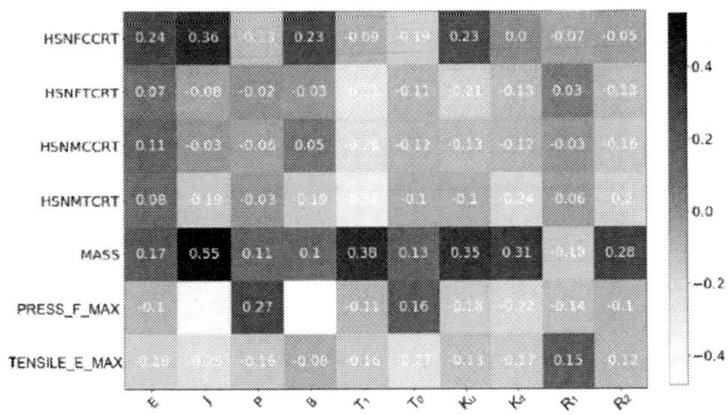

Figure 10. Correlation table.

3.2. *Responses*

The response values of the optimal design are shown in Table 4. For optimal design, the strain value is very close the boundary value and the other constraints are not, which mean the strain constraint is a key constraint. The maximum strain of each part is smaller than the strain constraint. The regions with large strain are the intersection and the connection between straight section and transition section (figure 11). Comparing to the initial design, the buckling load of the optimal design decreased by 66.71%, but it still meets the constraint.

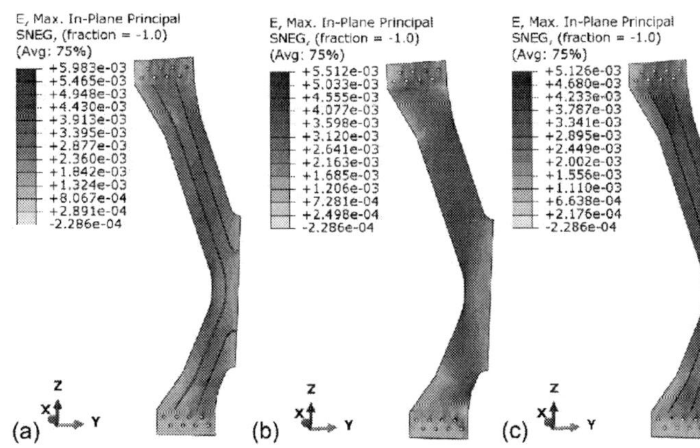

Figure 11. Strain contours under tensile load. (a) Outer hat stringer (b) Flat plate (c) Inner hat stringer.

3.3. Objective

The mass of the initial design and the optimal design (half of the X-shaped truss) is 7.838 Kg and 6.220 Kg, respectively. All of the variables except R_l are positively correlated with mass. Although the decrease of the R_l and the increase of the J induces the increase of the mass, which can reduce the strain under tensile load significantly. The decrease of the mass, largely because of the decrease of the thickness of the flat plate.

4. Conclusions

An optimization procedure of composite truss adapter was developed, integrating 3D CAD application and finite element analysis software to implement the automation of the geometric modelling and finite element analyzing. An optimal geometrical model of the X-shape truss subjected to multiple constraints with the minimum request of mass was obtained, which can be used as a reference in manufacturing. Based on the compare between initial design and the optimal design, it has been shown that the required carrying capacity in compressive load is easy to achieve. The main role of the optimization is to increase the load bearing capacity under tensile load. Also, it found that the tensile resistance can be significantly enhanced by widening the arm of the X-shaped truss. The optimal design can deliver up to 26.50% mass savings in comparison with initial design.

Acknowledgments
This work was financially supported by GK20182A040138 fund.

References

[1] Smith J P 2002 *38th AIAA/ASME/SAE/ASEE Joint Propulsion Conference & Exhibit* (Indianapolis)
[2] Du G, Xiao J, Zeng J and Peng C 2007 *Mater. Compos. Sinica.* ***24*** *46-50*
[3] Vasiliev V V and Razin A F 2006 *Compos. Struct. 76* 182-9
[4] Vasiliev V V, Barynin V A and Rasin A F 2001 *Compos. Struct.* **54** 361-70
[5] Belardi V G, Fanelli P and Vivio F 2018 *Compos. B Eng.* **139** 203-15
[6] Zheng Q, Jiang D, Huang C, Shang X and Ju S 2015 *Compos. Struct.* **131** 885-94
[7] Mohring H C, Eschelbacher S and Kimmelmann M 2018 *Procedia CIRP.* **77** 387-90
[8] Habibi H, Yari A, Hosseinzadeh S and Burkhani S 2016 *Int. J. Recent Adv. Mech. Eng.* **5** 61-9
[9] Vescovini R and Bisagni C J A 2012 *AIAA Journal* **50** 904-15
[10] Hosseinzadeh S, Yari A, Abbasi E and Absalan F 2014 *Int. J. Recent Adv. Mech. Eng.* **3** 11-26
[11] Yari A, Hosseinzadeh S and Galogahi M J I 2014 *Int. J. Recent Adv. Mech. Eng.* **3** 55-67
[12] Todoroki A and Ishikawa T 2004 *Compos. Struct.* **64** 349-57
[13] Kim S H and Na S W 1997 *Struct. Saf.* **19** 3-19
[14] Wei F T, Song L, Li Y and Liu P 2011 *IEEE.* 110-3
[15] Morris G M, Goodsell D S, Halliday R S, Huey R, Hart W E, Belew R K and Olson A J 1998 *J. Comput. Chem.* **19** 1639-62

Study on structural failure mechanism of explosive separation device based on the damage constitutive model

J M Wang[1], F Zhang[1], Y L Qin[1,2], Y Hong[1] and J F Jin[1]

[1]Beijing Institute of Space Long March Vehicle, Beijing 100076, PR China

E-mail: gxc8208@163.com

Abstract. The explosion separation device is the critical factor to ensure the effective separation of the payload and the launch vehicle. The rationality of the structural strength is usually assessed by the real explosion separation test. The separation of components in the deformation and damage is difficult to be obtained. The traditional method, based on the elasto-plastic constitutive, is also unable to simulate the damage of each component in the separation. The traditional strength check can be calculated by the elasto-plastic constitutive relation for large deformation, while the damage simulation can only be obtained by the progressive damage fracture. The metal material dynamic damage constitutive is applied based on the ABAQUES to simulate the damage and failure of the explosion separation. Then the finite element model of the explosion separation device can be established based on the elasto-plastic damage constitutive model, and the process of the ground test of the explosive separation device can be reproduced by the simulation of the corresponding FEM model. And then the weakness of the explosive separation device is finally determined by the calculation results. The experimental results are consistent with the simulation, which confirms the accuracy of the proposed method. The solution idea provides a reference basis for the product design and dynamic damage simulation.

1. Introduction

The explosion separation device is the critical factor to ensure the effective separation of the payload and the launch vehicle. The rationality of the structural strength is usually assessed by the real explosion separation test. The separation of components in deformation and damage can't be monitored, and the traditional general finite element analysis method, based on the elasto-plastic constitutive is also unable to simulate the explosion interconnected damage in the process of separation. At present, the research on explosive separation devices in engineering mainly focuses on the effect of pyrotechnics on explosive bolts. Four typical reagents were selected in Longfei D's research[1] to analyze the process of explosive bolts, and the separation of explosive bolts calculated by the self-contained plastic constitutive relation of ANSYS software is basically consistent with the engineering practice. The fracture characteristics of explosive bolts based on the ANSYS elasto-plastic follow-up model were studied in Hao Y's paper[2], and the optimal fracture structure of explosive bolts was obtained. The explicit dynamics method was used in Xixiong W's research[3] to predict the impact response of the explosive separation bolts, and the sensitive parameters are identified. Balden The hydrodynamic equations were used in V H's paper[4] to study the deformation and post-failure response of plates under blast loading. The explosive separation bolts were used in Hongda Z's research[5] to analyze the impact of the unlocking process, which revealed that the explosion shock and the strain energy shock are the sources of the entire separation shock. The potential damaging effects of high-order impacts were investigated in Jung-Ryul L's paper[6] generated by explosive separation devices on equipment. The numerical analysis method is used in Juho L's research[7] to analyze the phenomenon of separation reliability reduction caused by gap distance. The response and deformation of low carbon steel under impact loading were studied in Bonorchis D's paper[8] based on a set of independent local explosion load tests. The above researches are mainly studied based on the large deformation and elasto-plastic constitutive of mature finite element software without the damage constitutive, and the damage process of other protective structures caused by the explosive bolts in the separation process was not concerned.

In this paper, the structure deformation of the explosive separation device in the separation test is obtained by the finite element dynamic simulation. Firstly, the material damage constitutive relation is introduced into the material mechanics model based on the ABAQUES. Then the finite element model of the explosive separation device is established. And then the damage process of each components in the separation process is precisely analyzed. The process of the ground test of the explosive separation device can be reproduced, and the weakness of the explosive separation device is finally determined. The prediction results are consistent with the experiment, which proves the effectiveness of the proposed method. The solution idea provides a reference basis for the product design and dynamic damage simulation.

2. Dynamic damage constitutions of metallic materials

The stress failure under the external load of elasto-plastic metal materials is as shown in Figure 1, e.s. the elastic deformation $(\sigma < \sigma_s)$, the elasto-plastic hardening $(\sigma_s < \sigma < \sigma_b)$, and the stiffness degradation $(\sigma > \sigma_b)$. Permanent plastic deformation and damage will arise in the elastic-plastic hardening, and large deformation will arise accompanied by the damage degree increasing with the accumulation of permanent deformation. The material unloading stiffness will stay invariable before σ rising to σ_b for the first time, the material yield strength σ_s will be enlarged by the plastic deformation in accordance with the law of hardening, which is called the pre-damage. The material properties will be softened by continued increasing load, while the stress level σ is exceeding the strength limit σ_b for the first time, which eventually leading to fracture, as is called the post-damage. The damage critical level is defined while the stress level is rising to the strength limit σ_b for the first time.

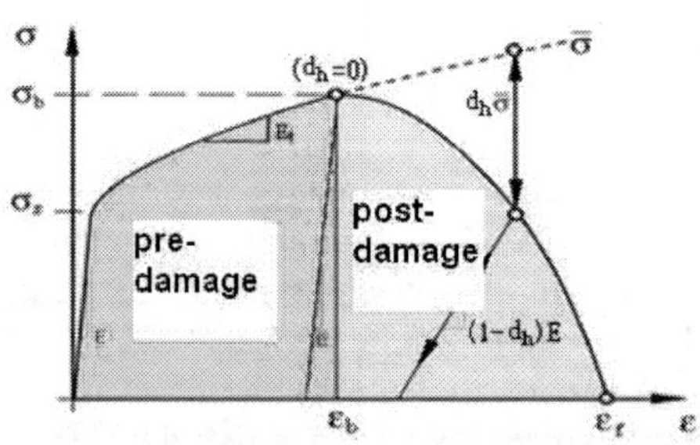

Figure 1. Constitutions of metallic materials.

According to the elasto-plastic theory, the hardening damage constitutive model of materials in the pre-damage is established, which can be expressed as:

$$\{d\sigma\} = [D]^{ep}\{d\varepsilon\} \tag{1}$$

Where $[D]^{ep}$ is the deformation stiffness of the material in the pre-damage, $\{d\varepsilon\}$ is the strain increment vector, and $\{d\sigma\}$ is the stress increment vector.

For metallic materials commonly used in structural design, the material hardening rate can be described by the yield surface equation as follows:

$$\sigma_{eq} - \left[\sigma_s + R\left(\varepsilon_{eq}^p\right)\right] = 0 \tag{2}$$

Where σ_{eq} is the Von Mises equivalent stress; $\sigma_y = \sigma_s + R\left(\varepsilon_{eq}^p\right)$ is hardening yield strength, σ_s is material static yield strength, p is material hardening function.

The differences between the material constitutions under dynamic load and static load are mainly reflected in the effect of strain rate on the hardening rate of metal materials. The influences of plastic hardening rate of different materials are different, which is not relative to the geometry characteristics. Comper_Symonds model and Johnson_Cook model [9-11] are two commonly used strain rate models in engineering. The independent influence function of the strain rate was introduced based on the static stiffness ratio as formula (2), and then the proportional relation of the dynamic hardening yield strength $\sigma_y^{'}$ and static infant flower yield strength σ_y were established.

Comper_Symonds model can be described as

$$Y = \frac{\sigma_y^{'}}{\sigma_y} = 1 + \left(\dot{\varepsilon}/D\right)^{\frac{1}{n}} \tag{3}$$

Johnson_Cook model(simplified) can be described as

$$Y = \frac{\sigma_y^{'}}{\sigma_y} = 1 + C\ln\left(\dot{\varepsilon}/\dot{\varepsilon}_0\right) \tag{4}$$

Where D , n , C is the strain rate correlation coefficient, ε_0 is the reference strain rate of Johnson_Cook. The static damage constitutive model of metal materials can be transformed into a dynamic solution while the static hardening yield strength σ_y in equation (2) is replaced by the dynamic hardening yield strength $\sigma_y^{'}$.

3. Research on structural damage mechanism of explosive separation device

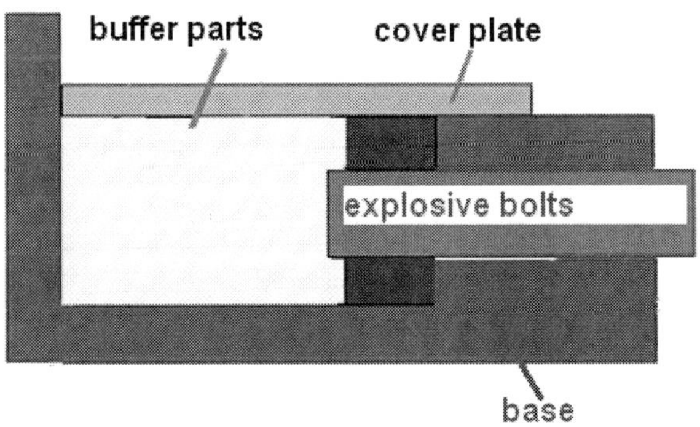

Figure 2. The explosion separation device.

The explosion separation device is mainly composed of explosive bolts and capturing assembly, as shown in Figure 2. The capturing assembly is composed of the buffer parts, the cover plate and the base. The buffer parts are used to slow down the impact of the explosive bolt on the cover plate and the base during the rapid separation. The cover plate and the base must be kept sealed before and after the separation to prevent the separation products from flying out to form residue. A gap of about 5 mm was discovered between the base and the cover plate after the explosion separation test, and the structural sealing was damaged. The redundancy will be presented in the form of the fragments from the buffer parts hitted by the explosive bolts, which poses the hidden danger to the safety of flight test. The explosion separation is unobservable and the damage of each part is difficult to analyze, then the finite element simulation is adopted to analyze the damage and deformation of each part in the separation, which can provide detailed exhibition for the separation and damage process.

The finite element analysis for explosion damage in the separation process is carried out based on the metal material damage constitutive. Firstly, the crash simulation model is established, the separation speed of the explosive bolt is determined, and then the damage process after separation of all components is simulated, the weakness of the components is determined.

The finite element model of explosive separation device is as shown in Figure 3. The elsto-plastic model is used to establish the base during the impact process in the finite element modelling. And the damage constitutive is used to model other components. The aluminium material damage constitutive is adopted to establish the model of the cover plate. The steel material dynamic damage constitutive is adopted to establish the explosive bolts during the transitory separation. The low-strength alloy steel dynamic damage constitutive is adopted to establish the fast-compressed cushioning parts. The material properties is as shown in Table 1. The shell elements are used in the buffer parts, and the hexahedral entity units are used in other structure. The cover plate and the base are connected with bolts, and the base is fixed to simulate real installation. The initial velocity of the explosive bolt is set to be 50 m/s based on the initial impulse of the separation.

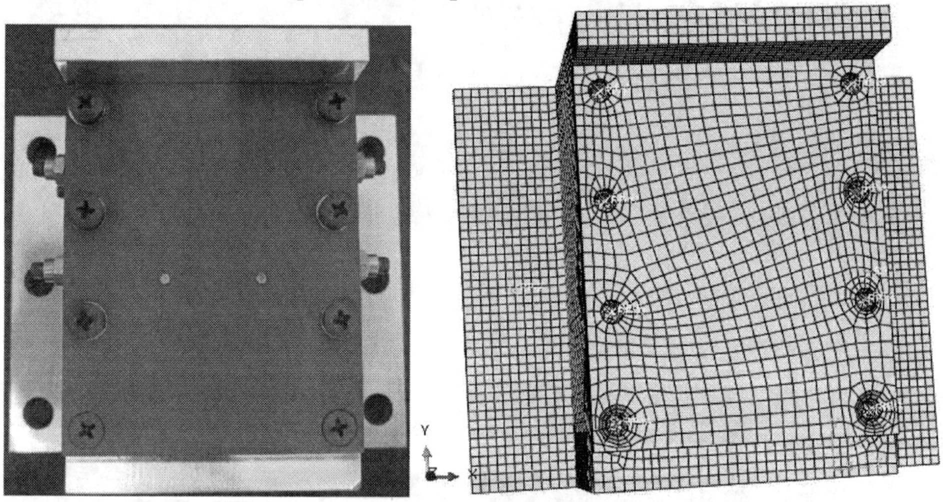

Figure 3. The explosive separation device and the finite element model.

Table 1. Material properties of explosive bolt box.

Parts	E/GPa	μ	σs/MPa	σb/MPa	Rou/g/cm^3
explosive bolt box	71	0.3	250	380	2.7
explosive bolts	200	0.3	885	1080	7.8
cover plate	68.6	0.3	117	315	2.7
buffer parts	200	0.3	300	615	7.8

The impact force curve of the explosion bolt to the bolt box is as shown in Figure 4. The impact force of the explosion bolt is shown as the purple curve and the impact of the explosive bolt to the buffer parts is shown as the blue curve in Figure 4. The impact force can be completely transmitted to the buffer parts on account of the good stiffness of the explosive bolt. It can be seen from the analysis that after the separation, the explosive bolt moved forward at a speed of 50 m/s, and the maximum impact force of the explosive bolt on the base is about 69.6 kN. The impact process can be described as the following course. Firstly, the movement of the shock buffer parts and the explosive bolt will be

IOP Conf. Series: Materials Science and Engineering **657** (2019) 012049 doi:10.1088/1757-899X/657/1/012049

continuous until the impact energy was completely exhausted. The large upward deformation will appear while the buffer parts was compressed, and then the large deformation will be discovered on the cover plate. Finally a large gap will be generated between the base and the cover plate. The deformation comparison of the explosion separation device between the ground test and the simulation results is as shown in Figure 5. The deformation comparison of the components is as shown in Figures 6-8. The actual deformation of the cover plate is 4.94 mm in the simulation, and the deformation is about 5 mm in the ground test, which confirms the accuracy and effectiveness of the simulation of the damage of the explosive separation device based on the metal dynamic damage constitutive. The simulation results are in good agreement with ground test experiment.

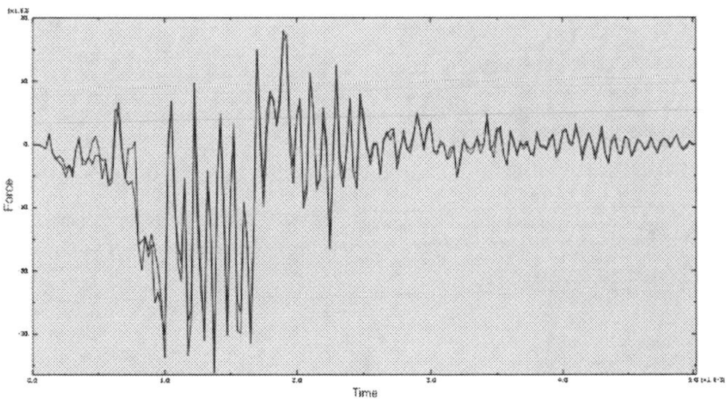

Figure 4. Impact force curve of explosion bolt on bolt box (maximum value is 2×34.8kN).

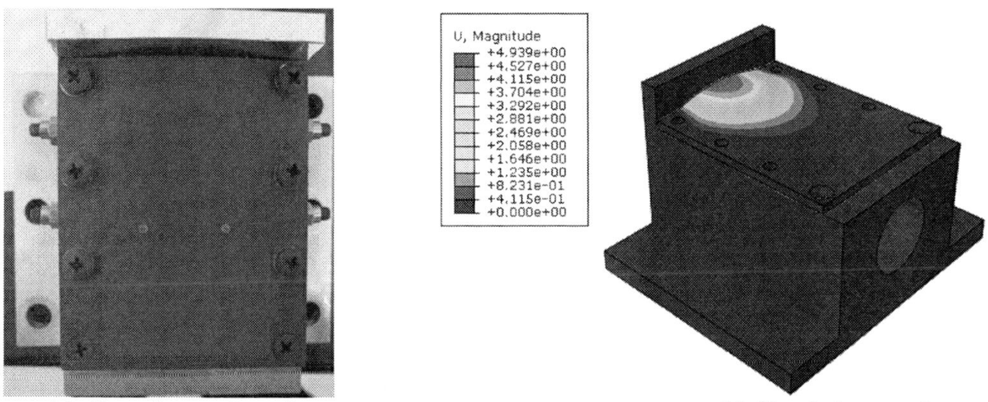

(a) Experimental results (b) Simulation results

Figure 5. Deformation of the bolt box after explosive separation.

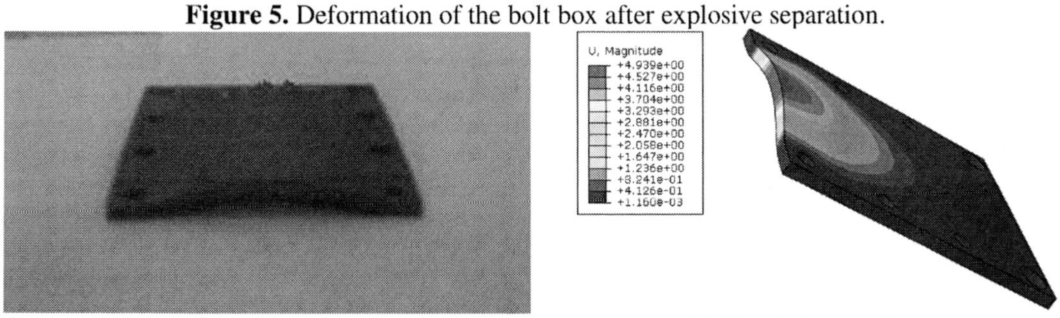

(a) Experimental results (b) Simulation results

Figure 6. Deformation of the cover plate after explosive separation.

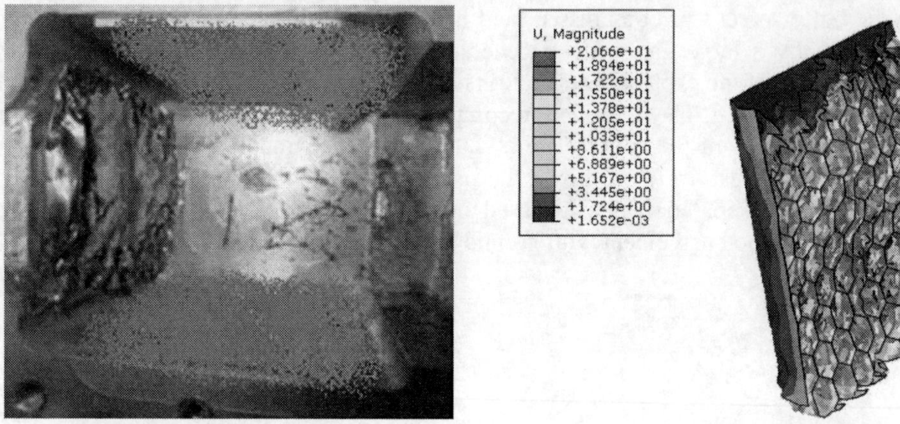

(a) Experimental results (b) Simulation results

Figure 7. Deformation of the buffer parts after explosive separation.

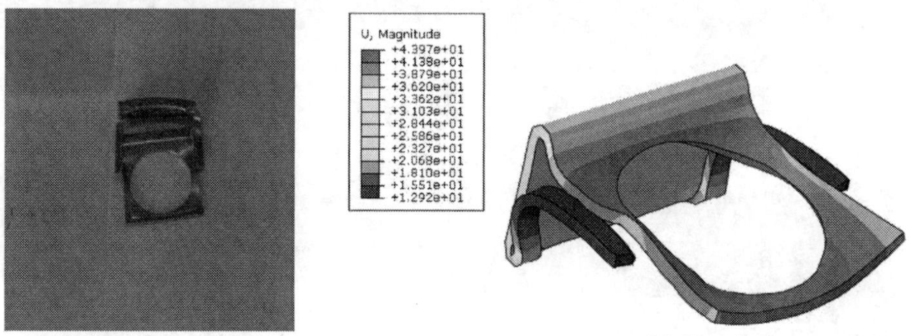

(a) Experimental results (b) Simulation results

Figure 8. Deformation of the corneal plate after explosive separation.

4. Conclusions

- The traditional method, based on the elasto-plastic constitutive, is unable to simulate the damage of each component in the separation. The traditional strength check can be calculated by the elasto-plastic constitutive relation for large deformation, while the damage simulation can only be obtained by the progressive damage fracture.
- The large deformation of the buffer parts is the immediate cause for the deformation of the cover plate. The structural improvement method was proposed to change the structural form of the buffer parts so that the deformation direction was changed from squeezing the cover plate to squeezing the base, which could effectively reduce the deformation of the cover plate and improve the structural sealing property.
- The damage of various components of the explosion separation device was accurately simulated in ABAQUES based on the dynamic damage constitutive of the metal materials. The research will be a great reference for the structural design improvement.

References

[1] Du L F, Ma Y H, Chen H N, Yang Sh B and Xia D X 2015 The simulation of action procedure of explosive bolt *Init. Pyrotech.* **3** 29-32

[2] Yang H 2017 Study on fracture characteristics of explosive bolts based on finite element model (Harbin, China: HIT) pp 19-35

[3] Wang X X, Qin Z Y, Ding J F and Chu F L 2017 Finite element modeling and pyroshock response analysis of separation nuts *Aerosp. Sci. Technol.* **68** 380-90
[4] Balden V.H and Nurick G.N 2005 Numerical simulation of the post-failure motion of steel plates subjected to blast loading *Int. J. Impact Eng.* **32** 14-34
[5] Zhao H D, Liu W, Ding J F, Sun Y, Li X and Liu Y Z 2018 Numerical study on separation shock characteristics of pyrotechnic separation nuts *Acta Astronaut.* **151** 893-903
[6] Lee J R, Chia Ch C and Kong Ch W 2012 Review of pyroshock wave measurement and simulation for space systems *Meas.* **45** 631-42
[7] Lee J, Han J H, Lee Y and Lee H 2014 Separation characteristics study of ridge-cut explosive bolts *Aerosp. Sci. Technol.* **39** 153-68
[8] Bonorchis D. and Nurick G.N 2009 The influence of boundary conditions on the loading of ectangular plates subjected to localised blast loading *Imp. Num. Sim.***36** 40-52
[9] Lemaitre J 1984 How to use damage mechanics *Nucl. Eng. Des.* **80** 233-45
[10] Qi Y J, Sun Q and Ma Y 2015 Dynamic mechanical properties and Johnson-Cook type constitutive equation of wood rubber shock absorber for vehicle *Scientia Silvae Sinicae* **51** 149-55
[11] Ji Y H 2009 Numerical simulation of Hard-body foreign object damage based on Johnson-Cook model (Jiangsu, China: NUAA) pp 7-28

Application of resonance demodulation method based on local characteristic-scale decomposition and spectral kurtosis in gearbox fault

R Cong[1,2], C H Li[2,3] and J Q Wu[2]

[1]School of Mechanical Engineering, Changzhou University, Changzhou 213164, China

[2]Mechanic science and engineering college, Northeast Petroleum University, Daqing 163318, China

E-mail: lichunhui0816@qq.com

Abstract. Resonance demodulation technology has been widely used for its advantages on magnification and selection of fault information. However, the selection of filter parameters is affected by the subjectivity, and the analysis results are easily affected by the noisy signals under actual working conditions. To solve this problem, a method combining relative entropy and correlation coefficient is proposed to analyze and select Intrinsic Scale Components obtained by Local Characteristic-scale Decomposition, remove false components, and to some extent, de-noise. Then, after reconstructing the remaining components, we use spectral kurtosis analysis to find out the most easily detected frequency bands of fault features and analyze them by resonance demodulation, aiming to diagnose gearbox failures. Finally, we compare this method with that based on the empirical mode decomposition. The research shows that the method proposed in this paper can effectively diagnose the gearbox fault and is better than the empirical mode decomposition.

1. Introduction

The vibration signal of the gearbox has the characteristic of amplitude modulation and frequency modulation, and to extract modulation information in the vibration signal by the demodulation method can diagnose fault. Resonance demodulation, also known as envelope analysis, is commonly used to demodulate at present, which has the advantages of magnifying and selecting fault information [1]. But the results of the analysis are easily affected by the noises in the signals under the actual working conditions; In addition, in practical applications, the filtering parameters of the demodulation method need to be chosen artificially, which means it will be greatly influenced by the subjectivity, leading to inaccurate analysis results. Therefore, a suitable method must be selected to make up for the defects of the resonance demodulation method. The concept of spectral kurtosis is proposed by Dwyer [2,3], and its essence is to use the time-frequency analysis to reflect the kurtosis size of original signal in each spectrum section. It can effectively show the non-stationarity and detect transient components in the noise signals. With its sensitivity to the fault pulse, it can clearly point out the frequency band of the transient components, providing the basis for the selection of filter parameters.

In recent years, many scholars have used both spectral kurtosis and signal decomposition method to make up for the shortcomings of the resonance demodulation method. Zhang and others [4] applied the wavelet de-noising and demodulation method to the motor fault diagnosis, succeeding in the compound fault diagnosis of motor. Cai and others [5] combined the empirical mode decomposition, (EMD) spectral kurtosis and resonance demodulation to diagnose the rolling bearing fault signal, thus overcoming the disadvantages of traditional envelope analysis while Zhou and others [6] did such diagnosis with the improved EMD algorithm and the resonance demodulation method. Cheng and others [7] used the Local Mean Decomposition (LMD) and spectral kurtosis analysis to get the best frequency band, and then used the resonance demodulation analysis to obtain the gearbox fault feature information. All these methods have achieved good results, but the effect of wavelet de-noising largely depends on the choice of basic functions and thresholds, which often requires designers to have rich experiences. EMD and LMD also have defects such as endpoint effect, frequency aliasing and large amount of computation.

In view of the above situation, this paper combines the LCD method, spectral kurtosis and demodulation analysis. LCD decomposition is proposed by Cheng, Yang and others [8] of Hunan University, based on the Intrinsic Time scale Decomposition (ITD). This method can divide a complex signal into a series of single component signals--Intrinsic Scale Component (ISC). Compared with other time-frequency decomposition methods, it has shorter operation time and more advantages in restraining endpoint effect and modal aliasing. Therefore, the combination of LCD method and spectral kurtosis to the gearbox fault diagnosis is more favorable to the accuracy of the diagnosis result.

However, as a new method, there are some defects in the LCD method, and the false component is one of them. As for the identification of false components, people ever used the naked eye relying on experience, but with the development of science and technology, it has been eliminated. At present, some research work has been done by scholars, and the correlation coefficient method [9] is more representative among them.

Correlation coefficient is the index to describe the correlation degree between variables. The value is [-1, 1], gamma >0, positive correlation, gamma <0, negative correlation, gamma =0, and the absolute value of gamma is greater, the correlation degree is higher. However, in some cases, the correlation coefficient between the ISC components is relatively close, which is easy to cause misjudgment.

In different cases, the selection of threshold is not the same. Ideally, the real components have absolute relationship with the original signal, while the false component is not. Actually, the problem of decomposition of the end effect will cause abnormal results in different degree, disturbing the choice of threshold for increasing the divergence value corresponding to the real components. So when dealing with different data, we need to adjust the threshold.

In this paper, we first calculate the relative entropy of each component and the original signal, and make the preliminary judgment of the degree of correlation component and the original signal according to the set threshold. Then we calculate the correlation coefficient, and make a further step to determine the correlation degree between component and the original signal, reducing the error caused by threshold set according to the experience and removing false component more effectively.

2. Study on the method of removing false components in LCD

The vibration signal of gearbox often contains a lot of noises. At the same time, there will be some false components in the process of signal decomposition. We must use appropriate criteria to analyze ISC components, so as to select reasonable components. The commonly used range degree is Euclidean distance, relative entropy, normalized energy difference, correlation coefficient and so on. In this paper, the relative entropy and correlation coefficient are used to remove the false components in the signal.

2.1. Relative entropy

The relative entropy is also called mutual entropy, cross entropy, Kullback entropy, and KL divergence and so on. P (x) and Q (x) are two probability distributions of the value of X, and the relative entropy of P to Q is [10]:

$$D(P\|Q) = \sum_{i=1}^{n} P(i) \ln \frac{P(i)}{Q(i)} \tag{1}$$

That is, calculating the mean value of the logarithmic difference between P and Q according to the probability P. Only if the sum of the probability P and Q is 1, and for any i to satisfy P (I) >0, Q (I) >0, formula (1) is defined. Any case of 0ln0 appears in formula (1), and its value is treated at 0.

The relative entropy has the following properties: (1) The relative entropy value is non-negative, i.e. D (P||Q) ≥ 0. (2) Although intuitively the relative entropy is a metric or a distance function, it is not actually a real measure or distance. Because the relative entropy is not symmetry, namely D (P||Q) = D (Q||P), when the relative entropy is used to analyze the similarity among data, it should be symmetric. That is:

$$D(P,Q)=\frac{1}{2}\times\left(D(P\parallel Q)+D(Q\parallel P)\right) \tag{2}$$

From above, we know that when the two random distributions are same, their relative entropy is zero. When the difference of two random distributions increases, their relative entropy will also increase.

2.2. *Correlation analysis*

Correlation analysis includes autocorrelation analysis and cross-correlation analysis. The autocorrelation function shows dependency relationships of signals between the value of one time and of the other time, and the cross-correlation function is a correlation statistic that expresses the dependence between two sets of data. The correlation function can describe the important information in the random signal. If the characteristic signal is a periodic signal, its correlation function is also periodic. When the characteristic signal is white noise, the autocorrelation function will be attenuated when the delay is large enough. The correlation analysis is used in this paper: the correlation between the components and the original signal is approximately equal to the autocorrelation of the components, and the correlation between the pseudo component and the original signal is very small. The cross correlation function is:

$$R_{xy}(\tau)=\lim_{T\to\infty}\frac{1}{T}\int_0^T x(t)y(t+\tau)dt \tag{3}$$

According to the above theory of relative entropy and correlation coefficient, the smaller the relative entropy is, the closer the component is to the original signal. The higher the correlation coefficient is, the greater the correlation between the component and the original signal. The key is to set a threshold for both relative entropy and correlation coefficient. When the relative entropy is less than the set threshold and the correlation coefficient is greater than the set threshold, it is the real component and vice versa. The relative entropy of this chapter is normalized, i.e. the real relative entropy of component tends to 0, but the actual ISC component is affected by the end effect. According to the experience of previous studies, the relative entropy threshold is set at 0.1; in light of that when the correlation coefficient of the two signals is less than 0.3, it can be considered that the two signals have no relativity, we can set the threshold of correlation coefficient at 0.3.

3. Estimation resonance demodulation method based on LCD and spectral kurtosis

Using LCD method, a complex signal can be decomposed into a series of ISC components. By analyzing the ISC components, a reasonable component can be selected for reconfiguration, so as to reduce the false components produced in the decomposition process and the influence of noises in the signals.

Then the spectral kurtosis of the signal is calculated to determine the filter parameters required by the resonance demodulation technology, and the definition of spectral kurtosis is as follows:

For the Wold-Cramer decomposition of a non-stationary signal, it is assumed that Y (T) is a system response and X (T) is a signal, then the expression of Y (T) is [9]:

$$Y(t)=\int_{-\infty}^{+\infty} e^{2\pi f i t}H(t,f;\varpi)dX(f) \tag{4}$$

In the formula, $H(t,f;\varpi)$ is a time-varying transfer function of a system. It is changed over time t and can be regarded as a complex envelope of the signal $Y(t)$.

$S_{2nY}(t,f)$ is the 2n order instantaneous moment of the signal $Y(t)$, and its expression is as follows:

$$S_{2nY}(t,f) \triangleq E\left\{\left|H(t,f)dX(f)\right|^{2n}\right\}/df = \left|dX(f)\right|^{2n} S_{2nX}(t,f) \tag{5}$$

In the formula, $S_{2nY}(t,f)$ is the instantaneous moment of the 2n order; E is the expected operator.

The expression of the 2n order spectral moment $S_{2nY}(t,f)$ for the definition of $Y(t)$ is as follows:

$$S_{2nY}(f) \triangleq E\left\{S_{2nY}(t,f)\right\} \tag{6}$$

The expression of the four order spectral cumulant $C_{4Y}(f)$ is as follows:

$$C_{4Y}(f) = S_{4Y}(f) - 2S_{2Y}^2(f), f \neq 0 \tag{7}$$

Therefore, the spectral kurtosis can be defined as:

$$K_Y(f) \triangleq \frac{C_{4Y}(f)}{S_{2Y}^2(f)} = \frac{S_{4Y}(f)}{S_{2Y}^2(f)} - 2 \tag{8}$$

In the actual gearbox fault diagnosis, the signals collected include the vibration signal and the noise signal, that is:

$$C(t) = x(t) + N(t) \tag{9}$$

Among them: $C(t)$ is the acquisition signal; $x(t)$ is the vibration signal of the gearbox; $N(t)$ is a noise signal.

The spectral kurtosis of the gearbox acquisition signal $C(t)$ can be expressed as:

$$K_x(f) = K_G(f)/\left[1+r(f)\right]^2 \tag{10}$$

$r(f)$ is a noise to signal ratio, and its value is equal to $S_N(f)/S_G(f)$, $S_N(f)$ and $S_G(f)$ are respectively the power spectral density of noise and gearbox vibration signals.

From formula (10), the spectral kurtosis value $K_x(f)$ of the acquisition signal is inversely proportional to the noise signal $r(f)$, and when the value of $r(f)$ is very small, the value of $K_x(f)$ is very large. Therefore, analyzing of each frequency band of the signal through spectral kurtosis and comparing the size of spectral kurtosis of each frequency band, we can find the best frequency band to get fault feature information.

To sum up, the steps of the resonance demodulation method based on LCD and spectral kurtosis are as follows:

The signal $C(t)$ is decomposed by LCD. When the correlation coefficient between the ISC component and the original signal is less than 0.1, the decomposition is stopped.

After calculating the relative entropy between the ISC component and the original signal, we calculate the correlation coefficients again, remove the false components and reconstruct the remaining components.

The spectral kurtosis of the reconstructed signal is calculated, and the spectral kurtosis of the reconstructed signal is obtained, and the filter parameters required by the resonance demodulation technology are determined.

The resonance demodulation analysis of the reconstructed signal is carried out, and the fault features of the gearbox are extracted and the fault type of the gear box is determined.

4. Simulation and example calculation

4.1. Simulation analysis

In order to verify the validity of the resonance demodulation method based on the LCD spectrum and kurtosis analysis, we construct the FM amplitude modulation simulation signal according to the characteristics of gearbox vibration frequency and amplitude, with meshing frequency at 385 Hz and the modulation frequency at 7 Hz. First of all, with the original termination condition of LCD method as the stopping basis, the signal is conducted by LCD decomposition. Then according to the relative entropy and the correlation coefficient of ISC component and simulation signal, we remove false component. The simulation signal is shown in formula (11).

$$x(t) = (1 + 2sin(14\pi t)) \times sin(770\pi t + 12sin(14\pi t)) \tag{11}$$

A white noise sequence with a mean value of 0 and a variance of 1 is added to X (T), and figure 1 is its time domain waveform.

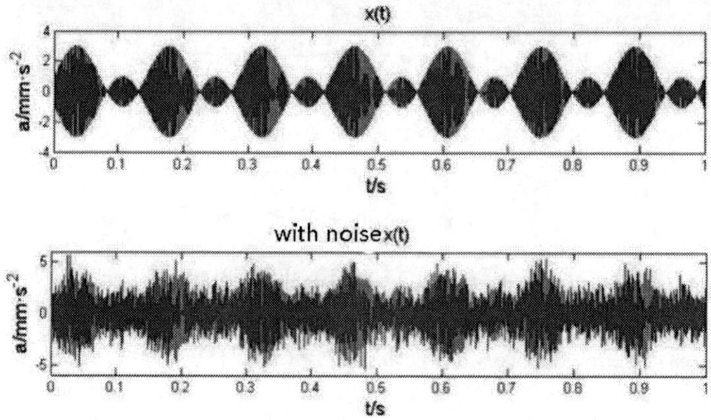

Figure 1. The time domain waveform of the simulation signal.

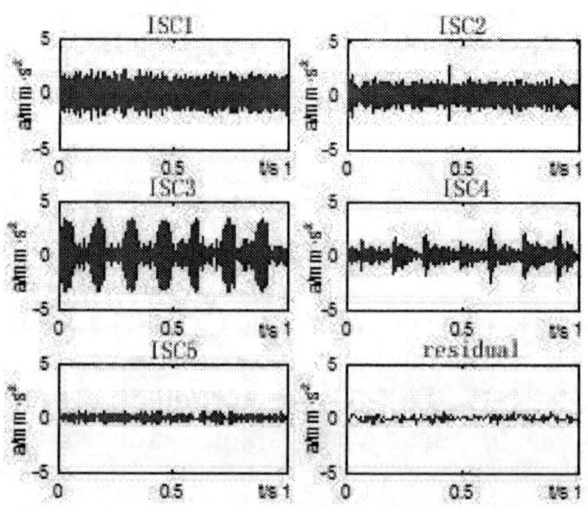

Figure 2. Results by using LCD of the simulation signal.

The 11 ISC components and residuals obtained after the LCD decomposition of the simulation signal are shown in figure 2. The relative entropy value of each ISC component and the simulation signal is calculated and the relative entropy is normalized. The results are shown in table 1. From table 1, we can see the relative entropy of each ISC component and the simulation signal, and draw a conclusion that the relationship between ISC3, ISC4 and ISC5 is closely related to the simulation signal, and the relative entropy is at least one order of magnitude different from other components. By calculating the

IOP Conf. Series: Materials Science and Engineering **657** (2019) 012050 doi:10.1088/1757-899X/657/1/012050

correlation coefficient between the two, the results are shown in table 1. It can be seen that the correlation coefficient of ISC5 and simulation signal is less than 0.3. It can be found that it is not related to the simulation signal, that is, it is a false component. The ISC3 and ISC4 are reconstructed, as shown in figure 3, and the correlation coefficient of the reconstructed signal and X (T) is calculated to be 0.9308. Therefore, we decompose the signal by LCD method, and remove the false components by relative entropy and correlation analysis, which can realize signal de-noising and retain the function of original signal basically.

Table 1. The Kullback-Leibler divergence and correlation coefficient between each ISC component and adding noise x(t) of the simulation signal.

	ISC1	ISC2	ISC3	ISC4	ISC5
Relative entropy	1	0.4182	0.0129	0.0222	0.1991
Correlation coefficient	0.1507	0.2213	0.6144	0.6639	0.0200

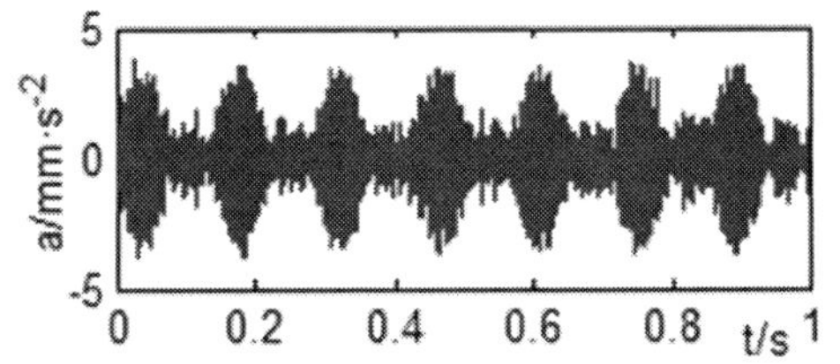

Figure 3. Time-domain waveform of the reconstructed signal.

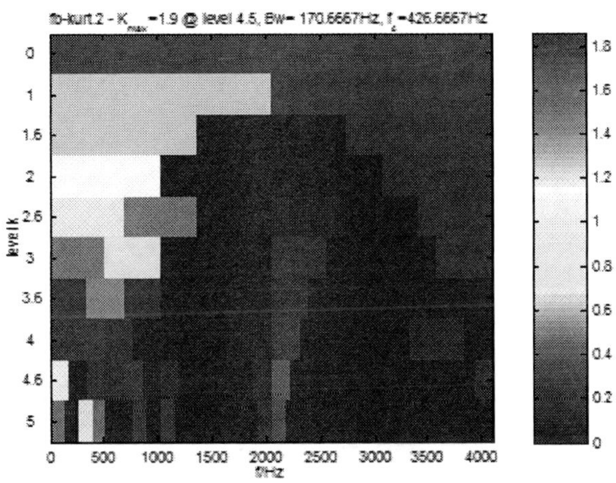

Figure 4. Spectral kurtosis of the simulation signal.

Next, we calculate the spectral kurtosis of reconstructed signal. Figure 4 is the spectral kurtosis diagram of reconstructed signal. The result shows that the highest spectral kurtosis is 1.8, the central frequency is 426.6667 Hz, and the bandwidth is 170.6667 Hz, that is, the signal to noise ratio of 0~170.6667 Hz is the largest zone. The selected filter parameters are used to filter the reconstructed signal, and then the signal after the filter is demodulated by resonance demodulation. The envelope spectrum is shown as figure 5. From figure 5, we can see clearly the frequency 7.362 Hz and its frequency doubling that is similar to the amplitude modulation frequency 7 Hz of the analog signal. We can conclude that the demodulation method based on LCD and spectral kurtosis effectively extract the modulation information of the simulation signal.

IOP Conf. Series: Materials Science and Engineering **657** (2019) 012050 doi:10.1088/1757-899X/657/1/012050

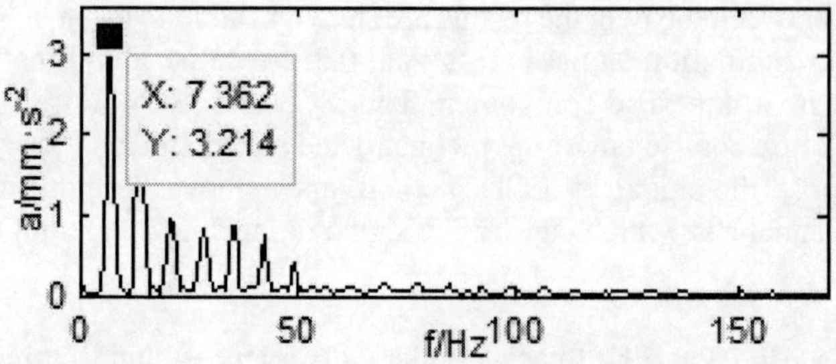

Figure 5. Envelope spectrum of the simulation signal.

4.2. Example analysis

In this paper, QPZZ-II rotary machinery vibration analysis and fault diagnosis experimental platform is used to simulate gearbox failure experiment. The experimental device is shown in figure 6. Two accelerometer sensors are used to collect signals, which are arranged in X direction and Y direction respectively. A total of experiments under normal gear state and three failure states were carried out. The data of gear crack and wear failure were analyzed in this paper. The time-domain and frequency domain waveform diagram of the fault data is shown in figure 7. Experimental parameters: the number of large gear teeth is 75, the number of small gear teeth is 55, the speed of the motor is 360 r/min when the signal is collected, the number of sampling points is 102400, and the sampling frequency is 10240 Hz. In order to make the effectiveness of the method more general, the data of small rotating frequency of 10 Hz is used for crack failure, while the wear fault is diagnosed by the data of small rotating frequency 6 Hz. The meshing frequency of them is 550 Hz and 330 Hz respectively.

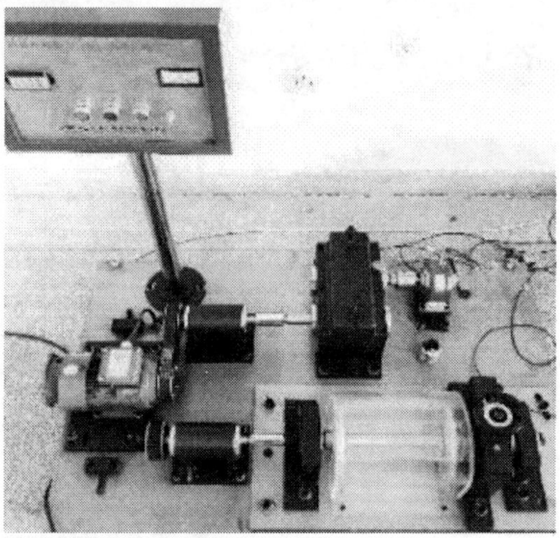

Figure 6. Experimental facility.

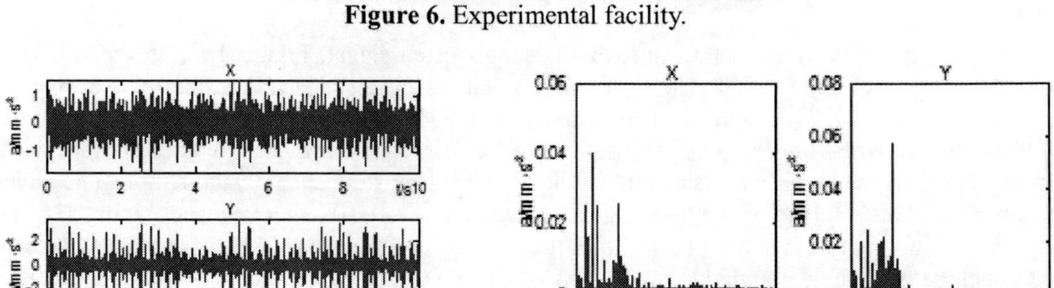

(a)

IOP Conf. Series: Materials Science and Engineering **657** (2019) 012050 doi:10.1088/1757-899X/657/1/012050

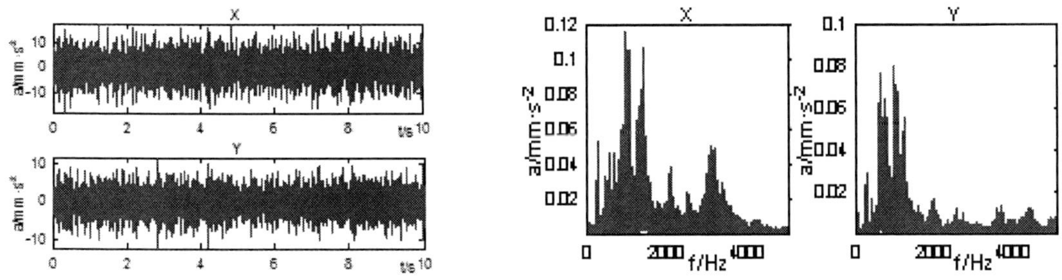

(b)

Figure 7. Time-domain waveform and frequency domain waveform of failure data. (a) crack fault and (b) wear fault.

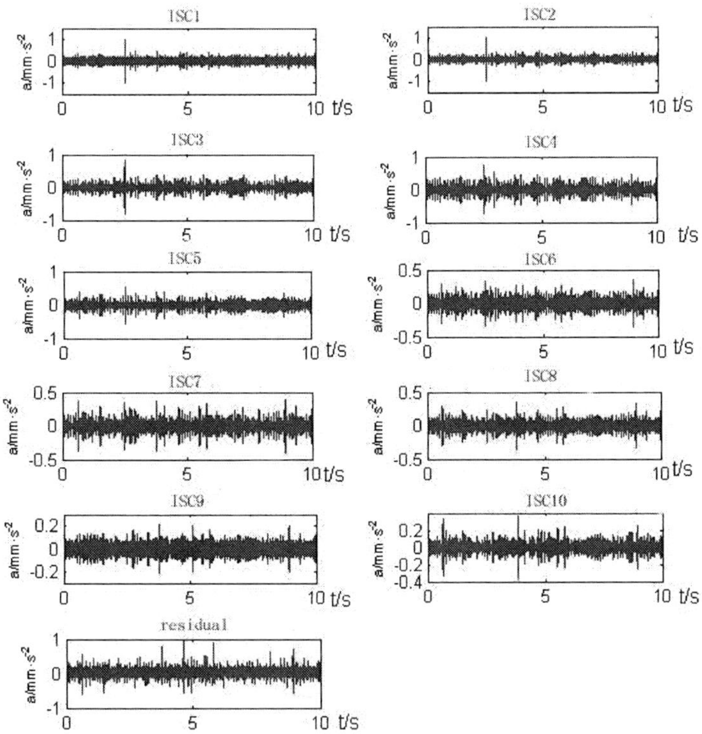

Figure 8. Results by using LCD of the crack fault signal.

The fault signals of gear with crack LCD decomposition results is shown in figure 8, and relative coefficient and calculation of each component and the original signal relative entropy is shown in table 2. According to the threshold of correlation coefficient, ISC1, ISC4, ISC5 component were selected to reconstruct, but the relative entropy of ISC1 exceeds the set threshold value, which means it is false component that should be removed. To reconstruct the components ISC4 and ISC5, as shown in figure 9, and to calculate the steep spectrum degree of signal reconstruction, we obtain spectral kurtosis, as shown in figure 10. We know that the spectral kurtosis maximum value is 485.8, the center frequency is 4048.1875 Hz, bandwidth is 426.125 Hz, namely (0~426.125 Hz) SNR maximum signal. The reconstructed signal is filtered by the above parameters, and then the signal after the filter is demodulated by resonance demodulation. The envelope spectrum is shown in figure 11. From the results of resonance demodulation analysis, we can see that there is a significant frequency impact component 9.312 Hz and its frequency doubling. This is approximate to crack fault gearbox vibration signal characteristic frequency 10Hz and frequency doubling, and the diagnostic error is 6.88%, within a reasonable range.

Table 2. The Kullback-Leibler divergence and correlation coefficient between each ISC component and adding noise x(t) of the crack fault signal.

	ISC1	ISC2	ISC3	ISC4	ISC5
Correlation coefficient	0.3804	0.2370	0.2093	0.3564	0.3271
Relative entropy	0.1220	0.0475	0	0.0123	0.0186
	ISC6	ISC7	ISC8	ISC9	ISC10
Correlation coefficient	0.2359	0.2355	0.2201	0.1505	0.1156
Relative entropy	0.1838	0.9842	0.7607	1	0.5521

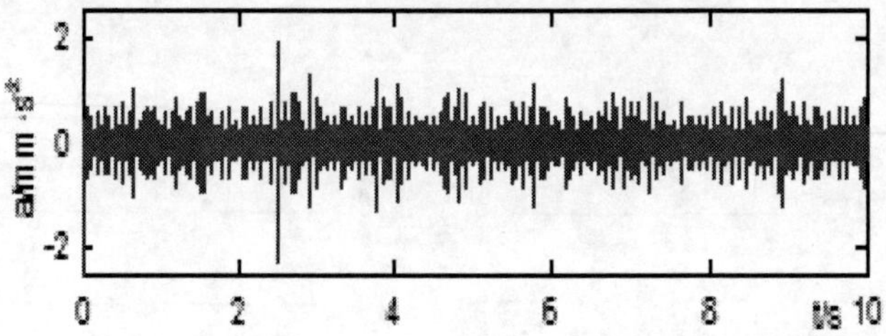

Figure 9. Time-domain waveform of the crack fault signal.

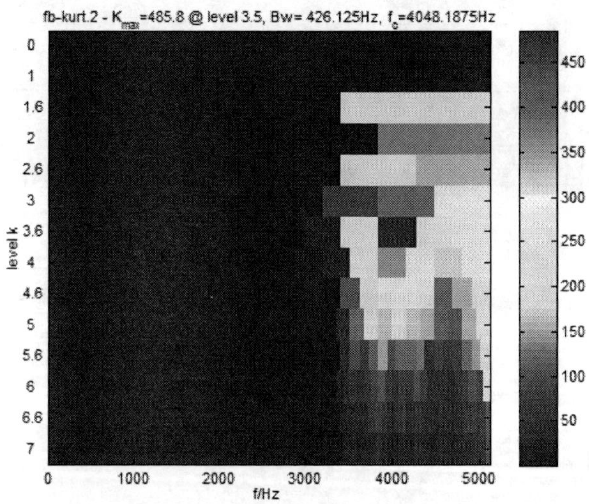

Figure 10. Spectral kurtosis of the crack fault signal.

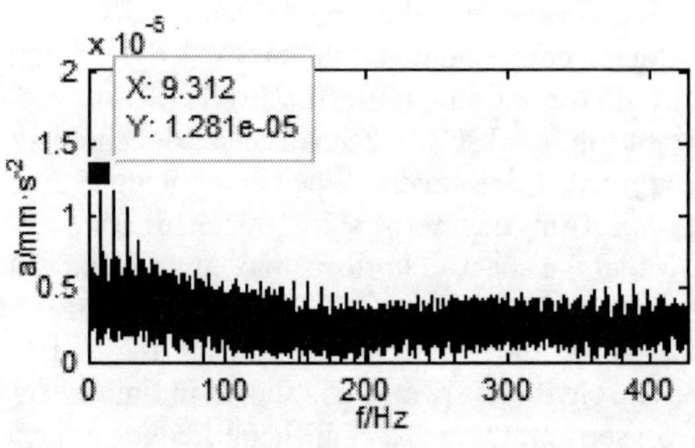

Figure 11. Envelope spectrum of the crack fault signal.

The wear of gear fault signal LCD decomposition results is shown in figure 12. The calculation of the relative entropy and correlation coefficient of each component and the original signal is shown in table 3. According to the set threshold of relative entropy and correlation coefficient, component of ISC1, ISC2, ISC3and ISC4 are selected to reconstruct, and the reconstructed signal is shown in figure 13. Spectrum of reconstructed signal was calculated, and spectral kurtosis is shown in figure 14. The spectral kurtosis maximum value is 15.5, the center frequency is 4474.3125 Hz, bandwidth is 1278.375 Hz, namely (0~1278.375 Hz) SNR maximum signal. The reconstructed signal is filtered by the above parameters, and then the signal after the filter is demodulated by resonance demodulation. The envelope spectrum is shown in figure 15. From the results of resonance demodulation analysis, we can see that there is a significant frequency impact component 304 Hz and its frequency doubling. It is similar to the characteristic frequency 330 Hz of the vibration signal of worn fault gearbox and its frequency doubling, and the diagnostic error is 7.88%.

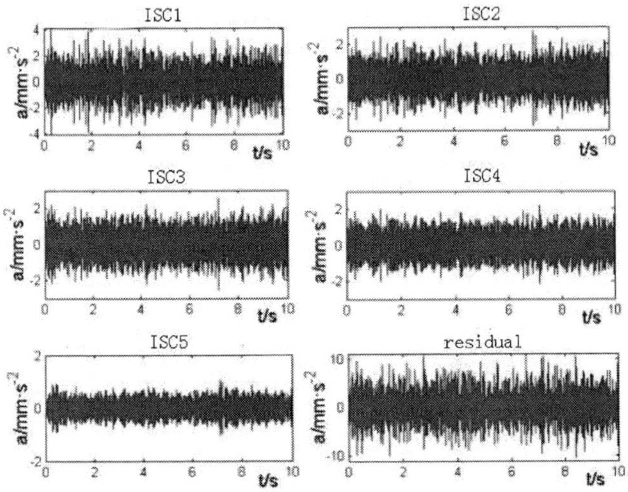

Figure 12. Results by using LCD of the wear fault signal.

Table 3. The Kullback-Leibler divergence and correlation coefficient between each ISC component and adding noise x(t) of the wear fault signal.

	ISC1	ISC2	ISC3	ISC4	ISC5
Relative entropy	0.4223	0.3331	0.4180	0.0292	1
Correlation coefficient	0.0374	0.0779	0	0.3937	0.1163

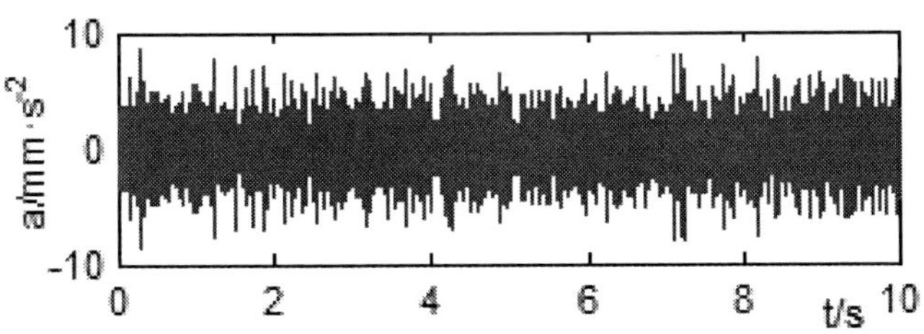

Figure 13. Time-domain waveform of the wear fault signal.

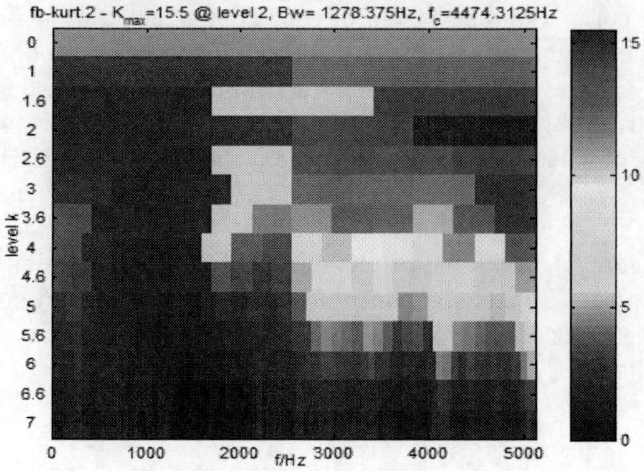

Figure 14. Spectral kurtosis of the wear fault signal.

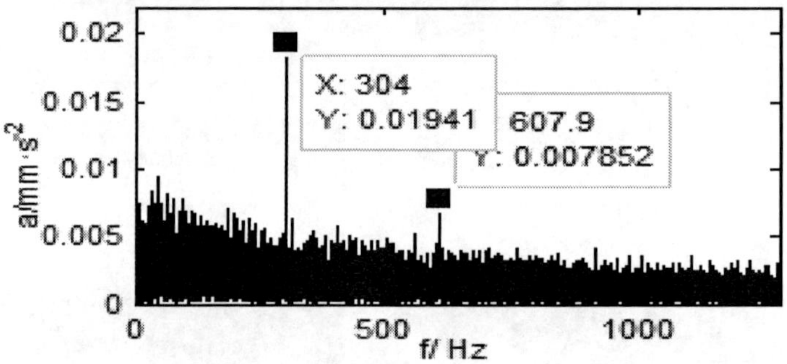

Figure 15. Envelope spectrum of the wear fault signal.

5. Conclusion

In this paper, the correlation coefficients and relative entropy are used to discriminate the components after LCD decomposition, and the false components are removed, and then the resonance demodulation analysis is carried out. By signal simulation, we can see the frequency 7.362 Hz and its frequency doubling that is close to the amplitude modulation frequency of the simulation signal 7 Hz.

Through analyzing the experimental data, there is a significant frequency impact component 304 Hz and its frequency doubling. It is similar to the characteristic frequency 330 Hz of the vibration signal of the worn fault gearbox, and the diagnostic error is 7.88%.

Through the above analysis, the correlation coefficient and relative entropy can be considered to have a good effect on the removal of false components after LCD decomposition.

Acknowledgment

This paper was supported by Foundation Project: National Natural Science Foundation, Project No. 51505079.

References

[1] Mcfadden P D and Smith J D 1984 Vibration monitoring of rolling element bearings by the high-frequency resonance technique- a review *Tribol. Int.* **17**(1) 3-10
[2] Dwyer R 1983 Detection of non-Gaussian signals by frequency domain Kurtosis estimation *IEEE Int. Conf. on ICASSP Acoustics, Speech, and Signal Processing* (Boston, USA) **8** 607-10
[3] Antoni J and Randall R B 2006 The spectral kurtosis: A useful tool for characterizing non-stationary signals *Mech. Syst. Signal Pr.* **20** 282-307
[4] Zhang X X and Liu Z X 2010 Application of resonance demodulation and wavelet denosing in

fault diagnosis of induction motors *Elect. Mach. Cont.* **14** 66-70
[5] Cai Y P, Li A H and Shi L S 2011 Roller bearing fault detection using improved envelope spectrum analysis based on emd and spectrum kurtosis *J. Vib. Shock* **30**(2) 167-72
[6] Zhi Z, Zhu Y S and Zhang Y Y 2013 Adaptive fault diagnosis of rolling bearings based on EEMD and demodulated resonance *J. Vib. Shock* **32** 76-80
[7] Cheng J S and Yang Y 2012 Application of spectral kurtosis approach based on local mean decomposition (lmd) in gear fault diagnosis *J. Vib. Shock* **31**(18) 20-3+54
[8] Cheng J S and Zheng J D 2012 A nonstationary signal analysis approach-the local characteristic-scale decomposition method *J. Vib. Eng.* **25**(2) 215-20
[9] Zhang J F and Huang Z C 2006 Discriminant approach to the machinery or sensor fault based on the k-l divergence *J. Mech. Strength* **28** 670-73
[10] Han Z H and Zhu X X 2012 A false component identification method of EMD based on kullback-leibler divergence *Proc. CSEE* **32** 112-17

IOP Conf. Series: Materials Science and Engineering **657** (2019) 012051 doi:10.1088/1757-899X/657/1/012051

Numerical analysis of mini-channel cooling applied in power battery

F X Huang[1], J Zhao[1,2], C Wang[1] and H Zhang[1]

[1]School of Mechanical Engineering, Guizhou University, Guiyang, 550025, China

E-mail: zhaojin9485@163.com

Abstract. Working in the extreme condition such as high temperature and discharge rate, causes the generated heat shapely increasing and rising of temperature of power battery. Further influence the cell performance. It is important to efficiently dissipate heat. In this paper, a mini-channel cooling system can be designed to cope with the problem. Moreover, creating numerical model to study different placement, ambient temperature and coolant inlet properties how to function on the system. The results express the system be placed at vertical direction can obtained a good temperature effect, only sacrificing few of pump energy than placed at horizontal. The results show the variation of ambient temperature has little effect on the temperature contribution of battery module, and the temperature of the battery pack is presented linear with the coolant inlet temperature. The temperature of battery pack is decrease and the uniformity is improved by increasing the inlet velocity of the coolant. In the end, increasing the height of the channel are a better scheme through the process of structure optimization.

Nomenclature

	Heat flow [W]	p	Pressure[pa]
λ_k	thermal conductivity [$W \cdot m^{-1} \cdot K^{-1}$]	ρ	Density[$Kg \cdot m^{-3}$]
A	area [m^2]	μ	dynamic viscosity coefficient
$\frac{\partial T}{\partial x}$	change rate of the temperature	E	total energy[J]
h	convective heat transfer coefficient [$W \cdot m^{-2} \cdot k^{-1}$]	u, v, w	Velocity of x, y, z direction[$m \cdot s^{-1}$]
Δt	temperature difference [$^\circ C$]	λ	loss coefficient.0
E	radiation force[$W \cdot m^{-2}$]	l	Length[mm]
ε	Emissivity(0□ 1)	d	diameter[mm]
σ_b	blackbody radiation constant[5.67×10^{-8} $W \cdot m^{-2} \cdot K^{-4}$]	v	Velocity[]
V	vector of velocity[]	ζ	local loss coefficient

1. Introduction

Nowadays, power battery used in electric vehicle is mainly lithium-ion battery. It has many remarkable advantages, such as high specific energy and power, low self-discharge rate, no memory effect and so on [1,2]. However, the working performance of battery cell is greatly affected by temperature. The optimum operating temperature range of lithium-ion cell is 20~45 ℃ [3], and the maximum temperature difference should be controlled in 5 ℃ [4].

With the increase of endurance mileage of electric vehicle as well as energy density of battery system, the heat generation is greatly risen. Therefore, more efficient heat dissipation way is needed.In the heat dissipation area, the common methods are air, liquid and phase change cooling [5-7]. Above all, using liquid to throw away heat is a bright method for high energy density batteries. Which is divided into direct and indirect two ways [8]. The indirect way uses the cold plate to separate the battery, the cold plate plays the key role in heat transfer, and the heat transfer coefficient is relatively increased.

Mini-channel can enhance heat transfer [9]. Qian Zhong [10] found that dividing the original channel of the chip cooler into two same small channels can effectively improve the cooling performance. Kandlikar [11] classified the flow channels according to the hydraulic diameter, and specified mini-channel that the hydraulic diameter range between 0.2 and 3 mm. In this paper, a novel mini-channel cooling system was proposed, the effects of some essential operate parameter, such as inlet velocity and temperature of the coolant as well as discharge rate of battery cell was investigated by means of computational fluid dynamic (CFD) analysis, the numerical simulations showed promising results. Finally, based on the purpose, ensuring the heat dissipation of the system and reducing the inlet and outlet pressure drop, the size optimization of the channel structure has been carried out.

2. Building mini-channel cooling system model

2.1. Numerical modeling

A commecial lithium-ion battery is chosen as researching object. The size of cell is 66x120x18 mm; density is 2100 $Kg{\cdot}m^{-3}$; specific heat capacity is 1400 $J{\cdot}Kg^{-1}{\cdot}K^{-1}$; thermal conductivity of X axis is 1 W / (m · K); and Y and Z axis is 5 W / (m · K).

Aluminum is chosen to be the material of the cold plate, and the thickness is 2 mm. a rectangular S-shaped flow channel, which is arranged symmetrically in the interior of every cold plate in the form of double channels. The dimensions of the channel are 1 mm (width) and 2 mm (hidth). Liquid water is used as coolant. Figure 1 shows the mini-channel battery cooling system established by Solidworks.

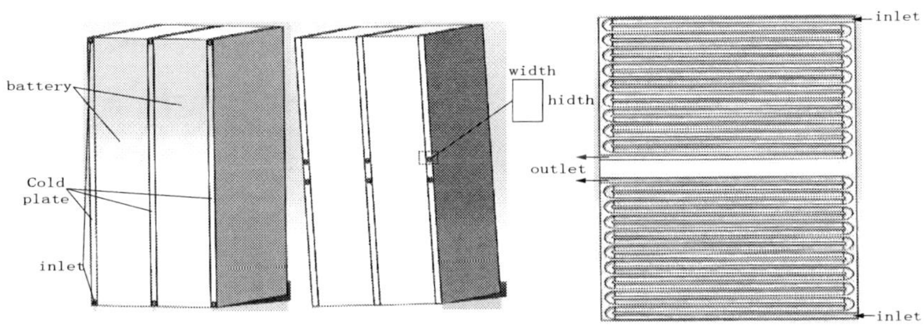

Figure 1. Mini-channel battery cooling system.

2.2. theoretical principle

2.2.1. Heat transfer.

- Heat conduction

$$\Phi = -\lambda_k A \frac{\partial T}{\partial x} \tag{1}$$

In above Φ, λ, A are the heat flux, the thermal conductivity of aluminum, the cross section area

IOP Conf. Series: Materials Science and Engineering **657** (2019) 012051 doi:10.1088/1757-899X/657/1/012051

perpendicular to the direction of the heat flow, $\frac{\partial T}{\partial x}$ is the change rate of the temperature with the direction of the direction, and the negative sign indicates that the direction of temperature transfer is the direction of the temperature decreasing.

- Thermal convection

$$\Phi = hA\Delta t \tag{2}$$

In above, h is the convective heat transfer coefficient, Δt represents the temperature difference between two objects of convection heat exchange.

- Thermal radiation

$$E = \varepsilon \delta_b T^4 \tag{3}$$

Where E represents the radiation force of the object, ε is the emissivity (also known as blackness), and its value is between 0 and 1, σ_b is a constant of black-body radiation with a value of 5.67×10^{-8}. The radiation heat transfer is very small, and its influence is ignored in this study.

2.2.2. Governing equations.

- The continuity equation

$$\nabla \cdot V = 0 \tag{4}$$

- The momentum conservation equation

$$\frac{\partial V}{\partial t} + (V \cdot \nabla)V = -\frac{\nabla p}{\rho} + \frac{\mu}{\rho}\nabla^2 V \tag{5}$$

- The energy conservation equation

$$\rho \cdot C_p \cdot \left(\frac{\partial E}{\partial t} + u \cdot \frac{\partial E}{\partial x} + v \cdot \frac{\partial E}{\partial y} + w \cdot \frac{\partial E}{\partial z}\right) = \lambda_k \cdot \left(\frac{\partial^2 E}{\partial x^2} + \frac{\partial^2 E}{\partial y^2} + \frac{\partial^2 E}{\partial z^2}\right) \tag{6}$$

In the fomula above, V represents the vector of velocity, p is pressure, ρ is density, μ represents the dynamic viscosity coefficient, E is the total energy. u , v and w represents the velocity of x y z direction respectively.

2.3. Computational grid

Before beginning finite element calculation, the model needs to be meshed, and the Mesh module brought with it by software ANSYS is used to generate the meshing of the model. The battery pack is automatic grid, the cooling plate is tetrahedral grid, and the fluid is generated mesh by sweep method. In the end, a total of 925465 nodes and 1963627 elements are obtained. Moreover, the average value of grid quality inspection parameter Element Quality is 0.845 and the average value of Skewness is 0.226. It is worthy to mention that the grid quality is good.

3. Placement mode and ambient temperature study

3.1. Effect of placement mode

For the cooling system, different placement can make the coolant has different flow direction. In order to study the influence of placement and abtaining a better assembly of battery and cold plate. Therefore, two kinds of models, horizontal and vertical, are established and analyzed by CFD simulation software-Fluent, seperately. The simulation conditions set as follows: the discharge rate is 1C, 3C and 5C,

IOP Conf. Series: Materials Science and Engineering **657** (2019) 012051 doi:10.1088/1757-899X/657/1/012051

respectively, the ambient temperature is 27 °C, at the inlet of coolant, the temperature is 20 °C and the velocity is 0.1 m·s^{-1}. Because the system will be installed in a narrow and tight space, the convective heat transfer coefficient between the system and the surrounding air is set as 1 W·m^{-1}·K^{-1}. The heat generation of each part of the battery pack is assumed uniform and the coolant is ideal fluid.

The simulation environment is set up in Fluent. Solver is pressure-based, steady, considering the influence of coolant gravity with the gravity acceleration value 9.8 m·s^{-2}. If Reynolds number in the mini-channel is up to 10^2 the flow can be treated as turbulent flow, so the low Reynolds number RNG model have been selected. Velocity inlet and pressure outlet boundary conditions have been applied to inlet and outlet in the whole process of numerical calculation. The results were listed in table 1.

Table 1. The simulation results of two placement.

Discharge rate	Ambient temperature	Placement mode	Maximum temperature	Minimum temperature	Pressure drop
1C	27°C	Horizontal	21.02°C	20.25°C	1582.97pa
3C	27°C	Horizontal	25.17°C	21.20°C	1582.97pa
5C	27°C	Horizontal	31.15°C	22.72°C	1582.97pa
1C	27°C	Vertical	20.99°C	20.32°C	1600.89pa
3C	27°C	Vertical	25.01°C	21.56°C	1600.89pa
5C	27°C	Vertical	30.08°C	23.35°C	1600.89pa

3.2. *Effect of ambient temperature*

In the natural convection cooling and the traditional air cooling, the battery pack working efficiency at charge and discharge process depends on the ambient temperature. In order to study the effect of ambient temperature for the battery pack in the mini-channel cooling system, especially higher discharge rate. At 5 C discharge rate, the same simulation conditions were set as section 3.1. The ambient temperature changes from 27 °C to 50 °C, including several temperature of 27 °C, 35 °C, 40 °C, 45 °C, 50 °C. It can simulate the transition of working cindition from room temperature to high temperature.

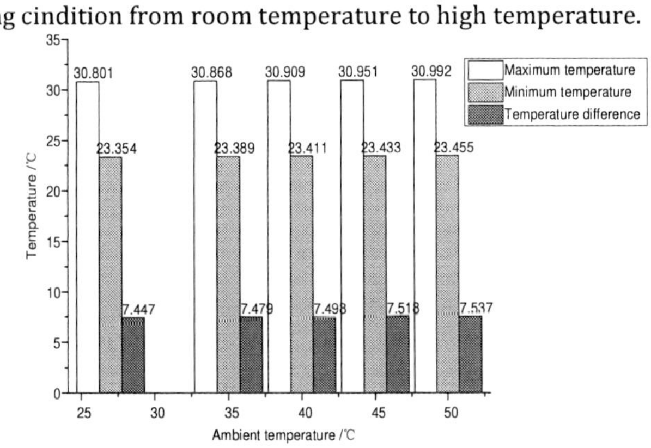

Figure 2. In 5C discharge rate, the temperature of the battery pack varies with the ambient temperature.

The variation of the maximum and the minimum temperature of the battery pack and the temperature difference between the battery pack with the ambient temperature are obtained. Figure 2 illustrates the numerical results that the maximum and the minimum temperature of the battery pack increase gradually with the increase of ambient temperature, but the change is relatively gentle. When the ambient temperature is 27 °C, the highest temperature of the battery pack is 30.801 °C and the lowest is 23.354 °C. However, while the ambient temperature rising 50 °C, the maximum temperature of the battery pack is kept at 30.992 °C and the minimum is kept at 23.455 °C, the maximum temperature is only increased by

0.191 °C and the minimum is only increased by 0.099 °C, meantime, the temperature difference is only increased by 0.09 °C. These results demonstrate that the mini-channel cooling system has strongly ability to adapt variable ambient temperature.

4. Study on inlet parameters of coolant

4.1. Different coolant inlet temperature

The inlet temperature of coolant is one of the important factors to determine the heat dissipation capacity of the system. Regarding the influence of the inlet temperature of the coolant. At 5 C discharge rate, the ambient temperature is 40 °C and the inlet velocity of the coolant is 0.1 $m \cdot s^{-1}$. The inlet temperature of the coolant range from 20 °C to 36 °C. Variation of battery pack temperature with coolant inlet temperature can be obtained on figure 3. The graph of result shows that the maximum and the minimum temperature of the battery pack present rising trend with the change of the inlet temperature of the coolant. After the inlet temperature reaches at 30 °C, the temperature of the battery pack exceeds its optimum temperature range. With the increasing of coolant inlet temperature where the temperature different began to decrease. However, the value is very small, which shows that the curve has no notably variation.

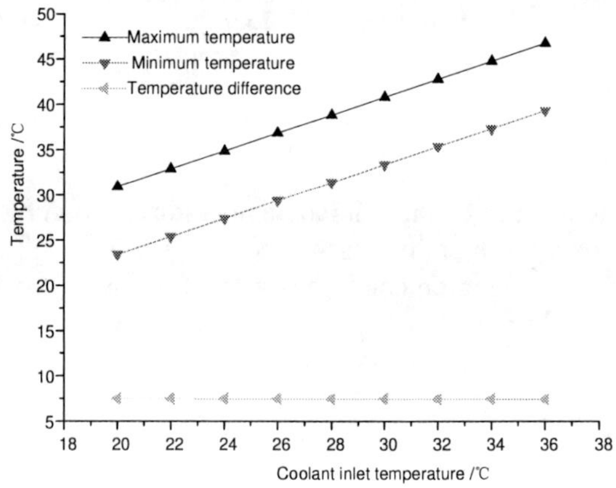

Figure 3. In 5C discharge rate, battery pack temperature varies with coolant inlet temperature.

4.2. Different coolant inlet velocity

The inlet velocity of coolant is another vital factor to effect heat dissipation performance of this system. With a view to the influence of inlet velocity, the discharge rate 5 C, the ambient temperature 40 °C and the inlet temperature of coolant 20 °C can be set up. The inlet velocity of coolant range between 0.1 and 0.5 $m \cdot s^{-1}$ with the variation gap is 0.05 $m \cdot s^{-1}$. At the end of the calculation, it is easily to knowledge the temperature distribution of battery pack and the pressure drop value of the coolant inlet and outlet.

It can be seen from figure 4 that the temperature of the area closely the cold plate on both sides of the cell is lower and the temperature in the middle is higher, the area of the high temperature in the cell decreases with the increase of the inlet velocity of the coolant, and the area of the low temperature increases with the increase of the inlet velocity of the coolant. Figure 5 shows that the highest and the lowest temperature of the battery pack decrease with the increase of the inlet velocity of the coolant. When the inlet velocity of coolant is less than 0.2 $m \cdot s^{-1}$, the maximum and minimum temperature of the battery pack get lower as the coolant inlet velocity increases. Once the coolant inlet velocity is increased to 0.2 $m \cdot s^{-1}$, the change becomes stable with the increase of the inlet velocity of coolant. The temperature difference between cells decreases with the increase of the inlet velocity of coolant. When the inlet velocity of coolant increases to 0.3 $m \cdot s^{-1}$, the inlet and outlet pressure of the coolant is 6004.45 pa, the temperature difference between batteries is less than 5 °C meeting the working requirement. From figure 6, it can be seen that the higher inlet velocity will lead to a higher pressure drop in the inlet and outlet of coolant.

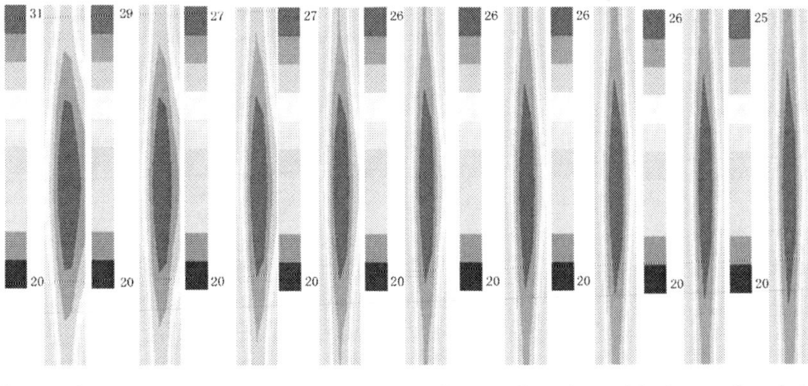

Figure 4. The cross-section temperature of the cell varies with the coolant inlet velocity, and the corresponding value are 0.1 0.15 0.2 0.25 0.3 0.35 0.4 0.45 0.5 $m{\cdot}s^{-1}$ respectively.

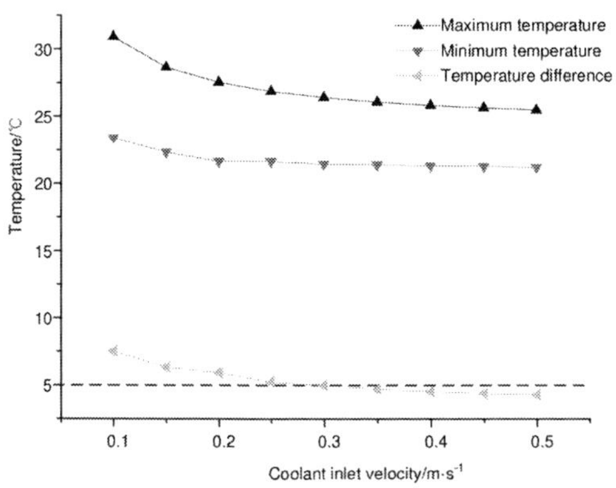

Figure 5. 5C discharge rate, the change of battery temperature with the inlet velocity of coolant.

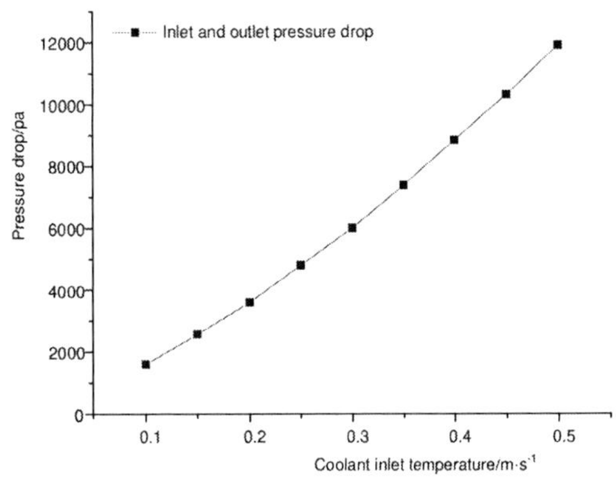

Figure 6. Variation of inlet and outlet pressure drop with inlet velocity of coolant.

5. Study on optimization of flow channel structure

According to the previous study, the increase of inlet velocity of coolant can efficiently decline maximum temperature and improve temperature uniformity of battery cells. At the same time, it will also significantly increase the inlet and outlet pressure drop of coolant. In order to reduce the inlet and outlet pressure drop of the coolant, the structure of the flow channel needs to be optimized. The pressure drop in the flow channel is mainly caused by the pressure loss along the way and local pressure loss.

The pressure loss along the way is defined as:

$$p_f = \lambda \frac{l}{d} \frac{\rho v^2}{2} \tag{7}$$

In above λ l d v are the loss coefficient along the way, the length of the channel, the hydraulic diameter of the channel, the coolant current velocity respectively.

The local pressure loss is defined as:

$$p_m = \zeta \frac{\rho v^2}{2} \tag{8}$$

Where ζ represents the local loss coefficient.

5.1. Channel height

Taking into account the influence for the height of flow channel and keeping other structure parameters constantly, the analysis is carried out to different flow channel model. The simulation conditions are as follows: discharge rate is 5 C; ambient temperature is 40°C; coolant inlet temperature is 20 °C; inlet velocity of coolant is 0.3 m·s^{-1}. A model study on the height of 2, 3, 4, 5 and 6 mm is established respectively.

Figure 7 presents that with the increase of the height of the flow channel, the trend of battery pack temperature is falling, and the maximum temperature difference between cells goes through a process of firstly increasing and then decreasing. Figure 8 expresses that with the increase of channel height, the inlet and outlet pressure drop of coolant declines, the height of coolant increases from 2 mm to 3 mm, and the inlet and outlet pressure drop of coolant decreases 2214.503 pa. With the increase of the height of the channel, the decrease of the inlet and outlet pressure drop becomes smaller, and the curvature of the curve decreases with the increase of the height of the channel. Properly increasing the height of the flow channel can reduce not only the flow pressure loss, but also reduce the process difficulty.

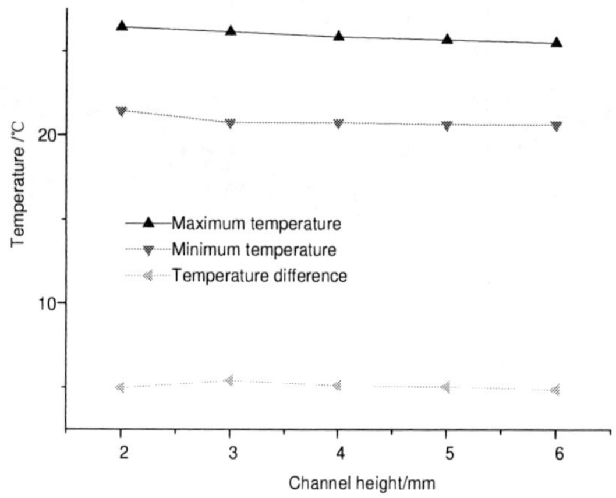

Figure 7. Variation of temperature of battery pack in different channel height.

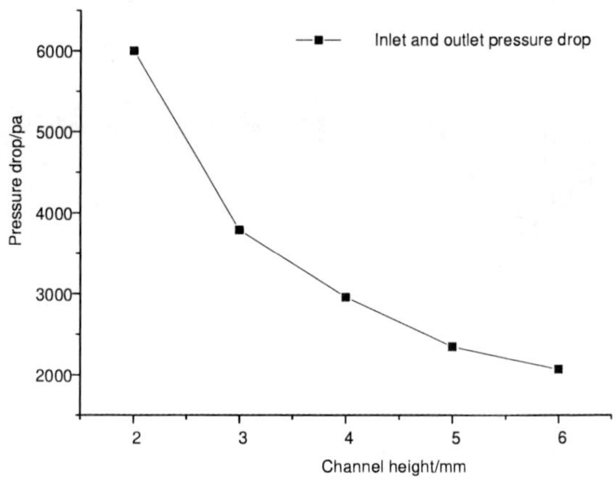

Figure 8. Variation of inlet and outlet pressure drop of coolant in different flow channel height.

5.2. Channel width

Exploring the influence of the flow channel width on the pressure loss, considering the analysis and study on the micro-channel, the width of the flow channel is selected to be 3 mm and the height of the flow channel range from 1 mm to 3 mm. The simulation conditions set are the same as Section 4.1.

Figure 9 depicts that the maximum and the minimum temperature of battery pack decreases with the increase of the width of the flow channel, and the maximum temperature difference also decreases. After the width of the flow channel is risen up to 1.5 mm, the temperature difference can make the battery pack obtained a good working performance. It can be seen from figure 10 shows increasing the width of the channel, the pressure drop of the coolant inlet and outlet decreases, the height of the channel increases from 1 mm to 1.5 mm, and the pressure drop of the inlet and outlet of the coolant decreases by 888.382 Pa. Continuing to increase the width of the flow channel, the pressure drop changing is smaller and smaller. Increasing the width of the channel can reduce the pressure drop, but at the same time, it also increases the thickness of the cold plate and the volume of the system.

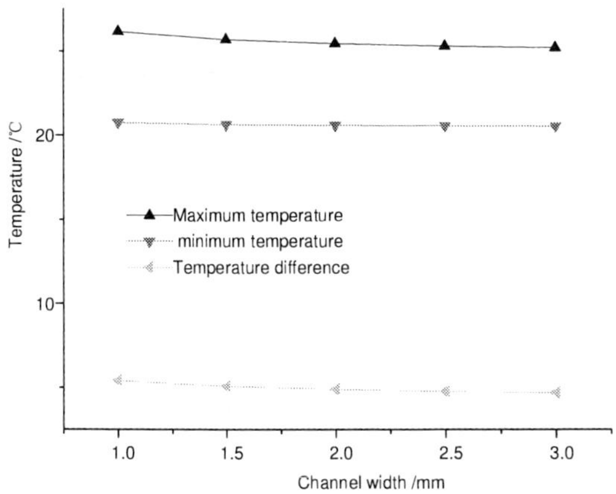

Figure 9. Variation of temperature of battery pack in different flow channel width.

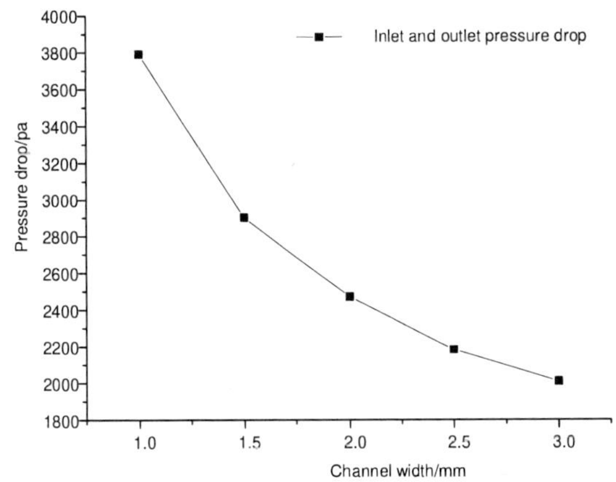

Figure 10. Variation of inlet and outlet pressure drop of coolant in different flow channel width.

6. Conclusions

Based on the research of heat dissipation of normal channel, a mini-channel cooling system applied in the power battery is designed and established in this paper, and the effects of system placement, ambient temperature, coolant inlet temperature and velocity on the temperature distribution of battery pack and the pressure drop at inlet and outlet of coolant are studied. Taking into account the pressure drop at the inlet and outlet of the coolant, the structure of the channel is optimized, and the influence of the width and the height of the channel can be studied. The following conclusions are obtained.

- Horizontal placement compared with vertical placement can make the battery pack obtain lower maximum temperature and higher uniformity, but the pressure drop at the inlet and outlet of coolant is relatively larger.
- The ambient temperature has little effect to the battery pack temperature. Even at 50 ℃ high temperature environment, the highest and the lowest temperature of the battery pack are only increased by 0.191 ℃ and 0.099 ℃ compared with the ambient temperature of 27 ℃.
- A linear ascending correlation existed on the temperature of battery pack and the coolant inlet temperature, but the temperature difference does not change obviously. The increase of inlet velocity of coolant can reduce the temperature of battery pack and improve the temperature uniformity among batteries, but at the same time, it also leads to a significant increase in the pressure drop at the inlet and outlet of coolant.
- With the increase of the width and height of the channel, the pressure drop at the inlet and outlet of the coolant decreases, and the value becomes smaller gradually. The increase of the height cannot increase the volume of the system, but can decline the manufacturing difficulty, and it is easier to be realized in practice.

References

[1] Liu X T, Chen Z H, Zhang C B and Wu J 2014 A novel temperature-compensated model for power Li-ion batteries with dual-particle-filter state of charge estimation *Appl. Energy* **123** 263-72

[2] Omar N, Monem M A, Firouz Y, Salminen J, Smekens J, Hegazy O *et al* 2014 Lithium iron phosphate based battery – assessment of the aging parameters and development of cycle life model *Appl. Energy* **113** 1575-85

[3] Yang N X, Zhang X W, Shang B B and Li G J 2016 Unbalanced discharging and aging due to temperature differences among the cells in a lithium-ion battery pack with parallel combination *J. Power Sources* **306** 733-41

[4] Chacko S and Chung Y M 2012 Thermal modelling of Li-ion polymer battery for electric vehicle

IOP Conf. Series: Materials Science and Engineering **657** (2019) 012051 doi:10.1088/1757-899X/657/1/012051

drive cycles *J. Power Sources* **213** 296-303

[5] Saw L H, Ye Y H, Tay A A O, Chong W T, Kuan S H and Yew M C 2016 Computational fluid dynamic and thermal analysis of Lithium-ion battery pack with air cooling *Appl. Energy* **177** 783-92

[6] Yang X H, Tan S C and Liu J 2016 Thermal management of Li-ion battery with liquid metal *Energy Convers. Manage.* **117** 577-85

[7] Shi S, Yu J Z, Xie Y Q *et al* 2017 Temperature rise characteristic of lithium battery integrated thermal management system combining phase change materials with air cooling *Journal of Beijing University of Aeronautics and Astronautics* **43** 1278-86

[8] Om N I, Zulkifli R and Gunnasegaran P 2018 Influence of the oblique fin arrangement on the fluid flow and thermal performance of liquid cold plate *Case Studies in Thermal Engineering* **12** 717-27

[9] Qian Z 2011 Heat transfer analysis of a chip cooler based on ANSYS *Cryo. & Supercond* **39** 44-6

[10] Zhang S, Liu X N, Yang Y J *et al* 2014 Numerical simulation of the flow boiling heat transfer based on micro — channel scale effect in cold — plate *Cryo. & Supercond* **2014** 55-8

[11] Kandlikar S G 2002 Fundamental issues related to flow boiling in minichannels and microchannels *Exp. Therm. Fluid Sci.* **26**(2-4) 389-407

Investigation of the 2.5D method in added resistance prediction of high speed trimaran

S M Wang[1] and W Y Duan[1,2]

[1]No. 145 Nantong Street, College of Shipbuilding Engineering, Harbin Engineering University, Harbin, China

E-mail: Wenyang_duan@126.com

Abstract. The numerical simulation of hydrodynamic performance of trimarans has attracted a significant attention recently. A series of added resistance model tests for a trimaran model at different speeds and regular waves are carried out. The variation trend of trimaran added resistance with wavelength to ship length ratio is analyzed for Fn=0.353 and Fn=0.471 under different outrigger layout. On these bases, the 2.5D method is applied to calculate the added resistance of the trimaran model in the experiment under various conditions and compare with the experimental results. By comparing with the experimental results, the scope of application of the 2.5D method in added resistance prediction is verified.

1. Introduction

As high-performance ships, trimarans have been widely applied in marine engineering nowadays. Since a trimaran usually consists of a main hull and two identical outriggers, it often has better rolling stability and lower wave resistance in some Froude numbers ranges. Also, the passenger accelerations of trimarans are usually smaller than that of catamarans in head-sea waves [1,2]. In recent years, the research on trimaran added resistance has been paid more and more attention under the background of the promotion of the concept of green ship. Trimaran added resistance refers to the difference between the resistance of a trimaran sailing in waves and that of a trimaran sailing at the same speed in calm water. At present, the research on the added resistance of trimaran is not mature enough compared to conventional mono-hull ships. This is because the configuration of the trimaran is quite unique (i.e. a main hull and two identical outriggers). Therefore, additional consideration should be given to hydrodynamic interference between the main hull and outriggers during the calculation of added resistance.

The earliest research on the ship added resistance can be traced back to 1942. Havelock *et al* [3] first proposed the method of obtaining heave force and pitching moment by integrating the wet surface pressure along the longitudinal direction to solve the added resistance of ships. The advantage of their work is that the calculation is relatively clear and simple, but the effect of diffraction force and viscosity is neglected which leads to inaccuracy of the results. Subsequently, Maruo [4] made up for the shortcomings of Havelock's work, considered the influence of incident, radiation and diffraction, and obtained the added resistance by using the distributed source method. Later, in 1972, Gerritsma and Beukelman [5] proposed the method of solving wave resistance enhancement by using radiation energy. The principle of this method is that the unsteady wave energy carried by the radiation waves generated by ship motions is considered to be obtained by the work done by added resistance. In this way, the average value of added resistance can be obtained by the radiation energy generated by ship moving in waves.

In the research work of ship added resistance in recent years, the solution methods of added resistance are more and more novel and varied. Representative examples include the Enhanced unified theory proposed by kashiwagi *et al* [6] in 2010 to solve the added resistance of ships. The significance of their work is to calculate the wave amplitude function required in the formula of added resistance proposed by Maruo, and discuss in detail the variation law of added resistance with traveling speed. Duan *et al* [7] also numerically predicted the hydrodynamic force and motion response of high-speed ships moving in regular waves by using the 2.5D theory. On this basis, a method for calculating added resistance of high-speed ships is recommended. In addition, Wan *et al* [8] also calculate and analyze the motion and added resistance of a Wigley III type model in head sea waves by the self-developed CFD solver (naoe-FOAM-SJTU). The numerical results show good agreement with the experimental results. In terms of the calculation of trimaran added resistance, Wu *et al* [9] analysis the influence of outrigger layout on trimaran's added resistance by CFD method. Although the numerical results agree well with the experimental result, the approach efficiency is not high enough. Therefore, this method is difficult to be used in trimaran outrigger layout analysis which requires large-scale computation. Besides, some major developments in the numerical modelling of wave propagation using the finite element method are also beneficial to the study of trimaran added resistance calculation [10,11].

In this paper, the model tests for a trimaran model at different speeds and regular waves are introduced. By measuring the resistance of the trimaran model in calm water and in waves at different speed, the variation law of trimaran added resistance with speed is analyzed. Based on the proposed model, the 2.5D method is applied to calculate the added resistance of the trimaran model according to the experimental conditions. By comparing the numerical results with the experimental results, the scope of application of the 2.5D method in added resistance prediction is verified.

2. Methodology

2.1. Trimaran added resistance model tests

In order to validate the application of the 2.5D method in trimaran added resistance prediction, the model tests of the trimaran under different speed and outrigger layout is performed in this paper. The model tests were conducted at Harbin Engineering University towing tank (108 m in length, 7 m in width and 3.5 m in depth). To measure the trimaran added resistance, a resistance sensor was set up on the trimaran model. The conditions of the trimaran model and the towing tank are shown in figure 1, the main parameters of this model are given in table 1.

Figure 1. Trimaran model and wave tank.

Table 1. Main parameters of the trimaran model.

Main feature	unit	value
Length overall	m	3.343
Breadth	m	0.641

IOP Conf. Series: Materials Science and Engineering **657** (2019) 012052 doi:10.1088/1757-899X/657/1/012052

Depth	m	0.249
Length waterline (main hull: L)	m	3.000
Beam waterline (main hull: B)	m	0.240
Drought (main hull)	m	0.122
Length waterline (side hull)	m	1.071
Beam waterline (side hull)	m	0.051
Drought (side hull)	m	0.043
Weight (whole ship)	kg	45.990

In the experiment we set 6 kinds of outrigger layout (see table 2), 2 transverse distance (*p/L*=0.096, 0.15) and 3 longitudinal distance (*a/L*=0.0, 0.26, 0.78), where a represents the longitudinal distance between main hull stern to outrigger stern and p represents the transverse distance between main hull center to outrigger center.

Table 2. Parameters of outrigger layout.

Serial number	*a/L*	*p/L*
L1T1	0.0	0.096
L1T2	0.0	0.15
L2T1	0.26	0.096
L2T2	0.26	0.15
L3T1	0.78	0.096
L3T2	0.78	0.15

The traveling speed we set in the experiment are $Fn = \frac{V}{\sqrt{gL}} = 0.353$ and 0.471 respectively. For the convenience of reference, the added resistance is dimensionless as $\frac{LR_a}{\rho g \varsigma^2 B^2}$ where R_a is the added resistance of the trimaran, ρ represent the density of water, g is gravity taking $g = 9.81 m/s^2$, ζ is the wave height, B is the breadth of the main hull. The added resistance of each condition is shown in the figures 2 and 3 (R_a is derived by the difference between resistance in waves and in calm water).

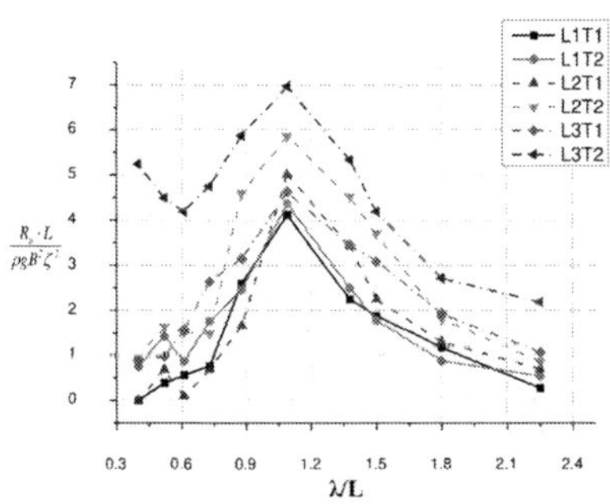

Figure 2. Added resistance of trimaran in different outrigger layout at *Fn*=0.353.

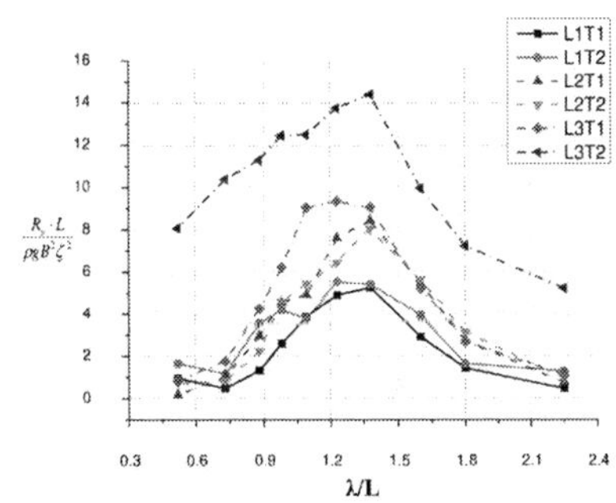

Figure 3. Added resistance of trimaran in different outrigger layout at *Fn*=0.471.

It can be obviously seen from figures 2 and 3 that L3T2 layout (*a/L*=0.26, *p/L*=0.15) has the highest value of added resistance in any traveling speed. For *Fn*=0.353, the wavelength to ship length ratio corresponding to the peak point of added resistance is around 1.1. While in the case of *Fn*=0.471, the wavelength to ship length ratio corresponding to the peak point of added resistance is around 1.3. In addition, for most cases, the added resistance will increase first then decrease with the increase of wave length to ship length ratio. To sum up, for both *Fn*=0.353 and *Fn*=0.471 and for most cases, it will lead to a smaller value of added resistance when the outriggers are placed near the stern of the main hull.

2.2. Theoretical calculation method of trimaran added resistance

High speed slender body potential flow theory is also named two dimensional and a half potential flow theory (2D+t theory or 2.5D theory). This method considers the free surface condition with forward speed and retains the assumption of two-dimensional flow field according to the geometry characteristics of slender body. It was first used by Faltinsen and Zhao [12] for the theoretical prediction of ships seakeeping performance and been widely applied since then. Based on the following assumptions, this method simplifies the calculation of ship motions. (1) Slender body assumption (the ship hull is assumed to be a slender body). (2) The steady wave-making potential is neglected. (3) There are no waves in front of the ship bow.

To predict added resistance of a trimaran, motions (including heave and pitch) of the trimaran must be accurately calculated. 2.5D method is adopted to calculate the vertical motion of the trimaran, the information of this method can be referred to reference [13] and will not be described here. According to the linear potential flow theory, the velocity potential of the flow field around the trimaran can be expressed as:

$$\Phi = \Phi_I + \Phi_B \tag{1}$$

In the frequency domain analysis, we have:

$$\Phi_I = Re(\phi_I e^{i\omega t}) \tag{2}$$

$$\Phi_B = Re[\phi_B e^{i\omega t}] \quad \phi_B = \textstyle\sum_{j=1}^{6} \eta_j \varphi_j + \varphi_D \tag{3}$$

$$\phi_I = \frac{iga}{\omega_0} e^{k_0 z} e^{-ik_0(x\cos\beta + y\sin\beta)} \tag{4}$$

where ϕ_I is the velocity potential of regular waves, ϕ_D is the diffraction potential. ϕ_j $(j = 1,2,\ldots 6)$ is the unit radiation potential. The steady drift force on a ship in waves can be expressed as a first order potential:

$$\vec{F}_m = -\frac{\rho}{2}[\iint_{S_B} (\Phi \frac{\partial}{\partial n} \nabla\Phi - \nabla\Phi \frac{\partial \Phi}{\partial n}) ds]_m \tag{5}$$

The velocity potential can be divided into the incident potential Φ_I and the disturbance potential Φ_B. It can be concluded that the added resistance of the ship sailing in waves can be expressed as five parts as follow:

$$R_a = F_3^1 + F_5^1 + F_3^2 + F_5^2 + F^3 \tag{6}$$

$$F_3^1 = Re\left[\frac{1}{2}\rho k_0 \eta_3 \int_L dx \int_{S_i} \omega_0 N_3 \phi_I{}^* dl\right] - Re\left[\frac{i}{2}\rho k_0 \eta_3 U \int_{S_1} N_3 \phi_I{}^* dl\right] \tag{7}$$

$$F_5^1 = Re\left[\frac{1}{2}\rho k_0 \eta_5 \int_L dx \int_{S_i} \omega_0 (-xN_3) \phi_I{}^* dl\right] - Re\left[\frac{i}{2}\rho k_0 \eta_5 U \int_{S_1} (-xN_3) \phi_I{}^* dl\right] \tag{8}$$

$$F_3^2 = Re\left[\frac{i}{2}\rho {k_0}^2 \eta_3 \int_L dx \int_{S_i} \phi_3 N_3 \phi_I{}^* dl\right] \tag{9}$$

$$F_5^2 = Re\left[\frac{i}{2}\rho {k_0}^2 \eta_5 \int_L dx \int_{S_i} \phi_5 N_3 \phi_I{}^* dl\right] \tag{10}$$

$$F^3 = Re\left[\frac{i}{2}\rho {k_0}^2 \int_L dx \int_{S_i} \phi_D N_3 \phi_I{}^* dl\right] \tag{11}$$

IOP Conf. Series: Materials Science and Engineering **657** (2019) 012052 doi:10.1088/1757-899X/657/1/012052

where S_i is the ith cross section (from the stern to the bow), N_3 is the vertical component of the normal vector in the transverse section. ϕ_3, ϕ_5 is the unit radiation potential of heave and pitch respectively. ϕ_D is the diffraction potential.

3. Results and discussion

For the convenience of reference, note that $R'_a = \frac{LR_a}{\rho g \varsigma^2 B^2}$ is the non-dimensional added resistance. The results obtained by experimental data (black symbols) are compared with the prediction results (red lines) and are shown in figures 4 and 5.

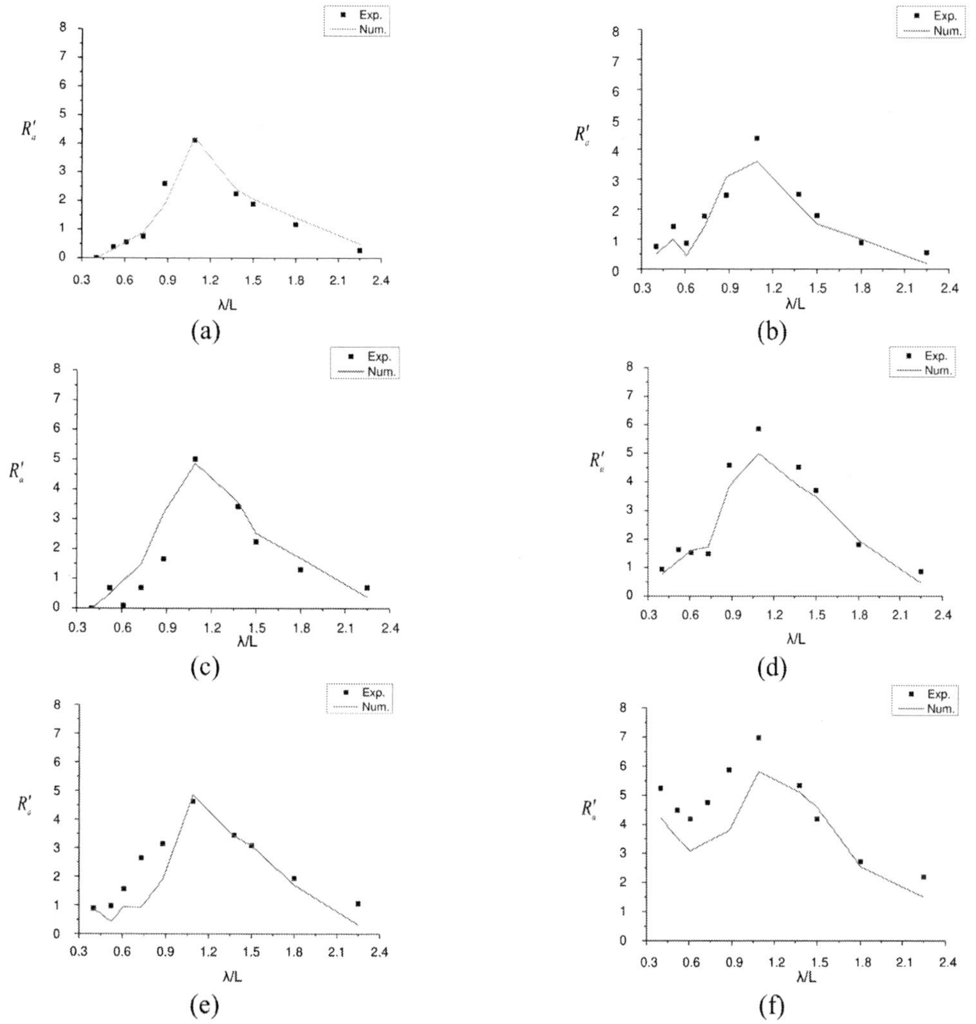

Figure 4. Added resistance of the trimaran at *Fn*=0.353. (a) *a/L*=0.0, *p/L*=0.096, (b) *a/L*=0.0, *p/L*=0.15, (c) *a/L*=0.26, *p/L*=0.096, (d) *a/L*=0.26, *p/L*=0.15, (e) *a/L*=0.78, *p/L*=0.096 and (f) *a/L*=0.78, *p/L*=0.15.

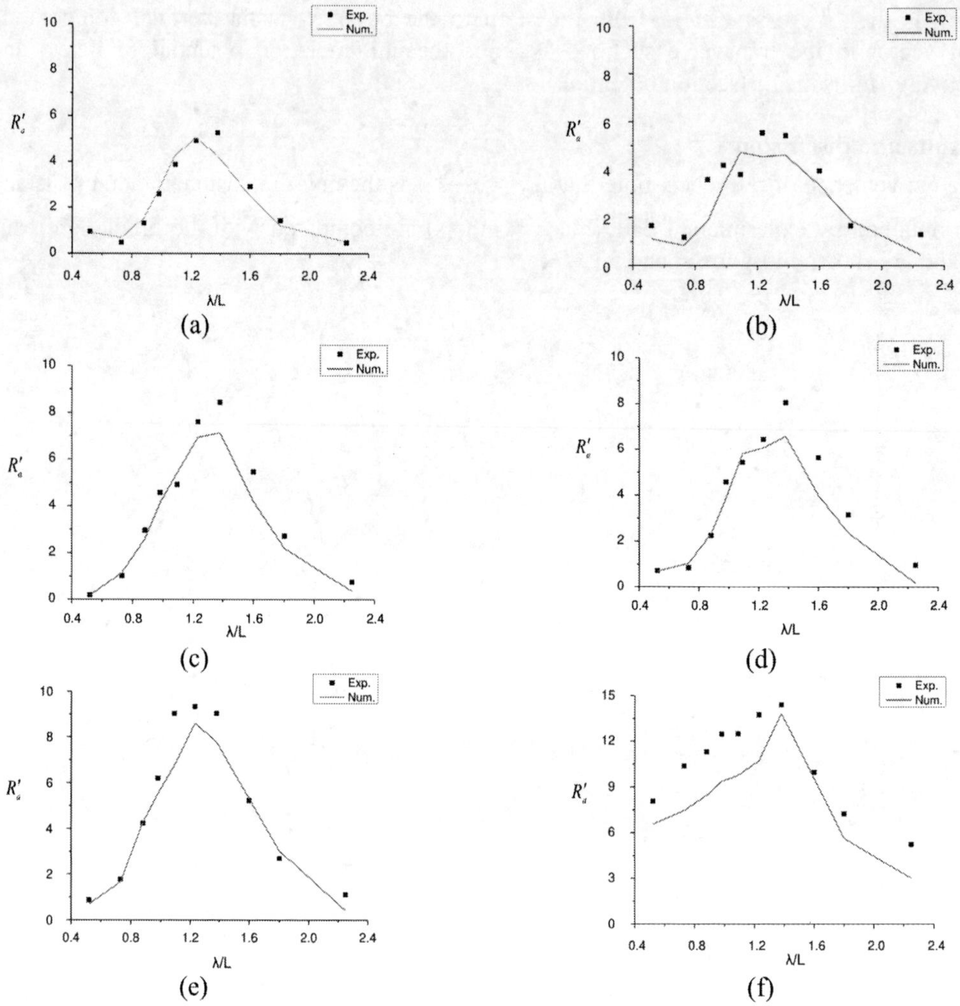

Figure 5. Added resistance of the trimaran at Fn=0.471. (a) a/L=0.0, p/L=0.096, (b) a/L=0.0, p/L=0.15, (c) a/L=0.26, p/L=0.096, (d) a/L=0.26, p/L=0.15, (e) a/L=0.78, p/L=0.096 and (f) a/L=0.78, p/L=0.15.

It can be observed that the results obtained by the proposed model are basically in good agreement with the test results, the average error of each group of results is distributed around 5% to 20%. Although the numerical prediction results in some individual conditions are far from the test results, the overall results show that the numerical prediction results can accurately reflect the variation tendency of added resistance curve.

For most conditions, the added resistance will increase first then decrease with the increase of wave length to ship length ratio. The wavelength to ship length ratio corresponding to the peak value of added resistance are mostly distributed between 1.0~1.2 for Fn=0.353 and 1.2~1.4 for Fn=0.471. With the increase of traveling speed, the wavelength to ship length ratio corresponding to the peak point of added resistance tends to increase. It should be noted that for both Fn=0.353 and Fn=0.471, the L3T2 layout (a/L=0.78, p/L=0.15) has the greatest value of added resistance. For L3T2 layout at Fn=0.471, the prediction results are quite different from the experimental results (figure 5(f)). This may due to the reason of slamming (the slamming phenomenon has been observed in the experiment).

4. Conclusion

This paper first introduced the added resistance model tests of a trimaran. Then a theoretical approach of trimaran added resistance prediction based on 2.5D method is described. Finally, the trimaran added resistance prediction method adopted in this paper is verified for *Fn*=0.353 and 0.471 by comparing the experimental results with the prediction results. The following conclusions can be obtained:

- In general, the prediction results are in good agreement with the experimental results. The predicted results can better reflect the trend of added resistance.
- In the case of the wavelength to ship length ratio more than 0.8, the added resistance will increase first and then decrease with the wavelength to ship length ratio for all outrigger layouts.
- The wavelength to ship length ratio corresponding to the peak value of added resistance are mostly distributed between 1.0~1.2 for *Fn*=0.353 and 1.2~1.4 for *Fn*=0.471.

Acknowledgment

This work was funded by the National Natural Science Foundation of China (projects of 51490671 and 51679043) to which the authors are most grateful.

References

[1] Davis M R and Holloway D S 2007 A comparison of the motions of trimarans, catamarans and monohulls *Aust. J. Mech. Eng.* **4** 183-95
[2] Lavroff J and Davis M R 2015 Slamming kinematics, impulse and energy transfer for wave-piercing catamarans *J. Ship. Res.* **59** 149-61
[3] Havelock T H 1942 The drifting force on a ship among waves *London Edinburgh Dublin Philos Mag. J. Sci.* **33** 467-75
[4] Hajime M 1957 The excess resistance of a ship in rough seas *Int. Shipbuild. Prog.* **4** 337-45
[5] Gerritsma J and Beukelman W 1972 Analysis of the resistance increase in waves of a fast cargo ship *Int. Shipbuild. Prog.* **19** 285-93
[6] Kashiwagi M, Ikeda T and Sasakawa T 2010 Effects of forward speed of a ship on added resistance in waves *Int. J. Offshore Polar* **20** 196-203
[7] Duan W Y, Wang R F, Ma W X and Ma S 2014 Numerical method investigation on added resistance of high speed ship in waves *Ship Mech.* **18** 1175-83 (in Chinese)
[8] Shen Z R, Ye H X and Wan D C 2012 Motion response and added resistance of ship in head waves based on RANS simulations *J. Hydrodyn. Ser. A* **27** 621-33 (in Chinese)
[9] Wu C S, Zhou D C, Gao L and Miao D C 2011 CFD computation of ship motions and added resistance for a high speed trimaran in regular head waves *Int. J. Nav. Arch. Ocean* **3** 105-10
[10] Jiang J, Mohamed M S, Seaid M and Li H 2018 Identifying the wavenumber for the inverse Helmholtz problem using an enriched finite element formulation *Comput. Method Appl. M.* **340** 615-29
[11] Drolia M, Mohamed M S, Laghrouche O, Seaid M and Trevelyan J 2017 Enriched finite elements for initial-value problem of transverse electromagnetic waves in time domain *Comput. Struct.* **182** 354-67
[12] Faltinsen O M and Umeda R Z 1991 The dynamics of ships numerical predictions of ship motions at high forward speed and discussion *Phil. Trans. Phys. Sci. Eng.* **334** 241-52
[13] Wang S M, Ma S and Duan W Y 2018 Seakeeping optimization of trimaran outrigger layout based on NSGA-II *Appl. Ocean Res.* **78** 110-22

IOP Conf. Series: Materials Science and Engineering **657** (2019) 012053 doi:10.1088/1757-899X/657/1/012053

Air compressor fault diagnosis based on lifting wavelet transform and probabilistic neural network

W S Yang[1,2], Y X Su[1] and Y P Chen[1]

[1]School of Automation, Wuhan University of Technology, Wuhan 430070, China

E-mail: 445926762@qq.com

Abstract. Due to the wide variety of faults, the components of air compressor fault condition vibration signal are complex. Fault features extracted by conventional time-frequency analysis and multi-resolution analysis are not comprehensive enough to reflect the fault condition of air compressor. To overcome this deficiency, this paper proposes an air compressor fault diagnosis algorithm based on lifting wavelet. Firstly, the vibration signal of the air compressor is decomposed by lifting wavelet; then the statistic such as the peak value and Kurtosis of the decomposition layer is calculated as the fault features. Finally, Probabilistic Neural Network is utilized to classify the fault state. In the experiment, the fault extraction methods such as Wavelet Packet Decomposition and Continuous Wavelet Transform are compared. The results indicate that the fault features extracted by lifting wavelet are more comprehensively to reflect the fault condition of air compressor while other fault character extraction methods need more fault characteristics; the fault diagnosis accuracy by utilizing proposed method is high and the training time is short, which is more suitable for the online fault diagnosis of the actual air compressor.

1. Introduction

In the air compressor health management system, fault diagnosis is generally performed by extracting fault characteristics from the air compressor vibration signal [1]. Therefore, the fault feature extraction in the vibration signal is the key to the high efficient fault diagnosis of the air compressor. Due to the wide variety of compressor failures and the rough working conditions, the vibration signal is doped with noise signal, stationary signal and non-stationary signal [2].

Conventional frequency analysis methods such as Discrete Fourier Transform and Fast Fourier transform can extract the signal frequency components and distribution. Since the air compressor fault vibration signal contains non-stationary signal, the fourier analysis has a poor resolution for the non-stationary signal with time-varying frequency, so the fault information extraction is not integrated [3].

The Wavelet Transform uses the wavelet function to convolve the signal based on the Fourier Transform, and can process the abrupt signal as well as the non-stationary signal. The Wavelet Transform essentially performs a series of correlation operations on each transform scale. When the signal characteristics are similar to those of the wavelet function, large wavelet coefficients will appear. Wavelet transform includes Continuous Wavelet Transform, Discrete Wavelet Transform and Wavelet Packet Transform [4]. When using Continuous Wavelet Transform to extract fault features, since many wavelet functions can be selected, and those can be evaluated by different verification criteria. Nishchal [3], Kalyan [4] used Shannon Entropy as the verification standard. In recent years, wavelet transform has been utilized as a popular method in the field of condition monitoring and fault diagnosis. Literature [16-18] utilized wavelet transform to extract air compressor vibration signal information for denoising and fault diagnosis.

Since the Continuous Wavelet Transform and the Discrete Wavelet Transform have poor resolution in the high frequency subband, if the mechanical vibration fault signal exists in the higher frequency subband, the use of wavelet transform will result in loss of fault features in some high frequency components. Wavelet Packet Transform can provide more detailed decomposition of high frequency subband, extract more medium and high frequency information signals, with better ability of localized time-frequency analysis [5].

The momentous difference between the Lifting Wavelet Transform and the classical wavelet transform is that the former does not use the Fourier transform, and all the operations are performed in the time domain, but the same time-frequency localization can be obtained [9]. Secondly, regardless of how the predictor and updater are chosen, the wavelet transform can always be fully reconstructed. The wavelet function and the scale function are determined by the predictor and the updater in lifting wavelet scheme. The selection of the predictor and updater is equivalent to the construction of wavelet functions in traditional wavelet analysis. Lifting wavelets have been used in gear and bearing fault diagnosis. Zhou [6] utilized the lifting wavelet to extract the fault features of the gearbox and gasoline engine. Jiao [7] utilized the lifting wavelet to extract the acoustic emission signal in the thermal leak diagnosis to determine the leakage frequency. Zhong [8] utilized the lifting wavelet to extract the turbopump fault. Compared with conventional wavelet analysis, for the same length data, the calculation of lifting wavelet is simpler, and in-place calculation is allowed, which consumes less memory and is more suitable for online processing of signals, adaptive and nonlinear transformation [11]. Lifting wavelet scheme has been applied in gear and bearing fault diagnosis [5-10], and less applied in air compressor fault diagnosis. To overcome the deficiency of difficulty in extracting fault features in air compressor fault diagnosis, lifting wavelet analysis combined with Probabilistic Neural Network was introduced to diagnose fault of the air compressor.

2. Lifting wavelet transform scheme

The Lifting Wavelet Transform consists of two processes of decomposition and reconstruction. The decomposition procedure includes three steps of split, prediction and update [9]; the reconstruction procedure is the inverse process of the decomposition.

Let $s = \{x(k), k \in Z\}$ be the data sequence. The decomposition procedure of the Lifting Wavelet Transform is as follows:

Split: The data sequence $s = \{x(k), k \in Z\}$ is decomposed into 2 sequences$x_o(k)$,$x_e(k)$ according to the equations (1) and (2):

$$x_o(k) = x(2k+1) \ \ k \in Z \tag{1}$$

$$x_e(k) = x(2k) \ \ k \in Z \tag{2}$$

Prediction: Using $x_e(k)$ to predict$x_o(k)$, the detail signal $d(k)$ is generated according to equation (3), where $P(\cdot)$ is a predictor.

$$d(k) = x_o(k) - P[x_e(k)] \ k \in Z \tag{3}$$

Update: The $x_e(k)$ is updated with the detail signal $d(k)$, and the approximation signal $c(k)$ is generated according to the equation (4),where $U(\cdot)$ is an updater.

$$c(k) = x_e(k) - U[d(k)] \ k \in Z \tag{4}$$

By iterating the above three steps, the detail signal and the approximation signal at different levels can be generated. Lifting wavelet reconstruction procedure is the inverse of the above process, and the reconstruction procedure is as shown in equation (5).

$$\begin{cases} x_e(k) = c(k) - U[d(k)] \ k \in Z \\ x_o(k) = d(d) + P[x_e(k)] \ k \in Z \end{cases} \tag{5}$$

IOP Conf. Series: Materials Science and Engineering **657** (2019) 012053 doi:10.1088/1757-899X/657/1/012053

From the decomposition of the lifting wavelet algorithm, it can be seen that the sequence is decomposed into a more compact representation, then different predictors and updaters are equivalent to select different filter banks in the wavelet decomposition. The lifting wavelet decomposition procedure are shown in figure 1. In this paper, the Interpolation Subdivision algorithm is used to construct the wavelet scale function and the wavelet function.

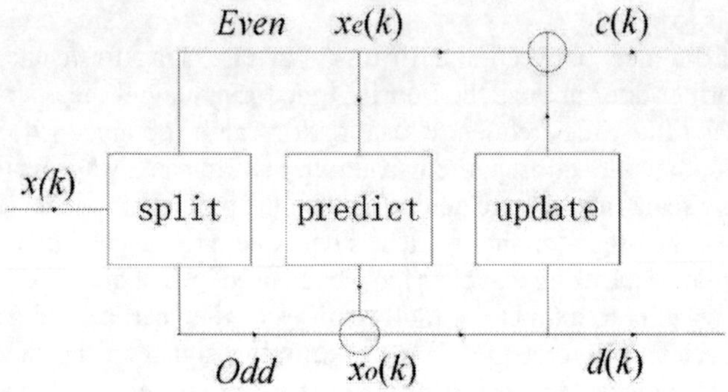

Figure 1. Decomposition procedure of lifting wavelet transform.

The Interpolation Subdivision algorithm constructs the wavelet scaling function as follows:

Step1: The detail coefficient is initialized to a 0 sequence, and the approximation coefficient is initialized to a δ sequence;

Step2: Reconstruct the signal by using equation (5);

Step3: Determine whether the wavelet scale function is smooth; if the scale function is not smooth, assign the signal value to the detail coefficient and get to Step2; otherwise the algorithm ends.

The Interpolation Subdivision algorithm constructs the wavelet function method similarly to the construction scale function method.

3. Probabilistic neural network

PNN (Probabilistic Neural Network) is a branch of a radial basis network and belong to a feedforward network, frequently used in classification and pattern recognition problems [13-15].

The theoretical basis of PNN is the Bayesian Minimum Risk Criterion (Bayesian Decision Theory). In the fault diagnosis classification, it can be described as: Let CA, CB be two types of failure modes. NA,NB is the number of samples for each failure mode, and N is the total number of samples. The fault feature samples are$X = (x_1, x_2, x_3, \cdots, x_n)$. The judgment is made using the equation (6).

$$\begin{aligned} & if\ h_A l_A f_A(X) > h_B l_B f_B(X),\ then\ X \in CA \\ & if\ h_A l_A f_A(X) < h_B l_B f_B(X),\ then\ X \in CB \end{aligned} \tag{6}$$

Where: h_A , h_B represents the prior probability of failure mode CA , CB , generally taking $h_A = N_A/N, h_B = N_B/N$;

l_A, l_B represents the cost factors of the misclassification; f_A, f_B represents the probability density functions, which can be generated by the Parzen method (7):

$$f_A = \frac{1}{(2\pi)^{P/2}\delta^P} \frac{1}{N_A} \sum_{i=1}^{N_a} exp\,[-\frac{(X - X_{ai})^T (X - X_{ai})}{2\delta^2}] \tag{7}$$

Where: X_{ai} represents the ith training vector of failure mode CA, and δ represents the smoothing parameter.

The PNN consists of an input layer, a mode layer, a summation layer and an output layer. The topology diagram is shown in figure 2.

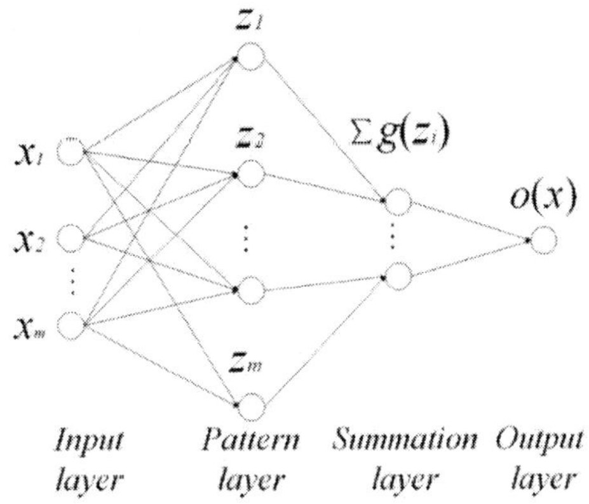

Figure 2. Probabilistic neural network structure.

The first layer is the Input layer. The number of Input layer neurons is consistent with the input sample dimension. The second layer is the Pattern layer. Each Mode layer node weights and sums the data transmitted from the Input layer, and passes those data to the Summation layer by Pattern layer through a nonlinear operation. The nonlinear operator uses equation (8):

$$g(z_j) = \exp((z_j - 1)/\delta^2) \tag{8}$$

The third layer is the Summation layer, and the data from the same type of Mode layer are accumulated to generate an estimated probability density function of the failure mode. The fourth layer is the Output layer, and the Output layer selects a neuron with the largest a posteriori probability density as the output of the entire system. Therefore, the PNN is equivalent to the Bayesian pattern classification method when the multivariate probability density function of the Gaussian kernel is used.

4. Proposed fault diagnosis scheme

The fault diagnosis scheme is divided into 3 steps.

The first step: pre-processing the acquired signal. The data set used in this paper is from [3], where the preprocessing process has been filtered, cropped, smoothed, and normalized. For details, please refer to literature [3].

The second step: fault feature extraction. In this paper, Daubechies-4 (db4) is used to perform 3-layer lifting wavelet decomposition on the pre-processed data. 5 statistics of peak value, mean, standard deviation, skewness and kurtosis are calculated in the third layer as a fault feature.

When the discrete Fourier transform is used, an 8-band signal energy spectrum is obtained as the fault feature.

When using continuous wavelet transform, the Morlet wavelet function is selected, and the signal is decomposed into 7 layers, and 5 statistics (same as lifting wavelet) in the 7th layer signal are taken as fault features.

When using the wavelet packet decomposition, the Daubechies-4 (db4) wavelet basis function is used to decompose the signal into 7 layers, and the energy of each node is calculated by the equation (9):

$$E_i = \sum_{n_i} x[n]^2 \tag{9}$$

In equation (9), $E_i$is energy at i^{th}node and n_i is the number of instances of i^{th}node.

A total of 254 node energies are obtained (excluding the first layer nodes). Therefore, 254 eigenvalues can be obtained by decomposing the signal using the wavelet packet decomposition.

The third step: classification of fault features, this paper uses Probabilistic Neural Networks to classify faults. 5 fault features are generated from lifting wavelet extraction. 8 fault features are generated from discrete Fourier transform, 5 fault features are generated from continuous wavelet transform, and 254 from wavelet packet transform.

5. Result and conclusion

The experimental data is divided into 8 categories, which include a healthy state, and 7 faulty states: LIV fault(LIV), leakage outlet valve (LOV) fault, non-return valve (NRV) fault, piston ring fault(Piston), flywheel fault (Flywheel),rider belt fault (Riderbelt)and bearing fault. Every category data set includes 225 24 Bit PCM format files at 50 kHz sampling rate.

The diagnosis accuracies is shown in table 1, and the time for training neural network is shown in table 2. Table 1 indicates that the fault diagnosis accuracy using the discrete Fourier analysis is the lowest (all 8 state), and the proposed algorithm has the highest accuracy. Among 8 fault conditions, Piston and Flywheel fault have the lowest accuracy. By utilizing Fast Fourier analysis for those 2 fault conditions, we can analysis their frequency distribution. Figure 3 is the raw signal frequency spectrum of these two faults. Figure 3 indicates that the vibration signals of both Piston and Flywheel faults contain 890 Hz and 2100 Hz frequencies. Therefore, it is difficult to distinguish the two types of faults by using the Discrete Fourier transform. Figure 4 is the 3rd decomposition layer frequency spectrum by using lifting wavelet transform. Figure 4 indicates that the lifting wavelet analysis can more fully extract the original signal fault features.

Table 1. Fault diagnosis accuracies.

Extraction Method	DFT	CWT	WPT	Lift Wavelet
Healthy	86%	92%	92%	94%
LOV	86%	94%	94%	96%
LIV	74%	96%	96%	97%
Piston	32%	72%	90%	89%
Flywheel	37%	80%	87%	90%
Bearing	77%	88%	88%	84%
NRV	75%	87%	87%	90%
Riderbelt	71%	79%	85%	92%

Table 2. PNN training time.

Extraction Method	Time (sec)
DFT	0.056
CWT	0.054
WPT	2.101
LiftWavelet	0.054

In order to further compare the extraction fault features ability between wavelet analysis and the lifting wavelet, the Wavelet Packet in the experiment selects 254 fault features, while the lifting wavelet selects only 5 fault features. From table 1, the fault diagnosis accuracy by using Wavelet Packet decomposition with 254 fault features can be close to the proposed algorithm. However, as can be seen from table 2, due to the large number of fault features, the training time increases accordingly.

Therefore, from the experiment, the proposed algorithm can have a higher fault diagnosis rate when selecting fewer fault features, so it is more suitable for online fault diagnosis.

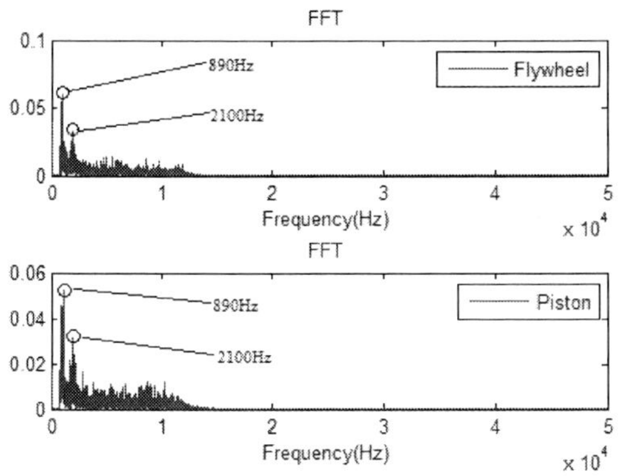

Figure 3. The raw signal frequency spectrum graph.

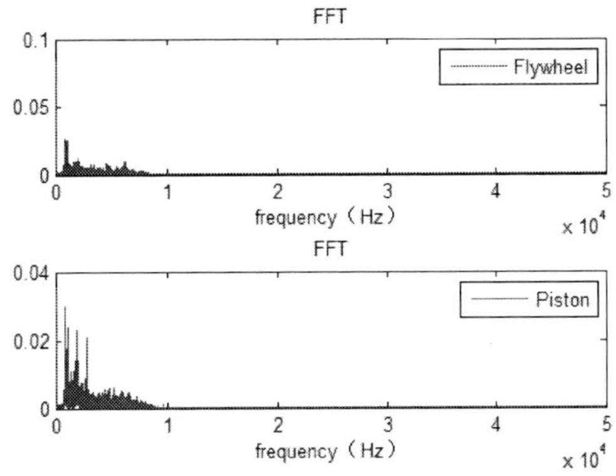

Figure 4. The frequency spectrum graph by utilizing Lifting wavelet transform.

References

[1] Goumas S K, Zervakis M E and Stavrakakis G S 2002 Classification of washing machines vibration signals using discrete wavelet analysis for feature extraction *IEEE T. Instrum. Meas.* **51** 497-508

[2] Mujun 2011 Fault diagnosis of air compressor based on RBF neural network *Int Conf Mechatronic Science* (Jilin, China)

[3] Verma N K, Sevakula R K, Dixit S and Salour A 2016 Intelligent condition based monitoring using acoustic signals for air compressors *IEEE T. Reliab.* **65** 291-309

[4] Bhavaraju K M, Kankar P K, Sharma S C and Harsha S P 2010 A comparative study on bearings faults classification by artificial neural networks and self-organizing maps using wavelets *Int. J. Eng. Sci. Tech.* **2** 1001-8

[5] Khan M A, Radwan S K and Rahman M A 2007 Real-time implementation of wavelet packet transform-based diagnosis and protection of three-phase induction motors *IEEE T. Energy Conver.* **22** 647-55

[6] Zhou R, Bao W, Li N, Huang X and Yu D 2010 Mechanical equipment fault diagnosis based on redundant second generation wavelet packet transform *Digit. Signal Process.* **20** 276-88

[7] Jiao Y L, Shi H and Wang X H 2016 Lifting wavelet denoising algorithm for acoustic emission signal *Int. Conf. Robots Intelligent Syst.* (Zhangjiajie, China)

[8] Zhong F, Li H, Wu Q and Hong T 2013 A fault detection algorithm for turbopump based on lifting wavelet and LMS *IEEE Int. Conf. Mecha. Automat.* (Takamatsu, Japan)

[9] Sweldens W 1996 The lifting scheme: a custom-design construction of biorthogonal wavelets *Appl. Comput. Harmonic Anal.* **3** 186-200

[10] Cusido R R O and Garcia 2008 Fault detection in induction machines by using power spectral density on the wavelet decompositions *IEEE T. Ind. Electron.* **55** 633-43

[11] Yang Z, Cai L, Gao L and Wang H 2012 Adaptive redundant lifting wavelet transform based on fitting for fault feature extraction of roller bearings *Sensors* **12** 4381-98

[12] Sweldens W 1998 The lifting scheme: A construction of second generation wavelets *SIAM J. Math. Anal.* **29** 511-46.

[13] Paydarnia H, Hajiaghasi S and Abbaszadeh K 2014 Improved structure of pnn using pca in transformer fault diagnostic *Arab. J. Sci. Eng.* **39** 4845-51

[14] Roy N and Bhattacharya K 2016 Signal analysis-based fault classification and estimation of fault location of an unbalanced network using s-transform and neural network *IEEJ T. Electrical and Electr. En.* **11** 401-9

[15] Zhang H Q, Zhao Y and Lian Y F 2012 Adaptive Probabilistic Neural Network fault-diagnosis algorithm for sins/gps integrated navigation system *J. Chinese Inert. Tech.* **20** 749-53

[16] Gong Y J, Fu Z, Xiang H Y, Zhang L and Meng C L 2010 Denoising vibration signals of food

refrigerant air compressor based on wavelet transform *Key Eng. Mater.* **440** 1037-41
[17] Verma N K, Gupta R, Sevakula R K and Salour A 2015 Signal transforms for feature extraction from vibration signal for air compressor monitoring *Tencon IEEE Region 10 Conf.* (Bangkok, Thailand)
[18] Weng P Y and Liu M K 2017 Roller bearing fault diagnosis based on wavelet packet decomposition and support vector machine *Int Conf Appl Syst Innovation* (Sapporo, Japan)

IOP Conf. Series: Materials Science and Engineering **657** (2019) 012054 doi:10.1088/1757-899X/657/1/012054

Investigations on the thermal behavior and associated thermal stresses for short pulse laser heating

D W Xue[1], J B Wang[1] and G Y Xu[1,2]

[1]School of Port and Transportation Engineering, Zhejiang Ocean University, Zhoushan 316022, Zhejiang, China

E-mail: xugy@zjou.edu.cn

Abstract. The fractional type Cattaneo heat conduction equation and the associated thermal stress equation are established for studying the thermal and related stress behavior of the short pulse laser heating. With the Laplace transform method, the analytical solution of the temperature distribution and thermal stress field are derived. Compared to the classical Fourier heat transfer model and the standard C-V one, the results of the present model show the thermal diffusion feature as well as the thermal wave behaviour, which is more realistic. Furthermore, the influences of the fractional order parameter on the thermal and related stress behavior are discussed. The thermal velocity, thermal variation rate and the peak value of the temperature and the thermal stress depend on the fractional order parameter. The fractional type Cattaneo heat model and associated thermal stress exhibit the non-local nature.

1. Introduction

Laser processing technology is in widespread application in manufacturing and modern medicine. In spite of many advantages, the non-uniform temperature field inside the material during laser processing will introduce extremely large temperature gradient, which leads to the generation of a great thermal stress. The clear understanding of the heat transfer process is of great importance to reveal the physical mechanism of the interaction between the pulse laser irradiation and the material.

The classical heat model, which is widely accepted, is based on the Fourier's law. However, evidences [1,2] show that the Fourier heat conduction model fails to characterize the cases such as the short pulse laser heating because it implies an unphysically infinite thermal propagation velocity. In fact, the short pulse laser refers to the laser with very short pulse duration time, generally on the order of nanoseconds or even femtoseconds and meanwhile with high-intensity energy power of heat flux. In such cases, the short laser beam will generate a great heat gradient and high heating rate. Therefore, as mentioned in the previous researches [3,4], the Fourier heat conduction model is a standard diffusion model of equilibrium thermodynamics and is not valid to describe the related problems.

Since the thermal wave effects have been observed in experiments by Peshkov [5], it has been widely accepted that the thermal signal transfers at a finite speed like waves in the case of extreme low temperature and short pulse laser heating, etc. Almost at the same time, modifications of Fourier model have been made to describe such transient heat transfer process. Among the various modifications, the most widely accepted one is the Cattaneo-Vernotte (C-V) model [6,7] which considers the lagging effects. Lots of investigations [8-10] have proved the wave nature of the heat behavior and shown the advantage of C-V model, because the lagging phenomenon and the faster thermal rising rate, which are different from the nature of Fourier model, are always existing [11].

However, there are also some investigations show the experimental temperature distribution deviates from that predicted by C-V model and indicate the limitation of C-V model [12-15]. Besides, other researchers have shown that the C-V model may introduce other non-physical reality, such as the local temperature lower than absolute zero [16], the violation to the second law of thermodynamics [17].

As a matter of fact, the above physical process described by the C-V equation is a typical anomalous diffusion process. From the macroscopic aspect, the motion of the large number of particles in anomalous diffusion process is not in accordance with the standard statistical distribution. And the mean square displacement of the particles is not in accordance with the standard linear behavior, and as is known, the mean square displacement of the particles takes the form $\langle x^2(t)\rangle \propto t^p$, where p refers to the anomalous diffusion exponent. For $p=1$, it is the normal diffusion, and $p\neq 1$ corresponds to the anomalous diffusion.

Recently, as pointed out in [18-20], there is an inherent relation between the fractional differential operator and the anomalous diffusion. As is known, the integer order differential operator is a local operator while the fractional order one is non-local [21]. This means that the next state of a system depends on all of its historical states rather than its current state, which is obviously more realistic and it is one of the most important reasons why the fractional calculus has the advantage to describe the phenomena mentioned above [21,22]. Lots of different physical phenomena have proved that the fractional order calculus is efficient in modeling the anomalous process, and it has shown [23-26] that the fractional differential equation overcomes the defect induced by the traditional Fourier equation and the C-V equation when described the anomalous diffusion process. So far, many investigations have been focused on the study of non-equilibrium heating process, but most of the investigations are concerning on the thermal response and the temperature field, there are only a few investigations concerning on the associated thermal stresses, including the outstanding work of Povstenko [27-30], Youssef [31], and Sherief [32]. For the short pulse laser heating process, there are little investigation on both the thermal propagation and temperature field as well as the associated thermal stresses with the fractional type Cattaneo heat equation.

Consequently, in the present study, the heat transfer process and the associated thermal stresses due to the interaction between the pulse laser irradiation are investigated. The fractional type Cattaneo heat equation and the thermal stress equation are derived first. With the Laplace transform method, the analytical solutions of above equations are then obtained. At last, an example is given to study the effects of the fractional parameters on the transient temperature and thermal stress field and the heat conduction mechanism during the anomalous diffusion process is discussed further.

2. Mathematical model

2.1. Fractional type Cattaneo heat transfer equation

Considering the metal material irradiating by a short pulse laser, the extensively used model is the well-known standard C-V model with the constitutive relation expressed as

$$q(r,t)+\tau\frac{\partial q}{\partial t}=-k\nabla T \tag{1}$$

The standard C-V model is in the form of integer order differential operator, and using the fractional Taylor series expansion, the fractional form of equation (1) is obtained

$$q(r,t)+\frac{\tau^p}{\Gamma(p+1)}\frac{\partial^p q}{\partial t^p}=-k\nabla T\ ,0<p<1 \tag{2}$$

Where p refers to the fractional order of differentiation, $\partial/\partial t^p$ stands for the Caputo's operator of fractional derivative.

The energy conservation equation is

$$\rho c_p \frac{\partial T}{\partial t} = -\nabla \cdot q + Q \tag{3}$$

One can get the fractional type Cattaneo heat transfer equation with equations (2) and (3) as:

$$\frac{\partial T}{\partial t} + \frac{\tau^p}{\Gamma(p+1)} \frac{\partial^{p+1} T}{\partial t^{p+1}} = a\nabla^2 T + \frac{1}{\rho c_p}(1 + \frac{\tau^p}{\Gamma(p+1)} \frac{\partial^p}{\partial t^p})Q, 0 \le p \le 1 \tag{4}$$

Where ρ is the material density and T is the temperature, c_p is the specific heat capacity. $a = k/(\rho c_p)$ stands for the thermal conductivity. Q is the heat resource. ∇^2 is the Laplace operator.

It is noted that as $\tau = 0$, equation (4) is changing in the form of the classical Fourier heat transfer equation and in the form of the standard C-V heat transfer equation with integer order as $p = 1$.

For the heating source of the short pulse laser, the non-Gaussian type laser with the following temporal profile is given:

$$I(t) = I_0 f(t) \tag{5}$$

where I_0 is the amplitude of laser peak power density, and $f(t)$ is expressed as:

$$f(t) = \exp(-\beta t) - \exp(-\mu t) \tag{6}$$

And it denotes the pulse laser intensity in which β and μ denote the laser pulse rise-time and the laser pulse fall-time parameter. The temporal variations of *f*(*t*) corresponding to different parameters of β and μ, and in the present paper, the parameters are chosen as $\beta = 0.5$, $\mu = 2.5$, shown in Figure.1, which is referred to [33,34].

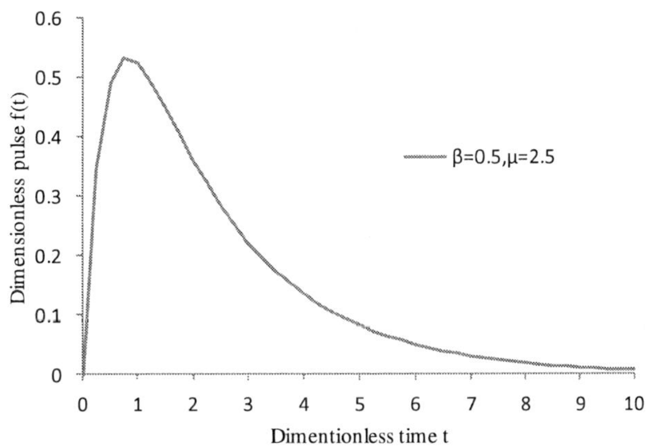

Figure 1. Dimensionless temporal distribution of the laser pulse intensity $f(t)$ with $\beta = 0.5$, $\mu = 2.5$.

Considering the surface reflection and absorption, the internal heat source is

$$Q = (1 - r_f) I(t) \delta \exp(-\delta x) \tag{7}$$

Where r_f and δ stand for the reflection coefficient, and the absorption coefficient.

Substituting the heat source (equation (7)) into the heat transfer equation (equation (4)), the one-dimensional fractional type Cattaneo heat equation is obtained.

$$\frac{1}{a}\left(\frac{\partial T}{\partial t}+\frac{\tau^p}{\Gamma(1+p)}\frac{\partial^{p+1}T}{\partial t^{p+1}}\right)=\frac{\partial^2 T}{\partial x^2}+\frac{\delta I_1 e^{-\delta x}}{k}\left[f(t)+\frac{\tau^p}{\Gamma(1+p)}\frac{d^p f}{dt^p}\right] \tag{8}$$

Where $I_1=(1-r_f)I_0$

The initial condition is

$$T(x,t)=T_0,\frac{\partial}{\partial t}T(x,t)=0,x>0,t=0 \tag{9}$$

Correspondingly, the boundary conditions are,

$$\frac{\partial}{\partial x}T(x,t)=0,x=0,t>0 \tag{10}$$

$$T(x,t)=T_0,x\to\infty,t=0 \tag{11}$$

2.2. Thermal stresses equation

The components of displacement for the half space with the traction free surface is as follows:

$$u_x=u(x,t),u_y=u_z=0 \tag{12}$$

Taking count into the stress-strain relation,

$$\varepsilon_{xx}=\partial u/\partial x,\varepsilon_{yy}=\varepsilon_{zz}=\varepsilon_{xy}=\varepsilon_{xz}=\varepsilon_{zy}=0 \tag{13}$$

One can get

$$\sigma_{xx}=(\lambda+2\kappa)\partial u/\partial x-(3\lambda+2\kappa)\alpha_T(T-T_0) \tag{14}$$

$$\sigma_{yy}=\lambda\,\partial u/\partial x-(3\lambda+2\kappa)\alpha_T(T-T_0) \tag{15}$$

Where $\lambda=\frac{E\nu}{(1+\nu)(1-2\nu)}$ is the Lame coefficient, E the elastic modulus, $\kappa=\frac{E}{2(1+\nu)}$ is the shear elastic modulus, ν is the Poisson ratio. The differential equation of one-dimensional thermal elastic motion is expressed as

$$\frac{\partial\sigma_{xx}}{\partial x}=\rho\frac{\partial^2 u}{\partial t^2} \tag{16}$$

Combining equations (14) and (16) with eliminating displacement u yields the differential equation of thermal stress

$$\frac{\partial^2\sigma_{xx}}{\partial x^2}-\frac{1}{V_e^2}\frac{\partial^2\sigma_{xx}}{\partial t^2}=\frac{1+\nu}{1-\nu}\rho\alpha_T\frac{\partial^2 T}{\partial t^2} \tag{17}$$

where $V_e=\sqrt{(\lambda+2\mu)/\rho}$ is the dilatation wave velocity.

The initial and boundary conditions are as follows

$$\sigma_{xx}=0,\ \frac{\partial\sigma_{xx}}{\partial t}=0,\ t=0,\ x>0 \tag{18}$$

$$\sigma_{xx} = 0\,,\ x = 0\,,\ t > 0 \tag{19}$$

$$\sigma_{xx} = 0\,,\ x \to \infty\,,\ t > 0 \tag{20}$$

3. Analytical solutions

For simplicity, the following non-dimensional quantities are introduced

$$T^* = k\delta(T - T_0)/I_1\,,t^* = a\delta^2 t\,,x^* = \delta x\,,\sigma_{xx}{}^* = \frac{1-\nu}{1+\nu}\frac{k\sigma_{xx}}{a^2\delta\rho\alpha_T I_1}\,,\tau^* = \alpha\delta^2\tau\,,(\beta,\mu)^* = (\beta,\mu)/\alpha\delta^2$$

3.1. Solution of the temperature field

Equations (8)-(11) will change into the following forms after dimensionalization (dropping the superscript for convenience).

$$\frac{\partial T}{\partial t} + \frac{\tau^p}{\Gamma(p+1)}\frac{\partial^{p+1}T}{\partial t^{p+1}} = \frac{\partial^2 T}{\partial x^2} + e^{-x}(1 + \frac{\tau^p}{\Gamma(p+1)}\frac{\partial^p}{\partial t^p})f(t) \tag{21}$$

$$T = 0\,,\ \frac{\partial T}{\partial t} = 0\,,\ x > 0\,,\ t = 0 \tag{22}$$

$$\frac{\partial T}{\partial x} = 0\,,\ x = 0\,,\ t > 0 \tag{23}$$

$$T = 0\,,\ x \to \infty\,, t > 0 \tag{24}$$

Equations (21)-(24) will change into the following form after the application of Laplace transform

$$\frac{d^2\overline{T}}{dx^2} - \left(1 + \tau^p s^p/\Gamma(1+p)\right)s\overline{T} = -\exp(-x)\left(1 + \tau^p s^p/\Gamma(1+p)\right)F(s) \tag{25}$$

$$\frac{d\overline{T}}{\partial x} = 0 \text{ at } x = 0 \tag{26}$$

$$\overline{T} = 0 \text{ at } x \to \infty \tag{27}$$

Where $\overline{T} = \int_0^\infty \exp(-st)T(x,t)dt\,, F(s) = \dfrac{1}{s+\beta} - \dfrac{1}{s+\mu}$

Laplace transformed solution of equation (27) combining equations (28) and (29) can be expressed

$$\overline{T}(x,s) = \overline{\phi}(x,s)F(s) \tag{28}$$

Where

$$\overline{\phi}(x,s) = \phi_1(s)[\exp(-x) - \phi_2(x,s)] \tag{29}$$

$$\overline{\phi}_1(s) = \frac{\left(1 + \tau^p s^p/\Gamma(1+p)\right)}{s\left(1 + \tau^p s^p/\Gamma(1+p)\right) - 1} \tag{30}$$

$$\overline{\phi_2}(x,s)=\frac{\exp\left(-x\sqrt{s\left(1+\tau^p s^p/\Gamma(1+p)\right)}\right)}{\sqrt{s\left(1+\tau^p s^p/\Gamma(1+p)\right)}} \tag{31}$$

The analytical solution of the equation (28) is

$$T(x,t)=\int_0^t \phi(x,\xi)f(t-\xi)d\xi \tag{32}$$

In which,

$$\phi(x,t)=\phi_1(t)\exp(-x)-\int_0^t \phi_1(\xi)\phi_2(x,t-\xi)d\xi \tag{33}$$

$$\phi_1(t)=\sum_{n=0}^{\infty}\frac{1}{n!}\left[\begin{array}{l}\tau^{-np}\Gamma^n(1+p)t^{n(1+p)}E_{p,n+1}^{(n)}\left(-\tau^{-p}\Gamma(1+p)t^p\right)+\\ \sum_{n=0}^{\infty}\tau^{-(n+1)p}\Gamma^{n+1}(1+p)t^{n(1+p)+p}E_{p,n+1+p}^{(n)}\left(-\tau^{-p}\Gamma(1+p)t^p\right)\end{array}\right] \tag{34}$$

According to inverse Laplace transform formula in [24]

$$\begin{aligned}\phi_2(x,t)&=L^{-1}\left[\frac{\exp\left(-x\sqrt{s\left(1+\tau^p s^p/\Gamma(1+p)\right)}\right)}{\sqrt{s\left(1+\tau^p s^p/\Gamma(1+p)\right)}}\right]=\\ &\frac{1}{\sqrt{\pi t}}\sum_{n=1}^{\infty}\Gamma^{n+1/2}(1+p)\frac{(-1)^n}{n!}\frac{t^{p(n+1/2)}}{\tau^{p(n+1/2)}}H_{0,2}^{2,0}\left[\frac{x^p\tau^p}{4t^{1+p}\Gamma(1+p)}\Big|_{(0,1),(n+1/2,1)}\right]\end{aligned} \tag{35}$$

where $E_{\alpha,\beta}^{(n)}(z)$ is two parameters Mittag-Leffler function [21]. The fractional derivative of $E_{\alpha,\beta}^{(n)}(z)$ is defined as

$$E_{\alpha,\beta}^{(n)}(z)\equiv\frac{d^n}{dz^n}E_{\alpha,\beta}(z)=\sum_{j=0}^{\infty}\frac{(j+n)!z^j}{j!\Gamma(\alpha j+\alpha n+\beta)} \tag{36}$$

It is noted that one can get the analytical solution of the standard C-V heat equation as $p=1$, and similarly, the analytical solution of the classical Fourier heat equation as $\tau=0$.

3.2. *Solution of the thermal stress field*

By introducing the above non-dimensional quantities into equations (17)-(20) and yields

$$\frac{\partial^2\sigma_{xx}}{\partial x^2}-\Lambda^2\frac{\partial^2\sigma_{xx}}{\partial t^2}=\frac{\partial^2 T}{\partial t^2} \tag{37}$$

$$\sigma_{xx}=0,\ \frac{\partial\sigma_{xx}}{\partial t}=0,\ x>0,\ t=0 \tag{38}$$

$$\sigma_{xx}=0,\ x\to\infty \text{ and } x=0,\ t>0 \tag{39}$$

where the dimensionless parameter $\Lambda=a\delta/V_e$.

Applying the Laplace transform to equations (37) and (39) yields

$$\frac{d^2\overline{\sigma}_{xx}}{dx^2} - \Lambda^2 s^2 \overline{\sigma}_{xx} = s^2 \overline{T} \tag{40}$$

$$\overline{\sigma}_{xx} = 0 \text{ at } x = 0 \text{ and } x \to \infty \tag{41}$$

where $\overline{\sigma}_{xx} = \int_0^\infty \exp(-st)\sigma_{xx}(x,t)dt$.

Solving equations (40) and (41) leads to the Laplace transformed solution

$$\overline{\sigma}_{xx} = \overline{\Sigma}_{xx} F(s) \tag{42}$$

Where

$$\overline{\Sigma}_{xx} = \overline{\phi}_1(s)\left[\overline{A}_1 \exp(-\Lambda s x) + \overline{A}_2 \exp(-x) + \overline{A}_3 \overline{\phi}_2(x,s)\right] \tag{43}$$

$$\overline{A}_1(s) = \frac{s^2}{\Lambda^2 s^2 - 1} + \frac{s}{\left(1 + \tau^p s^p / \Gamma(1+p)\right) - \Lambda^2 s} \frac{1}{\sqrt{s\left(1 + \tau^p s^p / \Gamma(1+p)\right)}} \tag{44}$$

$$\overline{A}_2(s) = -\frac{s^2}{\Lambda^2 s^2 - 1}, \overline{A}_3(s) = -\frac{s}{1 + \tau^p s^p / \Gamma(1+p) - \Lambda^2 s} \tag{45}$$

The analytical solution of dimensionless thermal stress is obtained

$$\sigma_{xx} = \int_0^t \Sigma_{xx}(x,\xi) f(t-\xi) d\xi \tag{46}$$

In which

$$\Sigma_{xx}(x,t) = \Sigma_{xx1}(x,t) + \Sigma_{xx2}(x,t) + \Sigma_{xx3}(x,t) \tag{47}$$

$$\Sigma_{xx1} = \frac{1}{\mathrm{A}^2}\left[\phi_1(t - \mathrm{A}x) + \frac{1}{\mathrm{A}}\int_0^{t-\mathrm{A}x} \phi_1(\xi)\sinh\left(\frac{t-\xi}{\mathrm{A}}\right)d\xi - \int_0^t \phi_1(t-\xi)\phi_6(\xi)\right] \tag{48}$$

$$\Sigma_{xx2} = -\frac{e^{-x}}{\mathrm{A}^2}\left[\phi_1(t) + \frac{1}{\mathrm{A}}\int_0^t \phi_1(\xi)\sinh\left(\frac{t-\xi}{\mathrm{A}}\right)d\xi\right] \tag{49}$$

$$\Sigma_{xx3} = -\frac{1}{\mathrm{A}^2}\left[\int_0^t \phi_1(t-\tau)\phi_2(x,\tau)d\tau + \frac{1}{\mathrm{A}^2}\int_0^t \phi_3(\tau)\phi_2(x,\tau)d\tau\right] \tag{50}$$

$$\phi_3(t) = \int_0^t \phi_4(t-\tau)\phi_1(\tau)d\tau \tag{51}$$

$$\phi_4(t) = \frac{\mathrm{A}^{-2m}\tau^{pm}}{m!} t^{m-pm}\left[\tau^p t^{-p} E_{1,1-p(m+1)}^{(m)}\left(\mathrm{A}^{-2}t\right) + E_{1,1-pm}^{(m)}\left(\mathrm{A}^{-2}t\right)\right] \tag{52}$$

$$\phi_5(t)=\sum_{n=0}^{\infty}\binom{-1/2}{n}\tau^{-p(n+1/2)}\frac{\Gamma^{(n+1/2)}(1+p)}{\Gamma(p(n+1/2)+1/2)}t^{p(n+1/2)-1/2} \tag{53}$$

$$\phi_6(t)=\int_0^t\phi_4(t-\tau)\phi_5(\tau)d\tau \tag{54}$$

4. Results and discussions

As mentioned above, equations (32) and (46) are the analytical solutions of the temperature and associated stress. To investigate the effect of the fractional order p on the thermal and associated stress behavior, the specific material copper (Cu) is chosen. The material parameters are listed in Table 1 and the corresponding constants are obtained as $\tau=4.25$, $\Lambda=1.66$, $V_e=4163m/s$.

Table 1. Thermal properties used in the present study.

δ (1/m)×10^{-7}	A (m^2/s)×10^4	C_e (J/kg.k)	ρ (kg/m^3)	k (W/m.K)	τ (s)×10^{11}	λ (N/m^2)×10^{-10}	μ (N/m^2)×10^{-10}
6.16	1.12	385	8930	385	1	7.76	3.86

4.1. The dimensionless temperature and thermal stress variation

Figure 2 presents the dimensionless temperature and thermal stress variations with time. It can be found from figure 2(a) that the temperature rising and falling rate are different for different positions. The temperature response at *x*=0 and 2 is obviously faster than that of *x*=4 and 6, and the farther away from the surface, the slower the temperature response, which indicates that the thermal wave propagates at a finite speed. Moreover, the peak value of the temperature decreases from *x*=0 to *x*=6 because of the decaying of the intensity of the pulse laser. Figure 2(b) gives the corresponding thermal stress fields. From figure 2(b), with the material surface(*x*=0) going through the stage of heating(0<*t*<5) and cooling(*t*>5), the inner part of the substrate material undergoes compression and tension. With the thermal wave signal transfers, the associated stress at different positions appears the similar distribution. That is to say, the thermal stress is first negative because of the compression and positive because of the tension, and the absolute peak thermal stress value for different positions occurs at the stage of heating. And the peak value takes place at different time for different positions as the thermal stress wave transfers.

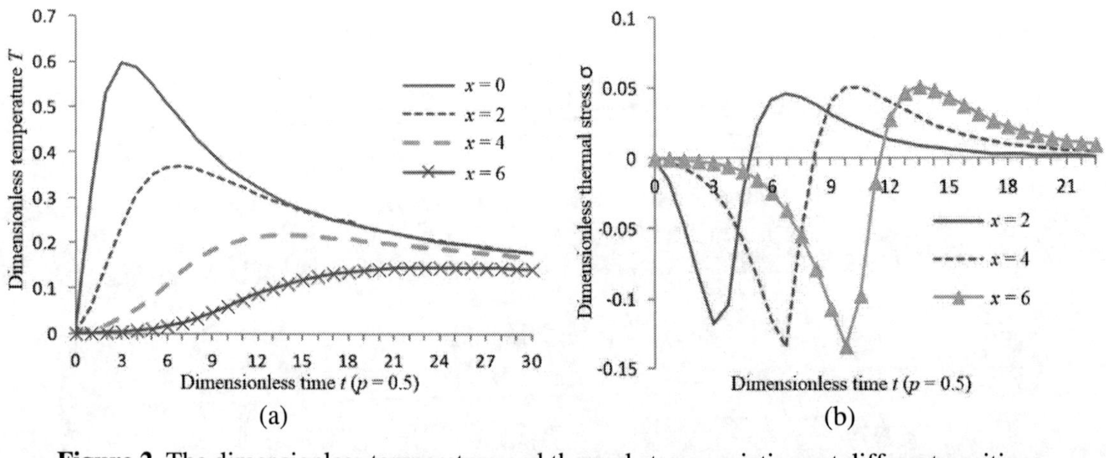

Figure 2. The dimensionless temperature and thermal stress variations at different positions.

Figure 3 shows the dimensionless temperature and associated stress distribution at different time. In figure 3(a), it is obvious that the heating area enlarges with heating time from *t*=2 to *t*=4. As mentioned above, the material surface (x=0) is heated by the laser source until t=5, the dimensionless temperature at t=6 is lower than that of t=4. It is noted that the heating and cooling process both exists at t=6. The non-synchronous heating and cooling is consistent with finite thermal wave velocity of the physical mechanism for non-equilibrium heat conduction. Figure 3(b) gives the corresponding thermal stress distribution at different time. It can be found that the all parts of the material is undergoing compressive deformation at t=2, while part of the material is in tension and part under compression at t=4 and t=6. As mentioned above, the heating stage leads to the compression and cooling generates the tension. The peak value of the stress is results from the large temperature gradient at different locations.

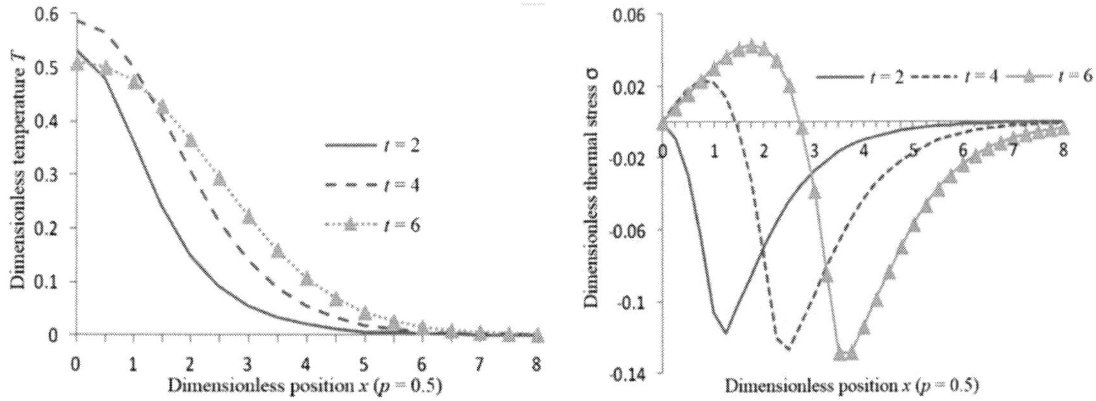

(a) (b)

Figure 3. The dimensionless temperature and associated stress distribution at different times.

4.2. The influences of the fractional order p on the temperature and associated stress field

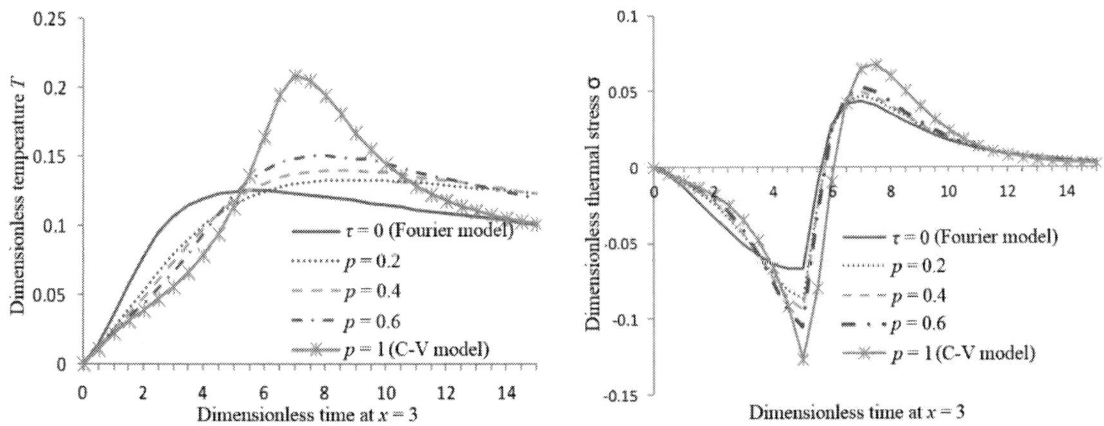

(a) (b)

Figure 4. The influences of the fractional order p on the temperature and associated stress variations.

Figure 4 shows the influences of the fractional order p on the dimensionless temperature and thermal stress variations at x=3. Figure 4(a) depicts the temperature variation. It is clear that the fractional order p influences the rising and falling rates of temperature directly. The larger the fractional order p, the faster the rising rate of the temperature. Moreover, the corresponding temperature ($0 < p < 1$) always lies between that of p=1 (the C-V model) and τ=0 (the Fourier model). From figure 4(b), the thermal stress variations are similar for the five different kinds of situations. But the peak value of the thermal stress gets larger with the fractional order gets bigger. The same as the temperature, the thermal stress predicted by fractional equation always lies between that of p=1 and τ=0.

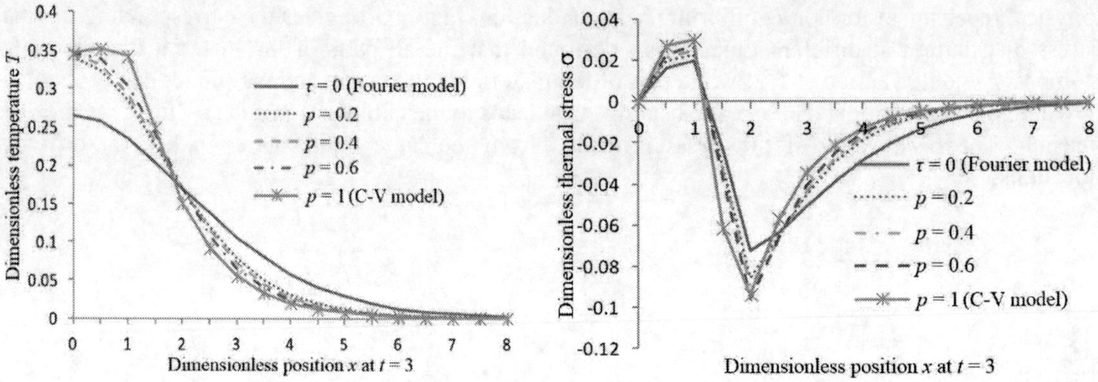

(a) (b)

Figure 5. The effects of the fractional order p on the temperature and thermal stress distributions.

Figure 5 shows the effects of the fractional order p on the dimensionless temperature and thermal stress distributions at t=3. Figure 5(a) represents the temperature distribution and figure 5(b) shows the corresponding thermal stress distribution. From figure 5(a), the temperature distribution from the fractional Cattaneo heat transfer equation stands between that of the classical Fourier and the standard C-V one. With p increasing, the surface temperature gets higher and the affected area is closer to the surface. Consequently, the larger the fractional order p is, the smaller the thermal wave velocity is. From figure 5(b), the thermal stress distribution is similar for the five different kinds of situations. But the peak value of thermal stress becomes smaller as the fractional order gets smaller.

5. Conclusion

The authors aim to introduce a mathematical model of fractional type Cattaneo heat conduction to reveal the heat transfer mechanism and the related thermal stress induced by short pulse laser heating, in which case, the thermal behavior exhibits both thermal diffusion and thermal wave. The heat transfer predicted by the present model shows the thermal behavior between thermal diffusion and the thermal wave feature. The non-local nature of the fractional type heat transfer model is more realistic.

The transient heat behavior is dependent on the fractional order parameter. Thermal wave mechanism is dominant as the fractional order parameter is larger, while thermal diffusion mechanism is dominant for the smaller one. With the fractional order parameter increasing, the thermal wave velocity decreases while the peak value of the temperature increases.

With the surface of the material undergoing the heating and cooling stage irradiated by the short pulse laser, a compressive stress develops first and is followed by a tensile stress. The associated thermal stress induced by transient heat behavior exhibits different features from the classical Fourier and standard C-V model. The peak value of the associated stress increases with the fractional order increasing.

References

[1] Xu Y S and Guo Z Y 1995 *Int. J. Heat Mass Tran.* **38** 2919-22
[2] Tang S W and Araki N 1996 *Heat Mass Tran.* **31** 359-63
[3] Joseph D D and Preziosi L 1989 *Rev. Mod. Phys.* **61** 41-73
[4] Wang X and Xu X 2001 *Appl. Phys. A* **73** 107-14
[5] Peshkov V 1944 *J. Phys.* **8** 381
[6] Cattaneo C 1958 *C. R. Acad. Sci.* **247** 431-3
[7] Vernotte P 1958 *C. R. Phys.* **246** 3154-5

[8] Hector L G, Kim W S and Ozisik M N 1992 *Int. J. Heat Mass Tran.* **35** 897-912
[9] Yilbas B S and Al-Aqeeli N 2009 *Opt. Laser Technol.* **41** 132-9
[10] Yilbas B S and Al-Dweik A Y 2012 *Physica B* **407** 4473-7
[11] Poletkin K V, Gurzadyan G G, Shang J and Kulish V 2012 *Appl Phys B* **107** 137-43
[12] Jiang F M, Liu D Y and Zhou J H 2003 *Microsc. Therm. Eng.* **6** 331-46
[13] Jou D, Criado-Sancho M and Casas-Vázquez J 2010 *J. Appl. Phys.* **107** 084302
[14] Auriault J L 2016 *Int. J. Eng. Sci.* **101** 45-9
[15] Maillet D 2019 *Int. J. Therm Sci.* **139** 424-32
[16] Bai C and Lavine A 1995 *J. Heat Trans.* **117** 256-63
[17] Rubin M 1992 *Int. J. Engrg. Sci.* **30** 1665-76
[18] Craiem D and Magin R 2010 *Phys. Bio.* **7** 013001
[19] Chaves A S 1998 *Phys. Lett. A* **239** 13-6
[20] Paradisi P, Cesari R, Mainardi F and Tampieri F 2001 *Physica A* **293** 130-42
[21] Podlubny L 1999 *Fractional Differential Equations* (New York: Academic Press)
[22] Mainardi F and Gorenflo R 2000 *J. Comput. Appl. Math.* **118** 283-99
[23] Hilfer R 2000 *Applications of Fractional Calculus in Physics* (Singapore: World Scientific)
[24] Qi H T, Xu H Y and Guo X W 2013 *Comput. Math. Appl.* **66** 824-31
[25] Mishra T N amd Rai K N 2016 *Appl. Math. Comput.* **273** 1006-17
[26] Xu G Y and Wang J B 2018 *Appl. Math. Mech.* 39 1465-76
[27] Povstenko Y Z 2005 *J. Therm. Stresses* **28** 83-102
[28] Povstenko Y Z 2008 *Q. J. Mech. Appl. Math.* **61** 523-47
[29] Povstenko Y Z 2011 *J. Therm. Stresses* **34** 97-114
[30] Povstenko Y Z 2015 *Appl. Math. Comput.* **257** 327-34
[31] Youssef H M 2010 *J. Heat Transfer* **132** 061301
[32] Sherief H H, El-Sayed A M A and El-Latief A M A 2010 *Int. J. Solids Struct.* **47** 269-75
[33] Yilbas B S 2002 *Int. J. Heat Mass Transf.* **45** 1993-2000
[34] Lam T T 2010 *Int. J. Therm. Sci.* **49** 1639-48

IOP Conf. Series: Materials Science and Engineering **657** (2019) 012055 doi:10.1088/1757-899X/657/1/012055

A feather precision measurement method

Y P Zhao[1,2] and J S Zhao[1,2]

[1]Department of Mechanical Engineering, Tsinghua University, Beijing 100084, P. R. China

[2]E-mail: zhaoyp16@mails.tsinghua.edu.cn; jingshanzhao@mail.tsinghua.edu.cn

Abstract. The methods of measurement of the microscopic and macroscopic morphology of feathers are discussed. A three-dimensional coordinate measuring machine and optical microscope are used together, that is, the macroscopic size of the feather is measured using the detection technology of the three-dimensional coordinate measuring machine, and the optical microscope is then used. The static accuracy error produced by the measurement result is (2.9064-4.0235) um, which is much smaller than the measurement error of 0.5 mm of the millimetre measuring ruler; The microscopic dimensions of the plume of the feather plume are measured by microscopy: the result is accurate to micron level, and the gradated scale is accurate to ± 0.5 mm. This study contributes to basic theoretical research in the fields of feather morphology, fractal analysis, and ecology.

1. Introduction

Natural selection gives birds colourful feathers (figures 1(a)-1(c)), and the flying behaviour of bird feathers has attracted more attention [1,2]. In the related research involving feathers, the range of measurement of feathers is an important reference index for judging bird species. Therefore, when the feather parameter is measured at large-scale, a three-dimensional coordinate measuring machine (micron-level) is used for precision measurement, and the measurement accuracy is much greater than that when using a measuring ruler (millimetre gradations). When a feather parameter is measured at small-scale, it is measured using an optical microscope (or scanning electron microscope (SEM)) for which the accuracy is of the order of microns. Use of a three-dimensional coordinate measuring machine with an optical microscope to overcome the defects of the measurement range and measure feather size parameters is proposed.

The three-dimensional coordinate measuring machine is a precision measuring device widely used in measuring the size of mechanical parts: it uses a laser-triggered measuring head for high-speed measurement. It has a large measuring range, fast response speed, and sensitive triggering. It is a micro-precision measuring tool.

Measurement of the size of the wings, feathers, and body of a bird is used to determine the ratio between the length of the wing and the fuselage, and is important in aircraft structural design. During three-dimensional (3-d) micro-surface morphology measurement [1], Jenni measured wing-length [3-5], Fl and F6-9 bending alignment in 5329 live birds of 51 species, using half-millimetre scale, where scale detection error is ± 0.5 mm [3]. The potential migration and differentiation of birds overwintering [6] maintains the natural phenomenon of species survival, not only indicating the importance of feathers for their survival, but also the bird niche (the ecological niche thereof) [7] determines the flight parameters of bird feathers. Access to food and escape from danger are thus predicated, and species survival is the result of natural selection. In addition, factors other than migration (dominance hypothesis, arrival time hypothesis, body-size hypothesis, Allen's rule, the resource partitioning hypothesis, and other potential limitations) also affect wing-length [8].

Figure 1. Natural selection gives birds colourful feathers (a) Embroidered bird (b) Pigeon (c) Parrot.

To analyse feather morphology, the measurement of feather leaf length currently involves use of a scale (to ± 0.5 mm) and blade area measurement using Adobe Photoshop 1 Version 5.0.2 (to ±0.1 mm^2) [9]. Studies have shown that the bird feather wingtip shape exhibits continuous pointedness [10], and the regional distribution of pointedness has important reference value for determining species distribution and mastering the ecological distribution thereof [11]. After collecting feather samples [12], statistical analysis (using SPSS™ software) [13-19] and binomial model methods [20,21] have been used. Other studies have shown that the wing beat cycle of bird flight begins with the wing reversed and the main feather rotates slightly to create a gap between successive feathers. The wing begins to move forward and upward, and lift is generated by the adjustment of the aerofoil [22]. According to research, there is a potential growth curve of bird feathers in the growth phase, that is, the growth of the feathers was described by Gompertz equations. It is a growth curve formed using non-linear regression [23,24]. In addition, studies have shown that feather morphology is related to bird behaviour and the prevailing environment, and shape variables are the most important component of morphological changes. Linear correlation measures are used to measure the shape and size of birds. Bird flight behaviour is characterised by habitat, dynamic flight (continuous tapping), flapping, gliding, and flying. The bird observes the surrounding environment from its habitat. When gliding, the wing stretches from the body in a fixed position without tapping, and the flight path is more or less straight. Therefore, when measuring the length of the wing, the wings on both sides should be fully extended. The power flight of the birds has two forms of continuous tapping and power glide therein [25,26].

Here. we proposed a precise measurement method for measuring the microscopic and macro-length dimensions of bird feathers, using an optical microscope and the high-precision measuring equipment on a three-dimensional coordinate measuring machine. The two are used together to measure various parameters thus providing references for researchers in morphology, fractal analysis, and ecology.

2. Materials and methods

2.1. Materials

The experimental materials were selected from white and black bird feathers and tail feathers (for example, goose feathers). They were immersed in, and disinfected with, a 75% alcohol solution, dried naturally, and used as needed. We applied double-sided adhesive tape to fix the feather position.

The experimental equipment included: a measuring ruler, an optical microscope, a FLY1086 model 3-d measuring machine, and a camera.

2.2. *Method*

2.2.1. Measuring the microscopic size of the feather-shaped plume. The knotted plume of the black feather was identified and fixed on the slide using double-sided adhesive tape. The specimen was placed on the optical microscope stage for microscopic observation, photography, and measurement.

The measurement methods of macro and micro-forms of feathers are shown in figure 2. The measurement of the macroscopic shape of the feather involves use of a measuring ruler (figure 2(A)). The microscopic morphology of the feather was imaged and filtered under an optical microscope (figure 2(B)). The micromorphological measurement method used on the knotted plume is shown in figures 2(C) and 2(D).

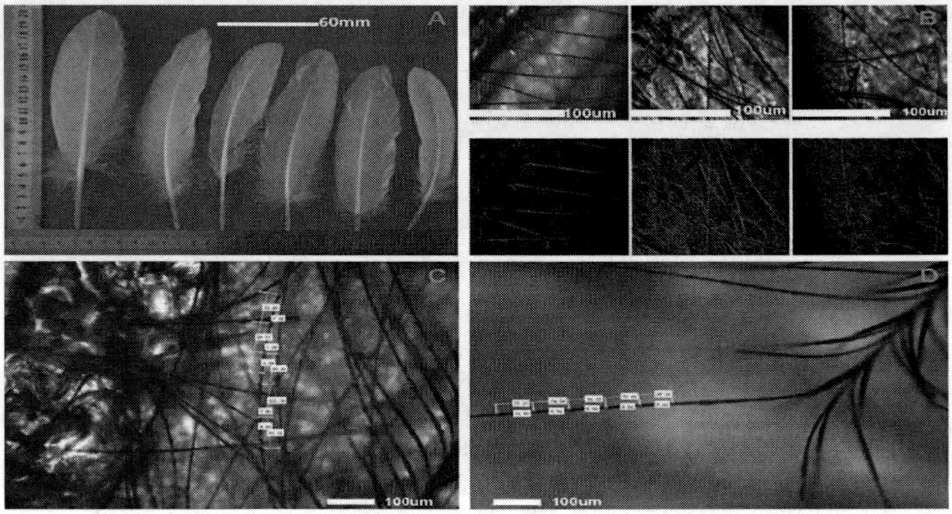

Figure 2. Measurement methods for macro and micro-forms of feathers. (a) Macroscopic morphology measurement, unit: mm. (b) Microscopic morphology under optical microscopy, unit: um. (c) Microscopic morphology measurement of knotted feathers. (d) Microscopic morphology measurement of knotted feathers.

2.2.2. Three-dimensional coordinate measuring machine measurement method. This experiment uses a FLY1086 model 3-d coordinate measuring machine as a high-precision measuring equipment, with a range of measurement of $X = 800$ mm, $Y = 1000$ mm, $Z = 600$ mm, and a resolution of 0.01 um was used. The chosen 3-d coordinate precision measuring machine includes: various bases, work benches, columns, guides, drive systems, measuring systems, electrical control cabinets, measuring heads, computer and control software, and printers. The measuring principle is follows: the measuring system used in the three-dimensional coordinate precision measuring machine is composed of a grating ruler and an infrared scanning system. During the measurement process, when the 3-d measuring head is in contact with the surface of the object to be measured, a sampling pulse is emitted, and the coordinates of the centre of the measuring head at this time are locked and saved. By measuring the surface several times, the spatial coordinate measurement head coordinate measuring machine equation can be obtained to determine the size and shape of the workpiece.

Two black feather tail feathers and one fly-feather were used, and they were washed and dried with 75% alcohol solution, and fixed on the high-level block of the 3-d coordinate measuring machine operating platform, and fixed with double-sided adhesive tape. Then the test data were checked on a point-by-point basis. The three-dimensional coordinate measuring machine and its precise measurement process for feathers are shown in figure 3. According to the error formula: the measured error value is:

$\Delta L = L_0$ (standard value) - $L(X)$ (measured value).

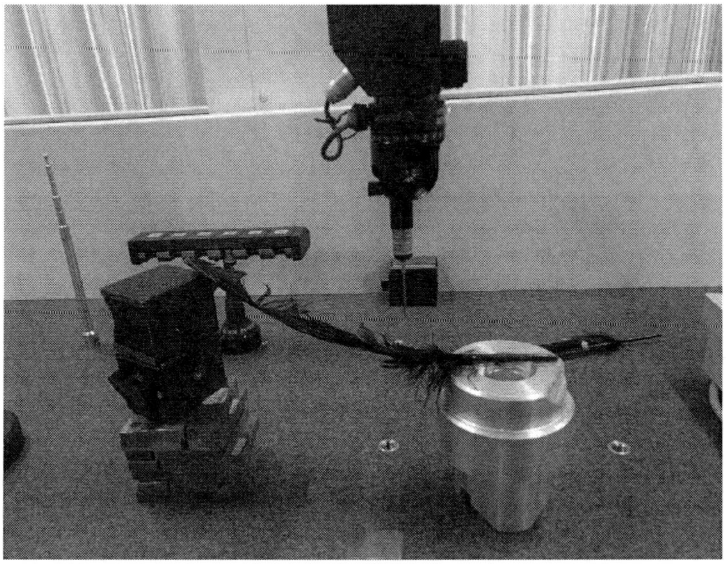

Figure 3. 3-d coordinate measuring machine and its precise measurement process.

It should be noted that the three-dimensional coordinate machine is suitable for the measurement of large feathers, and the precise measurement of small dimensions (for example, the plume) requires measurement by optical microscopy (measuring overall feather lengths required their combined use.

3. Results and discussion

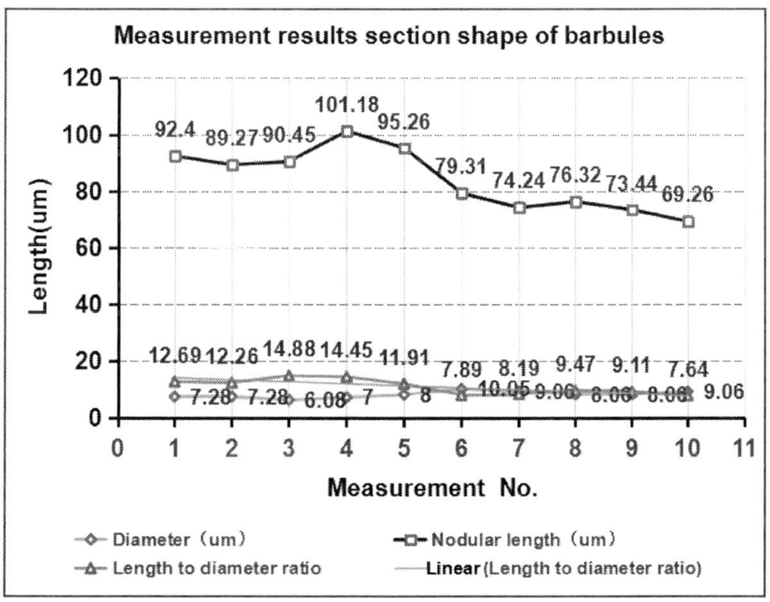

Figure 4. Analysis of measured results of knotted feather twigs.

From measured results of knotted feather twigs (figure 4), the diameter of the knotted plume is 6.08 um to 10.05 um, and the diameter of the curve (indicated by the thick line in the blue diamond in the figure) fluctuated less. The section length is 69.26 um to 101.18 um, and the curve shape formed by the section length (indicated by the black thick line in the figure) fluctuated to a significant extent. The aspect ratio is 7.64 to 14.88, and the curve formed by the aspect ratio (indicated by the thick line in the green triangle of the figure) tends to be linear (represented by the thin red line in the figure).

From the analysis of the distribution of the measurement results of the length and diameter of knotted feathers (figure 5), the diametral distribution of the knotted plumes is within 6.08 um to 10.05 um, and the section length is 69.26 um to 101.18 um, and the distribution thereof is concentrated centrally.

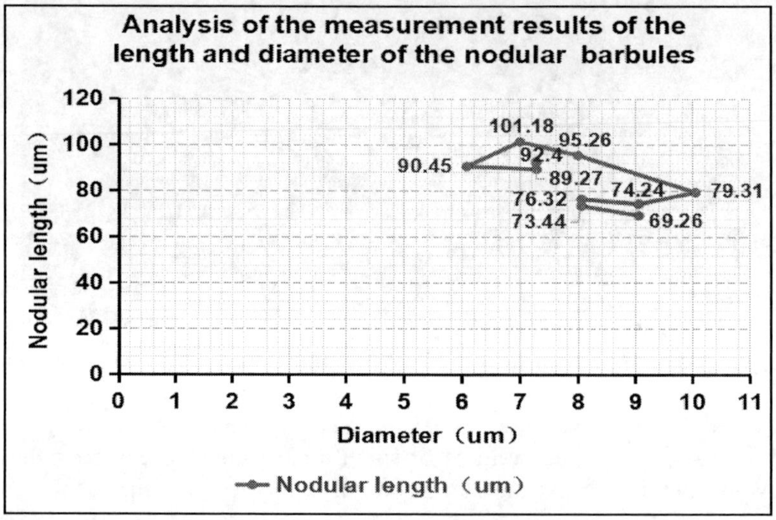

Figure 5. Analysis of the distribution range of the measured lengths and diameters of knotted plumes.

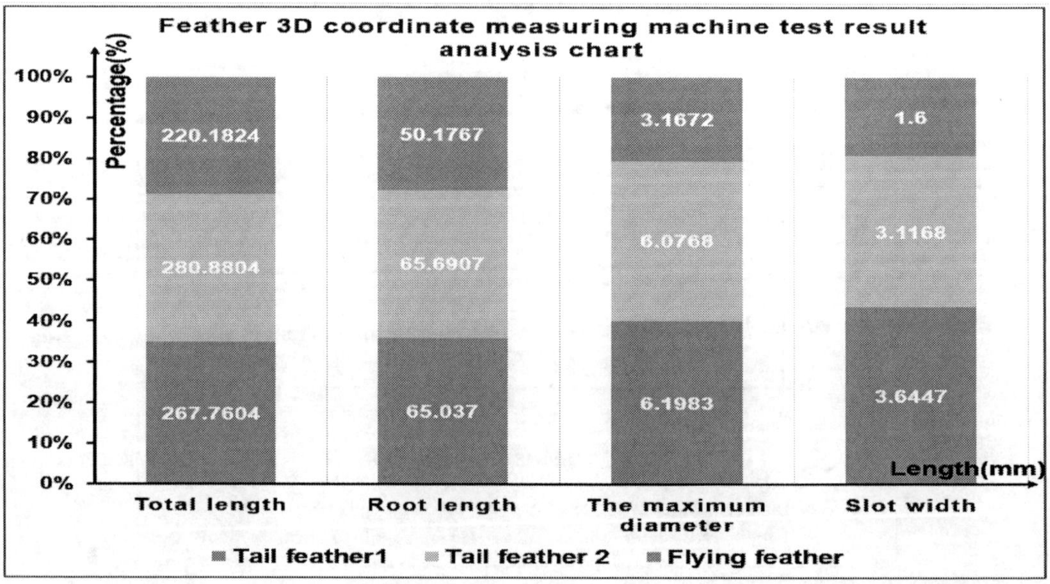

Figure 6. Analysis of the measured results of the feathers (3-d measurement coordinate machine).

The analysis of the measured results taken from the three-dimensional coordinate machine is shown in figure 6: the difference between the total length of the tail feathers 1 and 2 (red and yellow marking areas) of the same species, the diameter of the branching feathers, the length of the branch, and the aspect ratio were small, the total length between the tail feathers of the same species and the fly-feathers (green marking area), the diameter of the branch like plume, the length of the knot, and the ratio of the aspect ratios were quite different.

According to the user's manual for the FLY1086 three-dimensional coordinate measuring machine used in the experiment, the static accuracy error of the space measurement of the measuring machine is:

$$E \leqslant (2.9 + L/250)\ \mu\text{m}$$

The feather length measurement range of this example L is (1.600-280.8804) mm. The corresponding static accuracy error is (2.9064-4.0235) um. The error in the measured results is much less than the 0.5 mm error on the millimetre scale. The accuracy measurement error further indicates that the coordinate measuring machine is suitable for use in the case of measuring the macro-size of feathers in this case.

4. Conclusion

When measuring the micro and macro-length dimensions of the feather as a whole, it is recommended to use an optical microscope and a high-precision measuring device such as a three-dimensional coordinate measuring machine.

The total length of the tail feathers of the same species, the diameter of the branching plumes, the length of the branches, the length-to-diameter ratio differ little: the total length between the tail feathers and the flying feathers of the same species, the diameter of knotted feathers, and the length of the knot, differ significantly when expressed as aspect ratios.

The feather length measurement range of this example L is (1.600 to 280.8804) mm. The corresponding static accuracy error is (2.9064 to 4.0235) um. The error in the measured results is much less than the ± 0.5 mm error associated with the use of a millimetre gradated scale. The accuracy measurement error further indicates that the coordinate measuring machine is suitable for use when measuring the macro-size of feathers in this case.

The distribution of the length and diameter of the knotted feathers of the feathers is concentrated. The diameter distribution of the knotted plumes is 6.08 um-10.05 um, and the section length is 69.26 um-101.18 um.

Acknowledgment

This research was completed under the auspices of the National Natural Science Foundation of China (Grant no. 51575291). We thank the Optical Instrument Laboratory of the Department of Mechanical Engineering of Tsinghua University and the Precision Instrument Laboratory of the 3-d Coordinate Measuring Machine of Ningxia Institute of Business and Technology, Dr Ji Junjie, Associate Professor Aiping Wang, and Zhaodong Zhang for their enthusiasm and assistance.

References

[1] Harlander-Matauschek A, Isabel B, Corina L, Mirjana D and Werner B 2007 The relative preferences for wood shavings or feathers in high and low feather pecking birds *Appl. Anim. Behav. Sci.* **107** 78-87

[2] Vaquero-Alba I, Mcgowan A, Pincheira-Donoso D, Evans M R and Dall S R X 2016 A quantitative analysis of objective feather color assessment: measurements in the laboratory do not reflect true plumage color *The Auk.* **133** 325-37

[3] Sie Y D, Lin C Y and Chen S J 2018 3D surface morphology imaging of opaque microstructures via light-field microscopy *Sci. Rep.* **8** 1-13

[4] Xing L, McKellar R C, Wang M *et al* 2016 Mummified precocial bird wings in mid-Cretaceous Burmese amber *Nat. Commun.* **7** 1-7

[5] Harper D G C 1999 Feather mites, pectoral muscle condition, wing length and plumage coloration of passerines *Anim. Behav.* **58** 553-62

[6] Hargrove J, English S, Torr S J, Lord J, Haines L R, Van Schalkwyk C *et al* 2019 Wing length and host location in tsetse (glossina spp.): Implications for control using stationary baits *Parasite. Vector.* **12**(1) 1-13

[7] Hobson K A, Greenberg R, Van Wilgenburg S L *et al* 2010 Migratory connectivity in the rusty blackbird: isotopic evidence from feathers of historical and contemporary specimens *The Condor* **112** 778-88

[8] Sofía F -G, Antón P -R, Ivánde la H, Heather C P and Javier P-T 2015 Different space preferences and within-host competition promote niche partitioning between symbiotic feather mite species *Int. J. Parasitol.* **45** 655-62

[9] Nebel S, Rogers K G, Minton C D T *et al* 2013 Is geographical variation in the size of Australian shorebirds consistent with hypotheses on differential migration *Emu* **113** 99-111

[10] Jovani R, Serrano D, Frías O *et al* 2006 Shift in feather mite distribution during the molt of passerines: The case of barn swallows (Hirundo rustica) *Can. J. Zoo.* **84** 729-35

[11] Lockwood R and Rayner S J M V 1998 Avian wingtip shape reconsidered: Wingtip shape indices and morphological adaptations to migration *J. Avian Biol.* **29** 273-92

[12] Bianchi N, Ancora S, Fazio N D *et al* 2008 Cadmium, lead, and mercury levels in feathers of small passerine birds: Noninvasive sampling strategy. *Environ. Toxicol. Chem.* **27** 2064-70

[13] Rubtsov G A and Yakimenko V V 2012 The feather mite (Astigmata) fauna of some passerine birds (Passeriformes) in the south of Western Siberia *Entomol. Rev.* **92** 1020-31

[14] Hobson K A, Greenberg R, Van Wilgenburg S L *et al* 2010 Migratory connectivity in the rusty blackbird: Isotopic evidence from feathers of historical and contemporary specimens *The Condor* **112** 778-88

[15] Kooiker G 1986 Recommendations for analytical methods of heavy metals in bird eggs and feathers with atomic-absorption spectrophotometry (AAS) *J. Ornithol.* **127** 9-23

[16] Abdulreza M *et al* 2015 Mercury levels in avian feathers from different trophic levels of eight families collected from the northern region of Iran *Environ. Monit. Assess.* **187** 275

[17] Newton I 1999 An alternative approach to the measurement of seasonal trends in bird breeding success: a case study of the bullfinch Pyrrhula *J. Anim. Ecol.* **68** 698-707

[18] Murari S K, Frey F J, Frey B M *et al* 2005 Use of Pavo cristatus feather extract for the better management of snakebites: Neutralization of inflammatory reactions *J. Ethnopharmacol.* **99** 0-237

[19] Clark H R G 1992 Assessing avian diets using stable isotopes I: turnover of 13C in tissues *The Condor* **94** 181-8

[20] József R, Lajos R and Kiss B J 1997 Patterns in the distribution of avian lice (Phthiraptera: amblycera, ischnocera) *J. Avian Biol.* **28** 150-6

[21] Brown C R and Brown M B 2004 Group size and ectoparasitism affect daily survival probability in a colonial bird *Behav. Ecol. Sociobiol.* **56** 498-511

[22] Hedrick T L, Usherwood J R and Biewener A A 2004 Wing inertia and whole-body acceleration: An analysis of instantaneous aerodynamic force production in cockatiels (Nymphicus hollandicus) flying across a range of speeds *J. Exp. Biol.* **207** 1689-702

[23] Gous R, Moran E, Stilborn H *et al* 1999 Evaluation of the parameters needed to describe the overall growth, the chemical growth, and the growth of feathers and breast muscles of broilers *Poultry Sci.* **78** 812-21

[24] Mclennan J A, Dew L, Miles J *et al* 2004 Size matters: predation risk and juvenile growth in north island brown kiwi (Apteryx mantelli) *New Zeal. J. Ecol.* **28** 241-50

[25] Gamauf A 1998 Philippine birds of prey: interrelations among habitat, morphology, and behavior *Auk* **115** 713-26

[26] Gaston A J and Hipfner J M 2006 Body mass changes in brünnich's guillemots uria lomvia with age and breeding stage *J. Avian Biol.* **37** 101-9

IOP Conf. Series: Materials Science and Engineering **657** (2019) 012056 doi:10.1088/1757-899X/657/1/012056

Numerical simulation study on the effect of different mass flanges on strain growth of cylindrical shell

Y M Han[1], W B Gu[1,2], Y H Hu[1], J Q Liu[1] and Z Wang[1]

[1]College of Field Engineering, Army Engineering University of PLA, Nanjing, 210007, China

E-mail: guwenbin1@aliyun.com

Abstract. LS-DYNA software was used to establish the three-dimensional finite element simulation calculation model of cylindrical shell, and the effect of different mass flanges on the strain growth within the elastic response range of shell was analyzed and compared. The research found that: when the flange mass was changed, the radial and circumferential strain first peak value of shell basically did not change, while the axial first peak value changed greatly. The variation trend of strain growth response in radial direction and axial direction is basically the same, but the influence of axial strain growth is larger. Compared with the larger mass flange, the smaller mass flange may cause larger strain growth of the shell. Different flange mass only affects the time of strain growth and different frequency parts of excitation strain response, but hardly affects the appearance time of the "dangerous time domain". The longer the dangerous time domain is, the smaller the strain growth factor is.

1. Introduction

Explosive containers are special devices that limit explosive products and explosive energy in a certain confined space, and are widely used in civil and military fields. When the inside of the container is subjected to an impact load, a complex dynamic response is generated. Strain growth is a dynamic response of the vessel wall. It means that the maximum value of the strain does not appear at the first peak, but at the peak of the subsequent response. There are many factors that cause strain growth, which can be summarized into three aspects: a) resonance caused by explosive load [1-3]; b) disturbance caused by structure [4-6]; c) disturbance caused by modal superposition [7-15]. When the strain increases beyond the design allowable stress of the explosive container, the tank may be invalid, damaged, leaked, etc., posing a great threat to the safe use of the container. Therefore, the phenomenon of strain growth is the focus of the explosion container designer.

Since Buzukov [16] observed the cyclical growth of the strain of cylindrical explosive vessels in experiments in 1976, the strain growth phenomenon of explosive vessels has attracted the interest of many scholars. Karpp [4] once again observed the strain growth of the spherical container in the experiment, and established a two-dimensional axisymmetric spherical shell model, which simplified the flange into a mass, and verified that the flange will cause a spherical container large axial symmetric disturbance, this kind of disturbance that can cause severe strain growth, but did not give a more in-depth theoretical analysis; Li *et al* [11] simplified the explosion vessel into a two-dimensional ring for numerical simulation. By changing the partial mass of the ring, the effect of small structure disturbance on the ring strain growth was simulated. The results show that the perturbation of the small structure can accelerate the energy transformation from the breathing mode to the bending mode, thus aggravating the strain growth. In fact, these small structures can be nozzles, small openings, etc., but the effect of large structures (flanges, large openings, etc.) has not been studied in this paper. Hu *et al* [17] simplified the explosion vessel into a cylindrical shell, and analyzed the influence of flange radial length on strain growth through numerical simulation. It is believed that as the radial length of the flange increases, the bending disturbance of the various frequencies of the cylindrical shell is excited. This enhances the linear modal coupling response of the cylindrical shell, resulting in an increase in the strain growth factor. When more high frequency parts are introduced into the strain response, the strain growth time will also shorten.

As a necessary structure of the cylindrical container (as shown in figure 1), flanges play a safe role in confining explosive containers through bolt connections to prevent separation of tank body and end cover, leakage of toxic gases such as explosion products, etc. However, because flanges generally have a large mass, their different length, mass, position and thickness will greatly affect the dynamic response characteristics of the container after the impact of the explosion [17], which has the potential danger of inducing strain growth. However, there are currently few articles that systematically analyze the effect of flange on strain growth. Therefore, this paper will study the effects of different mass flanges on strain growth in terms of head peak, strain growth time and strain growth rate and so on.

Figure 1. Cylindrical explosive container with flange structure.

In this paper, the three-dimensional finite element model of the sliding-sliding boundary cylindrical shell is established and solved by LS-DYNA software. The influence of different mass flanges on the strain growth of the shell is analyzed. This can lay a foundation for the safe design of the explosion vessel, avoiding the dramatic impact of the flange on the tank strain growth, thereby enhancing the safety of the use of the explosion vessel.

2. Models and parameters

2.1. Finite element model

In order to investigate the effect of flange mass on the strain growth of cylindrical shells, under the premise of the same boundary conditions, this section designs four numerical simulations of different mass flanges, as shown in Table 1, and the model is shown in figure 2. The difference in flange mass is expressed in terms of different material densities.

In order to eliminate the adverse effects of the radial elongation of the flange, the following four sets of numerical models concentrate the mass of the flange on the cylindrical housing unit, and four models with four end flange thicknesses of 4.26 cm are provided. The four material densities are converted by the volume ratio of flanges with a material density of 7.83 g/cm^3 and the radial length of 1 cm, 3 cm, 5 cm and 7 cm, respectively.

Table 1. Numerical simulation of four different working conditions.

Variable	Condition 1	Condition 2	Condition 3	Condition 4
Density (g/cm^3)	16.18	33.99	53.27	74.01

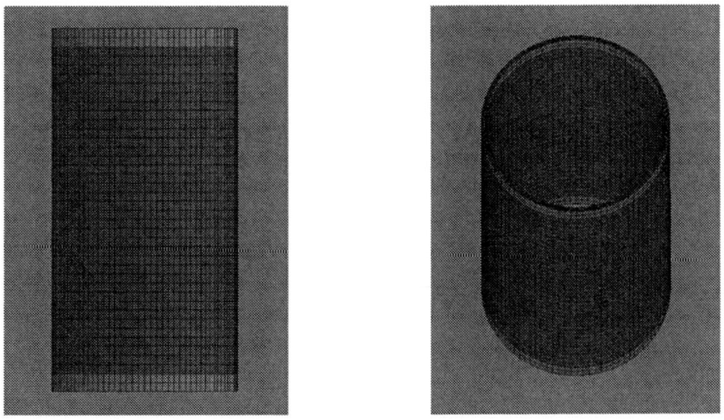

Figure 2. One of the cylindrical shell models with different mass flanges.

2.2. *Material parameters*

This paper is based on Dong's numerical model [15] and Zhu's experimental container [10] for numerical calculation. Referring to the modeling method of Hu [17], a cylindrical shell with an inner diameter of 21.28 cm, a thickness of 0.98 cm and a length of 85.2 cm was designed and divided into grids of approximately 0.16 cm, 1.45 cm and 2.13 cm in the radial, circumferential and axial directions, as shown in figure 2. Using the *MAT_ELASTIC material model, the parameters are shown in Table 2. Set the axial motion constraints at both ends of the housing.

Table 2. Material parameters of cylindrical housing.

Parameter	Young's modulus	Poisson's ratio	Density
Size	210 GPa	0.333	7.83 g/cm^3

The impact load on the container is reduced to a transient triangular wave pulse *P(t)*, where P_0 is the peak pressure and t_1 is the positive pressure action time:

$$\begin{aligned} P(t) &= P_0(1-\frac{t}{t_1}), \quad (0 \le t \le t_1) \\ P(t) &= 0, \quad (t > t_1) \end{aligned} \tag{1}$$

When the positive pressure action time of impact load is very small, the response of tank mainly depends on the impulse of load, and the waveform of load has little influence on it [3]. So set the pulse's positive pressure time to 50 microns, which controls the amount of the impulse by changing the peak size. The impulse of the load is half of the product of the peak value and the time of action, in Pa·S. The impulse acting on the inner surface of the cylindrical shell is set to be 300 Pa·S.

3. Effects of flange mass on strain growth of cylindrical shell

3.1. *Strain head peak*

It can be seen from Table 3 and Figure 3 that, under impact load, flanges of different mass have almost no effect on radial and circumferential strain head peak of shell centre torus, especially the circumferential strain head peak, whose variation range is only a few micro strains, but has a great effect on the axial strain head peak. It can be seen from Figure 3 that the axial strain head peak increases with the increase of flange mass. The reason is that: the flange is at both ends of the shell, when subjected to the impact load, the radial and circumferential elastic response of the flange initially affects only the two ends of the shell, while the axial tensile stress and compressive stress act on the axial direction of the entire shell, directly affecting the axial strain response of the shell, and this effect is proportional to the flange mass. Therefore, when considering the axial strain response, the effect of different flange mass is particularly important.

Table 3. Strain response characteristics of shells under different mass flanges.

	Density of flange g/cm^3	First peak compressive strain /με	Maximum compressive strain /με	Compressive strain growth factor	Compressive strain growth time
Radial	16.18	326.3	569.1	1.744	3860.9
	33.99	325.7	424.3	1.303	8656.5
	53.27	325.3	482.4	1.483	4101.5
	74.01	325.3	585.3	1.799	4364.0
Circumference	16.18	661.3	1048.0	1.585	3986.4
	33.99	659.2	791.0	1.200	7519.5
	53.27	656.5	904.8	1.378	3979.4
	74.01	653.9	1086.7	1.662	4240.0
Axial	16.18	37.4	184.1	4.925	590.5
	33.99	64.4	120.0	1.863	5811.9
	53.27	72.8	158.4	2.174	4182.4
	74.01	83.9	189.4	2.257	1707.0
	Density of flange g/cm^3	First peak tensile strain /με	Maximum tensile strain /με	Tensile strain growth factor	Tensile strain growth time
Radial	16.18	299.9	559.6	1.866	4236.5
	33.99	287.7	423.6	1.472	8278.5
	53.27	283.6	462.2	1.630	8525.4
	74.01	280.4	609.7	2.174	4240.0
Circumference	16.18	655.7	1047.2	1.597	3859.5
	33.99	655.7	789.4	1.204	7393.8
	53.27	655.7	914.4	1.395	4105.0
	74.01	655.7	1071.7	1.634	4366.5
Axial	16.18	78.4	192.2	2.452	2627.9
	33.99	114.2	117.9	1.033	6188.0
	53.27	130.2	173.4	1.331	1775.5
	74.01	146.2	250.1	1.710	1810.4

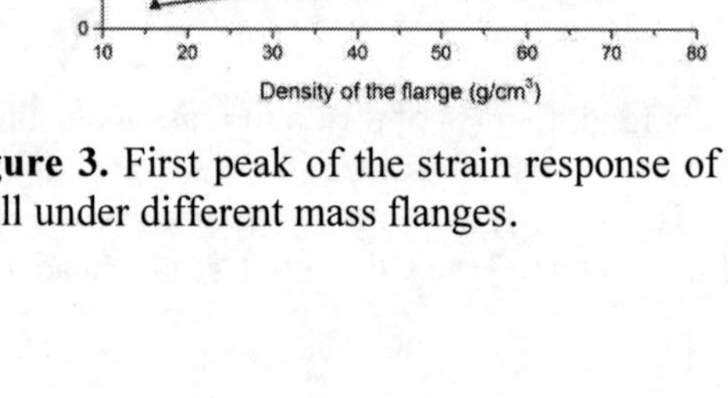

Figure 3. First peak of the strain response of the shell under different mass flanges.

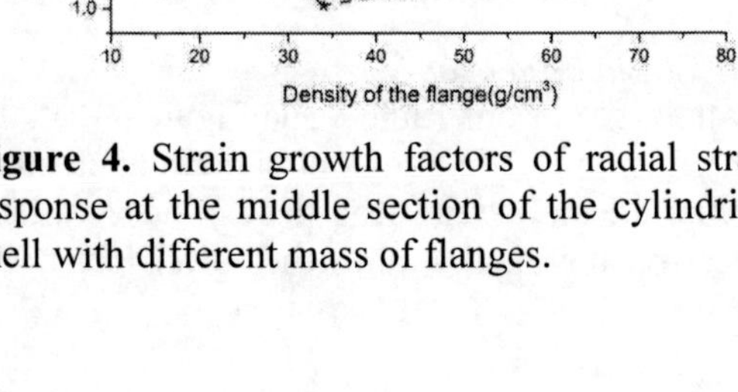

Figure 4. Strain growth factors of radial strain response at the middle section of the cylindrical shell with different mass of flanges.

3.2. Strain growth factor

As shown in figure 4, the strain growth factor shows the characteristics of high at both ends and low in the middle, and the change trend of radial and circumferential direction is basically the same, and the two circumferential lines basically overlap. For the axial direction, although the overall change trend is the same as the other two directions, the change range is large. This further confirms the point in 3.1. When the flange density is 16.18 g/cm^3, the axial tensile strain of the shell center torus is 2.5 times of the first peak value, and the compressive strain is up to 4.9 times, which is dangerous for the container, and may even cause the container failure or damage. It is indicated that the small mass flange may cause the larger strain increase of the shell. Therefore, when designing an explosive container, a reasonable selection of the mass of the structure can enhance the service life of the container.

3.3. Dangerous time domain

As can be seen from Table 3, the maximum radial and circumaxial strain values mainly concentrated around 4000 microseconds and 8500 microseconds, while the growth time of axial strain was distributed throughout the time domain. The maximum strain of the compressive strain and tensile strain of the middle section of the cylindrical shell with a flange density of 33.99 g/cm^3 and the tensile strain of the middle section of the cylindrical shell with a flange density of 53.27 g/cm^3 appears around 8500 μs. As can be seen from figure 5, a peak close to the maximum strain has appeared in the strain response curve of about 3000 μs. If the time period in which there are several strain peaks ≥ 90% of the maximum strain value is defined as the 'dangerous time domain' of the strain response (as shown in the dashed box in Figure 5(a)-5(d) of the four models), it can be considered that the mass of the flange has almost no influence on the starting time of the dangerous time domain. That is to say, the middle section of cylindrical shell with any flange mass appears higher radial strain growth at about 3000 μs in the condition. If the structure under high stress and strain will produce plastic deformation or damage, it has reached the relatively dangerous state about 3000 μs.

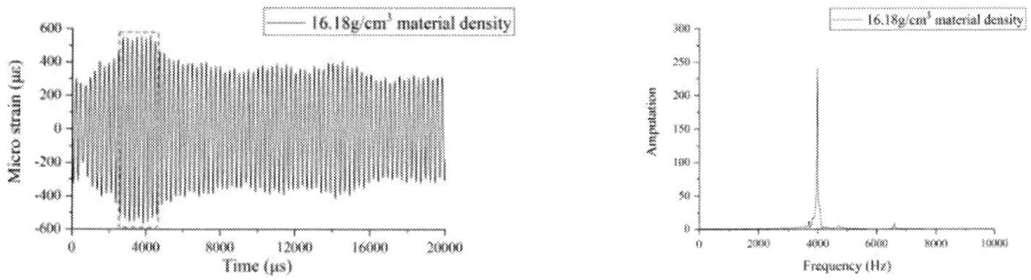

(a)

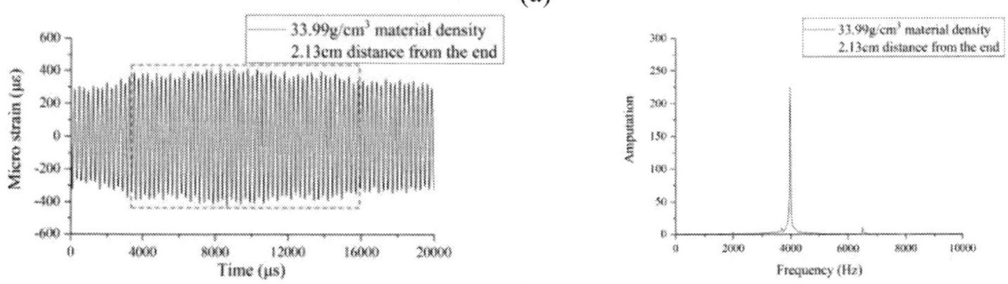

(b)

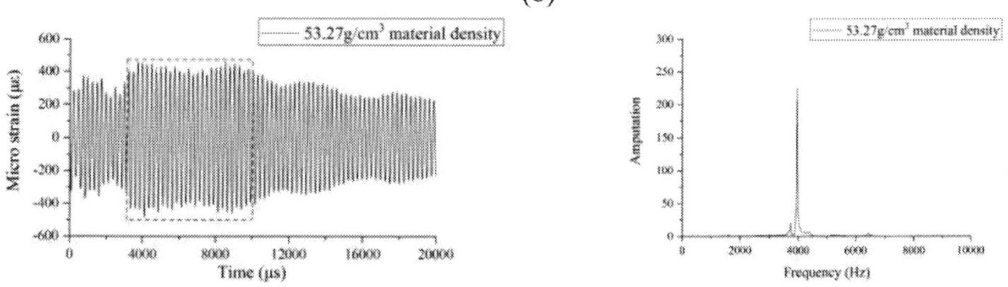

(c)

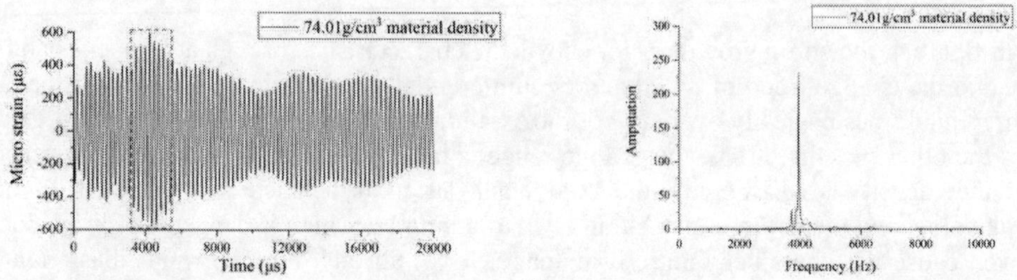

(d)

Figure 5. Radial strain response at the middle section of the cylindrical shell with different mass of flanges under 300 Pa·S transient impact load.

In the research process, it is found that both radial and circumferential directions have dangerous time domain, while axial direction has no such rule, so the radial direction is taken as an example to carry out the analysis. Table 4 lists the response time of the dangerous time domain caused by flanges of different mass. It can be obtained from the table that the dangerous time domain is the longest under the action of flange with a density of 33.99 g/cm^3. Although its strain growth factor is the smallest, it still cannot indicate that the effect of flange on shell under this mass is safe. The definition of fatigue damage is given in reference [18]: "the development process of local and permanent structural changes of materials that are subjected to disturbance load at a certain point and undergo crack or complete fracture after sufficient cyclic loading". The dangerous time domain with a long duration can also be regarded as a process of repeated loading with high stress, which may also lead to fatigue damage of the shell, which is unfavorable to the service life of the container. By comparing Table 3 and Table **4** and Figure 5, it can be concluded that, under the same impulse, the longer the continuous response time of the shell with high strain is, the smaller the maximum strain is. In other words, the longer the dangerous time domain is, the smaller the strain growth factor is.

Table 4. Dangerous time domain of radial strain.

Density of the flange (g/cm^3)	16.18	33.99	53.27	74.01
Dangerous time domain (μs)	2018.0	11093.9	6087.1	1278.0

3.4. Frequency characteristics

The flanges of the four shells are of different mass, but the spectrum diagram of their strain curves shows similar amplitudes and frequencies, that is, they all have the main peak of 4 kHz, but the flanges with a smaller mass will cause a small peak at about 6.5 khz, and the flanges with a larger mass will cause a small peak at about 3.7 khz. In the study or daily detection of explosion containment vessels, sensors or other devices need to be installed on them. If the vessels resonate with instruments, it is not conducive to measurement and detection. Therefore, when designing explosion containment vessels, the position of small frequency peak can be changed by adjusting the mass of flange to avoid resonance with the sensor or other instruments.

4. Conclusions

In this paper, the numerical calculation model of cylindrical shells is established to study the effect of different mass flanges on the strain growth of explosive vessels. The main conclusions are as follows:

- Under the same impact load, changing the mass of the flange has little effect on the radial and circumferential strain peaks of the central torus of the shell, but has a greater influence on the axial first strain peak, it increases with the increase of flange mass, so we must pay attention to it.
- When the mass of the flange is changed, the strain growth coefficient curves of the radial and circumferential directions of the shell center torus are very similar, and the circumferential

direction is basically identical, while the axial direction changes greatly. The effect of flange with different mass on axial strain growth reaches the maximum when the mass of four flanges is the smallest, which indicates that the flange with small mass may cause larger strain growth of shell.

- Flanges of different mass can affect the time of strain growth, but hardly affect the moment of occurrence of the dangerous time domain. The longer the danger time domain is, the smaller the strain growth factor is, and vice versa. For material fatigue damage, the continuous loading of high stress should not be ignored.
- Smaller flanges excite the high frequency portion of the strain response, while larger flanges excite the low frequency portion of the strain response. The mass of the flanges should be selected reasonably to avoid resonance between the response of different frequencies and other instruments on the vessels.

References

[1] Buzukov A A 1980 Forces produced by an explosion in an air-filled explosion chamber *Fiz Goreniya Vzryva* **16** 87-93
[2] Zhdan S A 1981 Dynamic load acting on the wall of an explosion chamber *Fiz Goreniya Vzryva* **17** 142-6
[3] Dong Q, Li Q M and Zheng J Y 2010 Interactive mechanisms between the internal blast loading and the dynamic elastic response of spherical containment vessels *Int. J. Impact Eng.* **37** 349-58
[4] Karpp R R, Duffey T A and Neal T R 1983 Response of containment vessels to explosive blast loading *J. Pressure Vessel Technol.* **105** 23-7
[5] Abakumov A I, Egunov V V, Ivanov A G *et al* 1984 Calculation and experiments on the deformation of explosion-chamber shells *Zh Prikl Mekh Tekh Fiz* **3** 127-30
[6] Dong Q, Li Q M, Zheng J Y *et al* 2010 Effects of structural perturbations on strain growth in containment vessels *J. Pressure Vessel Technol.* **132**(1) DOI:10.1115/1.4000372
[7] Duffey T A and Romero C 2003 Strain growth in spherical explosive chambers subjected to internal blast loading *Int. J. Impact Eng.* **28** 967-83
[8] Kornev V M *et al* 1979 Experimental investigation and analysis of the vibrations of the shell of an explosion chamber *Fiz Goreniya Vzryva* **15** 155-7
[9] Belov A I, Klapovskii V E, Kornilo V A *et al* 1984 Dynamics of a spherical shell under a nonsymmetric internal pulse loading *Fiz Goreniya Vzryva* **20** 71-4
[10] Zhu W H, Xue H L, Zhou G Q *et al* 1997 Dynamic response of cylindrical explosive chambers to internal blast loading produced by a concentrated charge *Int. J. Impact Eng.* **19** 831-45
[11] Li Q M, Dong Q and Zheng J Y 2008 Strain growth of the in-plane response in an elastic cylindrical shell *Int. J. Impact Eng.* **35** 1130-53
[12] Dong Q, Li Q M and Zheng J Y 2009 Investigation on the mechanisms of strain growth in cylindrical containment vessels subjected to internal blast loading *ASME PVP 2008* (Chicago,USA) **5** pp 211-20
[13] Liu W X, Zhang Q M *et al* 2017 Further research on mechanism of strain growth caused by superposition of different vibration modes *Int. J. Impact Eng.* **104** 1-12
[14] Dong Q, Li Q M and Zheng J Y 2010 Further study on strain growth in spherical containment vessels subjected to internal blast loading *Int. J. Impact Eng.* **37** 196-206
[15] Dong Q, Li Q M and Zheng J Y 2017 Strain growth in a finite-length cylindrical shell under internal pressure pulse *J. Pressure Vessel Technol.* **139**(2) DOI:10.1115/1.4035696
[16] Buzukov A A 1976 Characteristics of the behavior of the walls of explosion chambers under the action of pulsed loading *Fiz. Goreniya Vzryva* **12** 605-10
[17] Hu Y H, Gu W B, Liu J Q *et al* 2019 Numerical simulation of the effect of flange radial lengthon strain growth of cylindrical containment vessels *Vibroeng. Procedia* **25** 182-7
[18] ASTM 1979 American National Standard. *ANSI/ASTM*, E206-72

Experimental study on dynamic characteristics of wire-cable vibration isolator

H X Wang[1], D D Chen[1,2] and Y Y Ji[1]

[1]Mechanical Engineering college, Hubei University of Automotive Technology, 167 West Road of Motor City, Shiyan, Hubei 442002, China

E-mail: 8784145@163.com

Abstract. This study considers the dynamic experimental characteristics of the wire-cable vibration isolator. Orthogonal design scheme is adopted to reduce the test times. A wire-cable vibration isolator will be experimentally tested in the tension-compression mode to analyze its hysteretic behavior in more detail. Experimental data of restoring force and displacement are obtained through periodic test loading. The shape of hysteresis loops of vibration isolator under different excitation amplitudes, excitation frequencies and number of wire-cable loops have been obtained; the area of each hysteresis loop will be calculated by definite integral. The hysteresis loops under different conditions are studied, the results show that the wire-cable vibration isolator has excellent shock isolation characteristics; the isolator exhibits asymmetric hysteresis loops, which possess a hardening loading overlap in the loading curves. Finally, the key influence parameters of the dynamic characteristics are obtained, which establishes the foundation for designing the high performance wire-cable vibration isolator.

1. Introduction

Wire-cable isolator is a typical nonlinear hysteresis characteristics of vibration isolation device. In order to fully understand the cushioning performance of vibration isolation of the isolator, its dynamic characteristics is investigated experimentally, while would provide some exact experimental data for vibration isolator to model and identify parameter. Wu [1] took the common marine vibration isolator as the research object, analyzed the change of physical model of vibration isolator stiffness characteristics when ship tilts and rolls, and verified the theoretical research results by experiments. Wang [2] obtained the dynamic hysteresis characteristics of O-type wire-cable isolator in shearing, rolling and tension direction along with excitation amplitude and frequency by using periodic dynamic loading test method. In reference [3-5], the impact stiffness and damping of a rubber steel wire isolator were tested by falling hammer impact method. The acceleration vs. time curve of the loading mass was transformed into the hysteresis loop of the isolator to solve for the impact stiffness and the equivalent damping. To reduce the number of experiments, orthogonal design is designed firstly in this paper. Then, the shape and area of hysteresis loops of vibration isolator under different excitation amplitudes, excitation frequencies and number of wire-cable loops are obtained by means of data processing and numerical integration. Finally, the dynamic characteristics of hysteresis loops under different conditions would be studied.

2. Dynamic test of isolator

2.1. Test purpose and design scheme

In practical engineering applications, the isolator has always used in the dynamic environment, so, it is very important to study the isolator's dynamic characteristics in different conditions for design optimization of the wire-cable isolator. The test purpose would be to study the influence of displacement amplitude, frequency and number of wire-cable rings of isolator on the stiffness and damping of isolator. In order to reduce the number of tests, the specific implementation scheme of the test is designed as shown in table 1.

Table 1. Orthogonal test scheme.

Test number	Amplitude/mm	Frequency/Hz	Number of wire cable rings
1	6	3	6
2	6	4	6
3	6	5	6
4	6	6	6
5	6	7	6
6	6	8	6
7	6	9	6
8	1	5	6
9	2	5	6
10	3	5	6
11	4	5	6
12	5	5	6
13	6	5	4
14	6	5	5

2.2. Test setup

The setup for the quasi-static cyclic loading test was arranged on an electromagnetic vibration test bench (Model: MPA406/M232) as shown in figure 1. which can generate various types of excitation signals (sinusoidal wave, triangular wave signal, random signal, etc.). The test simplified schematic is shown in figure 1, and test installation drawing is shown in figure 2. The lower end of the vibration isolator is connected to the electromagnetic vibration table, the upper end is connected to the three-jaw chuck, the upper clamp is connected to the force sensor, and the three-jaw chuck is connected to the foundation through the frame. Considering that both the frame and the vibration table are fixed on the foundation, it can be approximately considered that the upper end of the vibration isolator is connected with the rigid body without displacement, and the lower end of the vibration isolator will be driven when the vibration table starts to vibrate. The laser displacement sensor is fixed on the frame, and the signal acquisition analyser is used to collect the signals of two force sensors and displacement sensors simultaneously, and transmits them to the computer for data processing. The excitation signal adopts sinusoidal displacement signal with A=1-6 mm and f=3-9 Hz.

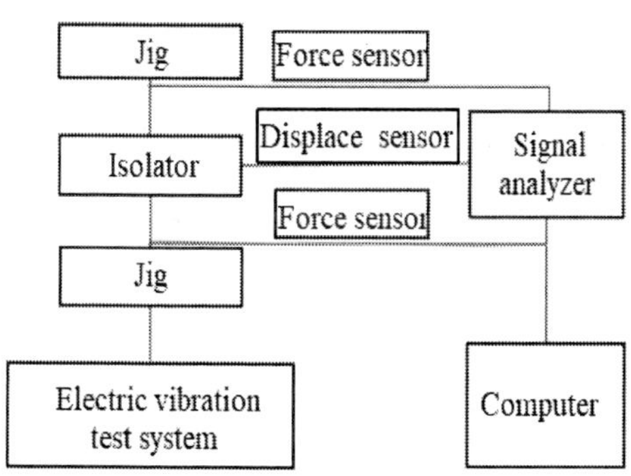

Figure 1. Test simplified schematic.

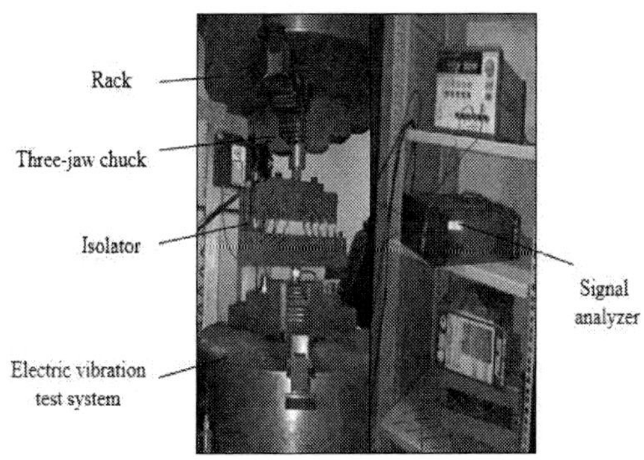

Figure 2. Test installation drawing.

3. Test data processing and analysis

3.1. Test data processing

In the test, deformation displacement excitation is adopted to the wire-cable isolator, which is used as the input signal and controlled by the closed loop system of the testing machine. Therefore, it can be considered that the excitation signal has no noise and can be used as the real deformation of the isolator. The restoring force of the vibration isolator is the output signal, which may be polluted by noise due to some reasons. Therefore, in order to more accurately display the dynamic hysteresis characteristics of the vibration isolator, it is necessary to conduct noise reduction processing on the collected restoring force data. Wavelet noise reduction [6] is a popular noise reduction method. However, this method is not suitable for wide range of white noise in practical application. For periodic vibration test, the measured restoring force is periodic signal, including multiple harmonics, which can be expressed as follows:

$$R(t) = \frac{R_0}{2} + \sum_{j=1}^{m} R_j \cos j\omega t + \sum_{j=1}^{m} R_j^* \sin j\omega t \tag{1}$$

Here $\{R\} = \{R_0 \ R_1 \ R_2 \ \cdots \ R_m \ R_1^* \ R_2^* \ \cdots \ R_m^*\}^T$ of $R(t)$, ω is the original excitation frequency, and *m* is the Fourier order, which can be selected according to the experimental data. Ko [7] use the frequency domain non-causal filtering method prosed by Ni [8] to conduct discrete Fourier series high-frequency noise deduction of the sampled data, obtain the harmonic component of the measurement restoration force, and then reconstruct the measurement periodic signal.

3.2. Test data analysis

The purpose of this test is to explore the influence of excitation frequency f, amplitude A, number N of wire-cable loops on the stiffness and damping characteristics of wire-cable isolator. Since the sampling frequency of displacement sensor and force sensor is 512 Hz, a large number of test data points are collected, and the data of each period are roughly same in the stable state, so the data of a steady-state period can be arbitrarily selected to analysis and research.

4. Analysis of test results

The collected experimental data are processed by 3.1 sections to obtain the hysteresis loop curves of wire-cable isolator at different frequencies, different loading amplitudes, different number of wire-cable rings. Through study, the influence of these three factors on dynamic characteristics of wire-cable isolator is analyzed.

4.1. Influence of amplitude A

The amplitude would influence the dynamic characteristics of the wire-cable isolator with certain parameters. For different amplitudes, the resilience of the elastic components of the isolator varies at the same frequency. The control variable method is used to study the dynamic stiffness and damping characteristics of wire-cable isolator under different amplitude displacement excitation. There are three parameters involved in this study, keeping the number *N* of wire-cable rings, frequency *f* unchanged, that is, the parameters *N*=6, frequency *f*=5 Hz are determined. The hysteresis curves, when excitation amplitude A=1-6 mm, are obtained as shown in figure 3. Meanwhile, in order to explore the changes of the area between hysteresis loops under different amplitude, select the test data with amplitude of 3 mm, and then use the nonlinear functions $R_u(x)$ and $R_l(x)$ to fit the data (upper and lower hysteresis restoring force) at the loading and unloading stages respectively, and obtain the nonlinear function expressions as follows:

$$R_u(x)=0.289x^5-0.7232x^4-1.764x^3+4.127x^2+49.08x+49.47 \tag{2}$$

$$R_1(x)=0.648x^5+1.432x^4-3.921x^3-3.573x^2+38.33x-75.1 \quad (3)$$

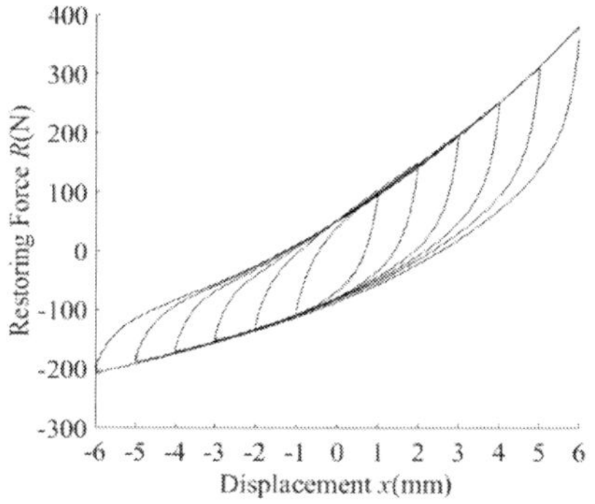

Figure 3. Hysteresis loop at different amplitude.

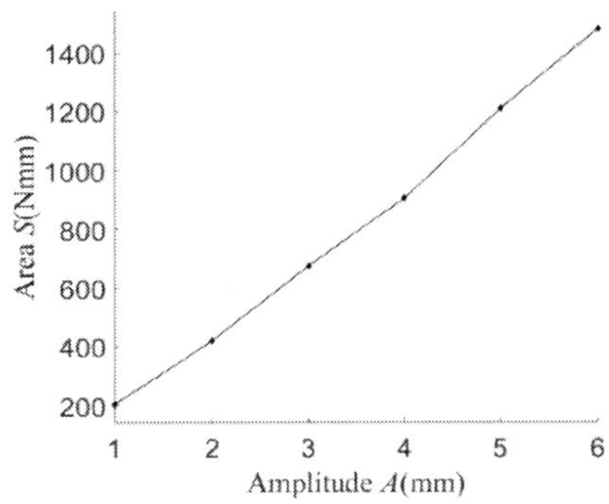

Figure 4. Area of each hysteresis loop.

Here, the difference ΔR between the upper and lower fitting curves is required, and then the area S enclosed by the difference curve and x=0 is calculated by definite integral, that is, the area of the hysteresis loop when A=3 mm, Similarly, the area of each hysteresis loop with different amplitudes can be obtained by using this method. The variation of the area S enclosed by the hysteresis curve under different amplitudes is shown in figure 4.

$$\Delta R=-0.359x^5-2.155x^4+2.175x^3+7.7x^2+10.75+124.57 \quad (4)$$

$$S=\int_{-3}^{3}\Delta R dx=676\text{Nmm} \quad (5)$$

It can be seen from figure 3, the stiffness of the vibration isolator gradually becomes harder when loading, while that of unloading gradually becomes softer when unloading, and the loading stiffness is larger than that of unloading, but the change of stiffness is more obvious when unloading. With the increase of amplitude, the inclination degree of each hysteresis loop of wire-cable isolator gradually flattens, the dynamic stiffness of wire-cable isolator increases gradually. It can be seen from the figure figure 4, with the increase of amplitude, the hysteresis area S significantly increases, the energy consumed by the elastic element also greatly increases, and the damping isolation performance is significantly enhanced.

4.2. *Influence of frequency f*

In this section, the influence rule of frequency factors on stiffness and damping of steel isolator is studied. The parameters N=6, A=6 mm are determined. The hysteresis curve of the isolator in the tension and pressure direction when frequency f=3-9 Hz is obtained, as shown in figure 5. The change of area S enclosed by the hysteresis curve at different frequencies is shown in figure 6.

According to figure 5, when the loading amplitudes are all 6 mm, the hysteresis curves are not very different. In other words, with the same excitation amplitude, the shape of hysteresis loops changes slightly with the change of excitation frequency, and the shapes of hysteresis loops are basically identical. As can be seen from figure 6, the area of hysteresis loop increases with the increase of loading excitation frequency, but the change is not obvious, which is consistent with the dynamic hysteresis and frequency independent characteristics of traditional wire-cable isolator.

IOP Conf. Series: Materials Science and Engineering **657** (2019) 012057 doi:10.1088/1757-899X/657/1/012057

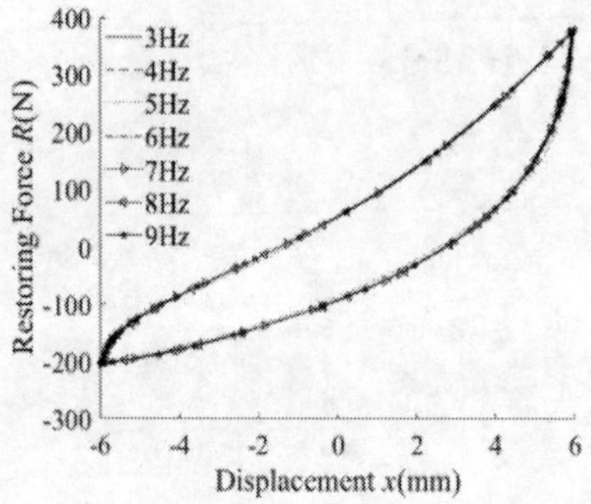

Figure 5. Hysteresis loop at different frequencies.

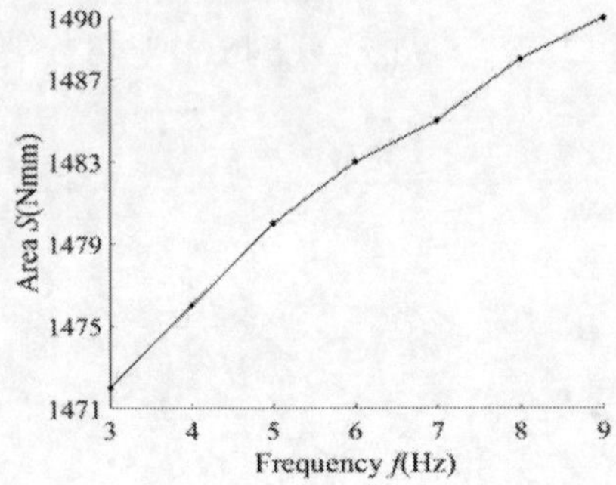

Figure 6. The area of each hysteresis loop.

4.3. *Influence of N*

Figure 7 shows the hysteresis curve of the elastic element when the amplitude A=6 mm, frequency f=5 Hz, and N=4, 5, 6. Figure 8 shows the change of the area S of hysteresis ring encircling when N changes.

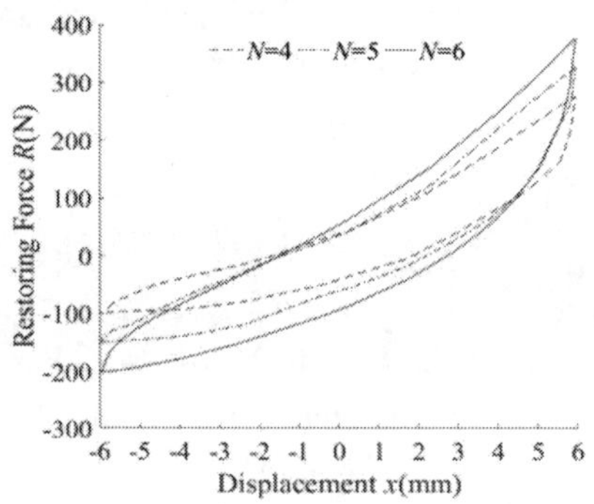

Figure 7. Hysteresis loop at different *N*.

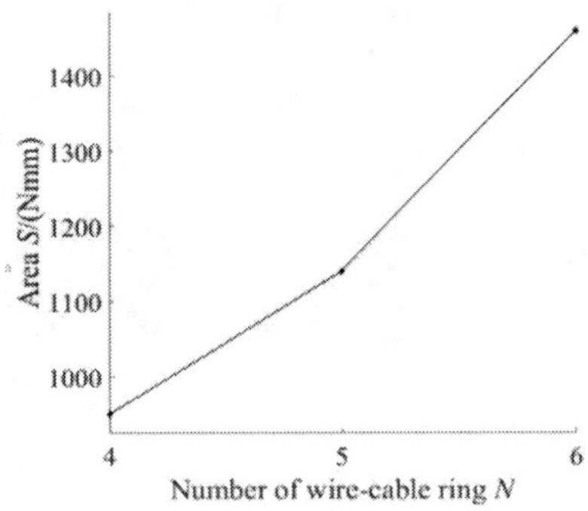

Figure 8. Area of each hysteresis loop.

As can be seen from figure 7 with the increase of *N*, the inclination degree of hysteresis loop obviously steepens, indicating that increasing *N* can improve the dynamic stiffness of the vibration isolator. According to figure 8, the area gradually increases with the increase of *N*, but it does not increase linearly. The main reason is that the wire-cable vibration isolator consists of a nonlinear system, and the superposition principle is only applicable to linear systems.

5. Conclusion

The dynamic characteristics of the wire-cable isolator are designed by orthogonal test. The dynamic restoring force hysteresis characteristics of the isolator under excitation displacement are obtained. The shape and area of hysteresis loops of vibration isolator under different excitation amplitudes, excitation frequencies and number of wire-cable loops are obtained by means of software processing and analysis data and numerical integration. The following conclusions are obtained by analyzing the test results:

With the increase of frequency, the shape of hysteresis loop is basically unchanged, which indicates that the dynamic performance of vibration isolator is independent of the loading frequency. With the increase of amplitude, the area of hysteresis loop gradually increases and the stiffness of vibration isolator decreases, the amplitude is negatively correlated with the stiffness and positively correlated with the damping. With the increase of the number of wire-cable loops, the dynamic stiffness of the vibration isolator increases, and the area of hysteresis loops increases in nonlinear relation. Therefore, increasing the number of wire-cable loops can improve the vibration isolation ability of the vibration isolator. The results showed that the excitation amplitudes and the number of wire-cable loops are the key influence parameters of the isolator's dynamic characteristics, which establishes the foundation for designing the high performance wire-cable vibration isolator.

Acknowledgments
This work was supported by the Natural Science Fund of Hubei Province (2017CFB741); the Foundation of Hubei Ministry of Education (B2016089); the Hubei Province Education and Science (2018GB030); the Doctoral Science Foundation, Hubei Automotive Industries Institute (Bk201406) and (Y2018202).

References

[1] Wu C, Shuai C G and Zhang S K 2018 The experimental study on stiffness of vibration isolator at different inclination in ship *Ship Sci. Technol.y* **40** 36-9
[2] Wang H X, Gong X S and Pan F 2016 Research on the dynamic behavior of O type wire-cable vibration isolator in three modes *Vibration Measurement and Diagnosis* **36** 1190-5
[3] Gong X S and Tang Y K 1999 New method for modeling of a nonlinear vibration system with hysteresis characteristics *J. Mech. Eng.* **3** 11-4
[4] Xie Y X, Chen Y and Che C D 2016 Experimental research on impact characteristics of a rubber steel wire isolator *Noise and Vibration Control* **36** 183-7
[5] Massa G D, Pagano S, Rocca E and Strano S 2013 Sensitive equipments on WRS-BTU isolators *Meccanic* **48** 1777-90
[6] Li M, Wu B and Zhang H M 2018 Research on estimation method of bending moment of underground diaphragm wall based on wavelet noise reduction *Tunnel Construction* **38** 120-5
[7] Ko J M, Ni Y Q and Tian Q L 1992 Hysteretic behavior and empirical modeling of a wire-cable vibration isolator *Int. J. Analyt. Exp. Model Anal.* **7** 111-27
[8] Ni Y Q, Ko J M and Wong C W 1999 Modelling and identification of a wire-cable vibration isolator via a cycle loading test Part 2: identification and response prediction *Proceedings of the Institution of Mechanical Engineers, Part I-Journal of Systems and Control Engineering* **213** 173-82

Dynamic effects of low-velocity impact on composite plates

R Wang[1], D D Chen[1], M Zhu[1] and S L Xing[1,2]

[1]College of Aerospace Science and Engineering, National University of Defense Technology, Chang Sha 410073, PR China

E-mail: happy_xing@nudt.edu.cn

Abstract. In low-velocity impact research on composites, quasi-static indentation experiments are typically used to make inner fractures. However, dynamic effects caused by impacts cannot be studied by quasi-static experiments. In this paper, the dynamic effects are discussed using simulation. Low-velocity impact processes were simulated through ABAQUS/Explicit software. Quasi-static processes were also simulated with ABAQUS/Explicit, in which very-low-speed displacement loadings were adopted to ensure that the kinetic energy could be ignored. The results show nearly no dynamic effect in the thickness direction, due to the thin nature of the composite plate and high-stress wave velocity in the through-the-thickness direction. In the radial direction, in contrast, apparent dynamic effects are observed, although the associated stress wave velocity is much higher. This may be attributed to the wave propagation from the impact site to the edge of the plate, reflecting repeatedly between the top and bottom surfaces of the plate.

1. Introduction

Carbon fiber composite laminates have been widely used in the field of engineering. The matrix cracks, interlaminar delamination, and fiber cracks can be easily induced in composite laminates when subjected to impact loading, which may lead to a serious degradation of load-carrying capacity of composite laminates. Therefore, to avoid catastrophic accidents, it is essential to carry out the damage assessment on composite structures. The stress condition of laminates under impact has not been defined.

Physical impact experiments are time-consuming and discrete compared with numerical simulation. Moreover, it is easier to observe stress changes using numerical models. However, theoretical analysis is an efficient tool for composite damage failure and strength prediction in impact experiments. After running the numbers in every possible combination, we can obtain a solution. Numerous studies have been undertaken to explore low-velocity impact by finite-element (FE) modelling, e.g., [1-6]. In the simulation model, a lot of delamination studies used cohesive elements to simulative interface absorption layer [4,7-9].

It is suggested that, re-expressed as "simulating with three-dimensional solid element models," this study attempts to implement in Abaqus/Explicit software prediction of the damage type of the composite laminate. In this paper, cohesive zone method is chosen over other delamination modelling approaches. between the plies of the plate is inset cohesive to simulate delamination. The numerical predictions coincided considerably well with the experimental curve, in, e.g., impact force-time and displacement-time history curves.

2. Research object

IOP Conf. Series: Materials Science and Engineering **657** (2019) 012058 doi:10.1088/1757-899X/657/1/012058

To study the dynamic effects of low-velocity impact on composite plates, multiple models were built and simulation models of stress presented to depict the difference between dynamic states and quasi-static. As shown in table 1, multiple models were built that determine the wave velocity and to contrast the stress charts of the quasi-static state with those of the dynamic state. A round plate is simple to observe dynamic effects. Therefore, a finite-element model was built as shown in figure 1. The numerical model was cut into a circular composite plate with different diameters to study the influence of the plate's length. As the plate's thickness affects quasi-static and dynamic states, models with different thicknesses were also built.

Table 1. Mechanical parameters of composite laminate.

	Diameter (mm)	Thickness (mm)
Case 1	50	2
Case 2	75	2
Case 3	100	2
Case 4	125	2
Case 5	75	5
Case 6	75	7.5
Case 7	75	10

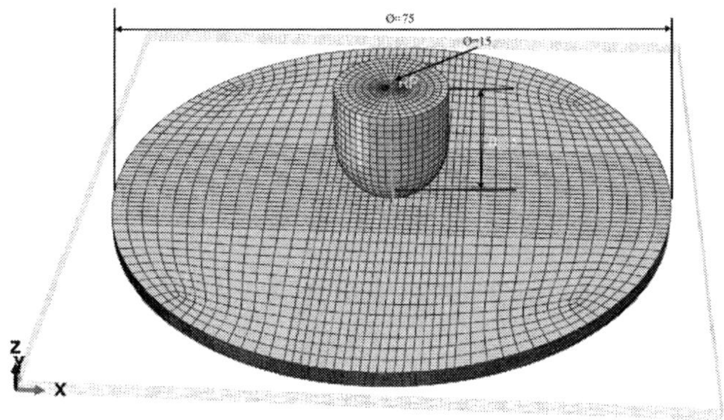

Figure 1. FE model.

3. Finite-element model

The laminate plate was manufactured with a carbon (Tenax HTS40 12 K 300)/epoxy (Cycom®977-2) composite. According to [10], the simulation model was cut into a circular plate modelled with a fully clamped edge and stacking sequence $[0/90]_{ns}$, which validated the feasibility of the FE modelling. The impactor material is analytical rigid, the diameter of which was 15 mm (see figure 1). From [7] and summarized in table 2, in the FE modelling, "general Contact" was used to define the contact schemes between the rigid impactor and the response of the impacted plate. Also, composite layer was contact by "general Contact", by means of a penalty function, in ABAQUS/Explicit. In addition, the tangential behavior was defined between the laminate surfaces with a friction coefficient μ=0.6. Between the rigid and the composite plate's top surface, the friction coefficient was set as μ=0.3

The hemispherical impactor of 1 kg was modelled with rigid elements (R3D4) and a hexahedral mesh and algorithm based on sweeping with an element size of 1 mm. All of the degrees of freedom were fixed with the exception of the z axis. The composite plate used C3D8R eight-node reduced integration elements for each ply. To reduce the computational cost, the FE model was split into different regions and different mesh sizes/density were built, with the impacted zone comprising 1 mm × 1 mm elements (as in figure 1), while other regions were cut into slices by 2 mm × 2 mm elements. Hourglass control was introduced to avoid unrealistic deformation of the simulation model. To simulate the delamination of the inter-laminate, eight-node three-dimensional cohesive elements

(COH3D8) were inserted between laminate faces, with a thickness of 0.0075 mm, and they were used to simulation model interface failure criteria. The max degradations were set to a value of 0.96 and when the elements failed the cohesive were removed from the finite element model until the failure criteria were satisfied.

Table 2. Mechanical parameters of composite laminate.

Density		1600 kg/m^3
Intra-laminar	Elastic properties	$E_1 = 153\,GPa; E_2 = E_3 = 10.3\,GPa$; $v_{12} = v_{13} = 0.3$; $v_{23} = 0.4$; $G_{12} = G_{13} = 6\,GPa$; $G_{23} = 3.7\,GPa$
	Strength (MPa)	$X^T = 2537$; $X_C = 1580$; $Y^T = 236$; $Y^C = 236$; $S_{12} = 90$; $S_{13} = 40$ S_{23}=40
	In-plate fracture toughness (kJ/m^2)	$G_{1C}^T = 91.6$; $G_{1C}^C = 79.9$; $G_{2C}^T = 0.22$; $G_{2C}^C = 1.1$; $G_S = 0.7$
Inter-laminar	Elastic properties	$Kn = 2000\,MPa$; $Ks = Kt = 1350\,MPa$
	Strength fracture energy	$Tn = 62.3\,MPa$; $Ts = Tt = 92.3\,MPa$ $Gn = 0.28\,N/mm$; $Gs = Gt = 0.79\,N/mm$

The simulation model was computed in the finite element code of Abaqus/Explicit by a Fortran subroutine (VUMAT). The velocity loading was used in dynamic simulation analysis with an initial velocity of 3.83 ms^{-1} and the quasi-static process used displacement loading. The loading curve is shown in figure 2.

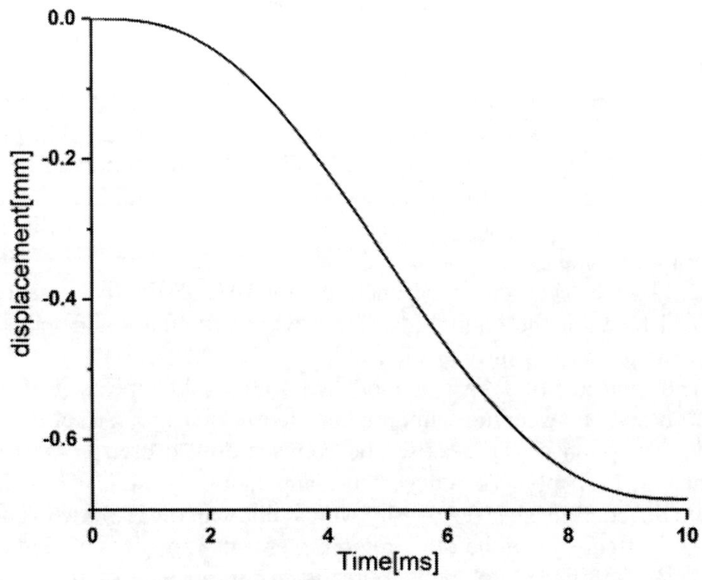

Figure 2. Spatial displacement along z axis.

IOP Conf. Series: Materials Science and Engineering **657** (2019) 012058 doi:10.1088/1757-899X/657/1/012058

4. Damage modelling

In general, FE models can be characterized by two main failure criteria: First, intra-laminar failure, due to fiber failure and matrix failure, matrix tensile and compressive damage or the damage of the fiber and matrix interface of the composite have disastrous consequences, and, second, inter-laminar failure and delamination that occurs between laminates. Composite damage mostly occurs in inter-laminar delamination with low-velocity impact.

4.1. Intra-laminar damage

Damage models can generally be studied by the number of failure criteria [11,12]. We used the Hashin criterion for the initiation criterion [13], which is one of the most widely used failure criteria for the modeling of initiation damage. The criterion has the following four different failure modes.

Fiber tensile failure ($\sigma_1 \geqslant 0$):

$$F_{ft} = \left(\frac{\sigma_1}{X_T}\right)^2 + \alpha\left(\frac{\tau_{12}}{S_{12}}\right)^2 + \alpha\left(\frac{\tau_{13}}{S_{13}}\right)^2 \geqslant 1 \quad (1)$$

Fiber compressive failure ($\sigma_1 < 0$):

$$F_{fc} = \left(\frac{\sigma_1}{X_C}\right)^2 \geqslant 1 \quad (2)$$

Matrix tensile failure ($\sigma_2 \geqslant 0$):

$$\left(\frac{\sigma_2}{Y_T}\right)^2 + \left(\frac{\tau_{12}}{S_{12}}\right)^2 \geqslant 1 \quad (3)$$

Matrix compressive failure ($\sigma_2 < 0$):

$$\left(\frac{\sigma_2}{2S_{12}}\right)^2 + \left(\frac{\sigma_2}{Y_c}\right)\left[\left(\frac{Y_C}{2S_{12}}\right)^2 - 1\right] + \left(\frac{\tau_{12}}{S_{12}}\right)^2 \geqslant 1 \quad (4)$$

In equation (1), σ_n denotes normal stress in the n direction, τ_{12} in-plane shear stress, and τ_{13} out-of-plane shear stress. X_T and X_C are the tensile and compressive strengths in the fiber direction, respectively; similarly, Y_T and Y_C are the tensile and compressive strengths in the transverse direction, respectively. S_{12} and S_{13} are the shear strengths in the fiber and transverse planes, respectively. The coefficient α in equation (1) is applied to determine the contribution to fiber tensile failure, which is uncontributed in this paper.

4.2. Delamination

In work by Camanho and Dávila, an inter-laminar failure criterion was proposed to judging damage for numerical simulation models [14]. The cohesive zone elements used mixed mode loading, which the model determined as damage modes I, II, and III simultaneously. The stress failure criterion used to estimate the delamination onset is given by:

$$\left(\frac{\sigma_n}{N}\right)^2 + \left(\frac{\sigma_s}{S}\right)^2 + \left(\frac{\sigma_t}{T}\right)^2 = 1 \quad (5)$$

where σ_i (i=n,s,t) stands for tractions in each pure mode in the normal n and shear directions s and t, respectively, while N, S, and T are the inter-laminar strength normal strength and two shear strengths, which can be calculated as:

$$\sigma_i = K_i \delta_i (i = n,s,t) \quad (6)$$

IOP Conf. Series: Materials Science and Engineering **657** (2019) 012058 doi:10.1088/1757-899X/657/1/012058

The Abaqus manual contains a detailed illustration regarding cohesive elements for the initiation criterion and failure criterion [15].

5. Results and discussion

The result of this experiment are similar to the simulation analysis results in [10]. During the impact process, the impact force-time, force-displacement, and energy-time history curves are captured. Additionally, the study of some impact dynamics problems was based on displacement load.

5.1. Validation of numerical model

The simulation prediction and experimental data were compared, including, e.g., impact force-time, force-displacement, and energy-time history curves under 7.4 J impact energies (see figure 3). The force-time history curves coincided, which upon initial contact reveal some oscillations in the time interval 0–0.5 ms. When the contact force approached the maximum value, strong variations of the force-time history curves occurred. The impactor velocity was reduced to zero and sprung back. The numerical simulation curve was higher than the experimental one and took a longer time to reach zero. This phenomenon may be because the damaged cohesive element being deleted, which thus produced contact forces between the delaminated plies.

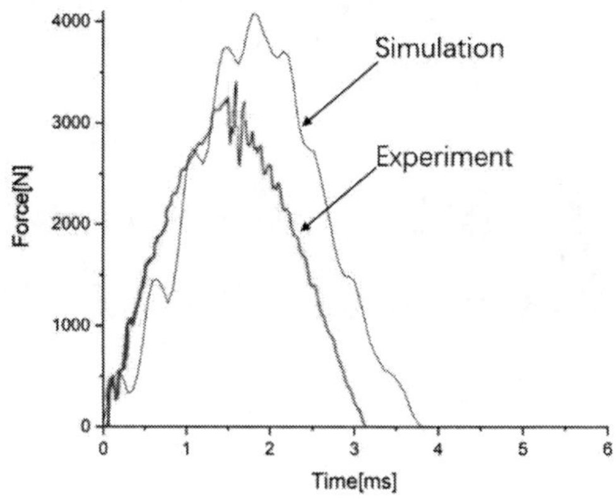

Figure 3. Numerical and experimental impact-force-time history curves.

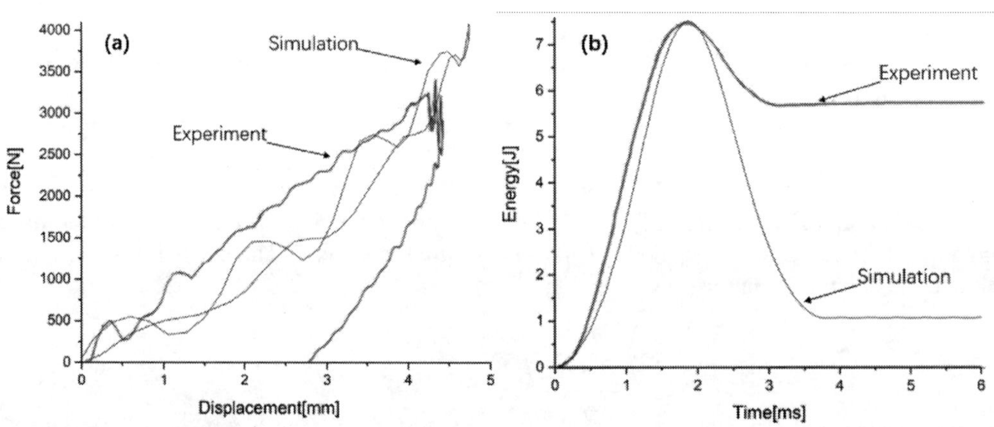

Figure 4. Experimental and numerical history curves: (a) force–displacement history curves and (b) energy-time history curves.

The experimental and numerical-simulation-predicted force-displacement curves are compared in figure 3. The numerical model takes more time than the experiment, as the time for maximum displacement was larger. This is because the finite element model spent longer time release energy than experiment, and the impacted plate can return to their initial position completely. In figure 4(a) the simulation curves evidence crossover, indicating that the elastic model must be symmetric so that the value increases and is recovered. In this process, it may be acceptable that the force-displacement curve while unloading exceeds the loading curve. In figure 4(b) are shown the energy-time curves, in which the simulation-predicted curves have released all of the energy. A linear spring model is used in this paper, in which low-energy impact cannot destroy the composite plies, so the laminate was fully recovered.

Figure 5 shows energy-time history curves for the quasi-static state. The internal energy extends far beyond the kinetic energy. This phenomenon is consistent with the quasi-static model.

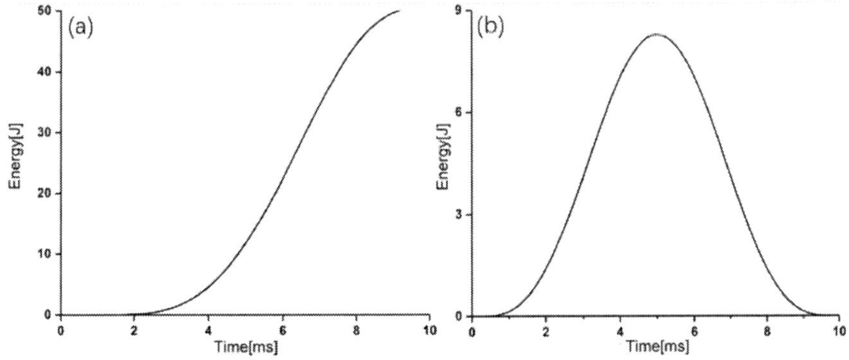

Figure 5. (a) Internal energy and (b) kinetic energy for entire model.

5.2. Analysis of dynamical effects

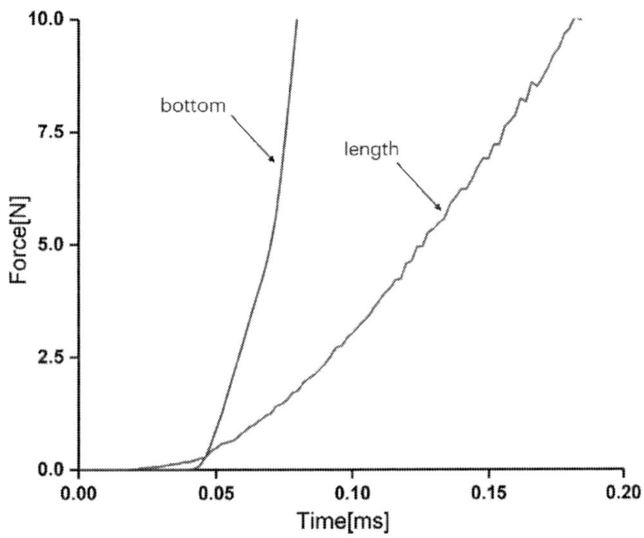

Figure 6. Plate stresses along radial direction and thickness.

The FEs were based on consideration of nonlinear dynamic analysis. In the dynamic model, there is a time difference between the side-point of the plate's top face and the plate's back face in the mid-point. To study the time difference, an FE model was built as impacted plate of thickness 10 mm and diameter 75 mm (table 1, Case 7). This case result is plotted in figure 6. The wave speed is the difference between the bottom and the radius. To further understand the difference, a variety of

IOP Conf. Series: Materials Science and Engineering **657** (2019) 012058 doi:10.1088/1757-899X/657/1/012058

thicknesses and ratios are modelled based on experimental design for a finite model; the composite plate size is presented in table 1. There is a clear difference between shock wave velocity along the thickness and along the length. The details are shown in tables 2 and 3. Apparently, propagation of shock waves in the thickness direction is slower than the orientation of plies in a laminate. The wave velocity is approximately 625 m/s along the z axis and the propagation speed is approximately 8000 m/s in the inter-laminate. This phenomenon may be because the speed of a shock wave is determined by Young's modulus and the density. In addition, the fiber modulus differs significantly from the resin modulus. In multilayer composites, the multilayer interfaces lead to the multiple reflections of acoustic waves. Thus, it can be seen that the propagation speed is slower in the thickness direction of shock waves.

Table 3. von Mises stress in plate along longitudinal direction.

Length (mm)	Dynamic (μs)	Quasi-static (μs)
25	4	1200
37.5	4	1200
50	6	1200
62.5	6	1200

Table 4. von Mises stresses in plate along thickness direction.

Thickness (mm)	Dynamic (μs)	Quasi-static (μs)
2	4	1300
5	6	1300
7.5	12	1300
10	16	1300

In the quasi-static model, the stress on each element arrives at the same time; the results are shown in tables 3 and 4. Displacement loading stresses the plate structure evenly. A composite plate is uniformly stress under quasi-static stress, and it was especially easy to draw this conclusion.

Table 5. von Mises stresses in plate (0.18 ms).

	Static	Dynamic	S/D
Case 1	Max 1.665×10^{2}	Max 1.984×10^{2}	0.839
	Min 1.487×10^{-1}	Min 4.279×10^{-2}	3.48
Case 2	Max 1.166×10^{2}	Max 2.072×10^{2}	0.563
	Min 4.676×10^{-2}	Min 1.120×10^{-2}	4.18
Case 3	Max 1.032×10^{2}	Max 2.067×10^{2}	0.499
	Min 4.325×10^{-2}	Min 1.128×10^{-2}	3.83
Case 4	Max 8.325	Max 2.057×10^{2}	0.405
	Min 3.180×10^{-2}	Min 2.971×10^{-2}	1.07
Case 5	Max 1.656×10^{2}	Max 2.759×10^{2}	0.6
	Min 4.249×10^{-2}	Min 1.372×10^{-2}	3.10
Case 6	Max 2.089×10^{2}	Max 3.137×10^{2}	0.666
	Min 2.919×10^{-2}	Min 1.029×10^{-2}	2.84
Case 7	Max 3.145×10^{2}	Max 4.127×10^{2}	0.762
	Min 3.874×10^{-2}	Min 5.970×10^{-2}	0.649

In the nonlinear dynamic analysis and dynamic stability analysis of impact under low energy, the loading mode will affect the distribution of stress. In figure 7, to facilitate observation, the difference between static and dynamic stress, three kinds of models are shown with the same loading distance. There is a clear difference between figures 7(c) and 7(d). The plate radius is an important factor for determining the composite stress state. The ratio of the maximum force and the minimum force is shown in table 5. From details presented in this table, a trend can be seen, namely, that the stress is more similar in both dynamic states and quasi-static when the plate is smaller. Upon changing the plate length, the stress will change significantly in the quasi-static state, while in the dynamic process, the stress remains the same with length variation. From the simulation chart in figure 7, a small influence of the variation of media thickness is observed, and both static and dynamic model stresses have increased. According to the results, although the shock wave is faster in an intra-laminate in the thickness direction, the dynamic stress effect is more visible in the length direction. One possible reason for this is that most of the shock-wave propagation undergoes multiple reflections inside the material, which leads to the length direction time difference being greater than that of the material thickness. The time difference is approximately an order of magnitude of 10 μs along the thickness direction. Minor effects occur with changing thickness.

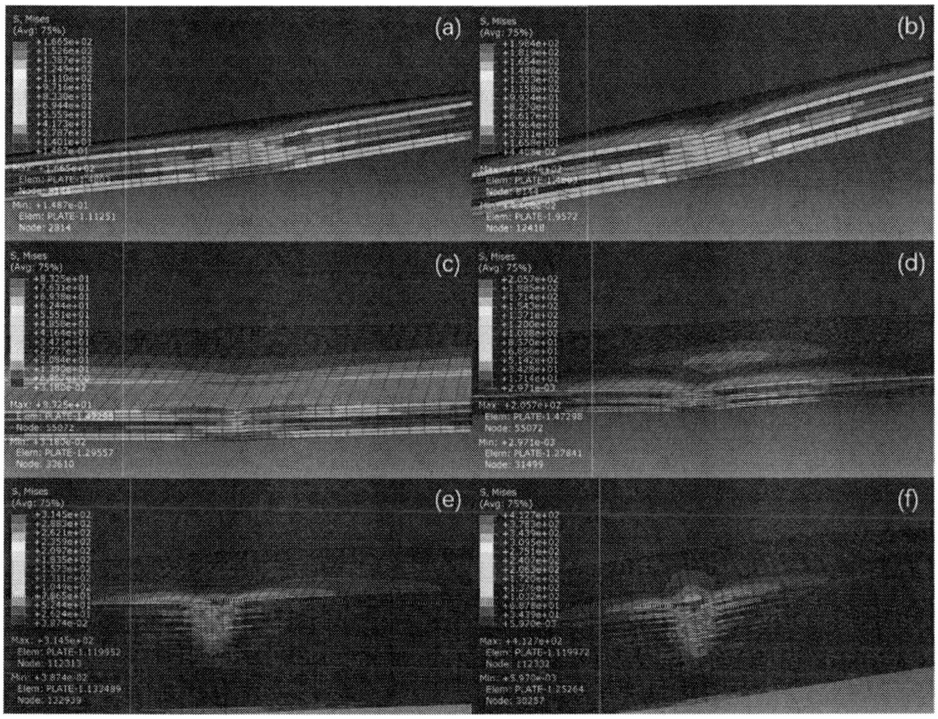

Figure 7. Plate stress deformation charts in quasi-static (a,c,e) and dynamic (b,d,f) models. FE models with (a,b) diameter 50 mm and thickness 2 mm, (c,d) diameter 125 mm thickness and 2 mm, and (e,f) diameter 75 mm and thickness 10 mm.

6. Conclusions

In this paper, first, a delamination analysis process for composite laminates under low-velocity impact was presented. Next, a calculation method was summarized and compared with quasi-static and dynamic models through corresponding finite-element analysis. Contrasting the velocity loading with distance loading, the FE model obviously changed with different model lengths. Ratio is an important parameter in analyzing dynamics effects. In multilayer composites, the multilayer interfaces lead to multiple reflections of acoustic waves. Hence, there is sufficient time difference along the length direction, rather than along the thickness direction, with the wave speed being slower.

Acknowledgments
The work was supported by the National Natural Science Foundation of China (Grant No. 51503223) and the Postdoctoral Science Foundation (Grant No. 2017M623378). We thank LetPub (www.letpub.com) for its linguistic assistance during the preparation of this manuscript.

References

[1] Soto, González E V, Maimí P, Martín de la Escalera F, Sainz de Aja J R and Alvarez E 2018 Low velocity impact and compression after impact simulation of thin ply laminates *Composites Part A* **109** 413-27
[2] Ghelli D and Minak G 2011 Low velocity impact and compression after impact tests on thin carbon/epoxy laminates *Composites Part B* **42** 2067-79
[3] Hongkarnjanakul N, Bouvet C and Rivallant S 2013 Validation of low velocity impact modelling on different stacking sequences of CFRP laminates and influence of fibre failure *Compos. Struct.* **106** 549-59
[4] Long S, Yao X and Zhang X 2015 Delamination prediction in composite laminates under low-velocity impact *Compos. Struct.* **132** 290-8
[5] Menna C, Asprone D, Caprino G, Lopresto V and Prota A 2011 Numerical simulation of impact tests on GFRP composite laminates *Int. J. Impact. Eng.* **38** 677-85
[6] Park H 2019 Investigation on low velocity impact behavior of sandwich composite and monolithic laminate plates using FEM analysis *Compos. Struct.* **220** 842-6
[7] Zhang C, Duodu E A and Gu J 2017 Finite element modeling of damage development in cross-ply composite laminates subjected to low velocity impact *Compos. Struct.* **173** 219-27
[8] Abir M R, Tay T E, Ridha M and Lee H P 2017 Modelling damage growth in composites subjected to impact and compression after impact *Compos. Struct.* **168** 13-25
[9] Thorsson S I, Waas A M and Rassaian M 2018 Low-velocity impact predictions of composite laminates using a continuum shell based modeling approach part A: Impact study *Int. J. Solids. Struct.* **155** 185-200
[10] Shi Y, Swait T and Soutis C 2012 Modelling damage evolution in composite laminates subjected to low velocity impact *Compos. Struct.* **94** 2902-13
[11] Hashin Z and Rotem A 1973 A fatigue failure criterion for fiber-reinforced materials *J. Compos Mater.* **7** 448-64.
[12] Hashin Z 1980 Failure criteria for uni-directional fibre composites *J. Appl. Mech.* **47** 329-34
[13] Murakami S 1988 Mechanical modeling of material damage *J. Appl. Mech.* **55** 280-6
[14] Camanho P P and Dávila C G 2002 Mixed-Mode decohesion finite elements for the simulation of delamination in composite materials *Tech. Rep. NASA/TM-2002- 21173*
[15] ABAQUS 2010 ABAQUS Version 6.10 Dessault systemes *Providence, RI*

IOP Conf. Series: Materials Science and Engineering **657** (2019) 012059 doi:10.1088/1757-899X/657/1/012059

Contact force analysis in a planar mechanism with translational clearance joint considering complex contact modes

M B Qian[1,2], Z Qin[2], S Z Yan[1,3] and L Zhang[1]

[1]State Key Laboratory of Tribology, Department of Mechanical Engineering, Tsinghua University, Beijing, 100084, P R China
[2]Department of Mechanical Engineering, Zhejiang A & F University, Hangzhou, 311300, China

E-mail: yansz@mail.tsinghua.edu.cn

Abstract. A translational joint model with clearance was established. A contact mode identification method was derived on the basis of the classification of contact modes. The positional relationships were classified into three different contact modes. The identification method was applied to a planar mechanism with translational joint. The influences of clearance size and velocity on dynamic characteristics between them were studied. The mean square error value of contact force can be used to evaluate the influence of the clearance on the dynamic characteristics of the mechanism.

1. Introduction

Clearance is an important factor affecting the dynamic performance of kinematic pairs because its existence is a necessary condition for relative motion between the connected components thereof [1-3]. A translational joint is a common motion pair: collisions between the translational joint sub-elements during the process of movement cause changes in dynamic characteristics due to the clearances therein [4-7].

At present, much research has been carried out on the subject of the effect of joint clearances on the dynamic characteristics of mechanical systems [8-11]. Translational joints with clearance are the research focus. By using a planar mechanism as an example, Jia *et al* [12,13] studied the influence of the clearance of the tripod ball sliding joint on dynamic characteristics of the mechanism, and a force constraint equation was established when there is a clearance in the kinematic pair. Song *et al* [14] proposed a dynamic analysis method for planar linkages, which considers the impact of translational joint with clearances. A virtual prototype model of a mechanism with translational joint with clearance was established based on the study of contact force and Coulomb friction force of the non-linear spring damper by Tong *et al* [15]. This model can reflect the collision and friction phenomenon between the sub-elements of translational joint with clearance.

To sum up, there is little research available on the motion state of translational joints and the contact mode between two elements when the clearance is considered. There is no systematic study of the motion characteristics of the two elements of the translational joint with clearance: however, the research in this field is important in any theoretical study and practical design of the dynamic behaviour of the mechanism. Here, we established a translational joint model with clearance and derived a contact-mode identification method on the basis of the classification of contact modes. To study the non-linear dynamic behaviour of mechanism with clearance, numerical calculation and simulation of a planar mechanism were conducted.

2. Translational joint with clearance model

Figure 1 shows a simplified model of a translational joint with clearance. The slider and the slide way are represented by *i* and *j* respectively. The local coordinate systems attached to *i* and *j* are O_i-$x_i y_i$ and O_j-$x_j y_j$, respectively. Points P, Q, R and S on the surface of the inner hole of the slideway refers to the geometric contact limits for the slider [16].

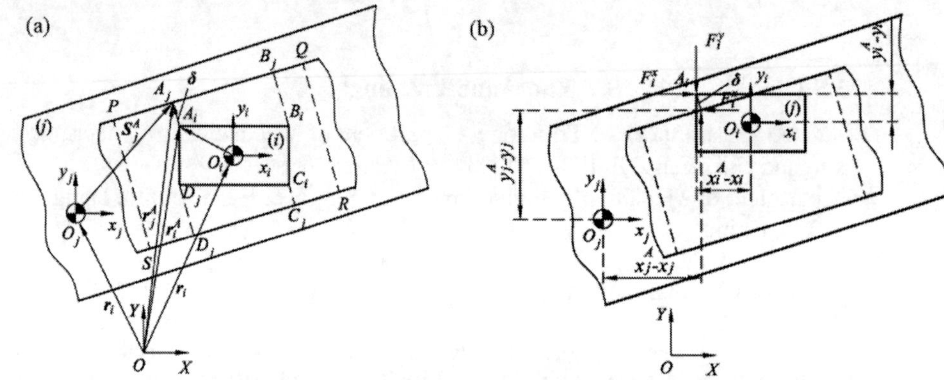

Figure 1. Two different contact conditions of translational joint with clearance (a) Free movement; (b) contact deformation.

Two different contact conditions of the slider and the slide way are shown in figure 1. Provided that the point A_j on the slide way is the potential contact point of the point A_i on the slider, A_j can be represented in the global coordinate system as [17]

$$\boldsymbol{r}_j^A = \boldsymbol{r}_j^P + \boldsymbol{t}^T \left(\boldsymbol{r}_i^A - \boldsymbol{r}_j^P\right)\boldsymbol{t} \tag{1}$$

$$\delta = \boldsymbol{r}_j^A - \boldsymbol{r}_i^A \tag{2}$$

When the slider and slideway are not in contact, vector δ represents aligned with vector *n*, otherwise it is opposite it: c is applied to represent the contact condition between *i* and *j*:

$$c = \begin{cases} 1, \delta^T \boldsymbol{n} > 0 \\ 0, \delta^T \boldsymbol{n} \le 0 \end{cases} \tag{3}$$

where δ denotes the piercing depth and its size is expressed as

$$\delta = \sqrt{\delta^T \delta} \tag{4}$$

The collision velocity $\dot{\delta}$ can be given as

$$\dot{\delta} = \dot{\boldsymbol{r}}_j + \dot{\boldsymbol{N}}_j \boldsymbol{s}_j^A - \dot{\boldsymbol{r}}_i - \dot{\boldsymbol{N}}_i \boldsymbol{s}_i^A \tag{5}$$

2.1. *The tangential friction*

The contact force between the sub-elements is mainly composed of normal contact force and tangential friction force in the contact region. The modified Coulomb friction model is applied to describe the tangential friction [18]:

$$F_T = -\mu(v_t) F_N \frac{v_t}{|v_t|} \tag{6}$$

Where $\mu(v_t)$ represents the effective dynamic friction coefficient, v_t is the relative sliding speed of the two bodies in the tangential direction.

The normal contact force can be analysed according to the different contact modes of the sub-elements (figure 2).

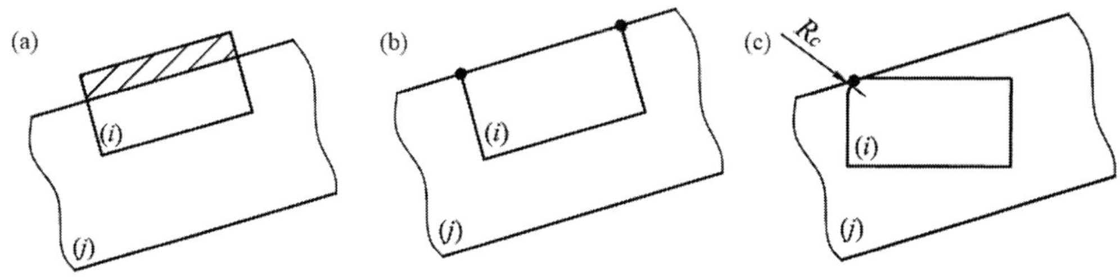

Figure 2. Three contact modes between *i* and *j*. (a) Surface contact; (b) line contact; (c) point contact.

2.2. *The normal contact force*

- When the slider is in surface contact with the slide way (figure 2(a)), a linear model is used to calculate the normal contact force on the contact surface [19]:

$$F_N = K\delta \tag{7}$$

where *K* is the generalised stiffness coefficient, which can be expressed as:

$$K = \frac{a}{0.475(\sigma_i + \sigma_j)} \tag{8}$$

$$\sigma_k = \frac{1-\mu_k^2}{E_k}, (k = i, j) \tag{9}$$

where a represents half of the perimeter of the contact surface, σ_i and σ_j are the material coefficients of the slider and the slide way, respectively, μ_k and E_k are the Poisson's ratio and elastic modulus of the material, respectively.

- When the slider is in line contact with the slide way (figure 2(b)), the contact is considered to be a cylinder-plane contact and the contact region can be regarded as a narrow rectangle, and a damping model is applied to calculate the normal contact force on the contact surface [20,21]:

$$F_N = Pl + C\dot{\delta}\delta \tag{10}$$

$$\delta = \frac{P(1-v^2)}{\pi E}\left(2\ln\frac{2r}{a} - \frac{v}{1-v}\right) \tag{11}$$

$$a = \left(\frac{4PR}{\pi E^*} \right)^{\frac{1}{2}} \tag{12}$$

where C represents the damping coefficient, and is generally 0.1% ~ 1% of the equivalent contact stiffness K, P denotes the pressure per unit length, $2a$ is the contact width, E is the elastic modulus of the material, r is the radius of the cylinder, and E^* and R are defined as

$$\frac{1}{E^*} = \frac{1-v_1^2}{E_1} + \frac{1-v_2^2}{E_2} \tag{13}$$

$$R = \frac{r_2 r_1}{r_2 - r_1} \tag{14}$$

- When the slider is in point contact with the slide way (figure 2(c)), the contact region can be idealised as an arc of radius R_c. The non-linear spring damping model is used to solve the normal contact force in this case [22]:

$$F_N = K\delta^e + C\dot{\delta}\delta \tag{15}$$

where e is the Hertz contact force index (generally taken as between 1.3 and 1.5 for metals).

3. Case study

In this section, a planar mechanism with translational clearance joint was studied (figure 3), and the geometric and inertial properties of the mechanism are listed in table 1. The difference in size between the slider and the slide way defines the size of the clearance. The crank is the driving link of the mechanism and rotates at a constant angular velocity of π rad/s. The step-size and time of simulation are 0.01 and 2 s respectively in the case study.

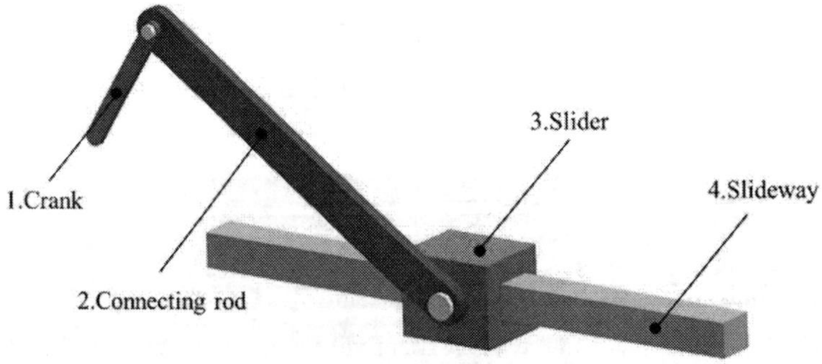

Figure 3. The mechanism with translational clearance joint.

Table 1. Geometric and inertial properties of the mechanism

Rods	Length (mm)	Width (mm)	Depth (mm)	Density (kg/m³)	Elastic modulus (N/m²)	Poisson's ratio
1	50	5	2.5	7801.0	2.07×10^{11}	0.29
2	142.2	14.2	7.1	7801.0	2.07×10^{11}	0.29
3	40	20	30	7801.0	2.07×10^{11}	0.29
4	250	20	30	7801.0	2.07×10^{11}	0.29

Based on the division of the contact modes of the sub-elements of the translational joint, a user sub-routine that can be dynamically linked with MSC.ADAMS was written in FORTRAN. Accordingly, the contact modes can be divided into three different types: point contact, line contact, and surface contact.

The influences of clearance size and velocity on the contact force between sub-elements were studied by changing the size of the inner hole of the slider while the width and depth of the slide way are fixed at 2 mm and 3 mm respectively. The parameters of the inner hole of the slider used in different cases are listed in table 2. The peak fluctuation, when the direction of contact force is positive, is taken as the research object, and the results are shown in figure 4.

Table 2. Parameters of the inner hole of the slider used in the four cases.

Cases	Length (mm)	Width (mm)	Depth (mm)	Clearance (mm)	Operation speed (rad/s)
1	40	2.2	3.2	0.1mm	π
2	40	2.1	3.1	0.5mm	π
3	40	2.2	3.2	0.1mm	1.22 π
4	40	2.2	3.2	0.1mm	2π

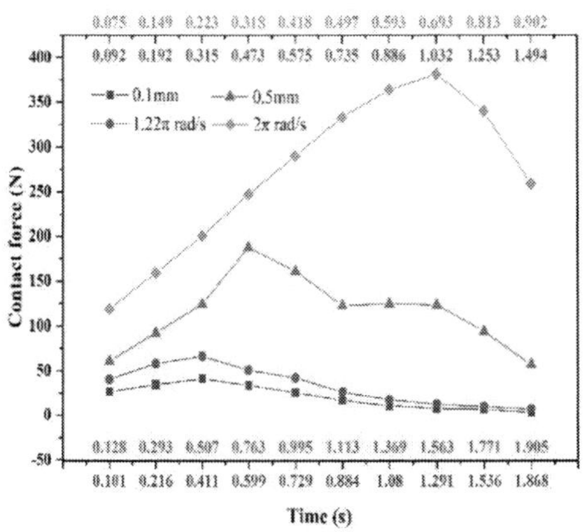

Figure 4. Contact force in different cases.

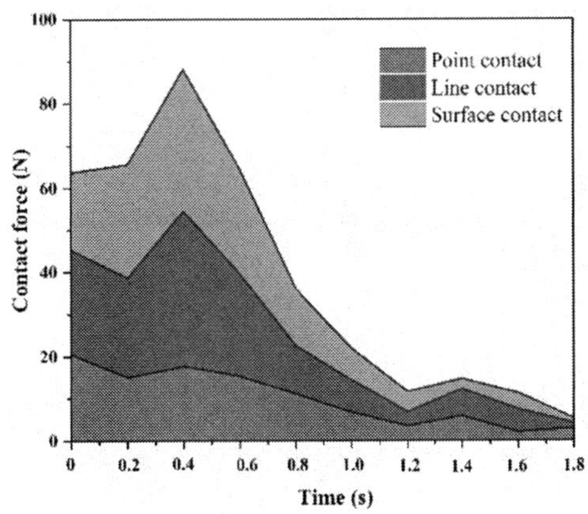

Figure 5. Contact force in different contact modes.

As can be seen from the figure 4, with the clearance size decreases or the speed increases, the contact force between sub-elements would increase. Meanwhile, the contact force was calculated according to different contact types, and we could assess the contact force under different contact modes. The result of the contact force in different contact modes in Case 1 is shown in figure 5. The simulation results are taken from different simulation intervals within one cycle.

The simulation results of the mechanism with 0.1 mm clearance size and π rad/s operation speed are divided according to different contact modes, as shown in figure 5. The simulation step-size and time of simulation are the same as the case study in figure 4. It is shown from figure 5 that, when the sub-elements are under different contact modes, the maximum contact force (F_{max}) in different modes during the simulation is different. This is because the magnitude of contact force is related to the positional relationship between sub-elements. We can also find that the contact force under different contact modes show irregular changes at the beginning of each simulation: however, the contact force continues to decrease after 0.5 s, which is consistent with the variation in contact force found previously.

This work can be employed for design optimization for translational joint design. The fluctuation amplitude and frequency of the contact force curve would change due to the existence of clearance. Therefore, the mean-square error value of contact force can be used to evaluate the influence of the clearance on the dynamic characteristics of the mechanism.

4. Results and discussion

We established a translational joint model with clearance and derived a contact mode identification method on the basis of the classification thereof. The positional relationships of the sub-elements were classified into three different contact modes: surface contact, line contact, and point contact. The contact force between sub-elements was calculated separately under each of the three different contact modes.

A contact-mode identification method was applied to a planar mechanism with translational joint. The influences of clearance size and velocity on the contact force between sub-elements were studied. when the sub-elements are under different modes of contact, the maximum contact force (F_{max}) in different modes during the simulation is different.

Acknowledgments

This work is supported by National Science Foundation of China under Contract Nos. 11872033, 51875531, and 51505432, and Natural Science Foundation of Beijing under Contract No. 3132017.

References

[1] Flores P, Ambrósio J, Claro J C P, Lankarani H M and Koshy C S 2006 A study on dynamics of mechanical systems including joints with clearance and lubrication *Mech. Mach. Theory* **41** 247-61

[2] Li J L, Huang H Z, Yan S Z and Yang Y 2017 Kinematic accuracy and dynamic performance of a simple planar space deployable mechanism with joint clearance considering parameter uncertainty *Acta. Astronaut.* **136** 34-45

[3] Xiang W W K, Yan, Wu J N and Robert X G 2014 Complexity evaluation of nonlinear dynamic behavior of mechanisms with clearance joints by using the fractal method *Proc. Inst. Mech. Eng, Part C J. Mech. Eng. Sci.* **228** 3482-3495

[4] Schwab L, Meijaard J P and Meijers P 2002 A comparison of revolute joint clearance models in the dynamic analysis of rigid and elastic mechanical systems *Mech. Mach. Theory* **37** 895-913

[5] Xiang W W K, Yan S Z and Wu J N 2015 A comprehensive method for joint wear prediction in planar mechanical systems with clearances considering complex contact conditions *Sci. China Tech. Sci.* **58** 86-96

[6] Farahanchi F and Shaw S W 1994 Chaotic and periodic dynamics of a slider-crank mechanism with slider clearance *J. Sound Vib.* **177** 307-24

[7] Olyaei and Ghazavi M R 2012 Stabilizing slider-crank mechanism with clearance joints *Mech. Mach. Theory* **53** 17-29

[8] Mukras S, Kim N H and Mauntler N A 2010 Analysis of planar multibody systems with revolute joint wear *Wear* **268** 643-52

[9] Bauchau A and Rodriguez J 2002 Modeling of joints with clearance in flexible multibody systems *Int. J. Solids Struct.* **39** 41-63

[10] Flores P, Koshy C S, Lankarani H M, Ambrósio J and Claro J C P 2011 Numerical and experimental investigation on multi-body systems with revolute clearance joints *Nonlinear Dyn.* **65** 383-98

[11] Flores P and Ambrósio J 2004 Revolute joints with clearance in multibody systems *Comput. Struct.* **82** 1359-69

[12] Jia X H, Ji L H, Jin D W and Zhang J C 2000 Modeling of tripod-ball sliding joint with clearance *J. Tsinghua Univ.* **40** 125-8

[13] Jia X H, Ji L H, Jin D W and Zhang J C 2000 Dynamic analysis of the crank slider mechanism including tripod-ball sliding joint with clearance *J. Mech. Sci. Technol.* **19** 698-700

[14] Song L and Yang J 2001 A method of planar linkage dynamic analysis with slider joint

impacting *Mach. Des. Res.* **17** 36-9

[15] Tong L and Wang H J 2007 Dynamic simulation of a crank-slider mechanism including sliding joint with clearance based on virtual prototyping *J. Chongqing Univ. Sci. Tech.* **9** 25-7

[16] Flores P, Ambrósio J and Claro J C 2008 Translational joints with clearance in rigid multibody systems *J. Comput. Nonlinear Dyn.* **3** 011007

[17] Flores P, Leine R and Glocker C 2010 Modeling and analysis of planar rigid multibody systems with translational clearance joints based on the non-smooth dynamics approach *Multibody Sys. Dyn.* **23** 165-90

[18] Flores P 2009 Modeling and simulation of wear in revolute clearance joints in multibody systems *Mech. Mach. Theory* **44** 1211-22

[19] Tian Q, Liu C, Machado M and Flores P 2011 A new model for dry and lubricated cylindrical joints with clearance in spatial flexible multibody systems *Nonlinear Dyn.* **64** 25-47

[20] Bai Z F and Zhao Y 2013 A hybrid contact force model of revolute joint with clearance for planar mechanical systems *Int. J. Nonlinear Mech.* **48** 15-36

[21] Lankarani H M and Nikravesh P E 1990 A contact force model with hysteresis damping for impact analysis of multibody systems *J. Mech. Des.* **112** 369-76

[22] Yigit A S, Ulsoy A G and Scott R A 1990 Spring-dashpot models for the dynamics of a radially rotating beam with impact *J. Sound Vib.* **142** 515-25

IOP Conf. Series: Materials Science and Engineering **657** (2019) 012060 doi:10.1088/1757-899X/657/1/012060

Research on low frequency electromagnetic environment of electric vehicle and human health

X W Dong[1,2] and M Lu[1]

[1]Key Laboratory of Opto-technology and Intelligent Control, Ministry of Education, Lanzhou Jiaotong University, Lanzhou 73007, China

E-mail: dxw007@126.com

Abstract. In order to study the influence of electromagnetic environment produced by low frequency current in power cables of electric vehicle on driver's and co-driver's tissues, especially in the central nervous system (CNS). The distribution of the magnetic induction intensity and the induction electric field intensity in human trunk and CNS of the driver's and co-driver's are calculated and analysed. The results are compared with the related standards of International Committee on Non-ionizing Radiation Protection (ICNIRP). The study shows that the induced fields in human trunk and CNS are much smaller than the occupational and public exposure limits defined by ICNIRP. Therefore, the low frequency current would be safe for human health to some extent.

1. Introduction

With the deterioration of global ecological environment and the worsening problem of energy and resource exhaustion, it has become a new competition focus in the automotive industry to vigorously promote the related technology research of electric vehicles. By 2020, Some organizations predict that, global sales of electric vehicles will reach more than 34.0 million, including 31% in Europe, 28% in China, 18% in the United States and 15% in Japan [1]. However, unlike traditional fuel-fired vehicles, the power system of electric vehicles usually requires hundreds of kilowatts of power [2]. The cabinet space of electric vehicles is relatively compact, and passengers are very close to electric power components such as high-power motors, inverters and high-voltage power cables. During the driving, passengers are inevitably exposed to the electromagnetic environment generated by these high-power electric equipments. If exposed to the electromagnetic environment for a long time, the human body may be discomfort, such as headache, dry skin and so on. Besides, it may also increase the probability of inducing diseases of nervous system and immune system [3].

In recent years, some researchers have measured and calculated the distribution of electromagnetic field in different electric vehicle compartments and obtain some conclusions about the characteristics of field distribution [4-6]. Especially, in the [7], the distribution of spatial field in different positions of ten electric car compartments which are popular in the Chinese market is measured, and the safety of electromagnetic exposure for child passengers is studied. Others have carried out relevant research on electromagnetic radiation of electric vehicle wireless charging system to human body [8,9]. However, there is few researches on human exposure to the low-frequency electromagnetic environment of power cables in electric vehicle.

When human body is exposed to electromagnetic field, it is impossible to obtain the distribution of physical quantities such as magnetic induction intensity and induction electric field intensity in various

IOP Conf. Series: Materials Science and Engineering **657** (2019) 012060 doi:10.1088/1757-899X/657/1/012060

tissues by external equipment. For the kind of problem, the general method is based on the principle of electromagnetic dosimetry, using the numerical calculation method to quantitatively analyze the distribution of electromagnetic field in human tissues [10]. The paper is based on the finite element method, the electromagnetic simulation software is used to model the electric vehicle body, power cables and passengers. The simulation model of electromagnetic environment of the electric vehicle power cables is established. The distribution of different induction fields in the trunk and CNS of driver's and co-driver's are calculated, and the results are compared with the standards [11] defined by International Committee on Non-ionizing Radiation Protection (ICNIRP), and evaluate the safety of electromagnetic exposure in the low-frequency electromagnetic environment. It has positive significance to enrich the research on electromagnetic radiation of electric vehicles and to eliminate the fear of electromagnetic exposure and occupational diseases.

2. Calculating model and dielectric parameters of human body

For the most electric vehicles, the power battery pack is usually located directly below or behind the car body. The power battery pack is connected with other electrical systems such as motors, inverters and other electric equipments by power cables. The general picture of typical electric vehicles is shown in figure 1.

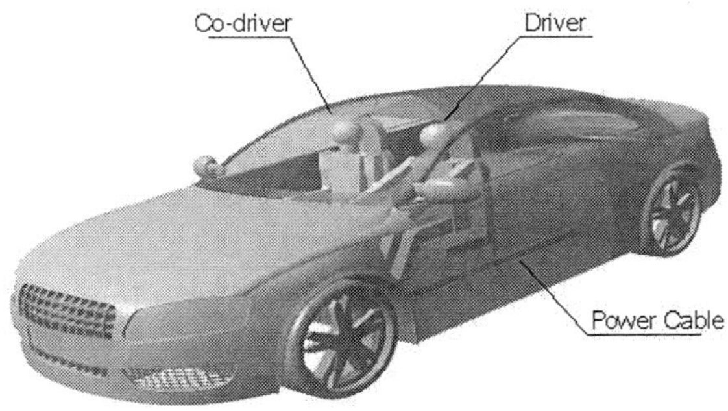

Figure 1. Overview of electric vehicles.

2.1. Calculating model

Using the Comsol Multiphysics as simulation platform, a simplified low-frequency electromagnetic environment model is established. As shown in figure 2, the simplified electromagnetic environment model includes car compartment, driver, co-driver, power cable, windows and so on.

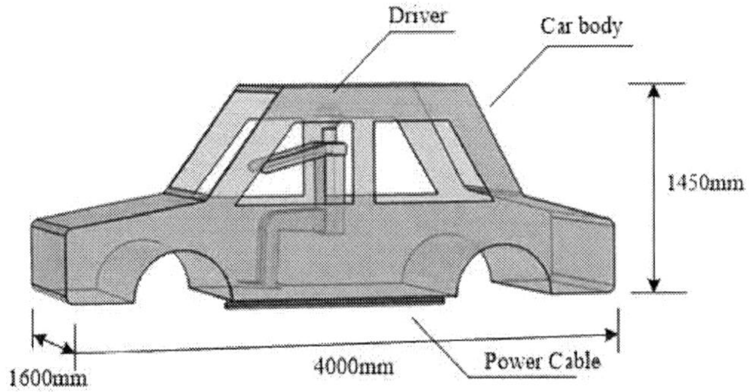

Figure 2. The model of the simple car with human body and electromagnetic power cable.

The human body model refers to the international standing posture of 1.75 m adult male and establishes the human sitting posture model in proportion, as shown in figure 3. For human head model, we also adopt a simple three-layer head model, which is commonly used in the related scientific research. In this model, the radius of scalp is 92 mm, the radius of skull is 85 mm, and the radius of brain is 80 mm [12].

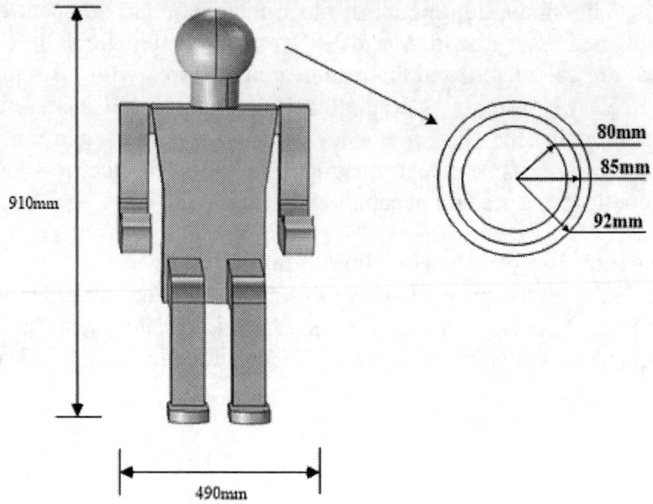

Figure 3. The sitting posture of human model and head dimensions.

2.2. *Human dielectric parameters*

The frequency-dependent tissue conductivity can be extracted from the internet database of Italian National Research Council [13,14]. For a simple human model, it is assumed that the human tissues is uniformity. The electrical conductivity and relative permittivity of the brain are taken as the average values of cerebrospinal fluid, white matter and gray matter; the electrical conductivity and relative permittivity of the trunk are taken as the average values of muscle, blood and bone. Thus, the approximate values of electrical conductivity and relative permittivity of the main tissues of the human body at 500 Hz are shown in table 1.

Table 1. Relative permittivity and conductivity of human main tissues at 500 Hz.

Biological tissues	Relative permittivity	Conductivity(S/m)
Scalp	34116.0	0.0006
Skull	22295.0	0.0813
Brain	153926.3	0.7192
Trunk	371684.8	0.3637

3. Calculating results

In the study, the AC/DC magnetic field module is used in Comsol Multiphysics software to calculate and analyze the magnetic induction intensity ($|B|$) and induced electric field intensity ($|E|$) in the trunk and CNS of driver's and co-driver's, when the current in the power cables of electric vehicles is 30 A in amplitude and 500 Hz in frequency. During the calculation, we use the platform of Sugon high performance cluster with 64 G memory and the calculating time is 3-4 hours. The results are as follows.

3.1. *Distribution of human magnetic induction intensity*

The distribution of magnetic induction intensity in the driver's trunk is shown in figure 4. The maximum magnetic induction intensity in the driver's trunk is 21.3 μT, in the area of the foot and ankle the magnetic induction intensity is larger, the magnetic induction intensity is relatively uniform in other parts of the trunk, and the value is much lower than that in the foot area. The magnetic induction intensity of human trunk in this position is much lower than the basic restriction of 1000 μT [11] of occupational exposure defined by ICNIRP.

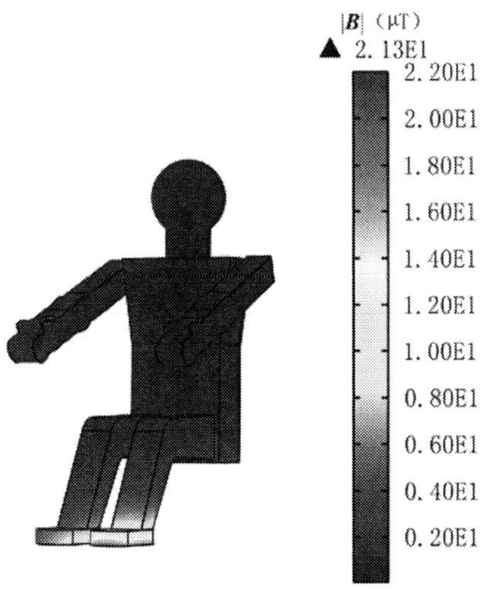

Figure 4. Distribution of magnetic induction intensity in driver's trunk.

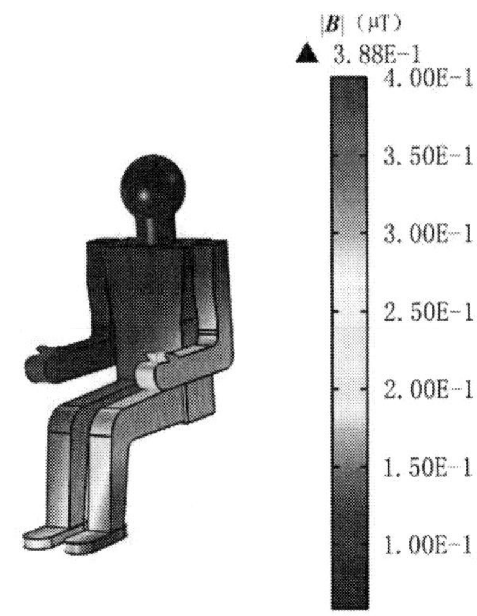

Figure 5. Distribution of magnetic induction intensity in co-driver's trunk.

Similarly, the distribution of magnetic induction intensity in co-driver's trunk is shown in figure 5. The maximum magnetic induction intensity in the co-driver's trunk is 0.388 μT, and the strongest area appears in the foot and the lateral area of the leg, which is directly related to the distance from the power cable. The magnetic induction intensity of human trunk in this position is also far lower than the basic restriction of 200 μT [11] of public exposure defined by ICNIRP.

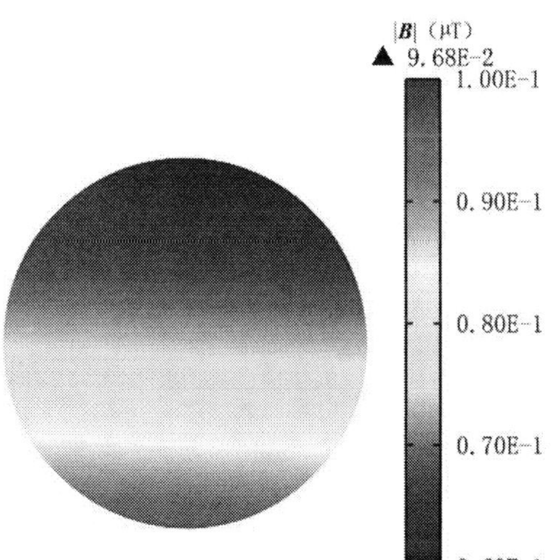

Figure 6. Distribution of magnetic induction intensity in driver's head.

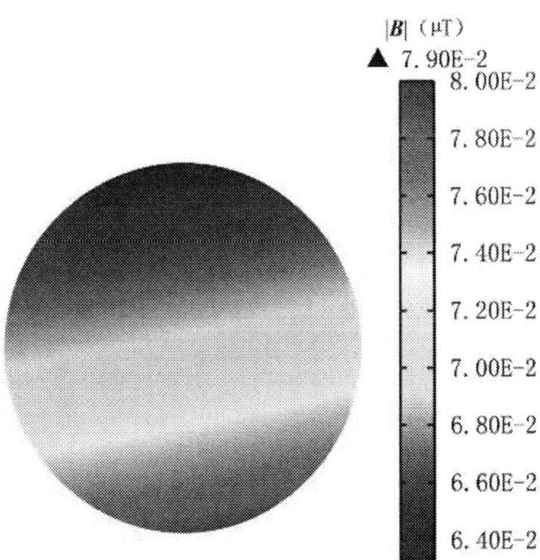

Figure 7. Distribution of magnetic induction intensity in co-driver's head.

The brain is the core of the CNS. Therefore, ICNIRP specifically defines the limits of electromagnetic field exposure in the head. The vertical section magnetic induction intensity of the driver's and co-driver's head is shown in figures 6 and 7, respectively. The maximum magnetic induction intensity of the driver's and the co-driver's head is 0.0968 μT and 0.079 μT, respectively. The closer the human head to the power cable, the greater the magnetic induction intensity. The magnetic induction intensity in the driver's and co-driver's head is much lower than the basic restriction of occupational exposure and public exposure defined by ICNIRP [11].

3.2. Distribution of human induced electric field intensity

According to the theory of electromagnetic field, the variable current produces a variable magnetic field, and the variable magnetic field would induce electric field. As shown in figure 8, the maximum induced electric field intensity in the driver's trunk is 16.7 mV/m, and the maximum induced electric field intensity is much lower than the basic restriction of 800 mV/m of the induced electric field intensity of occupational exposure defined by ICNIRP [11].

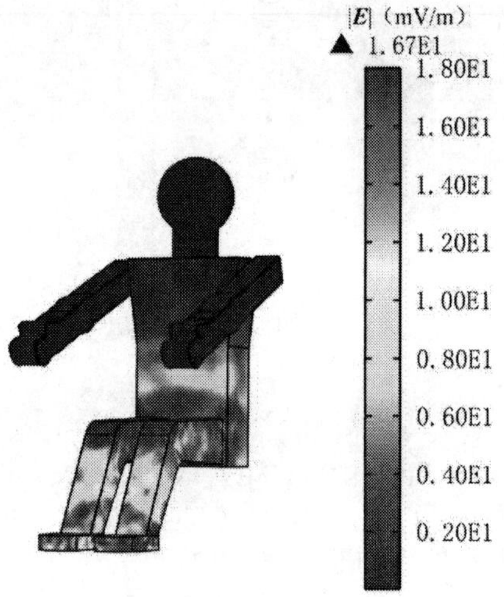

Figure 8. Distribution of induced electric field intensity in driver's trunk.

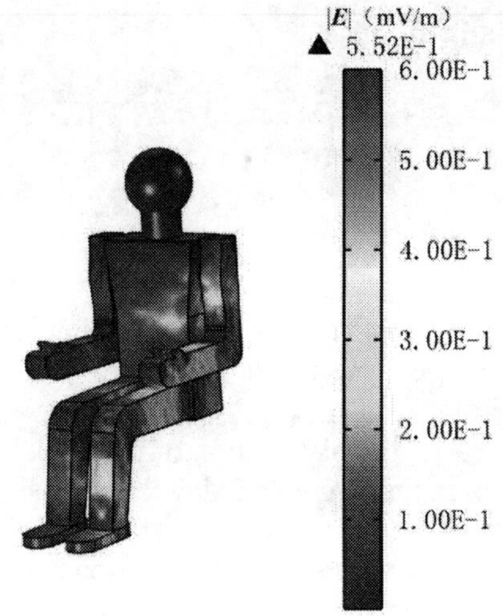

Figure 9. Distribution of induced electric field intensity in co-driver's trunk.

The induced electric field intensity in co-driver's trunk is shown in figure 9. The co-driver is much far from the power cable than the driver, so the value of induced electric field intensity in co-driver's trunk is much small. The maximum value of induced electric field intensity in the co-driver's trunk is 0.552 mV/m. The value is also far below the basic restriction of 400 mV/m of the induced electric field intensity for public exposure defined by ICNIRP [11]. From the figures 8 and 9, it can also be seen that the closer to the power cable, the greater the induced electric field intensity; conversely, the farther the distance, the smaller the induced electric field intensity.

For the induced electric field intensity in the CNS, the vertical sections of the head models of driver's and co-driver's are shown in figures 10 and 11 respectively. The maximum induced electric field intensity of driver's head is 1.78 mV/m, which is higher in skull and brain tissues, and the lowest in scalp area. The induced electric field intensity is lower than the basic restriction of 360 mV/m of occupational CNS exposure defined by ICNIRP [11] in driver's head. Similarly, the maximum induced electric field intensity in the co-driver's head is 0.0537 mV/m, which is concentrated on the side of the head near the power cable, and mainly concentrate in the skull and brain tissue. The maximum value is much lower than the basic restriction of 720 mV/m of public CNS exposure defined by ICNIRP [11].

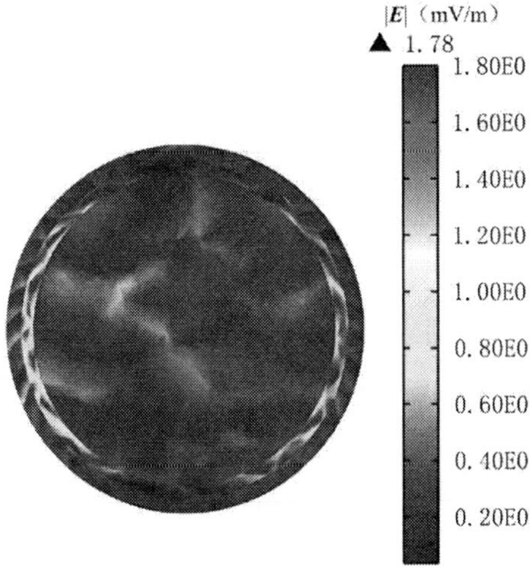

Figure 10. Induced electric field intensity of vertical section in driver's head.

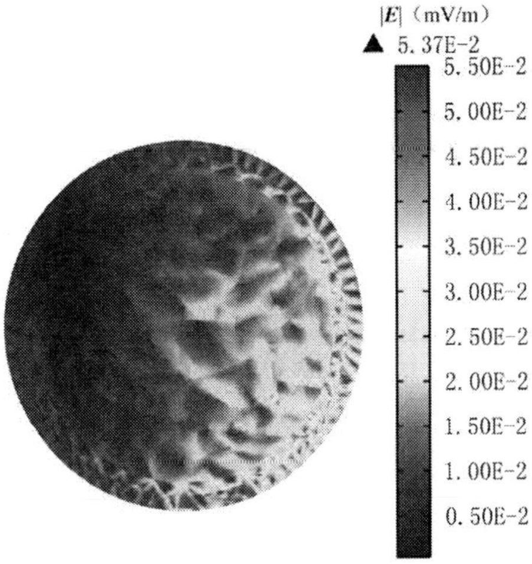

Figure 11. Induced electric field intensity of vertical section in co-driver's head.

4. Discussion

For study the electromagnetic exposure of human body in low-frequency electromagnetic environment of power cables in electric vehicles, the distribution of induced electromagnetic fields in human tissues, especially in the CNS of driver's and co-driver's is calculated. By establishing an equivalent electromagnetic environment in Comosol Multiphysics, the power cable with the current of 500 Hz in frequency and 30A in amplitude is modelled.

The numerical results show that under the action of the low frequency current, the magnitude and distribution characteristics of the magnetic induction intensity and the induced electric field intensity in different tissues of human body are directly related to the distance of the power cable. The closer the distance, the larger the values. The maximum values of magnetic induction intensity and induced electric field intensity in the trunk of driver's and co-driver's are much lower than the basic restriction of occupational exposure and public exposure defined by ICNIRP. Similarly, the induction fields in the CNS of driver's and co-driver's are also far below the basic restriction of occupational and public CNS exposure limits. Taking the values of magnetic induction intensity and induced electric field intensity as examples, the maximum magnetic induction intensity in driver's trunk reaches 2.13% of ICNIRP limits, and the corresponding value in co-driver's could only reach 0.19% of ICNIRP limits. The maximum induced electric field in the driver's CNS is only 0.49% of the ICNIRP limits, and the corresponding value in the co-driver's is only 0.07% of the ICNIRP limits.

Therefore, the electromagnetic environment in the cabinet of electrical vehicle would not damage the human tissues in a short time, and the study is similar to the electromagnetic wave emitted from the mobile phone and base station, the values are far below the ICNIRP limits. In fact, the result doesn't fully support the viewpoint that the radiation is not harmful to human health for long term exposure. It also needs to be validated by epidemiological methods for long-term follow-up of specific populations.

Acknowledgment

This research is financially supporting by National Natural Science Foundation of China (No. 51567015, No. 51867014), Opening Project of Key Laboratory of Opto-Technology and Intelligent Control, Ministry of Education (No. 2018015) and Lanzhou Jiaotong University Fund for Distinguished Young Scholars (No. 2016006).

References

[1] Wang F and Xia J 2016 *Safety Analysis and Design of Electric Vehicle Power Battery System*

(Beijing: Science Press) pp 18-26
[2] De Santiago J, Bernhoff H and Ekergard B 2012 *IEEE Trans. Veh. Technol.* **61** 475-84
[3] Wu T N and Qi D Y 2010 *Electromagnetic Field and Human Health* (Beijing: Posts and Telecommunications Press) pp 24-32
[4] Ruddle A R and Low L 2012 *IEEE Int. Symp. on Electromagnetic Compatibility* (Rome: IEEE)
[5] Andrea V, Alain F and Christof W 2015 *IEEE T Electromagn C* **57** 35-43
[6] Tell R A and Kavet R 2016 *Radiat Prot Dosim.* **172** 541-6
[7] Lin J, Lu M, Wu T, Yang L and Wu T N 2018 *Radlat. Prot. Dosim* **182** 190-9
[8] Xu G Z, Li C X and Zhao J 2017 *Trans. China Electrotech. Society.* **32** 152-7
[9] Chen C, Huang X L and Tan L L 2015 *Trans. China Electrotech. Society* **30** 61-7
[10] Lu M and Ueno S 2015 *IEEE T Magn.* **51** 5101204-1-4
[11] ICNIRP 2010 *Health Phys.* **99** 818-36
[12] Dong X W and Lu M 2019 *Auto. Engineering.* **41** 696-702
[13] Gabriel C, Gabriel S and Corthout E 1996 *Phys. Med. Biol* **41** 2231-49
[14] Gabriel S, Lau R W and Gabriel C 1996 *Phys. Med. Biol* **41** 2251-69

IOP Conf. Series: Materials Science and Engineering **657** (2019) 012061 doi:10.1088/1757-899X/657/1/012061

Estimation of aircraft magnetic field model coefficient under Geomagnetic Gradient Change

C J Li[1], B Zhang[2,3], T H Ouyang[1], Y Z Gong[1] and B Li[2]

[1]State Key Laboratory of Astronautic Dynamics, 710051, Xi'an, China
[2]Air and Missile Defense College, Air Force Engineering University,710051, Xi'an, China

E-mail: zhb8706@163.com

Abstract. Traditional airplane magnetic field models simplify geomagnetic field values as constants, overlooking changes to the model coefficient brought by gradually variational magnetic field which reduces the model accuracy. This paper puts forward an aircraft magnetic field model under geomagnetic gradient change. The geomagnetic gradient change is adopted the truncated singular value decomposition (TSVD) to solve multicollinearity in the model solving. It is proposed to use the coefficient obtained from the truncated singular value as initial conditions and utilize recursive damped least square (RDLS) to conduct adaptive adjustment to model coefficients aiming at the coefficient variation of induction field and Eddy Current Field caused by the geomagnetic field. The simulation outcome indicates that the method boasts a high magnetic compensation effect.

1. Introduction

The magnetic field of the aircraft and the field are induced by its motion exert an impact on magnetic anomaly detection of the aircraft. T-L model, the classical model of airborne magnetic interference, was first introduced by Tolles, Lawson and others [1,2] in 1943. As there is a certain correlation among variables of the model, the model displays strong multicollinearity which makes it very difficult to accurately estimate the model parameters. Tolles, Lawson did not put forward effective solutions to model solving and real-time compensation of the aircraft magnetic field in their paper. In 1961, Leliak invented a time-dependent method of magnetic interference between parameter estimation and aircraft maneuver based on T-L equation to make magnetic compensation [3]. His design has become a classic in the history of magnetic compensation and has been still a standard method of magnetic compensation till now. However, a excellent model solving has not been put forward. In 1979, Bickel put forward a Small-Signal magnetic compensation algorithm [4] which embodies the idea of a principal component estimation to a certain extent. It converts a parameter estimation of multivariate linear regression model with multicollinearity into an estimation of least square with the optimal solution under variance and has basically solved multicollinearity in the magnetic interference model at the expense of accuracy loss. Although the conventional solutions to multicollinearity, such as ridge estimate, principal components estimation, genetic algorithm [5-8], can evaluate parameters of a linear model, in case of ill-conditioned model linear equations, the deviation is quite big. An aircraft magnetic field compensation method based on neural network is proposed in Paper [9]. The method has slow convergence speed and is sometimes unstable.

At present, although there are various forms of aircraft magnetic interference equations, they still derive from the 16 Tolles-Lawson coefficient compensation models without changing physical meaning of the essence. The geomagnetic field is assumed to be a constant in models, and models are built for three kinds of magnetic fields (constant, induced and eddy current magnetic field) generated by aircraft maneuver. In the real aircraft flight, the geomagnetic field varies with time and location, which will reduce accuracy of the aircraft magnetic interference model and lead to compensation deviation during magnetic compensation. This paper transforms the T-L equation by integrating the geomagnetic gradient change with coefficients of the induction and eddy current magnetic field and establish airplane magnetic field model under gradually variational condition. Secondly, TSVD [10,11] is adopted to evaluate the T-L equation coefficients. In the end, considering the variation of the induced and eddy current magnetic field coefficients caused by the gradual change of the geomagnetic field, estimated values obtained from the TSVD is taken as the initial coefficient, and recursive damped least square (RDSL) [12,13] is used to conduct adaptive adjustment to coefficients of the induction and eddy current magnetic field.

2. Another section of my paper

Magnetic field detected by probe of a total field magnetometer [1,2]:

$$H_{Total} = H_{pd} + H_{id} + H_{ed} + He \tag{1}$$

Since in the direction cosine of the aircraft coordinate system, there is:

$$\cos^2 X + \cos^2 Y + \cos^2 Z = 1 \tag{2}$$

The geomagnetic field is

$$He(t) = He(t)\begin{pmatrix}\cos X & \cos Y & \cos Z\end{pmatrix}\begin{pmatrix}1 & 0 & 0\\ 0 & 1 & 0\\ 0 & 0 & 1\end{pmatrix}\begin{pmatrix}\cos X\\ \cos Y\\ \cos Z\end{pmatrix} \tag{3}$$

The induction field in the T-L model is:

$$H_{id} = \begin{pmatrix}\cos X & \cos Y & \cos Z\end{pmatrix}\begin{pmatrix}a_{11}He(t) & a_{12}He(t) & a_{13}He(t)\\ a_{21}He(t) & a_{22}He(t) & a_{23}He(t)\\ a_{31}He(t) & a_{32}He(t) & a_{33}He(t)\end{pmatrix}\begin{pmatrix}\cos X\\ \cos Y\\ \cos Z\end{pmatrix} \tag{4}$$

Then:

$$H_{id} = \begin{pmatrix}\cos X & \cos Y & \cos Z\end{pmatrix}\begin{pmatrix}(a_{11}+1)He(t) & a_{12}He(t) & a_{13}He(t)\\ a_{21}He(t) & (a_{22}+1)He(t) & a_{23}He(t)\\ a_{31}He(t) & a_{32}He(t) & (a_{33}+1)He(t)\end{pmatrix}\begin{pmatrix}\cos X\\ \cos Y\\ \cos Z\end{pmatrix} \tag{5}$$

$$\begin{aligned}H_{Total} &= H_{pd} + H_{ed} + H_{id} + He(t)\\ &= \begin{pmatrix}T & L & V\end{pmatrix}\begin{pmatrix}\cos X\\ \cos Y\\ \cos Z\end{pmatrix}\\ &+ \begin{pmatrix}\cos' X & \cos' Y & \cos' Z\end{pmatrix}\begin{pmatrix}b_{11}(t)' & b_{12}(t)' & b_{13}(t)'\\ b_{21}(t)' & b_{22}(t)' & b_{23}(t)'\\ b_{31}(t)' & b_{32}(t)' & b_{33}(t)'\end{pmatrix}\begin{pmatrix}\cos X\\ \cos Y\\ \cos Z\end{pmatrix}\\ &+ \begin{pmatrix}\cos X & \cos Y & \cos Z\end{pmatrix}\begin{pmatrix}a_{11}(t)' & a_{12}(t)' & a_{13}(t)'\\ a_{21}(t)' & a_{22}(t)' & a_{23}(t)'\\ a_{31}(t)' & a_{32}(t)' & a_{33}(t)'\end{pmatrix}\begin{pmatrix}\cos X\\ \cos Y\\ \cos Z\end{pmatrix}\end{aligned} \tag{6}$$

Among which,

$$b_{ij}(t)' = b_{ij}He(t), i = 1,2,3. j = 1,2,3$$

$$a_{ii}(t)' = (a_{ii} + 1)He(t), i = 1,2,3$$

$$a_{ij}(t)' = a_{ij}He(t), i = 1,2,3, j = 1,2,3, i \neq j$$

It can be seen that in equation (6), the gradually variational value of the geomagnetic field is integrated with the T-L model, and there is no need to discuss its size and change, and because it can be directly solved in equations. Therefore, there is no need to separate the magnetic interference generated by the air platform from the total field intensity signal. However, as $a_{ij}(t)'$ and $b_{ij}(t)'$ vary with the geomagnetic field, an algorithm should be sought out to carry out adaptive adjustment to it.

It is noted that H_{ed} is assumed to be constant in the derivation of T-L model eddy current field $He(t)$ [1,2]. There is

$$\overrightarrow{H}_{ed} = \frac{d\overrightarrow{He}}{dt}\mathbf{N} \tag{7}$$

among which $\mathbf{N}$ is the coefficient matrix that generates equivalent eddy current, and it is only essentially relevant to the nature of aircraft structure and material.

$$\begin{aligned}\frac{d\overrightarrow{He}}{dt} &= (u_1\frac{d\|\overrightarrow{He}\|}{dt} + \|\overrightarrow{He}\|\dot{u}_1 \quad u_2\frac{d\|\overrightarrow{He}\|}{dt} + \|\overrightarrow{He}\|\dot{u}_2 \quad u_3\frac{d\|\overrightarrow{He}\|}{dt} + \|\overrightarrow{He}\|\dot{u}_3) \\ &= (\|\overrightarrow{He}\|\dot{u}_1 \quad \|\overrightarrow{He}\|\dot{u}_2 \quad \|\overrightarrow{He}\|\dot{u}_3)\end{aligned} \tag{8}$$

In which $\begin{pmatrix}u_1 & u_2 & u_3\end{pmatrix}^T = \frac{\overrightarrow{He}}{\|\overrightarrow{He}\|}$ is direction vector of the geomagnetic field.

The model deviation resulting from geomagnetic field change Δ is:

$$\Delta = \Delta\overrightarrow{H}_{ed}\frac{\overrightarrow{He}}{\|\overrightarrow{He}\|} = (u_1\frac{d\|\overrightarrow{He}\|}{dt} \quad u_2\frac{d\|\overrightarrow{He}\|}{dt} \quad u_3\frac{d\|\overrightarrow{He}\|}{dt})N\frac{\overrightarrow{He}}{\|\overrightarrow{He}\|} \tag{9}$$

Since this paper assumes the geomagnetic field is varying gradually, Δ can be supposed 0, and at this moment the magnetic field model of the airplane is still the T-L equation. This assumption will definitely lead to a certain coefficient estimation deviation, but the simulation outcome at the end of this paper implies the effect of the magnetic compensation is still perfect.

3. Coefficient estimation of the field model

3.1. *TSVD parameter estimation algorithm*

Suppose [14]:

$$Kx = y + \varepsilon \tag{10}$$

among which $K \in R^{m\times n}$, and ε is a random error. The matrix K can be decomposed as:

$$K = U\Sigma V^T = \sum_{i=1}^{n} u_i \sigma_i v_i^T \tag{11}$$

among which u_i, v_i are column vectors of the matrix, and σ_i is the singular value of the matrix K. Then the least square solution of the equation can be written as:

$$x_{ls} = \sum_{i=1}^{n} \frac{u_i^T y}{\sigma_i} v_i \tag{12}$$

The ill-conditioned problems caused by the root and it is that when the singular value of operator K is too small or close to 0, it will bring large rounding error to the above equation. Currently there is no algorithm that can directly estimate x in the solution space of the operator K. In order to obtain a more accurate solution x, too small singular values should be truncated to create a stable solution space, and meanwhile make the space approximate to the original solution space as far as possible. The formula to evaluate x via TSVD is as follows:

$$x_{tsvd} = \sum_{i=1}^{k} \frac{u_i^T y}{\sigma_i} v_i (1 \le k < n) \tag{13}$$

among which, k is a truncated regularization parameter. In order to get a better estimation result, an appropriate parameter k is in need. If k is too big, a stable solution cannot be obtained whilst if too small, the approximate solution space will lose distortion. Here the generalized cross validation rule is applied [15,16]:

$$G(k) = (\frac{n}{n-k}) \left(\frac{\sum_{i=k+1}^{n} (u_i^T y)^2}{n-k} \right), k = 1, 2, ... \tag{14}$$

The k makes $G(k)$ minimum is optimal and x_{tsvd} can be got by introducing it to equation (13).

3.2. RDLS algorithm

Equation (6) can compensate for the magnetic interference caused by the geomagnetic field and the aircraft platform. Since the longer the duration ΔT is from calibration phase to compensation phase, the larger the gap between the estimated magnetic interference coefficient in the calibration phase and the actual magnetic interference coefficient in the compensation phase, finally the compensation effects deteriorates. Therefore, ΔT is needed to be reduced as much as possible. Here, suppose $\Delta T = 1$, namely, making use of the coefficient estimated from measured data from 0 to $t-1$ moment to compensate for the magnetic interference generated by the geomagnetic field and the aircraft platform during the t moment. The RDSL method is applied to estimate the magnetic interference coefficient in real time. The optimized criterion function for the algorithm is [17]:

$$\min J = \sum_{t=1}^{N} \beta^{N-t} [y(t) - \phi(t)\theta(t)]^2 + \mu \| \theta(t) - \theta(t-1) \|^2 \tag{15}$$

$y(t)$ is the output data for the magnetometer, $\theta(t)$ is a column vector of a magnetic interference coefficient to be estimated , $\phi(t)$ is a row vector constituted by direction cosine and its derivatives and $t = 1, 2, \cdots N$. The forgetting factor β is to emphasize the significance of "up-to-date" data in order to solve coefficient time-varying. The damping factor μ is to reduce adverse effect of noise on a real-time solution coefficient in the "up-to-date" data. The recursion formula for model coefficients is:

$$\begin{aligned} &\theta(t) = \theta(t-1) + \beta\mu P(t)[\theta(t-1) - \theta(t-2)] + P(t)\phi(t)[y(t) - \phi^T(t)\theta(t-1)] \\ &P(t)^{-1} = \mu(1-\beta)I + \beta P(t-1)^{-1} + \phi(t)\phi(t)^T \end{aligned} \tag{16}$$

The recursion formula for $P(t)$ is:

$$
\begin{aligned}
&P(t)=\frac{1}{\beta}[P'(t)-\frac{P'(t)\phi(t)\phi^T(t)P'(t)}{\beta+\phi^T(t)P'(t)\phi(t)}];\\
&P'(t)=P(t-1)-\sum_{i=1}^{n}\frac{P'_{i-1}(t-1)r_i r_i^T P'_{i-1}(t-1)\mu'}{1+r_i^T P'_{i-1}(t-1)r_i\mu'};\\
&P'_i(t-1)=P'_{i-1}(t-1)-\frac{P'_{i-1}(t-1)r_i r_i^T P'_{i-1}(t-1)\mu'}{1+r_i^T P'_{i-1}(t-1)r_i\mu'};\\
&P'_0(t-1)=P(t-1).
\end{aligned}
\tag{17}
$$

among which r_i is the successor vector of r_1, $r_1=[1,0,...,0]^T$, $\mu'=(1-\beta)\mu/\beta$.

Since the constant field coefficient does not differ with variation of the geomagnetic field, only the induction field and eddy field coefficient can be adjusted adaptively, and the initial coefficient can be obtained through TSVD algorithm based on generalized crossing validation.

4. Simulation analysis

The Algorithm simulation is semi-physical. Nonmagnetic revolving table is used to simulate aircraft rocking, pitching and yawing, and it gathers maneuver data and installs ferromagnetic objects to simulate the interference field on the platform. The geomagnetic gradient change in the experiment is shown in figure 1. Figure 2 shows the total field value in the maneuver process and magnetic compensation outcome of T-L equation based on TSVD parameter estimation. The TSVD compensation results based on multi-group experimental data are shown in figure 3, and the figure of merit [3,4] (FOM) of several groups are still relatively big, which implies that TSVD algorithm based on generalized cross validation has overcome ill-condition, however, the compensation effect is still not ideal. It mainly caused by the T-L model time variation.

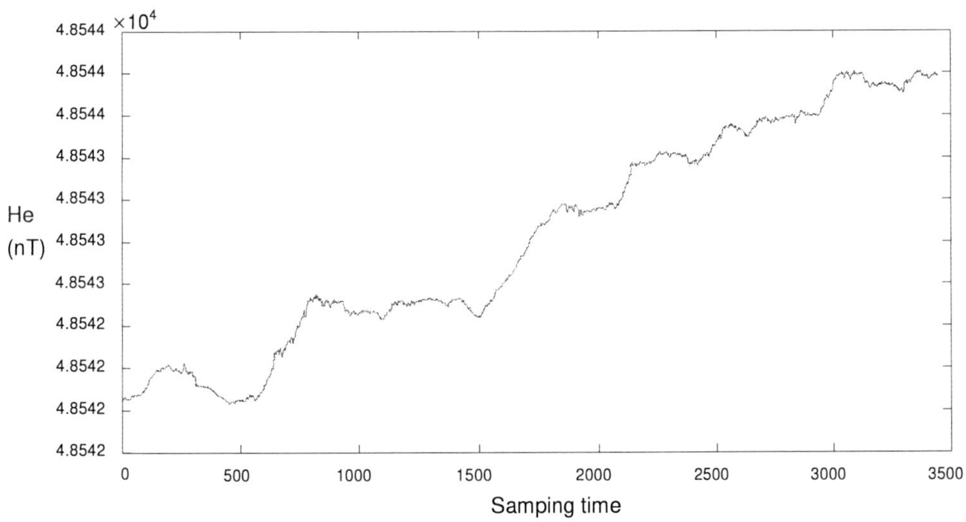

Figure 1. Gradually variational data of geomagnetic.

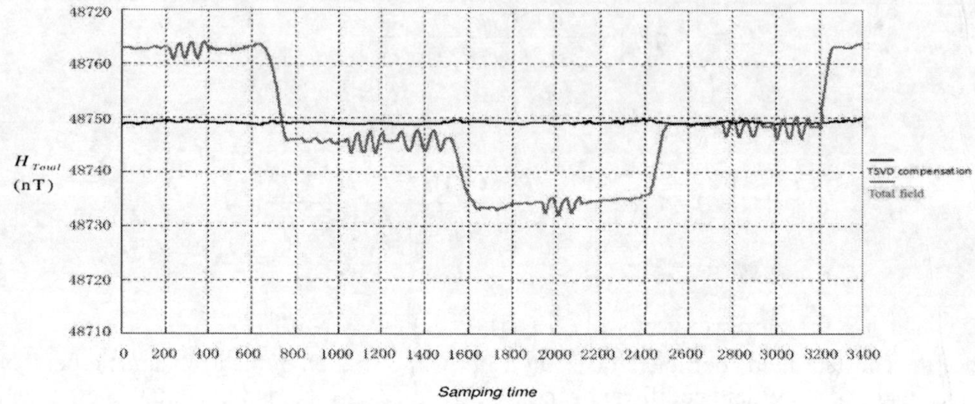

Figure 2. Magnetic compensation outcome of T-L equation based on TSVD parameter estimation.

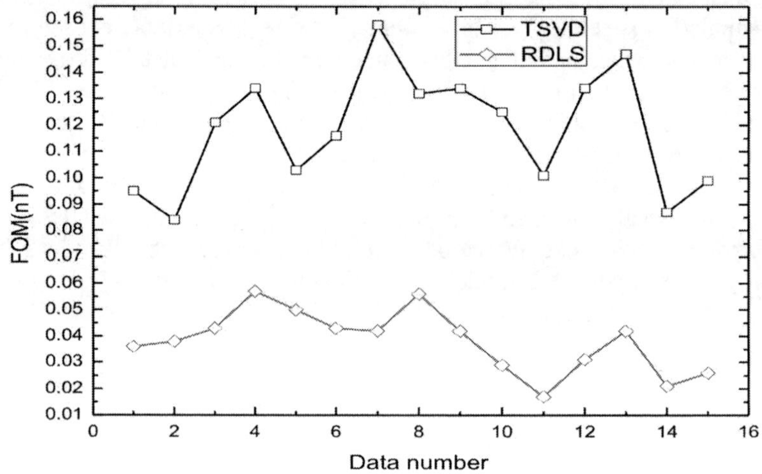

Figure 3. Multi-group compensation results of gradually variational geomagnetic field model based on RDLS and TSVD.

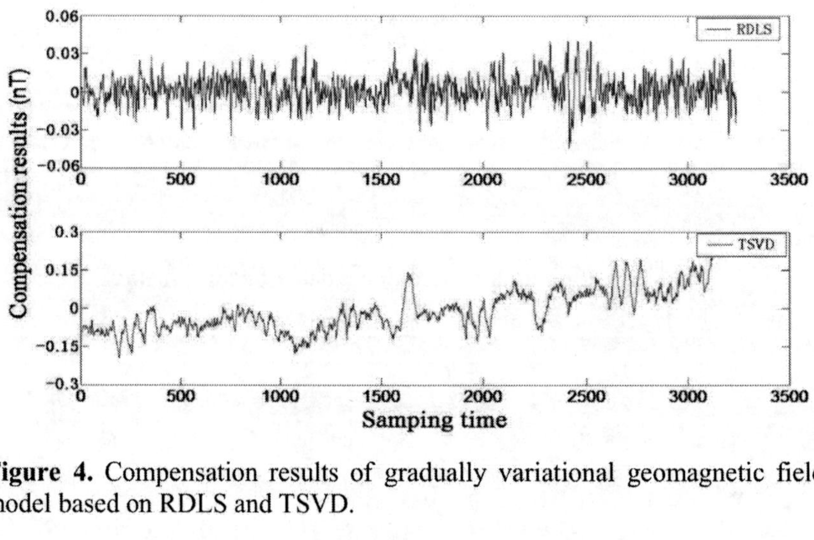

Figure 4. Compensation results of gradually variational geomagnetic field model based on RDLS and TSVD.

The solutions obtained from TSVD is taken as initial conditions and introduced to the RDLS algorithm in order to estimate model parameters of airplane magnetic field under the gradually variational magnetic field and adaptively update induction field and eddy field parameters. Figures 3 and 4 are the RDLS compensation results of the experimental data for this time and multi-groups comparing with TSVD. The method is more stable in compensation outcome with the peak to peak value lower than 0.1nT and performance better than TSVD algorithm. According to the data, the proposed adaptive compensation algorithm of the airplane magnetic field model under the geomagnetic gradient change can better track the variations of the geomagnetic field and adaptively adjust coefficients, and achieve a better compensation effect.

5. Conclusions

During the process of aircraft flight, the geomagnetic field varies with the change of time and location, which will generate deviation when being in use of the T-L model to carry out magnetic compensation. In this paper the T-L model is transformed by integrating the geomagnetic field with the induction and eddy current field coefficient, based on which adaptive magnetic interference compensation method of the airplane magnetic field model parameter estimation is designed under gradually variational geomagnetic field and. Simulation results indicate that the method boasts the better parameter estimation accuracy and a better magnetic compensation effect. However, in deriving the T-L equation, this method does not improve the assumption that the geomagnetic field is constant, which will inevitably affect the magnetic compensation accuracy. This is the direction of further research.

Acknowledgment

This research was financially supported by the Grenral financial Grant from the China Postdoctoral Science Foundation, grant No. 2017M623417.

References

[1] Tolles W E and Lawson J D 1950 *Magnetic Compensation of MAD Equipped Aircraft* (New York: Airborne Instruments Lab Inc)

[2] Tolles W E 1943 *Compensation of Induced Magnetic Fields in MAD Equipped Aircraft* (New York: Airborne Instruments Lab)

[3] Leliak P 1961 Identification and evaluation of magnetic field sources of magnetic airborne detector equipped aircraft IRE *Trans. Aerosp. Navig. Electron.* **9** 95-105

[4] Bickel S H 1979 Small signal compensation of magnetic fields resulting from aircraft maneuvers *IEEE Trans. on AES* **15** 515-25

[5] Zhu Z M, Chang J H and Li J S 2003 A method discussed to overcome the effects of the compound linear *Journal of Spacecraft TT & C Technology* **22** 7-12

[6] Praga-Alejo R J, Torres-Trevio L M and Pia-Monarrez M R 2008 Optimal determination of K constant of ridge regression using a simple genetic algorithm *Electronics, Robotics and Automotive Mechanics Conference* (Cuernavaca, Mexico) pp 39-44

[7] Hardwick C D 1984 Important design considerations airborne magnetic gradiometers *Geophysics* **11** 2004-18

[8] Sheinker A, Shkalim A and Salomonski N 2007 Processing of a scalar magnetometer signal contaminated by $1/f^{a}$ noise *Sensor. Actuat. A-P*hys **138** 105-11

[9] Williams and Peter M 1993 Aeromagnetic compensation using neural networks *Neural Comput. Appl.* **1** 207-14

[10] Hansen P C 1987 The truncated SVD as a method for regularization *BIT* **27** 534-53

[11] Xu P L 1998 Truncated SVD methods for discrete linear Ill-posed problems *Geophy J Int* **135** 505-14

[12] Chudamani R and Krishna V 2009 Real-time estimation of power system frequency using nonlinear least squares *IEEE T. Power De.* **24** 1021-8

[13] Pradhan A K 2005 Power system frequency estimation using least square technique *IEEE T. Power De.* **20** 1812-6

[14] Huang Q X, Liu D and Wang F 2007 Study on three-dimensional flame temperature distribution reconstruction based on truncated singular value decomposition *Acta Phys. Sin-Ch ED* **56** 6742-8

[15] Dong M, Jia X D, Chen P and Lv J 2011 Wavelet threshold de-noising of FOG dynamic signal based on generalized cross-validation *Journal of Information Engineering University* **12** 482-7

[16] Yang X H, Jin H Y and Jiao L C 2007 SAR speckle reduction based on generalized cross validation and cycle spinning *Journal of Electronics & Information Technology* **29** 1779-83

[17] Lin M Q, Chen Z Q and Yuan Z Z 2000 Recursive dampedleast square algorithm *Appl. Math. J. Chinese Univ* **15** 113-8

Study on parameter identification of vibration absorption system based on acceleration signal

T N Guo[1,2], T Yang[1] and Y C Gu[1]

[1]College of mechanical and electrical engineering, Beijing University of Technology, Beijing 100124, China

E-mail: guotn@163.com

Abstract. Based on the analysis and processing of acceleration signals, the time domain least square method and frequency domain analytic method are proposed to identify the parameters of vibration absorption system. The mathematical model of vibration reduction and Simulink simulation system are established, two parameter identification methods are validated by known system parameters and vibration signals; and then, based on the acceleration signal, the influence of signal processing on identification accuracy is analyzed. The two identification methods proposed in this paper can accurately identify the parameters of the vibration absorption system, Finally, the experimental results show that the two identification methods have good engineering applicability and scientific research applicability.

1. Introduction

Rubber shock absorber is widely used in military and civil fields. Its dynamic stiffness characteristics and other parameters are important basis for shock absorber design; the dynamic stiffness characteristic is the regular pattern that the dynamic stiffness of the system changes with the excitation frequency; through the dynamic stiffness characteristics and the working state of the designed shock absorber, the designed shock absorber can avoid the resonance zone and keep it in a good state of vibration reduction [1].

Wan *et al* [2] used polynomial fitting method to derive the identification process of dynamic stiffness and other parameters. He *et al* [3] studied the influence of vibration frequency, amplitude and preload on dynamic stiffness. Li *et al* [1] obtained the dynamic stiffness varying with frequency according to the static stiffness value and the formula of single degree of freedom. Overall, compared with the performance of shock absorber, the research on dynamic stiffness of shock absorber is less, and the research on parameter identification methods such as dynamic stiffness is less.

In this paper, two methods of identifying the parameters of vibration absorption system are proposed, which are the time domain least square method and frequency domain analytic method based on acceleration signal. The core of identification is to analyze the input signal and the output of the system to obtain the parameters of the vibration absorption system; the input and output include signals of acceleration, velocity and displacement. In vibration reduction experiments, the acceleration signal is easier to measure than the velocity signal, especially the displacement signal, acceleration sensor is selected in this experiment, the required signals of velocity and displacement are obtained by integrating the acceleration signals. Integration of acceleration signal is an important problem in identification process.

The integration of acceleration signal can be divided into time domain integration and frequency domain integration. Gu [4] of Tsinghua University verified the two integration methods through simulation and comparison; the research of Zhou [5] of the aeronautical industry helicopter Institute shows that the effect of frequency domain integration is better than that of time domain integration.

In this paper, the mathematic model of vibration absorption system and Simulink simulation system are established, and then two identification methods are validated by known system parameters, input signals and output signals of system, respectively; the acceleration signal is integrated in frequency domain to obtain the velocity and displacement signals which are substituted in the identification process to obtain the factors affecting the identification accuracy.

2. Least squares identification

2.1. Basic principles

Identification of parameters of vibration system by the time domain least square method is a process of data fitting [6]. After determining the mathematical model of the vibration system, the sum of squares of the difference between the left and right ends of the equation of the mathematical model is minimized according to the measured data, and the optimal system parameters are identified. Figure 1 is a mass-spring single-degree-of-freedom vibration system with damping.

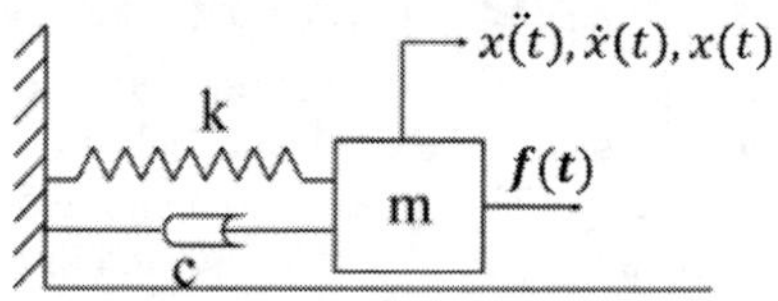

Figure 1. Excited vibration system.

In the picture m is the system mass, c and k are the system damping and stiffness parameters. There is the equation (1) for the harmonic excitation force f acting on the system.

$$f = m\ddot{x} + c\dot{x} + kx \tag{1}$$

In the formula, $\ddot{x}$, $\dot{x}$ and x are the signals of system vibration acceleration, velocity and displacement, respectively. Transform the equation into

$$f - m\ddot{x} = c_1\dot{x} + k_1 x \tag{2}$$

According to N sets of data of input and output collected in the simulation system, N covers at least one cycle of data, formula (2) makes N sets of difference on both sides of equation, and obtains formula (3).

$$\begin{cases} \Delta_1 = [f_1 - m\ddot{x}_1] - [c_1\dot{x}_1 + k_1 x_1] \\ \Delta_2 = [f_2 - m\ddot{x}_2] - [c_1\dot{x}_2 + k_1 x_2] \\ \dots \\ \Delta_n = [f_n - m\ddot{x}_n] - [c_1\dot{x}_n + k_1 x_n] \end{cases} \tag{3}$$

The best value of the parameter minimizes the sum of squares, so there is a formula (4).

$$\Delta = \Delta_1^2 + \Delta_2^2 + \dots + \Delta_n^2 = \min \tag{4}$$

When Δ takes the minimum value, the partial derivatives of k1, and c1 should be zero.

$$\begin{cases} \frac{\partial \Delta}{\partial c_1} = 2\Delta_1(-\dot{x}_1) + 2\Delta_2(-\dot{x}_2) + \ldots + 2\Delta_n(-\dot{x}_n) = 0 \\ \frac{\partial \Delta}{\partial k_1} = 2\Delta_1(-x_1) + 2\Delta_2(-x_2) + \ldots + 2\Delta_n(-x_n) = 0 \end{cases} \quad (5)$$

Formula (6) is converted into a matrix:

$$\begin{bmatrix} \dot{x}_1, \dot{x}_2, \ldots, \dot{x}_n \\ x_1, x_2, \ldots, x_n \end{bmatrix} \begin{bmatrix} \Delta_1 \\ \Delta_2 \\ \cdots \\ \Delta_n \end{bmatrix} = 0 \quad (6)$$

Set $A = \begin{bmatrix} \dot{x}_1, x_1 \\ \dot{x}_2, x_2 \\ \ldots, \ldots \\ \dot{x}_n, x_n \end{bmatrix}$, $F = \begin{bmatrix} f_1 - m\ddot{x}_1 \\ f_2 - m\ddot{x}_2 \\ \ldots \\ f_n - m\ddot{x}_n \end{bmatrix}$, $\Delta = \begin{bmatrix} \Delta_1 \\ \Delta_2 \\ \cdots \\ \Delta_n \end{bmatrix}$, $X = \begin{bmatrix} c_1 \\ k_1 \end{bmatrix}$, then formulas (3) and (6) can be expressed as formula (7).

$$\begin{cases} \Delta = F - AX \\ A^T \Delta = 0 \end{cases} \quad (7)$$

Solution (7) obtains the parameter matrix of vibration absorption system.

$$X = (A^T A)^{-1} A^T F \quad (8)$$

2.2. Simulation verification of least square identification

A mass-spring system with damping is built in Simulink, Setting the mass m, stiffness K and damping parameter C in vibration absorption system. The excitation force f (t) is input to the vibration absorption system to generate response signals of acceleration, velocity, displacement, and then the parameters k1 and c1 are identified. The accuracy of the identification method is obtained by comparing the errors between the identification parameters and the real parameters of the system. The verification process is shown in figure 2.

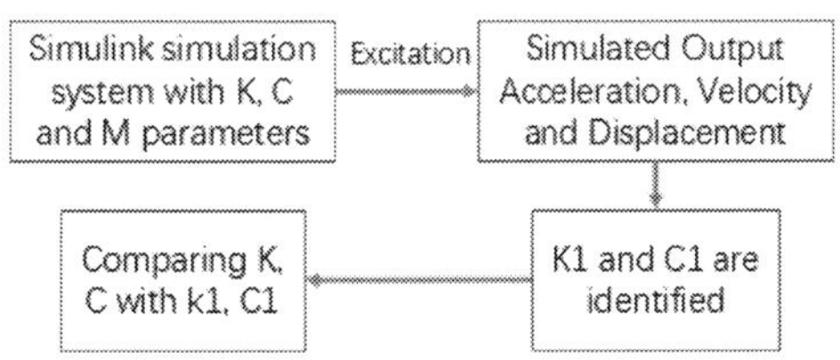

Figure 2. Verification process.

Set the system parameter mass m = 4, system stiffness k = 1000, damping ratio 0.1, that is, damping parameter C = 12.6491. After the excitation force is input, the generated signal is substituted into the identification process, and the parameters of the system are obtained as shown in table 1.

From table 1, it can be seen that the time domain least square method can accurately identify the parameters according to the known input signals and output signals of the system.

Table 1. Identification of parameters of vibration system.

Excitation frequency	sampling frequency	System stiffness k	Identification stiffness k1	System damping c	Identification of Damping c1
10hz	512hz	1000.000	1000.0000	12.6491	12.6491
50hz	1024hz		1000.0000		12.6491
100hz	2048hz		1000.0000		12.6491

2.3. Parameter identification based on the time domain least square method of acceleration signal

In this section, the signals of velocity and displacement are obtained by frequency domain integrating the acceleration signals, and then the parameters are identified.

Firstly, the simulated output acceleration $\ddot{x}$ is used as the basic data. Output velocity $\dot{x}_1$ and displacement x_1 are obtained by frequency domain integration. The simulation acceleration $\ddot{x}$, integral velocity $\dot{x}_1$ and integral displacement x_1 are substituted into the identification process, and the influence of the processing of acceleration signal on the system identification is obtained by comparing k1 and c1 with the real k and c of the system.

Frequency domain integration firstly transforms the time domain signal into Fourier transform, and then calculates the frequency domain signal of velocity and displacement. Finally, the inverse Fourier transform of the frequency domain signal of velocity and displacement is used to obtain the time domain signal of velocity and displacement.

Fourier component of acceleration time domain signal at any frequency can be expressed as formula (9).

$$\ddot{x}(t) = \ddot{X}e^{i\omega t} \tag{9}$$

The component of velocity signal can be obtained by integrating the component of acceleration signal, and the component of displacement signal can be obtained by integrating the component of velocity signal, as shown in formulas (10) and (11).

$$\dot{x}(t) = \int_0^t \ddot{X}e^{i\omega t}dt = \frac{\ddot{X}e^{i\omega t}}{i\omega} = \dot{X}e^{i\omega t} \tag{10}$$

$$x(t) = \int_0^t \frac{\ddot{X}e^{i\omega t}}{i\omega}dt = -\frac{\ddot{X}e^{i\omega t}}{\omega^2} = Xe^{i\omega t} \tag{11}$$

In the formula, $\ddot{x}(t)$, $\dot{x}(t)$ and $x(t)$ are the Fourier components of acceleration signal, velocity signal and displacement signal under frequency ω, respectively. Frequency ω is a variable, $\ddot{X}$, $\dot{X}$ and X are corresponding coefficients, i is imaginary.

Setting $\ddot{X}(\omega)$, $\dot{X}(\omega)$ and $X(\omega)$ as frequency domain signals of acceleration, velocity and displacement respectively, the conversion relations among signals of acceleration, velocity and displacement in frequency domain can be obtained from formulas (9), (10) and (11), as shown in formulas (12) and (13).

$$\dot{X}(\omega) = \frac{\ddot{X}(\omega)}{i\omega} \tag{12}$$

$$X(\omega) = -\frac{\ddot{X}(\omega)}{\omega^2} \tag{13}$$

The obtained time domain signals of velocity and displacement are substituted into the time domain least square method for system parameter identification. The results are shown in table 2.

Table 2. Identification of parameters of vibration system.

Excitation frequency	sampling frequency	System stiffness k	Identification stiffness k1	System damping c	Identification of Damping c1
10hz	512hz		1.0086e3		11.6854
50hz	1024hz	1000.000	1.5813e3	12.6491	12.0458
100hz	2048hz		2.1596e3		12.2947

From table 2, it can be seen that the accuracy of identification results of system parameters decreases with the increase of excitation frequency domain. The larger the sampling frequency is, the richer the information in the frequency domain integration is, the closer the obtained signal is to the real signal, and the more accurate the identification results will be. The identification results of the simulation experiments under 100 Hz input excitation at different sampling frequencies are shown in table 3.

Table 3. Identification of parameters of vibration system.

Excitation frequency	sampling frequency	System stiffness k	Identification stiffness k1	System damping c	Identification of Damping c1
100hz	256hz		5.5469e3		10.1388
100hz	2048hz	1000.000	2.1596e3	12.6491	12.2947
100hz	8192hz		1.2908e3		12.5767
100hz	32768hz		1.0657e3		12.6301

From the data in table 3, it can be seen that the larger the sampling frequency, the more accurate the identification results are. This is because the large sampling frequency substitutes more abundant signal information into the integration process, which makes the signal of frequency domain integration closer to the real output value of the system.

3. Identification of frequency domain analytic method

3.1. Basic principles

Frequency domain analytic method is the process of transforming time domain equation into frequency domain equation and finally obtaining analytic solution. Formula (1) is transformed by Laplace transform to obtain the frequency domain equation. Transform according to Laplace transform integral theorem:

$$F(s) = m\ddot{X}(s) + c_1 \frac{\ddot{X}(s)}{s} + k_1 \frac{\ddot{X}(s)}{s^2} \tag{14}$$

$$\frac{F(s)}{\ddot{X}(s)} = m + \frac{c_1}{s} + \frac{k_1}{s^2} \tag{15}$$

$\frac{F(s)}{\ddot{X}(s)}$ is the acceleration impedance of the system when replace s with $i\omega$,the ratio of the input signal of the system to the output signal of the system in complex form. It characterize the amplitude-frequency and phase-frequency characteristics of the input signal and the output signal of the system.

Set $P = \frac{F(s)}{\ddot{X}(s)} = P_R + P_I i$, P_R is its real part and P_I is its imaginary part. The modulus P is the

IOP Conf. Series: Materials Science and Engineering **657** (2019) 012062 doi:10.1088/1757-899X/657/1/012062

amplitude ratio of input excitation force signal to output acceleration signal. P is converted into angle, which is the phase difference between the input excitation force signal and the output acceleration signal [9].

The amplitude ratio and phase difference of signals of the input and output are calculated to obtain P [7], and the P-incorporation formula (15) and the formula (16) are obtained.

$$P_R + P_I i = \left(m - \frac{k_1}{\omega^2} \right) - \frac{c_1}{\omega} i \tag{16}$$

The numbers of formula (16) is complex on both sides of the equal sign, so the real parts on both sides are equal and the imaginary parts on both sides are equal. Solution formula (16) gets (17).

$$\begin{cases} k_1 = \omega^2 (m - P_R) \\ c_1 = -\omega P_I \end{cases} \tag{17}$$

3.2. Simulation and verification of frequency domain analytic method

The parameters k1 and c1 are identified from the output data of the system with known parameters k and c, and compared with the real values of k and c of the system. The results are shown in table 4.

Table 4. Identification of parameters of vibration system.

Excitation frequency	sampling frequency	System stiffness k	Identification stiffness k1	System damping c	Identification of Damping c1
10hz	512hz		1.0105e3		12.4102
50hz	1024hz	1000.000	1.5901e3	12.6491	12.0491
100hz	2048hz		2.2011e3		12.2985

As can be seen from table 4, there are errors in the identification results. The main factors affecting the identification results are P_R and P_I, which are determined by the amplitude ratio and phase difference of the input signal and the output signal of the system. In the process of calculating the amplitude ratio and phase difference of input and output signals, the accuracy is still affected by sampling frequency [7]. The results of the frequency domain analytic method under different sampling frequencies with simulation experiment of 100 Hz input excitation are shown in table 5.

Table 5. Identification of parameters of vibration system.

Excitation frequency	sampling frequency	System stiffness k	Identification stiffness k1	System damping c	Identification of Damping c1
100hz	512hz		5.6170e3		10.1428
100hz	2048hz	1000.000	2.2011e3	12.6491	12.2985
100hz	8192hz		1.3034e3		12.5795
100hz	32768hz		1.0760e3		12.6329

From the data in table 5, it can be seen that the frequency domain analytic method can accurately identify the system parameters, and the larger the sampling frequency, the more accurate the identification results are.

4. Examples verification

4.1. Experiments and identification processes

4.1.1. Introduction to experiments. In this section, a spacecraft electronic instrument box model is taken as an example for vibration absorption experiments. Vibration absorption system consisting of damper material and electronic instrument box model fixed on vibration frame. Figure 3 is a vibration absorption model and figure 4 is a vibration absorption experiment.

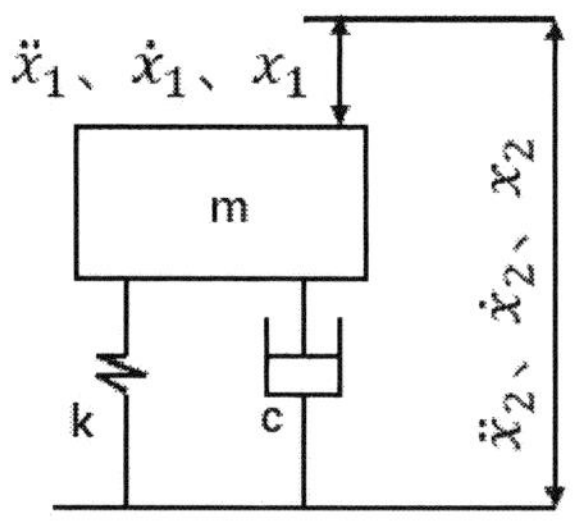

Figure 3. Vibration model.

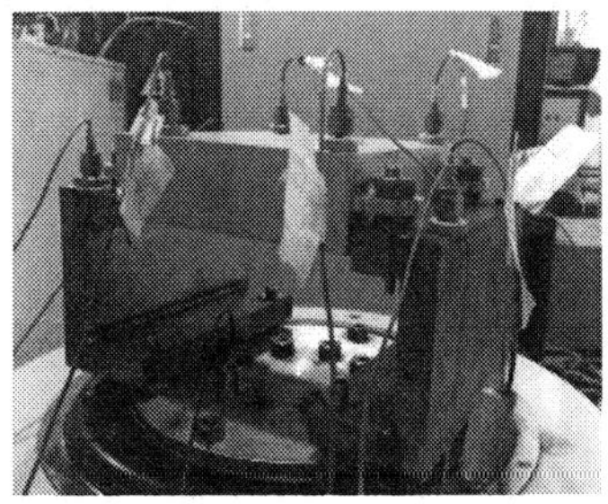

Figure 4. Experimental objects.

Formula (18) according to vibration reduction model.

$$m\ddot{x}_1 + c(\dot{x}_1 - \dot{x}_2) + k(x_1 - x_2) = 0 \tag{18}$$

$\ddot{x}_1$, $\dot{x}_1$ and x_1 are the signals of acceleration, velocity and displacement of the electronic instrument box model respectively; $\ddot{x}_2$, $\dot{x}_2$ and x_2 are signals of the acceleration, velocity and displacement of the shaking table respectively.

The shaking table is vibrated step by step in the range of 5-20 Hz with the step length of 1 HZ. Acceleration signals are collected in the experiment. Velocity and displacement signals are obtained by frequency domain integration.

4.1.2. Identification of experimental vibration system parameters by the time domain least square method. Formula (18) is converted to formula (19) when the parameters of the experimental system are identified by the time domain least square method.

$$m\ddot{x}_1 = c(\dot{x}_2 - \dot{x}_1) + k(x_2 - x_1) \tag{19}$$

4.1.3. Identification of experimental vibration system parameters by the frequency domain analytic method. When the parameters of the experimental system are identified by the frequency domain analytic method, formula (18) is transformed by Laplace transform integral theorem, and formula (20) is obtained.

$$[m + \frac{c}{s} + \frac{k}{s^2}]\ddot{X}_1 = [\frac{c}{s} + \frac{k}{s^2}]\ddot{X}_2 \tag{20}$$

Substituting $s = i\omega$ and $\frac{\ddot{X}_1(s)}{\ddot{X}_2(s)} = P_R + P_I i$ into formulas (20), (21) and (22) are obtained.

$$k = \frac{m\omega^2(P_R^2 + P_I^2 + P_R)}{P_R^2 - 2P_R + P_I^2 + 1} \tag{21}$$

$$c = \frac{-m\omega^2 P_R + P_R k - k}{\omega P_I} \tag{22}$$

4.2. Results and analysis

The dynamic stiffness and damping parameters of the experimental system identified by the two identification methods are shown in figures 5 and 6.

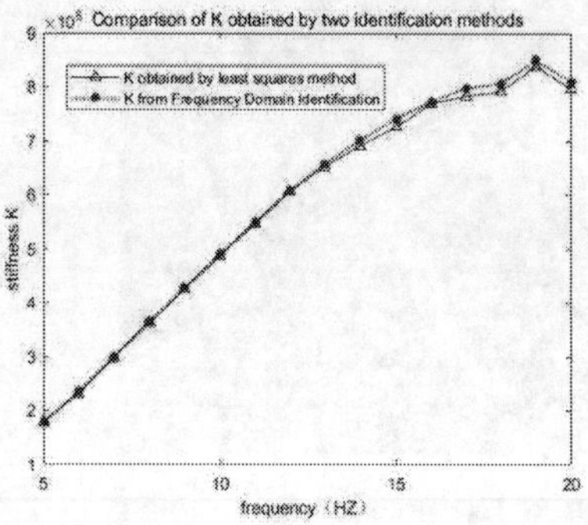

Figure 5. Dynamic stiffness of system.

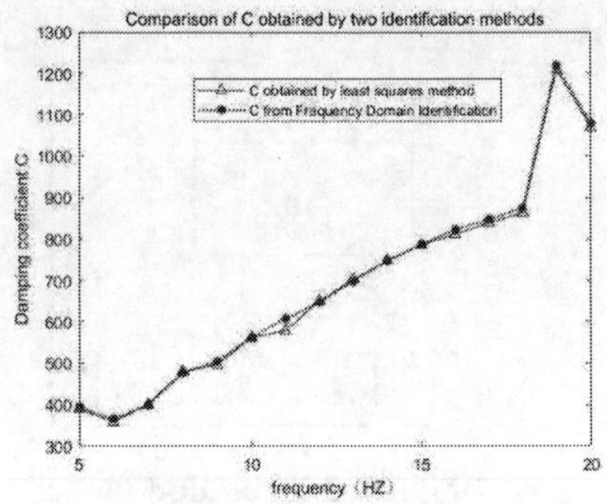

Figure 6. Dynamic damping of system.

From figures 5 and 6, it can be seen that the parameters of the experimental system identified by two different methods are in good agreement. The two methods are mutually corroborative and can confirm that the identification results are correct.

5. Conclusion

- The time domain least square method and the frequency domain analytic method proposed in this paper have good application of scientific research and engineering.
- Sampling frequency of signal is an important factor in parameter identification. The larger the sampling frequency, the more accurate the identification result is, but the calculation amount will increase accordingly.

References

[1] Li S Y, Zhang W G, Hou J Z *et al* 2017 Research on dynamic stiffness of vibration isolator *Acta Armamentarii* **38** 2274-9
[2] Wan Y Q, Fan L M and Y Qi Y 2007 Nonlinear characteristics a-nalysis for steel wire isolators *J. Vibration and Shock* **26** 46-9
[3] He F J and Lu X 2014 Experimental research on dynamic stiffness characteristics of a rubber isolator *Research and Exploration in Laboratory* **33** 42-4
[4] Gu M K and Lü Z H 2011 Identification of a mechanism's vibration velocity and displacement based on the acceleration measurement *Mech. Sci. Technol. Aerospace Eng.* **30** 522-6
[5] s*Engineer* **04** 147-9
[6] Qiu Y 2007 Application matlab in least square method with liner parameter *Sci. Mosaic* **05** 165-6
[7] Cheng J 1997 Research on phase difference detection algorithms for sinusoidal signals *Meas. Tech.* **10** 28-30

Dynamic modelling and analysis of super-high acceleration macro-micro-macro movement platform

L F Zhang[1,3], B Ma[2,3,4], J Wu[3], H Li[3] and G Q Li[3]

[1]School of Mechanical and Electrical Engineering, Henan University of Technology, Zhengzhou 450001, China
[2]Key Laboratory of Road Construction & Equipment of MOE, Chang'an University, Xi'an 710064, China
[3]WEIHUA Group Co., Ltd., Changyuan 453400, China

E-mail: m18681862425@163.com

Abstract. In microelectronics manufacturing industry, to obtain the high-acceleration of Marco-micro movement platform was very crucial and difficult. In this paper, a new macro-micro-macro movement platform is described. In addition, by the analysis of its driving principle, nine simplified dynamic models and dynamic equations under different motion conditions are established respectively. In different driving modes (single VCM driving mode, double VCMs reverse driving mode and double VCMs same driving mode, respectively), the displacement and velocity of floating stator, macro system and micro system was calculated, respectively. These results imply that the macro-micro-macro movement platform takes a periodic reciprocating motion and the max displacement and velocity of floating stator, macro system and micro system make different change rules, respectively. It concluded that the floating stator effectively isolating the vibration of macro-micro-macro movement platform.

1. Introduction

Microelectronics manufacturing industry has played vital role in the developing level of science and technology of a country [1,2]. In the microelectronics manufacturing field, the macro-micro movement platform is a vitally important equipment, and it is usually used to obtain the high-acceleration and super-precision positioning motion. However, the rapid enhancement of acceleration is difficult to be obtained, which cannot meet the fast development of high and new technology and the rapid rise of market demand. Therefore, it is very necessary to improve the acceleration of the macro-micro movement platform [3]. In many areas, the concept of "macro + micro" by single VCM and PZT has been widely used in the motion platform to effectively deal with the discrepancy of high-acceleration and super-precision positioning [4-7]. The macro-micro movement platform was established by the Harbin Institute of Technology, which can obtain the acceleration of 5 g and precision of ± 21 nm [8]. Heui etc has studied the research of macro-micro system by single VCM to obtain the repeat positioning accuracy of 19 nm and the acceleration of macro-micro system is increased by more than ten times [9]. Obviously, the single VCM and PSA are widely used in some macro-micro movement platforms to acquire the high acceleration [10-11]. But, the VCM usually provides a certainly-maximum output force. And the acceleration of motion platform cannot exceed the maximum acceleration to realize the ultra-high acceleration. Accordingly, the acceleration breakthrough is often at stagnant phase, and some main obstacles for further development of microelectronics manufacturing industry.

To achieve the high acceleration of macro-micro movement platform, the "macro + micro + macro" drive mode is established. Through the research on the macro-micro movement platform, the macro-micro-macro movement platform is proposed to explore the acceleration breakthrough. In this paper, the motion platform is realized by two VCMs and single PZT, and it is preliminarily built by according to related performance indexes and working conditions. Overall structure of the paper is arranged as follows: initially, a macro-micro-macro movement platform of super-high acceleration with double floating stator stage is introduced. Secondly, different driving modes are built, and dynamic theoretical analysis by different dynamic models of macro-micro-macro movement platform is established. In addition, some related performance indexes of macro-micro-macro movement platform are simulated. Finally, the conclusion is given.

2. Description of ultra-high acceleration macro-micro-macro movement platform

Using the concept of macro-micro-macro (MMM), a new movement platform is provided to obtain ultra-high acceleration. And the super-high acceleration of macro-micro-macro movement platform is obtained by the two same VCMs and floating stators and the random vibration is occurred, as shown in figure 1. Generally, two same VCMs are installed on the base, which join macro movement platform being installed on the output coils by connection arms. Two same floating stator platforms are applied in macro-micro-macro movement platform to reduce or isolate vibration between VCMs and bases, which consisted of damper, stopper, floating lock, spring, limiter, subbase, buffer device etc. peculiarly, in order to obtain different super-high acceleration, the different motion modes can adjust the quantity of output force. As the flexure hinge mechanism has some advantages of no friction, no clearance, high resolution, etc. and the PZT drive the flexure hinge mechanism can obtain an ultra-precision position. Through different VCMs driving control modes, the acceleration of macro-micro-macro movement platform can increase by more than ten times to achieve the goal of the super-high acceleration. By the coarse reading head reading and coarse grating ruler, the microscale localization will be more rapidly obtained. Simultaneously, the nanoscale localization is confirmed by fine reading head reading and fine grating ruler. The difference between microscale and nanoscale localization can be counted. In addition, the stator buffer device consisting of springs, damper and stopper and so on, can be used to reduce the vibration, and the PZT is used to reduce errors and achieve a super-precise positioning. Double stator buffer devices are located at the front and rear of two VCMs. The Macro-micro-macro movement platform is located at the base.

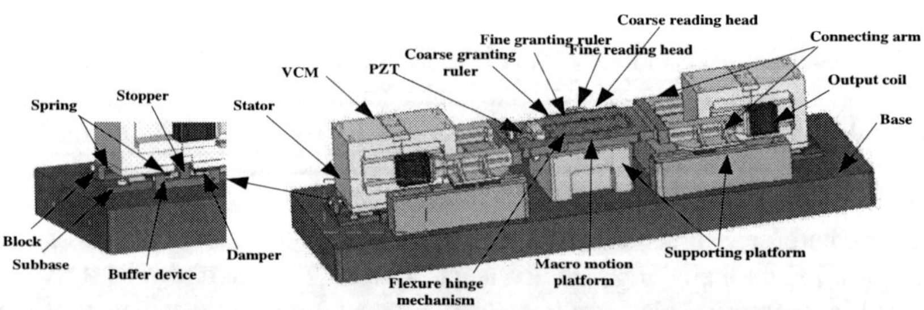

Figure 1. Structures of macro-micro-macro movement platform.

3. Dynamic theoretical analysis of macro-micro-macro movement platform

According to the macro-micro-macro movement platform proposed in this paper, it mainly includes macro system and micro system. By analyzing its transmission form, the macro-micro-macro movement platform is equivalent to a mass-spring-dashpot system. In addition, according to the structural characteristics of the macro-micro-macro movement platform, three motion conditions are listed: fixed stator mode, fixe-floating stator mode, and floating stator mode. And then, according to the driving mode of VCMs and PZT, it can be divided into nine motion conditions. Through the analysis of its driving principle, simplified dynamic models and dynamic equations under different motion conditions are established respectively.

3.1. Dynamic theoretical analysis of macro-micro-macro movement platform using fixed stator

In fixed stator mode, it can be a linear second-order system including three motion conditions: single VCM driving mode, double VCMs reverse driving mode, double VCMs same driving mode. Simplified dynamic models are obtained as shown in figures 2-4, respectively. Its system kinetic equations are established as shown in equations (1)-(3), respectively. From equation (1) to equation (9), m_0, m_1 and m_2 is the equivalent mass of floating stator, macro system and micro system, respectively; k_0, k_1 and k_2 is the equivalent stiffness of floating stator, macro system and micro system, respectively; c_0, c_1 and c_2 is the equivalent damping of floating stator, macro system and micro system, respectively; F_1, F_2, and F_3 is the input force of VCM_1, VCM_2 and PZT, respectively; and x_0, x_1 and x_2 is the displacement of floating stator, macro system and micro system, respectively.

$$\begin{cases} m_1\ddot{x}_1 + 2(c_1 + c_2)\dot{x}_1 + 2(k_1 + k_2)x_1 - 2c_2\dot{x}_2 - 2k_2x_2 = F_1 - F_2 \\ m_2\ddot{x}_2 + 2c_2(\dot{x}_2 - \dot{x}_1) + 2k_2(x_2 - x_1) = F_2 \end{cases} \quad (1)$$

$$\begin{cases} m_1\ddot{x}_1 + 2(c_1 + c_2)\dot{x}_1 + 2(k_1 + k_2)x_1 - 2c_2\dot{x}_2 - 2k_2x_2 = F_1 - F_3 - F_2 \\ m_2\ddot{x}_2 + 2c_2(\dot{x}_2 - \dot{x}_1) + 2k_2(x_2 - x_1) = F_2 \end{cases} \quad (2)$$

$$\begin{cases} m_1\ddot{x}_1 + 2(c_1 + c_2)\dot{x}_1 + 2(k_1 + k_2)x_1 - 2c_2\dot{x}_2 - 2k_2x_2 = F_1 + F_3 - F_2 \\ m_2\ddot{x}_2 + 2c_2(\dot{x}_2 - \dot{x}_1) + 2k_2(x_2 - x_1) = F_2 \end{cases} \quad (3)$$

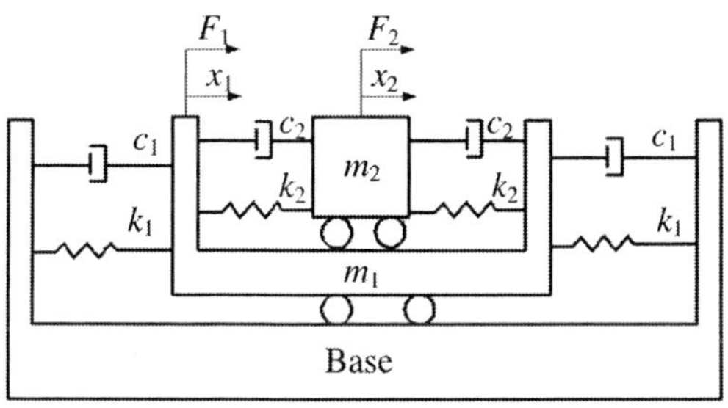

Figure 2. Simplified dynamic model of single VCM driving mode with fixed stator.

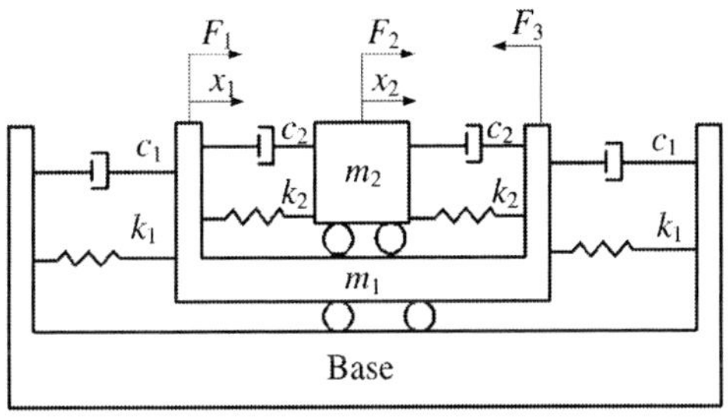

Figure 3. Simplified dynamic model of double VCMs reverse driving mode with fixed stator.

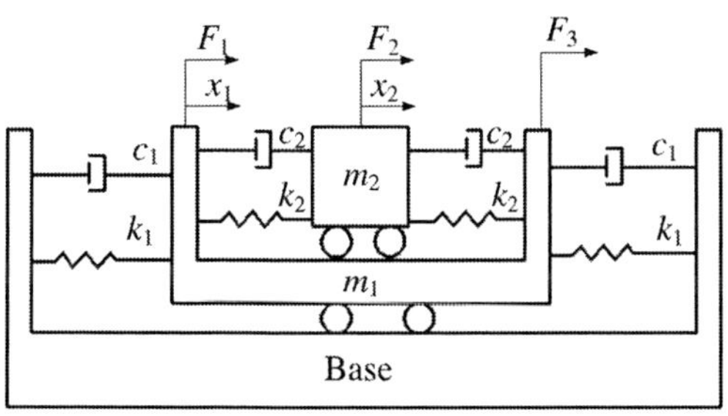

Figure 4. Simplified dynamic model of double VCMs same driving mode with fixed stator.

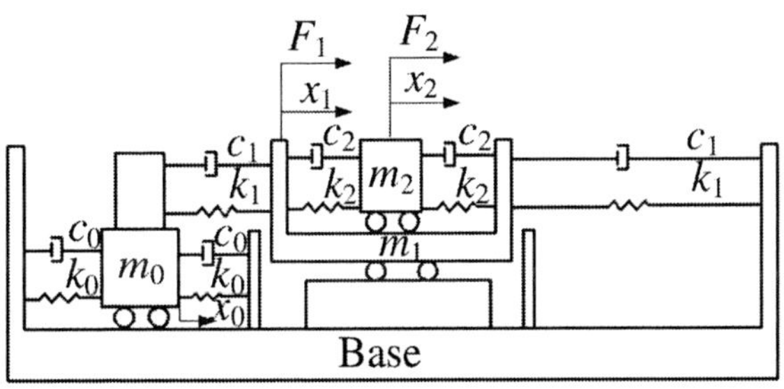

Figure 5. Simplified dynamic model of the single VCM driving mode with single fixed stator and single floating stator.

IOP Conf. Series: Materials Science and Engineering **657** (2019) 012063 doi:10.1088/1757-899X/657/1/012063

3.2. Dynamic theoretical analysis of macro-micro-macro movement platform using single fixed stator and single floating stator

In fixed-floating stator mode, it can be a linear second-order system including three motion conditions: single VCM driving mode, double VCMs reverse driving mode, double VCMs same driving mode. Simplified dynamic models are obtained as shown in figures 5-7, respectively. Its system kinetic equations are established as shown in equations (3)-(5), respectively.

$$\begin{cases} m_0\ddot{x}_0 + 2c_0\dot{x}_0 - c_1(\dot{x}_1 - \dot{x}_0) + 2k_0x_0 - k_1(x_1 - x_0) = -F_1 \\ m_1\ddot{x}_1 + c_1(2\dot{x}_1 - \dot{x}_0) - 2c_2(\dot{x}_2 - \dot{x}_1) + k_1(2x_1 - x_0) - 2k_2(x_2 - x_1) = F_1 - F_2 \\ m_2\ddot{x}_2 + 2c_2(\dot{x}_2 - \dot{x}_1) + 2k_2(x_2 - x_1) = F_2 \end{cases} \tag{4}$$

$$\begin{cases} m_0\ddot{x}_0 + 2c_0\dot{x}_0 - c_1(\dot{x}_1 - \dot{x}_0) + 2k_0x_0 - k_1(x_1 - x_0) = -F_1 - F_3 \\ m_1\ddot{x}_1 + c_1(2\dot{x}_1 - \dot{x}_0) - 2c_2(\dot{x}_2 - \dot{x}_1) + k_1(2x_1 - x_0) - 2k_2(x_2 - x_1) = F_1 - F_2 + F_3 \\ m_2\ddot{x}_2 + 2c_2(\dot{x}_2 - \dot{x}_1) + 2k_2(x_2 - x_1) = F_2 \end{cases} \tag{5}$$

$$\begin{cases} m_0\ddot{x}_0 + 2c_0\dot{x}_0 - c_1(\dot{x}_1 - \dot{x}_0) + 2k_0x_0 - k_1(x_1 - x_0) = -F_1 + F_3 \\ m_1\ddot{x}_1 + c_1(2\dot{x}_1 - \dot{x}_0) - 2c_2(\dot{x}_2 - \dot{x}_1) + k_1(2x_1 - x_0) - 2k_2(x_2 - x_1) = F_1 - F_2 - F_3 \\ m_2\ddot{x}_2 + 2c_2(\dot{x}_2 - \dot{x}_1) + 2k_2(x_2 - x_1) = F_2 \end{cases} \tag{6}$$

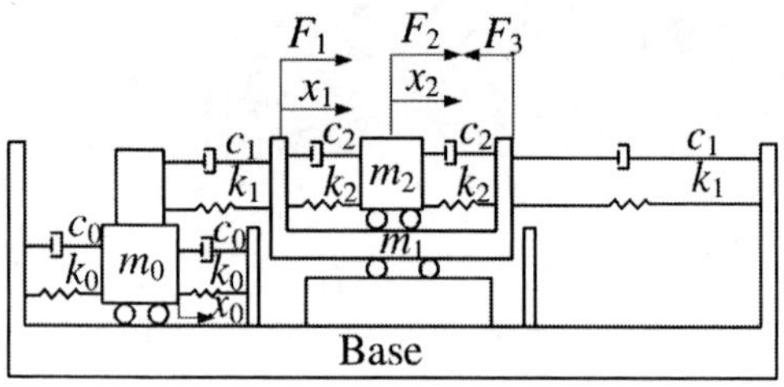

Figure 6. Simplified dynamic model of the double VCMs same driving mode with single fixed stator and single floating stator.

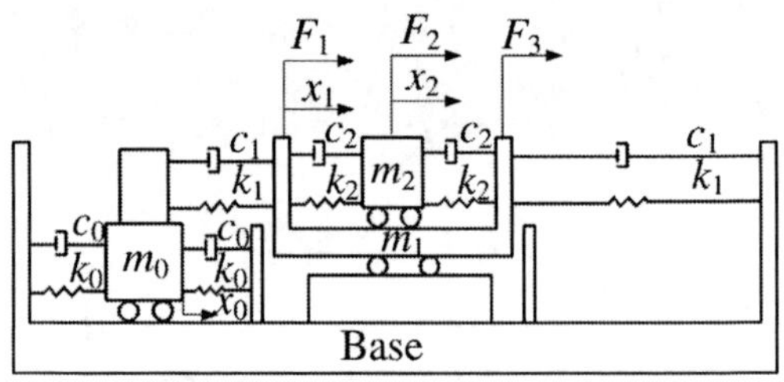

Figure 7. Simplified dynamic model of the double VCMs reverse driving mode with single fixed stator and single floating stator.

3.3. Dynamic theoretical analysis of macro-micro-macro movement platform using double floating stator

In floating stator mode, it can be a linear second-order system including three motion conditions: single VCM driving mode, double VCMs reverse driving mode, double VCMs same driving mode. Simplified dynamic models are obtained as shown in figures 8-10, respectively. Its system kinetic equations are established as shown in equations (7)-(9), respectively.

$$\begin{cases} 2m_0\ddot{x}_0 + 4c_0\dot{x}_0 - 2c_1(\dot{x}_1 - \dot{x}_0) + 4k_0x_0 - 2k_1(x_1 - x_0) = -F_1 \\ m_1\ddot{x}_1 + 2c_1(\dot{x}_1 - \dot{x}_0) - 2c_2(\dot{x}_2 - \dot{x}_1) + 2k_1(x_1 - x_0) - 2k_2(x_2 - x_1) = F_1 - F_2 \\ m_2\ddot{x}_2 + 2c_2(\dot{x}_2 - \dot{x}_1) + 2k_2(x_2 - x_1) = F_2 \end{cases} \tag{7}$$

$$\begin{cases} 2m_0\ddot{x}_0 + 4c_0\dot{x}_0 - 2c_1(\dot{x}_1 - \dot{x}_0) + 4k_0x_0 - 2k_1(x_1 - x_0) = -F_1 - F_3 \\ m_1\ddot{x}_1 + 2c_1(\dot{x}_1 - \dot{x}_0) - 2c_2(\dot{x}_2 - \dot{x}_1) + 2k_1(x_1 - x_0) - 2k_2(x_2 - x_1) = F_1 + F_3 - F_2 \\ m_2\ddot{x}_2 + 2c_2(\dot{x}_2 - \dot{x}_1) + 2k_2(x_2 - x_1) = F_2 \end{cases} \tag{8}$$

$$\begin{cases} 2m_0\ddot{x}_0 + 4c_0\dot{x}_0 - 2c_1(\dot{x}_1 - \dot{x}_0) + 4k_0x_0 - 2k_1(x_1 - x_0) = -F_1 + F_3 \\ m_1\ddot{x}_1 + 2c_1(\dot{x}_1 - \dot{x}_0) - 2c_2(\dot{x}_2 - \dot{x}_1) + 2k_1(x_1 - x_0) - 2k_2(x_2 - x_1) = F_1 - F_3 - F_2 \\ m_2\ddot{x}_2 + 2c_2(\dot{x}_2 - \dot{x}_1) + 2k_2(x_2 - x_1) = F_2 \end{cases} \tag{9}$$

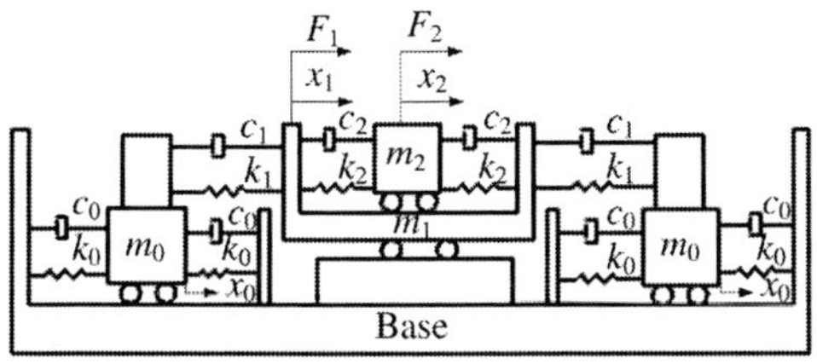

Figure 8. Simplified dynamic model of the single VCM driving mode with double floating stator.

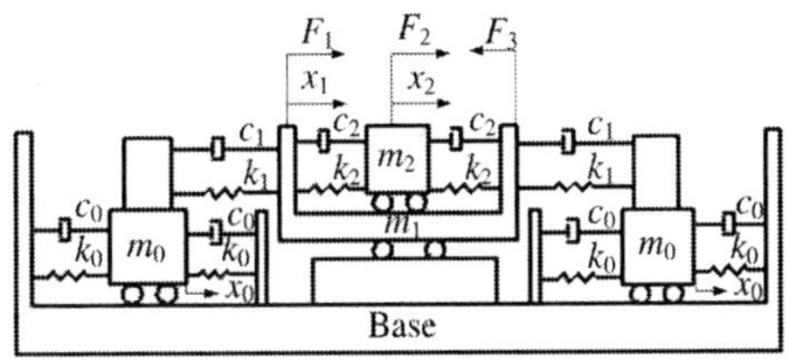

Figure 9. Simplified dynamic model of the double VCMs same driving mode with double floating stator.

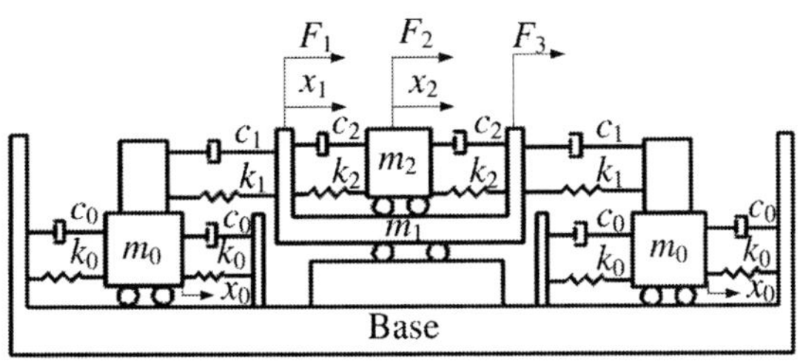

Figure 10. Simplified dynamic model of the double VCMs reverse driving mode with double floating stator.

4. Results and discussion

The reference values of kinetic parameters are shown in table 1. Equations (1)-(9) are converted to the matrix equation as follows:

$$\mathbf{M}\ddot{X} + \mathbf{C}\dot{X} + \mathbf{K}X = \mathbf{F} \tag{10}$$

Table 1. The values of dynamic parameters.

m_0(kg)	k_0(N/m)	c_0(N·s/m)	m_1(kg)	k_1(N/m)	c_1(N·s/m)	m_2(kg)	k_2(N/m)	c_2(N·s/m)
13	2.8×10^7	0.02	0.942	6.6×10^7	0.05	1.227	2.8×10^7	0.02

Defined **F=0** and **C=0**, the equation (10) can be converted by equation (11):

$$\mathbf{M}\ddot{X} + +\mathbf{K}X = \mathbf{0} \tag{11}$$

So, in different motor mode, by the values of table 1 and equation (11), the natural frequency of movement platform can be calculated. In fixed stator mode, the system's natural frequency are 596 Hz and 1825 Hz respectively, and in fixed-floating stator mode, the system's natural frequency are 941 Hz, 941 Hz and 2371 Hz respectively, and in floating stator mode, the system's natural frequency are 886 Hz, 886 Hz and 2345 Hz respectively. Therefore, they are far less than the driver frequency of PZT (20 kHz). The results indicate that the motion platform cannot appear the resonance, and the system of motion platform can be regard as a rigid body motion system.

Identified F_1=300 N, F_2=3000 N and F_3=100 N, respectively, the displacement and velocity of

IOP Conf. Series: Materials Science and Engineering **657** (2019) 012063 doi:10.1088/1757-899X/657/1/012063

macro-micro-macro motion platform are obtained by MATLAB. In different driving modes (single VCM driving mode, double VCMs reverse driving mode and double VCMs same driving mode, respectively), the displacement and velocity of floating stator, macro system and micro system are shown in figures 11-13, respectively. These results imply that the system takes a periodic reciprocating motion. In different VCM driving modes (single VCM driving mode, double VCMs reverse driving mode and double VCMs same driving mode, respectively) with fixed stator, the max velocity of macro system and micro system nearly reached at 1.1988 m/s and 1.9054 m/s, respectively. In addition, the max displacement and max velocity of macro system and micro system gradually increases by different VCM driving modes. In different VCM driving modes (single VCM driving mode, double VCMs reverse driving mode and double VCMs same driving mode, respectively) with fixed-floating stator, the max velocity of floating stator, macro system and micro system nearly reached at 0.5249 m/s, 1.1002 m/s and 1.8315 m/s, respectively.

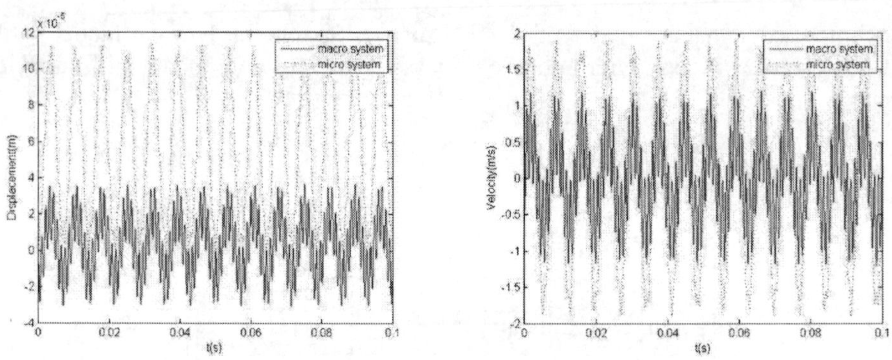

(a) (b)

Figure 11. The displacement (a) and velocity (b) of driving modes with fixed stator.

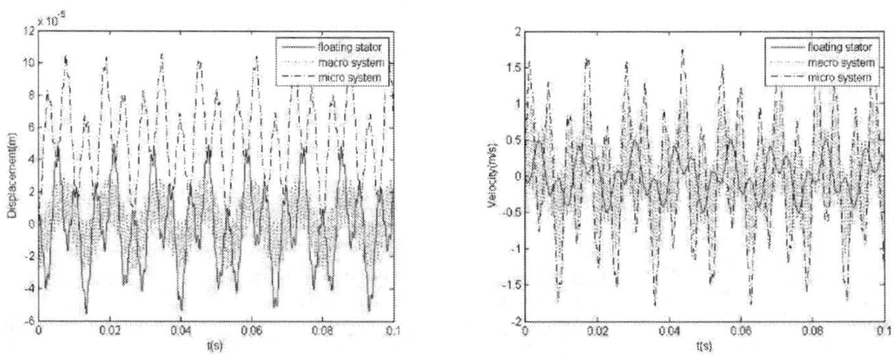

(a) (b)

Figure 12. The displacement (a) and velocity (b) of driving modes with single fixed stator and single floating stator.

Meanwhile, by different VCM driving modes, the max displacement of floating stator and macro system gradually decreases firstly and then increases, reversely, the max displacement of micro system gradually increases firstly and then decreases, and the max velocity of floating stator, macro system and micro system gradually increases firstly and then decreases. In different VCM driving modes (single VCM driving mode, double VCMs reverse driving mode and double VCMs same driving mode, respectively) with floating stator, the max velocity of floating stator, macro system and micro system nearly reached at 0.5676 m/s, 0.9412 m/s and 1.8488 m/s, respectively.

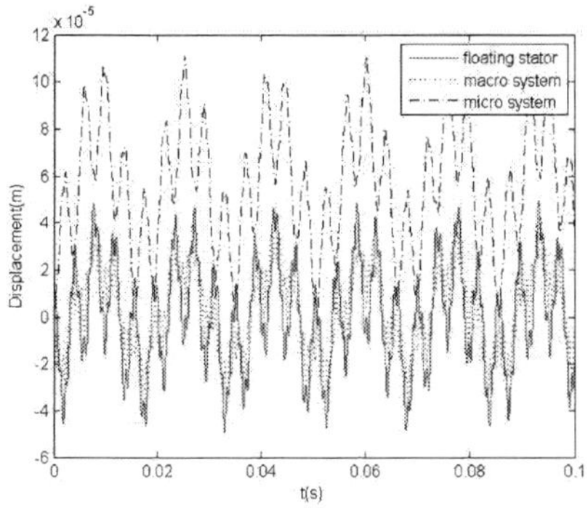

(a)

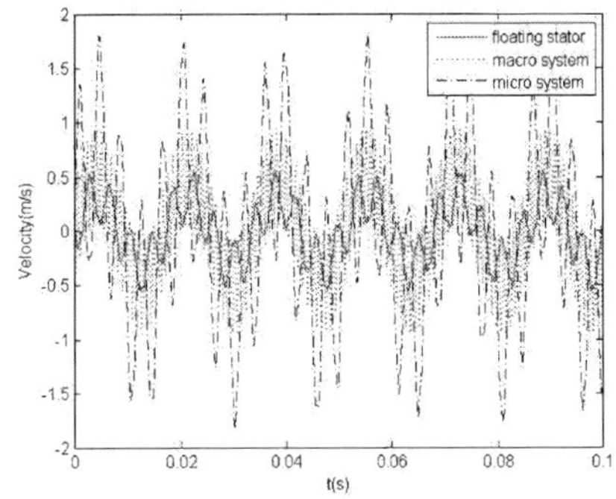

(b)

Figure 13. The displacement (a) and velocity (b) of driving modes with double floating stator.

By different VCM driving modes, the max displacement of floating stator and micro system gradually decreases at firstly and then increases; reversely, and the max displacement of macro system gradually decreases. In addition, the max velocity of floating stator and micro system gradually increases firstly and then decreases. However, the max velocity of macro system gradually decreases firstly and then increases. By compared with the max velocity of floating stator, macro system and micro system, respectively, it indicated that floating stator nearly weaken the vibration of macro-micro-macro movement platform in same driving modes.

5. Conclusion

In this paper, a new macro-micro-macro movement platform is provided. Its different dynamic models were obtained by different VCM driving modes. By the simulation results, it can be concluded that the macro-micro-macro movement platform takes a periodic reciprocating motion and the floating stator nearly isolates the vibration of macro-micro-macro movement platform.

Acknowledgments

This article was funded by the project as follows: the National Natural Science Foundation of China (Grant No. 51705132), the National Natural Science Foundation of China (Grant No. 51705217), the Science and Technology Department of Henan Province Natural Science Project (Grant No. 172102210215) and the Education Department of Henan Province Natural Science Project (Grant No. 17A460008).

References

[1] Wang Y, Ma H, Yang J and Wang K 2017 Industry 4.0: A way from mass customization to mass personalization production *Adv. Manuf.* **5** 11-320

[2] Zhang L, Gao J and Chen X 2017 A rapid dynamic positioning method for settling time reduction through a macro-micro composite stage with high positioning accuracy *IEEE IEEE Trans. Ind. Electron.* **99** 1

[3] Zhang L, Li X, Fang J and Long Z 2017 Multi-objective positioning design optimization of flexure hinge mechanism considering thermal-mechanical coupling deformation and natural frequency *Adv. Mech. Eng.* **9** 1-17

[4] Elfizy A, Bone G and Elbestawi M 2005 Design and control of a dual-stage feed drive *Int. J. Manch. Tool. Manuf.* **45** 153-65

[5] Fang J, Long Z, Wang M, Zhang L and Dai X 2016 Multi–mode sliding mode control for precision linear stage based on fixed or floating stator *Rev. Sci. Instrum.* **87** 1-10

[6] Fang J W, Zhang L F, Long Z L and Wang M Y 2018 A fuzzy adaptive sliding mode control for

the precision position of piezo-actuated nano positioning stage *Int. J. Precis. Eng. Manuf.* **19** 1447-56

[7] Wang H 2013 Research on high performance positioning stage with floating stator (Harbin, China: Harbin Institute of Technology)

[8] Heui J, Dong S and Jong H 2001 Ultra-precision positioning system for servo motor-piezo actuator using the dual loop and digital filter implementation *Int. J. Manch. Tool. Manuf.* **41** 51-63

[9] Dong W, Tang J and EIDeeb Y 2008 Design of dual-stage actuation system for high precision optical manufacturing *Proc. SPIE* **6928** 1-7

[10] Ha J, Kung Y, Hu S and Fung R 2006 Optimal design of a micro-positioning Scott-Russell mechanism by Taguchi method *Sens. Actuators* **125** 565-72

[11] Matsubara A, Fujita T, Kono D *et al* 2007 Nano positioning drive with piezoelectric actuator integrated into support bearing unit of ball screw *Proceedings of the 35th International MATADOR Conference* (Manchester) **2007** 323-6

IOP Conf. Series: Materials Science and Engineering **657** (2019) 012064 doi:10.1088/1757-899X/657/1/012064

Research on seismic absorption performance of adjacent structures with additional tuned mass dampers

Q X Sun[1,2]

[1]Jiangsu University of Science and Technology, Zhenjiang City, Jiangsu Province, China

E-mail: 1624667415@qq.com

Abstract. In order to investigate the seismic performance of adjacent structures with additional tuned mass dampers, this article takes two adjacent single degree of freedom systems as an example, and takes the transfer rate of displacement response of adjacent structures as performance index, analyses the relationship between damping parameters and structural parameters of adjacent structures with additional tuned mass dampers. The numerical relationship between the structural parameters of adjacent structures with tuned mass dampers ,the parameters of tuned mass dampers, and Specific Performance of the adjacent structures in Earthquakes has been find out.The optimum structural parameters are obtained by theoretical deduction. Taking displacement response curve as an index, the reliability of the optimum structural parameters is verified.

1. Introduction

In recent years, the world's total population has gradually increased, the proportion of urban population has gradually expanded, and the per capita possession area has gradually decreased. As a result, urban buildings have become more and more dense, and many adjacent structures have been formed due to the small space of buildings. When an earthquake occurs, these adjacent structures may collide with each other, leading to structural damage and collapse. In the Loma Prieta earthquake that occurred in 1989, more than 200 of the 500 buildings surveyed collided [1]. In the 2011 Christchurch Earthquake, the structural damage caused by the collision was Investigated, according to the survey results, more than 6% of the structures were severely damaged by collisions [2]. The collision between adjacent buildings caused by earthquakes has resulted in huge loss of life and property. Therefore, it is very meaningful to study the ways to prevent these buildings from colliding with each other and to improve the seismic performance of these buildings. Klein [3], an American, first proposed the idea of connecting adjacent structures to improve the seismic capacity of the structure. Later, he and his colleagues studied the control reaction of the connected structure connected by the energy dissipating component and the semi-active control device under the action of earthquake. SAAnagnostopoulos [4] studied the collision of two adjacent single-degree-of-freedom structures under earthquake action, and proposed the installation of viscoelastic dampers at the collision. Westermo [5] proposed the position in the same plane of adjacent structures. With articulated connecting rods, a number of subsequent studies have been conducted to reduce structural dynamic response by joining adjacent structures.. Sun [6] simplified the adjacent structure to a two-degree-of-freedom system with viscous dampers connected by selecting the frequency ratio and mass ratio of adjacent structures as parameters, and the relationship between the mean square value of the structural displacement response and the connected damping parameters.. Chen and Pei [7] made a more indepth research on the basis of their predecessors, and discussed the energy dissipation and damping of the adjacent structures of two adjacent multi-layer steel frame structures connected with viscous dampers and metal dampers. In recent years, domestic and foreign scholars have analyzed the selected research models by adding numerical simulation and experimental research methods by adding damping devices between adjacent structures, and analyzing the rationality of damping control methods from different levels. However, there are few researches have been done on the addition of uned mass dampers between adjacent structures. The purpose of this paper is to find out the numerical relationship between the structural parameters of adjacent structures with tuned mass dampers, the parameters of tuned mass dampers, and Specific Performance of the adjacent structures in Earthquakes. And to find out the structural parameters of adjacent structures most suitable for installing tuned mass dampers.

2. Numerical model

The Mechanical model of adjacent structures of additional tuned mass dampers is shown in figure 1.

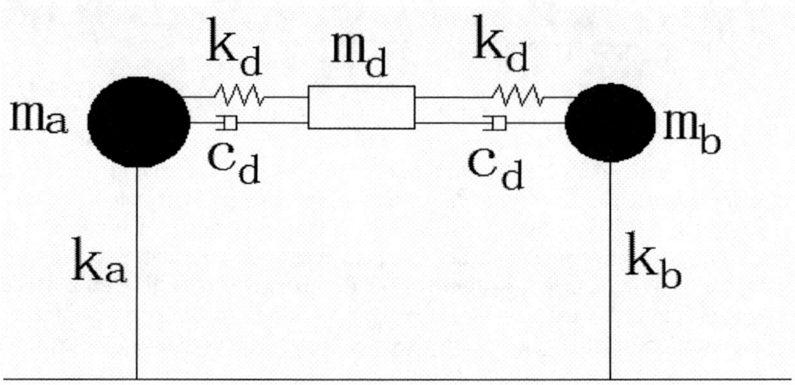

Figure 1. Mechanical model.

The vibration differential equation corresponding to figure 1 is:

$$m_a\ddot{x}_a + k_a x_a + c_d(\dot{x}_a - \dot{x}_d) + k_d(x_a - x_d) = -m_a\ddot{y} \tag{1}$$

$$m_b\ddot{x}_b + k_b x_b + c_d(\dot{x}_b - \dot{x}_d) + k_d(x_b - x_d) = -m_b\ddot{y} \tag{2}$$

$$m_d\ddot{x}_d + c_d(\dot{x}_d - \dot{x}_a) + k_d(x_d - x_a) + c_d(\dot{x}_d - \dot{x}_b) + k_d(x_d - x_b) = -m_d\ddot{y} \tag{3}$$

In the formula, x_a, x_b, x_d are the displacements of A, B and D relative to the ground, and Y is the the displacement of the ground. Suppose the input ground motion is a simple harmonic, let $y = Y \cdot e^{i\omega t}$, the displacement response of structure A and B under earthquake action is $x_a = X_a \cdot e^{i\omega t}$,$x_b = X_b \cdot e^{i\omega t}$,$x_d = X_d \cdot e^{i\omega t}$. Y, X_a, X_b, X_d represent the corresponding amplitudes of each vibration. Substitute Y, X_a, X_b, X_d into expressions (1), (2) and (3), by combining the same items, it is concluded that:

$$\frac{X_a}{Y}(-m_a\omega^2 + k_a + ic_d\omega + k_d) + \frac{X_d}{Y}(-ic_d\omega - k_d) = m_a\omega^2$$

$$\frac{X_b}{Y}(-m_b\omega^2 + k_b + ic_d\omega + k_d) + \frac{X_d}{Y}(-ic_d\omega - k_d) = m_b\omega^2$$

$$\frac{X_a}{Y}(-ic_d\omega - k_d) + \frac{X_b}{Y}(-i\omega c_d - k_d) + \frac{X_d}{Y}(-m_d\omega^2 + 2ic_d\omega + 2k_d) = m_d\omega^2$$

IOP Conf. Series: Materials Science and Engineering **657** (2019) 012064 doi:10.1088/1757-899X/657/1/012064

By substituting $\omega_a = (k_a/m_a)^{1/2}$, $\omega_b = (k_b/m_b)^{1/2}$, $\omega_d = (k_d/m_d)^{1/2}$ into the above formulas, we can get the result that:

$$\left|\frac{X_a}{Y}\right| = \left|\frac{GC^2 + BED + HBC - C^2E}{ABD - C^2B - AC^2}\right| \tag{4}$$

$$\left|\frac{X_b}{Y}\right| = \left|\frac{EC^2 + AGD + HAC - C^2G}{ABD - C^2B - C^2A}\right| \tag{5}$$

$$\left|\frac{X_d}{Y}\right| = \left|\frac{ABH + BCE + ACG}{ABD - BC^2 - AC^2}\right| \tag{6}$$

In the formula: $A = -m_a\omega^2 + \omega_a{}^2 m_a + ic_d\omega + \omega_d{}^2 m_d$, $B = -m_b\omega^2 + \omega_b{}^2 m_b + ic_d\omega + \omega_d{}^2 m_d$, $C = ic_d\omega + \omega_d{}^2 m_d$, $D = -m_d\omega^2 + 2ic_d\omega + 2\omega_d{}^2 m_d$, $E = m_a\omega^2$, $G = m_b\omega^2$, $H = m_d\omega^2$.

Verify the displacement transfer formula:

It is assumed that the characteristic parameters of the structure are: $k_a = 60, m_a = 2, k_b = 40, m_b = 1, m_d = 0.2(m_a + m_b)/2 = 0.3$ [8].

The natural vibration period and the natural vibration frequency corresponding to the structures A and B are:

$$T_a = \frac{2\pi}{\omega_a} = 1.14s,\ \omega_a = \sqrt{k_a/m_a} = \frac{5.48rad}{s},$$

$$1\ T_b = \frac{2\pi}{\omega_b} = 0.99s,\ \omega_b = \sqrt{k_b/m_b} = 6.32rad/s.$$

When the damper parameter $k_d = 0$, the displacement transfer function curves corresponding to the structure A, B are shown in Fig. 2 and Fig. 3.

Set two buildings A and B as the main structure, and the dampers of the mass D and the connecting mass and dampers are substructures. When $c_d = 0$ (the substructure damping is 0), it means that the substructure is not attached to the main structure. At this time, A and B are two independent structures. When A and B are subjected to external excitation, they are equivalent to the traditional structures without damping device. When $c_d \to \infty$ of the substructure, it means that the substructure is completely fixed to the main structure (A, B, D are connected as a whole). If A and B are externally excited, they are equivalent to the traditional structure without damping device. In the data and trend of figures 4 and 5, equations (4) and (5) are reasonable.

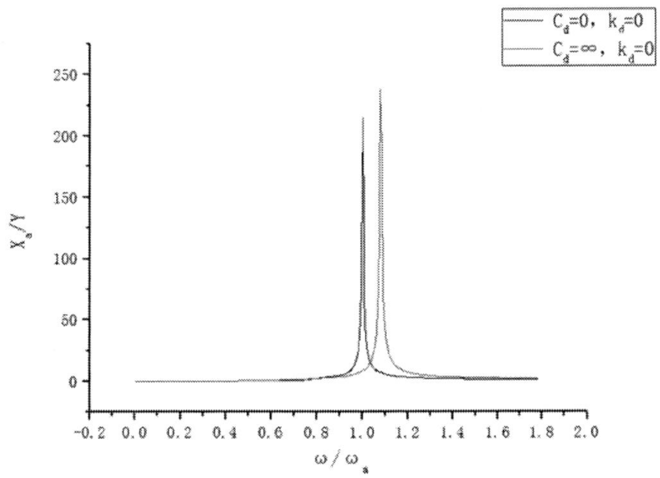

Figure 2. Displacement transfer function curve of structure A at $k_d = 0$.

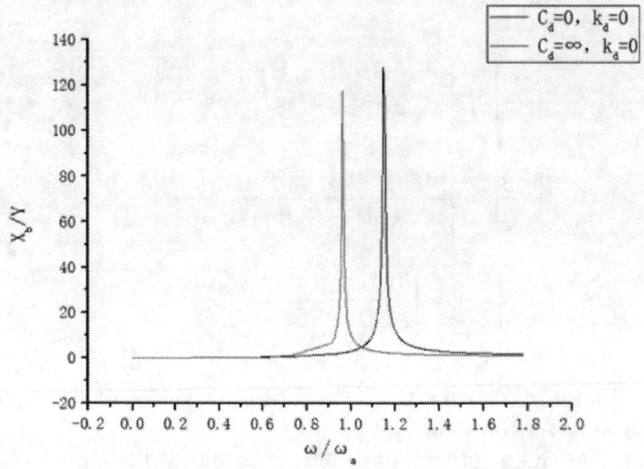

Figure 3. Displacement transfer function curve of structure B at $k_d = 0$.

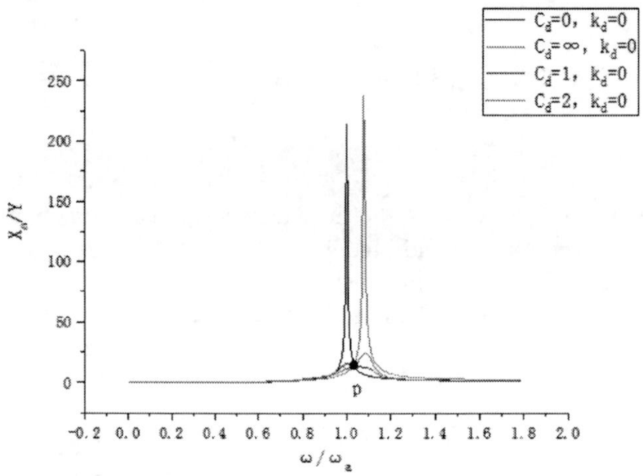

Figure 4. Displacement transfer function curve of structure A at $k_d = 0$.

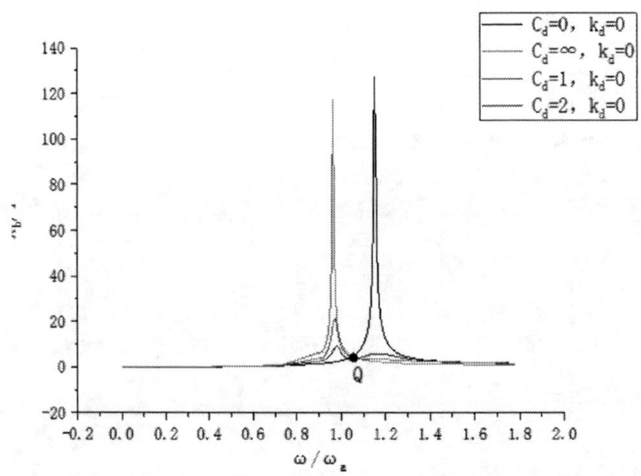

Figure 5. Displacement transfer function curve of structure B at $k_d = 0$.

IOP Conf. Series: Materials Science and Engineering 657 (2019) 012064 doi:10.1088/1757-899X/657/1/012064

3. Calculation and analysis of case model

It is still assumed that the characteristic parameters to the structure are:

$$k_a = 60,, k_b = 40.$$

$$m_a = \ 2, m_b = 1, m_d = 0.2(m_a + m_b)/2{=}0.3 \text{ [8]}.$$

The natural vibration period and the natural vibration frequency corresponding to the structures A and B are:

$$T_a = \frac{2\pi}{\omega_a} = 1.14s, \ \omega_a = \sqrt{k_a/m_a} = \frac{5.48rad}{s},$$

$$1T_b = \frac{2\pi}{\omega_b} = 0.99s, \ \omega_b = \sqrt{k_b/m_b} = 6.32rad/s.$$

The displacement transmission curves of the two buildings A and B when $k_d = 0$ are shown in Fig. 4 and Fig. 5.

From figures 4 and 5, it can be seen that with the increase of damping coefficient, the amplitude of displacement transfer function decreases first and then increases. The displacement transfer function corresponding to system A and B passes exceeds a certain point when the damping coefficient c_d changes and is recorded as P and Q points respectively. According to the fixed-point theory, these fixed-point heights should be equal in seismic design.

When the damping parameter $c_d = 0$, $k_d = 0$, the displacement transfer function of corresponding structure A and B can be obtained from formula (4) and formula (5) as follows:

$$\left|\frac{X_a}{Y}\right| = \omega^2/|\omega^2 - {\omega_a}^2|, \ \left|\frac{X_b}{Y}\right| = \omega^2/|\omega^2 - {\omega_b}^2| \qquad (7)$$

When the damping parameter $c_d = \infty$, k_d is constant, the displacement transfer function of corresponding structure A and B can be obtained from formula (4) and formula (5) as follows:

$$\left|\frac{X_a}{Y}\right| = \omega^2/|\omega^2 - {\omega_\infty}^2|, \ \left|\frac{X_b}{Y}\right| = \omega^2/|\omega^2 - {\omega_\infty}^2| \qquad (8)$$

In the formula: $\omega_\infty = \sqrt{(k_a + k_b)/(m_a + m_b + m_d)}$.

When the damping parameter c_d is constant and $k_d = \infty$, the displacement transfer function of corresponding structure A and B can be obtained from formula (4) and formula (5) as follows:

$$\left|\frac{X_a}{Y}\right| = \omega^2/|\omega^2 - {\omega_\infty}^2|, \ \left|\frac{X_b}{Y}\right| = \omega^2/|\omega^2 - {\omega_\infty}^2| \qquad (9)$$

Assuming that P and Q are two fixed points on the displacement transfer function curve of structure A and B, the corresponding ω to P and Q should satisfy the equation (10). The ω value corresponding to the fixed point solved by the equation (10) should satisfy the formula (11). By substituting the corresponding ω at the fixed point into equation (8), the displacement transfer coefficients at the fixed points of A and B structures can be obtained as formula (12) and formula (13) from equation (12).

Formula (13) shows that: $\frac{k_a}{k_b} = \frac{-J \pm \sqrt{J^2 - 4I^2}}{2I}$

In the formula: $I{=}m_a m_b$, $J{=}{m_a}^2 + {m_b}^2 + 2m_a m_d + 2m_b m_d + {m_d}^2$

$$\begin{cases} \left|\frac{X_a}{Y}\right|_{c_{d=0}} = \left|\frac{X_a}{Y}\right|_{c_{d=\infty}}, & \omega = \omega_P \\ \left|\frac{X_b}{Y}\right|_{c_{d=0}} = \left|\frac{X_b}{Y}\right|_{c_{d=\infty}}, & \omega = \omega_Q \end{cases} \qquad (10)$$

$$\begin{cases} \omega_P = \sqrt{(\omega_\infty{}^2 + \omega_a{}^2)/2} \\ \omega_Q = \sqrt{(\omega_\infty{}^2 + \omega_b{}^2)/2} \end{cases} \tag{11}$$

$$\begin{cases} \left|\dfrac{X_a}{Y}\right|_{\omega=\omega_P} = |(\omega_\infty{}^2 + \omega_a{}^2)/(\omega_\infty{}^2 - \omega_a{}^2)| \\ \left|\dfrac{X_b}{Y}\right|_{\omega=\omega_Q} = |(\omega_\infty{}^2 + \omega_b{}^2)/(\omega_\infty{}^2 - \omega_b{}^2)| \end{cases} \tag{12}$$

$$\left|\frac{(k_a + k_b)m_a + k_a(m_a + m_b + m_d)}{(k_a + k_b)m_a - k_a(m_a + m_b + m_d)}\right| = \left|\frac{(k_a + k_b)m_a + k_a(m_a + m_b + m_d)}{(k_a + k_b)m_a - k_a(m_a + m_b + m_d)}\right| \tag{13}$$

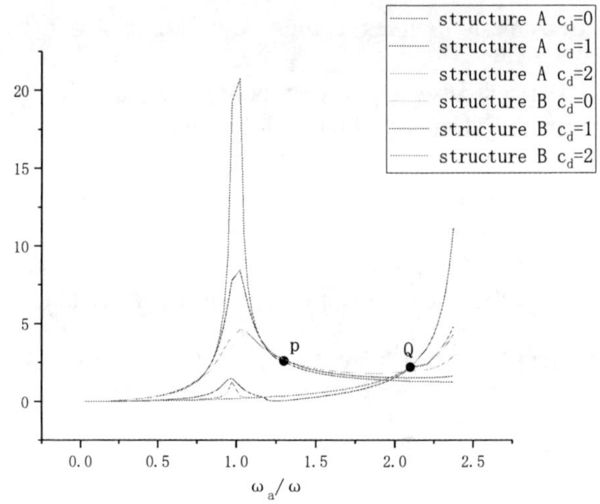

Figure 6. Displacement transfer function curves of structures A and B at $k_d = 0$.

The displacement transfer curves of A and B buildings under the satisfactoriness formula (13) are shown in figure 6. It can be seen from the same height of the fixed point that formula (13) is reasonable.

4. Conclusion

- In this paper, the numerical relationship between the structural parameters of adjacent structures with tuned mass dampers, the parameters of tuned mass dampers, and Specific Performance of the adjacent structures in Earthquakes has been find out by theoretical derivation, and the formula (4), (5) are verified by numerical simulation.
- The structural parameters of the adjacent structure which is most suitable for installing the tuned mass damper are obtained by theoretical derivation, and the result is verified by the displacement transfer curve (ie, formula (13)).

References

[1] Kasai K and Maison B F 1997 Building pounding damageduring the 1989 Loma Prieta earthquake *Eng. Struct.* **19** 195-207
[2] Cole G L, Dhakal R P and Turner F M 2012 Building pounding damage observed in the 2011 Christchurch earthquake *Earthq. Eng. Struct. Dynam.* **41** 893

[3] Inaudi J A and Kelly J M 1993 Hybrid isolation systems for equipment protection *Earthq. Eng. Struct. Dynam.* **22** 297-313

[4] Stavros A A 1988 Pounding of buildings in series during earthquakes *Earthq. Eng. Struct. Dynam.* **16** 443-5

[5] Bruce D W 1989 The dynamics of interstructural connection to prevent pounding *Earthq. Eng. Struct. Dynam.* **18** 687-99

[6] Sun H S and Dai S J 2009 Seismic optimization analysis of viscous dampers connecting adjacent structures *7th National Academic Conference on Safety and Protection of Engineering Structures* (Ningbo University, Zhejiang, China) pp 401-6

[7] Chen C Q and Pei X S 2018 Study on the seismic performance of adjacent structures with additional dampers *J. Jiangsu Univ. Sci. Technol. (Natur. Sci. Edit.)* **2018** 444-8

[8] Zhang M 2007 *Seismic Analysis and Damping Control of Building Structures* (Chengdu: Southwest Jiaotong University Press)

IOP Conf. Series: Materials Science and Engineering **657** (2019) 012065 doi:10.1088/1757-899X/657/1/012065

Effective elastic properties of doubly periodic array of functionally graded inclusions by an iterative FE-BE coupling method

Z Y Liu[1,2], D Li[1], Y C Xie[1], D Wang[1] and J G Zhang[1]

[1]Beijing Institute of Spacecraft Environment Engineering, Beijing 100094, China

E-mail: liuzhaoyan2011@163.com

Abstract. This paper is devoted to the study of the effective elastic properties of doubly periodic array of functionally graded inclusion problems by an iterative FE-BE coupling method. This method has some advantages compared with boundary element method (BEM), e.g. (1) inclusions can be isotropic, orthotropic, anisotropic or functionally graded materials; (2) various shaped inclusions can be easily solved; (3) the fundamental solutions for functionally graded inclusion materials are not needed. Some numerical examples are used to validate the applicability and reliability of the adopted scheme.

1. Introduction

Determining the elastic effective properties of a composite material has been the subject of numerous works [1]. Functionally graded materials (FGM) which were first reported [2] in the 1980s are a new generation of composite materials characterized by a continuously varying property. The concept of FGM is to take advantage of certain desirable features of the continuously and functionally varying volume composition of constituent particles and optimize the distribution of material properties such as strength, hardness, thermal resistance, etc, so that one can expect that functionally graded inclusion composites, which is formed when graded inclusions are embedded randomly in a host medium, will be applied in various fields such as mechanical, civil, medicine and so on.

A review about functionally graded materials and their successful implementation in a wide range of areas is available in [3]. The effect of the interphase zones around inclusions on effective elastic properties of the heterogeneous materials has been studied [4]. There are some works addressing the response of functionally graded composite systems with different geometries such as hollow cylinders, coatings on substrate and sandwich panels. In these researches, some benchmark results for the accuracy of numerical solutions have been obtained [5], which can be used to examine numerical solution accuracy from finite element method (FEM), BEM or other numerical methods.

Periodic or approximately periodic microstructures can be found in many natural or manmade materials. For this, micromechanics for solids with periodic microstructures has attracted the attention of many scholars. In principle, few exact solutions exist, except when the microstructure of the composite is known completely, for example, periodic arrays of spheres [6] or fiber-reinforced composites in the case of specific geometries [7]. Unfortunately, in most situations, this is not the case. Numerical methods, including finite element method [8] and boundary element method [9] etc., are naturally used to calculate the effective properties. Maxwell's scheme [10] has been developed in several works of Mogilevskaya *et al* [11] for the case of a material reinforced with parallel infinite

IOP Conf. Series: Materials Science and Engineering **657** (2019) 012065 doi:10.1088/1757-899X/657/1/012065

fibers and Kushch *et al* [12] for a material with spherical particles. Also, Mogilevskaya and Crouch [13] combine Maxwell's methodology with the BEM for evaluating the effective properties of composite, porous, and microcracked isotropic materials with periodic or random structure.

Dong [14] adopted an effective symmetric-iterative FE-BE coupling method to investigate elastic problems and obtained series of ideal results by comparing with analytical or exist results. In the present paper, this FE-BE coupling method is further developed and combined with unit cell models to evaluate the effective elastic properties of functionally graded inclusion problems with doubly periodic array.

Following [15], a rectangular cell containing single inclusion is cut from doubly periodic inclusion problems. The boundary condition on the rectangular cell is available for two loading systems, i.e. the tension and the in-plane shear. Then, the iterative FE-BE coupling method [16] is utilized to calculate those models. In calculation, a small stiffness matrix from BE domain is taken as a corresponding loading matrix in FE region. We first obtain displacements of the common interface nodes by iteratively calculating the coupled equations. After that, stress and displacement responses of the medium can be acquired by adopting the BE method. Numerical results are obtained for three examples. These results are compared with those from ABAQUS to demonstrate the validity of the adopted method.

2. Statement of the problem

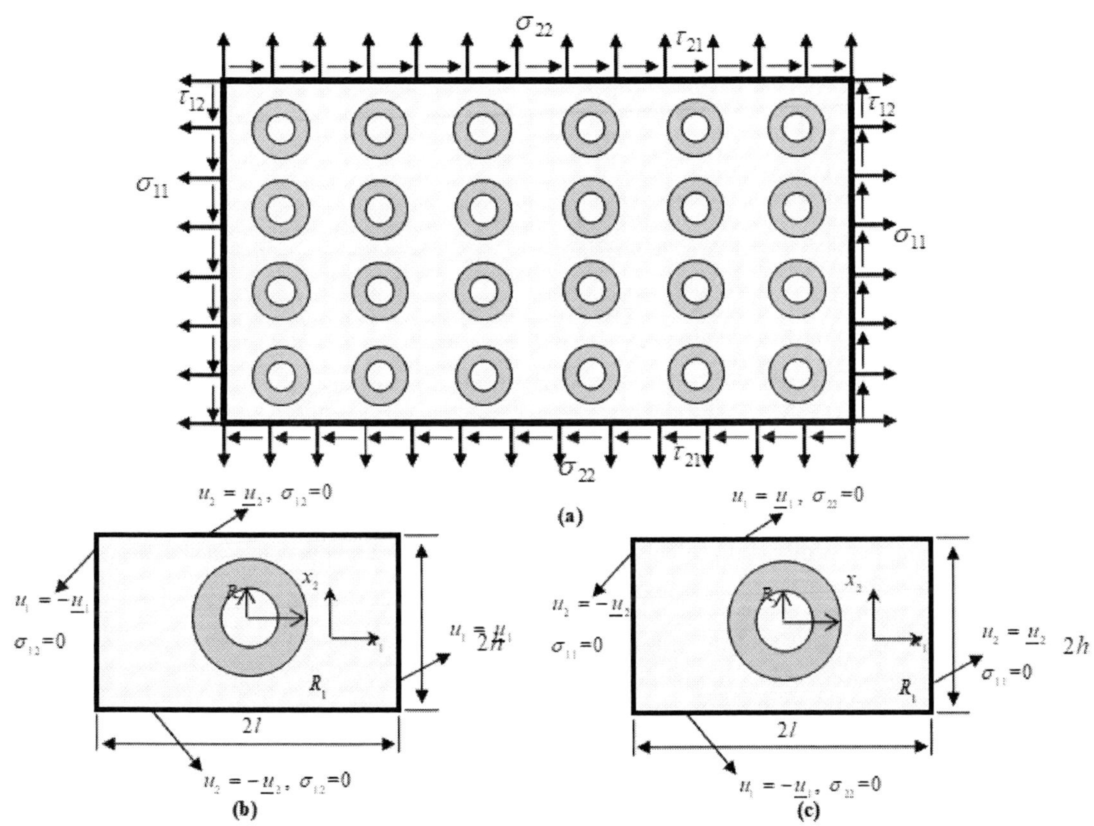

Figure 1. (a) Doubly periodic inclusion problems; (b) rectangular cell containing single inclusion and its boundary condition for tension loading; (c) rectangular cell containing single inclusion and its boundary condition for shear force.

We refer to an infinite isotropic elastic matrix containing functionally graded inclusions with three

IOP Conf. Series: Materials Science and Engineering **657** (2019) 012065 doi:10.1088/1757-899X/657/1/012065

different cross sections (circular, elliptic or hollow cylinder). Inclusions are arranged in a regular lattice determined by two families of parallel lines, respectively parallel to the x_1 axis and x_2 axis, as sketched in figure 1(a). This geometrical structure subject to remote tension and in-plane shear forces can be interpreted as a two dimensional array of unit cells, developing periodically along the x_1 and x_2 directions. The geometry and the boundary conditions of the unit cell, i.e. the microstructure of the composite, can be displayed in figures 1(b) and 1(c). The cell sides measure $2l$ and $2h$ respectively, and the inclusion outer and inner radius are R_1 and R_2.

The materials of the matrix medium are assumed to be linearly elastic and isotropic, while for the functionally graded materials of inclusions, Young's modulus depends only on the radial direction and Poisson's ratio is assumed to be constant. To this end, a family of problems is introduced, indexed by parameters n scaling the microstructure. When solving these problems, we consider the following variation law cases.

- circular inclusions

$$E(r) = E_o\left(\frac{r}{R}\right)^n \quad , \; 0 \le r \le R \tag{1}$$

- elliptic inclusions

$$E(x, y) = E_o\left(\frac{x^2}{c_1^2} + \frac{y^2}{c_2^2}\right)^n \quad , \; 0 \le x \le c_1, 0 \le y \le c_2 \tag{2}$$

- hollow cylinder inclusions

$$E(r) = E_o\left(\frac{r}{R_2}\right)^n \quad , \; R_1 \le r \le R_2 \tag{3}$$

where n is the inhomogeneity parameter that can be assigned to be an arbitrary value between -5 and 5, and the homogenization limit is obtained by letting n go to zero. E_0 is the elastic modulus of the matrix medium. In equation (1) R is the radius of circular inclusions. c_1 and c_2 in equation (2) are respectively the x semi-major axis and y semi-major axis of elliptic inclusions.

The main objective of this work is to obtain the effective elastic properties of doubly periodic functionally graded inclusion problems by using the FE-BE coupling method. A medium containing doubly periodic array of inclusions of various shapes can be considered as a homogeneous orthotropic medium. The detailed formula for calculating the effective elastic properties can be found in paper [19]. Here I won't repeat.

3. Formulation of iterative FE-BE coupling algorithm

Consider a linear elastic solid of domain Ω enclosed by a boundary Γ. The solid can be regarded as two decomposed portions Ω_F and Ω_B which are modeled by FEM and BEM, respectively. For the FEM, the algebraic assembled element equations are given by

$$\begin{bmatrix} \mathbf{K}_{ff} & \mathbf{K}_{fc} \\ \mathbf{K}_{cf} & \mathbf{K}_{cc} \end{bmatrix} \begin{Bmatrix} \mathbf{U}_f \\ \mathbf{U}_f^c \end{Bmatrix} = \begin{Bmatrix} \mathbf{F}_f \\ \mathbf{F}_f^c \end{Bmatrix} \tag{4}$$

Where $\mathbf{U}_f$ and $\mathbf{U}_f^c$ denote the vectors of non-interface and interface displacements in the FE region, $\mathbf{F}_f$ and $\mathbf{F}_f^c$ are the associated loading vectors, respectively. $\mathbf{K}_{ff}$, $\mathbf{K}_{fc}$, $\mathbf{K}_{cf}$ and $\mathbf{K}_{cc}$ are respectively sub-matrices of the global stiffness matrix of the FE region related to the vectors $\mathbf{U}_f$ and $\mathbf{U}_f^c$.

Neglecting body forces and after the process of boundary dicretization, the BEM partitioned equations are obtained

$$\begin{bmatrix} \mathbf{A}_{bb} & \mathbf{A}_{bc} \\ \mathbf{0} & \mathbf{A}_{cc} \end{bmatrix} \begin{Bmatrix} \mathbf{X}_b \\ \mathbf{T}_b^c \end{Bmatrix} = \begin{Bmatrix} \mathbf{f}_b \\ \mathbf{f}_b^c \end{Bmatrix} + \begin{bmatrix} \mathbf{A}_b \\ \mathbf{A}_c \end{bmatrix} \{\mathbf{U}_b^c\} \tag{5}$$

Where $\mathbf{X}_b$ is the vector of non-interface in the BE region; $\mathbf{U}_b^c$ and $\mathbf{T}_b^c$ are the interface displacements and tractions, respectively; $\mathbf{f}_b$ and $\mathbf{f}_b^c$ are two vectors which contains the contribution of the known values.

It is understood that the governing equations in the FE domain involve nodal displacements and forces; in contrast, the primary unknowns in the BE domain are displacements and tractions. To facilitate the coupling, the relations between tractions and forces need to be established. After this, the displacement compatibility and force equilibrium along the common interface are adopted. Finally, the FE-BE coupling system of equations are obtained as follows

$$\begin{bmatrix} \mathbf{K}_{ff} & \mathbf{K}_{fc} \\ \mathbf{K}_{cf} & \mathbf{K}_{cc} + \mathbf{K}_l \end{bmatrix} \begin{Bmatrix} \mathbf{U}_f \\ \mathbf{U}_f^c \end{Bmatrix}^{n+1} = \begin{Bmatrix} \mathbf{F}_f \\ -\overline{\mathbf{F}}_b^c \end{Bmatrix} + \begin{Bmatrix} \mathbf{0} \\ -\mathbf{K}_b^c + \mathbf{K}_l \end{Bmatrix} \{\mathbf{U}_f^c\}^n \tag{6}$$

Where $\mathbf{K}_l = (1/2)\left(\left[\mathbf{K}_b^c\right]^T + \left[\mathbf{K}_b^c\right]\right)$, $\overline{\mathbf{F}}_b^c = [\mathbf{M}][\mathbf{A}_{cc}]^{-1}\{\mathbf{f}_b^c\}$ in which $\mathbf{M}$ is the converting matrix from tractions to nodal forces on the BE side of the FE-BE interface.

4. Numerical examples

In this section, an example is given to demonstrate the feasibility and effectiveness of the present scheme by comparing results with the ABAQUS. In order to ensure the numerical accuracy of the present method, some convergence tests have to be done.

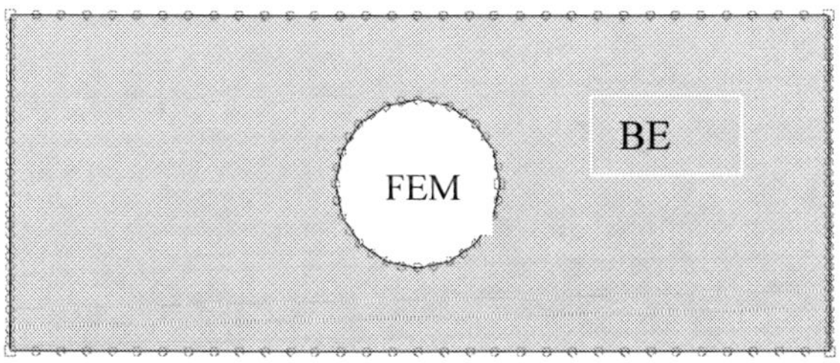

(a)

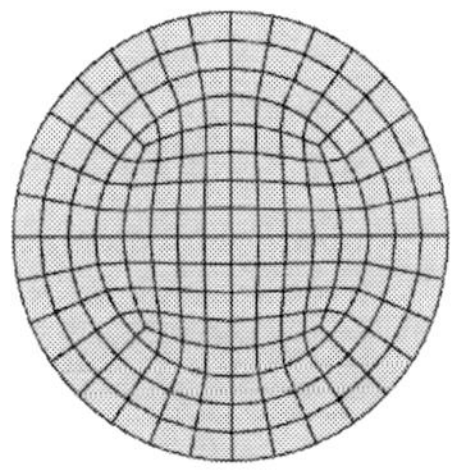

(b)

Figure 2. The rectangular cell with one circular inclusion: (a) node distribution of the medium in BEM region; (b) element distribution of the inclusion in FEM region.

As shown in figure 2, this example is studied for doubly periodic circular inclusion whose radius is taken as $R = 10$. The elastic parameters of the medium are taken as $E_0 = 1$ and $v_0 = 0.3$, while for inclusions, they possess elastic modulus E_i that vary continuously as a function of position referring to equation (1), and the Poisson's ration v_i is assuming to be the same as medium's. The cell boundary lengths are respectively taken as $2l$ and $2h$ (see figures 1(b) and 1(c)).

In numerical implementation, each edge of the outer rectangular cell for various cases is divided into 32 quadratic boundary elements. The BE node distribution of this region is shown in figure 2(a).

The numbers of finite elements in the inclusion are chosen as 20, 56, 96, 192 and 256, respectively.

The corresponding numbers of the interface elements between the inclusion and the matrix are 8, 12, 24, 32 and 32, respectively. The convergence results for different values of effective properties are displayed in figures 3 and 4. With the increase of the number of finite elements, one can find that the convergence results have been obtained for 192 finite elements within the inclusion together with 32 interface elements. Therefore, we will choose this type of subdivision, the FE distribution of which is depicted in figure 2(b), to carry out further numerical analysis.

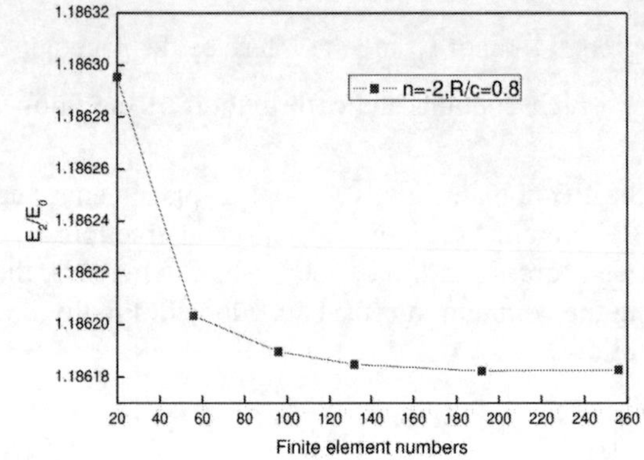

Figure 3. Normalized effective elastic modulus E_2/E_0 with increasing finite element numbers ($n=-2, R/c=0.8$).

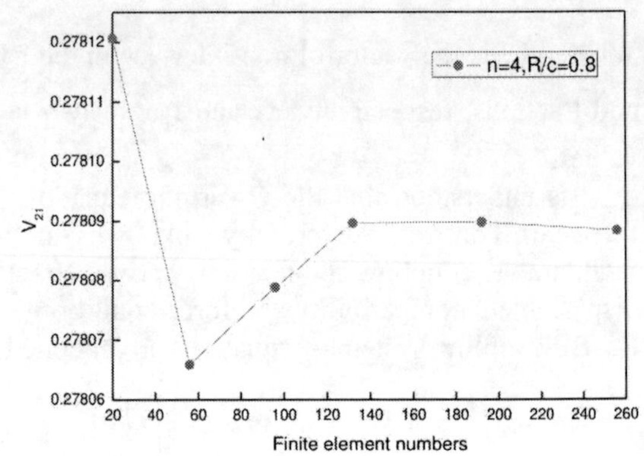

Figure 4. Effective Poisson's ratio v_{21} with increasing finite element numbers ($n=4, R/c=0.8$).

In ABAQUS model, an element type of CPS8R (one 8-node biquadratic plane stress quadrilateral, reduced integration) is used. The whole model is divided into 5261 elements, 609 of which are belonged to the inclusion region. The medium and the inclusion are tied together.

By using the coupling method, we plot the effective elastic properties for $R/c=0.4\,(c=\min(l,h))$ and different values of n in figures 5-8. Here we keep the radius of inclusion be $R=10$, so h and l are changed according to the relationship of R/c and $h/l=0.4$.

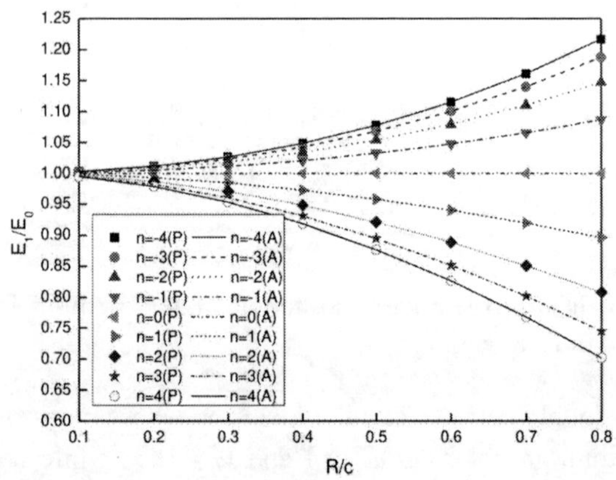

Figure 5. Effective elastic modulus along x_1 direction of doubly periodic circular inclusions for various n and c; A: ABAQUS, P: the present method.

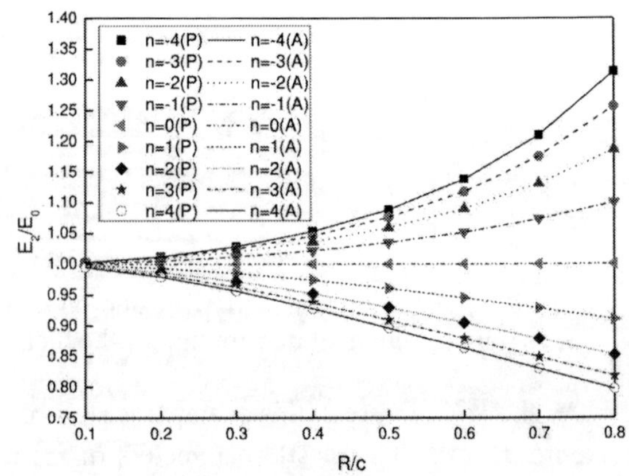

Figure 6. Effective elastic modulus along x_2 direction of doubly periodic circular inclusions for various n and c; A: ABAQUS, P: the present method.

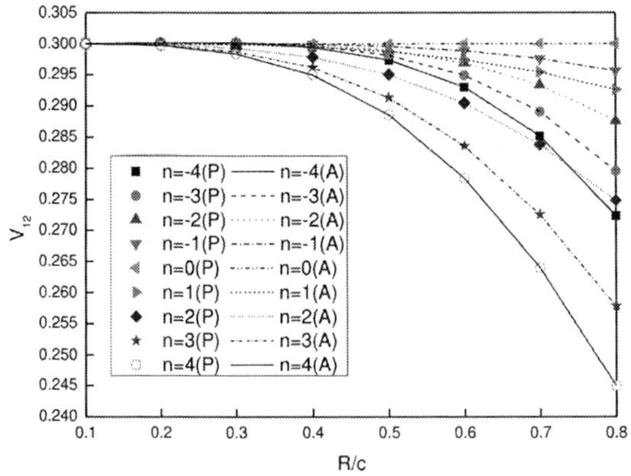

Figure 7. Effective Poisson's ratio v_{12} of doubly periodic circular inclusions for various n and c ; A: ABAQUS, P: the present method.

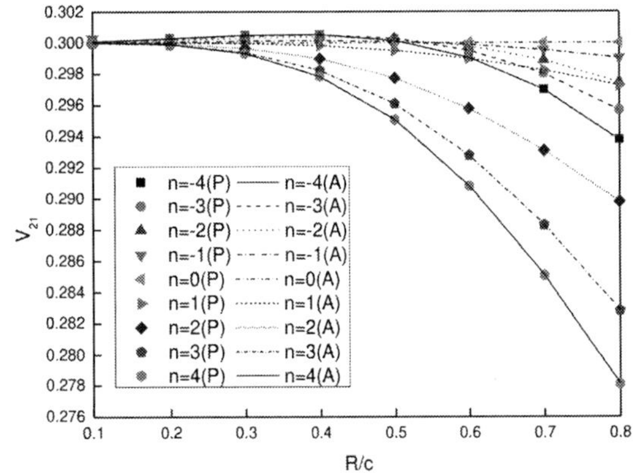

Figure 8. Effective Poisson's ratio v_{21} of doubly periodic circular inclusions for various n and c ; A: ABAQUS, P: the present method.

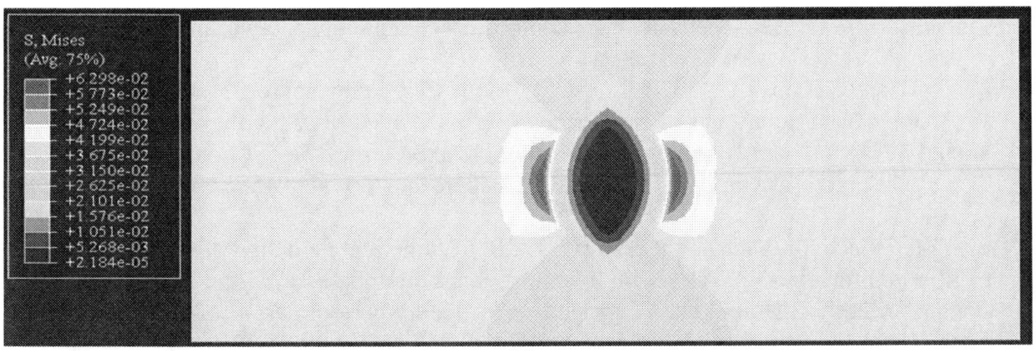

(a)

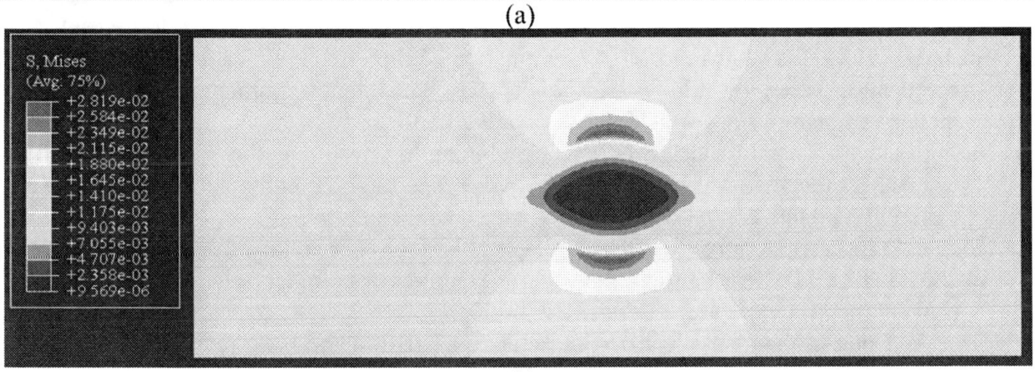

(b)

Figure 9. Mises stress contours of doubly periodic circular inclusions for $n = 4$ and $R / c = 0.4$.

It can be seen that the results from the present method are in excellent agreement with ABAQUS's. Besides, we can observe for different values of n that the harder the inclusion becomes, the larger the effective elastic moduli E_1 , E_2 and G_{12} . The bigger the inclusion becomes, i.e. R / c is increased, the larger the effective elastic moduli E_1 and E_2 and G_{12} . One can also notice that when $E_i / E_0 = 1$, i.e. the homogeneous medium, the normalized effective elastic and shear moduli reduce to 1, respectively, as expected. Effective Poisson's ratios v_{12} and v_{21} are shown in figures (7) and (8), respectively. By checking the results (see figures (5)-(8)), one can find that the relationship of $E_1 \nu_{21} / (E_2 \nu_{12}) = 1$ has been well approximated. Figure 9 depicts the plot contours of Mises stress with $n = 4$ and $R / c = 0.4$. We can observe the "obvious layer structures" which shows the outstanding merit of functionally graded materials, i.e. their properties can be tailored via the design of the material gradients. Its significance, for the development of engineering components, relies on the fact that properly tuning the fibre grading profile allows to modulate the material stress state, thus enhancing material performance and durability.

References

[1] Hashin Z 1983 Analysis of composite materials-A survey *J. Appl. Mech.* **50** 481-505
[2] Koizumi M 1997 FGM activities in Japan *Compos. Part B-Eng.* **28** 1-4
[3] Birman V and Byrd L W 2007 Modeling and analysis of functionally graded materials and structures *Appl. Mech. Rev.* **60** 195-216
[4] Lutz M P and Zimmerman R W 1996 Effect of the interphase zone on the bulk modulus of a particular composite *J. Appl. Mech.* **63** 855-61
[5] Batra R C 2011 Material tailoring and universal relations for axysimmetric deformations of functionally graded rubberlike cylinders and spheres *Math. Mech. Solids* **16** 729-38
[6] Sangani A 1987 Elastic coefficients of composites containing spherical inclusions in a periodic array *J. Mech.Phys. Solids* **35** 1-21
[7] Kantor Y and Bergman D 1982 Elastostatic resonances. a new approach to the calculation of the effective elastic constants of composites *J. Mech. Phys. Solids* **30** 355-76
[8] Michel J C, Moulinec H and Suquet P 1999 Effective properties of composite materials with periodic microstructure: A computational approach *Comput. Meth. Appl. Mech. Eng.* **172** 109-43
[9] Wang J, Crouch S L and Mogilevskaya S G 2006 Numerical modeling of the elastic behavior of fiber-reinforced composites with inhomogeneous interphases *Compos. Sci. Technol.* **66** 1-18
[10] Dong C Y 2006 Effective elastic properties of doubly periodic array of inclusions of various shapes by the boundary element method *Int. J. Solids Struct.s* **43** 7919-38
[11] Maxwell J C 1873 *A Treatise on Electricity and Magnetism* (Oxford: Clarendon Press)
[12] Kushch V I, Mogilevskaya *et al* 2013 Evaluation of the effective elastic moduli of particulate composites based on Maxwell's concept of equivalent inhomogeneity: microstructure-induced anisotropy *J. Mech. Mater. Struct.* **8** 283-303
[13] Mogilevskaya S G and Crouch S L 2013 Combining Maxwell's methodology with the BEM for evaluating the two-dimensional effective properties of composite and micro-cracked materials *Comput. Mech.* **51** 377-89
[14] Dong C Y 2001 An iterative FE-BE coupling method for elastostatics *Comput. Struct.* **79** 293-9
[15] Ma J M and Le M F 1992 A new method for coupling of boundary element method and finite element method *Appl. Math. Model.* **12** 43-6
[16] Liu Z Y, Dong C Y and Bai Y 2015 An iterative FE–BE method and rectangular cell model for effective elastic properties of doubly periodic anisotropic inclusion composites *Compos. Struct.* **124** 253-62

IOP Conf. Series: Materials Science and Engineering **657** (2019) 012066 doi:10.1088/1757-899X/657/1/012066

Investigation on characteristics of arc extinguishing time of a new lighting protection gap for transmission lines

R B Yan[1,2], J F Wang[1], Z J Li[1] and SS Huang[1]

[1]College of Electrical Engineering, Guangxi University, Nanning 530004, Guangxi Zhuang Autonomous Region, China

E-mail: 956657229@qq.com

Abstract. The present research envisages the development of a novel arc extinguishing lightning protection gap in pursuit of reducing the lightning trip-out rate and accident rate. In the developed setup, the jetting stream arc extinction lightning protection gap device was used for transmission line. When this transmission line was struck by lightning and the lightning surge was going to break down the gap, the arc extinguishing lightning protection gap immediately jetted out sufficient electronegative gas flow that quickly cut off the arc and effectively prevented the insulator from being damaged by the arc. Furthermore, when the line was attacked by lightning, the generated overvoltage instantaneously broke the gap that was parallel to the insulator string, so that the lightning current was oriented to the earth from the gap. Simultaneously, it triggered the gas generator and generated a high speed stream, then extinguished the frequency continuous current arc rapidly, thus effectively reducing the rate of line trip to ensure the reliability of power supply. We used a high-speed camera and ordinary camera that automatically captured the process of the jetting stream arc in the extinction lightning protection gap device, while extinguishing the 1kA power frequency arc. The image of the high-speed camera indicated that the power frequency arc of 1kA was quenched in 10 ms before the relay protection action of the test loop, which made that the overhead line lightning flashover but the circuit breaker did not trip. Thus, a reduction in the rate of lightning trip was observed by the developed setup, and it has an immense potential for application at the transmission lines of China.

1. Introduction

The southern part of China is located in the subtropical region, with a strong lightning activity. The annual average number of thunderstorm days in Guangxi is more than 80. In recent years, the overhead line of the power grid has been caused by lightning strikes, which affects the safe operation of the power grid. On an average, the 35 kV and below overhead lines of Guangxi Power Grid face more than 100 lightning strikes per year, resulting in a large number of line blackouts and a significant increase of the workload for repair and maintenance lines, thereby seriously affecting the reliability of power supply.

There are two main concepts employed in the protection from lightning strikes at home and abroad *viz.* the "intercept-type" and the "scatter-type". At present, the developed countries like Japan, Germany, and France have generally adopted the "scatter-type" concept [1].

In the recent years, significant attempts were made by China for developing new lightning protection products based on concept of "scatter-type" device. Until now, China has developed the arcing horn and parallel gap which fit for different voltage levels, and some a few of these have been installed and are operational in actual line [2-11].

Ordinary insulator strings in parallel gap is known as arcing horn. The core idea is to allow the overhead transmission lines have a certain rate of lightning trip. The parallel gap fulfills the purpose of the priority breakdown of lightning by use metal electrodes to short circuit the insulator strings dry arcing distance. Through this the lightning energy is released by the discharge gap and eases the power frequency arc. Although it has a lightning flashover, auto reclosing can succeed, so it has no permanent fault. However, this method faces the following difficulties:

- This gap is adopted at the expense of the lightning trip rate with a low accident rate. It cannot self-extinguish the power frequency continuous current arc, which causes an increase in the lightning trip rate and the circuit breaker also gets a load to cut off short circuiting the arc that requires time. So, the overall reliability of power grid supply declines.
- When the auto reclosing is unable to move or it does not exist, the power frequency continuous current arc cannot be self-extinguished and continue to burn the gap. The useful time will shorten even more.

Therefore, this paper raises a new arc extinction lightning protection gap, which impulses a discharge path by the normal clearance with the function of fixed position lightning. Also, it can sense the lightning breakdown signal to trigger the stream generator in a short time. The stream generator device can jet high-speed airflow to extinguish the arc, so the transmission and distribution system cannot trip during the lightning flashover. Moreover, reduction in the lightning trip rate and protection of the insulator strings and transmission lines improves the power grid reliability.

The prototype of the arc extinction lightning protection gap developed in the present work successfully accomplished the interrupter test verification of 1kA, 5kA, 10kA short circuit arc in a large capacity testing laboratory of Xi'an high-voltage apparatus research institute.

2. Rapid quenching experiment of the arc by extinction lightning protection gap

2.1. Experimental

This jetting stream arc extinction lightning protection gap device was used for transmission line. When the line was attacked by lightning, the generated overvoltage led to a priority breakdown of the gap which was parallel to the insulator string. Hence, the lightning current was oriented to the earth from the gap. Simultaneously, it triggered the gas generator and generated a high speed stream that extinguished the frequency continuous current arc rapidly, thus effectively reducing the rate of line trip to ensure the reliability of power supply. During recent years, the theoretical basis this lightning protection gap has been extensively studied [12-17]. The test run in several power supply bureaus of Guangxi Power Grid has achieved good operating results.

This device included: wire connector, high voltage electrode, ground electrode, gas generator, signal pickup assembly, and earth connector. The arc extinction lightning protection gap as shown as Figure 1.

The whole device through wire connector-1 and earth connector 5 tied with the insulator string, simulated the real situation of the clearance installed in the overhead line insulator string. During the test, the high voltage electrode-2 was connected to the generator outlet conductor, and the ground electrode-3 was connected with earth metal. Since, the parallel clearance located the path of lightning impulse flashover, the clearance between the high-voltage electrode and the ground electrode short circuited by 0.8 mm fuse, to ensure the burning of the frequency continuous current arc in the ends of gap after the lightning stroke. The signal pickup assembly-6 triggered the gas generator-4 after the detection of the current signal. The gas generator-4 generated a high speed stream to extinguish the arc. If the working succeeds, the arc gets extinguished in 0 - 15 ms.

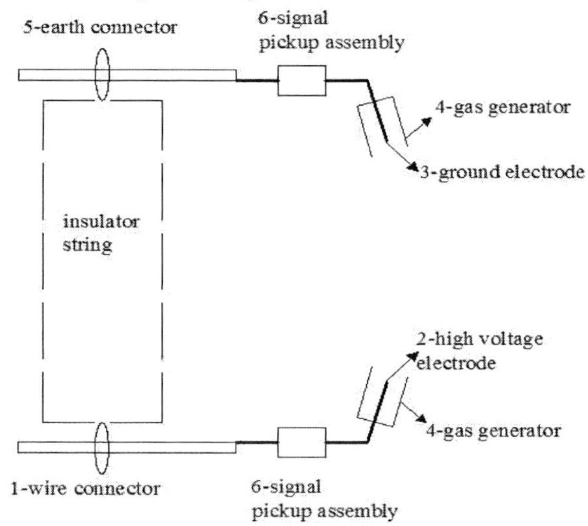

Figure 1. Schematic drawing of the arc-quenching lightning protection gap

2.2. Test for rapid extinguishing of the power frequency arc

The test for the rapid extinguishing of the arc by the lightning protection gap was conducted in the large capacity testing laboratory of the Xi'an high-voltage apparatus research institute. A voltage of 12 KV and 50 Hz power frequency current was provided using a short circuit generator to adjust reactor.

The gap sample was pre-commissioned in the order of 200 ms without the replacement of any other components, except the gas generating device, and the current RMS value was 1 kA, 5 kA, 10 kA for 3 arc-extinguishing tests. Therefore, when the effective value of the arc flowing through the gap was 1 kA, the current waveform was collected, and the arc was extinguished at 7.5 ms, and the waveform is as shown in Figure 2 and 3.

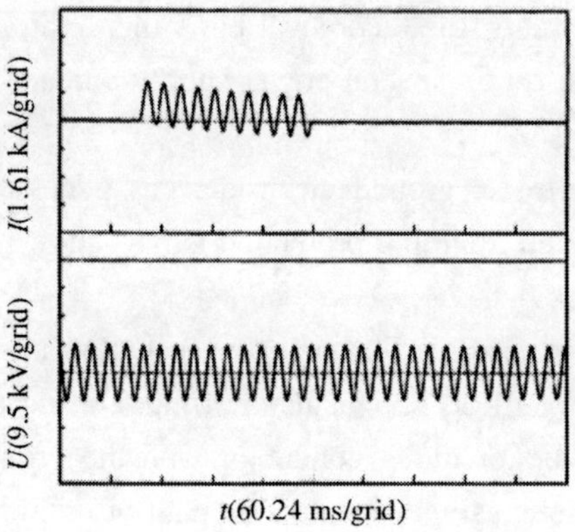

Figure 2. Voltage and current waveforms of 1kA arc before the rapid arc-quenching test

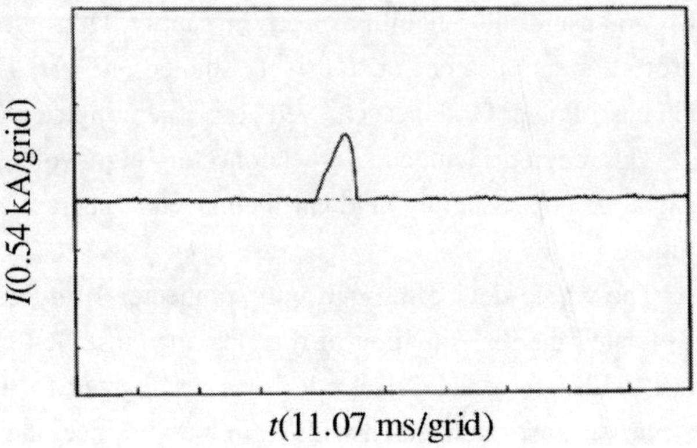

Figure 3. Voltage and current waveforms in the rapid arc-quenching test

In addition, in the arc-extinguishing experiment conducted at 5 kA and 10 kA, because the sample was sensitive, the arc was extinguished, the current was in ampere level, and the measurement system failed to collect the current waveform. After the test, the appearance of the sample gas generating device was intact, and the gas generating device operated successfully, which was judged from the successful extinguishing of the sample. The pre-commissioning and extinction waveforms of the 10 kA arc are shown in Figure 4 and 5. The pre-commissioning and extinction waveforms of the 10 kA arc are shown in Figure 6 and 7.

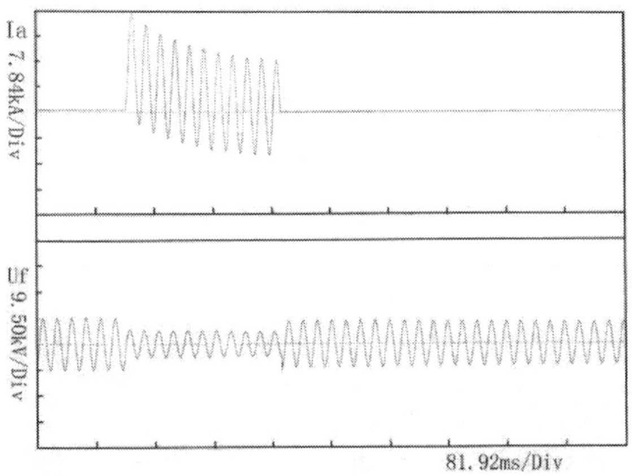

Figure 4. Voltage and current waveforms of 10 kA arc before the rapid arc-quenching test

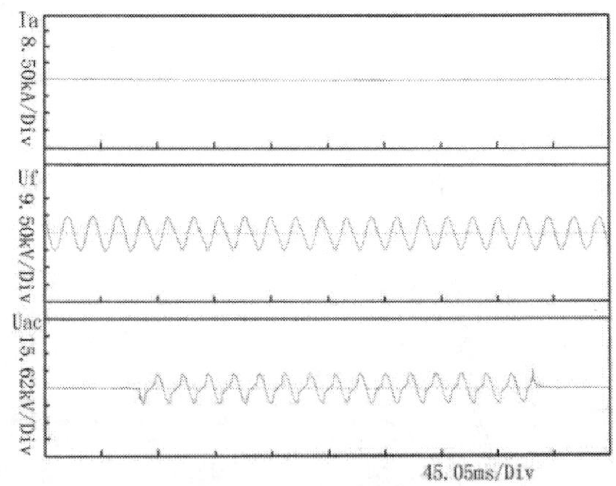

Figure 5. Voltage and current waveforms in the rapid arc-quenching test

IOP Conf. Series: Materials Science and Engineering **657** (2019) 012066 doi:10.1088/1757-899X/657/1/012066

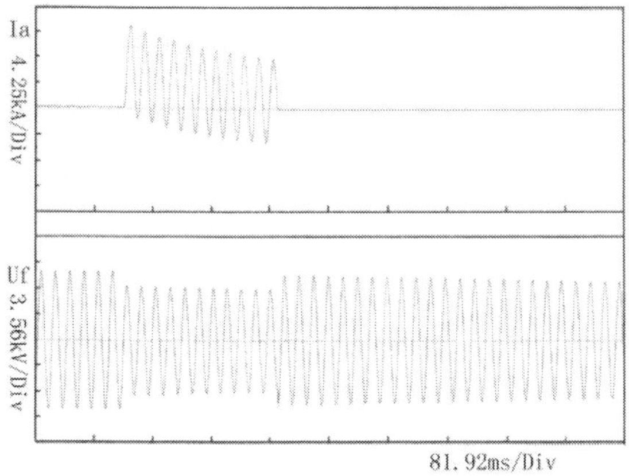

Figure 6. Voltage and current waveforms of 5kA arc before the rapid arc-quenching test

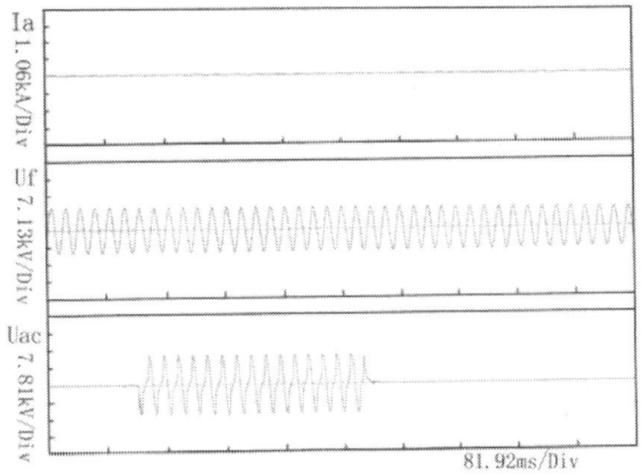

Figure 7. Voltage and current waveforms in the rapid arc-quenching test

At the test site, the ordinary camera was used to record the extinguishing of the power frequency arc by the lightning protection gap, and the visual effect of the extinguishing the power frequency arc in the gap is shown in Figure 8.

Figure 9 shows the microscopic process of the high-speed airflow captured by a high-speed camera at a speed of 6 000 s/s. It shows the guidance of the arc-free lightning protection gap gas generating device with the gap current. On its activation, the high-speed airflow acted on the power-frequency freewheeling arc to extinguish it.

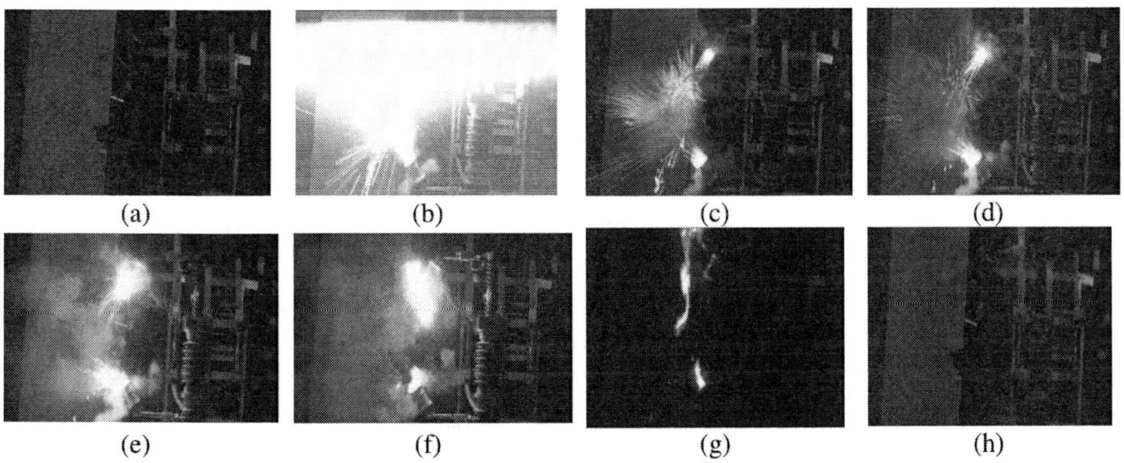

Figure 8. Arc screens extinguish process using ordinary camera: (a) Before the device act; (b) Switch was closed and device in operation; (c) Airflow expands the arc; (d) Arc cracks began to appear; (e) Arc cracks was obvious; (f) Arc began to cease; (g) Arc continues to weaken; (h) Extinguishing of arc

3. Analysis of extinguishing of the rapid frequency continuous arc by the lightning protection gap

The jetting stream lightning protection gap was through the acquisition of current pulse signal and then instantly triggered the gas generator. After 200 μs of the acquisition signal, a high speed stream was generated to role on the earlier arc and strongly interposed the clearance of the built arc process.

The moment when the frequency continuous current arc just rises (the extinguished instantaneous value is only a few percent of frequency current amplitude), it gets extinguished out. The success of this device in extinguishing the arc is because of "fast rule strong", and does not impacted by the current steady-state magnitude and closing phase angle.

The arc-extinguishing lightning protection device was already tested for its response time for generating gas before the studies at the Xi'an High-voltage Electrical Apparatus Research Institute. In this test, only a small arc was used as a trigger power source. Figure 10 shows the measured jet airflow response time, which is about 200 μs. The test results were consistent with the assumptions. The rapidly generated jet airflow was the main reason for the rapid extinguishing of the arc-frequency lightning protection power frequency arc. It strongly impacted the arc and interrupted the arcing process when the arc current did not yet develop into to a steady state.

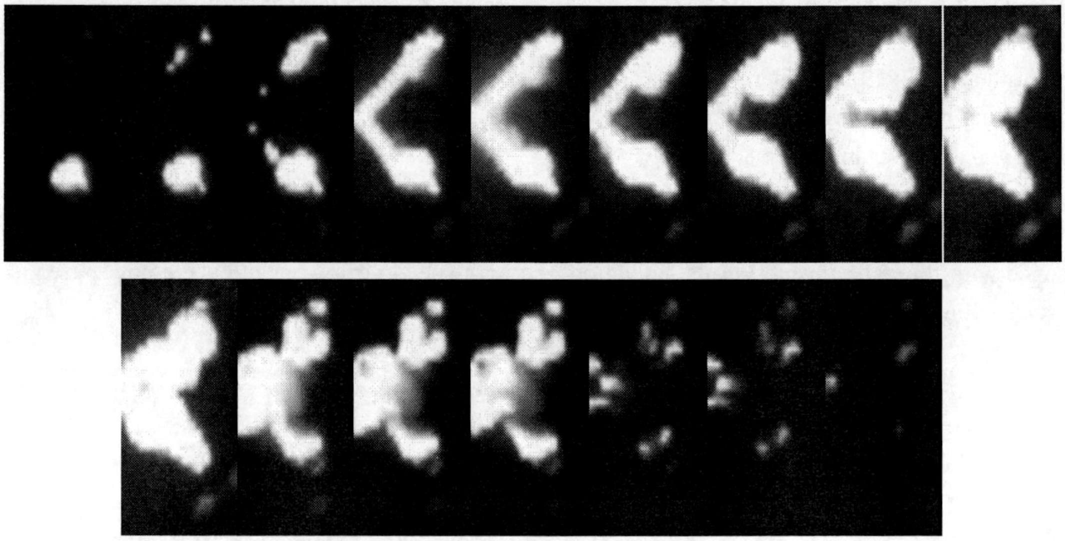

Figure 9. Arc extinguishing process captured by the high-speed camera

In the present work, we tested the arc-proof lightning protection gap for 3 times, and the steady-state amplitude was 1, 5, 10 kA in the 3 arc-extinguishing tests. The short-circuit arc waveform with 7.5 ms was collected at 1 kA, and the other 2 times. The reason is that the occurrence time of the gas flow of the gas generating device was related to the current power flowing through the signal collecting device. Once the current power met the requirements, the gas generating device was triggered to generate a high-speed air flow. It was expected that the power frequency arc of 5 kA, 10 kA triggered the gas generating device in a shorter time than the 1 kA power frequency arc, as the arc of the high-speed air current was faster for 1 kA power frequency arc. Hence, the arc current in these two tests (5 and 10 kA) and the existence time was extremely short, which was outside the scope of the system test, resulting in the system not collecting the relevant waveform.

The test result of Xi'an high-voltage apparatus research institute was consistent with our envisaged results. This is the reason why the arc extinction lightning protection gap frequency current arc was extinguished rapidly, when the arc current was just created, and did not develop into a steady state due to the strong gas stream that interrupted its built-arc process.

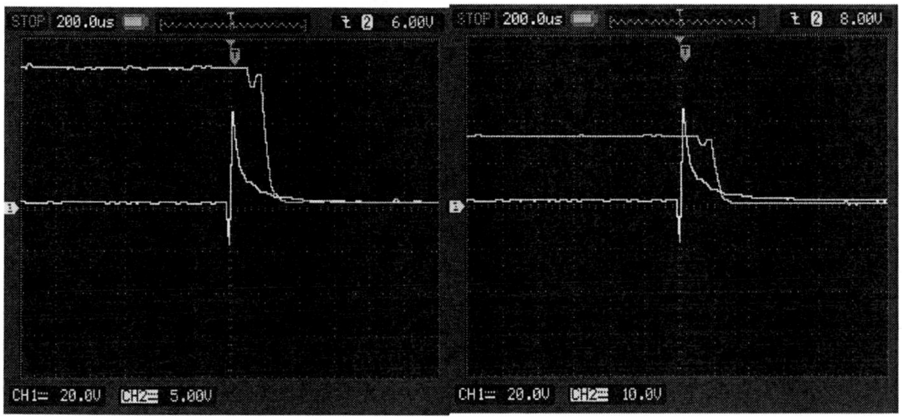

Figure 10. Measured jet airflow response time (yellow trigger pulse, blue for the air to produce time; response time ~ 0.2ms)

During the test, we performed the arc extinction lightning protection gap test with expected commission steady-state amplitude of 1kA. It was only collected short arc waveform in 1kA level, whose duration was 7.5 ms. The occurrence time of gas generator device and the jetting stream have a relationship with short current power which pass through signal acquisition device. Once the short circuit current power met the requirements, it triggered the gas generator device to generate a high speed gas stream. Since, the short circuit power was proportional to the power frequency arc, thus, the greater the power frequency electric arc, the shorter the trigger time was needed to generate the gas by the gas-generator. If the expected frequency current arc in the test was 5 kA or 10 kA, the trigger time of the gas generator device would be earlier than that needed for 1 kA. The high-speed gas generated by the gas generator affected the power frequency electric arc depending on the time needed to trigger, which was more conducive to rapid arc extinguishing.

The triggering of the gas generating device is shown in Figure 11. t10, t5, and t1 correspond to the triggering time of the gas generating device when the power frequency short-circuit current amplitude was 10, 5, 1 kA, and $t10 < t5 < t1$. When the device was actually installed on the transmission line, the gas generating device was directly triggered by the lightning pulse, which was earlier than the triggering of the short-circuit current in the test, and the airflow was applied to the instantaneous value of the short-circuiting arc, so that the probability of the device cutting off the arc was more high. The time for the lightning current to reach the peak was about 1~5 μs, and the required grooming time was extremely short, much less than 100 μs. The inherent reaction time of the arc extinguishing device ensured that the lightning current was smoothly introduced into the earth.

IOP Conf. Series: Materials Science and Engineering **657** (2019) 012066 doi:10.1088/1757-899X/657/1/012066

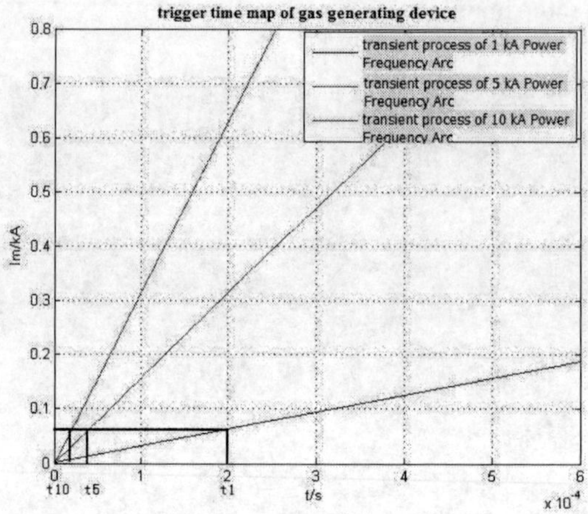

Figure 11. Trigger time map of gas generating device under different amplitude of current

The process of extinguishing the arc from the gap can be further analyzed from the image recorded by the high-speed camera in Figure 8. The essence of power frequency arc is the plasma. The plasma gap arc exhibits the force of contracting to the center of the arc column under the action of its own magnetic field force [18], and the temperature is very high. Therefore, the shape is fine and the color is blue. We observed that shape of the light generated by the arc extinguishing device was relatively thick and the color is white, so the thick white point at both ends of the gap was the light when the arc extinguishing device exploded. It can be seen from Figure 7 that the gas generating device already operated when the gap was just turned on and the gap arc was not completely formed. The high-speed airflow of the explosive jet followed the power-frequency freewheeling arc and rapidly elongated the freewheeling arc that caused the arc diffusion coefficient to increase rapidly, and the deionization process quickly returned the gas to a neutral insulating state. At the same time, the gas carried a lot of energy through heat conduction and convection, so that the arc temperature was suddenly reduced and the thermal ionization could not be maintained. The final gap arc was quickly extinguished because the insulation recovery strength was greater than the breakdown recovery voltage, and the energy supply was cut off.

4. Conclusion

- The nature of power frequency arc was the plasma, which exited for a very short time. The high-speed camera was used to study the arc shape and the arc development process, which recorded the arc quenching process automatically and clearly.
- The speed of gas generator device in response to the lightning pulse jetting stream and the speed and pressure were the main reasons of the extinguishing of the power frequency continuous current arc. Furthermore, shorter response time of gas generator device, and a higher speed and pressure of the jetting stream can lead to faster extinguishing of the arc.
- Because of its speed, the arc extinction lightning protection gap extinguished the arc. Although, the ends of frequency continuous current arc were burned, but the instantaneous value of extinguishing the arc was small and with a very short time, so it did not exhibit a

great influence on its clearance.

References

[1] He J L,Zeng R and Chen S M 2009 Lightning protection study of transmission line,part III:Protection measures*High Volt.Eng.***35**2917-23

[2] Liu G S,Zheng Y L and Ran X B2007 Analysis on operational experience of grounding electrode line of mawo side in tian-guang HVDC transmission line with arcing horn*Power Syst. Technol.***31**37-9

[3] Lin F C, Zhan H M, Gong D W*et al*2003Experi-mental research on arcing horn of insulators in ultra-high towers *High Volt. Eng.***29** 21-2

[4] Yang Q,Tan W,Sima W X*et al*2011Voltage equalizing characteristics of composite insulator with parallel gap lightning protection device employed*High Volt. Eng.***37** 1910-5

[5] Chen W J, Sun Z Y andLi G F2006Development of parallel gap lightning protection device for 110 kV and 220 kV overhead lines *Power Syst. Technol.***30**70-5

[6] Chen X Y,Yin C R and Yang T 2011Design and operation on parallelgap lightning protection device ofcomposite insulators for 110 kV overhead transmission line *Electric Power Construction***32**57-61

[7] Terzija and KoglinH 2002Long arc in free air: Laboratory testing,modeling, simulation and model parameters estimation*IEE Proc. Gener. Transm. Distrib.***49** 319-25

[8] Li M G, Lin S Q andXu F2011Study and application of arc ignition parallel gap in 35kV overhead transmission line *Guangxi Electric Power***2011**1-5

[9] Li Q X, Lin S Q, Li M G andXu F 2011Study on the arc ignition parallel gap which used in composite insulator string of 35 kV overhead transmission lines*Insul. Surg. Arres.***2011**16-20

[10] Gu S Q, He J L, Chen W J andXu G Z2006Magnetic force computation for the electric arc of parallel gap lightning protection device on overhead transmission lines*Proc. Chin. Soc. Elect. Eng.***26**140-5

[11] Ge D,FengH Q and Yuan L H2008Experimental research on large power frequency current of parallel gap of insulator strings *High Voltage Engineering***34**1499-503

[12] Wang J F,Huang Z D and Wang Y L2012A dynamic chain arc model for coupling of explosion airflow field *Proc. Chin. Soc. Elect. Eng.***32**154-60

[13] Nagy I E, Matti L*et al* 2007Modeling and experimental verification of high impedance arcing fault in medium voltage networks*IEEE Trans. On Dielectrics and Electrical Insulation***14**375-83

[14] Wang J F,Huang W,Qu Z X and Liu R2012Jet stream interrupter gap lightning protection device for 35 kV overhead transmission lines *High Volt. Appar.***2012**57-61

[15] Lin S Q,Lin H and Wang J F2012Study on model of constructing arc about parallel gap lightning protection device with arc-extinguished by high-speed airflow*Insul. Surg. Arres.***2012**57-61

[16] Wang J F, Qu Z X,Su H Y,Huang W and Liu R2012 An approach to suppress lighting overvoltage of 35 kV overhead transmission lines by air-blast arc-quenching gap*Power Syst. Technol.***2012**

[17] Wang J F,Tang J,Huang Z D and Pan K N2011Arc extinguishing mechanism of jet stream

interrupter gap lightning protection device*Power Syst. Technol.***35**228-31
[18] Dai D S, Song Y L, Zhang H andZhu Y F 2002 Study on arc force in plasma welding*Transactions of The China Welding Institution***23**51-4

IOP Conf. Series: Materials Science and Engineering **657** (2019) 012067 doi:10.1088/1757-899X/657/1/012067

Simulation on temperature field of initial segment of electromagnetic rail launch

S Y Tong[1,2,3], G Feng[1], Z M Lian[1] and J S Cheng[2]

[1]Air and Missile Defense College, Air Force Engineering University, Xi'an, China
[2]Institute of Electrical Engineering, Chinese Academy of Sciences, Beijing, China

E-mail: 504806552@qq.com

Abstract. The temperature field analysis of electromagnetic rail launcher is an important basis for studying the launching stability and life. Aiming at the rail groove erosion phenomenon, the launching initial segment is selected as the research site. The heat generation mechanism of Joule heat and friction heat is analyzed, the heat source model is established, the three-dimensional finite element simulation model is established and the position, velocity and current waveform of the launching initial section are determined according to the circuit simulation, the finite element simulation results show that the heat in the initial stage of launching mainly concentrates on the contact surface between armature and rail. The friction heat is the main factor affecting the temperature rise of the contact surface, and the uneven distribution of the current density will lead to the uneven temperature distribution of the contact surface.

1. Introduction

Electromagnetic rail launcher is a new type of kinetic energy weapon which uses electromagnetic energy to accelerate the load to ultra-high speed. It has a broad application prospect in the military field for it can break through the limit of traditional propellant launching and achieve controllable thrust and velocity [1-3]. During the launching process, the pulsed power supply provides MA-level current in several ms, and the armature will generate thousands of times the gravitational acceleration in the strong magnetic field. Keeping good sliding contact between armature and rail is of great significance to the stability and life of the launcher under these extreme conditions [4].

Temperature is one of the important indicators for describing the contact state. The influence factors of mechanical characteristics on the contact surface between armature and rail include pre-tightening force during armature assembly, electromagnetic force during current conduction and thermal stress due to friction heat and Joule heat [5]. The temperature of the contact surface rises sharply due to the accumulation of heat during the launch process, which leads to significant changes in the mechanical properties of materials, such as material softening, melting, and metal oxidation, which will result in the instability of the contact state [6]. The heat accumulation makes the contact condition worse, and the stability and life of the device are significantly reduced especially under the condition of multiple continuous launches [7].

In 2002, Institution of Advanced Technology (ITA, in America) began to focus on the study of the electrical contact characteristics of armature rail and claimed that the armature transition was solved in the range of 2-3 km/s speed [3]. The problem of rail anti-ablation under discontinuous launching was solved by 2013 [8]. Due to the late start of domestic research, rail life has become one of the difficulties restricting experimental research. In recent years, some domestic scholars have carried out some research on the thermophysical characteristics and thermal management of electromagnetic rail launcher. Lin [9] has used numerical simulation to study the transient temperature field under Joule heat source. Wang [10] has analyzed the temporal and spatial distribution of contact heat on the calculation model of contact pressure and contact resistance. Lin [11] has established a two-dimensional model of rail transient temperature for simulating degree effect and compared with experiment. But they adopted more idealized conditions and did not analyze specific experimental phenomena. Existing launching tests show that the rail edge grooves are prone to occur in the initial stage of launching. The groove pits deepen continuously with multiple launches. The most probable reason is that local high temperature causes material softening and stress concentration [3,6]. Thus in this paper, a heat source model and a three-dimensional finite element simulation model will be established to simulate the temperature field of the rail and armature in the initial stage of launching, which will provide a reference for the study of the stability and life of the electromagnetic rail launcher, and corresponding optimization methods to solve the problem of groove corrosion as well.

2. Establishment of heat source model

In the process of electromagnetic launching, the generation of heat source includes two aspects: Joule heat generated by armature current and friction heat generated by relative sliding of armature. Both of them act on armature and rail at the same time with the characteristics of time accumulation and space distribution.

2.1. Joule heat

The current, provided by pulse power supply, flows through rail and armature to form a closed circuit. Energy is consumed in the form of heat on resistance. Considering the imperfect contact condition, the resistance can be divided into resistance of armature and rail body and contact resistance [10,12]. The body resistance generates heat in the conductor while the contact resistance generates heat on the contact surface. The total resistance can be described as

$$R = R_a + R_r + 2R_c \tag{1}$$

where R is the total resistance of armature rail; R_a, R_r are the resistance of armature and rail body; R_c is the contact resistance of one side between armature and rail. Contact resistance is generally composed of shrinkage resistance and film resistance. The oxide film on metal surface is destroyed during armature assembly, thus the film resistance can be neglected. Contact resistance is

$$R_c = R_s + R_b \tag{2}$$

$$R_s = \frac{\rho_a + \rho_r}{4na} \tag{3}$$

where ρ_a, ρ_r are the resistivity of armature and rail materials, n is the number of micro contact spots on the contact surface which can be estimated at 10 per 4 square millimetres [13], a is the average radius of contact spots. The actual contact area of contact surface A_c is usually much smaller than the ideal contact area, which can be derived from the following formula

$$A_c = \frac{F_c}{\xi H_a} \tag{4}$$

$$A_c = n\pi a^2 \tag{5}$$

where F_c represents the contact pressure on one side contact surface; H_a is the hardness of the softer side of contact material, generally the armature material, takes the value of 150 HB; ξ is the elastic

IOP Conf. Series: Materials Science and Engineering **657** (2019) 012067 doi:10.1088/1757-899X/657/1/012067

deformation correction coefficient between 0 and 1, which decreases with the increase of the contact pressure, takes 0.1. Eliminating the conductive speckle radius a which is difficult to determine, the thermal power of the contact resistance can be expressed by combining the above formulas as

$$P_c(t) = I^2(t)R_c(t) = \frac{(\rho_a + \rho_r)}{4}\left(\frac{\pi\xi H_a}{nF_c(t)}\right)^{1/2} \cdot I^2(t) \tag{6}$$

2.2. Frictional heat

Friction heat is generated on the contact surface, and the velocity of launching initial stage is small. However, due to the larger current and the larger contact pressure between armature and rail, the influence of friction heat should be considered. The frictional heat generation power on one side between armature and rail can be described as follows:

$$P_f(t) = \mu F_c(t)v(t) \tag{7}$$

where v is the relative sliding speed of amature; μ is the sliding friction coefficient, as the movement proceeds, its value will decrease continuously.

3. Establishment of simulation model

3.1. Structural model and material parameters

According to the actual structure of the electromagnetic rail launcher, a three-dimensional finite element model is established after corresponding simplification, which is shown in figure 1. The rail is 10 mm high, 20 mm wide, 1.8 m long, 16 mm length between two rails, 2 mm high and 10 mm wide for rail bump, 25 mm long and 10 mm wide for U-type armature, and 5 mm round corner for shoulder.

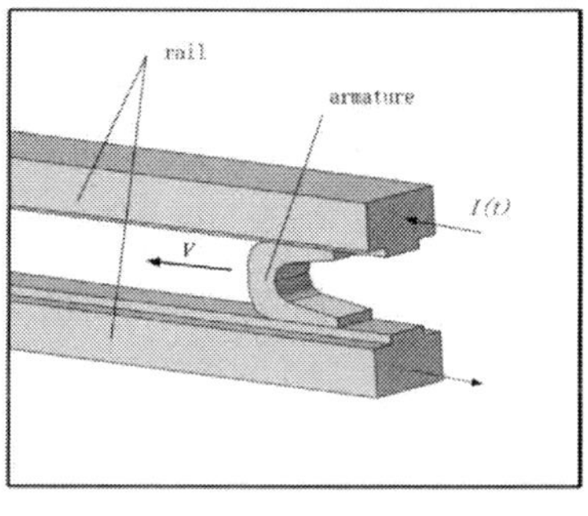

Figure 1. Finite element model of electromagnetic rail launcher.

The armature material is aluminium alloy and the rail is copper alloy. Some important parameters are shown in table 1. Among them, ρ is the material density, μ_r is the relative permeability, σ is the conductivity, k is the thermal conductivity and C_p is the constant pressure heat capacity.

Table 1. Some important parameters of materials.

	$\rho/(kg/m^3)$	μ_r	$\sigma/(S/m)$	$k/(W/(m\cdot K))$	$C_p/(J/(kg\cdot K))$
armature	2689	1	3.8×107	238	900
rail	8933	1	5.8×107	400	385

3.2. Circuit model

The circuit model is used to obtain a more realistic current waveform. The circuit model is shown in figure 2. Among them, pulse capacitor C1 provides current, R2 and L2 are equivalent resistance and inductance of power supply and circuit, R1 and L1 are wave-modulated resistance and inductance, and resistance R3 and inductance L3 of launcher on the right side actually show linear variation. The calculation formulas are as

$$R_3 = R_0 + R'x(t) \tag{8}$$

$$L_3 = L_0 + L'x(t) \tag{9}$$

where R_0 and L_0 are the resistance and inductance of the armature rail at the initial launching position, R' and L' are the resistance gradient and the inductance gradient, $x(t)$ is the armature displacement. The initial position of armature launching is 15 mm away from the tail of rail. The approximation is obtained by finite element analysis that R_0 is 2.25 $u\Omega$, L_0 is 8.58 nH, R' is 0.15 $m\Omega / m$, L' is 0.57 $\mu H / m$.

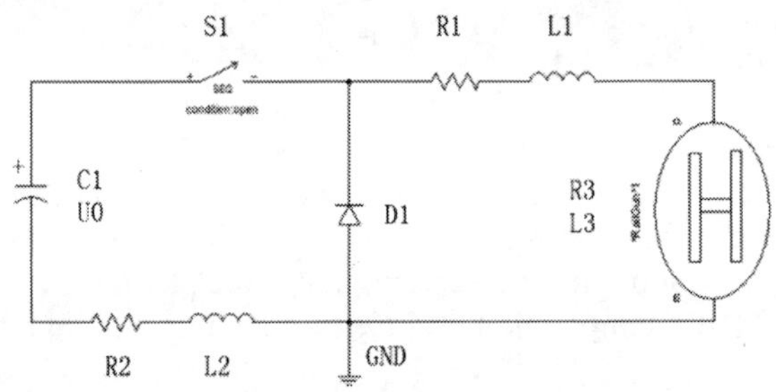

Figure 2. Circuit model.

For unknown variables $x(t)$, it is necessary to bring the armature kinematics equation into the solution. Ignoring the effect of air drag, the kinetic equation of armature at time t is

$$F_z = ma(t) = m\frac{\mathrm{d}v(t)}{\mathrm{d}t} = m\frac{\mathrm{d}^2 x(t)}{\mathrm{d}t^2} = \frac{1}{2}L'I^2(t) - 2\mu F_c \tag{10}$$

where F_z is the resultant force in the direction of armature motion; $a(t)$ is the armature acceleration; $v(t)$ is the armature speed; μ is the friction coefficient of contact area, which is constant 0.15. The contact pressure of the central rail F_c can be calculated by finite element simulation or estimated by the following formula [7]:

$$F_c(t) = F_0 + \frac{1}{4}\alpha L'I^2(\mathrm{t})\cos\theta \tag{11}$$

where F_0 is the pre-pressure of the interference assembly, which can be regarded as constant value when the friction loss of material is neglected. According to Marshall formula of "1g/1A", F_0 is 1kN; α is the coefficient of transforming electromagnetic expansion force of armature arm into electromagnetic extrusion force, taken as 0.5; $\cos\theta$ is the cosine of the angle between armature tail wing and rail, taken as 0.956.

Assuming that the rail is infinitely long and the maximum time is 5 ms for transient circuit analysis, the circuit waveform curve, armature velocity curve and displacement curve can be obtained. As is shown in figures 3-5, the current reaches a peak value of 200 kA at 0.23 ms. This paper sets the armature to move 1.65 m before leaving the guide rail. The exit time of the armature is about 1.68 ms, and the exit speed is about 1733 m/s.

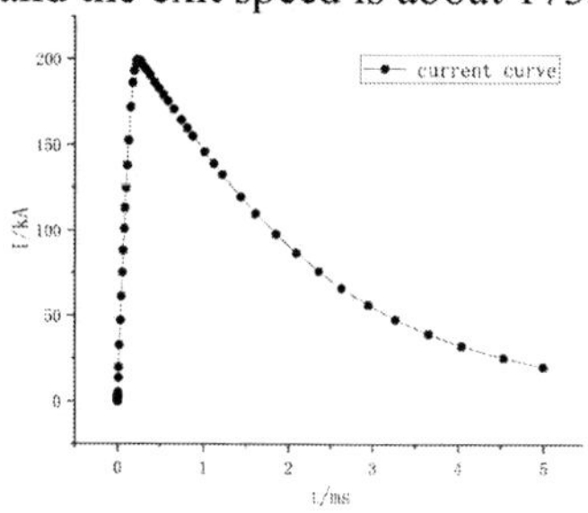

Figure 3. Current waveform.

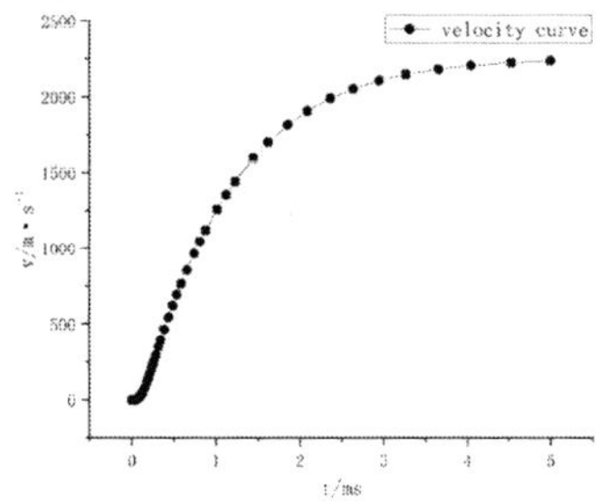

Figure 4. Armature velocity curve.

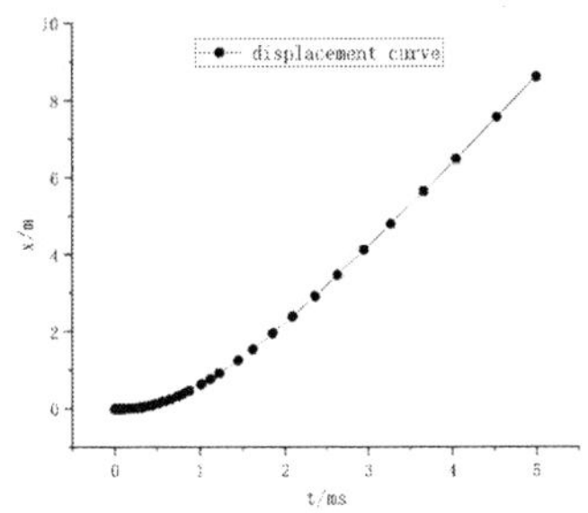

Figure 5. Armature displacement curve.

3.3. Initial segment simulation model

It can be seen from the experimental phenomena of rail groove corrosion in the initial stage of launch that most of the groove corrosion occurs in the time and position where the loading current is large and the contact time is long [4,6]. In this paper, the position of armature ,about two armature lengths (51 mm), is selected as the fixed research position. The time is 0.35 ms, the speed is 409 m/s and the current position is in the descending edge, reaching 192 kA.

Considering the skin effect of current, the armature is fixed, the Joule heat source of contact resistance and the friction heat source are generated at the interface of contact surface. The heat source added to the contact surface is taken as a constant. The contact resistance calculated by the above formula is about 3.313 $\mu\Omega$ respectively, and the peak heat generating power is 1.325×10^5 W and 2.285×10^5 W. Using transient electromagnetic-thermal coupling simulation, considering the worst heat generation, the model is set as heat insulation, the ambient temperature is 20°C, and the total time is 0.35 ms. The current density and temperature field distribution on armature and rail are investigated.

4. Simulation results and analysis

4.1. Current density

Current density is an important factor to measure Joule heat. In the simulation, different sections are selected to investigate the current density at different time, as shown in figures 6 and 7.

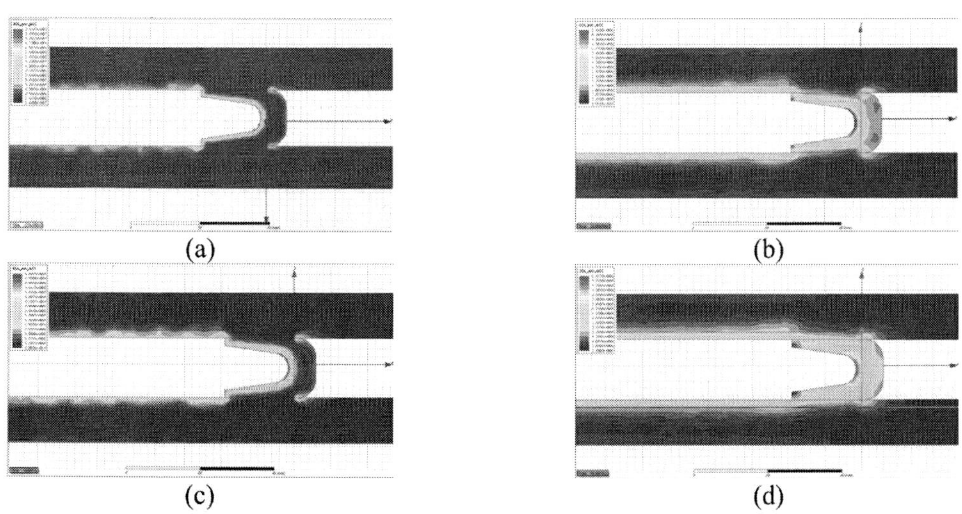

Figure 6. Current density diagram of side section. (a) 0.025 ms, (b) 0.1 ms, (c) 0.225 ms and (d) 0.35 ms.

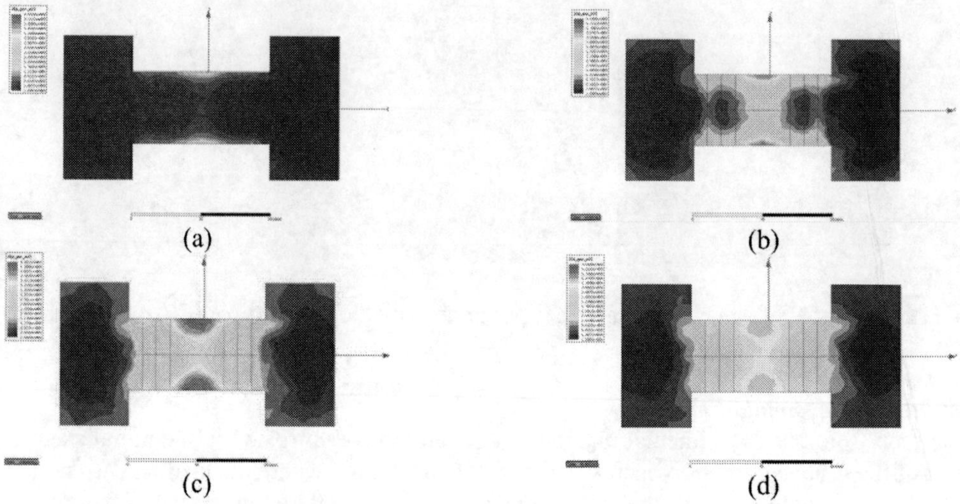

Figure 7. Current density diagram of longitudinal section. (a) 0.025 ms, (b) 0.1 ms, (c) 0.225 ms and (d) 0.35 ms.

From the cross section of figure 6, it can be seen that the skin effect is obvious at the moment of 0.025 ms when the current is just passed in. The maximum current density is located at the neck of the armature, followed by the junction between the armature tail and the rail, and the current in the rail is concentrated in the convex platform. With the passage of time, the current gradually permeates into the material. The current density near the peak time of current curve (0.225 ms) reaches 5.663×109 A/m^2, but the maximum current density appears before the peak time of current current, reaching 5.998×109 A/m^2.

From the longitudinal section of figure 7, it can be seen that the current initially gathers at the center of edge sides of the armature, and then diffuses to the whole section. The current on the contact surface gathers at the end of armature at edge sides, where the heat generating power is high and the material is easy to soften due to local high temperature, which is close to the groove location of armature and rail in the experiments.

4.2. Temperature distribution

The location and diffusion path of temperature can be visually understood by temperature distribution map. After 0.35 ms, the temperature distribution of armature and rails ,investigated by transient thermal simulation method, is shown in figures 8 and 9.

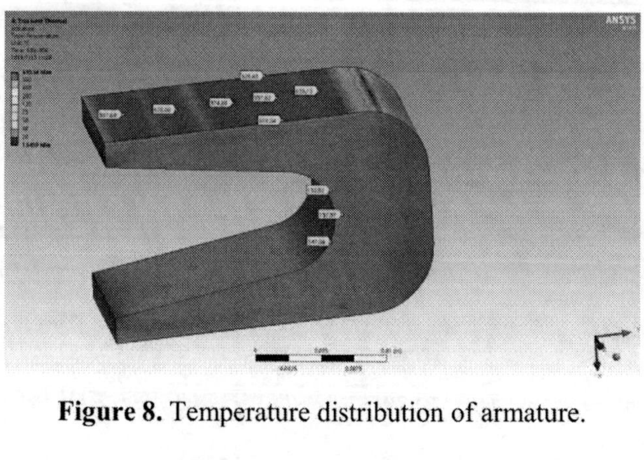

Figure 8. Temperature distribution of armature.

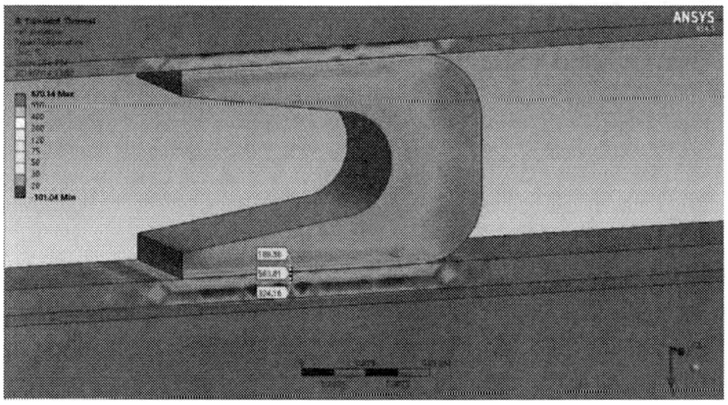

Figure 9. Temperature distribution of rail.

As shown in figure 8, the highest temperature of the armature appears on the contact surface. In the initial stage of launch, contact heat (including contact resistance Joule heat and friction heat) is generated rapidly on the contact surface. Due to the limited heat transfer capacity of materials in a short time, the instantaneous high temperature on the contact surface reaches 670°C, which exceeds the melting point of armature material. This is consistent with the aluminum layer detected on the inner surface of rail in experiments. Because of the skin effect of the current, the temperature of the tail and side edges of the armature on the contact surface is slightly higher than that of other parts. In addition, the temperature concentration occurs at the inner corner of the armature neck, which corresponds to the current density concentration in the upper text.

Because of the armature movement, the rail thermal contact area is larger during the process of launch. The actual temperature value of the rail should be lower than that of the simulation, but the distribution is similar. As can be seen from figure 9, the heat generation of the rail is concentrated in the thin layer on the surface of the convex platform, and thermal diffusion occurs both in the axial and radial direction of the launch. When the Joule heat source of electromagnetic field is coupled into the thermal field, the inconsistent finite element meshing results in abnormal temperature in small number of units, but does not affect the overall solution.

4.3. Optimization analysis

From the point of view of heat generation, Joule heat plays an important role both inside and on the contact surface between armature and rail, but the friction heat on the contact surface of armature rail is more significant at the initial stage of launch. Local high temperature on the contact surface aggravates the stress concentration, which leads to groove corrosion on the rail. When the armature melts locally, the electrical contact performance decreases, which will aggravate the groove corrosion on the rail. Through the simulation analysis in this paper, the following aspects can be considered to reduce the thermal effect in the initial stage of launch:

- Material aspect: adopting armature material with high melting point, rail material with high strength and lubricant with good heat transfer performance on the contact surface.
- Structural aspects: carrying out fillet treatment at the end of armature tail fin, changing the section shape of contact surface (concave or convex) and reasonably controlling the armature interference.
- Cooling passage can be set in the rail to accelerate heat conduction.

5. Conclusion

In this paper, the launching initial section is selected as the research site for the rail groove erosion phenomenon. The heat generation mechanism of Joule heat and friction heat is analyzed, the corresponding heat source model is established, and the three-dimensional finite element simulation model based on circuit simulation is established. The main conclusions are as follows:

- In the initial stage of launch, the heat is concentrated on the contact surface, and the heat transfer effect in the conductor is not obvious.
- Under hundreds of kA peak current loading, friction heat is the main factor that causes the temperature rise of the contact surface at the initial stage of launch, which is also the basic cause of groove corrosion..
- The uneven distribution of current density will lead to uneven temperature distribution on the contact surface, accelerating the formation of groove pits on edge of contact surface.

In order to obtain more reliable conclusions, the follow-up study will start from the coupling of force field on the contact surface. The non-uniform distribution of contact force makes the current and heat more concentrated, especially at the edge of the contact area.

6. Acknowledgements

I thank members of my lab, L Xiong, L Wang, and many other colleagues for their discussions on finite element simulation and helpful comments on this manuscript. The work described in this paper was fully supported by a grant from Defense Industrial Technology Development Program(JCKY2018130C008).

References

[1] Wang Y and Xiao F 1995 *The Principle of Electrical Gun* (Beijing: National Defence Industry Press) (In Chinese)

[2] Ma W M and Lu J Y 2016 Electromagnetic launch technology *J. Natl. Univ. Def. Technol.* **38** 1-5

[3] Qiao Z M, Lei B, Lv Q G *et al* 2016 Analysis of the key technologies and development tendency of electromagnetic railguns *J. Gun Launch Control* **37** 91-5

[4] Cao R G, Su M, Li S Z *et al* 2016 Summarize and analysis of rail-gun's grooving phenomenon with large pulse current and high-velocity sliding electrical contact *High Vol. Eng.* **42** 2822-9

[5] Zhang Y S, Lu J Y, Tan S *et al* 2018 Analysis of force load on the rail in rapid-fire mode of electromagnetic rail launcher *ACTA Arm. II* **39** 618-24

[6] Lu J Y and Sun W 2016 Preliminary research on erosion technique of rail electromagnetic launch system *J. Naval Univ. Eng.* **28** 7-10

[7] Zhang Y S, Lu J Y, Tan S *et al* 2016 Research on distribution characteristic of thermal energy in continuous electromagnetic rail launch *J. Naval Univ. Eng.* **28** 94-9

[8] Wild B, Schupper C, Alouahabi F *et al* 2015 The influence of the rail material on the multishot performance of the rapid fire railgun *IEEE Trans. Plasma Sci.* **43** 2095-9

[9] Lin Q H and Li B M 2017 Numerical simulation of transient temperature field in the electromagnetic railgun *J. Eng. Thermophys.* **38** 149-54

[10] Wang Z H, Wan M and Li X J 2016 Characteristic of temporal and spatial distribution of railgun contact heat *Chinese J. High Pressure Phys.* **30** 511-6

[11] Lin L S, Zhao Y and Yuan W Q 2016 Thermal effect analysis of railgun tests under transient conditions *High Vol. Eng.* **42** 2864-9

[12] Guo F Y and Chen Z H 2008 *Electrical Contact Theory and Its Application Technology* (Beijing: China Electric Power Press)

[13] Jin L W, Lei B and Zhang Q 2012 Heat generation mechanism analysis and numerical simulation of temperature field in electromagnetic railgun *J. Gun Launch Control* **4** 9-12

IOP Conf. Series: Materials Science and Engineering **657** (2019) 012068 doi:10.1088/1757-899X/657/1/012068

Simulation design method of free boundary simulator for high precision spacecraft on orbit

M H Liu[1,3], H Cao[2] and H Y Gao[1]

[1]Beijing Institute of Spacecraft Environment Engineering, Beijing, 100094, China
[2]China Academy of Space Technology, Beijing, 100094, China

E-mail: liuminghui.china@163.com

Abstract. The reproduction of on-orbit free boundary is one of the key technologies for highly accurate ground micro-vibration test of high precision optical spacecraft. This paper focuses on the simulation design method of the free boundary simulator based on finite element method (FEM). The preliminary design is modeled by one-dimensional spring elements, which is effective to optimize the critical design parameters. Then the detailed design model is established to verify the capability of strength, the stability and the modal performance of the free boundary simulator. The actual simulator is manufactured according to the simulation design. And the test of the actual simulator indicates that, the simulation results match well with the experimental results. The development of the free boundary simulator of spacecraft can be instructed effectively by the simulation method in this paper.

1. Introduction

The micro-vibration test is an important approach to evaluate the effect of the moving mechanism on the spacecraft with high precision optical payloads in orbit, and to obtain the structural response transmission characteristics from the vibration source to the payloads, to improve the imaging quality of optical payloads on the spacecraft. One of the key technologies of the micro-vibration test is the reproduction of the "free-free" boundary condition of the spacecraft on orbit.

Several methods have been proposed to reproduce free boundary condition for spacecraft during micro-vibration test. Such as air bags were used to give flight-like boundary conditions for Solar Dynamic Observatory [1]. With the goal of getting as close as possible to a spacecraft free-free condition, the Artemis satellite was mounted on top of a system of pneumatic isolators. The six resonant modes of the system were kept below 2.8 Hz [2]. A ground based test adapter was developed to safely support a large payload and isolate it from ground borne vibration during system level stability testing. The finite element model based simulation was used for predicting the isolation system rigid body modes [3]. The simulation is an effective way for the design of the free boundary simulator, however, the simulation method is not comprehensive presented in the literature.

This paper introduces the two-step design method of the free boundary simulator based on FEM. Firstly the preliminary design is modeled by spring elements, which is effective to optimize the critical design parameters. Then the detailed design model is established to verify the strength capability, the stability and the modal performance of the free boundary simulator. The actual simulator is manufactured according to the simulation design. And the test of the actual simulator indicates that, the simulation results match well with the experimental results. It is concluded that the development of the free boundary simulator of spacecraft can be instructed effectively by the simulation method in this paper.

2. Preliminary simulation of on-orbit free boundary simulator

The spacecraft is in a free state without any external constraints during its flight on orbit due to the lack of gravity. By the theory of structural dynamics, the spacecraft possesses six rigid body modes of zero frequency. However, external constraints have to be applied to hold the spacecraft at rest due to the gravity of the earth on ground. The goal of on-orbit free boundary reproduction on ground is to make the spacecraft possess six quasi rigid body modes after the external constraints applied. Therefore, the free boundary simulator shown in figure 1 is proposed. The difficulties of the design are heavy payload bearing, high stability, and low suspension frequencies. Effective simulation method is of necessary to accomplish the design.

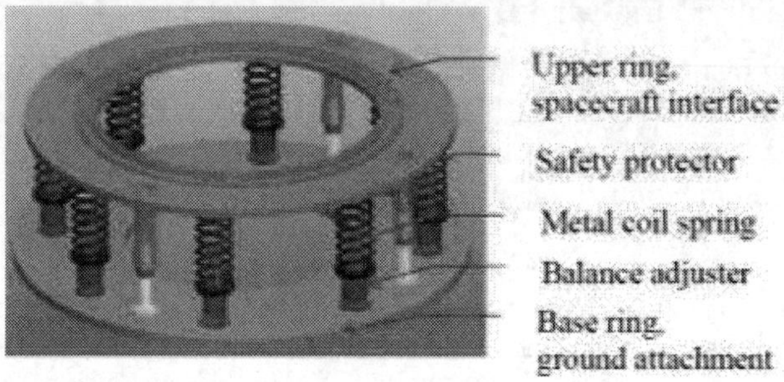

Figure 1. The sketch of the on-orbit free boundary simulator.

2.1. Modeling of preliminary design

In the stage of preliminary design, the emphasis is to simulate the mechanical parameters of the metal coil spring based on the configuration of the free boundary simulator, so that the requirements of load-bearing and suspension frequencies are satisfied. For the convenience of analysis and optimization of the stiffness parameters of the coil springs, the one-dimensional spring element of NASTRAN is used to simulate the coil springs. And the lumped mass element is accepted to represent the spacecraft with mass and moment of inertia. The spacecraft and the springs are connected by RBE2 elements to simplify the bolt connection. The combined model of the spacecraft and the free boundary simulator is illustrated in figure 2.

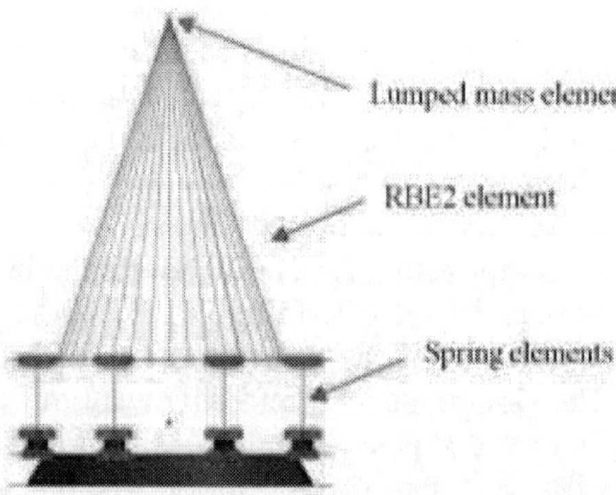

Figure 2. The preliminary model of spacecraft and free boundary simulator.

2.2. Sensitivity analysis of design parameters

A group of initial stiffness parameters of the springs are set for the preliminary design, as shown in table 1. Modal analysis results show that the suspension frequencies are high. The bounce modal frequency is 2.6 Hz, and the second rocking modal frequency is 3.2 Hz. The initial parameters do not satisfy the design requirements. Modification of the initial parameters should carry on.

Table 1. The initial stiffness parameters.

Stiffness parameters	Parameter value
Lateral stiffness(K_X, N/m)	40000
Lateral stiffness(K_Y, N/m)	40000
Axial stiffness(K_Z, N/m)	80000
Bending stiffness(K_RX, Nm/rad)	1000
Bending stiffness(K_RY, Nm/rad)	1000
Torsional stiffness(K_RZ, Nm/rad)	3000

As seen that there are six stiffness parameters for the spring element in the preliminary design. In order to achieve high efficiency of design, the sensitivity analysis is performed, so that the most sensitive parameters to the suspension frequencies can be identified. The six stiffness parameters are defined as design variables, and the first six modal frequencies of the system are set as response objective. The sensitive matrix is obtained by sensitivity analysis, which is in table 2. It shows that, the first rocking modes are mainly affected by lateral stiffness and axial stiffness, the torsion mode is basically affected by lateral stiffness, and the bounce mode and the second rocking modes are mainly affected by axial stiffness. It is also seen that the bending and torsional stiffness of the spring have weak influence on the suspension modes. The sensitivity results are used to instruct the optimization of the parameters.

Table 2. The sensitivity matrix of stiffness parameters.

Mode	K_X	K_Y	K_Z	K_RX	K_RY	K_RZ
First rocking (Y)	0.00	6.23	3.01	0.03	0.00	0.00
First rocking (X)	6.89	0.00	2.70	0.00	0.02	0.00
Torsion	5.83	8.24	0.00	0.00	0.00	1.69
Bounce	0.00	0.00	28.67	0.00	0.00	0.00
Second rocking (X)	8.82	0.00	23.36	0.00	0.11	0.00
Second rocking (Y)	0.00	14.65	30.60	0.23	0.00	0.00

2.3. Results of preliminary design

Based on the sensitivity analysis results, more targeted selection of parameters are made for the design optimization of suspension frequencies. The optimized parameters of stiffness are listed in table 3. It shows that obvious modification is produced for lateral and axial stiffness. And the optimized suspension frequencies are shown in table 4. The shapes of the first and second rocking mode are illustrated in figure 3. The suspension frequencies satisfy the design requirements. The preliminary design is accepted for detailed design.

Table 3. The optimized parameters of stiffness.

Stiffness parameters	Parameter value
Lateral stiffness(K_X, N/m)	24000
Lateral stiffness(K_Y, N/m)	24000
Axial stiffness(K_Z, N/m)	48800
Bending stiffness(K_RX, Nm/rad)	1000
Bending stiffness(K_RY, Nm/rad)	1000
Torsional stiffness(K_RZ, Nm/rad)	3000

Table 4. The suspension frequencies of preliminary design.

Mode order	Frequency (Hz)	Mode shape
1	0.43	First rocking (Y)
2	0.48	First rocking (X)
3	0.8	Torsion
4	2.0	Bounce
5	2.1	Second rocking (X)
6	2.5	Second rocking (Y)

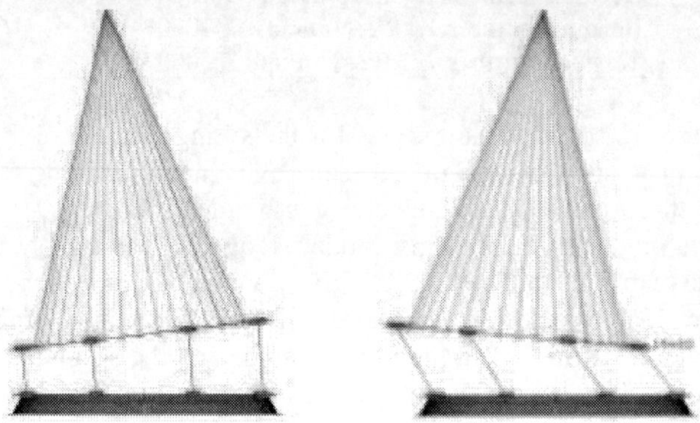

Figure 3. The shapes of the first and second rocking mode.

3. Detailed simulation and test validation of on-orbit free boundary simulator

3.1. Modeling of detailed design

Instructed by the preliminary design simulation results, the particular model of coil spring can be selected, and the diameter of the thread of the spring, the diameter of the coil, the height of the spring, and the pitch of the spring are all determined. The detailed design model is established. The beam element is used to present the coil spring. The spacecraft can be modelled by lumped mass element as same as in preliminary design. And if a detailed spacecraft model is given, the lumped mass element can be replaced for more detailed simulation. The mechanical interface structure is represented by solid element. The spacecraft and the interface are connected by RBE2 elements as same as in preliminary design to simplify the bolt connection. The combined model of the spacecraft and the free boundary simulator is illustrated in figure 4.

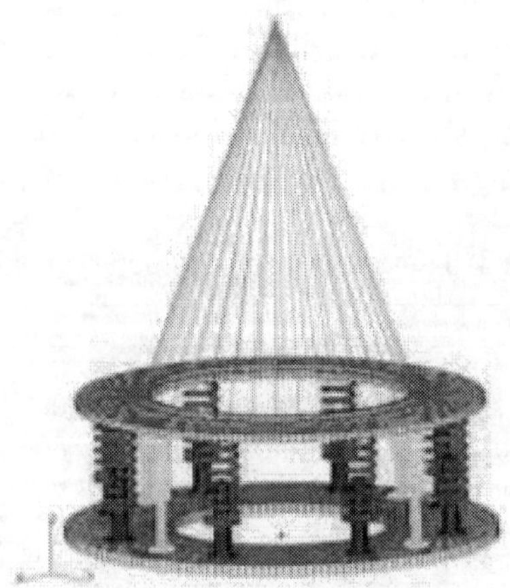

Figure 4. The detailed model of spacecraft and free boundary simulator.

IOP Conf. Series: Materials Science and Engineering **657** (2019) 012068 doi:10.1088/1757-899X/657/1/012068

3.2. Analysis of the design

Comprehensive analyses of static strength, stability, and mode are performed by using the detailed model. The deadweight of the spacecraft is loaded on the model with fixed base ring of the free boundary simulator as boundary condition. The maximum displacement of the system is 71.8 mm shown in figure 5(a), which is the linear deformation of the spring. And the maximum stress of the structure is 36.6 MPa, located at the supporting structure of the spring, as presented in figure 5(b). The deformation satisfies the operating requirements of spring, and the stress satisfies the structure strength requirements.

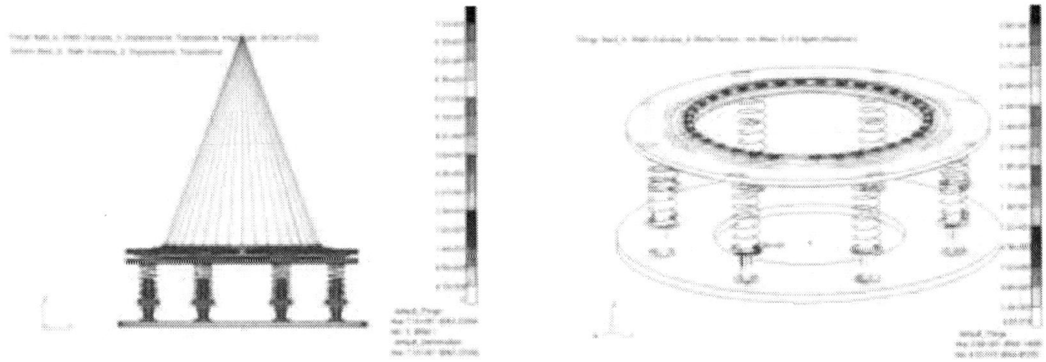

(a) (b)

Figure 5. The static analysis results of the detailed model. (a) static deformation (mm) and (b) static stress (MPa).

The stability of the free boundary simulator loaded by spacecraft deadweight is shown in figure 6. The first buckling mode shape of the system is toppling deformation of the springs. And the assurance factor is 1.6 which is satisfying for the engineering application.

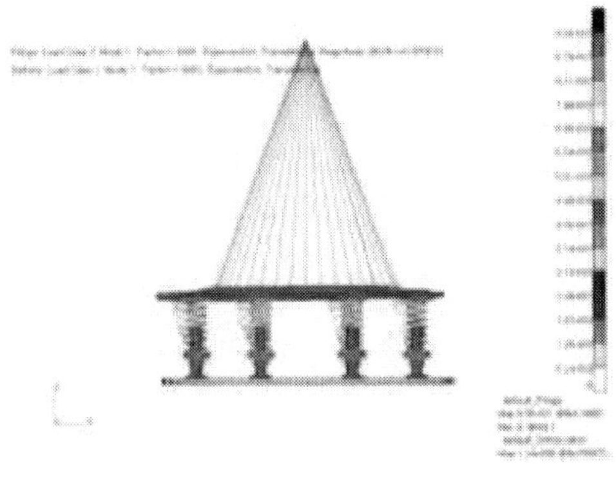

Figure 6. The first buckling mode deformation.

Table 5. The suspension frequencies of detailed design.

Mode order	Frequency (Hz)	Mode shape
1	0.45	First rocking (Y)
2	0.48	First rocking (X)
3	0.8	Torsion
4	2.0	Bounce
5	2.06	Second rocking (X)
6	2.08	Second rocking (Y)

Modal analysis results of the system are shown in table 5. The highest suspension of the system is 2.08 Hz. The mode shapes of the first and second rocking mode are presented in figure 7. The suspension of the system satisfies the design requirements.

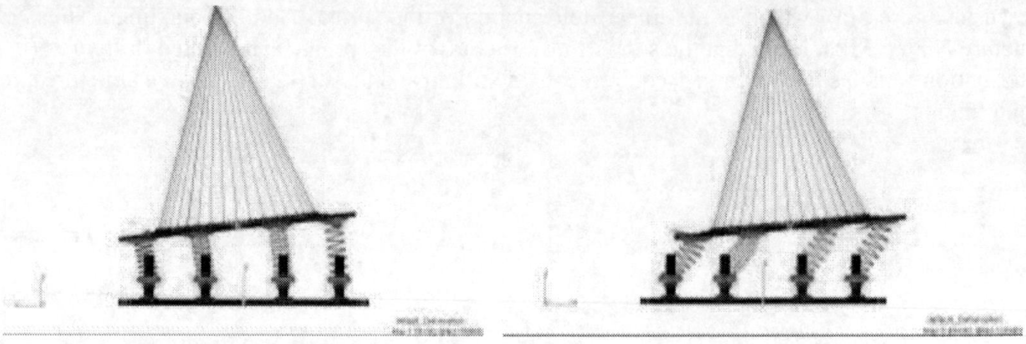

Figure 7. The shapes of the first and second rocking mode.

3.3. Test validation

Based on the simulation of the detailed design, the free boundary simulator is manufactured, as shown in figure 8. The base ring of the simulator is fixed on the test hall floor. And the upper ring is used to attach spacecraft.

Figure 8. The manufactured free boundary simulator.

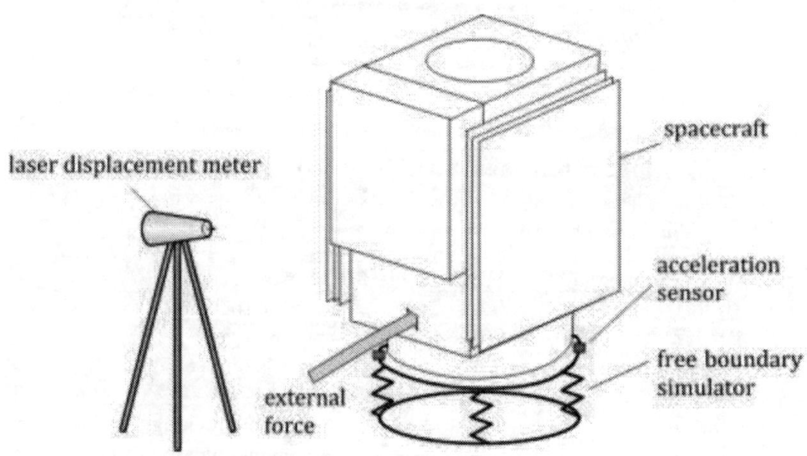

Figure 9. The test configuration of the system.

The test configuration is illustrated in figure 9. Suspension frequencies measurement is performed by application of external force and torque along the X, Y and Z direction of the spacecraft to make the spacecraft deviate from the static balance position. Then the applied force is removed to make the system freely oscillate, and the free oscillation response of the system is measured by acceleration sensor and laser displacement meter. The oscillation period is extracted by analyzing the measured response curves to obtain the suspension frequencies.

The test results are listed in table 6. It is found that there are certain errors between the simulation and test by comparing tables 5 and 6. One of the reasons is that the lumped mass model is used in the simulation, which is different than the real spacecraft. However the simulation results match well with the experimental results. It is satisfied for engineering application.

Table 6. The tested suspension frequencies.

Mode order	Frequency (Hz)	Mode shape
1	0.35	First rocking (Y)
2	0.37	First rocking (X)
3	0.91	Torsion
4	1.98	Bounce
5	1.99	Second rocking (X)
6	1.99	Second rocking (Y)

4. Conclusion

A simulation method of the free boundary simulator based on FEM is proposed for the design of on-orbit free boundary simulator of spacecraft. The method basically consists of two steps. The first step is to simulate the design by one-dimensional spring elements, with focuses on the identification of sensitive design parameters for effective optimization. The second step sets up the detailed simulation model to analysis the accurate static and dynamic performances of the free boundary simulator, for the purpose of engineering application. The simulation is validated by the test of the actual simulator. It is concluded that the proposed method provides effective instruction to the design of the on-orbit free boundary simulator of spacecraft.

References

[1] Liu K C, Thomas K, Peiman M, Pete M, Crrl B and William H 2008 Jitter Test Program and On-Orbit Mitigation Strategies for Solar Dynamic Observatory

[2] Marchante E M and Munoz L 1997 Artemis satellite micro-vibrations testing and analysis activities 1997 *48th International Astronautical Congress* (6-10 October, Turin, Italy)

[3] Pendleton S C, Basile J P, Academia G M and Kidney S L 2009 Payload isolation system for ground based stability testing: Design and test validation *50th AIAA/ASME/ASCE/AHS/ASC Structures, Structural Dynamics, and Materials Conf.* (4-7 May 2009 Palm Springs California)

IOP Conf. Series: Materials Science and Engineering **657** (2019) 012069 doi:10.1088/1757-899X/657/1/012069

Numerical study on the ideal interference of aircraft riveting joints

S Sun[1,2,3]

[1]AVIC China Aero-Polytechnology Establishment, No.7 Jingshun Road, Beijing, China
[2]Key Laboratory of Quality Infrastructure Efficacy Research, AQSIQ, Beijing, China

E-mail: Sunsheng369@126.com

Abstract. The interference is the main parameter affecting the fatigue lifetime of the interference-fit joints. In this paper, the numerical simulation method of the interference-fit riveting process was studied for the object of the aluminium alloy joint commonly used in aircraft. On this basis, the influence of different interferences on the joint life was calculated and determined, which can provide a basis for determining the ideal riveting interference. The numerical simulation results show that the model built in this paper can well describe the whole forming process of the rivet including the overall upsetting and local upsetting. With the increase of the interference, the fatigue lifetime of the joint increases first and then decreases. For the joint structure given in this paper, when the interference is between 2.9% and 3.23%, a higher life increment can be obtained, which is basically consistent with the ideal interference obtained by experiments.

1. Introduction

Due to its characteristics such as wide application range, stable quality and low cost, the riveting is still the most important connection method for metal structures in current aircraft manufacturing. The life and reliability of the joint play a vital role in aviation safety. Because of the high cost and long cycle of the life test for the riveting joint, the optimization design method based on numerical simulation method has been gradually applied. Langrand, Blanchot, Repetto *et al* used different methods to establish the finite element analysis model of the riveting process, which provided ideas for the numerical simulation method of the strength of the riveting joint [1-3]. Chakherlo carried out simulated fatigue tests on the riveted structure and obtained the relationship between the interference and the fatigue life [4]. Zeng studied the effect of residual stress due to interference-fit on the fatigue behavior [5]. Abdelal investigated the dimensional growth of aircraft panels while being riveted with stiffeners based on numerical and experimental methods, he found that a higher degree of predictability of the riveting process are obtained using explicit FE models [6]. Li studied the influence of riveting dies with different angles on the electromagnetic riveting quality and provided ideas for the application of numerical simulation in the research of electromagnetic riveting technology [7]. Moreover, the structural analysis methods of composite joints were studied by Cao, Dean and Yoo [8-10].

On the basis of previous studies, the effect of the interference on the fatigue lifetime of the electromagnetic riveting joint was studied by means of numerical simulation method in this paper, which can help to select the reasonable interference in the engineering application of interference-fit

IOP Conf. Series: Materials Science and Engineering **657** (2019) 012069 doi:10.1088/1757-899X/657/1/012069

strengthening technology.

2. Numerical simulation of the forming process of the interference fit riveting joint

Compared with ordinary riveting, electromagnetic riveting technology can achieve relatively uniform interference fit when the clearance between the rivet and hole is relatively large or the interlayer thickness is relatively large. Therefore, the electromagnetic riveting joint is taken as the research object in this paper.

The electromagnetic riveting joint consists of riveting die, rivet and sandwich plates, and their geometric relationship is shown in figure 1. In this structure, the rivet diameter is 4 mm, rivet length is 9.5 mm, width of the sandwich plate is 24 mm and the thickness of each sandwich plate is 3 mm. The riveting joint is an axisymmetric structure as a whole. For improving the calculation efficiency, a 1/2 symmetry model is established by ABAQUS as shown in figure 2. Because the rigidity of the riveting die is much larger than that of the rivet and sandwich plates, it is treated as a rigid body, that is, its deformation is not considered in the analysis process. The R3D4 elements are used for the die's mesh generation, and the rivet and sandwich plates are divided by the C3D8RT elements which can consider the temperature.

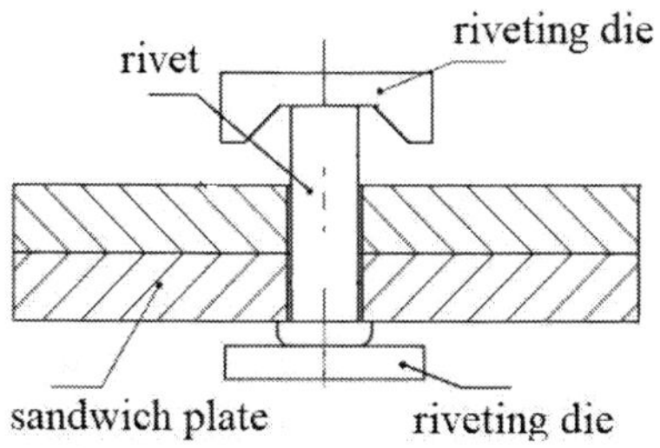

Figure 1. Structure schematic of the riveting joint.

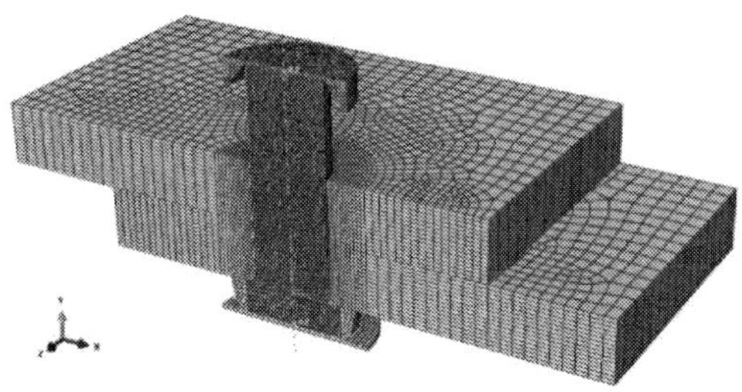

Figure 2. The 1/2 symmetry model of the riveting joint.

As a rigid body, the deformation of the riveting die does not be considered. So only the parameters of the rivet and the sandwich plates need to be set. In this paper, the material of the rivet is 2A10 aluminum alloy, and the sandwich plates uses 2024 aluminum alloy. The material performance parameters are shown in table 1.

Table 1. Performance index of the materials.

Material	Elasticity modulus (GPa)	Poisson's ratio	Density ($kg \cdot m^{-3}$)	Thermoconductivity ($W \cdot m^{-1} \cdot K^{-1}$)	Specific heat ($J \cdot kg^{-1} \cdot K^{-1}$)	Thermal expansivity ($m \cdot m^{-1} \cdot {}^{\circ}C^{-1}$)
2024	71	0.33	2750	154	880	2.56e-5
2A10	73.1	0.33	2770	121	875	2.47e-5

The electromagnetic riveting usually produces 30% -50% strain at the instant of several hundred microseconds to milliseconds, and its strain rate is several magnitudes higher than that of the ordinary riveting. In order to consider the influence of the loading rate and temperature rise on the flow properties of the materials, the Johnson-Cook thermo-viscoplastic constitutive model of the two materials needs to be fitted.

$$\sigma = (A + B\varepsilon^n)\left[1 + Cln\varepsilon_p^*\right]\left[1 - T^{*m}\right] \tag{1}$$

$$\varepsilon_p^* = \varepsilon_p / \varepsilon_0 \tag{2}$$

$$T^* = (T - T_0)/(T_m - T_0) \tag{3}$$

Among these equations, ε_p is the strain rate, $\varepsilon_0$is the reference strain rate, T_0 and T_m are the room temperature and melting temperature, and C and m are the strain rate sensitive coefficient and thermal softening coefficient. The other five parameters A, B, C, m, n are the intrinsic parameters of the material. A, B and n are the strain strengthening coefficients, C is the strain rate strengthening coefficient, and m is the temperature softening coefficient. The values of these five parameters can be calculated [11] according to the data from the quasi-static test and Hopkinson pressure bar test. The room temperature T_0 is set at 20°C. The JC constitutive model parameters of the materials are shown in table 2.

Table 2. The JC constitutive model parameters of the materials.

Material	A(MPa)	B(MPa)	C	n	m	T_m(°C)
2024	71	0.33	2750	154	880	2.56e-5
2A10	73.1	0.33	2770	121	875	2.47e-5

In order to guarantee the accuracy of the riveting force, the riveting force curve was measured directly through the experimental method [11]. The riveting force is loaded on the surface of the riveting die, and the plastic deformation process of the rivet in the riveting process includes two parts: the overall upsetting and the local free upsetting. The forming process of the upsetting head of the rivet is shown in figure 3. When t=0.35 ms (figure 3(d)), the rivet head has been basically formed. At this time, the riveting force has been rapidly attenuated, and the force is not enough to continue forming, and a small amount of rebound will occur in the area of the upsetting head.

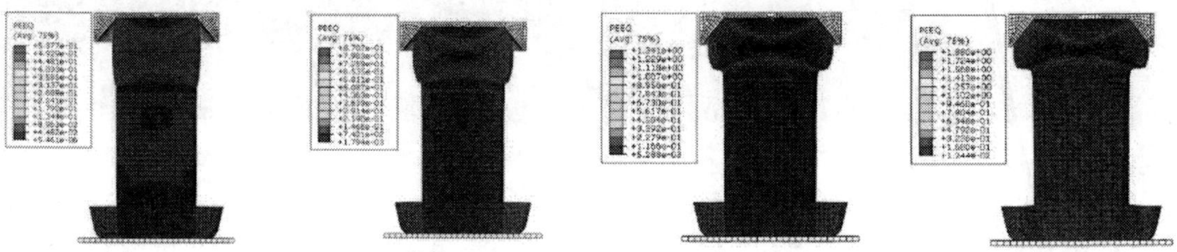

(a) (b) (c) (d)

Figure 3. The forming process of the upsetting head of the rivet.

Finally, the riveting joint after forming is calculated as shown in figure 4.

Figure 4. The final forming result of the joint.

3. Calculation of the fatigue lifetime of the interference fit riveting joint

The fatigue failure of the riveting joint is mainly manifested by the tension failure of the sandwich plates. The fatigue failure of the sandwich plates occurred earlier than that of the rivet at the stress concentration position. Therefore, the fatigue lifetime of the sandwich plates is chosen to evaluate the fatigue strength of the riveting joint.

3.1. Analysis of residual stress at the edge of sandwich plates hole

The residual pressure stress is generated on the surface of the part and hole by the interference fit connection to reduce the mean stress of spectrum load at the local fatigue risk point, which has a significant gain on the fatigue lifetime of the structure. Therefore, the condition of residual stress at the edge of sandwich plates hole was analyzed firstly in this paper. According to figure 5, three node paths are established. The path direction extends from the edge of the hole to the edge of the plate width side.

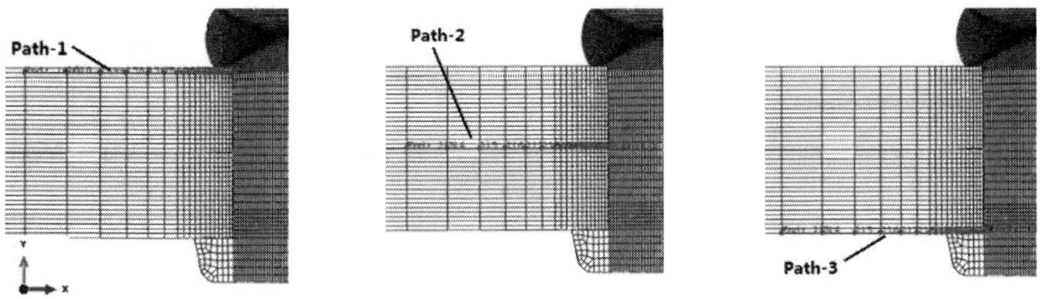

(a) (b) (c)

Figure 5. Schematic of the three node paths.

After the riveting die is unloaded and the rivet rod finishes the rebound process, the residual stresses in the x-direction of each node in the three paths are extracted. The path distance is taken as the abscissa and the residual stress is taken as the ordinate to make figure 6. It can be observed from figure 6 that the residual pressure stress is at the edge of the sandwich plates hole in the interference fit connection. It can reduce the average stress of the spectrum load when subjected to the tension-tension cyclic external load, which has a beneficial effect on improving fatigue lifetime of the riveting joint. When the distance from the edge of the hole is far (more than 6 mm), the residual stress of each path approaches to 0. Observe the path-1 curve, at the position of 1 mm from the edge of the hole, the residual stress on the side of the upsetting head becomes the tensile stress, and the generation of the residual tensile stress will have an adverse effect on the fatigue strength of the area. Based on the above analysis, the riveting interference fit has an obvious strengthening effect on the edge of the hole, but has little effect on the area far from the hole, and even may have an adverse effect on some areas.

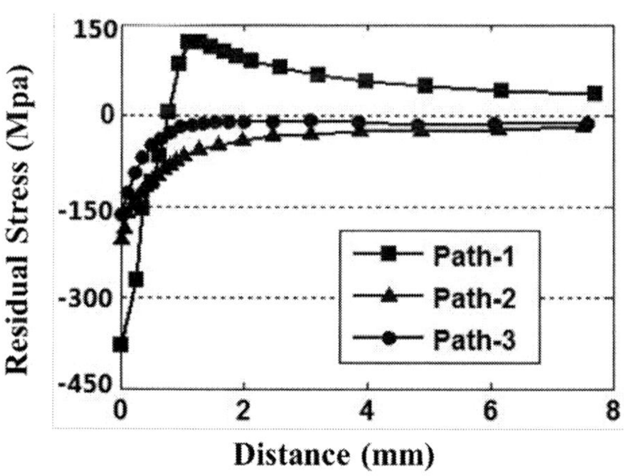

Figure 6. The residual stresses (x-direction) of each node.

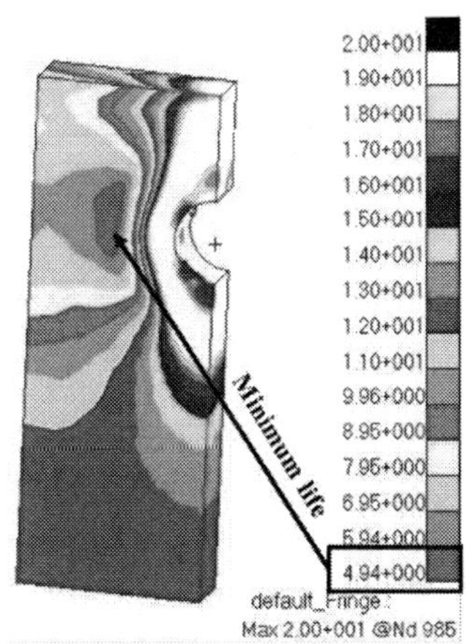

Figure 7. The minimum fatigue life of the sandwich plate.

3.2. Calculation of the fatigue lifetime of the sandwich plate

The stress results of the sandwich plate are substituted into the fatigue life analysis software MSC. Fatigue to calculate the fatigue lifetime of the sandwich plate. Take the tension-tension load with the stress ratio of 0.1, the stress level is 162 Mpa (about 60% of the shear strength of the joint), the length direction of the plate is chosen as the loading direction, and the load type is a sinusoidal wave. The life cloud-chart of the joint is calculated as shown in figure 7. The minimum life value is 86500 cycles (the logarithmic life is 4.94). The minimum life is not at the edge of the hole, but at a certain distance. According to the residual stress analysis results in Section 3.1, this is due to the effect of the interference fit, which produces the residual pressure stress at the edge of the hole and the residual tensile stress at the periphery of the hole.

4. Effect of the interference on fatigue lifetime of the riveting joint

4.1. Control of the interference of the riveting joint model

In order to study the influence of different interferences on the strength of the joint, it is first necessary to obtain electromagnetic riveting joint models with different interferences. The main factors affecting the interference of the joint include the clearance between the rivet and hole, the extension length of the rivet and the inclination angle of the riveting die. Compared with the clearance between the rivet and hole and the extension length, it is more convenient to control the inclination angle of the riveting die as a changing factor in the numerical simulation. Therefore, the models with different interferences are obtained by changing the inclination angle of the riveting die.

In order to obtain the electromagnetic riveting joint model with the interference in the range of 0.5%~4.5%, the numerical simulation of the riveting process was carried out by selecting the riveting dies with 180° (flat punch), 120°, 90°, 70°, 60° and 45° respectively. The calculation results of the interference respectively are 0.96%, 1.97%, 2.75%, 2.9%, 3.23%, 3.9%. It can be seen from the results that as the inclination angle of the riveting die decreases, the overall interference increases gradually.

4.2. Effect of the interference on the residual stress of the sandwich plate

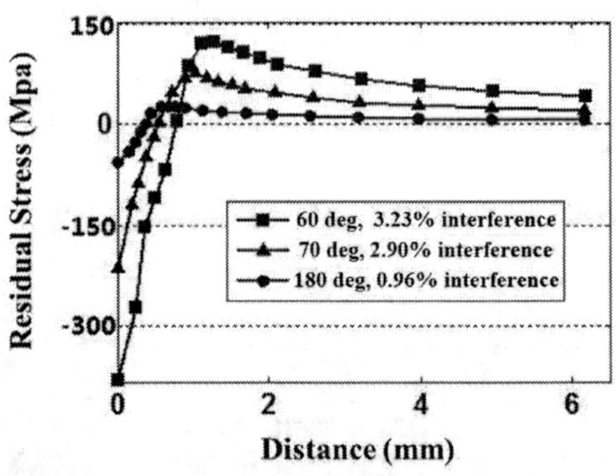

Figure 8. The Variation curve of Residual stress with different interferences.

According to the residual stress analysis method in Section 3.1, the X-direction residual stresses of the models with different interferences are analysed as shown in figure 8. It is observed from the figure that the residual pressure stress can be significantly increased with the increase of the interference, which is also the reason why the interference has a great influence on the fatigue lifetime of the structure. Comparing with the three curves, it is observed that as the interference increases, the effect range of the residual pressure stress increases, which has a positive effect on the improvement of the fatigue strength, but at the same time, it can be seen that the peak value of the residual tensile stress generated in the area far from the hole also increases as the interference increases, which will have an adverse effect on the life of the joint. Therefore, it can be inferred that for the interference fit riveting joint, it is not always better for larger amount of interference, but there is an optimal interference, when the interference is close to the optimal interference, the joint has the maximum fatigue life.

4.3. Determination of the ideal interference

The fatigue lifetime of the models with different interferences are analysed according to the life calculation method of the joint in Section 3.2. The fatigue lifetime of the aluminium alloy sandwich plate is calculated respectively when the inclination angle of the riveting die is 180°, 90°, 70°, 60°, and 45°. The results obtained are shown in table 3. In the table, the "Life increment" indicates the ratio of the fatigue life of riveting plates with different interferences to that of the riveting plate with 0.96% interference.

Table 3. Fatigue life of the sandwich plate with different interferences.

Inclination angle(deg)	Interference(%)	Fatigue life(cycles)	Life increment
180°	0.96%	2×10^4	1
90°	2.5%	4.17×10^5	2.085
70°	2.9%	7.13×10^5	3.565
60°	3.23%	8.65×10^4	4.325

From the data in table 3, the interference is taken as the abscissa and the fatigue life is taken as the ordinate, and the fatigue life curve is plotted as the interference changes, as shown in figure 9.

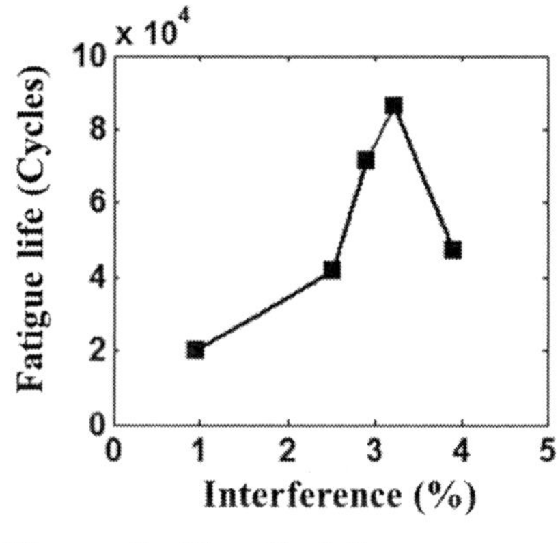

Figure 9. The Variation curve of fatigue life along with interferences.

It can be seen from figure 9 that as the interference increases, fatigue lifetime of the joint increases first and then decreases. When the interference is about 3%, the joint has a maximum fatigue life. Based on the analysis results of the residual stress, when the interference is less than 1%, the residual pressure stress generated by the interference fit is very small, and its maximum residual stress is only 15% of the maximum value when the interference is 3.23%, so the improvement effect of too small interference on the fatigue life is not obvious. When the interference reaches 3.9%, fatigue life has a significant downward trend. By analyzing the residual stress, when the interference increases, the effect range of the residual pressure stress will also increase, but at the same time, the peak value of the tensile stress generated in the area far from the hole will also increase. From figure 8, when the interference increases from 2.9% to 3.23%, the peak value of the residual tensile stress increases by 50%. Therefore, excessive interference will have an adverse effect on the fatigue strength of the structure. In summary, when the interference is between 2.9% and 3.23%, the joint can obtain a higher life increment.

4.4. Comparison with experimental research conclusions

The optimum interference of aviation aluminium alloy joints with the same thickness of 6mm was studied by experimental method, and the recommended optimum interference was 3.1%, which is basically consistent with the conclusion of 2.9%~3.23% ideal interference given in this paper [12].

5. Conclusions

In this paper, we studied the relationship between interference and fatigue life of riveting joints by means of numerical simulation method, which can provide a reference for the selection of the interference in engineering applications. The main conclusions are as follows:

The numerical simulation of the forming process of the electromagnetic riveting is carried out. The finite element model established by the Johnson-Cook material constitutive model can completely describe the whole forming process of the rivet, including the overall upsetting and local upsetting.

Based on MSC. Fatigue, the calculation of the fatigue lifetime of the interference fit riveting joint is carried out. When the interference is large, the position with the minimum fatigue life does not appear at the edge of the hole, but at a certain distance. This is due to the effect of the interference fit, which produces the residual pressure stress at the edge of the hole and the residual tensile stress at the periphery of the hole.

The effect of the interference on residual stress and fatigue lifetime of riveting joint was analyzed. As the amount of interference increases, the fatigue lifetime increases first and then decreases. For the structural parameters given in this paper, when the interference is between 2.9% and 3.23%, a higher life increment can be obtained.

References

[1] Langrand B, Deletombe E, Markiewicz E and Drazétic P 2001 Riveted joint modeling for numerical analysis of airframe crashworthiness *Fintet. Elem. Anal. Des* **38** 21-44

[2] Blanchot V and Daidie A 2006 Riveted assembly modeling: Study and numerical characterization of a riveting process *J. Mater. Process. Tech* **180** 13-5

[3] Repetto E A, Radovitzky R, Ortiz M, Lundquist R C and Sandstrom D R 1999 A finite element study of electromagnetic riveting *J. Manuf. Sci. Eng* **121** 61-8

[4] Chakherlou T N and Vogwell J 2010 A novel method of cold expansion which creates near-uniform compressive tangential residual stress around a fastener hole *Fatig. Fract. Eng. Mater. Struct* **27** 343-51

[5] Zeng C, Tian W and Liao W H 2016 The effect of residual stress due to interference fit on the fatigue behavior of a fastener hole with edge cracks *Eng. Fail. Anal* **66** 72-87

[6] Abdelal G F, Georgiou G, Cooper J, Robotham A, Levers A and Lunt P 2015 Numerical and experimental investigation of aircraft panel deformations during riveting process *J. Manuf. Sci. Eng* **137** 011009

[7] Li Y H, Cao Z Q, Zhang Q L and Feng D G 2013 Study on the effect of riveting quality with different angles of the forming cup of riveting die *Acta Aeronoutica et Astronautica Sinica* **34** 426-33 (in Chinese)

[8] Cao Z and Cardew-Hall M 2006 Interference-fit riveting technique in fiber composite laminates *Aerosp. Sci. Technol* **10** 327-30

[9] Dean A, Sahraee S, Reinoso J and Rolfes R 2016 Finite deformation model for short fiber reinforced composites: Application to hybrid metal-composite clinching joints *Compos. Struct* **151** 162-71

[10] Yoo S Y, Kim C H, Kweon, J H and Choi J H 2016 The structural analysis and strength evaluation of the rivet nut joint for composite repair *Compos. Struct* **136** 662-8

[11] Milani A S, Dabboussi W, Nemes J A and Abeyaratne R C 2009 An improved multi-objective identification of Johnson–Cook material parameters *Int. J. Impact. Eng* **36** 294-302

[12] Guo L Z and Xing W Z 2000 Experimental study on optimumi nterference-fit level of interference fit *New Technology & New Process* **1** 14-6 (in Chinese)

IOP Conf. Series: Materials Science and Engineering **657** (2019) 012070 doi:10.1088/1757-899X/657/1/012070

Numerical simulation of heavy truck's aerodynamic noise

J Liu[1,3], C Zhuang[1], S Liu[1] and H M Zhou[2]

[1]Jiangsu XCMG Construction Machinery Research Institute Ltd., 26 Tuolanshan Road, XUZ 221004, China
[2]Xuzhou XCMG Automobile Manufacturing Co. Ltd., NO. 19 Zhujiang East Road, High and New Technology Industry Development Zones, XUZ 221100, China

E-mail: 65863166@qq.com

Abstract. A 3D numerical simulation is built on heavy truck including cab, frame, and tires. And vehicle exterior flow field in steady state is computed by computational fluid dynamics software. In the computation process, the distribution of sound power levels on the surface of vehicle body as well as the connection between sound power level and vehicle speed are obtained from aerodynamic noise source, which is using broadband noise source models. Taking the steady flow field results as original values, the exterior flow field computations in transient state are performed by large eddy simulation. Based on the transient fluid field, the near field aerodynamic noise from cab accessories is predicted using FW-H model, and the comparison of the simulation results and the test results is carried out. The results show that the aerodynamic noise sources are from the upwind side with a great pressure gradient. Attentions should be paid more on rearview mirror, diversion trench and sun visor for Aerodynamic noise optimization of heavy truck. The maximum sound power level of aerodynamic noise in vehicle body accessories is quadratic to vehicle speed. And the aerodynamic noise of heavy trucks is analysed by numerical simulation, in which noise distribution law is consistent with the test. Therefore, the numerical simulation can be adopted to provide the guidance for the design of vehicle body with low noise.

1. Introduction

The noise level of heavy truck is one of important performance parameters. It causes environmental noise pollution and affects driving comfort. At present, it has been an important indicator for measuring the overall quality of vehicle. Vehicle noise is mainly composed of engine noise, transmission noise, tires noise, intake and exhaust noise, plate vibration noise, aerodynamic noise and so on. Significant achievements have been made in reducing engine noise, tire noise, intake and exhaust noise after years of effort by researchers. As speed-increasing and sealing property-improving, aerodynamic noise at high speed has become the crucial part of vehicle noise.

At present, in the terms of passenger vehicle and high-speed train, the aerodynamic noise has been studied widely by specialists and scholars, and remarkable achievements have been made in reducing aerodynamic noise. However, there are few studies on the aerodynamic noise of heavy trucks. The main reasons are that the aerodynamic noise from traditional heavy trucks is relatively less due to its low speed, and most noises are from the power transmission system because of the powerful engines. According the internal and international market situation, the higher the speed of heavy trucks, the greater the proportion of aerodynamic noise. Research shows that, the aerodynamic noise increases with the speed to the sixth power, which means for each double of speed, aerodynamic noise will increase by about 18 db. As speed is below 60 km/h, it can obviously affect internal vehicle. In 1972, as Boeheim in Germany tested aerodynamic noise on 15 kinds of cars, he found that when vehicle speed reached over 113 km/h, the inside noise generated by aerodynamic noise from exterior flow field was 62-78db (A), and when the speed arrived 177 km/h, the inside noise was up to 72-87db (A). Meanwhile, for the flourish development of new energy vehicles in recent years, its market share in the future will take higher. As early as 1999, as Norman et al from Ford Motor Company in the United States proposed, although the noise of electric vehicles was less than that of traditional internal-combustion engine vehicles, there were more component noises and wind noises in the absence of engine masking effect. Therefore, the study of aerodynamic noise on heavy trucks is significant.

Tests and simulation are the main methods of aerodynamic noise research. The tests include real vehicle test and acoustic wind tunnel test. Common simulations include numerical computation method that is used to study the noises generated by fluid and solid boundary interaction based on computational aeroacoustics (CAA). In 1952, Lighthill put forward the Lighthill equation describing fluid motion and sounds based on Navier-Stokes equations (N-S equations for short) that describes fluid flowing. In 1955, based on the theory of Lighthill, Curle noticed the influence on a stationary solid surface. In 1969, Ffowc Williams-Hawkings deepened the theory of Curle to the noise problem caused by the interaction between the moving wall and the fluid, and then proposed the Ffowc Williams-Hawkings equation (FW-H equation for short).

2. Theory basis

2.1. Proudman equation

There is no obvious frequency band for turbulence in many engineering applications, and acoustic energy is distributed continuously in frequency over a wide band, which is called broadband noise problem. In this problem, Proudman equation is widely used. On the basis of acoustic analogy theory of Lighthill, Proudman ignored the differential of delay time and replaced it to equivalent synchronous covariance, in order to derive a radiated sound power expression of isotropic turbulence per unit volume at lower Mach number and higher Reynolds number. After that, Lilly dealt with the delay time differential and derived a new expression of radiated sound power. Both of these expressions can calculate the radiative sound power P_A (in unit W/m^3) of isotropic turbulence per unit volume.

$$P_A = \alpha\rho_0\left(\frac{u^3}{l}\right)\frac{u^5}{a_0^5} \tag{1}$$

Type (1) can be written in forms of k and ε.

$$P_A = \alpha_\varepsilon\rho_0\varepsilon M_t^5 \tag{2}$$

Sound power level is defined as

$$L_p = 10\lg(P_A/P_r) \tag{3}$$

Type: u is turbulence speed, l is turbulence scale, a_0 is sound speed, α is model constant, $M_t = \sqrt{2k}/a_0$, $P_r = 10^{-12}W/m^3$ is reference acoustic power.

In the type, α_ε takes $\alpha_\varepsilon = 0.1$ based on isotropic turbulence computation of Sarker and Hussaini. In the calculation, exterior vehicle steady flow field is computed by standard k-ε turbulence model first to obtain the turbulent kinetic energy and turbulent dissipation rate of each node in the flow field. Then sound power level distribution is got after calculating the noise power at a node of the volume element by Proundman equation.

2.2. Fowcs Williams-Hawkings equation

The noise caused by the running vehicle and air interaction accords with FW-H equation:

IOP Conf. Series: Materials Science and Engineering **657** (2019) 012070 doi:10.1088/1757-899X/657/1/012070

$$\frac{\partial^2 \rho}{\partial t^2} - c_0^2 \frac{\partial^2 \rho}{\partial x_i^2} = \frac{\partial^2 T_{ij}}{\partial x_i \partial x_j} - \frac{\partial}{\partial x_i}\left[(p_i \delta_{ij} - \tau_{ij})\frac{\partial f}{\partial x_j}\delta(f)\right] + \frac{\partial}{\partial t}\left[\rho_0 u_i \frac{\partial f}{\partial x_i}\delta(f)\right] \quad (4)$$

Including: $T_{ij} = (P_{ij} + \rho u_i u_j) - (c_0^2 \rho + p_0)\delta_{ij} = \rho u_i u_j - \tau_{ij} + (p - c_0^2 \rho)\delta_{ij}$, u_i means the fluid velocity component in x_i direction, $\delta(f)$ for Dirac-delta function, δ_{ij} for Kronecker delta.

On right side of equation (4), the first one is Lighthill sound source, which is quadrupole sound source, because T_{ij} is Lighthill stress tensor that includes velocity variables. The second one is dipole source caused by pressure pulsation. Third one is monopole source caused by surface acceleration.

3. Numerical model

3D geometrical model is built on a heavy truck as research object including cab, frame, axle, tires, etc. As various shapes and sizes of containers can be hung behind the head of motor tractor, only the traction head and chassis are analyzed in this model. The geometric model of vehicle body was divided into triangular grid. For the complex model, using triangular grid can improve mesh quality as well as retain the original geometric features. To perfectly seize aerodynamic noise of vehicle body, there are many detail features retaining including cab doorknob, rearview mirror and so on, which conduct mesh encryption. The final body gird model is shown as figure 1.

Figure 1. Surface mesh model.

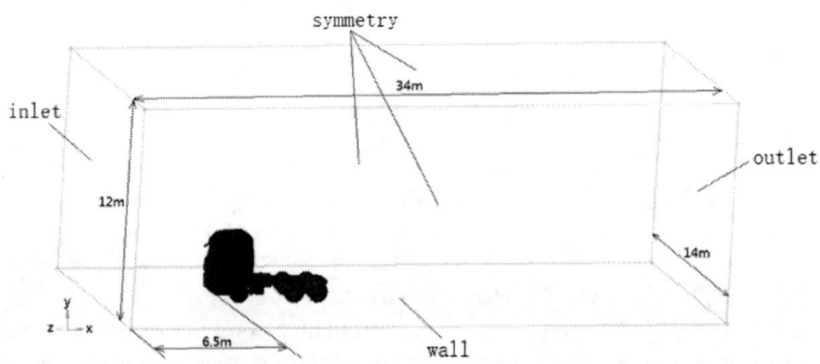

Figure 2. Computational domain and boundary conditions.

For the vehicle is running on the plat ground, the length, width and height of the computational domain are respectively 34 m, 14 m and 12 m according the vehicle body size. The vehicle head is 6.5 m away from the inlet. The schematic of computational domain and boundary conditions is shown as figure 2. Meanwhile the type of inlet is velocity-inlet. The velocity of inlet is equal to vehicle speed. The outlet is taken as pressure outlet, the ground as wall boundary, the top surface and two sides as symmetry boundary, and the surface of vehicle body as impenetrable solid wall. Finally, the unstructured tetrahedral grid of global computing domain is generated in accordance with the body surface and each boundary surfaces. The minimum grid size is 2 mm, the maximum grid is 180 mm, and the total grid number is more than ten millions.

4. Surface noise source analysis

4.1. Surface sound power level distribution

It can rapidly assess the noise intensity caused by moving surfaces in fluid by using broadband noise sources model. All source terms only need the typical RANS equation solution, for transient flow solution is unnecessary. Therefore, there is less calculation in broadband noise sources model than that in FW-H equation. Surface sound power level can be obtained using broadband noise sources model. Based on the k-ε two equation turbulence model, the steady exterior flow field of a vehicle at a constant speed on the plat ground is calculated, and the sound power level distribution of vehicle body surface is obtained by using broadband noise model.

When the vehicle speed is 100 km/h, the static pressure, speed and sound power level distribution of body surface is obtained by computation, shown as figure 3. The results show that the sound power level is higher in the location where surface speed and pressure gradient are greater. Regardless of containers, higher sound power level is mainly from the upwind side of rearview mirror, sun shield, diversion trench, bumper, tire and other parts. Considering the effect of aerodynamic noise on driver's ears, the aerodynamic noises caused by bumper, tire and other places far from door and glass are difficult to be heard. The rearview mirror, sun shield and diversion trench are closer to door and glass gap, so that the noises may easily come into the vehicle from these gaps due to poor sealing. As a result, it influences the parotic noises in the vehicle.

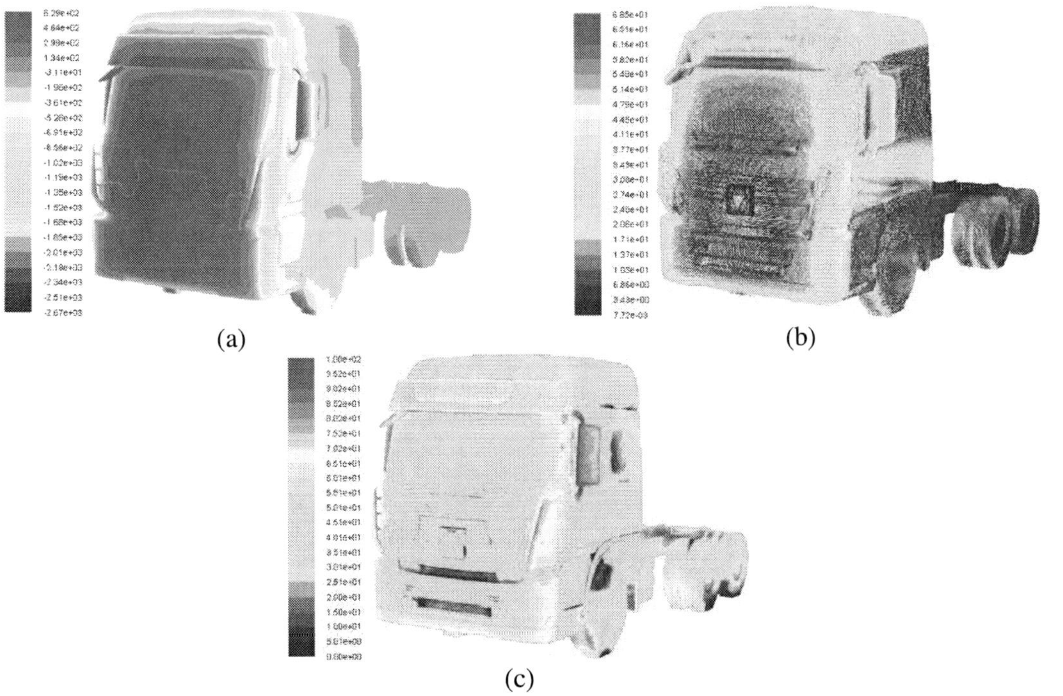

Figure 3. Flow field parameter distributions. (a) Static pressure distribution, (b) Velocity distribution and (c) Sound power level distribution.

IOP Conf. Series: Materials Science and Engineering **657** (2019) 012070 doi:10.1088/1757-899X/657/1/012070

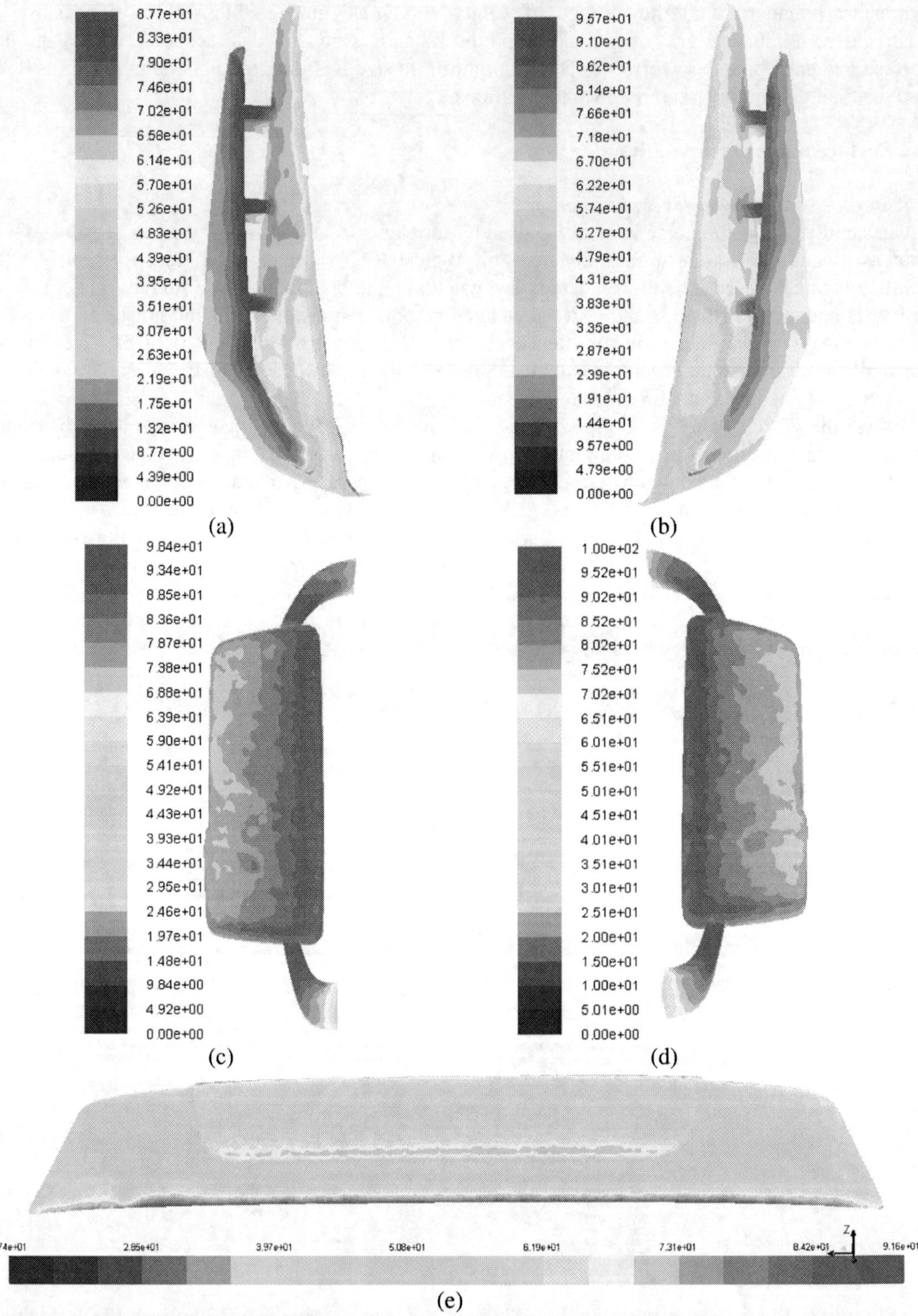

Figure 4. Sound power level distribution of main parts. (a) Sound power level distribution of right diversion trench, (b) Sound power level distribution of left diversion trench, (c) Sound power level distribution of right mirror, (d) Sound power level distribution of left mirror and (e) Power level distribution of sun shield.

The distribution of sound power level in rearview mirror, diversion trench and sun shield is shown as figure 4. It can be seen that higher sound power level of rearview mirrors is generated at the root of connecting rod, close to the mirror body and edge of door. Higher sound power level of sun shield is generated at the lower edge. Higher sound power level of left and right diversion trenches appears at leading edge of the out plates and fins. All of them are the positions where high static pressure transforms into low static pressure, and low wind speed changes into high wind speed on the upwind side, that is, the location where there is greater static pressure and speed gradient places. Maximum sound power level of each accessory is shown in table 1. Ultimately, the biggest noise comes from rearview mirrors, and the secondary noise is from diversion trenches and sun shield is followed. The asymmetry of left and right accessories as well as errors in simulation result in inconsistent distribution of sound power level at rearview mirrors and diversion trenches.

Table 1. Maximum sound power level of main positions.

Position	Maximum Sound Power Level
Left Mirror	100.23 dB
Right Mirror	98.36 dB
Left Diversion Trench	95.75 dB
Right Diversion Trench	87.74 dB
Sun Shield	91.6 dB

4.2. Surface acoustic power at different speeds

When a vehicle is traveling at constant speeds, the maximum sound power level curve of major components is as shown in figure 5. From that, the sound power level of each components sequence in same. The highest is left mirror, following right mirror, left diversion trench, sun shield, right diversion trench. The maximum sound power level of each parts increases with speed improving, and is quadratic to speed.

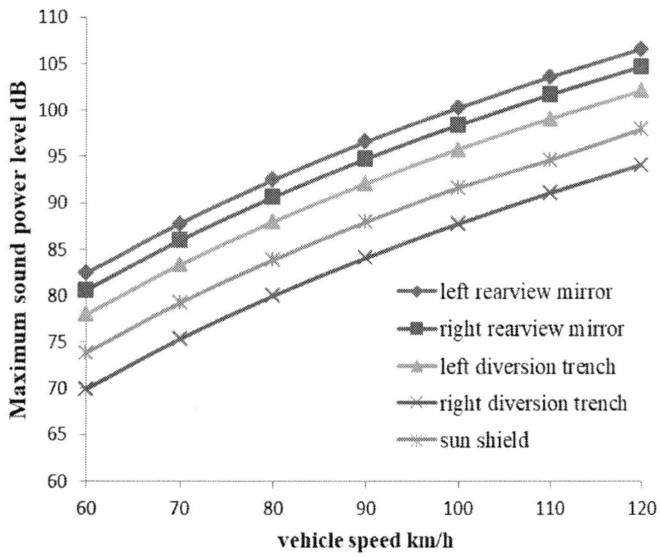

Figure 5. Maximum sound power level of each position at various vehicle speeds.

5. Near-field aerodynamic noise analysis

Taking the steady flow field results as original values, the transient flow field computations are performed by large eddy simulation (LES), and FW-H equation is solved at the same time. All vehicle surfaces are taken as aerodynamic noise sources, and noise receivers are arranged near key positions such as rearview mirrors and sun shield, shown as figure 6. Set time step to 0.00033 s, and sound pressure result of each receiver with the highest frequency of 1500 Hz are computed. To verify the accuracy of simulation results, it is tested on a real vehicle with the same noise receivers, shown as figure 7.

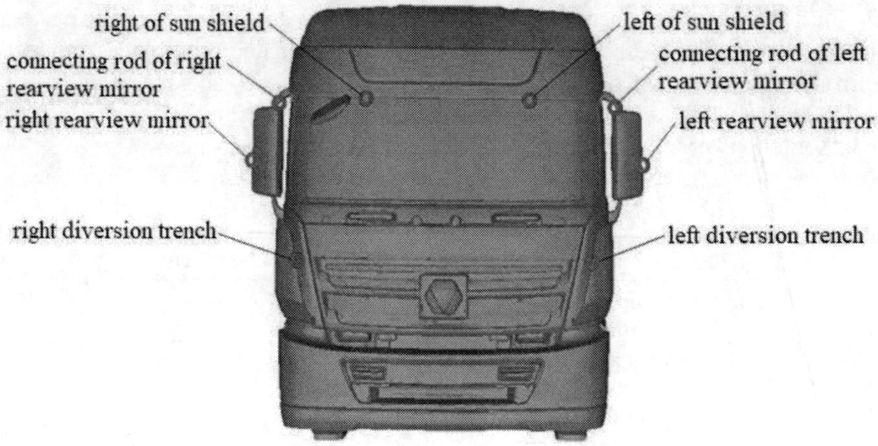

Figure 6. Sound receivers schematic.

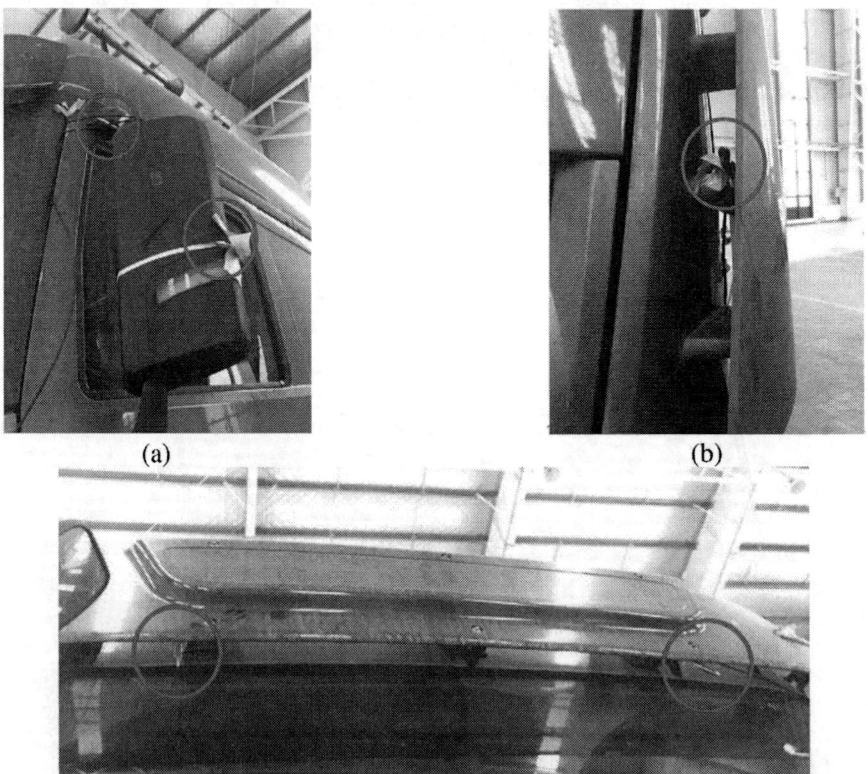

Figure 7. Microphones layout. (a) Microphones on rearview mirror, (b) Microphones on diversion trench and (c) Microphones on sun shield.

When the vehicle speed is 100 km/h, the simulation and test results comparison for A-weighted sound pressure level is shown as figure 8. It can be seen that test sound pressure level of each receiver is about 20 dB (A), larger than that in simulation. On the one hand, there is the simulation error. On the other hand, there is only aerodynamic sound source in the simulation. While the vehicle is traveling in test, other sound sources, such as power transmission noise and tires noise, contribute more to measurement points, which results in the test results are generally larger than the simulation results. However, the near field sound pressure level of rearview mirrors, diversion trenches and sun shield are in same sequence, which is basically consistent with the sequencing results in steady state simulation. The maximum noise is from connecting rod of left rearview mirror. Therefore, the simulation model can reflect the aerodynamic noise distribution of vehicles, which also can be adopted for qualitative analysis and research of aerodynamic noise, so as to provide reference and guidance for vehicle body design with lower noise.

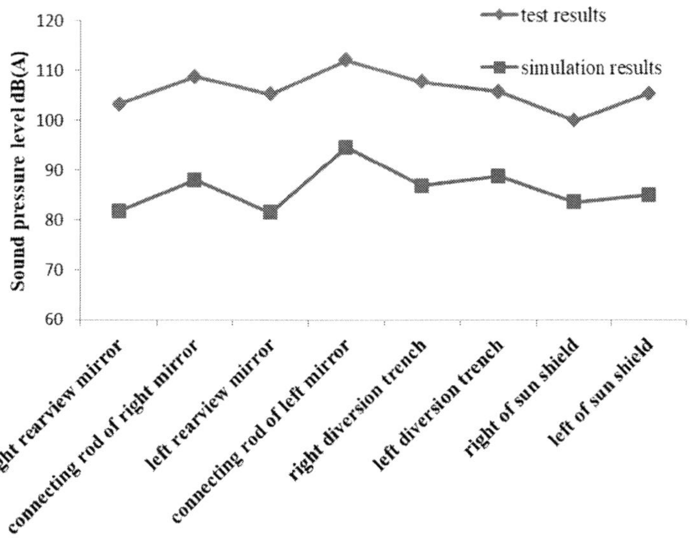

Figure 8. Comparison of sound pressure level between test and simulation.

6. Conclusion

- As vehicle is running, the greater aerodynamic noise is from the big pressure gradient at upwind side, that is, the positions where high static pressure changes to low static pressure, and low wind speed changes to high wind speed at the upwind side.
- The greater aerodynamic noises for heavy truck, which has a huge effect on inside vehicle, are mainly the positons of rearview mirrors, diversion trenches and sun shield, needing wind noise optimization. Meanwhile, the rearview mirror generates the largest aerodynamic noise, so it is necessary to focus on the connecting place between the rearview mirror rod and mirror body, as well as its upwind edge.
- The maximum sound power level of each parts increases with speed improving, and is quadratic to speed.
- Based on the analysis of aerodynamic noise of heavy truck with numerical computation method, the distribution rules consistent with tests can be obtained. Therefore, the simulation results can help the optimization of vehicle body accessories and provide the guidance of body design for low noise.

Acknowledgement
This work has been completed under the auspices of the Six Talent Peaks Project in Jiangsu Province (Grant No. JXQC-043).

References

[1] Yang B 2008 Research on automobile exterior aerodynamics noise (ChangChun: JinLin University)
[2] Yu L, Huang H, Shao G T and Lin X Q 2016 Aerodynamic optimization analysis of the heavy duty truck based on PowerFLOW *Auto Sci-Tech* **251** 38-45
[3] Hammache M, Michaelian M and Browand F 2001 Aerodynamic forces on truck models including two trucks in tandem *California PATH Research Report*
[4] Chilbule C, Upadhyay A and Mukkamala Y 2014 Analyzing the profile modification of truck-trailer to prune the aerodynamic drag and its repercussion on fuel consumption *12th Global Congress on Manufacturing and Management (Vellore)* pp 1208-19
[5] Chowdhury H, Loganathan B and Mustary I 2016 Effect of various deflectors on drag reduction for trucks *1st Int Conf. Energy and Power (Melbourne)* pp 561-6
[6] Wang Y G, Yang C, Yang Z G and Li Q L 2014 Analysis of aerodynamics noise of vehicle's external surface *Tech. Acoust.* **33** 50-5
[7] Ma D Y 2004 *Modern Acoustics Theory* (Beijing: Science Press) pp 293-8
[8] Zhang J, Sun B C, Guo T, Fang J and Zhao W Z 2015 Research on aerodynamic noise radiated from whole body surface of high-speed train and its distribution *J. Chin. Rail. Soc.* **37** 10-7
[9] Luo Z M, Gu Z Q, Zong Y Q, Liu L G and Jiang C M 2016 Wind buffeting noise analysis and control for vehicle under crosswind *Acta Aerodyn Sin.* **34** 468-75
[10] Huang S, Liang X F and Yang M Z 2012 Numerical Simulation of aerodynamic noise and noise reduction of high-speed train connection section *Acta Aerodyn Sin.* **30** 254-9
[11] Zheng Z Y 2012 A study on the numerical simulation of high-speed vehicle's external aerodynamic acoustic field (ChengDu: Southwest Jiaotong University)
[12] Proudman I 1952 The generation of noise by isotropic turbulence *Proc. R. Soc. London, Ser. A , Mathematical and Physical Sciences* **214** 119-32
[13] Buchheim R, Marketzke J and Piatek R 1985 The control of aerodynamic parameters influence vehicle dynamics *SAE Technical Paper Series* 850279
[14] Sarkar S and Hussaini M Y 1993 Computation of the sound generated by isotropic turbulence (Washington, D.C.: NASAt Report) pp 74-93

IOP Conf. Series: Materials Science and Engineering **657** (2019) 012071 doi:10.1088/1757-899X/657/1/012071

Numerical simulation of dendritic evolution based on an improved cellular automaton model involving solute field

J J Wang[1,2], H J Meng[1], J Yang[1] and Z Xie[1]

[1]School of Information Science and Engineering, Northeastern University, Shenyang 110819, China

E-mail: neuwangjing@163.com

Abstract. This paper studied the single crystal growth of Fe-0.64C alloy under a fixed cooling rate, proposed an improved cellular automaton model. The influence of solute field on dendrite growth is especially considered into the model. And a new solute distribution scheme based on Fick's first law and solute conservation was proposed from the microscopic scale. In addition, the present work improved the acquisition algorithm, which takes the growth length of the interface cell to the neighbouring cell as the determination condition of the capture. A solid fraction calculation method based on the solidification area of interface cell to total area is also proposed. The results show that this model can not only dynamically display the formation process of secondary dendrites and high-order dendrites, but also simulate the distribution of solute inside the dendrites and at the interface front.

1. Introduction

The morphology of the dendrites and the microsegregation formed during the redistribution of the solute play a decisive role in the mechanical properties of the metal. In the traditional process, the trial and error method is used to study the growth of dendrites. Although this method is intuitive, it has blindness, and the workload is large, which wastes manpower and material resources. Otherwise, using computer technology to numerically simulate the dendrite evolution process can avoid a large number of experiments and reduce the research cost. Therefore, since the 1980s, studying the dendritic growth mechanism and using numerical simulation to reproduce the dendrite evolution process has been a hot topic in the field of materials science. The existing research methods are divided into three categories: deterministic methods, phase field methods, and random methods [1-3]. The deterministic model does not conform to the randomness of the nucleation and growth process of the actual solidification process; the phase field method can only simulate small area solidification due to complicated calculation; the MC method in the random method does not consider the details of macroscopic and microscopic transmission, so it lacks a certain physical background; the CA method not only combines the advantages of PF model and MC model, but also has a solid physical basis, which can reflect the influence of solute concentration and undercooling on the solidification process.

Rappaze *et al* [4,5] first used the CA method in the field of materials science and established a two-dimensional model of dendrites growth from a mesoscopic scale, then optimized the model to establish a three-dimensional model [6]. Based on this, Nastac [7] proposed the Neumann rule, established a fully coupled model of two-dimensional dendrite growth, and randomized the dendrite nucleation process. Zhu *et al* [8] added the influence of solute redistribution and curvature on the basis of the traditional CA model, improved the phenomenon that the dendrite arm growth deviated from the preferred growth direction, and applied this model to the evolution process of three-dimensional dendrites [9,10]. Zhu and Stefanescu [11] proposed a new calculation method of dendrite growth kinetics based on the interface solute balance principle, which obtained secondary and tertiary branches of dendrites. Yin and Wang *et al* [12,13] considered the influence of flow field on dendrite growth, constructed a lattice Boltzmann (LB-CM) model and a cellular automata lattice Boltzmann (LB-CA), and simulated the phenomenon of forward deflection of dendrites. Artemev *et al* [14] proposed an improved method for the solute partitioning process. Nakagawa, Chen *et al* [15-17] used a modified eccentricity algorithm to attenuate the anisotropy brought by the mesh, and simulated the dendritic morphology under different preferred orientations. Krane, Shi *et al* [18,19] also proposed improvements to attenuate the anisotropy. Luo *et al* [20] simulated the dendritic growth of Fe-C alloy under convection. In the past two decades, the research on the numerical model of dendrite growth has made great progress, and the accuracy of the model is getting higher and higher, but the mesh anisotropy existing in the modeling process cannot be completely eliminated, and the solute redistribution model during dendrite growth process also lacks systematic and complete analysis.

The actual casting process is very complicated, the liquid metal is cooled under different boundary conditions, then the crystal is driven to nucleate and grow after the melt receiving undercooling. In order to simplify the model and provide the dendrites growth environment, the following assumptions are made on the temperature of dendrites growth environment: the initial undercooling is 1°C, and the cooling rate is 20°C/s. Based on the above assumptions, this paper proposed a new solute distribution scheme, improved the acquisition algorithm and solid fraction calculation method of the cellular automaton model, finally simulated the evolution of dendrites during solidification of a Fe-0.64C alloy with a size of 4 mm×4 mm.

2. Improved cellular automaton model

The present work used an improved cellular automaton model to simulate the evolution of dendrites with temperature field and solute field. In order to simulate secondary and high-order dendrites, the calculation area is divided into 400×400 cells with a size of 10 μm; each cell is given characteristic parameters such as temperature, concentration, preferred orientation and state. Since the dendritic morphology has a four-fold symmetry, the preferred orientation $\theta \in [-45°, 45°]$, and each cell will undergo a transition of liquid ($f_s = 0$), interface ($0 < f_s < 1$), solid ($f_s = 1$). All cells are initialized to a liquid state, and a nucleation grain with a preferred orientation $0°$ is placed at the central of the calculation region. The process of grain growth in the undercooling melt is accompanied by the discharge of solute, the solute is enriched at the front edge of the solid-liquid interface, which lowers the melting point temperature and reduces the undercooling of the melt, eventually leading to a decrease in the growth rate of dendrites. The cooling rate in the dendrite growth process was assumed to be a fixed value. The new solute distribution model is based on Fick's first law and mass conservation, which makes a comprehensive analysis of the excess solute emissions at the interface front. In addition, the present model proposed new calculation methods for the capture and solid fraction, see details in Section 3. The mechanism of dendritic growth and solute transport in the model are introduced as follows.

2.1. Growth Kinetics equation

In the actual casting process, the undercooling is affected by the diffusion of solute, heat and curvature of the interface. The total undercooling of the interface is a combination of various undercooling:

$$\Delta T = \Delta T_K - \Delta T_r - \Delta T_c + \Delta T_T + \Delta T_P \tag{1}$$

ΔT_K, ΔT_r, ΔT_T, ΔT_c and ΔT_p are kinetic, curvature, solute, thermal and pressure undercooling, respectively. During the growth of dendrites, the kinetic undercooling is generally in the order of 0.01 K~0.05 K, and the effect of pressure undercooling on the equilibrium temperature is also very small.

Therefore, in order to simplify the model, this paper ignores the two kinds of undercooling and established kinetics model:

$$\Delta T = \Delta T_T - \Delta T_r - \Delta T_c \tag{2}$$

The KGT [21] model was used to establish the relationship between the growth speed of the dendritic tip and the undercooling:

$$v = \alpha \Delta T^2 + \beta \Delta T^3 \tag{3}$$

Growth coefficients α and β can be obtained by experimental regression.

2.2. Solute distribution model

It can be seen from the equation (2) that the solute undercooling is an important component of the total undercooling of the melt and it has a great influence on the growth of the dendrites. The traditional CA model ignores the solute field, which inevitably affects the morphology of the dendrites. Therefore, this model specifically incorporates the influence of the solute field. Since the diffusion coefficient of the solid is 3 to 4 orders of magnitude smaller than that of the liquid, the solid solute diffusion is ignored, and only the diffusion of the solute in the liquid and the interface cell is considered. Since the diffusion coefficient of the solid phase is 3 to 4 orders of magnitude smaller than that of the liquid phase, the solid phase solute diffusion is ignored, and only the diffusion of the solute in the liquid phase and the interface cell is considered.

The solute diffusion model is based on Fick's first law, as in equation (4):

$$J = \frac{dm}{Adt} = -D_i \left(\frac{\partial C}{\partial x} \right) \tag{4}$$

J stands for solute diffusion flux; D_i indicates solute diffusion coefficient, i is cell state with two values: L and I (liquid and interface). D_i can be calculated according to equation (5):

$$D_i = f_s D_s + f_l D_l \tag{5}$$

f_s、f_l represents solid fraction and liquid fraction; D_s、D_l indicates the solute diffusion coefficient of solid and liquid, and it can be obtained according to the empirical formula, which take values 1×10^{-9} and 1×10^{-7} respectively in this paper.

Equation (6) is the solute transport equation established based on mass balance:

$$\frac{\partial C}{\partial t} \Delta x \Delta y \Delta t = J_{i-1,j} - J_{i+1,j} + J_{i,j-1} - J_{i,j+1} \tag{6}$$

Interface solute redistribution follows local conservation:

$$C_s^* = C_l^* \tag{7}$$

The equilibrium solute concentration in the interface can be calculated by equation (8) [22]:

$$C_l^* = C_0 + \frac{T^* - T_l^{eq}}{m_l} + \frac{\Gamma K}{m_l} \tag{8}$$

C_0, T^*, T_l^{eq}, K, Γ and m_l indicates initial solute concentration, interface temperature, initial liquidus temperature coefficient, interface curvature and liquidus slope, respectively.

3. Numerical methods

3.1. Solution of the solute field

When the solid fraction of the cell increases by Δf_s, the amount of excess solute can be obtained by equation (9), which is derived from equation (7).

$$dC = (1-k)C_l^t \Delta f_s \tag{9}$$

If the remaining liquid of the current cell can completely receive the excess solute, the excess solute will be completely discharged to the remaining liquid, and the compositions of liquid and solid are updated by the equations (10) and (11), where N represents the number of steps iterated from the start of solidification.

$$C_l^{t+\Delta t} = \frac{C_l^t\left(1-f_s^t\right) - kC_l^t \Delta f_s^t}{1-f_s^t} \tag{10}$$

$$C_s^{t+\Delta t} = \frac{\sum_{n=1}^{N} k\Delta f_s(n) C_l(n)}{f_s^t} \tag{11}$$

Otherwise: when $C_l^{t+\Delta t} > C_l^*$, let $C_l^{t+\Delta t} = C_l^*$ and $dC' = (C_l^* - C_l^{t+\Delta t})\Delta f_s$; if $f_s^{t+\Delta t} = 1$, then $C_l^{t+\Delta t} = 0$ and $dC' = dC$.

While $dC' > 0$, if there are no liquid neighbours, dC' will be enriched at the grain boundaries to form grain boundary segregation; otherwise, they will be reasonably distributed into the liquid neighbours. As shown in figure 1, in the case of the solute distribution is isotropic, dC' will be evenly distributed in the circle, resulting in a difference in the amount of solute between the nearest neighbour and the next nearest neighbour ,which can be represented by $\Delta C^n : \Delta C^{nn} = S^n : S^{nn} = 1:0.52$. S^n and S^{nn} are the area of the nearest neighbour and the next nearest neighbour in the circle.

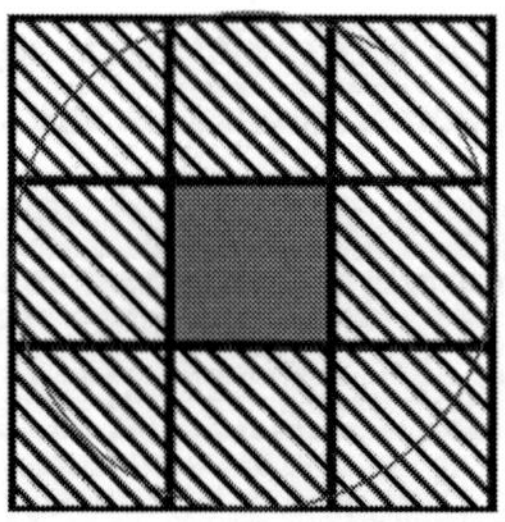

Figure 1. Solute distribution diagram (interface cell, liquid cell).

Solute diffusion in liquid and interface is solved by cellular automaton method.

3.2. Growth and capture algorithms

As shown in figure 2, the shape of the grain is represented by an eccentric square, θ indicates its preferred orientation, the half-diagonal length is calculated by the equation (12):

$$L(t) = \int_{t_0}^{t} v\left(\Delta T(\tau)\right) d\tau \tag{12}$$

v is the growth speed of the dendrite tip.

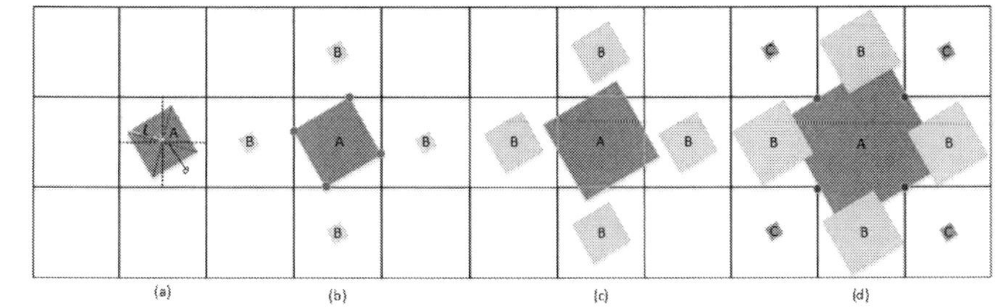

Figure 2. Dendrite evolution process in present model.

When the four vertices of the eccentric square contact the nearest neighbours B_s as shown in figure 2(b), B_s can be captured as the interface cell, so as C_s in figure 2(d).

All the captured cells inherit the growth parameters of A. When all neighbours are captured, cell A is completely solidified. Solid fraction of the interface cell at time *t* can be calculated by equation (13):

$$f_s(t) = S(\mathrm{t})/A \tag{13}$$

A is the total area of the cell; $S(\mathrm{t})$ represents the solidified area of the cell.

4. Results and discussion

Figure 3 displays the evolution of dendrites simulated by the present model, which exhibits a clear dendritic morphology as well as secondary and high-order dendrites. (a), (b) and (c) correspond to the morphology of dendrites at time $t_a = 0.2s$, $t_b = 0.6s$ and $t_c = 0.66s$, respectively.

Figure 3. Dendrite evolution process in present model.

The result of the model without considering the solute field is shown in figure 4. It can only simulate the outline of dendrite growth, and cannot fine dendritic structure. (a), (b), and (c) correspond to the morphology of dendrites at time $t_a = 0.1s$, $t_b = 0.3s$ and $t_c = 0.34s$, respectively.

It can be seen from the figure 3(a) that the nucleated crystal first grows along the primary dendrite arm. As the solidification continues, the dendrites continuously extend into the melt, and the solute is enriched at the interface front; the interface of the dendritic arm loses equilibrium under the action of solute diffusion and surface tension, then the secondary dendrite arm shown in figure 3(b) is generated. Solute will also be enriched between the secondary dendrite arms, breaking the balance of the secondary dendrite arm interface, forming tertiary dendritic arms. And there will be more high-order dendritic arms before the end of solidification, eventually forming a dendritic crystal. If the influence of the solute field is neglected, as shown in the figure 4, the dendritic interface always advances steadily and maintains the state of the flat interface, so no branching is formed and the actual dendritic morphology cannot be simulated.

(a) (b) (c)

Figure 4. Dendrite evolution process in previous model.

In addition, the model can also simulate the distribution of solute in the interior of the dendrite and the front of the interface. Figure 5 shows the distribution of solute corresponding to the morphology of the dendrites in figure 3. It can be seen that the solute is enriched at the interface and it is not uniform inside the dendrite.

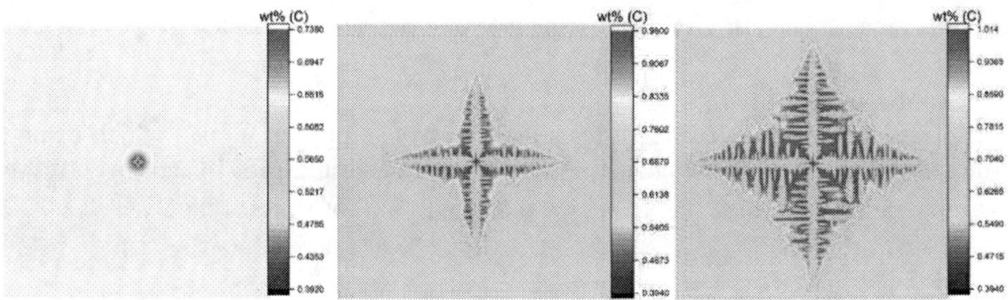

(a) (b) (c)

Figure 5. Solute distribution near dendrites.

The distribution of solute along the direction of the primary dendrites is shown in figure 6.

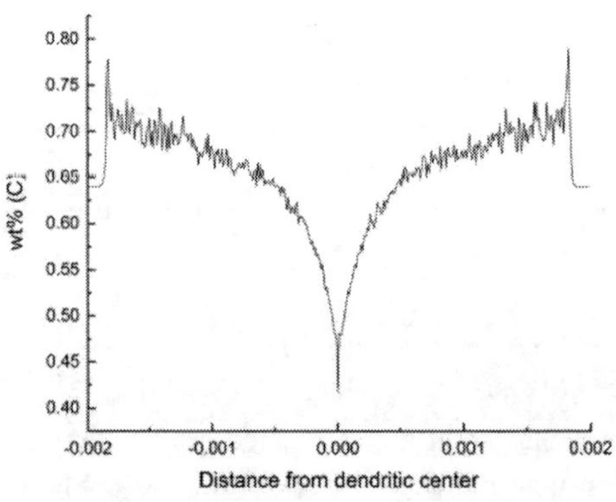

Figure 6. Solute distribution of primary dendrite arms.

The amplitude of the fluctuation between the two peaks of the curve represents the degree of segregation of the solute in the dendrites. The peaks on both sides represent the enrichment of the solute at the interface front. From figure 6, the distribution of the solute inside and at the front of the dendrite can be seen more clearly.

5. Conclusion

The present work improved the cellular automaton model and proposed a new solute distribution scheme, eventually completed the dynamic display of the dendrite evolution. Compared with the traditional model:

- The model can simulate the secondary and higher dendritic arms of dendrites;
- It is able to simulate the distribution of solute inside the dendrite and at the front of the interface.

Acknowledgments

The authors thank the financial support of National Natural Science Foundation of China (61703084).

References

[1] Stefanescu D M 1999 *Acta Mater.* **47** 4253-62
[2] Ludwig A 2002 *Int. J. Thermophys* **23** 1131-46
[3] Mat M D and Liegbusi O J 2002 *Int. J. Heat Mass Transf.* **45** 279-89
[4] Rappaz M and Gandin C 1993 *Acta Mater.* **41** 345-60
[5] Gandin C A and Rappan M 1994 *Acta Mater.* **42** 2233-46
[6] Gandin C A, Desbiolles J L, Rappaz M and Thevoz P 1999 *Metall. Mater. Trans.* A. **30** 3153-65
[7] Zhu M F, Kim J M and Hong C P 2001 *ISIJ Int.* **41** 992-8
[8] Beltran-Sanchez L and Stefanescu D M 2003 *Metall. Mater. Trans.* A. **35** 2471-85
[9] Zhu M F and Hong C P 2004 *Metall. Mater. Trans.* A. **35** 1555-63
[10] Zhu M F, Lee S Y and Hong C P 2004 *Phys. Rev.* E. **69** 061610
[11] Zhu M F and Stefanescu D 2007 *Acta Mater.* **55** 1741-55
[12] Yin H and Felicelli S D 2010 *Acta Materialia* **58** 1455-65
[13] Yin H, Felicelli S D and Wang L 2011 *Acta Mater.* **59** 3124-36
[14] Artemev A and Goldak J A 1997 *Can. Metall. Q.* **36** 57-64
[15] Nakagawa M, Natsume Y and Ohsasa K 2006 *ISIJ Int.* **46** 909-13
[16] Chen R, Xu Q Y and Liu B C 2014 *J. Mater. Sci. Technol.* **30** 1311-20
[17] Wang W, Lee P D and Mclean M 2003 *Acta Mater.* **51** 2971-87
[18] Shi Y F, Zhang Y, Xu Q Y *et al* 2012 *IOP Conf. Ser.-Mat. Sci.* **33** 012112
[19] Krane M J M, Johnson D R and Raghavan S 2009 *Appl. Math. Model.* **33** 2234-47
[20] Luo S, Wang W L and Zhu M Y 2018 *Int. J. Heat Mass Transf.* **116** 940-50
[21] Kurz W, Giovanola B and Trivedi R 1986 *Acta Metall.* **34** 823-30
[22] Luo S and Zhu M Y 2013 *Comput. Mater. Sci.* **71** 10-8

The rolling contact research of three dimensional wheel-rail based on finite element analysis

J H Tian[1,2], G Q Jing[1], X X Lu[1], K Xiao[1] and H R Zhang[1]

[1]CAE Analysis Room for Engineering Application, School of Mechatronic Engineering, Xi'an Technological University, Xi'an 710032 China

E-mail: 540673737@qq.com

Abstract. In this paper, wheel-rail contact model was established and analyzed, some useful results were obtained. The results show that In the early stage of wheel-rail contact, the wheel-rail contact pressure increases gradually and the contact point shape changes from regular polygon to irregular polygon in the process of wheel approaching rail joint. The equivalent stress and elastic deformation trend of the rail are basically the same, both of which decrease rapidly and then slowly increase. The plastic deformation of the rail gradually increases and accumulates, and the stress concentration area occurs at the side of the rail joint. When the wheel starts rolling, the equivalent stress of the wheel gradually decreases at first, and then increase. The peak value of the maximum equivalent stress occurs at the joint where the wheel reaches the rail.

1. Introduction

With the increase of railway load-carrying capacity and speed, the research on wheel-rail contact is becoming more and more important [1].

The research methods of wheel-rail contact mainly include Hertz's analytical method [2], Kalker's programs, and Fastsim and CONTACT [3]. Document [4] used the finite element method to analyze the influence of axle weight, wheel transverse motion, elastic modulus and poisson's ratio on the contact stress of wheel-rail. Document [5] compared the influence of different profile of wheel on the contact stress of wheel-rail under the static contact condition. Document [6-10] have used this method to obtain the optimized solutions for different wheel-rail interaction problems.

In this paper, three dimensional wheel-rail rolling contact model was established to analyze wheel-rail contact problems. The distribution of the wheel-rail contact pressure, equivalent stress, plastic strain and elastic strain during wheel rolling were obtained.

2. Establishment of the wheel-rail contact model

A commonly used 840 mm diameter elastic wheel and corresponding rail were created. The x-axis was the axial direction of the wheel, the y-axis was the vertical direction along the rail, and the z-axis was the running direction of the wheel. The slope of 1:20 was utilized, since it could be adjusted easily with sleepers' slope. The rail's length was 500 millimeter to simulate the initial stage of wheel-rail contact. It could be seen in figure 1. Friction coefficient 0.3 was used in this model.

The 3D geometric model was meshed with SOLID185, that was 8-noded hex elements as shown in figure 2. Where bi-linear kinematic Hardening elastic-plastic material was used for the wheel-rail rolling contact model , and the material properties were in table 1. It could give the stress-strain

IOP Conf. Series: Materials Science and Engineering **657** (2019) 012072 doi:10.1088/1757-899X/657/1/012072

relation of the material as shown in figure 3.

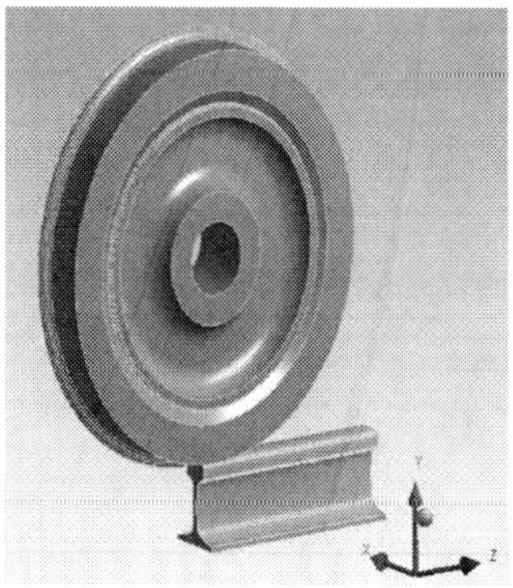

Figure 1. Geometric model.

Figure 2. The wheel-rail finite element contact model.

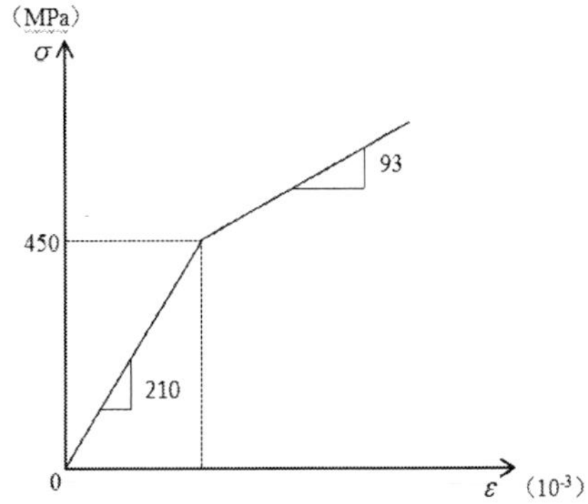

Figure 3. Stress-strain relation of the material.

Table 1. The properties of the materials.

Material	Modulus of elasticity (GPa)	Yield stress (MPa)	Modulus of plasticity (GPa)	Poisson's ratio
Steel	210	450	93	0.3

Two load steps were applied for realizing wheel rolling on rail. In the first load step, the translation degree of the x, z direction was fixed, the translation degree of the y direction was free and the load F 10.5 t was applied in the negative y direction at pilot node. In the second load step, the rotational degree of the x direction was free and translation displacement of the z direction was 400 mm for wheel. All freedom degrees at the bottom of the rail were constrained. The final wheel-rail rolling contact model can be seen from figure 2.

According to the basic Hertz contact theory, the theory of the normal contact pressure for three-dimensional problem can be expressed as:

$$P(x,y) = P_0\left(1-\frac{x^2}{a^2}-\frac{y^2}{b^2}\right)^{1/2} \quad (1)$$

Where $P(x,y)$ is the contact pressure at the contact point and P_0 is the maximum contact pressure of the wheel and rail. a represents the major semi axis of the contact point and b represents the minor semi axis of the contact point.

$$P_0 = \frac{3Q}{2\pi ab} \quad (2)$$

where the Q is the axle load.

3. Results and discussions

Extended Lagrange algorithm was used, the calculation time was set at 4 seconds, and the load steps were divided into 2000 substeps and contact pressure, stress and deformation were obtained. Total displacement for both wheel and rail at three different times was shown in figure 4. It can be seen from figure 4, with the increase of time, the wheel rolled from the left end of the rail to the right. when t=0.6 s, the maximum equivalent displacement is 117.671 at the top of the wheel rim. From 0.6 s to 4 s, The maximum equivalent displacement is 363.965, 606.324, 803.918 successively. the maximum equivalent displacement of the wheel-rail gradually increases, and the maximum value is always at the end of the wheel rim and gradually moves towards to the rolling direction of the wheel, this simulation is similar to the real wheel-rail rolling.

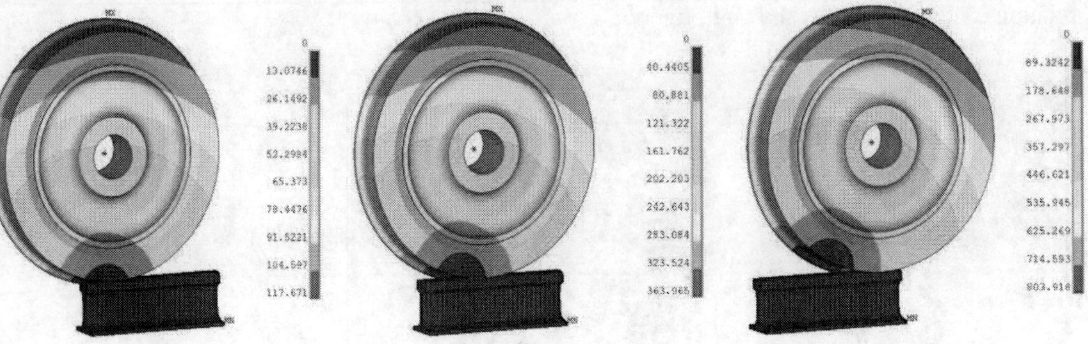

Figure 4. Total displacement of the wheel on the rail at three different times. (a) t=0.6s, (b) t=1.8s and (c) t=4s.

3.1. Discussion of the contact pressure

Figure 5 shows the pressure diagram of wheel-rail contact, the maximum contact pressure occurs at the contact spot, it's position is marked with MX. The maximum contact pressure value is 1029.57 MPa. When the Q is 10.5t, a is 8.1 mm, b is 6 mm. According to the equation (2), the wheel-rail contact pressure is 1024 MPa. The numerical result of the wheel-rail contact pressure is consistent with the basic Hertz contact theory, the error between them is 0.5%, which verifies the accuracy of the wheel-rail contact simulation model.

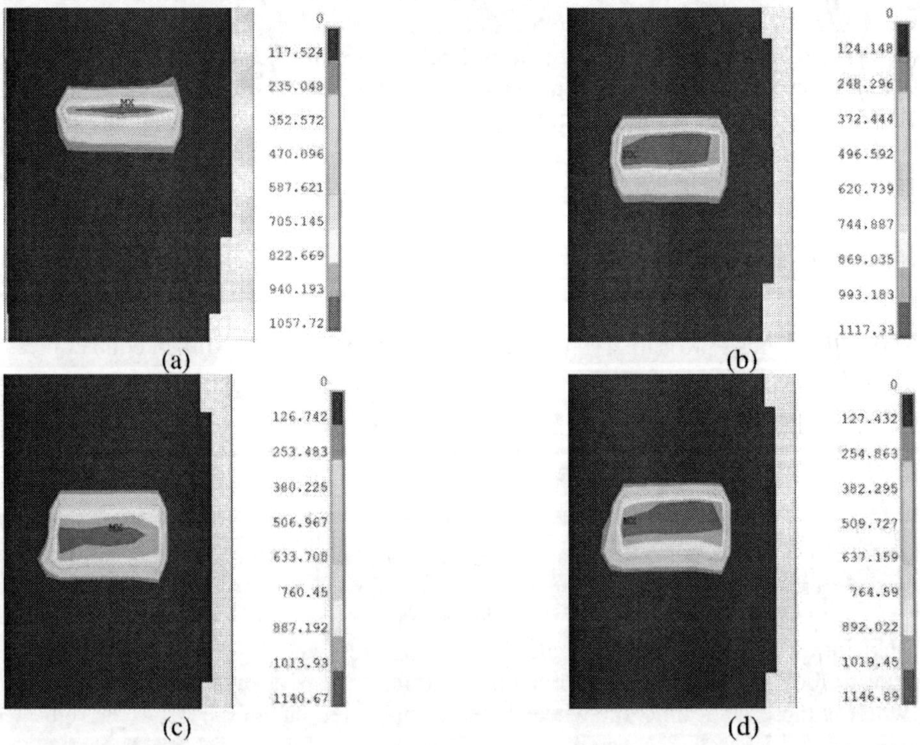

Figure 5. Contact pressure distribution of the wheel-rail. (a) t=0.6s, (b) t=1.8s, (c) t=3s and (d) t=4s.

It can also be seen from figure 5, with the wheel rolling, the area of the wheel-rail contact spot is almost constant, but the shape of the contact spot has a slight change at different time. The initial regular polygon shape gradually evolves into an irregular polygon, this is due to the corrugation effect of the rail.

The curve of the wheel-rail contact pressure at 0.6s, 1.8s, 3s, 4s is presented in figure 6. It can be seen from figure 6, with the wheel rolling, the contact pressure rises dramatically at the beginning, and wheel gets closer to the rail joint, the wheel-rail contact pressure increase slowly, after 3s, the maximal contact pressure is stable around 1140 MPa. Therefore, the maximum value for wheel-rail rolling model at 4s is 89.19 MPa higher than that at initial time.

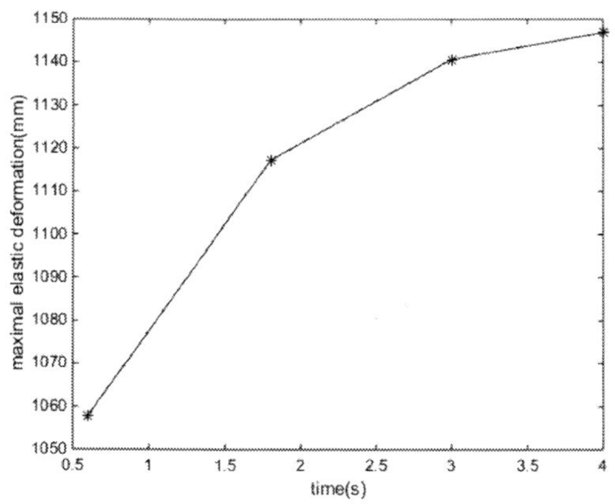

Figure 6. The contact pressure curve of the wheel-rail at different time.

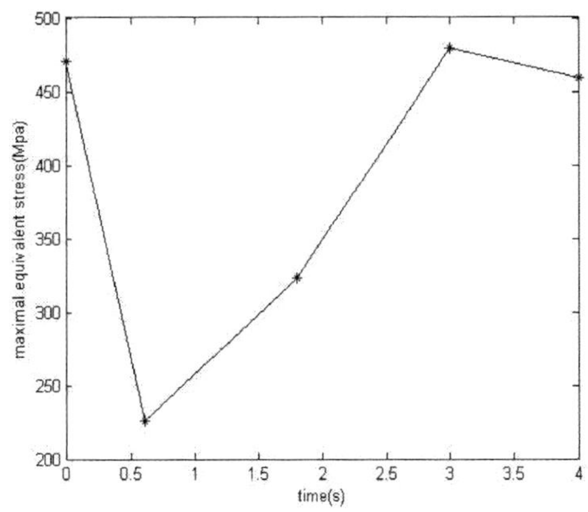

Figure 7. The maximum equivalent stress distribution curve of the rail.

3.2. Discussion of the equivalent stress and elastic deformation for rail and wheel

Distribution curve of the rail maximum equivalent stress at different time is shown in figure 7, it can be seen that with the increase of time, the equivalent stress of the rail decreases sharply at beginning, and then increases relatively slowly. Till 3s, The wheel and track deformation tend to stabilize the maximum equivalent stress reaches the peak. and then it stable at 460 MPa. Till 4s, the wheel and rail contact point is close to the track edge, as the difference of bearing structure between track edge and middle rail, so the maximum equivalent stress distribution curve of the rail decreases.

Equivalent stress distribution of the rail at different time is plotted in figure 8. With the wheel rolling, the equivalent stress area moves toward to the rolling direction of the wheel, and the equivalent stress influence area gradually diffuse from the wheel-rail contact area to the surrounding area. It also can be seen that when the wheel is on the rail joint at 0s or 4s, the distribution of the equivalent stress is the approximate semicircle area connecting the rail upper and lower surface, and extending to the rail joint, which is easy to cause stress concentration and crush rail joint.

The elastic deformation distribution is roughly consistent with the equivalent stress distribution of the rail.

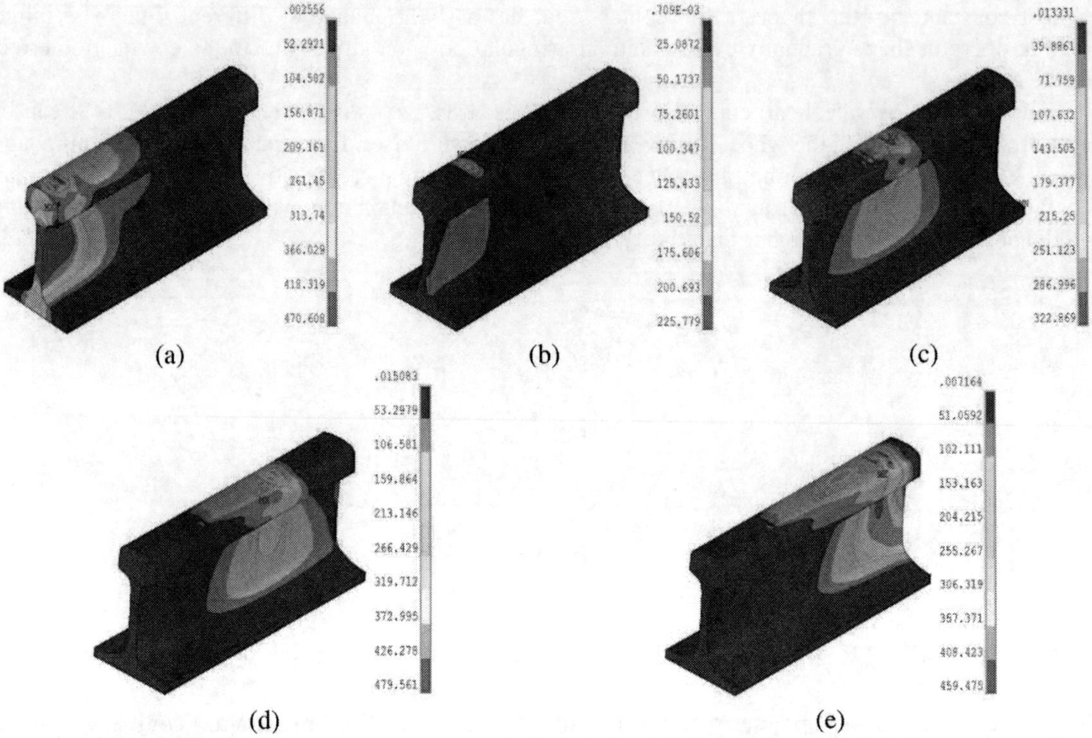

Figure 8. Equivalent stress of the rail. (a) t=0s, (b) t=0.6s, (c) t=1.8s, (d) t=3s and (e) t=4s.

The equivalent stress distribution and the maximum equivalent stress distribution curve are plotted in figures 9 and 10. As can be seen from figure 9, with the wheel rolling, the maximum equivalent stress is 434 MPa, is lower than the yield limit of wheel material, so the plastic deformation of the wheel is not occur. The shape of the contact spot is relatively elliptic and the equivalent stress distribution is adjusted with the Hertz contact theory. It also can be seen from figure 10, as the wheel rolling, the maximal equivalent stress of the wheel gradually decreases at first, and then increase. When the wheel reaches to the rail joint, the maximum equivalent stress reaches the peak.

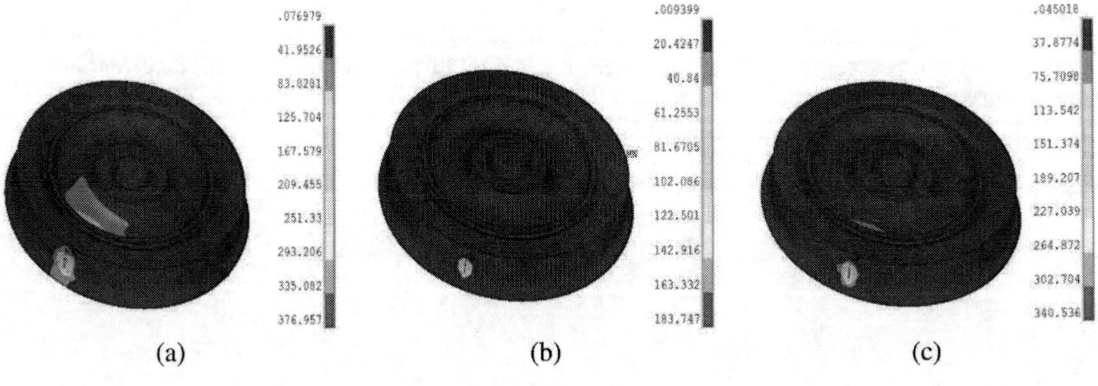

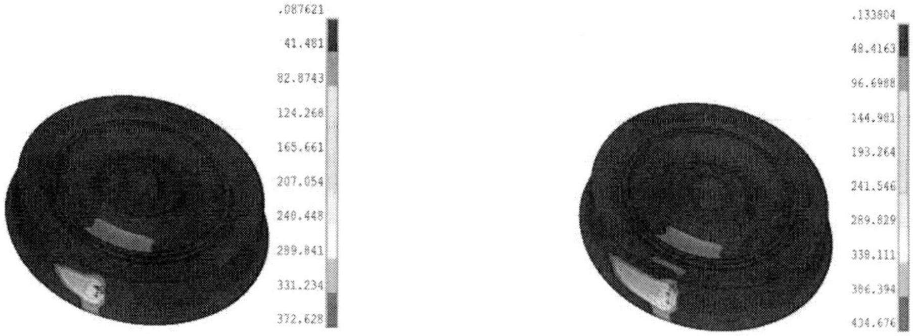

(d) (e)

Figure 9. The equivalent stress distribution of the wheel. (a) t=0s, (b) t=0.6s, (c) t=1.8s, (d) t=3s and (e) t=4s.

The elastic deformation distribution and the maximum elastic deformation distribution curve of the wheel are shown in figures 11 and figure 12, It can be seen from figure 12, the maximal elastic deformation occurs at wheel-rail contact spot and its maximal elastic deformation fluctuates 0.002 mm with the wheel rolling. It also can be seen from figure 11, with the increase of time, the equivalent elastic deformation of the wheel decreases dramatically at first, and then increases relatively sharply. When the time is 4s, it reaches the peak.

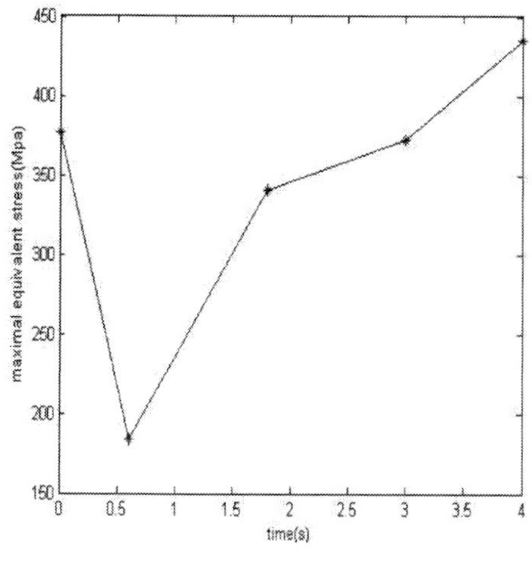

Figure 10. The maximal equivalent stress distribution curve of the wheel,

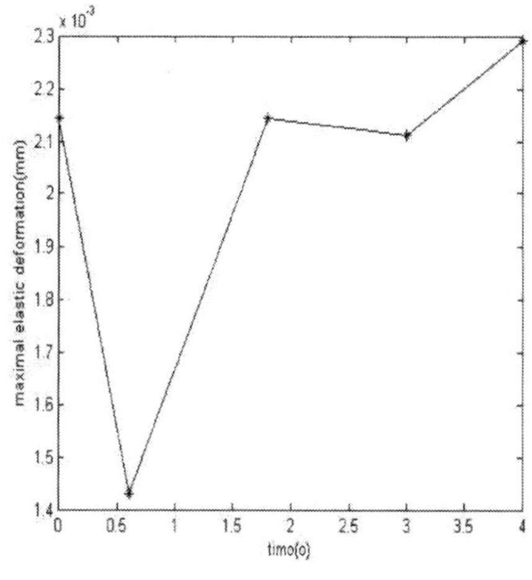

Figure 11. The maximal elastic deformation distribution curve of the wheel.

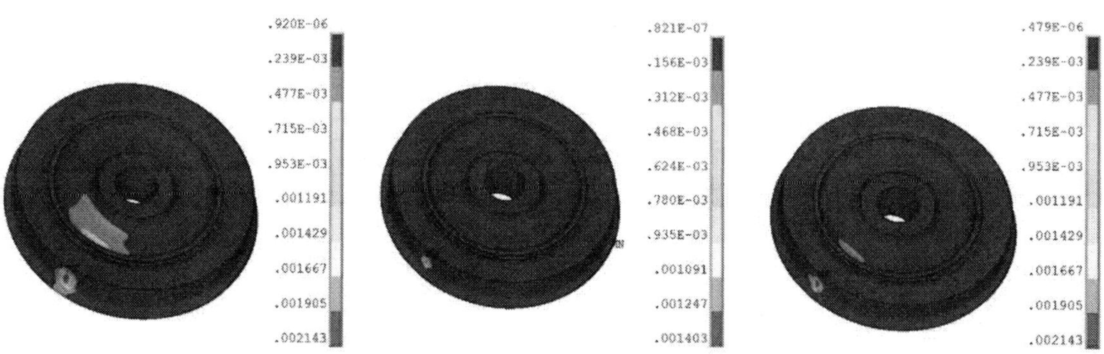

(a) (b) (c)

IOP Conf. Series: Materials Science and Engineering **657** (2019) 012072 doi:10.1088/1757-899X/657/1/012072

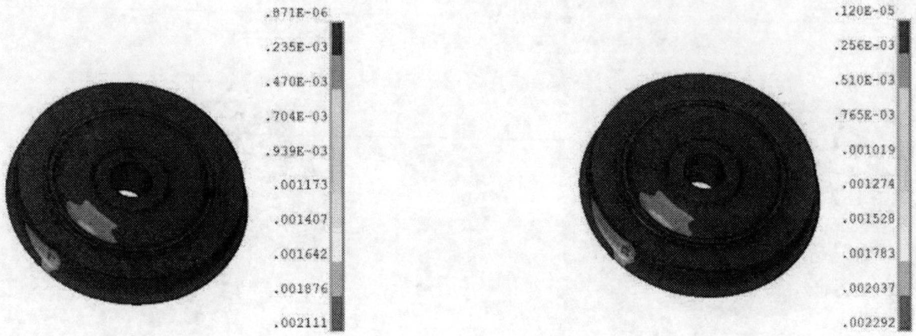

(d) (e)

Figure 12. The elastic deformation distribution of the wheel. (a) t=0s, (b) t=0.6s, (c) t=1.8s, (d) t=3s and (e) t=4s.

3.3. Discussion of the plastic deformation of rail

When the wheel rolling only rail has plastic deformation and the plastic deformation occur at 0s, 3s, 4s. Equivalent plastic deformation for the rail is shown in figure 13, with the wheel rolling, the plastic deformation for the rail occurs on the surface of rail. As the wheel rolling slowly forward, the plastic deformation area of the rail increases. At 0s, the plastic deformation area occurs at the side of the rail joint, this trends to lead to resulting in crushing of the rail joint. Hence, accumulation of these plastic deformation with the time will result in permanent deformation and lead to its corrugation.

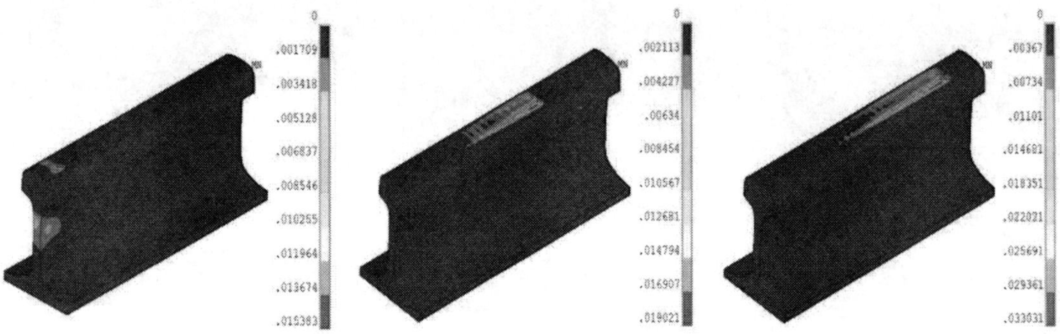

Figure 13. The plastic deformation distribution of the rail.

4. Conclusions

In this study, the rolling contact of wheel-rail system is used to investigate the contact pressure, equivalent stress, elastic deformation and plastic deformation. Some useful results are obtained.

As the wheel rolling, the wheel-rail contact pressure gradually increases, and the initial regular polygon gradually evolves into an irregular polygon. With the change of time, the equivalent stress and elastic deformation trend of the rail are basically the same, both of which decrease rapidly and then gradually increase, and reach the peak value when the time is 3s. For the wheel, when the wheel reaches to the rail joint, the maximum equivalent stress reaches the peak. With the increase of time, the plastic deformation of the rail gradually increases and accumulates, and the stress concentration area occurs at the side of the rail joint.

Acknowledgments

The research is supported by the National Natural Science Foundation of China (No. 11302159), the Principal Foundation of Xi'an Technological University (No. XAGDXJJ18004), and Basic Research Plan of Natural Science in Shannxi Province (No. 2018JM5099).

IOP Conf. Series: Materials Science and Engineering **657** (2019) 012072 doi:10.1088/1757-899X/657/1/012072

References

[1] Ma C H, Shi S L and Wu Y P 2006 Finite element analysis of the contact stress of wheel and rail *Journal of Shijiazhuang Railway University (Natural Science Edition)* **19** 32-5
[2] Johnson K L and *Contact Mechanics* 1985 (Cambridge, UK: Cambridge University Press)
[3] Kalker J J and Johnson K L 1990 *Three-Dimensional Elastic Bodies in Rolling Contact* (TH Delft, The Netherlands: Kluwer Academic Pub)
[4] Guo J 2017 Research on wheel and rail contact conditions and material parameters of high-speed trains based on ANSYS East China Jiaotong University
[5] Sharma S K and Kumar A 2016 Dynamics analysis of wheel rail contact using FEA *Procedia Eng.* **144** 1119-28
[6] Gu S, Yang X, Zhou S *et al* 2016 An innovative contact partition model for wheel/rail normal contact *Wear* **366-367** 38-48
[7] Liu Y, Liu L and Mahadevan S 2007 Analysis of subsurface crack propagation under rolling contact loading in railroad wheels using FEM *Eng. Fract. Mech.* **74** 2659-74
[8] Farjoo M, Daniel W, Bellette P *et al* 2013 Field statistical and finite element analysis of rail squats *Eng. Fract. Mech.* **109** 117-29
[9] Wen Z F, Jin X S and Zhang W H 2003 Finite element analysis of contact-impact of wheel/rail at rail gap *Tribol.* **23** 240-4
[10] Wen Z F, Jin X S and Zhang W H 2003 Elastic-plastic finite element analysis of three dimensional contact-impact at rail joint *Chin. J. Mech. Eng.* **16** 411-6

IOP Conf. Series: Materials Science and Engineering **657** (2019) 012073 doi:10.1088/1757-899X/657/1/012073

Application of transient nonlinear thermal analysis on quenching of hot die steel

J B Qi[1,2], K Kang[1], Z Y Hu[1], Z L Jin[1] and H P Ren[1]

[1]Institute of Materials and Metallurgy, Inner Mongolia University of Science and Technology, Baotou Inner Mongolia 014010, China

E-mail: qjbqjb1980@126.com

Abstract. Hot die steels contact with high temperature metals, and possess well thermal conductivity, wear resistance and red hardness. The carbon content should be controlled in the range of 0.3wt.% to 0.6wt.%, and added alloy elements to improve hardenability. The complex composition of the alloy affects its thermal and physical properties. For billets with large volume, it is difficult to measure the temperature of all parts. Through numerical simulation, the temperature field distribution in quenching process of hollow steel pipe with a length of 7 m is obtained by adopting the transient nonlinear thermal analysis algorithm of general finite element software. The results show that the steel temperature decreases to the target temperature of 19°C, it takes about 10 to 12 minutes for the thin wall and 18 to 20 minutes for the thick wall. It is about 4 minutes to descend martensite transition temperature of 265°C at the thick end. Steel temperature decreases to medium temperature, with the wall thickness increases, time slightly longer. The characteristics and phase changing latent heat of hot die steel quenching process are taken into account, and the corresponding treatment is carried out when setting thermophysical parameters. The accurate simulation results are obtained and have certain significance for the research of related materials heat treatment.

1. Introduction

Temperature field simulations about thermal and physical properties of materials mainly include specific heat capacity, phase change latent heat, thermal conductivity and density. The thermophysical parameters of materials are a function of temperature. So in the iterative solution, the parameters varying with temperature can be calculated by interpolation and employed as the approximate value of the relevant parameters at the next moment [1].

Due to the dramatic temperature changes of the quenching process and the temperature measurement complexity of the large workpiece, numerical simulation technologies on quenching process always carry out a large amount of research. The contents covered the phase change of steel quenching [2,3], thermal elastic-plastic [4], phase transformation plasticity and correlation of internal stress with martensite transformation [5], performance and the residual stress after quenching [6], surface heat transfer and boundary conditions [7], the nonlinear thermal analysis of quenching process, temperature and stress field coupling [8], etc.

When metals and alloys are quenched in water, spindle oil and other liquid media, the surface heat transfer coefficient always increases to the maximum value and then decreases as the quenching temperature decreases. In the early stage of quenching, the temperature difference between the workpiece and the medium is significant. When the hot workpiece is suddenly put into the quenching medium, the medium around the workpiece is heated and vaporized. There is a layer of steam film between the workpiece and the quenching medium. The steam film with poor thermal conductivity leads to a small surface heat transfer coefficient. As the temperature of the workpiece decreases and the quenching time continues, the steam film and bubbles gradually break. The heat transfer between the specimen and the quenching liquid becomes fierce, and the surface comprehensive heat transfer coefficient and cooling speed rapidly increase to the maximum. When the workpiece temperature is low, the surface heat transfer coefficient decreases rapidly with the decrease of temperature difference between quenching specimen and medium.

2. Numerical simulation methods

The analysis involving phase transitions is always transient analysis, and the load in the transient analysis varies with time. In order to accurately describe the quenching process, it is necessary to comprehensively consider the relationship about thermal properties, phase change latent heat, surface heat transfer coefficient, temperature and time for quenching cooling such transient problems. The matrix form of transient nonlinear thermal analysis can be expressed as:

$$(C(T))\{\dot{T}\}+(K(T))\{T\}=\{Q(T,t)\} \tag{1}$$

Where $(C(T))\{\dot{T}\}$ is the heat storage phase, $C(T)$ is the heat capacity matrix, $K(T)$ is the heat conduction matrix, $Q(T,t)$ is the heat flux density of heat source contained in the object, T and t are the temperature and the quenching cooling time respectively.

In the quenching process accompanied by phase change, enthalpy can be defined. The heat storage property, namely phase change latent heat, can be calculated and superimposed into the specific heat capacity matrix. Since the simulation process includes air cooling before quenching, boundary conditions and material properties that change with temperature. Such as surface convection coefficient, thermal conductivity coefficient, enthalpy, etc., which have typical characteristics of nonlinear thermal analysis. These parameters are applied according to the temperature function. The ingredients of hot die steel are shown in Table 1.

Table 1. Composition analysis of materials (mass fraction, %).

C	Si	Mn	Cr	Ni	Mo	V
0.35~0.45	0.80~1.10	0.80~1.10	1.30~1.50	2.50~3.50	0.40~0.60	0.20~0.60

The density is calculated according to the influence coefficient of elements in the steel [9].

$$\rho=\rho_0+x\cdot V\rho \tag{2}$$

Where ρ_0 is the density of pure iron, x is the mass fraction of elements contained in steel, $V\rho$ is the effect to the steel density of increasing the element content by 1 wt.%. The values of $V\rho$ are listed in Table 2.

Table 2. Effect of steel density increasing 1wt.% element content.

Element	C	Si	Mn	Cr	Ni	Mo	V
$V\rho$(g/cm^3)	-0.06	-0.035	-0.0034	-0.006	0.011	0.024	-0.018

Hot die steel quenching process:

910°C solid solution ⟶ air cooling 2 minutes ⟶ quenching in 19°C flowing water for 40 minutes.

Thermal conductivity and specific heat capacity are inherent properties of materials, which are related to the composition, density, temperature factors of materials. In this paper, the thermal conductivity and specific heat capacity vary between 30-35 W/m•K and 480~570 J/kg•K, respectively.

The overall fluctuation range is small. Firstly, the emissivity is set as 1 on the basis of black body in the simulation process, and then it is determined according to the oxidation condition of material surface when the temperature is measured. According to the ratio between the measured temperature of the infrared thermometer and the material surface, 0.93°C is higher than 600°C and 0.85°C is lower than 600°C

Surface heat transfer coefficient for air cooling after solid solution treatment, according to the empirical value [9] combined with the calculation. The air at 20°C is naturally convection on the surface of the steel tube, and the heat transfer coefficient can be obtained about $70W/m^2 \bullet K$. All material parameters used for calculation are shown in Table 3. With the change of temperature, the values of thermal conductivity coefficient and specific heat capacity change rarely, so the average values are taken. However, in the quenching process, due to the strong heat exchange between the material surface and the medium, the heat transfer coefficient is not stable. Currently, there is no reliable measurement method, and the calculated value is only given according to the three stages of the quenching process [10].

Table 3. Parameters of materials.

Density/(kg/m^3)	Thermal conductivity coefficient/(W/m·K)	Specific heat capacity/(J/kg·K)	Surface heat transfer coefficient/(W/m^2·K)	Stefan-Boltzmann constant/(w/(m^2 ·k^4))
7830	32	520	Flowing water: 2000 Gas: 70	5.67e-8

3. Analysis of air cooling temperature field

The eight-node three-dimensional hexahedron element was selected as the unit entity type, and the three-dimensional thermal element was selected as the surface effect element. There were 37,624 elements in the calculation material, and the time step was 10s. The mesh is divided by tetrahedron or hexahedron mapping pattern. Figure 1 shows the temperature distribution cloud diagram during air cooling of hot die steel tube normalizing. Different colours represent the corresponding temperature.

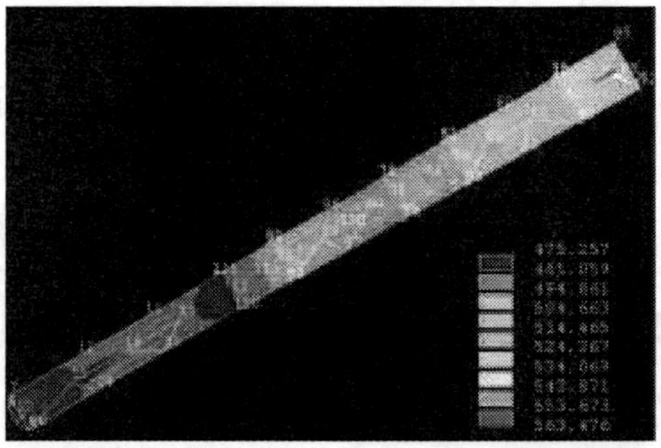

Figure 1. Temperature distribution cloud diagram during air cooling of hot die steel tube normalizing.

Figure 2 to 4 show the temperature changes of nodes at different positions in tube. Results show that after 2 min of air cooling, the 125 mm thin-walled temperature of node 6 and node 15 decreases from 1183 K to 1093.23 K and 1121.65 K, respectively. With 155 mm wall thickness, node 98 and node 99 temperatures are 1116.87 K and 1126.00 K, respectively. The node 62 and node 45 temperatures of 200 mm thick-walled are 1127.30 K and 1106.12 K, respectively. The closer it is to the end of the pipe, the faster temperature will drop. This is mainly due to additional heat dissipation from the end surface of the pipe. In the early stage, the temperature drop of billet is fast, and then gradually slows down. The absolute value of the slope of temperature change curve decreases with the time elapse.

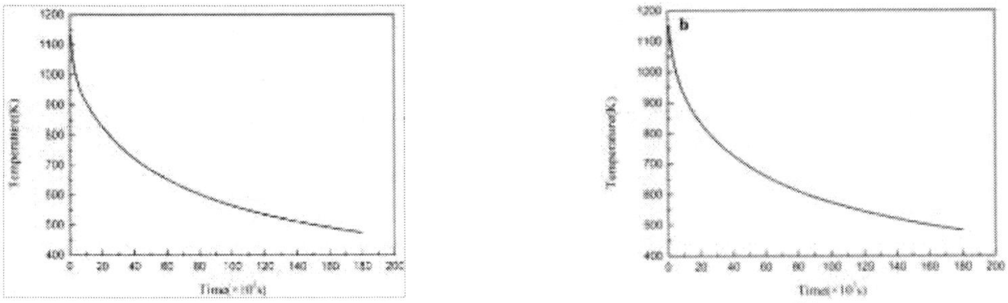

(a) (b)

Figure 2. Curve of temperature change at 125mm wall thickness. (a) Node 6 and (b) Node 15.

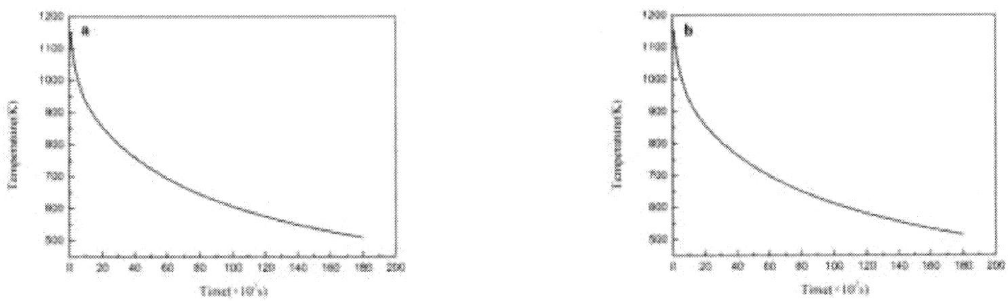

Figure 3. Curve of temperature change at 155 mm wall thickness. (a) Node 98 and (b) Node 99.

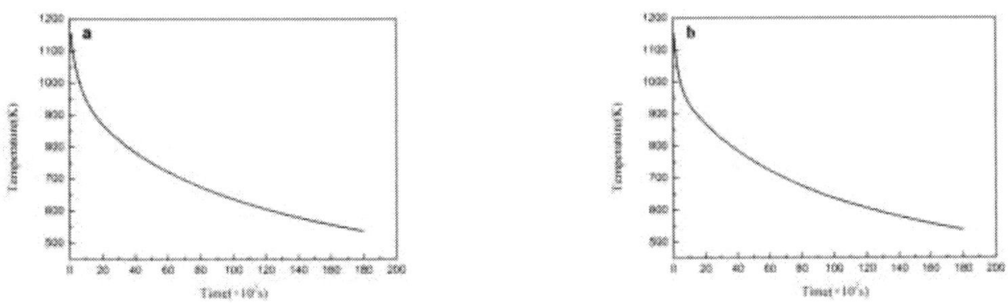

Figure 4. Curve of temperature change at 200 mm wall thickness. (a) Node 62 and (b) Node 45.

4. Temperature field analysis of quenching process

Quenching can be divided into three stages: membrane boiling, transition boiling and nucleation boiling [7].

Heat transfer coefficient is very important for quenching process. Figure 5 shows the application of heat transfer loads on the inner and outer surfaces of steel tubes. As a parameter of the boundary condition, it has noteworthy influence on the simulation accuracy. Therefore, in quenching heat analysis, it is necessary to obtain an accurate relationship curve between surface temperature and heat transfer coefficient of materials. However, it is difficult to measure surface heat transfer coefficient in practice. In this paper, the method combining empirical value and theoretical calculation is used to estimate the surface heat transfer coefficient. The quenching medium is flowing water.

IOP Conf. Series: Materials Science and Engineering **657** (2019) 012073 doi:10.1088/1757-899X/657/1/012073

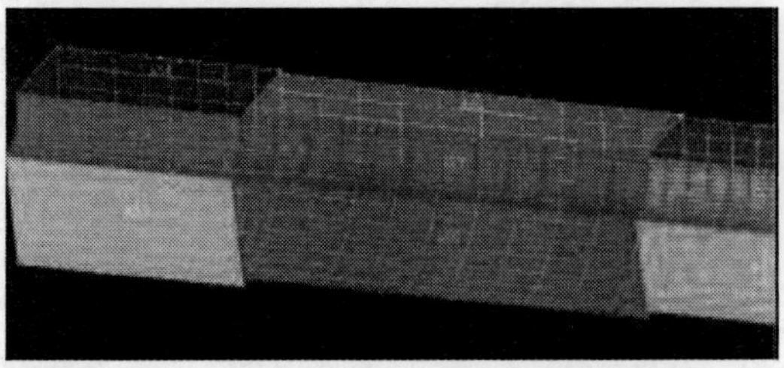

Figure 5. Heat transfer loads on the inner and outer surfaces of steel tubes.

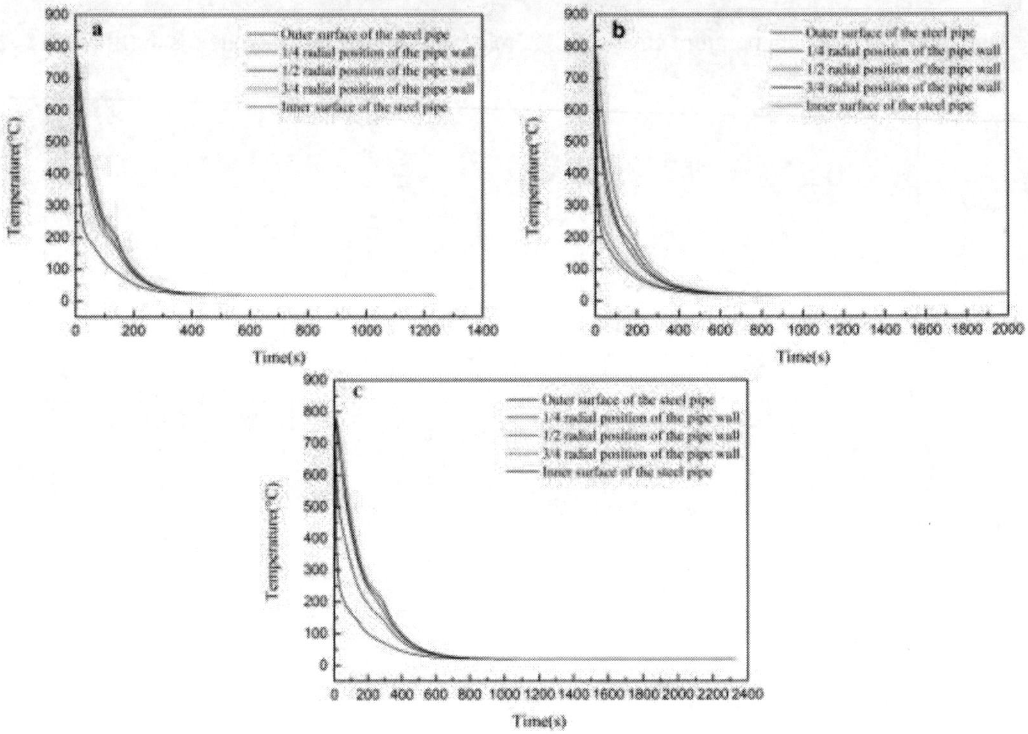

Figure 6. Temperature distribution curve in the quenching process of hot die steel pipe. (a) the thinnest wall, (b) the mid-wall, (c) the thickest wall.

The latent heat release of quenching specimens can be described by using the equivalent heat capacity method at constant pressure. Comparatively accurate physical parameters are obtained by combining theoretical calculation with actual measurement. As can be seen in figure 6, near the martensite phase transformation temperature, the slope of temperature distribution curve are changed. The overall trend is with lower temperature, its absolute value decrease and then increase. Phase change latent heat release can lead to slow cooling speed near the martensitic transformation point of material, and increased cooling speed until phase change latent heat released in full. The faster temperature drops, the earlier this kind of phenomenon appears, the lower corresponding temperature shows.

5. Conclusions

The steel pipe is quenched to the target temperature of 19°C, it takes about 10 to 12 minutes for the thin wall, 18 to 20 minutes for the thick wall, and 14 to 16 minutes for the longitudinal center of the pipe. The radial centers of hot die steel pipe wall quenching to martensite transition temperature, the

IOP Conf. Series: Materials Science and Engineering **657** (2019) 012073 doi:10.1088/1757-899X/657/1/012073

thick wall and thin wall take about 233s and 123s, respectively. With the pipe wall thickness increasing, time slightly longer in the process of steel quenching to medium temperature.

References

[1] Zhou C, Liao D M, Chen T *et al* 2013 Numerical calculation and program development of temperature field in air cooling process based on finite element method *Foundry* **62** 850-4

[2] Fernendes F, Denis S and Simon A 1985 Mathematical model coupling phase transformation and temperature evolution during quenching of steel *Mater. Sci. Technol.* **1** 838-44

[3] Ning Q H, Dong J Y and Li Y G 2018 Finite element simulation of water spray quenching for large size NM450 steel plate *Heat Treat. Met.* **43** 221-6

[4] Rammerstorfer F G, Fischer D F and Mitter W 1981 On thermo-elastic-plastic analysis of heat treatment processes including creep and phase changes *Comput. Struct.* **1981** 771-9

[5] Denis S, Egautier, Simon A *et al* 1985 Stress-phase-transformation interactions basic principles, modelling, and calculation of internal stresses *Mater. Sci. Technol.* **1** 805-15

[6] Wang L, Li Z L, Yang D F *et al* 2015 Finite element simulation and experimental study on residual stress of atomized gas quenching *Hot Work Techol.* **44** 175-8

[7] Xie J B 2003 Numerical simulation and application research of metal and alloy quenching in different media (Kunming: Kunming University of Science and Technology)

[8] Zhang L Q, Wang J, Luo X M *et al* 2017 Study on coupling simulation of flow field, temperature field, structure field and stress field in heat treatment process *Heat Treat. Met.* **42** 181-6

[9] Tan Z and Guo G W 1994 Thermophysical Properties of Engineering Alloys (Beijing: Metallurgical Industry Press)

[10] Kopun R and Skerget L 2012 Experimental and numerical investigation of boiling flow in a vertical pipe with phase change using a multi-fluid modelling approach *Adv. Fluid Mech. IX* **74** 219-30

IOP Conf. Series: Materials Science and Engineering **657** (2019) 012074 doi:10.1088/1757-899X/657/1/012074

Effects of herringbone grooves on aerodynamic journal bearings based on finite element method

S Y Gao[1,2,4], S Yang[2,3], Y G Shi[2], P Sun[2] and H Chen[2]

[1]The State Key Laboratory of Mechanical Transmissions, Chongqing University, 400044, Chongqing, PR China
[2]Institute of Advanced Manufacturing Technology, Hefei Institutes of Physical Science, Chinese Academy of Sciences, Changzhou, 213164, Jiangsu, PR China
[3]State Key Lab of Digital Manufacturing Equipment and Technology, Wuhan, 430074, Hubei, PR China

E-mail: sygao@iamt.ac.cn

Abstract. Aerodynamic journal bearings are commonly used in high speed machines, which have weaknesses of low steady-state performance specifications. It is of great significance to study the effects of herringbone grooves on the characteristics of aerodynamic journal bearings. This paper investigates effects of herringbone grooves on functions of aerodynamic journal bearings by numerically solving the Reynolds equation based on improved finite element method. The obtained results show that the herringbone grooved aerodynamic journal bearings have a critical speed and they should run at a speed of greater than the critical speed. Ultra-high speeds and low eccentricity ratios are the suitable working conditions of the herringbone grooved aerodynamic journal bearings. The performance of aerodynamic journal bearings is enhanced with advancing groove numbers, and an optimal groove depth is existed with respect to the herringbone grooved aerodynamic journal bearing.

1. Introduction

Aerodynamic bearings have outstanding characteristics of almost no wear, few heating, no external pressurized air supply and clean operation, which are extensively applied in various engineering industries including high speed spindle, micro gas turbine engine, inertial gyroscope, laser scanner, cryogenic expander, et al. However, aerodynamic bearings have weaknesses on load capacity, stiffness and so on. Herringbone grooves had been reported to improve performance of the aerodynamic bearings.

Large quantity of papers had been published to investigate herringbone groove effects on the aerodynamic bearing. Air journal bearings with herringbone grooves were analysed by narrow groove theory (NGT) [1-3] which had an assumption of infinite groove number. However, the obtained pressure profiles had no saw-toothed ripples in the groove boarders and the smoothed pressure filed was acquired. Furthermore, the Reynolds equation was numerically calculated by finite difference method (FDM) [4, 5] or finite element method (FEM) [6-8] to analyse the effects of herringbone grooves on the aerodynamic bearing. Based on the aforementioned analyses, many papers had been published to investigate the herringbone grooved air bearings, however, few efforts had been done to comprehensively analyse herringbone groove effects on the aerodynamic journal bearing (AJB) under ultra-high-speed condition based on FEM. Therefore, this paper will numerically solve the Reynolds

IOP Conf. Series: Materials Science and Engineering **657** (2019) 012074 doi:10.1088/1757-899X/657/1/012074

equation to study herringbone groove effects on the AJB under ultra-high-speed condition with the improved FEM.

2. Herringbone grooved aerodynamic journal bearing (HGAJB)

Figure 1 shows a typical geometry structure of a HGAJB, which has a smooth bearing bush and a herringbone grooved journal. Two groups of herringbone grooves are evenly incised around the journal surface, which are symmetric to the plane a-a of the bearing. The outer ends of the bearing bush and the herringbone grooves are aligned.

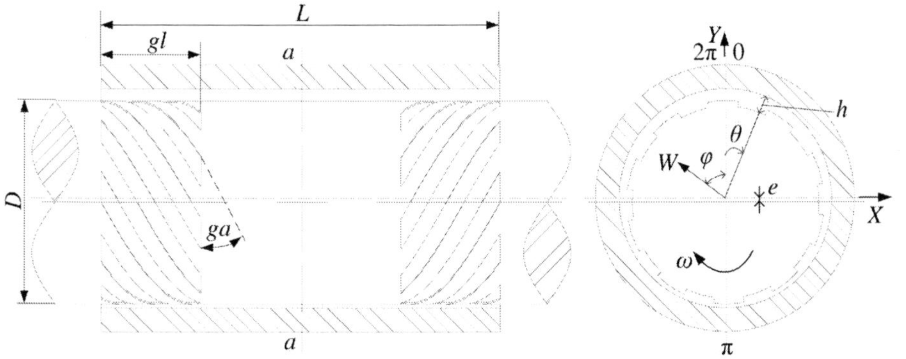

Figure 1. Schematic of a HGAJB.

Figure 2 illustrates an unfolded plane of the air film in a HGAJB. *L* is axial length of the bearing, *D* is diameter of the bearing bore, *gl* denotes axial herringbone groove length, *ga* denotes the herringbone groove angle, *gd* denotes the herringbone groove depth, *gw* is the width of a herringbone groove, and *gt* is the overall width of a pair of groove-ridge, *gn* denotes groove number of one groove group. Besides, g_{lr}, determined by $gl/(L/2)$, denotes the ratio of groove length; g_{wr}, defined by gw/gt, denotes the ratio of groove width.

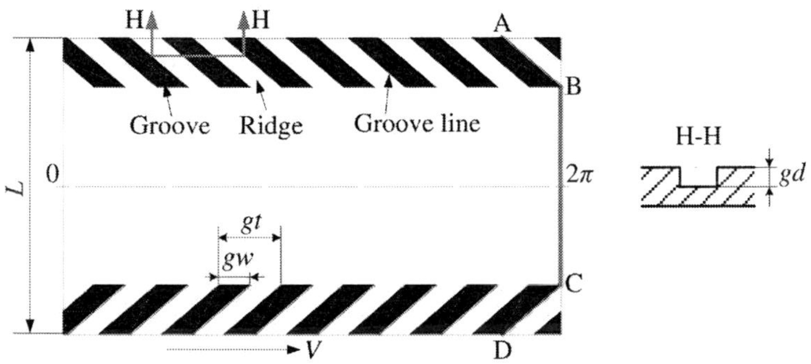

Figure 2. The unfolded air film with herringbone grooves.

3. Calculation model

3.1. Mathematical equations

Inertial effect is ignored due to the negligible mass of the lubricant gas. Besides, the current HGAJB model incorporates the following assumptions:

(a) There are no geometric errors of the bearing bore and the journal.
(b) There is no tilt for the displacement of the journal.
(c) The gas viscosity is considered be constant.
(d) Air film pressures are virtually invariable along air film thickness direction.

The dimensionless Reynolds equation for the aerodynamic journal bearing is depicted in equation (1) [9, 10]

$$\frac{\partial}{\partial \bar{x}}\left(\bar{h}^{3}\frac{\partial \bar{p}^{2}}{\partial \bar{x}}\right)+\frac{\partial}{\partial \bar{z}}\left(\bar{h}^{3}\frac{\partial \bar{p}^{2}}{\partial \bar{z}}\right)=\Lambda_{x}\frac{\partial(\bar{h}\bar{p})}{\partial \bar{x}}+\Lambda_{z}\frac{\partial(\bar{h}\bar{p})}{\partial \bar{z}} \tag{1}$$

where Λ_x and Λ_z denote the dimensionless bearing numbers, $\bar{x}$ and $\bar{z}$ denote the dimensionless Cartesian coordinates, $\bar{h}$ denotes the dimensionless air film thickness, $\bar{p}$ denotes the dimensionless air film pressure.

In this study, aforementioned dimensionless Reynolds equation has the following boundary conditions.

1. Pressure at the outer border of the bearing: $p=p_a$, p_a is ambient pressure;
2. Pressure at the symmetric boundary: $\frac{\partial p}{\partial n}=0$, n is the normal direction of the symmetric boundary.

The numerical solution of the Reynolds equation is calculated by FEM with the triangular element, and the finite element formula of the Reynolds equation for aerodynamic journal bearings is deduced as follows

$$\begin{gathered}\sum_{e\in\Delta i}\int_{\Delta e}\left(\mathbf{N}^{eT}\mathbf{h}^{e}\right)^{3}d\bar{x}d\bar{z}\left(c_{i}\mathbf{c}^{eT}+b_{i}\mathbf{b}^{eT}\right)\mathbf{f}^{e}\frac{1}{(2\Delta e)^{2}}\\ -\sum_{e\in\Delta i}\lambda\left(c_{i}\bar{u}+b_{i}\bar{\omega}\right)\mathbf{h}^{eT}\int_{\Delta e}\mathbf{N}^{e}\mathbf{N}^{eT}d\bar{x}d\bar{z}\frac{1}{2\Delta e}\left(\mathbf{f}^{1/2}\right)^{e}=0\\ (i=1,2,\ldots,q)\end{gathered} \tag{2}$$

where $\mathbf{N}^e$ is the shape function, $\mathbf{h}^e$ is the column matrix of the node thickness, c_i and b_i are the coefficients of the interpolation function, $\mathbf{c}^e$ and $\mathbf{b}^e$ are the coefficient matrixes of the interpolation function, $\mathbf{f}^e$ is the column matrix of dimensionless node pressure square, $\bar{u}$ and $\bar{\omega}$ are the dimensionless velocities, λ is the bearing coefficient.

Iteration method is applied to solve the finite element formula and the unknown nodes' pressures are obtained. Furthermore, properties of the aerodynamic journal bearing, such as load capacity, attitude angle and stiffness, can be calculated (the detailed mathematical equations are published in reference [9]).

3.2. Computational domain

In this study, the computational domain of aerodynamic journal bearing is determined by half of the fluid zone which is symmetric to plane a-a depicted in figure 1. The air film of the AJB is unfolded into a plane along the direction of A-B-C-D illustrated in figure 2. Figure 3 depicts the meshed computational domain with triangle elements and boundary conditions. Plane *k-k* locates at middle position in the AJB along the axial direction, and plane *m-m* is positioned at the inside end of the groove. It should be noted that the groove node thicknesses are superposed by the ridge node thicknesses and the groove depth. Thickness discontinuity of the air film locates in the transition boundaries between the ridge zone and the groove zone. In this study, the node thickness depends on the corresponding elements to which the node belongs [11].

Grid independence test is conducted to determine the computation grid size. Scheme A: 5 segments are uniformly meshed in each groove zone and ridge zone along peripheral direction. Scheme B: 10 segments are uniformly meshed in each groove zone and ridge zone along peripheral direction. Identical grids are meshed for schemes A and B in the axial direction. Geometrical parameters and working conditions of the bearings for schemes A and B are also identical. Difference of the calculated load capacities between scheme A and scheme B is 0.23%, manifesting that the performance parameters of the aerodynamic journal bearings are almost identical calculated by the two meshing schemes. Less grid quantity needs less computational time, so meshing scheme A is used to determine the grid size.

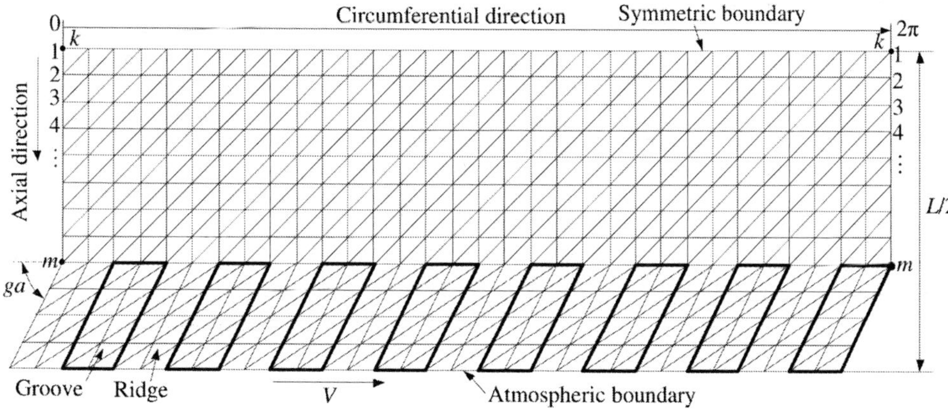

Figure 3. Schematic diagram of grids and boundary conditions for the HGAJB.

3.3. *Experimental verification*

It is exceedingly hard to test the effects of the herringbone groove on the property specifications of the AJB. Gao et al. [11] employed a round-about way to verify the improved FEM simulation method for analysing the effects of herringbone grooves on the property behaviors of ultra-high-speed aerostatic journal bearings. Equation (1) of reference [11] for aerostatic bearings can be transformed into equation (1) of this paper for aerodynamic bearings by ignoring the mass flow factor. Hence, the proposed calculational method of this paper is also indirectly verified based on the reference [11].

4. Results and discussions

There are many geometrical parameters of herringbone grooves, and the parameters of groove number and groove depth are selected as the variables to research herringbone groove effects on performance of aerodynamic journal bearing under various running conditions. An analysing procedure is developed by MATLAB. The nonlinear equations are numerical solved by the 'fsolve' function with the termination tolerance of 1e-20. Pressure distribution, load capacity, attitude angle and air film stiffness of aerodynamic journal bearings are thoroughly studied at various speeds (n) and eccentricity ratios (ε).

In this paper, the principle size parameters of the bearing are defined as follows: L=34.813mm, D=19.01mm, h_m=10μm, where h_m is the air film thickness with no grooves and no eccentricity. The groove angle, ga, is fixed at 30°, the groove length ratio, g_{lr}, is 1/3 and the groove width ratio, g_{wr}, is 1/2. The atmospheric pressure (p_a) is 101,325Pa, and the reference air film pressure (p_s) is 0.6MPa. The atmospheric air density (ρ_a) is 1.205kg/m^3, and the atmospheric air dynamic viscosity (η) is 1.81e-5Pa·s.

4.1. *Groove number*

In this section, the groove depth, gd, is fixed at 10μm. Figure 4 depicts three-dimensional dimensionless air film pressure profiles for variable groove numbers under conditions of n of 200,000r/min and ε of 0.3. It can be seen from figure 4 that major changes of air film pressures are occurred in the groove-ridge zones, presenting a zigzag profile of the air film pressure. Herringbone grooves pump outside gas into the bearing film, leading to the increased pressure of the air film, shown in figure 4 (a) and (b). Furthermore, pressure fluctuation amplitudes of the air film in adjacent groove zones and ridge zones ease up with enhancing groove numbers.

IOP Conf. Series: Materials Science and Engineering **657** (2019) 012074 doi:10.1088/1757-899X/657/1/012074

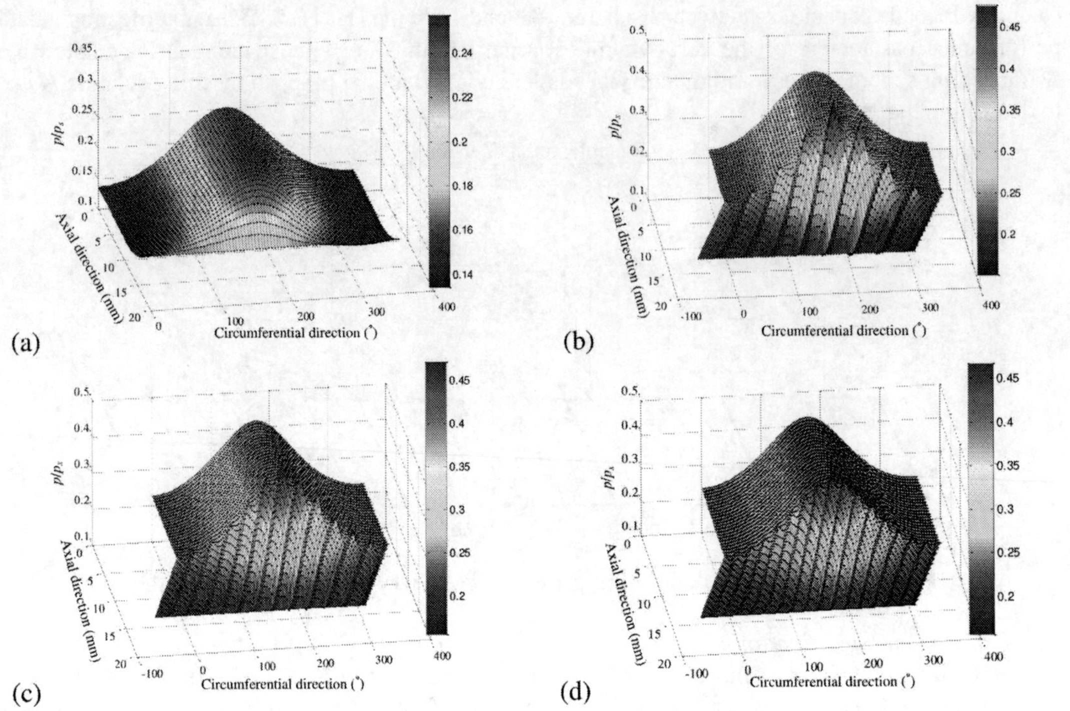

Figure 4. Dimensionless pressure profiles of the aerodynamic bearing films with variable groove numbers. (a) groove number is 0. (b) groove number is 8. (c) groove number is 12. (d) groove number is 16.

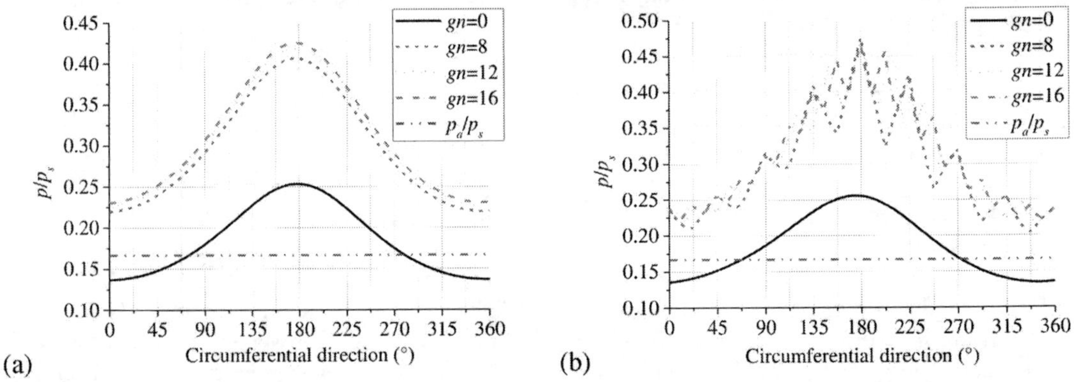

Figure 5. Dimensionless air film pressures corresponding to diverse groove numbers. (a) Plane *k-k*. (b) Plane *m-m*.

Dimensionless pressures of the air film at cross-sections *k-k* and *m-m* with respect to diverse groove numbers are illustrated in figure 5. Figure 5 depicts that herringbone grooves significantly increase air film pressures, and air film pressure increases with enhancing groove numbers. However, the increasing amplitude of the air film pressure is not obvious from *gn* of 8 to 16. It can be also found that the negative pressure appears when *gn* of 0 and disappears when *gn* of 8 or more. When there is no groove in the aerodynamic journal bearing, the external air is pumped into the air film in negative pressure zone, and the inner air is pumped out of the air film in positive pressure zone. When the herringbone groove number reaches 8 or more, only positive pressure appears in the bearing film, which presses the inner gas out of the air gap meanwhile the herringbone grooves pump the external gas into bearing film.

Figure 6 depicts property data of the AJB corresponding to diverse groove numbers under conditions of different eccentricities and speeds. The load capacity is reduced by herringbone grooves at speed of lower than 50,000r/min, as shown in figure 6(a). However, progressive enhancement of the aerodynamic effect resulted from herringbone grooves occurs corresponding to enhancing speeds. The load capacities with herringbone grooves are larger than those with non-grooves at speed of greater than 50,000r/min. Hence, the HGAJB has a critical speed which should be lower than its running speed. Besides, the grooved load capacity increases with enhancing groove numbers. The load capacity is increased by 107.25% from *gn* of 0 to 16 at speed of 400,000r/min and eccentricity of 0.3.

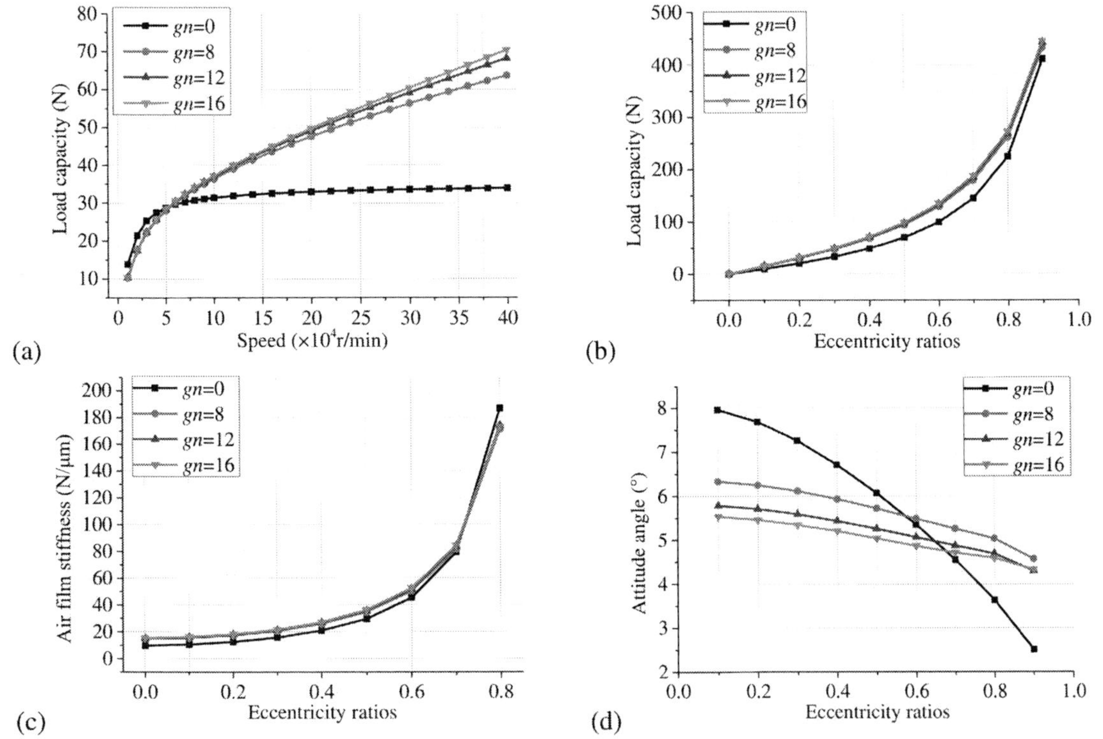

Figure 6. Performance parameters of the AJB corresponding to diverse groove numbers. (a) ε=0.3. (b) n=200,000r/min. (c) n=200,000r/min. (d) n=200,000r/min.

Figure 6 (b) illustrates that the load capacities with herringbone grooves are larger than those with non-grooves with various eccentricities and n of 200,000r/min. Figure 6 (c) and (d) present that herringbone grooves can improve air film stiffness with eccentricity ratio of lower than 0.7 and the attitude angle is declined with ε of less than 0.5 at n=200,000r/min. However, air film stiffnesses are decreased and attitude angles are increased by herringbone grooves with $\varepsilon > 0.7$ at n=200,000r/min. Besides, load capacities and air film stiffnesses increase with enhancing groove numbers, and attitude angles decrease with enhancing groove numbers. Based on aforementioned analysis results, ultra-high speeds and low eccentricity ratios are optimum operating conditions for herringbone grooved aerodynamic journal bearings. The performance parameters of HGAJB are advanced with enhancing groove numbers.

4.2. Groove depth

In this section, the groove number, *gn*, is fixed at 16. Figure 7 depicts property data of the AJB corresponding to diverse groove depths under conditions of different eccentricities and speeds. Figure 7 (a) illustrates that the load capacity with *gd* of 20μm increases with increasing speeds and is greater than those of other groove depths when the speed reaches 240,000r/min or more. The load capacity with *gd* of 5μm is larger than those of other groove depths at speed of $n \leq 40{,}000$r/min and becomes the lowest at speed of $n>200{,}000$r/min. The load capacity with ε of 0.3 for various groove depths has the following relation $W_{gd=20\mu m} > W_{gd=10\mu m} > W_{gd=30\mu m} > W_{gd=5\mu m}$ at speed of $n \geq 240{,}000$r/min.

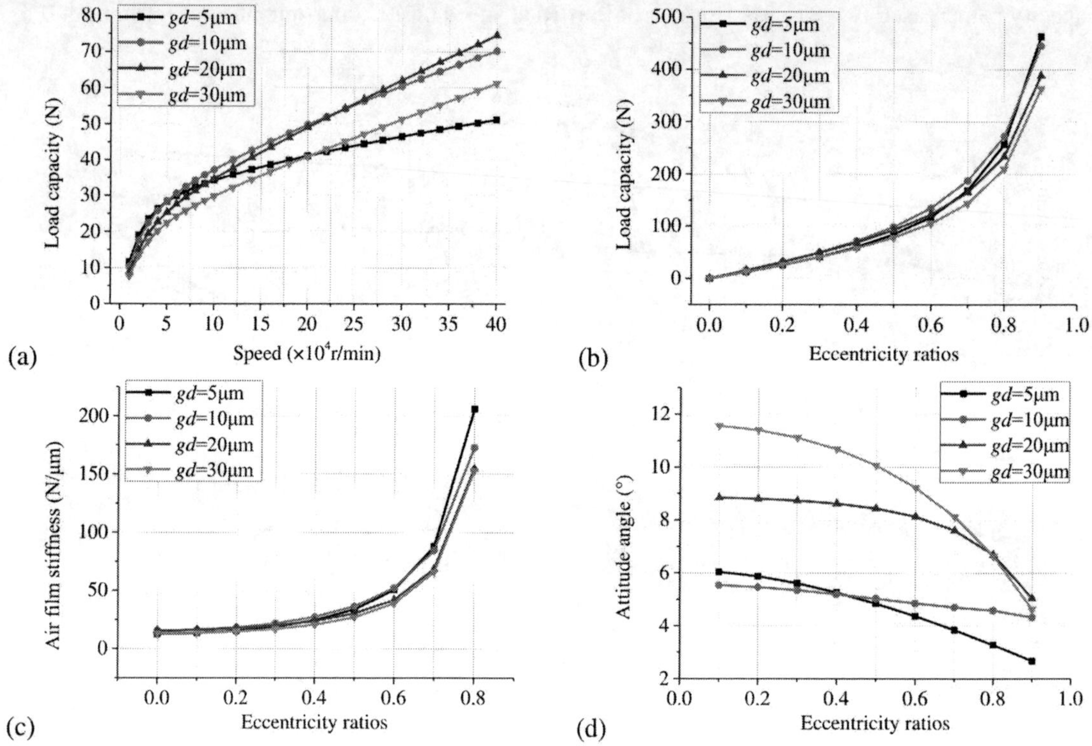

Figure 7. Performance parameters of the AJB corresponding to diverse groove depths. (a) ε=0.3. (b) n=200,000r/min. (c) n=200,000r/min. (d) n=200,000r/min.

Figure 7 (b) describes that the load capacity with *gd* of 10μm is greater than others at $0.2 \leq \varepsilon \leq 0.8$ and n=200,000r/min. Figure 7 (c) presents that the air film stiffness with *gd* of 10μm is greater than others at $0.1 \leq \varepsilon \leq 0.6$ and n=200,000r/min, and the air film stiffness with *gd* of 5μm becomes the greatest one at $\varepsilon > 0.6$ and n=200,000r/min. Figure 7 (d) depicts that the attitude angles decrease with increasing eccentricity ratios for various groove depths at n=200,000r/min, and the attitude angle with *gd* of 30μm has the highest value at $\varepsilon < 0.8$, and the attitude angle with *gd* of 5μm becomes the smallest one at $\varepsilon > 0.4$. In conclusion, too big or too small groove depth is hard to improve performance characteristics of HGAJB, and an optimal groove depth exists in the HGAJB with ultra-high speeds and low eccentricity ratios.

5. Conclusions

Effects of herringbone grooves on properties of the AJB are investigated by numerically solving the Reynolds equation based on FEM, and air film pressure, bearing load capacity, bearing attitude angle and air film stiffness are obtained under different operating conditions. Based on the aforementioned analysis results, the conclusions are as follows

(1) The HGAJB has a critical speed which should be lower than its running speed.

(2) Ultra-high speeds and low eccentricity ratios are optimum operating conditions for herringbone grooved aerodynamic journal bearings. The performance parameters of HGAJB are advanced with enhancing groove numbers.

(3) Too big or too small groove depth is hard to improve performance characteristics of HGAJB, and an optimal groove depth exists in the HGAJB with ultra-high speeds and low eccentricity ratios.

Acknowledgements

The authors thank for the funding support for this research by the State Key Laboratory of Mechanical Transmissions (Grant No. SKLMT-KFKT-201801), the National Natural Science Foundation of China (Grant No. 51705501, 51705500), the CASHIPS Director's Fund (Grant No. YZJJ201622) and the State Key Lab of Digital Manufacturing Equipment and Technology in China (Grant No. DMETKF2017017).

References

[1] Vohr JH and Chow CY 1965 Characteristics of herringbone-grooved, gas-lubricated journal bearings *J. Basic. Eng.* **87** 568-76

[2] Schiffmann J and Favrat D 2010 The effect of real gas on the properties of herringbone grooved journal bearings *Tribol. Int.* **43** 1602-14

[3] Miyanaga N and Tomioka J 2016 Effect of support stiffness and damping on stability characteristics of herringbone-grooved aerodynamic journal bearings mounted on viscoelastic supports *Tribol. Int.* **100** 195-203

[4] Chen SK, Chou HC and Kang Y 2012 Stability analysis of hydrodynamic bearing with herringbone grooved sleeve *Tribol. Int.* **55** 15-28

[5] Han Y, Xiong S, Wang J and Wang QJ 2016 A new singularity treatment approach for journal-bearing mixed lubrication modeled by the finite difference method with a herringbone mesh *J. Tribol.-T. ASME.* **138** 011704-1-10

[6] Zirkelback N and Andres LS 1998 Finite element analysis of herringbone groove journal bearings: a parametric study *J. Tribol.-T. ASME.* **120** 234-40

[7] Faria MTC 2001 Some performance characteristics of high speed gas lubricated herringbone groove journal bearings *JSME. Int. J. Ser. C. Mech. Syst. Mach. Elem. Manuf.* **44** 775-81

[8] Jang GH and Yoon JW 2002 Nonlinear dynamic analysis of a hydrodynamic journal bearing considering the effect of a rotating or stationary herringbone groove *J. Tribol.-T. ASME.* **124** 297-304

[9] Liu T, Liu Y H and Chen S J 1990 *Aerostatic lubrication* (Harbin: Press of Harbin Institute of Technology)

[10] Gao S, Cheng K, Chen S, Ding H and Fu H 2017 Computational design and analysis of aerostatic journal bearings with application to ultra-high speed spindles *P. I. Mech. Eng. C.-J. Mec.* **231** 1205-20

[11] Gao S, Shi Y, Xu L, Chen H and Cheng K 2019 Investigation on influences of herringbone grooves for the aerostatic journal bearings applied to ultra-high-speed spindles *P. I. Mech. Eng. C.-J. Mec.* **233** 5795-812

How to guarantee the results of investigations with limited statistical information

K Voinov[1,3] and A Esbulatova[2]

[1]National Research University of Information Technologies, Mechanics and Optics (University ITMO) Saint-Petersburg, Russia
[2]West-Kazakhstan Agrarian-technical University, Russia

E-mail: forstar@mail.ru

Abstract. What kind of problems do researchers usually have after investigations connected with different physical processes (e.g., temperature, deterioration, pressure, vibration, and so forth)? In many mathematical papers, textbooks and reference books it is suggested that 340-360 statistical data are needed to obtain the right conclusion by experimentation [1-3]. In these cases, much money (and time, days and even months) will be required. Usually, researchers try to restrict the amount of statistical data to 3, 5, 10 or 30 data points. But in this case, they do not have enough information to determine the trustworthiness of their deductions. Our task is the following: to solve this problem and demonstrate the fact that if there are even only three points of observation, we can still obtain a convincing response/result. It is rather difficult to overestimate the significance of this achievement in practice.

1. Introduction

Many people visit shooting ranges to shoot at a target such as can be seen in figure 1. A target has several circles/circumferences; if a bullet hits the centre of the target, this is considered to be excellent (10) and any other situation is not as good. If the shot was in the "milk" (outside the biggest diameter in the target), then this is the worst case.

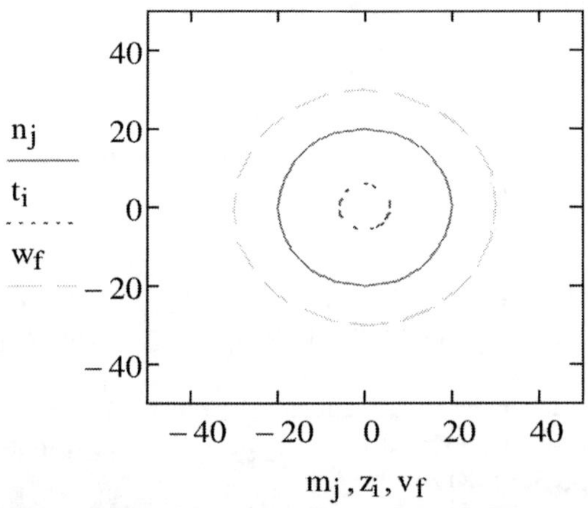

Figure 1. A target which has four zones, for example.

Building on this idea, the technical aspect will be considered, which is linked with probability in technology, products and with our lives. Suppose that any parts, objects, products, systems are to be manufactured. Any purchaser wants to obtain them in the demanded limits in terms of size, quality and reliability. Namely, all diameters and properties must be achieved within the bounds of tolerance (maybe, ±1 mm or % for chemical elements or compositions). Consequently, if it will be more or less than the installed limits, the buyer can get spoilage parts/products/wares. So, there is the following problem: how often must a producer check these objects in plants/factories/works/manufacture? In our approach, by making very small selections periodically from the general aggregate (only 3 or 5 positions) we can effectively control the quality of all objects and products (parts, systems, butter, cream, kefir, cottage cheese, and so on). By using this effective method, the law of distribution for spoilage objects does not need to be determined. The right result will be obtained if the target is created, for example, with several circumferences using a particular computer program in MathCad [4-7]:

$$j := 1..37 \qquad i := 1..37 \qquad n_j := 20\cos(10\cdot j\cdot h) \quad m_j := 20\sin(10\cdot j\cdot h)$$

$$f := 1..37 \qquad h := 0.01745 \qquad t_i := \frac{30\cos(20\cdot i\cdot h)}{5} \quad z_i := \frac{30\sin(20\cdot i\cdot h)}{5}$$

$$w_f := 30\cos(15\cdot f\cdot h) \quad v_f := 30\sin(15\cdot f\cdot h)$$

Here, there are *i*, *f* and *j* which are indexes of the variable symbols, and *h* which is a constant to transfer degrees into radians.

There are four levels (figure 1, table 1) to estimate our technical task connected with the quality and precision for the product. These will correlate with the next levels of the confidence coefficient/probability, namely: practically 100% for the center/smallest circle, 98% for the inner zone in the next circle; 95% for the zone in the third circle. Other circles are not shown here, but they must correspond to the next probabilities which will be less than 95%. Some results of the calculations are below.

Table 1. The calculated meanings.

$n_j =$	$m_j =$	$t_i =$	$z_i =$
19.696	3.473	5.638	2.052
18.794	6.84	4.596	3.857
17.321	10	3	5.196
15.321	12.856	1.042	5.909
12.856	15.321	-1.042	5.909
10	17.32	-3	5.196
6.841	18.794	-4.596	3.857
3.473	19.696	-5.638	2.052
$5.265\cdot10^{-4}$	20	-6	$3.159\cdot10^{-4}$
-3.472	19.696	-5.638	-2.052
-6.84	18.794	-4.597	-3.856
-9.999	17.321	-3	-5.196
-12.855	15.321	-1.042	-5.909
-15.32	12.856	1.041	-5.909
-17.32	10.001	3	-5.196
...	...	...	...

These are the definitions for the smallest diameter (t_i) and the next is a little larger (n_j). The full 37 values are not shown here for brevity.

2. Instruction of the statistical data (3 points with lines) and two circles with these points and lines

Suppose that there is a tolerance for a part (diameter is 1 mm) and for among the group. Then, the average radius for the inner circumference *R* must be 0.5 mm which is essential for each pattern. For another physical value (perhaps how far a train is running between 100-130 km) we must draw further circles with R_1 = 50 km and R_2 = 65 km. If the points of our observation are between these two circles, this indicates that there is no problem with this term of operation for the system. This is the same approach that has to be used for any system/object/product.

IOP Conf. Series: Materials Science and Engineering **657** (2019) 012075 doi:10.1088/1757-899X/657/1/012075

Here, we depict three conventional dots/points as the results of our measurement using a PC and MathCad (see figure 2). All dots are linked with straight lines in order to see them correctly on the screen. Following this, we put them on the circles (figure 3).

$i := 1..2 \qquad j := 1..2 \qquad k := 1..2 \qquad h := 0.01745 \qquad g := 1..37$

$a1_i := \begin{pmatrix} 3 \\ 1 \end{pmatrix} \quad a2_i := \begin{pmatrix} 1.5 \\ 6 \end{pmatrix} \quad b1_j := \begin{pmatrix} 3 \\ 5 \end{pmatrix} \quad b2_j := \begin{pmatrix} 1.5 \\ 7.5 \end{pmatrix} \quad c1_k := \begin{pmatrix} 5 \\ 1 \end{pmatrix} \quad c2_k := \begin{pmatrix} 7.5 \\ 6 \end{pmatrix}$

$$f_g := 20\cos(10 \cdot g \cdot h) \qquad v_g := \frac{30\cos(20 \cdot g \cdot h)}{5}$$

$$t_g := 20\sin(10 \cdot g \cdot h) \qquad z_g := \frac{30\sin(20 \cdot g \cdot h)}{5}$$

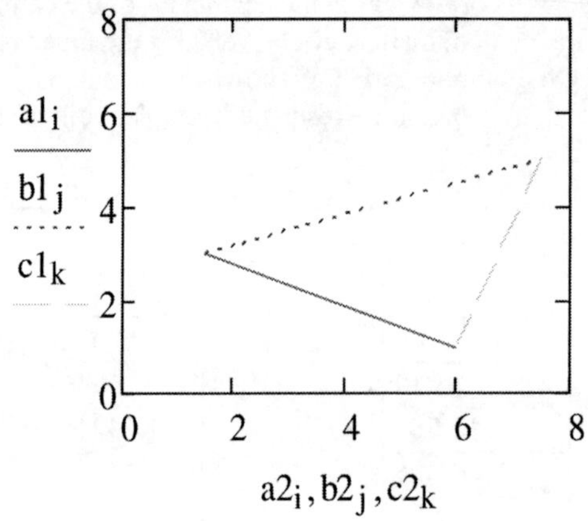

Figure 2. Three points from our measurements as a result after investigations are joined in a triangle.

Finally, these three points are placed on the picture (figure 3) with circles which determine the confidence of the applied probability.

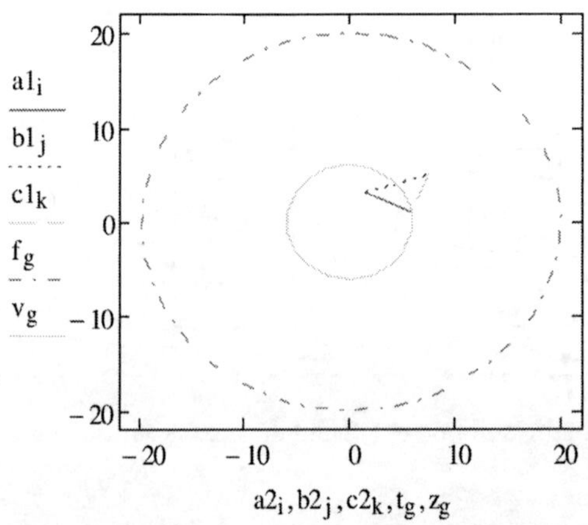

Figure 3. Version of the state for three points with connected straight lines.

If we should want to have the situation where all points are in the smallest circle, this depicted real situation is not suitable and the applied technology used in manufacture was not good. This is needed to be improved as soon as possible. If we require that the points of observation fall into the second (large) circle, then this is achievable. In this case we must not correct our technology to produce parts or products.

IOP Conf. Series: Materials Science and Engineering **657** (2019) 012075 doi:10.1088/1757-899X/657/1/012075

3. Instruction of the tolerances borders in a view as two parallel lines, as a quadrant and a rectangle which make the correlation with the definite tolerances

Projecting the new parts, we set different tolerances for these (for the external diameter, for the inner opening or for the length). The tolerances can be positive, negative or ±.

Projecting the new parts, we set different tolerances for these (for the external diameter, for the inner opening or for the length). The tolerances can be positive, negative or ±.

If there is a negative tolerance for the part 20-18.6 = 1.4 mm, this will be considered below in MathCad. In this case, the external diameter for the part is 20.0 mm (maximum) and the minimum is 18.6 mm. This shows that the quality is excellent (see figure 4).

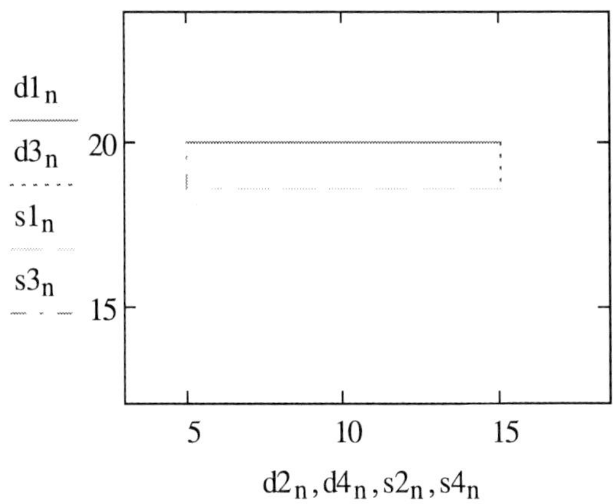

Figure 4. The range between the two lines (dl_n and $s1_n$) is 1.4 mm and the accuracy is 100%.

Here, we will not construct circumferences as it will be enough to build a rectangle or square. Examples for the rectangle are given below.

$$n := 1..2$$

$$d1_n := \begin{pmatrix}20\\20\end{pmatrix} \quad d2_n := \begin{pmatrix}5\\15\end{pmatrix} \quad d3_n := \begin{pmatrix}20\\18.6\end{pmatrix} \quad d4_n := \begin{pmatrix}15\\15\end{pmatrix}$$

$$s1_n := \begin{pmatrix}18.6\\18.6\end{pmatrix} \quad s2_n := \begin{pmatrix}15\\5\end{pmatrix} \quad s3_n := \begin{pmatrix}18.6\\20\end{pmatrix} \quad s4_n := \begin{pmatrix}5\\5\end{pmatrix}$$

Thus, if the real diameter sizes for our parts are between these horizontal lines, it means that both the quality and the precision are the highest. So, in this case we have the maximum guarantee probability (100%) for quality.

If we want to have several levels for our probability in accordance with standard demands for the wares/products, these can be obtained using the next approach, for example (see figure 5). Each square or rectangle or the diameters of circles must correlate to the different level of the appointed quality and probability (%).

The vertical border between the two horizontal lines is 20 and 18.6 (mm), but the second inner square has borders on the 2% worse and, finally, the third square has two horizontal lines on the 5% worse (the worst result in 95% only). Consequently, the second and the third squares have tolerances which are more than we can get in the range 20.0…18.6. Furthermore, we can apply this common rule in the way that we have made it above.

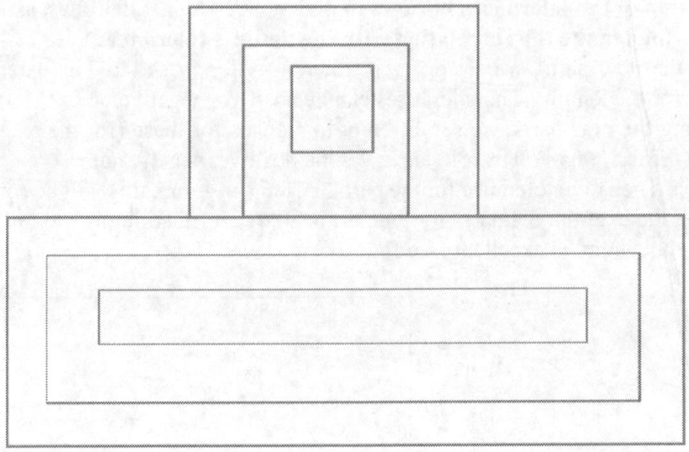

Figure 5. Three possible levels in the square and in the rectangle of the quality for the products. (For example, in the center 100%, for the middle 98%, and for the outer 95%.)

4. Ractical example

Suppose that there are two openings in our design with diameter 18 mm which have mobile contacts with two shafts. Each shaft is put into the hole. Thus, we have obtained two pairs of friction. The guaranteed term of operation without repair must be 10 months for each pair. The statistical data for wear and tear is as follows: $r1_j$ for the first opening and $r2_j$ for the second. The permissible range for deterioration is shown in figure 4 (horizontal lines 18.6-20.0 mm) for each hole. Writing the computer program and drawing the common chart, we can analyze the reliability of these pairs of friction (figure 6) [8-9]. During this experiment (period t_j = 0–10 months) we fixed n = 6 points/dimensions ($r1_j$ and $r2_j$) of the wear for each opening.

$j := 1..6$ $n := 6$ $m := 1..2$

$r1_j :=$	$r2_j :=$	$t_j :=$
18	18	0
18.2	18.8	2
18.6	19	4
19.2	19.5	6
19.3	22	8
19.8	23	10

$d1_m :=$	$d2_m :=$	$d3_m :=$	$d4_m :=$
20	5	20	15
20	15	18.6	15

$s1_m :=$	$s2_m :=$	$s3_m :=$	$s4_m :=$	$f1_m :=$	$f2_m :=$
18.6	15	18.6	5	10	10
18.6	5	20	5	23.8	10

The result of this investigation is as follows: the wear and tear for hole number one ($r1_j$) is very good over the whole period of operation; the deterioration ($r2_j$) for hole number two is not good. Consequently, the material and/or lubricant for the part which has the second hole should be improved.

Such technical examples can also be used for the equipment which we use to produce different dairy products.

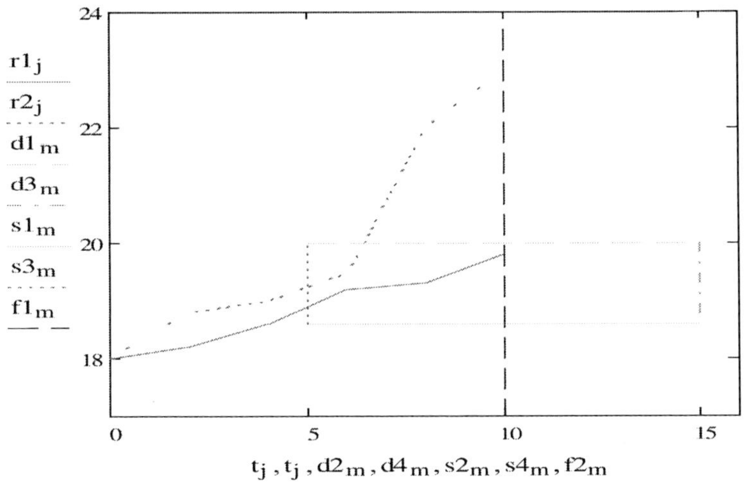

Figure 6. Two curves of wear for the two holes and the permissible range for their operation (lines $d1_m$ and $s1_m$) and the limit of the term of operation (vertical dotted line).

5. Results and Conclusion

A new and effective method is suggested in this paper. Namely:

- Rather small amounts of parts or objects can be investigated if the final information is in a narrow sphere/field with different levels of probability (about 100% guarantee);
- Suggested computer programs to help obtain clear results (charts);
- Shown practical examples that can be applied to estimate the real term of operation for different systems;
- Even when using from three to five points of observation, the right conclusion about the quality and reliability can be obtained with different levels of probability and trustworthiness for all the results of the general aggregate;
- And also very importantly, the real law of distribution is not needed for our statistical data using only extremely small initial information.

6. Short discussion

A review of the relevant publications connected with statistical information showed that there have not been such decisions as suggested in this article. The expounded ways permit researchers to correctly use new methods in their investigations utilizing small amounts of statistical data to obtain the right answer/conclusions about parts/objects/technology processes. If the points were concentrated very closely together and in the limit of the standard tolerance, it shows that there is no need to make any changes or improvements. This information can be chosen by means of casual selection: namely, a consideration of what kind of points will be needed to attain an acceptable probability connected with the quality and reliability.

References

[1] Dlin A M 1958 *Mathematical Statistics in Engineering* (M.: Soviet Science) p 466
[2] Xydson D 1967 *Statistics to physicists* (M.: Peace) p 217
[3] Johnson N and Lyon F 1980 *Data Processing Methods* (M.: Peace) p 236
[4] Kir'yanov D 2001 *MathCAD 2001, self-instructor* (SPb.: BXV-Petersburg) p 544
[5] D'akonov V 2001 *MathCAD 8/2000. Special reference book* (SPb.: PITER) p 592
[6] Voinov K N 2015 *Problems and decisions in the questions of the wear and tear* (St.-Petersburg: Nestor-History) p 500

[7] Kudryavtse E M 2001 *Mathcad 2000 Pro.* (M.: DMK Press) p 576
[8] Voinov K N 2016 *International Encyclopaedia Tribology: Computer Programmes and Examples to Calculate in Tribology and in Tribo-Technics* (St.-Petersburg: Nestor-History) p 240
[9] Voinov K N 2016 *International Encyclopaedia Tribology: Computer Programmes and Examples to Calculate in Tribology and in Tribo-Technics* (St.-Petersburg: Nestor-History) p 272

AUTHOR INDEX

AUTHOR INDEX

AUTHOR INDEX